Principles of

Unit Operations

John Wiley & Sons, Inc.

NEW YORK • LONDON • SYDNEY

Alan S. Foust

Leonard A. Wenzel

Curtis W. Clump

Louis Maus

L. Bryce Andersen

Department of Chemical Engineering
Lehigh University
Bethlehem, Pennsylvania

Principles of

Unit Operations

Library of Congress Catalog Card Number: 60–6454

ISBN 0 471 26896 8

Preface

The treatment of unit operations given in this book emphasizes the scientific principles upon which the operations are based, and groups those with similar physical bases so that they may be considered together. The development usually begins with an analysis of the physical behavior of a system and the establishment of a simplified physical model. A basic mathematical relation is written using the model and is solved. The resulting general expression is then applied to the specific unit operations. In order to maintain a clarity of presentation at an elementary level, refinements of the physical models and the resulting elaborate mathematics necessary for rigorous treatment of complex situations are generally omitted, and, in order to emphasize underlying similarities among the various unit operations, descriptions of equipment and specialized calculation methods are condensed. However, since visualization of equipment helps to add significance to the theoretical treatment and since the young engineer should be familiar with major equipment types, the important pieces of processing equipment are shown in line drawings and photographs and are discussed briefly. The more important of the specialized calculation methods necessary for process design are considered after the underlying principles have been fully developed.

The traditional concept of unit operations has been a major factor in the phenomenal success of chemical engineers and chemical engineering in the last fifty years. We believe that the unification presented here is the next logical step in the evolution of the concept of unit operations. This treatment is offered in the belief that it is more efficient in teaching, more economical in time, more adequate in its presentation of the fundamentals, and more effective in training toward the definition and solution of broad problems in chemical processing. This book should serve as a basis for advanced work in the more specialized theory and practice of the individual unit operations. The engineer educated in this approach may not be as immediately adept in the manipulation of a given specialized procedure of calculation, but he should be firmer in his understanding of the fundamental principles, more aware of the similarities among many of the unit operations, and more flexible and original in his solution of new processing problems. In short, he should be more readily adaptable to change and progress.

The continuing rapid extensions of knowledge of unit operations has created a serious problem in the coverage of this information within the time properly allotted in an undergraduate program. The increasing number of process steps that might be included as unit operations and the breadth of their applicability dictate that instruction be systematized and be made more adaptable to newer operations. This treatment, with its unification of the principles of similar operations, makes it possible to maintain a realistic balance between the unit operations and other vital facets of chemical engineering education.

Many of the formalized calculations that have occupied chemical engineers in the past will very shortly be done by electronic computers. Multicomponent distillation calculations and multiple-effect evaporator calculations already have been programmed for machine computation. Work is going forward in the use of computers to design chemical reactors, to determine the dynamics of a system during start-up, to predict and optimize the response of systems to automatic control, and to help in many other applications. The use of a computer to do in minutes what previously took man-months of engineering time has opened up several avenues of development. One avenue is that complex engineering problems, which previously have been solved only crudely and perhaps only qualitatively, can now be answered with high accuracy. Many of these applications have been in the field of process economics, as for example in optimizing the products from a petroleum refinery. Another avenue may be that rigorous but cumbersome calculation procedures may become preferable to approximate, short-cut methods. This has already happened to some extent in the field of multicomponent distillation. For work of this sort to be effective, the chemical engineer with thorough understanding of the mechanism of the process must cooperate with the mathematician who understands the mathematical possibilities and limitations of the computer. Thus, the increasing need is for a chemical engineer with firm mastery of the fundamental characteristics of process operations and in addition with the mathematical background necessary to attack the problem of describing these operations by a mathematical model. Traditional training in the details of a calculation method is thus important only as mental discipline.

With the increasing understanding of the fundamental principles, it is possible to classify the unit operations into groups based upon similar principles. In this book two major groups are considered: the stage operations and the rate operations. The stage operations are considered by using a generalized model which is applied to all of the mass-transfer operations. The rate operations are introduced with a thorough coverage of the principles of molecular and turbulent transport. After the fundamental similarities of each group of operations have been considered, the principles are applied to the analysis of the more common operations in each group.

A generalized treatment of the mass-transfer stage operations is presented in Part I. A generalized method of calculation based upon the physical model of an equilibrium stage is developed without regard to the nature of the particular phases in contact. Specific examples are taken from the various mass-transfer operations. The stage operations have been placed first because they are based on simple stoichiometric and equilibrium concepts and thus follow logically from the stoichiometry course which usually precedes the unit operations course. Full coverage of Part I requires at least two semester hours.

The fundamental principles of the rate operations are developed in Part II. Included are the operations in which a property of a phase diffuses or is transferred under the influence of a potential gradient. Molecular and turbulent transport of heat, mass, and momentum are considered in detail. Turbulence is explained at the junior-year level without the rigorous and abstruse concepts of a more complete advanced consideration. Full coverage of all the material in Part II requires at least three semester hours. Our experience is that coverage is possible in three semester hours only if sections of Chapter 13 are covered superficially and then reviewed when the corresponding operations are studied in Part III.

In Part III the principles introduced in Parts I and II are applied to the calculations involved in process design of equipment for the various operations. Our objective is to make the transition from principle to practice without obscuring the principles with an excess of practical details and special methods. Because Part III is dependent on Part II and to a minor extent on Part I, we recommend that Part III be introduced only after a thorough coverage of the other parts. All the material in Part III may be covered in four semester hours. The order of presentation of the major topics in Part III is flexible, and it may be rearranged according to the wishes of the instructor. For example, momentum transfer

(Chapters 20 through 22) may be covered before heat and mass transfer (Chapters 15 through 19). With this flexibility the instructor is free to omit subjects at his discretion.

Appendix A discusses dimensions, units, and dimensional analysis. These subjects are included in an appendix to avoid disruption of the major development of the principles. Knowledge of dimensions and units as discussed in Appendix A is necessary for Part II. The concepts of dimensional analysis are fundamental to the principles developed in Chapter 13. If these topics have not been studied earlier, they should be introduced when needed. Supplementary material beyond that given in this appendix may be introduced if the instructor desires.

Appendix B deals with the measurement and description of small particles. This material is relevant to any operation involving the presence of a particulate solid phase and especially to the operations discussed in Chapters 18, 19, and 22. Other aspects of small particle technology are omitted because they are so frequently available in specialized courses.

The entire book may be covered in nine or ten semester hours. Judicious selection of material and abbreviated consideration of subjects considered by the instructor to be of more limited utility makes an eight-hour course quite feasible.

Part II might serve as a nucleus for a basic three-hour course in transport operations for all engineers, with selected applications drawn from Part III. Such a basic course is in agreement with the recommendations of the American Society for Engineering Education for a more unified and fundamental coverage of heat, mass, and momentum transfer.

This book is the result of several years of teaching the unified approach. The preliminary draft has been used for over two years as the textbook for junior students in the chemical engineering curriculum at Lehigh University. The material has undergone several revisions based upon the experiences gained from using the preliminary draft. Every effort has been made to insure that the material included in this book can be taught successfully to junior-year engineering students. Our experience has been that the generalized approach is briefly more bewildering to the average student than is the traditional approach; however, after a short period, the material becomes clear, and the student ultimately gains a greater understanding of the unit operations.

The integration of unit operations with the important fields of kinetics, thermodynamics, and economics can be accomplished in a subsequent design course. The unit operations are among the most important tools of the chemical engineer, but they must not be allowed to crowd out other important subjects in a chemical engineering curriculum. Although human relations is seldom formally taught, it is no less important to the chemical engineer than is his background in the physical sciences and economics. Fortunate indeed is the young engineer whose education has been broad enough to impress upon him the importance of all three aspects of the triad of engineering: physical science, economics, and human relations.

We wish to express our appreciation to the administration of Lehigh University for its cooperation in the testing of this book. We also wish to thank the several classes of students on whom the developing versions of this book were tested. Their forbearance, cooperation, and suggestions have been of great help in preparing the final version.

ALAN S. FOUST
LEONARD A. WENZEL
CURTIS W. CLUMP
LOUIS MAUS
L. BRYCE ANDERSEN

Bethlehem, Pennsylvania
December 1959

Contents

This modern chemical plant produces a large number of organic chemicals starting with raw materials derived from petroleum. Ethyl chloride and ethyl alcohol are produced from ethylene. From propylene come allyl chloride, epichlorohydrin, glycerin, isopropyl alcohol, dimethyl ketone, methyl isobutyl carbinol, methyl isobutyl ketone, diacetone alcohol, and hexylene glycol. Butylene is the starting material for secondary butyl alcohol and methyl ethyl ketone. Many of these products are in turn starting materials for the synthesis of numerous other organic compounds.

Design, construction, and operation of plants such as this require the close cooperation of technical specialists in all fields of engineering. All of the major areas of chemical engineering, including the Unit Operations, are also utilized in such a complex installation to maintain most profitable balance of products with markets. (*Shell Chemical Corporation*)

chapter 1

Unit Operations in Chemical Engineering

Chemical engineering is defined as ". . . the application of the principles of the physical sciences, together with the principles of economics and human relations, to fields that pertain directly to processes and process equipment in which matter is treated to effect a change in state, energy content, or composition . . ." (1).* This very vague definition is intentionally broad and indefinite as to the extent of the field. It is probably as satisfactory a definition as any practicing chemical engineer would give to cover his work. It should be noted that considerable emphasis is placed on the process and process equipment. The work of many chemical engineers would better be called process engineering. The process may be any collection of steps involving changes in chemical composition or involving certain physical changes in material being prepared, processed, separated, or purified. The work of many chemical engineers involves choosing the appropriate steps in the appropriate order to formulate a process for accomplishing a chemical manufacturing operation, a separation, or a purification. Since each of the steps constituting a process is subject to variations, the process engineer must also specify the exact conditions under which each step is to be carried out.

As the process evolves and equipment must be designed, the work of the chemical engineer merges with that of the mechanical and civil engineer. Since the transfer of primary responsibility from the process engineer to the mechanical engineer can take place satisfactorily at various stages of the design, it is impossible to define a fixed extent to which the responsibility should be called that of a chemical engineer or a stage at which the mechanical engineer should take over responsibility for equipment.

Obviously, the physical sciences referred to in the definition are primarily chemistry and physics. The processes normally expected to be the responsibility of the chemical engineer almost always involve at least one and sometimes more chemical reactions. For this reason, the process engineer must understand the chemistry of all reactions involved. In most processes being carried out on a large scale, however, the chemistry has been previously worked out, and the physical changes incident to preparation and purification of the reaction mixtures demand considerably more study than does the chemical reaction. Frequent application of the principles of physics and of physical chemistry are required in the processing steps which produce physical changes, such as vaporization, condensation, or crystallization. As a process evolves into a plant and the work merges with that of mechanical designers, the science of mechanics becomes increasingly important. The chemical engineers who specialize in equipment must have thorough and extensive grounding in mechanics of materials.

Since all the engineer's work must be quantitative, mathematics is the fundamental tool of the engineer in all his work. Throughout the development of a chemical process, from the initial conception of the chemical reaction in a laboratory, which must be expressed in a quantitative equation, through the material and energy balances, to the prediction of size of equipment necessary for a given plant, mathematical expressions of all variables are constantly used. In the economic study to determine the most profitable operating conditions and in accounting for the sales receipts and distribution of income to profits and costs, including replacement of the plant, mathematical calculations are universal.

The existence or contemplation of a process implies that a material is to be produced for which customers will pay. It must be delivered in a quantity, of a quality,

---

* References are collected at the end of each chapter and parenthetical numbers refer to the References at the end of the chapter.

and at a price which are acceptable to the customer. Simultaneously, it must pay for materials, labor, and equipment used in the manufacture and return a profit over and above all costs. Many materials produced by the chemical industry are planned and plants built before the real market potential has been developed. For a completely new product some estimate of the size of the market must be made, and the plant scaled in proportion.

The human-relations aspect of engineering practice is not usually emphasized in undergraduate training because of the great quantity of technical information and techniques which the student must learn. That this may be a fallacious course is implied by the fact that failures of young engineers because of personnel problems are at least five times as frequent as failures because of inadequate technical training. All engineers must realize that the industry in which they are working requires team effort of all personnel. Valuable information can be obtained from operators of limited educational background who have observed similar processes. The man who has "lived" with an operation has probably observed actions and effects and has learned methods of detailed control that cannot be approached by formal theory alone. The best engineering job can be done only with proper regard for all available facts regardless of their source. A new process or the technical improvement of an existing one which is designed without due regard for the operators is usually destined to failure. The start-up of a new plant or the installation of a technical change is likely to be much smoother and the cost of it much less if the operating personnel understand the objectives and are convinced of their soundness.

SOME BASIC CONCEPTS

Before attempting to describe the operations which comprise a chemical process, it is necessary to introduce several basic concepts which must be understood before a description of the operations is meaningful.

Equilibrium. There exists for all combinations of phases a condition of zero net interchange of *properties* (usually mass or energy in chemical processing) called equilibrium. For all such combinations not at equilibrium, the difference in concentration of some property between that in the existing condition and that which would exist at the equilibrium condition is a driving force, or a potential difference, tending to alter the system toward the equilibrium condition. The tendency of thermal energy to flow from a region of high concentration—hot body—to a region of low concentration—cold body—is universally familiar. Similarly, the tendency of electrical energy to flow from a region of high potential to one of low potential in accordance with Ohm's law ($I = E/R$) is well known. The tendency of acetic acid to flow from an acetic acid–water solution into an ether

phase in contact with it is less widely known. The description of this equilibrium is considerably more complicated than the statement of equality of temperatures which describes the equilibrium of energy of molecules. Material will flow from a region of high concentration (activity) to one of low concentration (activity), just as heat and electricity flow from high- to low-concentration regions in the situations mentioned above.

The expression of the equilibrium condition is familiar to all in connection with electrical and thermal energy. The concentration of such energy is expressed directly as a voltage potential or a temperature. Accordingly, two bodies at the same electrical potential, or at the same temperature, will be in equilibrium with regard to that particular kind of energy. For the equilibrium between a liquid and its vapor, the vapor-pressure curve is reasonably familiar. The curve expresses in pressure units the concentration of vapor which is in equilibrium with the pure liquid when both are at a specified temperature. In case of a liquid mixture, equilibrium must exist between the liquid phase and the vapor phase in regard to each and every constituent present. For a binary mixture, the relation is a relatively simple one describing the concentration or partial pressure of each constituent in the vapor phase which is in equilibrium with a liquid of one particular composition at the specified temperature. Obviously, the vapor will be of different composition when it is in equilibrium with different liquid mixtures. The expressions for equilibrium in multicomponent mixtures between the liquid phase and its vapor or between two liquid phases having partial solubilities become more involved. In every case the condition must be satisfied that the potential for each constituent is identical in all equilibrium phases of a particular system.

Driving Force. When two substances or phases not at equilibrium are brought into contact, there is a tendency for a change to take place which will result in an approach toward the equilibrium condition. The difference between the existing condition and the equilibrium condition is the driving force causing this change. The difference can be expressed in terms of concentrations of the various properties of substances. For example, if liquid water of low energy concentration, i.e., low temperature, is brought in contact with water vapor of high energy concentration; that is, at high temperature, energy will be transferred from the vapor phase to the liquid phase until the energy concentration is the same in both phases. In this particular case, if the amount of liquid is large in comparison with the vapor, both phases become one by the condensation of the vapor as its energy is transferred to the cold water. The final mixture will be an increased amount of liquid water at a higher temperature than initially and a decreased amount of water vapor. This combination reaches equilibrium very quickly, at a temperature such that the vapor pressure of the water equals the pressure of the vapor phase. A similar line of reasoning can be followed in the case of two electrical condensers charged to different concentrations, i.e., voltage. If they are brought into electrical contact, the electrical energy will flow from the region of higher concentration to that of lower. Both condensers will be charged to the same voltage when equilibrium is reached.

Separations. Obviously, the separation of a solution, or other physically homogeneous mixture, requires preferential transfer of a constituent to a second phase which may be physically separated from the residual mixture. Illustrations are the dehumidification of air by condensing or by freezing a part of the moisture, or the use of a liquid solvent which is insoluble in the unextracted material. Any two phases which exhibit

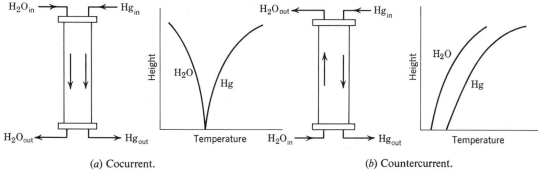

(a) Cocurrent.   (b) Countercurrent.

Figure 1.1. Flow and temperature in a contractor.

A less familiar type of driving force exists when a solution of acetic acid and water is brought in contact with isopropyl ether. The three materials will usually separate into two liquid phases, each containing some quantity of all three components. The concentration of each of the three substances in each of the two phases must be known to describe the equilibrium condition. If two phases which are not in equilibrium are brought together, a transfer analogous to that for electrical and thermal energy will occur. The result will be a transfer of isopropyl ether into the water-acid phase and the transfer of both water and acid into the ether phase until the potential of each constituent is identical in the two phases. There is no convenient and simple expression for the chemical potential; hence, the amount per unit volume, or concentration, of mass in such a phase is commonly so designated. Mass concentration is not a rigorous definition, but the more accurate and more complex functions of activity, fugacity, and free energy demand more knowledge of physical chemistry than is expected at this time. In the preceding example the mass concentration of a component is different in each phase at equilibrium.

In all cases discussed above the potential (concentration) of an existing substance or mixture when compared with the potential at the equilibrium condition yields a difference in potential which is a driving force, tending to change the conditions of the system toward the equilibrium. The driving forces, or differences in the potential of energy or matter, will tend to produce a change at a rate which is directly proportional to the difference from the equilibrium potential. The rate at which the system changes toward equilibrium is one of the major topics to be covered in this book.

preferential distribution of constituents and which may be easily separated may be involved in a separation operation. Two solid phases may be very difficultly separated; a liquid and a gas or solid usually may be easily separated; two liquids of approximately equal density and no interfacial tension may resist all practicable separation means short of altering one of the phases.

Flow Patterns. In many of the operations for transferring energy of material from one phase to another, it is necessary to bring two streams into contact to permit a change toward equilibrium of energy or of material, or both. The transfer may be accomplished with both streams flowing in the same direction, i.e., *cocurrent flow.* If cocurrent flow is used, the limit in amount of transfer which can occur is firmly set by the equilibrium conditions which will be reached between the two streams being contacted. If, however, the two streams being contacted are made to flow in opposite directions, transfer of material or energy in considerably greater amounts is possible. Such a flow pattern is known as *countercurrent flow.*

As an illustration, if a stream of hot mercury and a stream of cold water are allowed to reach thermal equilibrium, the temperature attained can be predicted by a heat balance which recognizes the relative quantities of the streams, their initial temperatures, and their heat capacities. If the streams flow simultaneously from the same inlet point to the same outlet point, the equilibrium temperature is definite, and the path is as indicated in Figure 1.1a. If the streams are made to flow in opposite directions, as by letting the mercury flow downward through an upflowing stream of water, it is possible for the entering hot-mercury stream to raise the temperature of the leaving cool-water stream to a temperature above that to which the mercury stream is lowered as it leaves the contacting equipment, as indicated in Figure 1.1*b*. The counterflow principle is used in many chemical-engineering operations in order to permit greater transfer of a property than would be indicated merely by the attainment of a single equilibrium between the leaving streams.

Continuous and Batch Operation. In the majority of chemical processing operations, it is more economical to maintain continuous and steady operation of equipment, with a minimum of disturbances and shutdowns. This is not always practical in some small-scale operations, in operations where extremely corrosive conditions force frequent repairs, and in others for various specific reasons. Because of the greater productivity of continuously operating equipment and the resultant lower unit cost, it is usually advantageous to operate equipment continuously. This means that time is not a variable in the analysis of such a process, except during the rather brief start-up and shutdown periods. The time rate of transfer or of reaction is important in fixing the necessary size and capacity of equipment, but the performance is expected to be the same today, tomorrow, or next year if the operating conditions remain the same. Conditions are not constant throughout a system at any time, but those at a particular point are constant with time.

When small quantities of material are to be processed, it is often more convenient to charge the entire quantity of material to the equipment, process it in place, and remove the products. This is called a batch operation.

An operation which is variant with time is spoken of as a *transient* or *unsteady state*, in contrast with that spoken of as *steady state*, in which conditions are invariant with time. The quenching of a steel part for heat treating and the freezing of ice cubes in a domestic refrigerator are illustrations of unsteady-state operations. In *batch operations*, almost the entire cycle is a start-up transient and a shutdown transient. In a *continuous operation*, the time during which the start-up transient exists may be extremely small in comparison with the steady-state operation. Analysis of transient or batch operations is usually more complex than in steady-state operation. Because of the greater simplicity and the wide occurrence throughout chemical processing of steady-state operations, the introductory treatment is in terms of conditions which do not vary with time. Analysis of a transient operation is different from the steady state only in the introduction of the additional variable of time. This variable complicates the analysis but does not fundamentally change it.

UNIT OPERATIONS

Chemical processes may consist of widely varying sequences of steps, the principles of which are independent of the material being operated upon and of other characteristics of the particular system. In the design of a process, each step to be used can be studied individually if the steps are recognized. Some of the steps are chemical reactions, whereas others are physical changes. The versatility of chemical engineering originates in training to the practice of breaking up a complex process into individual physical steps, called unit operations, and into the chemical reactions. The unit-operations concept in chemical engineering is based on the philosophy that the widely varying sequences of steps can be reduced to simple operations or reactions, which are identical in fundamentals regardless of the material being processed. This principle, which became obvious to the pioneers during the development of the American chemical industry, was first clearly presented by A. D. Little in 1915:

"Any chemical process, on whatever scale conducted, may be resolved into a coordinated series of what may be termed 'unit actions,' as pulverizing, mixing, heating, roasting, absorbing, condensing, lixiviating, precipitating, crystallizing, filtering, dissolving, electrolyzing and so on. The number of these basic unit operations is not very large and relatively few of them are involved in any particular process. The complexity of chemical engineering results from the variety of conditions as to temperature, pressure, etc., under which the unit actions must be carried out in different processes and from the limitations as to materials of construction and design of apparatus imposed by the physical and chemical character of the reacting substances." (2)

The original listing of the unit operations quoted above names twelve actions, not all of which are considered unit operations. Additional ones have been designated since then, at a modest rate over the years but recently at an accelerating rate. Fluid flow, heat transfer, distillation, humidification, gas absorption, sedimentation, classification, agitation, and centrifugation have long been recognized. In recent years ion exchange, adsorption, gaseous diffusion, fluidization, thermal diffusion, hypersorption, chromatography, and others have also been proposed for designation as unit operations.

In general the term *unit operations* has been restricted to those operations in which the changes are essentially physical. This is not universally true, because the term gas absorption is used appropriately for the operation of removing one gas from a mixture whether the removal is accomplished by physical solution or by chemical reaction with the solvent. Very frequently chemical changes occur in a material being distilled or heated. In such cases the physical operation is the primary concern, and, if a chemical change occurs simultaneously, it is usually handled by a modification of the physical properties of the material.

The typical chemical manufacturing operation involves a few chemical steps which are probably straightforward and well understood. Rather extensive equipment and operations are usually needed for refining or further preparing the usually complex mixture for use as an end product. The result is that the work of the typical process engineer is much more concerned with physical changes than with chemical reactions. The importance of the chemical reactions must not be overlooked because of the economic importance of small improvements in percentage yield from chemical reactions. In many cases a relatively small percentage improvement in yield may economically justify considerably more extensive processing operations and equipment.

All unit operations are based upon principles of science which are translated into industrial applications in various fields of engineering. The flow of fluids, for instance, has been studied extensively in theory under the name of hydrodynamics or fluid mechanics. It has been an important part of the work of civil engineers under the name of hydraulics and is of major importance in sanitary engineering. Problems of water supply and control have been met by every civilization.

Heat transfer has been the subject of many theoretical investigations by physicists and mathematicians; it has played a major part in the generation of power from fuels, as developed by mechanical engineers. Dissipation of heat in electrical equipment is a major limitation on the power output of such machinery. Pyrometallurgy and the heat treatment of materials of construction and tools represent additional major applications.

Throughout industry, one finds examples of most of the unit operations in applications which are in the province of other engineering fields. However, the chemical engineer must carry out many unit operations on materials of widely varying physical and chemical properties under extremes of conditions such as temperature and pressure. The unit operations which are used to separate mixtures into more or less pure substances are unique to chemical engineering. The materials being processed may be naturally occurring mixtures, or they may be the products of chemical reactions, which virtually never yield a pure substance.

INTEGRATION OF THE UNIT OPERATIONS

The traditional presentation of the unit operations has been the collection of appropriate theoretical and practical information about each unit operation as a package. In previous textbooks each operation has been presented rather independently from the others. It is seldom obvious in introductory presentations that several of the unit operations overlap in their foundations and are quite intricately related to each other. The interrelations become more obvious in monographs on several of the unit operations because of the impossibility of presenting the theory on any one of the operations completely without regard for the influence of others. Specifically,

heat transfer in a flowing system cannot be completely presented without consideration of the fluid mechanics; mass transfer cannot be divorced from heat transfer and fluid mechanics.

With increasing information has come broader recognition of the basic similarities. Conversely, recognition and exploitation of the similarities has contributed to a broader understanding of each operation. It now appears that the compartmentalization of information by unit operation leads to unnecessary repetition and waste of time and that study of basic principles common to a group of the operations will lead to a better understanding of all of them.

This book presents under single headings those operations having similar fundamentals, using generalized nomenclature and concepts. This presentation has been found to result in time economy in learning and is believed to contribute a greater breadth of understanding of all the operations when the interrelations are understood.

Analysis of the Unit Operations. The unit operations might be analyzed and grouped using any one of three possible methods. A unit operation may be analyzed using a simple physical model which reproduces the action of the operation; it might be analyzed by considering the equipment used for the operation; or it might be analyzed starting with a mathematical expression which describes the action and which is tested using experimental process data.

The physical model is established by careful study of the basic physical mechanism. The model is then applied to a real situation either through mathematical expression or by physical description. Since the model is idealized, some corrections are necessary in its application to real operations. This approach develops an understanding of the basic similarities among the principles of the various unit operations.

The grouping could be made in terms of those operations which are accomplished in similar equipment or in which a similar function occurs. For most of the operations, the art preceded a scientific understanding, and equipment was built and operated on the basis of woefully incomplete basic knowledge. Some improvements and refinements came, as expected, purely from the art and the equipment. Grouping on the basis of equipment and its functioning exposes one to the risk of only perpetuating the mistakes of the past. A thorough understanding of the basic operation seems much more likely to yield improvements in operations.

The operations could also be grouped in the light of similarity of the basic mathematical formulation of the operation. This method of grouping is unsatisfactory because of the perversity of molecules in their disregard of mathematics. Because of the nonlinearities involved and because boundary conditions of one phase usually respond to changes occurring in an adjacent phase, it is frequently impossible to formulate the boundary conditions for solution of a mathematical expression in manageable terms.

Each of the three modes of grouping could be used as a basis. The physical model of the fundamental operation is the most satisfactory approach and is used in this presentation. Wherever possible, the physical model is described mathematically, and the performance is expressed in mathematical relations derived from the fundamental principles. This formulation gives the best basis for understanding and refining those operations in which the art is ahead of theory. This is true in spite of the fact that the models are oversimplified and that the mathematical formulation of the behavior of the model cannot be transposed perfectly into an expression of the behavior of the prototype.

It should be obvious that there is no universal criterion dictating a particular choice of method of analysis and that all contributing factors should be recognized in deciding upon a particular mode. Any grouping requires some arbitrary choice and always leaves one with some of the operations which fit poorly into the general scheme. Such operations must be studied individually.

Two Major Physical Models. One widely applicable model for unit operations is a device in which two streams, or phases, are brought together, allowed to reach equilibrium, then separated and withdrawn. Since it is assumed that the leaving streams are at equilibrium, this model is called an *equilibrium stage*. Evaluation of the changes in the streams which must be accomplished to attain equilibrium establishes a measure of ultimate performance. Real equipment is evaluated by expressing the changes accomplished in it as a fraction or percentage of the changes that would occur in an equilibrium stage. In another possible model for transfer of a property between two streams we visualize the carriers of the property, evaluate their number and rate of migration, and arrive at an expression of the *rate of transfer* between the two streams in continuous contact. This rate of transfer multiplied by the *time of contact* yields an expression for the amount of transfer accomplished. The equilibrium-stage model may be expressed mathematically in a finite-difference equation relating entering concentrations of any property with the equilibrium concentrations of the property in the leaving streams. Graphical techniques frequently can be used more conveniently than the finite-difference equation. The mathematical expression for the rate-of-transfer model is a differential equation which can sometimes be integrated rigorously but more frequently must be handled in terms of average conditions. Since a large number of chemical processing operations are actually carried out either stagewise or in continuous contact, these two models are widely applicable for the analysis of unit operations.

Most of the unit operations can be studied on either of the two bases. Many of them are carried out sometimes in continuous-contact equipment and sometimes in stagewise equipment. In some operations the advantage of one or the other mode of analysis may be obvious. In many others, the choice is dictated by availability of the necessary data and constants. Equilibrium data are a part of the stock in trade of the physical chemist and are available for a large number of substances under various conditions. To some extent, the convenience in analysis is related to the work of earlier investigators, in that their results may have been interpreted in a fashion which makes one or the other analysis more convenient. Choice of one method of analysis does not necessarily restrict the actual operation to the same model.

The Stage Operations. Operations in which stage contacting is frequently used will be considered first. The model is the device in which the two incoming streams interact to attain equilibrium between the streams as they leave the stage. The model is known as an *equilibrium stage*, and is assumed always to yield two product streams in equilibrium with each other. The generalized treatment does not require a specification of the property being transferred or of the nature of the phases being contacted. The practical analysis is based on the fraction of transfer accomplished in the actual stage as compared to the equilibrium stage. Presentation will be in as completely general terms as possible, without regard to the particular nature of the phases in a particular case.

Stage contacting may be illustrated using the mercury and water streams discussed earlier. As shown in Figure 1.1, the mercury and water streams are in continuous contact, and heat is transferred continuously from the hot stream to the cold stream. For stage contacting the equipment is modified as described below. If the hot-mercury and cold-water streams used above as an illustration are mixed intimately and then fed to a settler where the phases are separated, the outflowing streams will be at practically the same temperature. The equilibrium temperature can be predicted by a material and an energy balance. Suppose now that two mixer-settlers are provided, one of which receives the hot mercury and the other of which receives the cold water. The mercury leaving the warmer mixer-settler flows to the cooler mixer-settler, and the water leaving the cooler mixer-settler flows to the warmer mixer-settler. The two mixer-settlers will accomplish the transfer of more heat than the one. If the number of mixer-settlers is increased to n, even more energy can be taken from the mercury. In this case, the mercury would be passed through them in the order 1, 2, 3, . . . , n, and the water n, . . . , 3, 2, 1. The introduction of additional stages decreases the heat transferred per stage because the potential difference from equilibrium becomes less, but the total transfer is increased. No one would do this particular operation in the manner described, but many transfer operations use stage contacting. Stage contacting is a common way of extracting one component from a liquid mixture by preferential solution of that compound or group of compounds such as in the removal of sludge-forming components from lubricating oils.

The Rate Operations. The unit operations involving continuous contacting depend upon the rate of transfer, and are therefore called *rate operations*. The transfer of a large number of properties of a material—such as electrical, magnetic, thermal, mass, and momentum concentrations—follows the same basic mathematical expression of rate of transfer as a function of concentration gradient,

$$\frac{\partial \Gamma}{\partial \theta} = \delta \frac{\partial^2 \Gamma}{\partial x^2} \qquad (1.1)$$

where Γ = concentration of the property to be transferred

θ = time

x = distance measured in direction of transport

δ = proportionality constant for a system

This equation is frequently called the diffusion equation. It is a general expression which reduces to Ohm's law for electrical flow for specified conditions. The broad study of electrical and magnetic transport is the important "field theory" of the electrical engineer. These two phenomena follow well-established laws and involve relatively constant proportionality factors (such as δ in the above equation). Since boundary conditions can usually be evaluated, analytical solutions are frequently possible for engineering calculations. Chemical substances are less well behaved mathematically, and the proportionality "constants" are seldom really constant. The boundary conditions are more elusive; hence, chemical engineers are seldom in position to apply mathematically elegant and rigorous solutions of the diffusion equation. In order to arrive at a solution of the diffusion equation, it may be simplified into a finite-increment equation for average conditions rather than solved as a differential.

In the simplest cases, quite unusual in chemical processing, the rate of transport is constant with time *and* position within the system. The driving force may be assumed to be constant and distributed over a path of fixed length and of constant area. The physical properties of the path may be constant so that the proportionality factor δ may be assumed to be constant. These assumptions are the simplifications that have been introduced in arriving at Ohm's law in its form as usually presented in introductory physics courses. The counterpart in chemical transport becomes

$$\text{Rate of transport} = \frac{\text{driving force/unit distance}}{\text{resistance/unit of path area}} \qquad (1.2)$$

Since chemical substances seldom fit nice mathematical equations and since chemical equilibrium is constantly upsetting neat formulations of boundary conditions which would permit rigorous solutions of Equation 1.1, various averages and approximations must be used in arriving at an answer in an economical length of time. The simplifications usually approach Equation 1.2 much more closely than the rigorous diffusion equation.

For the rate operations, analysis must be based upon the driving force causing a change, the time during which a driving force is allowed to act, and the quantity of material upon which it acts. The diffusion equation above expresses the transient behavior of a large number of properties under the influence of a driving force for transport of the property. In chemical engineering, mass, momentum, and thermal energy are the three properties whose transport is the most frequently involved. As mentioned above, it is universal that these three properties, along with a number of others in which chemical engineers are less frequently concerned, tend to flow from regions of high concentration to regions of low concentration. Accurate prediction of the amount of the property which flows from a donor region (source) to a receiver region (sink) can be made if the driving force, the area of the path, and the unit resistivity of the path (the proportionality constant used in Equation 1.1) are accurately known. Throughout the study of the rate operations, the importance of a clear understanding of the meaning of concentration cannot be overstressed. In every case, the concentration expresses the amount of property per unit volume of the phase being processed. The amount being transferred can usually be expressed in some absolute unit measuring that quantity such as Btu's, or pound moles. It can also be expressed in terms of the decrease in concentration of the property in a known amount of phase having a known capacity for this property. For example, a quantity of energy leaving a system as heat can be expressed in terms of the number of Btu's or calories of energy. It can also be expressed in terms of the decrease of temperature of a known amount of the phase. These generalizations will become more meaningful as different operations are analyzed and the transported quantities are expressed in terms of the various possible units. Since the basic principles of transport are identical for the three properties, an analysis will be offered in completely general terms before specification of the particular property in specific operations.

Unsteady-State Operation. The diffusion equation, Equation 1.1, is applicable to a change which is a function of time.

GENERAL CONSIDERATIONS

The understanding of the basic physical principles of an operation and the formulation of these principles into a mathematical expression are the first requirements for applying the principles of the unit operations. In engineering practice, however, numerical values must always be incorporated, and a practical answer obtained.

The same problem may be met by the design engineer in specifying equipment, by the operating engineer in checking the performance of installed equipment, or by any engineer in seeking improvement in quality or quantity. It is therefore necessary that mathematical and/or graphical techniques be available which will permit the prediction of any unknown answer for a particular system regardless of whether the unknown is a composition, quantity, temperature, or number of stages required to accomplish a specified amount of enrichment of any chosen property.

Although this book is devoted exclusively to the principles of the unit operations of chemical engineering, it should be emphasized to the prospective chemical engineer that unit operations are only one sector of chemical engineering. The real objective is the engineering of the most economical process. The unit operations are techniques in arriving at this process, but they must not be allowed to crowd out of consideration the other important scientific principles which must be recognized.

The best process can be designed only with proper regard for the basic chemistry, kinetics, and thermodynamics, with adequate recognition of the limitations imposed by materials of construction and auxiliaries to the plant. The equipment design will involve work by engineers trained in disciplines normally not covered by the chemical engineer. The ultimate object of the engineering is the accumulation of a profit from the operation. The largest return of profit, after all costs are accounted for, demands the full exploitation of all the technical factors involved, favorable human relations within the producing team, and accurate knowledge of the amount of product which can be sold for the maximum eventual profit.

REFERENCES

1. Constitution of the American Institute of Chemical Engineers.
2. Little, A. D., *Report to the Corporation of M.I.T.*, as quoted in *Silver Anniversary Volume*, p. 7, AIChE, 1933.

part I

Stage Operations

The industrial unit operations which effect a transfer of mass from one phase to another by discontinuous contacting of the two phases can be classified as *stage operations*. The basic calculations required to design equipment for the various stage operations are based on many identical concepts. This section will consider the general concepts involved in the design of multistage equipment. Where the various stage operations differ, each will be discussed in detail, but wherever possible the operations will be considered in a unified manner. The details of calculations will be illustrated with examples from specific stage operations. A general graphical method of calculation for steady-state stage operations will be developed in Chapters 2 through 6. In Chapter 7, simplified graphical and analytical methods will be considered. Chapter 8 is devoted to the evaluation of stage operations at unsteady state.

A complete tabulation of the general notation and its application to specific stage operations is given at the end of Part I. The reader is urged to study it carefully as he reads the general development.

chapter 2

Mass-Transfer Operations

Many operations in the chemical process industry involve the transfer of mass from one phase to another. Usually one component of the phase will transfer to a greater extent than another, thereby causing a separation of the components of the mixture. For example, crude petroleum can be separated into several components by mass transfer between a liquid and a vapor phase. The chemical engineer is concerned with the distribution of components between two phases at equilibrium and with the rate of transfer of the components from one phase to another.

The *rate* of mass transfer must be considered in the design of equipment where the two phases are in continuous contact and mass is interchanged continuously between the phases. This subject will be considered in a later section. In many mass-transfer operations, equipment is designed to give discontinuous contact of phases in a series of stages. The initial calculations involved in equipment design or in the evaluation of performance of existing equipment are based on relatively simple stoichiometric and equilibrium relationships.

A *stage* may be defined as a unit of equipment in which two dissimilar phases are brought into intimate contact with each other and then are mechanically separated. During the contact various diffusing components of the mixture redistribute themselves between the phases. The resultant two phases have approached equilibrium and therefore have compositions different from the initial phases. By successive contact and separation of dissimilar phases (a multistage operation) large changes in the compositions of the phases are possible. In an *equilibrium stage* the two phases are well mixed for a time sufficient to allow establishment of thermodynamic equilibrium between the phases leaving the stage. At equilibrium no further net change of composition of the phases is possible for a given set of operating conditions. In actual industrial equipment it is usually not practical to allow sufficient time with thorough mixing to attain equilibrium. Therefore, an *actual stage* does not accomplish as large a change in composition as an equilibrium stage. The *stage efficiency* is defined as the ratio of a composition change in an actual stage to that in an equilibrium stage. Stage efficiencies for industrial equipment range between a few per cent and about 100 per cent. Since an equilibrium stage gives the greatest composition change possible for a given set of operating conditions, it is also referred to as an ideal, or theoretical, stage.

The calculation of equipment requirements for industrial multistage operations usually involves the determination of the number of equilibrium stages followed by the application of stage efficiency to give the number of actual stages required. An equilibrium stage can be represented schematically:

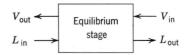

Two phases, V_{in} and L_{in}, are mixed and allowed to come to equilibrium. The phases are then mechanically separated and leave the stage as V_{out} and L_{out}, which are in equilibrium with one another. An everyday example of a single-stage mass-transfer operation is the vacuum coffee maker. Here, the hot water (V_{in}) and the ground coffee (L_{in}) are contacted to distribute the soluble constituents of the coffee between the liquid and solid. If given sufficient time, the dissolved coffee would come to equilibrium with that in the grounds. The coffee solution (V_{out}) is then mechanically separated (by pouring, for example) from the grounds (L_{out}). In actual coffee making, the time of contact with the

Plan view of top plate

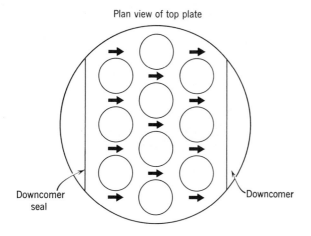

Downcomer seal

Downcomer

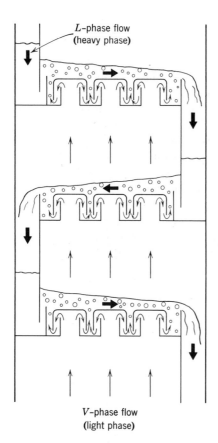

L-phase flow
(heavy phase)

V-phase flow
(light phase)

Figure 2.1. Crossflow bubble-cap plates for contacting two fluid phases. The V-phase flow is indicated by the light arrows. The L-phase flow is shown by heavy arrows. The bubble caps disperse the V-phase in the L-phase. They are designed to minimize leakage of the L-phase back through the V-phase channels.

resultant approach to equilibrium and the ratio of water to coffee (V_{in}/L_{in}) determine whether the coffee is "weak" or "strong."

Equipment for Stage Operations. Equipment for stage operations varies greatly in size and in construction details, but there are many fundamental similarities.

In general, each stage of the equipment mixes the incoming two phases thoroughly so that the material can be transferred as rapidly as possible from one phase to the other. Each stage then must separate the resultant two phases as completely as possible and pass them on to the adjacent stages. Some industrial equipment may consist of a single stage, but more often multistage units are employed with countercurrent flow of the two phases. Multistage operation permits greater changes in the compositions of the two phases than can be accomplished in one stage.

The names usually attached to the various mass-transfer operations evolved before the similarities among the stage operations were fully understood. The primary difference between the various stage operations is the nature of the two phases involved in each operation. In *distillation* a vapor phase contacts a liquid phase, and mass is transferred both from the liquid to the vapor and

Figure 2.2. Exploded view of a bubble cap. The riser shown at the bottom of the picture is attached to the plate. Gas flows up through the riser into the cap and out of the vertical slots, where it is dispersed in the liquid phase which covers slots in the bubble cap. (Vulcan Copper and Supply Co.)

from the vapor to the liquid. The liquid and vapor generally contain the same components but in different relative quantities. The liquid is at its bubble point* and the vapor in equilibrium is at its dew point*. Mass is transferred simultaneously from the liquid by vaporization and from the vapor by condensation. The net

In the most simple case of gas absorption none of the liquid absorbent vaporizes, and the gas contains only one constituent that will dissolve to any extent. For example, ammonia is absorbed from an air–ammonia mixture by liquid water at room temperature. Ammonia is soluble in water, but air is almost insoluble. The

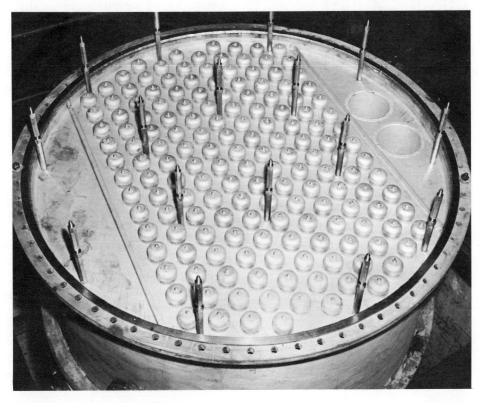

Figure 2.3. Bubble-cap plate with crossflow. The sheet-metal dams or weirs that run along the left and right sides of the plate maintain the liquid level high enough to cover the slots in the bubble caps. The liquid flows downward from the plate above on to the left side of plate pictured. It then flows over the weir and across the plate to the weir and downcomers at the right side. In this case there are two circular downcomers which deliver the liquid to the next plate below. The inlet weir maintains a liquid level covering the lower end of the downcomer, so that the gas cannot shortcircuit up through the downcomer. The rods projecting from the plate are for support of the next plate above. (Vulcan Copper and Supply Co.)

effect is an increase in concentration of the more volatile component* in the vapor and an increase in concentration of the less volatile component* in the liquid. Vaporization and condensation involve the latent heats of vaporization of the components, and, therefore, heat effects must often be considered in distillation calculations. Liquids with different vapor pressures at the same temperature can be separated by distillation. For example, crude oil can be separated into a number of fractions such as light gases, naphtha, gasoline, kerosene, fuel oils, lubricating oils, and asphalt.

The mass-transfer operation which is called *gas absorption* involves the transfer of a soluble component of a gas phase into a relatively nonvolatile liquid absorbent.

water will not vaporize to an appreciable extent at room temperature. Therefore, only ammonia will be transferred from the gas phase to the water. As ammonia is transferred to the liquid phase its concentration will increase until the dissolved ammonia is in equilibrium with that in the gas phase. When equilibrium is reached, no more ammonia will be transferred, since there is no concentration driving force for mass transfer.

Often the heat effects of absorption are small. The liquid absorbent is below its bubble point, and the gas phase is well above its dew point. A further difference between distillation and gas absorption is that the liquid and gas phases usually do not contain all of the same components.

Stripping, or *desorption*, is the opposite of absorption. In this case, the soluble gas is transferred from the liquid

---

* Defined in Chapter 3.

Plan view of top plate

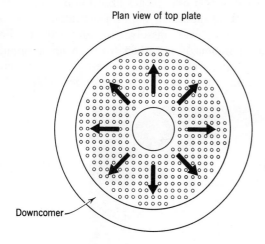

Downcomer

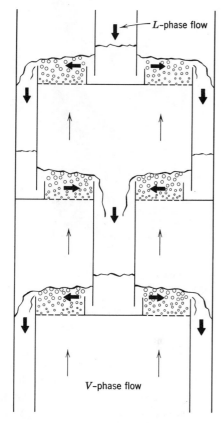

L-phase flow

V-phase flow

Figure 2.4. Disk-and-doughnut perforated plates for contacting two fluid phases. The light arrows indicate the V-phase flow, and the heavy arrows show the L-phase flow. The V-phase is dispersed in the L-phase on each plate by passage through the small holes in the plate.

to the gas phase, because the concentration in the liquid is greater than that in equilibrium with the gas and the concentration driving force is opposite to that for absorption. For example, ammonia can be stripped from an aqueous solution by bubbling fresh air through the solution. Since the entering air contains no

ammonia and the liquid does, transfer will be from the liquid to the gas.

Equipment used for multistage contacting of the liquid and gas in distillation and gas absorption includes bubble-cap and perforated-plate columns. A three-stage section of a typical bubble-cap plate column is shown in Figure 2.1. Such plates are usually circular and may vary from a few inches to as much as 30 feet in diameter, depending on the quantity to be processed. The gas or vapor (V-phase) flows upward through the plate and is finely dispersed in the liquid by the bubble-cap. The construction of the bubble-cap shown in Figure 2.2 minimizes the leakage of the liquid through the gas channels. The liquid flows across the plate and then downward to the next plate. For this reason, it is referred to as a crossflow plate. A typical industrial bubble-cap plate with crossflow is shown in Figure 2.3.

The perforated plate serves the same purpose as a bubble-cap plate. The gas flows upward through small holes in the plate, which disperse the gas as fine bubbles in the liquid on the perforated plate, as shown in Figure 2.4. The liquid flows radially from the center of the top plate to the circumference, then downward to the plate below. On the lower plate the liquid flows from the circumference to the center downspout and then down to the next lower plate. Thus, the downcomers from plate to plate will alternate between the center and the circumference, as shown. This arrangement is commonly called the "disk-and-doughnut" column.

Many liquid flow patterns are used in bubble-cap and perforated-plate columns. For example, crossflow is used with perforated plates, as shown in Figure 2.5, and the disk-and-doughnut arrangement can be applied to bubble-cap plates. Figure 2.6 shows a bubble-cap plate with a radial-flow pattern. A split-flow pattern is shown in Figures 2.7 and 2.8. Both a bubble-cap plate and a perforated plate act as a single stage. The stage efficiency will depend upon the flow characteristics on the plate and upon the transport properties of the materials being processed.

Many other types of stage-contact equipment are used to produce the desired large interfacial area between phases. Industrial distillation and absorption towers, which are assemblies of individual plates, are pictured in Chapters 6 and 7.

In *liquid-liquid extraction* the two phases involved are both liquids. In the simple case, a solute is removed from one phase (the raffinate) by solution into the other phase (the extract). In most cases the situation is further complicated by partial mutual solubility of the two solvents. Where the two liquid phases are easily separable, perforated-plate columns similar to that of Figure 2.4 may be used. Where the liquid phases have nearly equal densities or tend to emulsify, centrifugal extractors are employed. Another common type of equipment for liquid extraction is the mixer settler (Figure 2.9a). In this apparatus the two liquid phases are thoroughly mixed by an impeller or by a mixing jet. The phases are then separated by gravity or by centrifugal force. The individual stages can be grouped vertically in a column with the impellers on a common shaft, or they can be arranged horizontally. Figure 2.9b shows a typical mixer-settler arrangement with impeller mixing and horizontal arrangement. The units have a small vertical gradient, so that one of the fluid phases may flow by gravity from stage to stage while the other is pumped in the opposite direction. Mixer settlers are used in the petroleum industry and in the extraction and purification of uranium in the nuclear energy industry. A typical industrial mixer settler is shown in Figure 2.10.

In *solid-liquid extraction* a solid solute is dissolved from an insoluble residue into a liquid-solvent phase. A simple example of a one-stage solid-liquid extraction process is the coffee pot mentioned earlier. Here the soluble coffee is dissolved from the insoluble grounds by the solvent, water; and the resultant solution is separated from the grounds.

In many cases solid-liquid extraction is not a true interphase mass-transfer process. Often all the solute enters already dissolved or is rapidly dissolved in the

Figure 2.6. Bubble-cap plates with a radial-flow pattern. Parts of two plates are pictured. On the plate resting in a vertical position the liquid is supplied at six points around the circumference, as indicated by the inlet weirs. The liquid flows from the circumference into the downcomer at the center, where it flows to the plate below. A portion of the next plate above is shown in the left foreground before installation. On this plate the liquid flows from the center to the six downcomers at the circumference (two of which are visible). The spacing between plates is indicated approximately by the height of the downcomers. (Vulcan Copper and Supply Co.)

first few stages. The remaining stages merely split the resultant solution into two parts: (1) that adhering to the insoluble residue and (2) that flowing out of the stage as extract. The concentration of solute in the two completely miscible liquids changes from stage to stage, but there is no interphase mass transfer after the solute is completely dissolved. Even though true interphase transfer does not occur, the calculation of stages can be carried out in the same manner as that for the other stage operations if the "equilibrium" is specially defined. This will be discussed later.

Since a solid does not flow as do the fluids in the operations previously discussed, the equipment for solid-liquid extraction is different from that for distillation, absorption, and liquid-liquid extraction. Some means must be provided to move the solid phase countercurrent to the liquid. For example, the rotating-blade extractor (Figure 2.11) gives stagewise contact between the finely divided solid and a liquid solvent. This unit may be used to extract oil from ground cottonseeds with hexane

Figure 2.5. Perforated plate with crossflow. When the plate pictured is installed, the liquid will flow from the plate above into the space shown at the top of the plate. From there it flows across the plate to the outlet weir and into the three downcomers to the plate below. The four grooves running the length of the plate are for structural strength. (Vulcan Copper and Supply Co.)

Figure 2.7. Split flow on a bubble-cap plate. The liquid flows in both directions from the center to the downcomers on each side. On the plate below the one pictured, the liquid flow is from each side to a central downcomer. (Vulcan Copper and Supply Co.)

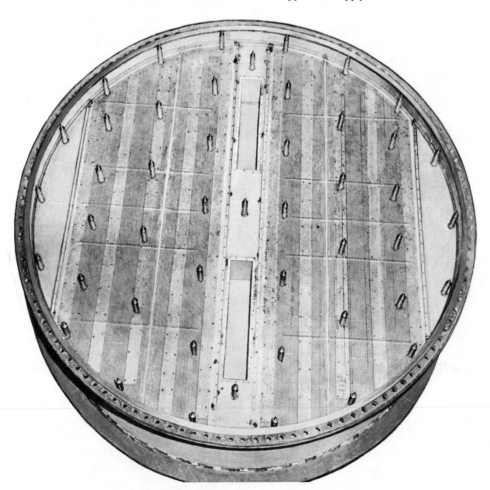

Figure 2.8. Split flow on a perforated plate. The liquid flow is from the left and right sides to the two downcomers at the center. (Vulcan Copper and Supply Co.)

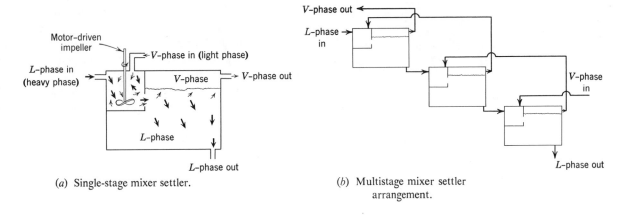

(a) Single-stage mixer settler.

(b) Multistage mixer settler arrangement.

Figure 2.9. Mixer settlers for liquid-liquid extraction. The two entering phases are thoroughly mixed by the impeller. The mixture flows into the settling tank, where the two phases are allowed to separate under the influence of gravity.

Figure 2.10. An industrial mixer-settler unit for liquid extraction. The unit is used to extract undesirable components from a lubricating-oil distillate using nitrobenzene as a solvent. Five stages are used with simple countercurrent flow of the two phases, similar to the arrangement in Figure 2.9b. The five horizontal tanks pictured are the settler units. The mixer units are out of sight beneath the platform at the far end of the settler tanks. The mixer for each stage consists of tank approximately 4 ft in diameter and 8 ft high, equipped with a motor-driven impeller for mixing. The mixed phases are fed to the settling tanks, each of which is 7¾ ft in diameter and 33 ft long.
(Atlantic Refining Co.)

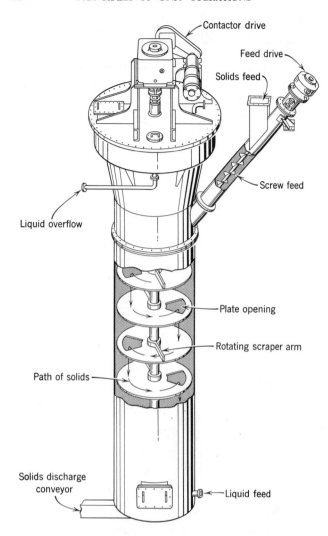

Figure 2.11. Rotating-blade column for countercurrent contact of a liquid and a solid phase. The finely divided solid is fed into the top of the column by a screw conveyor. It is deposited on the top plate, where it has contact with the upward-flowing liquid. The slowly rotating scraper arms push the solid through the opening in the plate, where it falls to the plate below. The leached solids are discharged from the column by a liquid-tight conveyor at the base.

The units are usually built with at least twenty plates, spaced not more than 12 in. apart. The solids normally take between 30 and 45 min to pass through the tower. The columns may be built in any size from 1 ft to 7 ft in diameter, and may handle up to 1000 tons/day of solids. A number of applications with appropriate feed streams are listed below.

Operation	Liquid Feed	Solids Feed
Ore Leaching	water, acid, alkali	ore
Dissolving	solvent	slow-to-dissolve solids
Decolorization	sugar syrup	charcoal
Ion Exchange	solution	ion-exchange resin
Adsorption	solution	adsorbent granules
Oil-seed extraction	hexane or other solvent	oil-seed flakes

(Allis-Chalmers Manufacturing Co.)

as a solvent and to extract peanut oil, linseed oil, and soybean oil. The ground seeds are fed to the top plate, which is rotating as shown. After one rotation the seeds will be scraped through a slot in the plate by the stationary scraper and will fall to the plate below, where the process is repeated. The solvent is supplied at the bottom of the tower. It flows upward through the slot in each plate, contacting the ground seeds in a stagewise manner. The rotating-blade column can be used for many solid-liquid contact operations, as indicated on Figure 2.11. A typical industrial installation is shown in Figure 2.12.

For large-scale operations, such as the leaching of copper ores, large open agitated tanks are used in an arrangement similar to that of Figure 2.9b. The solid phase may be moved from stage to stage mechanically, or it may be mixed with sufficient liquid to make a slurry which will flow to the next stage. In some cases the solid is left in the same tank, and successively less concentrated solutions are contacted with it.

Figure 2.12. Aureomycin extraction column. At one stage in the manufacturing process, the antibiotic Aureomycin must be extracted from a solid cake coming from a continuous filter. Most of the Aureomycin is extracted with a liquid solvent in large tanks. The cake is then fed to the 31-stage rotating-blade extractor shown in this figure. Here the remaining Aureomycin is extracted with fresh solvent. The construction of the column is similar to that diagrammed in Figure 2.11. (Allis-Chalmers Manufacturing Co.)

Adsorption involves the transfer of mass from either a gas or a liquid to the surface of a solid. The fluid does not dissolve in the solid in the usual sense; it adheres to the surface. Thus, adsorption is not a true interphase transfer operation. However, an equilibrium is attained between the adsorbed fluid and that remaining in the bulk fluid phase, so that stage calculations can be made in the usual manner. Adsorption is used in the oil industry to separate compounds not easily separable by the operations discussed previously. Multistage countercurrent adsorption has been investigated in recent years to separate a mixture of hydrocarbon vapors into its constituents in a standard bubble-cap column (1). The unit employed a very finely divided charcoal adsorbent which would flow like a liquid when agitated by the gas flowing upward through it.

There are a number of other mass-transfer operations. However, the principal ones in which stagewise contacting is used have been discussed. Operations involving continuous contact will be discussed in later sections when the rate of transfer is considered.

REFERENCES

1. Etherington, L.D., et al., *Chem. Eng. Prog.*, **52**, 274 (1956).

chapter 3

Phase Relationships

The choice of a mass-transfer operation for use in separating a mixture depends upon the phase characteristics, equilibrium relationships, and chemical properties of the material to be processed and upon the economics of the possible separating methods. For example, the soluble constituents of coffee cannot be separated from the ground bean by distillation because of the solid phase present, the temperature instability of the soluble portion, and the probable difference between the soluble material and any distilled portion. Similarly, penicillin cannot be removed from its fermentation broth by distillation because of the sensitivity of the penicillin to heat. Here liquid-liquid extraction is used. Liquid extraction has also been used in the petroleum industry for removal of aromatic compounds from paraffinic compounds boiling in the same temperature range.

Usually the phase relationships and chemical characteristics of the process material allow separation by any one of several methods. In such cases the choice of method is based upon economic evaluation of each alternative. For example, separation of ammonia (boiling point, −33.4°C) from water (boiling point, 100°C) is very easily accomplished by distillation or by stripping. On the other hand, the separation of propane (boiling point, −42.2°C) from propylene (boiling point, −47°C) is more difficult. However, the separation may be accomplished by distillation, stripping, or liquid-liquid extraction. Economic analyses for both these systems have shown distillation to be preferable. Distillation also has the advantage of not adding a third component to the system. In absorption, adsorption, and extraction it is necessary to add a third component (such as an absorbent oil, an adsorbent charcoal, or an organic solvent) which must later be removed to give pure products.

Other unit operations can be used for difficult separations. For example, *m*-xylene and *p*-xylene have boiling points of 138.8°C and 138.5°C respectively, thereby making separation by distillation very difficult. However their melting points (*m*-xylene, −53.6°C; *p*-xylene, 13.2°C) are sufficiently different to permit separation by fractional crystallization.

An understanding of the various phase-equilibrium relationships is necessary in the selection of the mass-transfer operation most appropriate for a given separation. The variation of phase equilibria with temperature and pressure is often an important consideration. Furthermore, the number of stages of equipment required to produce a desired change in phase composition depends upon the distribution of components between phases at equilibrium. Such equilibrium data are determined experimentally. In some cases, the equilibrium relationship can be predicted from theory or from empirical relationships. For example, Raoult's law can be used to describe certain vapor-liquid equilibria.

The Phase Rule. Before the phase relationships for specific systems are discussed, the general problem of phase equilibria will be considered. The *phase rule*, first stated by J. Willard Gibbs, is a useful tool in the consideration of phase equilibria. The phase rule is derived by considering the number of variables in a system together with the number of equations relating them. The rule may be stated as

$$V = C + 2 - P \qquad (3.1)$$

where $V =$ the number of intensive variables that can be varied independently

$C =$ the number of components in the system

$P =$ the number of phases in the system

A *phase* may be defined as a physically distinct and homogeneous portion of a system. A phase may be either a solid, a liquid, or a gas. Several solid and liquid phases may coexist; but, since gases are totally miscible with each other, there can be only one gas phase. An *intensive variable* is independent of the total quantity of the phase. For example, the temperature, pressure, and composition of a phase are intensive variables. On the other hand, the total volume of the phase depends upon the quantity and is, therefore, an *extensive variable*. Similarly, the over-all composition of a system of several phases depends upon the extent of each phase and is an extensive variable. For phase equilibria the number of *components* of a phase may be defined as the least number of chemical species necessary to prepare the phase.

The phase rule is useful in predicting the number of intensive variables that may be varied independently in any system. For example, consider liquid water. Since it is pure, $C = 1$. With the single liquid phase, $P = 1$, and, by Equation 3.1, $V = 2$. It is therefore possible to vary both the temperature and pressure of a single pure liquid phase. Consideration of an equilibrium mixture of water and steam shows that $C = 1$, $P = 2$, and $V = 1$. Thus, only one intensive property can be varied independently. If the temperature is specified, the pressure is automatically set.

If the phase rule is applied to a vapor-liquid equilibrium mixture of ethanol and water, it is found that $C = 2$, $P = 2$, and $V = 2$. Thus two variables may be set. The two may be any combination of the temperature, pressure, and phase concentration. For example, if the composition and pressure of a phase are specified, the temperature is set. Further application of the phase rule will be useful in considering specific systems discussed in the remainder of this chapter.

GAS-LIQUID EQUILIBRIA

Calculations for distillation and gas absorption require knowledge of gas-liquid equilibria. A simple expression of vapor-liquid equilibrium is Raoult's law.

$$p_a = P_a x_a \tag{3.2}$$

or

$$y_a = \frac{p_a}{P} = \frac{P_a}{P} x_a \tag{3.3}$$

where x_a = mole fraction of component a in liquid

y_a = mole fraction of component a in vapor

p_a = partial pressure of component a in vapor

P_a = vapor pressure of component a at the given temperature

P = total pressure

These equations indicate that the vapor evolved from a liquid mixture will be a mixture of the same components as the liquid. The vapor will normally be richer in the component having the higher vapor pressure at the temperature of the vaporization. A pure component will never be evolved from a liquid mixture, although in the limiting case the vapor pressure of one component may be so low as to make the component practically nonvolatile.

Raoult's law is accurate only in predicting vapor-liquid equilibria for an ideal solution in equilibrium with an ideal gas mixture. Solutions that show negligible deviation from ideality include those whose components have similar structure and physical properties, such as benzene–toluene, propane–butane, and methanol–ethanol. Raoult's law shows that the compositions in an equilibrium mixture depend upon the total pressure of the system and upon the vapor pressures of the components. The vapor pressures vary with temperature but not with composition or total pressure.

In certain systems, where Raoult's law does not apply, phase compositions can be predicted by

$$y_a = K_a x_a \tag{3.4}$$

where K_a is an experimentally determined constant. When K_a is independent of composition and dependent only upon temperature, Equation 3.4 expresses Henry's law.

In a binary (two-component) system the component with the higher vapor pressure at a given temperature is referred to as the "more volatile component," whereas that with the lower vapor pressure is called the "less volatile component." By convention, the composition of a binary mixture will be expressed as the concentration of the more volatile component.

For binary systems where a is the more volatile component and b is the less volatile component, assuming Raoult's law is valid,

$$\frac{y_a}{x_a} \frac{x_b}{y_b} = \frac{P_a}{P_b} \tag{3.5}$$

Since $y_b = 1 - y_a$ and $x_b = 1 - x_a$,

$$\left[\frac{y_a}{x_a}\right]\left[\frac{1 - x_a}{1 - y_a}\right] = \frac{P_a}{P_b} = \alpha_{ab} \tag{3.6}$$

Equation 3.6 defines the *relative volatility* (α_{ab}) of component a relative to b. For systems which do not follow Raoult's law, the relative volatility is defined as $\alpha_{ab} = K_a/K_b$. The relative volatility is constant when either Henry's or Raoult's law holds. In other cases it varies with composition. Figure 3.1 shows vapor-liquid composition data for two systems where the relative volatility is constant and also for two systems where the relative volatility varies with composition. The systems shown in Figure 3.1 are at constant pressure, but the temperature varies with composition. The variation of temperatures is shown in Figures 3.2 and 3.3.

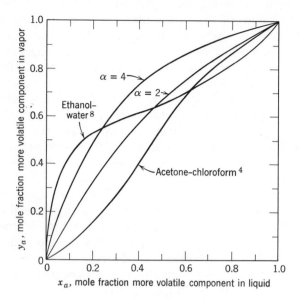

Figure 3.1. Typical vapor-liquid equilibrium curves at 1 atm total pressure.

Illustration 3.1. A mixture of butane and pentane is at equilibrium at 3 atm pressure and 100°F. Calculate the compositions of the liquid and vapor: (*a*) using Raoult's law, (*b*) using experimental values of K_a and Equation 3.4.

SOLUTION. (*a*) Vapor-pressure data are found in tables of physical data (Appendix D):

Vapor pressure of pentane at 100°F = 830 mm Hg
Vapor pressure of butane at 100°F = 2650 mm Hg
Total pressure: $P = 3 \times 760 = 2280$ mm Hg

For butane: $y_B = \dfrac{P_B}{P} x_B = \dfrac{2650}{2280} x_B = 1.16 x_B$

For pentane: $y_P = \dfrac{P_P}{P} x_P = \dfrac{830}{2280} x_P = 0.363 x_P$

Since only butane and pentane are present in the liquid and vapor, $x_B + x_P = 1$, and $y_B + y_P = 1$. There are now four equations and four unknowns. Solving,

$$y_B + y_P = 1 = 1.16 x_B + 0.363 x_P$$
$$1 = 1.16 x_B + 0.363(1 - x_B), \text{ etc.}$$
$$x_B = 0.80 \quad y_B = 0.93$$
$$x_P = 0.20 \quad y_P = 0.07$$

(*b*) Experimental values of K at 3 atm and 100°F are for butane $K_B = 1.15$ and for pentane $K_P = 0.36$ (Appendix D).

Therefore, $y_B = 1.15 x_B \quad y_P = 0.36 x_P$ and as before,

$$x_B = 0.81 \quad y_B = 0.93$$
$$x_P = 0.19 \quad y_P = 0.07$$

The small difference between the values in parts (*a*) and (*b*) is within the precision with which the charts for vapor pressure and K can be read. Therefore, mixtures of butane and pentane follow Raoult's law at 3 atm pressure. In general, deviations from Raoult's law are greater at higher pressures.

Illustration 3.2. What is the relative volatility of butane to pentane at 100°F and 465 psia?

SOLUTION. Assuming Raoult's law holds at elevated pressure, from Equation 3.5,

$$\alpha_{B-P} = \frac{2650}{830} = 3.2.$$

Using experimental values of K at 465 psia (Appendix D): $K_B = 0.24$, $K_P = 0.085$.

$$\alpha_{B-P} = \frac{0.24}{0.085} = 2.8$$

This indicates that at 465 psia butane–pentane mixtures deviate from Raoult's law.

Raoult's law can be used to calculate the dew point and bubble point of ideal mixtures. A mixture does not boil at a single temperature for a constant total pressure, in contrast to the behavior of a pure liquid. The temperature at which a liquid mixture of a given composition begins to vaporize as the temperature is increased is called the "bubble point." Conversely, the temperature at which a vapor mixture first begins to condense on cooling is called the "dew point." For a pure liquid the bubble point and dew point are identical and equal to the boiling point, since a pure component vaporizes or condenses at one temperature.

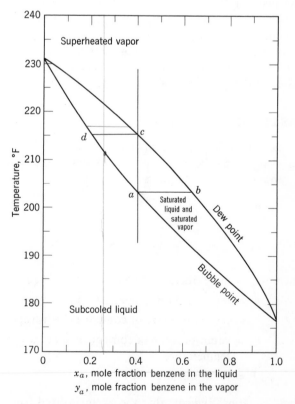

Figure 3.2. Temperature-composition diagram for liquid-vapor equilibrium of benzene and toluene at 1 atm.

Application of the phase rule to the vaporization of single-component liquids shows that there can be one independent variable. Thus, if the over-all pressure has been specified, the temperature of the vaporization is set. On the other hand, in the vaporization of a two-component liquid, there are two independent variables. In this case, after the pressure is specified either the temperature or the composition may be varied independently. Since the liquid composition will vary as vaporization proceeds, the temperature must then vary. A *boiling range* results, extending between the bubble point and the dew point.

Illustration 3.3. Calculate the dew point of a gaseous mixture containing 20 mole percent benzene, 30 mole percent toluene, and 50 mole percent *o*-xylene at 1 atm total pressure.

SOLUTION. At this moderate pressure for these similar compounds Raoult's law will be sufficiently accurate. The vapor pressures of the three components depend upon temperature, which is to be determined. A trial-and-error procedure is necessary, since the vapor pressures cannot be expressed as simple mathematical functions of temperature.

For benzene: $y_B = \dfrac{P_B}{760} x_B = 0.2;\quad x_B = \dfrac{152}{P_B}$

For toluene: $y_T = \dfrac{P_T}{760} x_T = 0.3;\quad x_T = \dfrac{228}{P_T}$

For xylene: $y_X = \dfrac{P_X}{760} x_X = 0.5;\quad x_X = \dfrac{380}{P_X}$

At the correct temperature $x_B + x_T + x_X = 1$;

$$\frac{152}{P_B} + \frac{228}{P_T} + \frac{380}{P_X} = 1$$

The left side of this equation must equal 1. If a temperature is assumed, vapor pressures at this temperature may be found from tables, and the left side of the equation may be calculated. If the left side does not equal 1, a new temperature is assumed.

Assume $T = 100°F$: $P_B = 165$ mm Hg; $P_T = 49$ mm Hg; $P_X = 15$ mm Hg.

$$\frac{152}{165} + \frac{228}{49} + \frac{380}{15} = 30.9 > 1$$

Therefore, the first trial of $T = 100°F$ is too low.

Assume $T = 270°F$: $P_B = 3000$ mm Hg; $P_T = 1550$ mm Hg; $P_X = 540$ mm Hg.

$$\frac{152}{3000} + \frac{228}{1550} + \frac{380}{540} = 0.90 < 1$$

Therefore, the temperature is less than $270°F$.

Assume $T = 263°F$: $P_B = 2600$ mm Hg; $P_T = 1180$ mm Hg; $P_X = 510$ mm Hg.

$$\frac{152}{2600} + \frac{228}{1180} + \frac{380}{510} = 0.997 \doteq 1.0$$

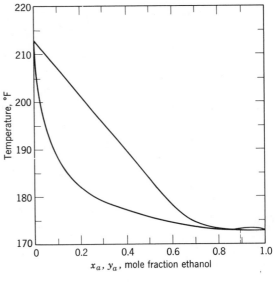

(a) Ethanol–water (8).

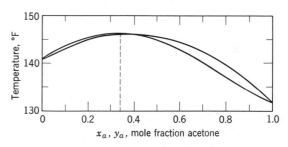

(b) Acetone–chloroform (4).

Figure 3.3. Temperature-composition diagram for mixtures forming an azeotrope at 1 atm.

The check at $263°F$ is sufficiently close to consider this the dew point of the mixture.

At the dew point, the first liquid formed has a composition such that the pressure exerted by each component of the liquid is equal to the partial pressure of the component in the vapor. The composition of the first equilibrium liquid formed as condensation begins at $263°F$ is

$$x_B = \frac{152}{2600} = 0.059 \doteq 0.06$$

$$x_T = \frac{228}{1180} = 0.193 \doteq 0.19$$

$$x_X = \frac{380}{510} = 0.745 \doteq \underline{0.75}$$
$$\underline{\overline{0.997}}$$

The dew point and bubble point of a binary mixture are functions of its composition, as shown in Figure 3.2 for the system benzene–toluene. The curves are calculated from vapor-pressure data, assuming Raoult's law is valid for this system. The dew point and bubble point become identical at $x_a, y_a = 0$ and $x_a, y_a = 1.0$, since these compositions represent pure toluene and pure

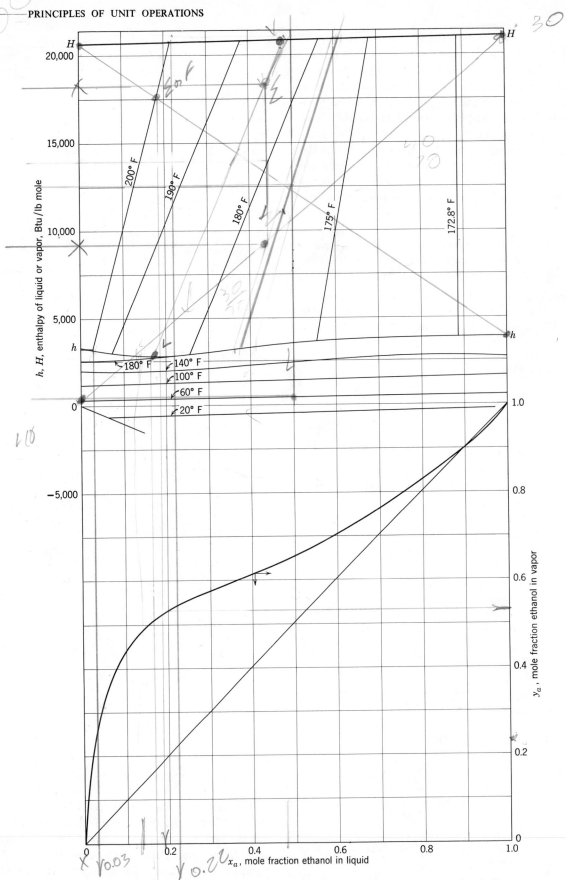

Figure 3.4. Enthalpy-composition diagram for ethanol–water mixtures at 1 atm pressure (2). Reference state: liquid water at 32°F, liquid ethanol at 32°F.

benzene respectively. The single values of temperatures at these compositions are the boiling points of the pure components.

If a liquid mixture containing 0.4 mole fraction benzene is heated (as represented by moving up the vertical line on Figure 3.2), it will start to vaporize at its bubble point (point a), which is 203.5°F. The first equilibrium vapor formed is also at 203.5°F, and its composition (b) is 0.625 mole fraction benzene. As the temperature is raised (from a to c) more liquid evaporates until the last liquid vaporizes at 215°F (d). During the vaporization process the liquid composition varies from 0.4 to 0.215 and the vapor from 0.625 to 0.4 mole fraction benzene, but the over-all composition of the liquid-vapor mixture remains constant at 0.4 mole fraction benzene.

The region below the bubble-point curve represents the liquid phase. A liquid at a temperature below its bubble point may be referred to as a *subcooled liquid*, whereas one at its bubble point is called a *saturated liquid*. Similarly, a vapor at its dew point is called a *saturated vapor*. Between the bubble-point and dew-point curves is a region of two phases where both a saturated liquid and a saturated vapor coexist. The relative quantities of liquid and vapor in the two-phase region can be calculated by a material balance (see Chapter 4). Above the dew-point curve lies the region of *superheated vapors*.

Nonideal solutions have temperature-composition curves substantially different from those of Figure 3.2. For example, Figure 3.3 presents two nonideal systems, ethanol–water, and acetone–chloroform. Each system forms an azeotrope. An *azeotrope* is a mixture which has an equilibrium vapor of the same composition as the liquid. Such a characteristic is undesirable if a separation of the mixture into its components by distillation is necessary, since no composition richer than the azeotrope can be attained. At the azeotropic composition the dew point and bubble point are equal, and the mixture vaporizes at a single temperature. This conclusion can be reached by applying the phase rule. In a binary liquid-vapor system there can be two independent variables. Usually the pressure is specified and so there remains one variable. The definition of the azeotrope sets the compositions of the two phases as equal to each other, and this restriction eliminates the final independent variable. Therefore, the temperature is set at an unique value. For example, in the ethanol–water system the equilibrium liquid and vapor compositions become equal at 0.894 mole fraction ethanol. Since the dew point and bubble point are also equal at this composition, the mixture will vaporize at a single temperature, as does a pure liquid. For this reason, azeotropes are often called "constant-boiling mixtures." The azeotrope boils at a temperature lower than the boiling point of either pure ethanol or pure water. Systems of this type are referred

to as "minimum-boiling mixtures." On the other hand, the system acetone–chloroform has a maximum-boiling azeotrope at 0.34 mole fraction acetone (Figure 3.3b).

Certain distillation calculations require knowledge of the variation of the enthalpies of the liquid and vapor with composition. Enthalpy-concentration data are given for ethanol–water in Figure 3.4 and for ammonia–water in Figure 3.5. The molar enthalpy is plotted on the vertical axis, and the mole fraction of the more volatile component on the horizontal axis. Any point on the diagram defines both the enthalpy and the composition of a binary mixture.

The enthalpy of a binary mixture is relative to some arbitrary reference condition chosen for the pure components. For example, in Figure 3.4 the enthalpies of pure liquid ethanol at 32°F and pure liquid water at 32°F were chosen equal to zero. Mixtures of ethanol and water may have enthalpies different from zero at 32°F because of enthalpy changes on solution and dilution. The changes are small for the ethanol–water system. Liquid mixtures at temperatures other than 32°F will have enthalpies different from zero because of the enthalpy change on heating, as determined by the heat capacity and the temperature change. If a mixture is vaporized, its enthalpy will increase by an amount equal to the latent heat of vaporization. All enthalpy changes are considered in constructing an enthalpy-composition diagram for a binary system from experimental data on heat capacities, heats of solution, and heats of vaporization.

On Figure 3.4, the curve labeled *H* represents vapors at the dew point (saturated vapors), and the curve labeled *h* represents liquids at the bubble point (saturated liquids). The vertical distance between the curves is therefore the latent heat of vaporization of a mixture. Between the two curves is a liquid-vapor two-phase region. To determine points on the saturated vapor and liquid curves which are in equilibrium with each other, use is made of the equilibrium curve on the lower part of the diagram. A value of the vapor composition, y_a, may be chosen on the vapor enthalpy curve (*H*). From the equilibrium curve, the value of x_a in equilibrium with the chosen y_a may be determined. The value of x_a may then be plotted on the liquid enthalpy curve (*h*). A *tie line* then can be drawn connecting the two points which are in equilibrium. The temperature-labeled lines in the two-phase region are tie lines.

If the equilibrium temperature is given, the vapor and liquid compositions are obtained from a temperature-composition plot, such as Figure 3.3a for the ethanol–water system. The region below the saturated-liquid curve represents subcooled liquids. Lines of constant temperature (isotherms) are given in this region. In the ethanol–water system the heats of solution and dilution are small, so that the isotherms are nearly horizontal

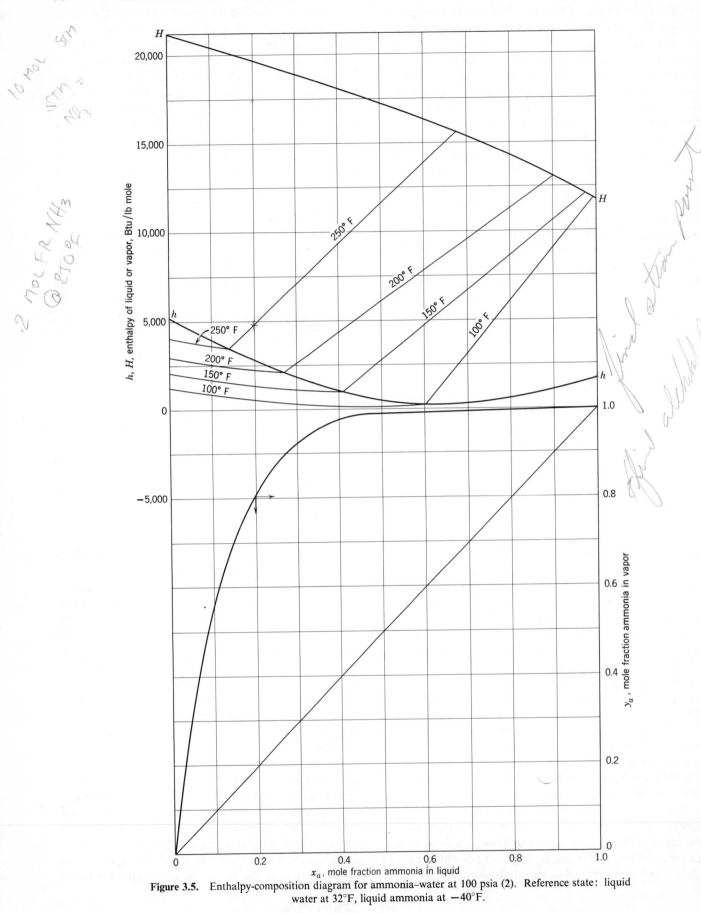

Figure 3.5. Enthalpy-composition diagram for ammonia–water at 100 psia (2). Reference state: liquid water at 32°F, liquid ammonia at −40°F.

straight lines, indicating constant enthalpy. Above the saturated-vapor curve is the region of superheated vapors.

Illustration 3.4. What is the composition of a vapor in equilibrium with a saturated liquid of 0.5 mole fraction ethanol, 0.5 mole fraction water?

SOLUTION. On Figure 3.6, a vertical line is drawn downward from the liquid enthalpy curve at $x_a = 0.5$. The equilibrium value of y_a is determined by the intersection of the vertical with the equilibrium curve and may be read from the right-hand vertical axis as $y_a = 0.66$. To transpose this value to the horizontal axis, a horizontal is drawn from the equilibrium curve to the diagonal, and then a vertical is drawn to the saturated-vapor enthalpy curve. The value on the H curve may be connected with the original value on the h curve by a tie line. The temperature of the equilibrium mixture can be determined by reference to Figure 3.3a. It is approximately 176°F.

Illustration 3.5. Determine the compositions of the phases present under the following conditions:
   (*a*) Ethanol–water at 177°F and 1 atm.
   (*b*) Ammonia–water at 1 atm, $x_a = 0.3$.
   (*c*) Ethanol–water at 100°F and 1 atm.
   (*d*) Ammonia–water at 100 psia, $H = 10,000$, over-all mole fraction $NH_3 = 0.5$.
   (*e*) Ammonia–water at 100 psia, $H = 20,000$, over-all mole fraction ammonia $= 0.4$.

SOLUTIONS. (*a*) The compositions are most easily determined from Figure 3.3a. A horizontal at 177°F intersects the bubble-point and dew-point lines at the equilibrium compositions, $x_a = 0.42$, $y_a = 0.63$.
   (*b*) From the equilibrium curve of Figure 3.5, at $x_a = 0.3$, $y_a = 0.93$.
   (*c*) Since the composition is not given, the mixture cannot be located exactly on Figure 3.4. However, at 100°F *all* mixtures of ethanol and water are subcooled liquids, as indicated by the 100°F isotherm. Therefore, only a liquid phase will be present at 100°F and 1 atm, and there is no equilibrium vapor.
   (*d*) The point ($H = 10,000$ Btu/lb mole, $z_a = 0.5$) is located on Figure 3.5. Since the point lies in the two-phase region, there will be a liquid and vapor in equilibrium. It is necessary to determine a tie line through the point. A first approximation is made by drawing a tie line through the point with a slope intermediate between those of the 250 and 200°F isotherms. The values of x_a and y_a obtained by the intersections of the tie line with the enthalpy curves are checked on the equilibrium diagram in the lower part of Figure 3.5. The correct values are $x_a = 0.17$, $y_a = 0.74$.
   (*e*) The point ($H = 20,000$ Btu/lb mole, $z_a = 0.4$) is located on Figure 3.5. Since the point is in the superheated-vapor region, there will be only a vapor phase of composition $y_a = 0.4$.

Gas-absorption calculations often require data on equilibrium between the gas dissolved in a liquid phase and the gas phase. The equilibria often can be expressed

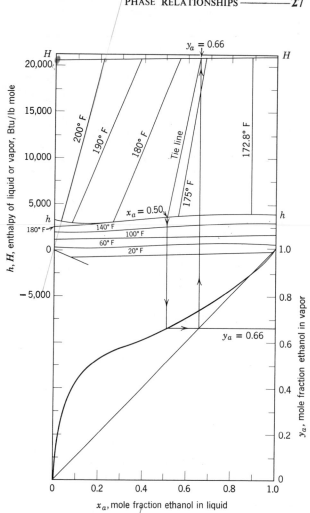

Figure 3.6. Solution to Illustration 3.4.

by Henry's law or Raoult's law for the solute. Where the equations are inaccurate, data may be tabulated or plotted on an equilibrium diagram. For example, in the absorption of ammonia from an air–ammonia mixture into water, the slight solubility of air may be neglected, and the concentration of ammonia in the liquid may be tabulated as a function of the partial pressure of ammonia in the gas (Appendix D).

LIQUID-LIQUID EQUILIBRIA

In liquid extraction one component of a solution is transferred to another liquid phase which is relatively insoluble in the first solution. In the most simple case the solute is partitioned between two insoluble liquid phases. Equilibrium data for this case can be recorded as weight ratios of solute to solvent in each of the phases at equilibrium. An example is the partition of uranyl nitrate between nitric acid solution and an organic solvent, as shown in Figure 3.7.

In many cases the two solvents are partially soluble in each other, and in addition the concentration of the solute

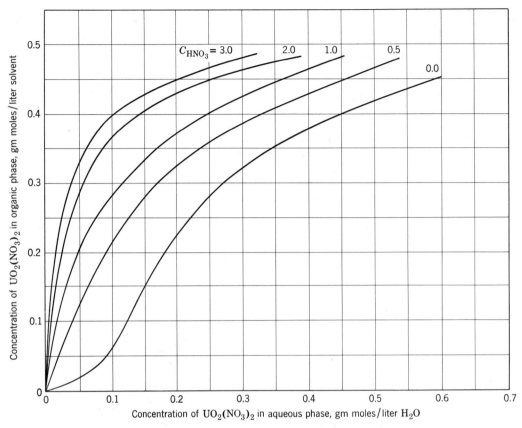

Figure 3.7. Distribution of uranyl nitrate between aqueous nitric acid solutions and an organic solvent at 25°C (3). Organic solvent: 30 volume percent tributyl phosphate in kerosene c_{HNO_3} = gm moles nitric acid per liter of water.

may influence the mutual solubility of the solvents. In such cases equilibrium data must be plotted on a three-component diagram. Application of the phase rule to two-phase, three-component systems shows that there can be three independent variables. In this case there are four possible variables: temperature, pressure, and the concentrations of two components. The concentration of the third component is not independent since it is set by stipulating the other two. However, the concentrations of the two components are not entirely independent because it was stipulated that the two phases are in equilibrium. Therefore, if temperature, pressure, and one concentration are specified, the other two concentrations are set.

Data for ternary systems are often reported on equilateral-triangular diagrams; however, for engineering calculations a right triangle is more convenient for reporting the data, since ordinary rectangular-coordinate graph paper may be used in constructing the diagram. If the calculation involves only relatively low solute concentrations, only that part of the diagram should be plotted, with an expanded horizontal scale for greater graphical accuracy.

The ternary system isopropyl ether–acetic acid–water

(Figure 3.8) is representative of systems in which one pair of liquids is partially miscible and the other two pairs are completely miscible. In Figure 3.8 each vertex of the triangle represents a pure component: the right angle, pure water; the upper vertex, pure isopropyl ether; and the right-hand vertex, pure acetic acid. In general, the right angle represents a pure phase consisting of the major component of the raffinate which is not extracted (b)*; the upper vertex, pure solvent used in extraction (c); and the right-hand vertex, pure solute (a) to be extracted from the raffinate by the solvent. The horizontal boundary of the triangle represents mixtures of solute and unextracted raffinate component, with no solvent present. The vertical boundary represents binary mixtures of the unextracted raffinate component and solvent. Since $x_a + x_b + x_c = 1$, it is necessary to plot the concentrations of only two components. The third can always be found by difference. On the right-triangular diagram, a mixture is most easily plotted by the mass-fraction solute (x_a or y_a) and the mass-fraction solvent (x_c or y_c). The mass fraction of unextracted raffinate component (x_b or y_b) need not be plotted. The diagonal side of the triangle

─────────────────────

* Hereafter called the *unextracted raffinate component.*

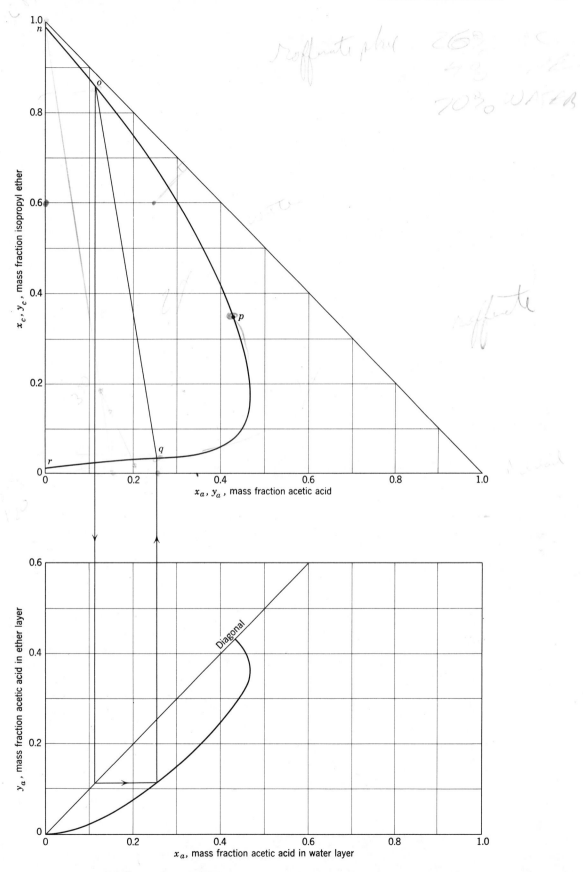

Figure 3.8. Isopropyl ether–acetic acid–water system at 20°C. (7)

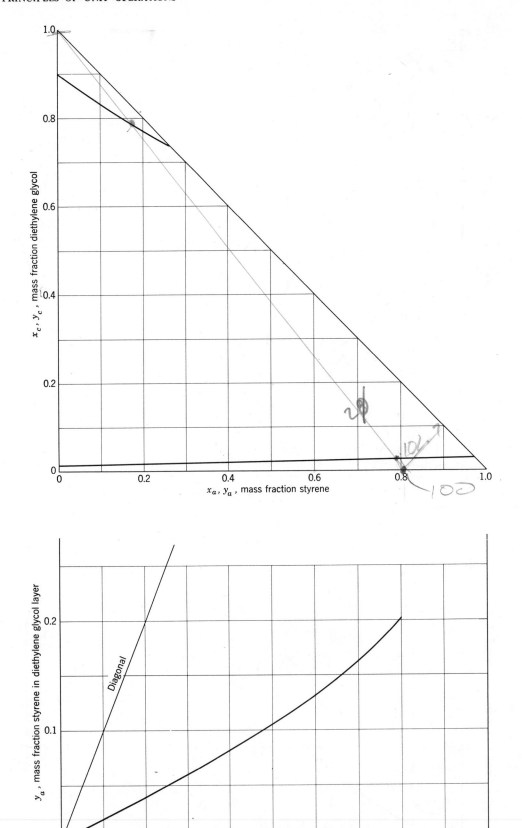

Figure 3.9. Diethylene glycol–styrene–ethylbenzene system at 25°C, 1 atm. (1)

represents mixtures where $x_b = 0$ or $y_b = 0$. The coordinate axis for component b actually is a perpendicular to the diagonal side of the triangle from the right angle. Since the third coordinate axis is not used, it is omitted.

Any point within the triangle represents a three-component mixture. Certain over-all compositions exist as one liquid phase, whereas others may split into two liquid phases. The curve *nopqr* (Figure 3.8) separates the two-phase region (to the left) from the one-phase region (to the right) and is called the *phase envelope*. Compositions of phases in equilibrium must lie on the phase envelope. A mixture on the lower part of the envelope (such as *q*) contains a high concentration of the unextracted component and is called the *raffinate phase*. A mixture on the upper part of the envelope (such as *o*) is rich in the solvent and is referred to as the *extract*. Compositions of phases which are in equilibrium may be connected by a tie line, such as *o–q*. Alternatively, equilibrium compositions may be determined by use of the equilibrium curve on the lower part of the diagram. At the *plait point* (*p*) the tie line has been reduced to a point where the extract and raffinate are identical. Note that the equilibrium curve ends at the plait point at $x_a = y_a$.

The system diethylene glycol–styrene–ethylbenzene is representative of systems in which two of the three pairs of liquids are partially soluble (Figure 3.9). In this system the extract and raffinate phases are represented by independent curves, and there is no plait point. The line for the extract phase on the diagram is inconveniently short for stage calculations. This situation can be remedied by replotting the data on a *solvent-free* basis (Figure 3.10). On the horizontal axis is plotted

$$X_a, Y_a, \frac{\text{mass of solute}}{\text{mass of solute} + \text{unextracted raffinate component}}$$

On the vertical axis is plotted

$$X_c, Y_c, \frac{\text{mass of solvent}}{\text{mass of solute} + \text{unextracted raffinate component}}$$

With these definitions, $X_a + X_b = 1$ and X_c may have any value. Similarly, $Y_a + Y_b = 1$, and Y_c may have any value. For example, for pure solvent Y_c is infinite.

Illustration 3.6. Calculate the data of Figure 3.10 from that given in Figure 3.9.

SOLUTION. On the extract curve of Figure 3.9, at $y_a = 0.10$, $y_c = 0.83$. Therefore, $y_b = 0.07$. Basis: 1 lb of this extract.

$$Y_a = \frac{\text{mass solute}}{\text{mass solute} + \text{unextracted component}}$$

$$= \frac{0.10}{0.10 + 0.07} = 0.59$$

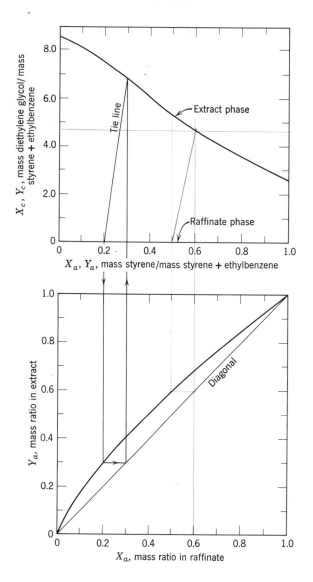

Figure 3.10. Diethylene glycol–styrene–ethylbenzene system at 25°C on solvent-free coordinates.

$$Y_c = \frac{\text{mass solvent}}{\text{mass solute} + \text{unextracted component}}$$

$$= \frac{0.83}{0.10 + 0.07} = 4.88$$

The point ($Y_a = 0.59$, $Y_c = 4.88$) may then be plotted as a point on the extract curve of Figure 3.10. Additional points may be calculated in this manner, and the extract curve may then be plotted. On the raffinate curve of Figure 3.9 at $x_a = 0.8$, $x_c = 0.022$ and $x_b = 0.178$. Therefore,

$$X_a = \frac{0.8}{0.8 + 0.178} = 0.817; \quad X_c = \frac{0.022}{0.8 + 0.178} = 0.0225$$

This point may be plotted on the raffinate curve of Figure 3.10. Additional points on the raffinate curve may be plotted in the same way. Since the solvent concentration of the raffinate is low, the curve nearly coincides with $X_c = 0$ for this system.

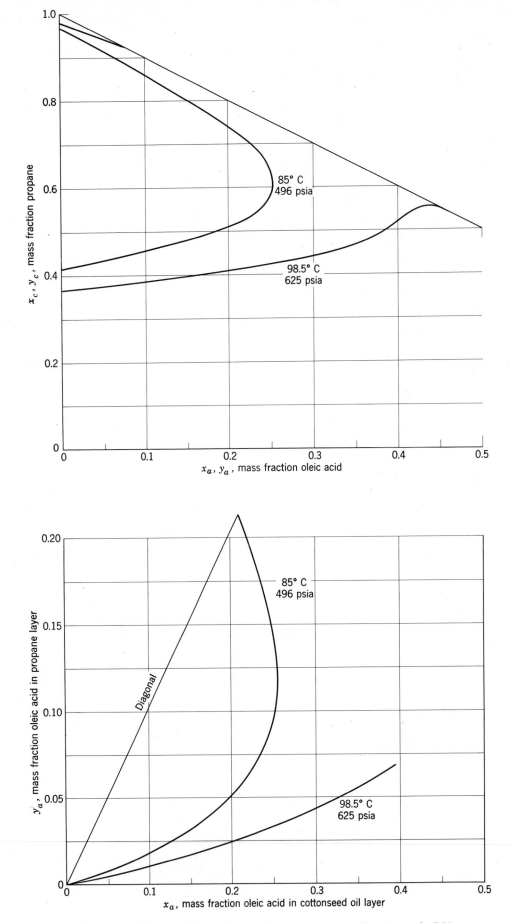

Figure 3.11. Ternary equilibrium diagram for propane–oleic acid–cottonseed oil (5).

32

The propane–oleic acid–cottonseed oil system (Figure 3.11) demonstrates how the solubility characteristics of a system may vary with temperature. In this case an increase in temperature decreases mutual solubilities, contrary to the more common case where an increase in temperature causes an increase in solubility. Systems which form two phases at lower temperatures may be completely miscible at a higher temperature. Figure 3.11 also demonstrates the use of a partial triangle and a change of scale which are often desirable for accuracy in graphical calculations. Since none of the data for the system are beyond $x_a = 0.5$, the diagram beyond this point is eliminated and the remaining diagram is increased in scale.

SOLID-FLUID EQUILIBRIA

Adsorption has been observed to occur by both physical and chemical mechanisms. Physical adsorption occurs when the intermolecular forces of attraction between the fluid molecules and the solid surface are greater than the attractive forces between molecules of the fluid itself. The molecules of fluid adhere to the surface of the solid adsorbent, and an equilibrium is established between the adsorbed fluid and that remaining in the fluid phase.

Experimental physical adsorption isotherms for a number of pure hydrocarbon vapors on silica gel are given in Figure 3.12. Adsorption of a mixture of gases

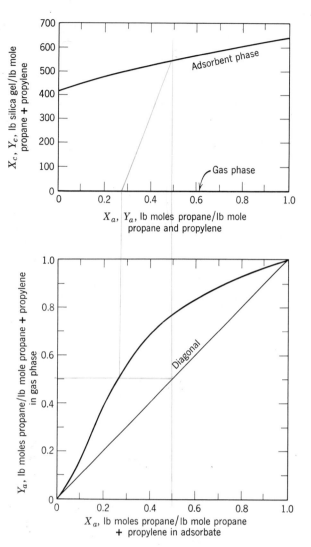

Figure 3.13. Simultaneous adsorption of propane and propylene on silica gel at 25°C (6).

results in an adsorbate of composition different from that in the gas. Therefore, it is possible to separate gaseous mixtures by selective adsorption. The equilibrium compositions of gas phase and adsorbate for propane-propylene mixtures are given in Figure 3.13. This equilibrium diagram closely resembles those for distillation (Figure 3.4) and liquid extraction (Figure 3.10). An analogy can be drawn among the adsorbent in adsorption, the enthalpy in distillation, and the solvent in extraction. However, it should be noted that the gas phase (V) occurs at $Y_c = 0$, and the solid phase (L) has values of X_c greater than zero, in contrast to an analogous diagram for liquid extraction, such as Figure 3.9.

Data for physical adsorption can often be expressed by an empirical equation. Common equations for adsorption isotherms include

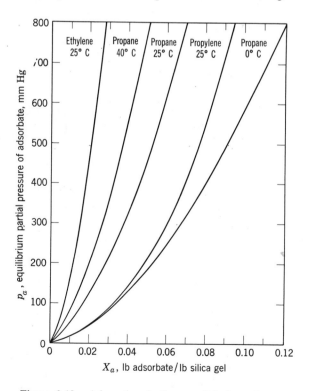

Figure 3.12. Adsorption isotherms of hydrocarbons on silica gel (6).

Freundlich: $\qquad X = k_1 p^{1/n}$ \hfill (3.7)

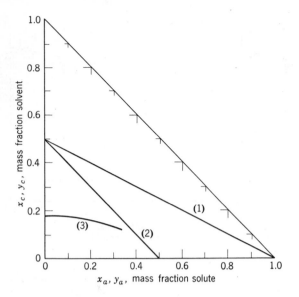

Figure 3.14. Underflow compositions for solid-liquid extraction.

Curve 1. Constant ratio of $\dfrac{\text{lb adhering solvent}}{\text{lb inerts}} = 1.0$

Curve 2. Constant ratio of $\dfrac{\text{lb adhering solution}}{\text{lb inerts}} = 1.0$

Curve 3. Experimentally determined curve for halibut livers and ether(10)

Langmuir: $X = \dfrac{k_2 p}{1 + k_3 p}$ (3.8)

$= k_2 p$ for small values of p

where $X = \dfrac{\text{weight of adsorbate}}{\text{unit weight of adsorbent}}$

p = partial pressure of adsorbed gas in the gas phase

n, k_1, k_2, k_3 = empirical constants

Chemical adsorption, or chemisorption, involves chemical interaction between the adsorbed fluid and the adsorbent solid. In many cases the adsorption is irreversible, and it is difficult to separate the adsorbate from the adsorbent. For this reason only physical adsorption lends itself to continuous stagewise operation.

Adsorption is employed to process both liquids and gases industrially. Common adsorbents include silica gel, charcoal, alumina, and various natural clays.

Ion exchange is an important operation which resembles chemical adsorption. The solid ion-exchange resin is manufactured to include an ion which will be replaced by a particular component ion of the fluid phase. A common example of ion exchange is the "zeolite" water softener, where sodium ions in the resin are exchanged for calcium ions in the water. Since industrial ion-exchange operations are batch or continuous-contact rather than stagewise, they will be considered in a later chapter.

In solid-liquid extraction a liquid solvent is used to dissolve a soluble solid from an insoluble solid. Some of the resultant solution adheres to the insoluble residue. An "equilibrium stage" can be defined as one where the liquid adhering to the solids leaving has the same composition as the liquid-extract phase leaving. It is then necessary to determine experimentally the mass of adhering liquid per unit mass of insoluble solids. Such data can be plotted on a ternary diagram, as shown in Figure 3.14. The resulting curves are referred to as "underflow loci" and do not represent true equilibrium conditions. Included are curves for constant ratios of solvent to inerts and solution to inerts and an experimentally determined curve for halibut livers with ether as the solvent. The underflow may be considered a mixture of inert solids and solution. The right-angle vertex represents pure inert solids and the hypotenuse of the ternary diagram represents clear solutions. Any line joining a clear-solution composition on the hypotenuse with the right angle intersects the underflow locus at the underflow composition in equilibrium with the clear solution. Therefore, the portion of the line from the hypotenuse to the underflow locus is an equilibrium tie line. The underflow locus determines the relative quantity of solution which adheres to the inert solids. The data of Figure 3.13 are representative of underflow loci. When experimental data are available, they should be used. Where data are limited, curves such as 1 and 2 may be used as approximations.

REFERENCES

1. Boobar, M. G., et al., *Ind. Eng. Chem.*, **43**, 2922 (1951).
2. Brown, G. G., et al., *Unit Operations*, John Wiley and Sons, 1950.
3. Codding, J. W., W. O. Haas, and F. K. Hermann, *Ind. Eng. Chem.*, **50**, 145 (1958).
4. Fordyce, C. R., and D. R. Simonson, *Ind. Eng. Chem.*, **41**, 104 (1949).
5. Hixson, A. W., and J. B. Bocklemann, *Trans. Am. Inst. Chem. Eng.*, **38**, 891 (1942).
6. Lewis, W. K., et al., *J. Am. Chem. Soc.*, **72**, 1153 (1950).
7. Othmer, D. F., R. E. White, and E. Trueger, *Ind. Eng. Chem.*, **33**, 1240 (1941).
8. Otsuki, H., and F. C. Williams, *Chem. Eng. Progr. Symposium Ser. Vol.* 49, No. 6, 55 (1953).
9. Perry, J. H., ed., *Chemical Engineers' Handbook*, 3rd ed., McGraw-Hill Book Co., 1950.
10. Ravenscroft, E. A., *Ind. Eng. Chem.*, **28**, 851 (1936).

PROBLEMS

3.1. Calculate the dew point and bubble point of the following mixture at a total pressure of 2 atm:

Hexane 40 mole percent
Heptane 25 mole percent
Octane 35 mole percent

3.2. An equilibrium mixture of liquid and vapor has an over-all composition of 0.55 mole fraction benzene and 0.45 mole fraction toluene at 1 atm total pressure and 205°F.

(*a*) Calculate the compositions of liquid and vapor and check the values with Figure 3.2.

(*b*) What percentage of the vapor will condense if the total pressure is increased to 2 atm?

3.3. Determine the composition of the phases specified for each of the following vapor–liquid equilibria:

(*a*) The vapor phase in equilibrium with a liquid of 30 mole percent ethanol and 70 mole percent water at 1 atm.

(*b*) The vapor phase in equilibrium with a liquid of 40 mole percent *a* and 60 mole percent *b*, where the relativity volatility of *a* to *b* is 2.7.

(*c*) The liquid and vapor phases for a mixture of acetone and chloroform at a temperature of 140°F and at total pressure of 1 atm.

(*d*) The liquid and vapor phases for a mixture of ammonia and water with an over-all composition of 40 mole percent ammonia at 200°F and 100 psia.

3.4. A vapor containing 25 mole percent propane and 75 mole percent ethane is compressed at 30°F.

(*a*) What pressure must be applied to condense one-half of the vapor (on a molar basis)?

(*b*) What is the composition of the liquid formed in (*a*) if the final liquid and vapor are in equilibrium?

(*c*) What would occur if the vapor were compressed to 400 psia? Vapor pressures at 30°F: ethane, 350 psia; propane, 67 psia.

3.5. A liquid mixture consists of 25 mole percent benzene and 75 mole percent toluene. The mixture is heated in a closed container to a final temperature of 217°F and pressure of 1 atm. (*a*) Determine the composition of each phase in the container at the final conditions. (*b*) What percentage of the liquid was vaporized?

3.6. Calculate the relative volatility of ethanol to water for a mixture in which the liquid phase contains 0.1 mole fraction ethanol. Repeat the calculation for liquids of 0.2, 0.4, 0.6, 0.7, 0.8, 0.894, and 0.95 mole fraction ethanol.

3.7. 1 lb mole of a mixture of 40 mole percent ethanol, 60 mole percent water is at 20°C and 1 atm.

(*a*) What is its enthalpy and physical state?

(*b*) How much heat must be added to it to completely vaporize it to a saturated vapor?

(*c*) What will be the temperature of the saturated vapor of part (*b*)?

3.8. 3 lb moles of an ammonia–water vapor exist at 250°F and 100 psia. As the vapor is cooled, droplets of condensate form immediately.

(*a*) What is the composition of the vapor?

(*b*) How much heat must be removed to cool the mixture to 200°F?

(*c*) What will be the composition of the phases present at 200°F?

(*d*) Estimate the temperature to which the vapor must be cooled to assure complete condensation.

3.9. Which of the following mixtures will form two phases, and what are the compositions of the phases?

(*a*) 25 mass percent acetic acid, 60 mass percent isopropyl ether, and 15 mass percent water at 20°C.

(*b*) 25 mass percent acetic acid, 70 mass percent isopropyl ether, and 5 mass percent water.

(*c*) 30 mass percent styrene and 70 mass percent ethylbenzene at 25°C.

(*d*) 20 mass percent styrene and 80 mass percent ethylbenzene.

(*e*) 30 mass percent styrene and 70 mass percent diethylene glycol.

(*f*) 30 mass percent diethylene glycol and 70 mass percent ethylbenzene.

(*g*) 45 mass percent diethylene glycol, 20 mass percent styrene, and 35 mass percent ethylbenzene.

(*h*) 30 mass percent oleic acid, 50 mass percent propane, and 20 mass percent cottonseed oil at 85°C and 496 psia.

(*i*) Same composition as in part (*h*) but 98.5°C and 625 psia.

3.10. Determine the composition of the phases specified for each of the following phase equilibria:

(*a*) The organic solvent phase in equilibrium with an aqueous solution of uranyl nitrate of concentration 0.4 gm mole $UO_2(NO_3)_2$ per liter H_2O at 25°C.

(*b*) The two liquid phases in a mixture with an over-all composition of 0.25 mass fraction acetic acid, 0.30 mass fraction isopropyl ether, 0.45 mass fraction water at 20°C.

(*c*) The diethylene glycol-rich liquid phase in equilibrium with an ethylbenzene-rich liquid phase containing 0.6 lb styrene per pound styrene + ethylbenzene at 25°C.

(*d*) The silica-gel phase in equilibrium with a gas containing 0.6 mole fraction propane, 0.4 mole fraction propylene.

(*e*) An underflow sludge in equilibrium with a solution containing 20 mass percent solute and 80 mass percent solvent. The sludge retains 2 lb of solution per pound of insoluble material.

3.11. Construct a temperature-composition diagram similar to Figure 3.2 for the system hexane-octane. Using the diagram, solve the following problem:

A vapor containing 43 per cent hexane and 57 per cent octane is cooled. At what temperature does condensation begin? What is the composition of the first liquid formed? At what temperature does the last vapor condense? What was the composition of the last vapor?

3.12. Convert the propane–oleic acid–cottonseed oil ternary diagram at 98.5°C, 625 psia (Figure 3.11) to solvent-free coordinates. On the diagram determine the composition of the two phases present in a mixture whose overall composition is 10 mass per cent oleic acid, 60 mass per cent propane, and 30 mass per cent cottonseed oil.

3.13. Plot the following underflow loci by calculating five points on each line:

(*a*) Constant ratio of pounds of adhering solvent to pounds of inerts = 0.68.

(*b*) Constant ratio of pounds of adhering solution to pounds of inerts = 0.68.

The aqueous solution contains HNO_3 @ conc of 61.9 m/liter

chapter 4

Equilibrium Stage Calculations

The calculation of the effluent streams from a single equilibrium stage can be accomplished numerically or graphically. In many cases, particularly in multistage operations, graphical calculations are more convenient and timesaving. The mathematical manipulations employed are the same in either case. The following discussion applies generally to calculations on ternary, enthalpy-concentration, and solvent-free diagrams, although the notation is for the ternary diagram.

SINGLE EQUILIBRIUM STAGE

Usually the quantity and composition of the two streams entering an equilibrium stage are known (Figure 4.1). If the two phases are thoroughly mixed together,

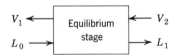

Figure 4.1. Single equilibrium stage.

a mixture Σ results. Material balances for this operation give

Total material balance:	$L_0 + V_2 = \Sigma$	(4.1)
Component a balance:	$L_0 x_{a0} + V_2 y_{a2} = \Sigma z_a$	(4.2)
Component b balance:	$L_0 x_{b0} + V_2 y_{b2} = \Sigma z_b$	(4.3)

where L_0 and V_2 are the total mass, or moles, of the streams; x_0 and y_2 are the respective mass, or mole, fraction; and Σ and z are the total mass and the mass or mole fraction, respectively, of the mixture. An equation for component c may also be written, but it would not be an independent equation since $x_c = 1 - x_a - x_b$. The quantity (Σ) and composition (z) of the sum can be determined by the material balances. If the sum lies in

the two-phase region of the system, it will split into two phases (L_1 and V_1). The composition of the phases is determined from the equilibrium relationship, and the quantity of each phase may be determined by

$$\Sigma = L_1 + V_1 \qquad (4.4)$$
$$\Sigma z_a = L_1 x_{a1} + V_1 y_{a1} \qquad (4.5)$$
$$\Sigma z_b = L_1 x_{b1} + V_1 y_{b1} \qquad (4.6)$$

The concentrations of the phases leaving an equilibrium stage depend not only on the equilibrium relationships of the system but also on the relative quantities of the two phases entering the stage. That is, the L_0/V_2 ratio will determine the location of the sigma point, and the tie line through the sigma point determines the exit concentrations.

Illustration 4.1. What is the composition of the phases leaving an equilibrium stage where the following two phases are fed to the stage?

L-phase: 100 lb acetic acid, 20 lb isopropyl ether, 80 lb water

V-phase: 15 lb acetic acid, 82 lb isopropyl ether, 3 lb water

SOLUTION. From the data given

$$L_0 = 200, \quad x_{a0} = 0.5, \quad x_{c0} = 0.1, \quad x_{b0} = 0.4$$
$$V_2 = 100, \quad y_{a2} = 0.15, \quad y_{c2} = 0.82, \quad y_{b2} = 0.03$$

By Equation 4.1, $\quad \Sigma = L_0 + V_2 = 200 + 100 = 300$

and by Equation 4.2, $z_a = \dfrac{(200)(0.5) + 100(0.15)}{300} = 0.383$

Similarly $\quad z_c = \dfrac{(200)(0.1) + 100(0.82)}{300} = 0.34$

and $z_b = 1 - 0.383 - 0.34 = 0.277$. This composition can be located in the two-phase region of the ternary diagram

(Figure 4.2). The mixture will split into two phases, as indicated by the tie line through z. The compositions of the phases leaving the stage may be read from the diagram $x_{a1} = 0.45$, $x_{b1} = 0.45$ $x_{c1} = 0.10$; $y_{a1} = 0.32$, $y_{b1} = 0.11$, $y_{c1} = 0.57$.

The quantities of L_1 and V_1 are determined by Equations 4.4 and 4.5.

$$300 = L_1 + V_1$$

$$(300)(0.383) = L_1(0.45) + V_1(0.32)$$

Solving, $L_1 = 146$ lb $\quad V_1 = 154$ lb

Graphical calculations are often more rapid than numerical calculations. Probably the most important graphical technique is the determination of the composition of a mixture resulting from the addition of two mixtures. The derivation which follows will develop the equations which justify the graphical addition of mixtures using phase diagrams. It will be shown that the composition of the mixture resulting from the addition of two mixtures will lie on a straight line between the compositions of the original two mixtures. This is called the *graphical-addition rule*. Then a relationship determining the exact location of the resultant mixture on the straight line will be developed. The relationship is the *inverse lever-arm rule*. The derivations are based on the material-balance equations (Equations 4.1, 4.2, 4.3) and utilize simple geometrical relationships.

If Σ in Equation 4.2 is replaced by the expression of Equation 4.1,

$$L_0 x_{a0} + V_2 y_{a2} = (L_0 + V_2) z_a$$

Solving for z_a gives

$$z_a = \frac{L_0 x_{a0}}{L_0 + V_2} + \frac{V_2 y_{a2}}{L_0 + V_2} = \frac{\dfrac{L_0}{V_2}}{\dfrac{L_0}{V_2} + 1} x_{a0} + \frac{1}{\dfrac{L_0}{V_2} + 1} y_{a2}$$

and

$$z_a \left[\frac{L_0}{V_2} + 1 \right] = \frac{L_0}{V_2} x_{a0} + y_{a2}$$

Finally,

$$\frac{L_0}{V_2} = \frac{y_{a2} - z_a}{z_a - x_{a0}} \qquad (4.7)$$

Equation 4.7 relates the composition of the resultant mixture to the compositions and masses of the original two mixtures. The compositions are expressed in terms of the concentration of component a.

Similarly, combining Equations 4.1 and 4.3 gives for component b,

$$\frac{L_0}{V_2} = \left(\frac{y_{b2} - z_b}{z_b - x_{b0}} \right) \qquad (4.8)$$

Combining Equations 4.7 and 4.8 gives

$$\frac{y_{a2} - z_a}{z_a - x_{a0}} = \frac{y_{b2} - z_b}{z_b - x_{b0}} \qquad (4.9)$$

which will be used in the following geometrical argument.

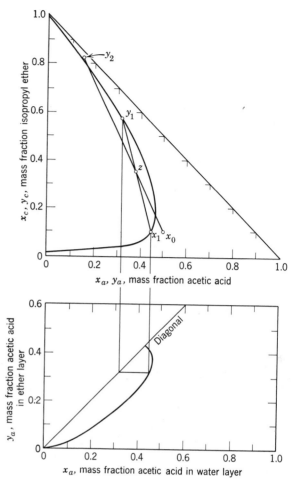

Figure 4.2. Solutions to Illustrations 4.1 and 4.2.

The compositions representing L_0, V_2, and Σ can be plotted on a ternary diagram, as shown in Figure 4.3. The composition z of Σ is shown in Figure 4.3 to be at some point *not* on a straight line between x and y, since first it must be shown that angle α is equal to angle β before plotting x, y, and z on a straight line is justified. It will be shown that, if $L_0 + V_2 = \Sigma$, the composition of Σ must lie on a straight line between the compositions of L_0 and V_2. This is proved by showing that angle α = angle β. The tangent of α is $(z_b - y_{b2})/(z_a - y_{a2})$, and the tangent of β is $(x_{b0} - z_b)/(x_{a0} - z_a)$. Therefore, from Equation 4.9, $\tan \alpha = \tan \beta$, and point (z_a, z_b) must lie on a straight line between points (x_a, x_b) and (y_a, y_b), as shown in Figure 4.4. The exact location of (z_a, z_b) is determined by the relative quantities of the L and V phases. Equations 4.7 and 4.8 each define the location. They are both expressions of the *inverse lever-arm rule*. The rule can also be stated as follows: If a mixture of L_0 is added to another mixture V_2, the composition of the resultant mixture Σ will lie on a straight line between the compositions of L and V such that the ratio L_0/V_2 is equal to the distance from V_2 to Σ

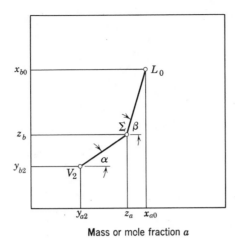

Figure 4.3. Proof of the addition rule.

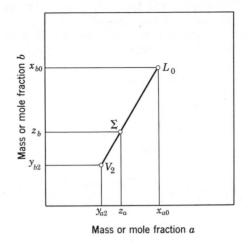

Figure 4.4. Graphical addition.

divided by the distance from L_0 to Σ. That is,

$$\frac{L_0}{V_2} = \frac{\overline{zy_2}}{\overline{x_0z}} = \frac{z - y_2}{x_0 - z} \qquad (4.10)$$

The distances between L_0, V_2, and Σ may be measured along the line connecting them, but it is more convenient to use one of the coordinate scales as indicated by Equation 4.10. The inverse feature of the lever-arm rule is indicated in its definition. For example, a very large quantity of phase L_0 added to a small quantity of phase V_2 would produce a mixture Σ of a composition nearly the same as L_0. Therefore, the composition z of Σ would lie very close to the composition x of L_0 but on the straight line between x and y. Alternate forms of the lever-arm rule can be derived.

$$\frac{L_0}{\Sigma} = \frac{z - y_2}{x_0 - y_2} \qquad (4.11)$$

and

$$\frac{V_2}{\Sigma} = \frac{x_0 - z}{x_0 - y_2} \qquad (4.12)$$

These equations are particularly useful in locating the point z. Equations 4.10, 4.11, and 4.12 are written without component subscripts. They apply to each of the three components.

Illustration 4.2. Calculate Illustration 4.1 using graphical methods.

SOLUTION. The compositions of L_0 and V_2 given in Illustration 4.1 are plotted on Figure 4.2. The composition z of the mixture Σ must lie on the straight line between x_0 and y_2. Its exact location is determined by Equation 4.10.

$$\frac{z - y_2}{x_0 - z} = \frac{200}{100} = 2$$

Therefore the distance from y_2 to z is twice the distance from x_0 to z, and the point z may be located, as shown. The same conclusion results from the application of Equation 4.11.

$$\frac{z - y_2}{x_0 - y_2} = \frac{2}{3}$$

Therefore, the point z lies two-thirds of the distance from y_2 to x_0, and from Figure 4.2, $z_a = 0.38$, $z_b = 0.34$, $z_c = 0.28$. These values check Illustration 4.1 within the accuracy of the graph.

To find the quantities of L_1 and V_1 the inverse lever-arm rule is applied to the tie line through z which connects x_1 and y_1 on the phase envelope. Equation 4.10 may be rewritten for the splitting of mixture Σ into two phases.

$$\frac{L_1}{V_1} = \frac{z - y_1}{x_1 - z}$$

The right-hand side of this equation can be evaluated by measuring the distances, or more easily by substituting the coordinates. For a,

$$\frac{L_1}{V_1} = \frac{0.38 - 0.32}{0.45 - 0.38} = 0.86$$

or for b,

$$\frac{L_1}{V_1} = \frac{0.34 - 0.57}{0.10 - 0.34} = 0.96$$

The error of graphical reading causes the answers to disagree. Probably the value for b is more accurate, since the differences are larger and less sensitive to a graphical error. Therefore

$$\frac{L_1}{V_1} = 0.96$$

$$L_1 + V_1 = \Sigma = 300$$

and

$$V_1 = \frac{300}{1.96} = 153 \text{ lb}$$

$$L_1 = 300 - 153 = 147 \text{ lb}$$

These results check with Illustration 4.1.

Subtraction of mixtures can also be carried out graphically. The difference (Δ) can be expressed by

$$\Delta = L - V \text{ for the total mass} \qquad (4.13)$$

$$\Delta x_\Delta = Lx - Vy \text{ for any of the components} \qquad (4.14)$$

If Equation 4.13 is rearranged, $\Delta + V = L$, it becomes clear that graphical subtraction is equivalent to graphical addition and that the inverse lever-arm rule applies. In this case the composition of the difference, x_Δ, lies on a straight line drawn through x and y but beyond x from y. An enthalpy balance similar to Equation 4.14 may also be written. Examination of Figure 4.5 will show the similarity between graphical addition and subtraction.

Illustration 4.3. A saturated liquid containing 50 mole percent ethanol and 50 mole percent water is heated in a closed tank. Liquid is vaporized until the temperature of the remaining liquid reaches 180°F at 1 atm pressure. (*a*) How much of the original liquid was vaporized? (*b*) How much heat per pound mole was added to the system to vaporize the liquid?

SOLUTION: (*a*) Since the liquid was vaporized in a closed tank, the vapor formed is still in contact with the liquid, and equilibrium conditions will exist. Therefore, the liquid is at its bubble point, which is 180°F. Its composition must lie on the saturated-liquid enthalpy curve of Figure 4.6 at 180°F. Thus, the remaining liquid contains 26.5 mole percent ethanol.

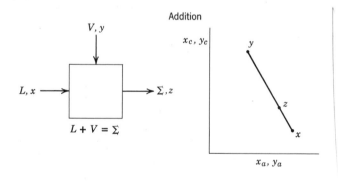

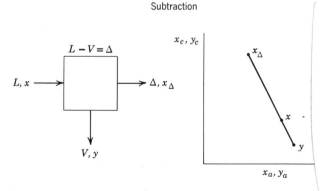

Figure 4.5. Comparison of graphical addition and subtraction.

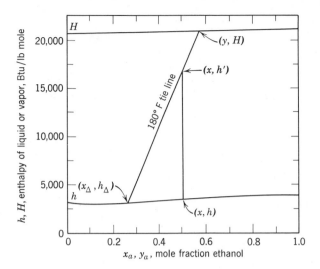

Figure 4.6. Solution to Illustration 4.3.

The vapor must lie on the saturated-vapor curve at 180°F, so that the mole fraction ethanol in the vapor is 0.565. To calculate the quantity vaporized, let the original liquid $= L$, the final liquid $= \Delta$, and the final vapor $= V$. That is, the vapor V is subtracted from the liquid L to give Δ, the remaining liquid.

$$\Delta = L - V$$

By the inverse lever-arm rule

$$\frac{V}{L} = \frac{\text{length of line } \overline{xx_\Delta}}{\text{length of line } \overline{yx_\Delta}} = 0.78$$

or

$$\frac{V}{L} = \frac{x - x_\Delta}{y - x_\Delta} = \frac{0.50 - 0.265}{0.565 - 0.265} = 0.78$$

Therefore, 78 mole percent of the original liquid was vaporized.

(*b*) Basis: 1 lb mole of original liquid

Initial enthalpy $= Lh$ (at $x = 0.5$)

Final enthalpy $= \Delta h_\Delta + VH$

Heat added to the system $= Q$

From Figure 4.6,

at $x = 0.5$, saturated liquid: $h = 3500$ Btu/lb mole;

at $x = 0.265$, saturated liquid: $h_\Delta = 3100$ Btu/lb mole;

at $y = 0.565$, saturated vapor, $H = 21,000$ Btu/lb mole

An enthalpy balance gives

$$Lh + Q = \Delta h_\Delta + VH \qquad (4.15)$$

Therefore

$$Q = (0.22)(3100) + (0.78)(20,900) - (1)(3500)$$

$$= 13,500 \text{ Btu/lb mole original liquid}$$

A direct graphical method can be used.

If Equation 4.15 is rewritten,

$$h + \frac{Q}{L} = \frac{\Delta}{L} h_\Delta + \frac{V}{L} H \qquad (4.15a)$$

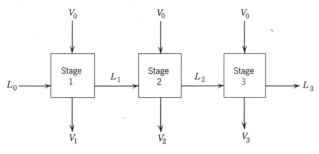

Figure 4.7. Multiple-stage contacting.

be contacted with fresh solvent again to remove additional solute. Such an arrangement using three stages is shown in Figure 4.7.

Illustration 4.4. One thousand pounds of a roasted copper ore containing 10 per cent $CuSO_4$, 85 per cent insoluble gangue and 5 per cent moisture is to be extracted by washing it three times with 2000-lb batches of fresh water. It has been found that the solids retain 0.8 lb of solution per pound of gangue. What is the composition of the final underflow sludge after three washings?

SOLUTION. The underflow locus is located on Figure 4.8 from the following two points:

At $x_a = 0$, $x_c = \dfrac{0.8}{1.8} = 0.445$, $x_b = \dfrac{1}{1.8} = 0.555$

At $x_a = 0.1$, $x_c = \dfrac{0.8}{1.8} - 0.1 = 0.345$, $x_b = \dfrac{1}{1.8} = 0.555$

The process is represented by Figure 4.7. The compositions of the incoming ore (x_0) and the fresh solvent (y_0) are located on Figure 4.8

$$x_{a0} = 0.10, \quad x_{c0} = 0.05; \qquad y_{a0} = 0.0, \quad y_{c0} = 1.0.$$

In the first stage L_0 and V_0 are mixed, and L_1 and V_1 leave the stage in equilibrium. For leaching, equilibrium occurs

The sum $(\Delta h_\Delta/L + VH/L)$ lies on the 180°F tie line at $x = 0.5$, where $h' = 17{,}000$. Therefore, $Q/L = h' - h$, the vertical distance at $x = 0.5$ from the saturated liquid curve to the 180°F tie line.

Since $L = 1$ lb mole,

$$Q = h' - h = 17{,}000 - 3500 = 13{,}500 \text{ Btu/lb mole}$$

MULTIPLE EQUILIBRIUM STAGES

Numerical and graphical calculations can be applied to a series of equilibrium stages. For example, the underflow leaving a stage in solid-liquid extraction may

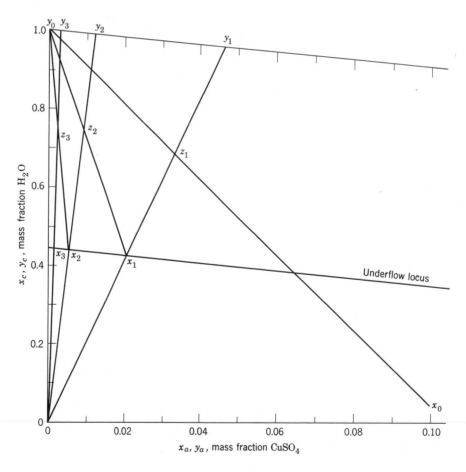

Figure 4.8. Solution to Illustration 4.4.

when the solution adhering to the gangue has the same composition as the solution leaving in the extract. Therefore, the underflow consists of gangue, which is represented by the right angle, and solution, whose composition lies on the diagonal, since it contains no solids. Since $L_0 = 1000$ and $V_0 = 2000$, the composition of $\Sigma_1 = L_0 + V_0$ is located by the inverse lever-arm rule. Now, $\Sigma_1 = L_1 + V_1$, and L_1 consists of insoluble gangue and solution of composition y_1. Therefore, the mixture $L_1 + V_1$ can be considered to consist of gangue and solution. Since the gangue is represented by the right angle ($x_a = 0$, $x_b = 1.0$, $x_c = 0.0$), a straight line through $x_b = 1.0$ and z_1 will locate y_1 where it intersects the diagonal.

Since $L_1 = \Sigma_1 - V_1$, the composition x_1 of L_1 is located at the intersection of the underflow locus with the straight line through z_Σ and y_1. In the second stage L_1 is mixed with V_0 and the procedure outlined above is repeated.

$$\Sigma_2 = L_1 + V_0 = L_2 + V_2, \quad \text{etc.}$$

However, first it is necessary to calculate L_1 by a balance around the first stage.

$$\frac{L_1}{V_1} = \frac{\overline{y_1 z_1}}{\overline{z_1 x_1}} = \frac{0.0465 - 0.0333}{0.0333 - 0.0210} = 1.07$$

and, since $L_1 + V_1 = 3000$ lb, $L_1 = 1550$ lb and $V_1 = 1450$ lb. For the second stage Σ_2 is located by the inverse lever-arm rule with $L_1 = 1550$ and $V_0 = 2000$. Then the quantity of L_2 is determined.

$$\frac{L_2}{V_2} = \frac{\overline{y_2 z_2}}{\overline{z_2 x_2}} = \frac{0.99 - 0.74}{0.74 - 0.44} = 0.83$$

Since $L_2 + V_2 = 3550$, $L_2 = 1615$ lb, $V_2 = 1935$ lb. For stage 3, $\Sigma_3 = L_2 + V_0 = L_3 + V_3$, etc. From Figure 4.8, the composition of the underflow from the third stage is 0.16 per cent $CuSO_4$. The mass of the underflow (L_3) is 1555 lb. Therefore, the percentage $CuSO_4$ not recovered is

$$\frac{(0.0016)(1555)}{(0.10)(1000)} = 2.5 \text{ per cent}$$

To recover the copper sulfate from the product solutions in Illustration 4.4, the large quantity of water would have to be evaporated. The quantity of water required for a fresh wash to each stage is large, and the resultant solutions are dilute. An arrangement which gives a more concentrated extract and requires less solvent for a given recovery of solute is discussed in the next chapter.

PROBLEMS

4.1. Calculate graphically the composition of the final mixture resulting from the following processes. Use a triangular diagram where H_2O is plotted on the vertical axis, salt is plotted on the horizontal axis, and sand is plotted on the diagonal axis.

(a) 100 lb of water is evaporated from 1000 lb of the following mixture: H_2O, 35 per cent; salt, 20 per cent; sand, 45 per cent.

(b) 2000 lb of water, 5000 lb of salt, and 3000 lb of sand are mixed together.

(c) The following two mixtures are mixed together and then 250 lb of sand is strained out and 150 lb of salt is added. Initial mixtures:

1000 lb: 40 per cent sand, 10 per cent H_2O, 50 per cent salt
200 lb: 10 per cent sand, 70 per cent H_2O, 20 per cent salt

(d) 300 lb of the solution in the following mixture is poured off of the insoluble sand.

1000 lb: 35 per cent sand, 55 per cent H_2O, 10 per cent salt

(e) 245 lb of a mixture (70 per cent sand, 20 per cent H_2O, 10 per cent salt) is removed from the following mixture:

400 lb: 40 per cent sand, 10 per cent H_2O, 50 per cent salt.

4.2. The following two mixtures are to be added together:

125 lb: $x_a = 0.52$, $x_b = 0.36$, $x_c = 0.12$
82 lb: $x_a = 0.25$, $x_b = 0.05$, $x_c = 0.70$

Determine the composition of the mixture graphically using the inverse lever-arm rule:

(a) By measuring along the line between the two compositions.

(b) By using the x_a coordinate scale for measurement.

(c) By using the x_c coordinate scale for measurement.

4.3. The following two phases are mixed together:

200 lb: 20 per cent oleic acid, 80 per cent cottonseed oil
100 lb: pure propane

Determine graphically the composition of the mixture:

(a) Using Equation 4.10.

(b) Using Equation 4.11.

(c) Using Equation 4.12.

4.4. Calculate graphically the composition of the mixture resulting when the following three solutions are mixed.

250 lb: 15 per cent acetic acid, 85 per cent water
50 lb: pure isopropyl ether
100 lb: 20 per cent acetic acid, 78 per cent water, 2 per cent isopropyl ether

4.5. 40 lb moles of water at 1 atm pressure and 60°F is mixed with 30 lb moles of ethanol at its dew point at 1 atm. The resultant mixture is heated to 185°F at 1 atm.

(a) How much heat was added to the mixture?

(b) What are the compositions and quantities of the resultant mixtures?

4.6. The following pairs of phases are brought into contact and are allowed to come to equilibrium in a single stage. Calculate graphically the mass and composition of the resulting two phases.

(a) 15 lb moles of liquid: 0.65 mole fraction ethanol, 0.35 mole fraction water at 140°F and 1 atm.

25 lb moles of vapor: an ethanol–water mixture at its dew point of 200°F and 1 atm.

(b) 250 lb of pure liquid propane at 98.5°C and 625 psia.

100 lb of a liquid: 30 mass percent oleic acid, 70 mass percent cottonseed oil at 98.5°C and 625 psia.

(c) 400 tons of a roasted copper ore containing 12 mass percent $CuSO_4$ and 88 mass percent insoluble inerts.

1000 tons of water.

(The inerts will retain 1.3 tons of water per ton of inerts. Of course, some $CuSO_4$ is dissolved in the 1.3 tons of water.)

(d) 1 lb mole of a gas of 40 per cent propane, 60 per cent propylene at 25°C.

300 lb silica gel at 25°C.

4.7. Derive Equations 4.11 and 4.12, and show with a sketch the utility of the equations in graphical calculations.

4.8. Calculate the minimum quantity of diethylene glycol which must be added to 100 lb of a solution of 80 mass percent styrene and 20 mass percent ethylbenzene to produce two phases. How much more diethylene glycol must be added to produce one phase again?

4.9. A seashore sand contains 85 per cent insoluble sand, 12 per cent salt, and 3 per cent water. 1000 lb of this mixture is to be extracted so that after drying it will contain only 0.2 per cent salt. How many washings with 2000-lb batches of pure water are required to give the desired purity? The sand retains 0.5 lb of water per pound of insoluble sand.

4.10. 1000 lb of a solution of 35 mass percent acetic acid in water is to be extracted with 2000 lb of pure isopropyl ether. Calculate the percentage removal of the acetic acid from the aqueous phase for each of the following proposals.

(*a*) The ether is split into four 500-lb fractions. The aqueous solution is mixed with one 500-lb fraction, the ether phase is separated, and the remaining aqueous phase is washed with the second 500-lb ether fraction. The process is continued until all four fractions are used.

(*b*) The same as part (*a*) but the ether fractions are successively 1000, 500, 250, 250 lb.

(*c*) The same as part (*a*) but only three washes of 667 lb are used.

(*d*) A single wash of 2000 lb of ether is used.

4.11. How much propane must be added to 1 lb of pure cottonseed oil to give two phases? How much more propane must be added before the mixture returns to one phase?

4.12. 1 lb mole of a gas containing 50 mole percent propane and 50 mole percent propylene is allowed to come to equilibrium with 300 lb of silica gel at 25°C and 1 atm. The resultant gas is withdrawn and contacted with a fresh batch of 300 lb of silica gel. What is the final composition of the gas after three contacts with successive batches of 300 lb of silica gel?

4.13. Illustration 4.4 is solved correctly within the accuracy of the graphical method. However, on logical grounds the underflow leaving each stage should be a constant (i.e., $L_1 = L_2 = L_3$). Without using a graph evaluate the logically correct value of the underflow from each stage.

chapter 5

Countercurrent Multistage Operations

Since the two phases leaving an equilibrium stage are in equilibrium, no separation of valuable constituents greater than that at equilibrium is possible in one stage. It is possible to increase the recovery of a valuable constituent by contacting the phase with fresh solvent in several equilibrium stages, as discussed in Chapter 4. However, the use of a fresh solvent feed to each stage requires large quantities of solvent and produces a very dilute solution of solute in solvent. To conserve solvent and to produce a more highly concentrated extract product, countercurrent multistage contacting of phases is employed. An example of industrial countercurrent multistage contacting is shown in Figure 5.1.

In countercurrent multistage contact the two phases enter at opposite ends of a series, or cascade, of equilibrium stages (Figure 5.2). The phases flow in directions counter to each other. In this way the concentration of solute in the V-phase product can be increased, and a high recovery of solute is possible with the use of a smaller quantity of solvent. This discussion has used the terms of extraction, but the principles of countercurrent multistage contact apply to all stage operations.

Usually one of two questions is answered by calculations for multistage equipment.

1. How many equilibrium stages are required to obtain the desired product separation or recovery?
2. What product recovery can be obtained with existing equipment of a known number of equilibrium stages?

In answering either question both the equilibrium relationships and the relative quantities of the two phases must be considered.

THE OVER-ALL MATERIAL BALANCE

In Figure 5.2 L and V refer to the mass of each phase; x and y refer to the composition of phases L and V respectively; and h and H refer to the concentration of any conserved property in phases L and V, respectively. The mass may be expressed in pounds or pound moles, and the composition as mass or mole fraction. The subscripts 0, 1, 2, etc., identify the stage *from* which a stream is flowing. N refers to the *last* stage in a cascade, whereas n is used to designate *any* stage in the cascade.

The conserved property most frequently of interest in stage operations is the enthalpy of the phases. Therefore, h and H will refer the enthalpy per unit mass of the phase. In the following development the equations involving composition and enthalpy will both be written. It will be seen that the equations are of identical form and that it would be possible to write a single general set of equations which could be used to express the conservation of either mass or enthalpy.

An over-all material and enthalpy balance around the entire cascade gives

Total material balance:

$$L_0 + V_{N+1} = L_N + V_1 = \Sigma \qquad (5.1)$$

Component balance:

$$L_0 x_0 + V_{N+1} y_{N+1} = L_N x_N + V_1 y_1 = \Sigma z_\Sigma \qquad (5.2)$$

Enthalpy balance:

$$L_0 h_0 + V_{N+1} H_{N+1} = L_N h_N + V_1 H_1 = \Sigma h_\Sigma \qquad (5.3)$$

These equations state that the material and enthalpy entering the cascade (in streams L_0 and V_{N+1}) must equal the material and enthalpy leaving the cascade (in streams L_N and V_1).

Equations 5.1, 5.2, and 5.3 also define the sum or sigma point, which is useful in calculating balances around the entire cascade. Substitution for Σ in Equation 5.2 from Equation 5.1 gives

$$z_\Sigma = \frac{L_0 x_0 + V_{N+1} y_{N+1}}{L_0 + V_{N+1}} = \frac{L_N x_N + V_1 y_1}{L_N + V_1} \qquad (5.4)$$

Figure 5.1. Liquid-liquid extraction applied to the purification of gasoline. The three towers pictured in the foreground are used to extract undesirable mercaptans from a gasoline fraction of crude oil to produce a low-sulfur gasoline. The three towers are operated in parallel. Raw gasoline entering at the bottom of each tower is contacted countercurrently with caustic methanol in the lower section to remove the mercaptans. In the upper section of each tower the gasoline is contacted with a caustic solution to recover methanol which has dissolved in the gasoline. The methanol is fed to each tower between the upper and lower sections, and the caustic solution is supplied at the top of the tower. The columns are $3\frac{1}{2}$ ft, 4 ft, and 5 ft in diameter and 63 ft high. Each column has nine perforated plates. Auxiliary equipment includes a stripping tower where the methanol is removed from the caustic solution by steam stripping and a distillation tower where the methanol is recovered from an aqueous solution. (Atlantic Refining Company.)

For the enthalpy coordinate of the sigma point, combination of Equations 5.1 and 5.3 gives

$$h_\Sigma = \frac{L_0 h_0 + V_{N+1} H_{n+1}}{L_0 + V_{N+1}} = \frac{L_N h_N + V_1 H_1}{L_N + V_1} \quad (5.5)$$

The equations are simply analytical expressions of the addition rule developed in Chapter 4. Examination of the group on the right side of Equation 5.4 shows that it is simply the sum of the masses of one component

divided by the sum of the total masses of the two streams L_N and V_1. Similarly, Equation 5.5 gives the sum of the enthalpies of the two streams divided by the total masses of the streams. Calculations utilizing Equations 5.4 and 5.5 are often more easily accomplished graphically. The procedure is shown in Figure 5.3. For example, perhaps the mass and composition of L_0 and V_{N+1} are known. It is then possible to locate z_Σ, as shown. Then, if x_N is known, y_1 can be determined by drawing a straight line between x_N and z_Σ and extending it to intersect the extract curve. The masses of L_N and V_1 can be determined by applying the inverse lever-arm rule to the line $\overline{x_N z_\Sigma y_1}$.

STAGE-TO-STAGE CALCULATIONS

In one type of calculation for countercurrent flow enough information is available to calculate the over-all material balance, as outlined above. After an over-all material balance has been used to determine the end conditions of the cascade, stage-to-stage calculations may be made to determine the number of equilibrium stages required to give the desired end conditions.

Calculation may proceed from either end of the cascade. If, for example, y_1 and H_1 are known, use of the equilibrium relationship will give x_1 and h_1. It is now necessary to develop an interrelationship between x_1, h_1 and y_2, H_2. The two streams L_1 and V_2 are passing each other between stages 1 and 2. In order to calculate the quantity and composition of V_2 from those of L_1 the new concept of *net flow* must be introduced. Since there is no accumulation of a constituent, a property, or a total quantity in the cascade, the net flow is constant throughout the cascade. The net flow is a fictitious stream. It is the difference between two streams which are not in reality subtracted one from the other.

Net flow to the *right* in the cascade shown in Figure 5.2 is defined as the difference between the flow to the right and the flow to the left, or

Total net flow:
$$\Delta = L_0 - V_1 = L_n - V_{n+1} = L_N - V_{N+1} \quad (5.6)$$

Component net flow:
$$\Delta x_\Delta = L_0 x_0 - V_1 y_1 = L_n x_n - V_{n+1} y_{n+1}$$
$$= L_N x_N - V_{N+1} y_{N+1} \quad (5.7)$$

Property net flow:
$$\Delta h_\Delta = L_0 h_0 - V_1 H_1 = L_n h_n - V_{n+1} H_{n+1}$$
$$= L_N h_N - V_{N+1} H_{N+1} \quad (5.8)$$

The total net flow (Δ) of material at the left end of the cascade is equal to that material flowing to the right (L_0) minus that flowing to the left (V_1). Similarly, at the right end of the cascade, the total net flow (Δ) is equal to $L_N - V_{N+1}$. By Equation 5.1

$$L_0 - V_1 = L_N - V_{N+1}$$

Therefore, the total net flow at any stage in the cascade is constant. If the mass of V_{N+1} is greater than L_N, Δ is negative, and the net flow is to the left in Figure 5.2. Since a net flow to the right is defined as a positive quantity, an actual net flow of any component to the left has a negative numerical value.

Equations 5.6, 5.7, and 5.8 also define the net flow at any stage n in the cascade. It can be seen that the component net flow is constant through all stages at steady state, so that the delta point (x_Δ) is valid at any point in the cascade. Therefore, since $\Delta = L_1 - V_2$ and $\Delta x_\Delta = L_1 x_1 - V_2 y_2$, the value of y_2 lies on the

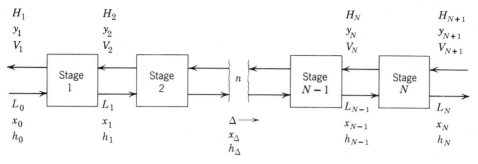

Figure 5.2. Countercurrent multistage contact.

Equation 5.7 defines the net flow of one component. For example, if copper sulfate is to be extracted from a roasted ore (L_0) by water (V_{N+1}), there will be a net flow of insoluble gangue and copper sulfate to the right and a net flow of water to the left. The term x_Δ is a fictitious concentration of Δ which would result if V were actually subtracted from L. Combining Equations 5.6 and 5.7 gives

$$x_\Delta = \frac{\Delta x_\Delta}{\Delta} = \frac{L_0 x_0 - V_1 y_1}{L_0 - V_1} = \frac{L_n x_n - V_{n+1} y_{n+1}}{L_n - V_{n+1}}$$

$$= \frac{L_N x_N - V_{N+1} y_{N+1}}{L_N - V_{N+1}}$$

$$= \frac{\text{net flow of component}}{\text{total net flow}} \quad (5.9)$$

The numerators of Equation 5.9 are the net flow of any component, and the denominators are the total net flow. Therefore, if the net flow of a component is in a direction opposite to the total net flow, x_Δ will have a negative value for that component.

Application of the graphical-addition rule to Equations 5.6 and 5.7 for the entire cascade shows that x_Δ should lie on the straight line $\overline{y_1 x_0}$ extended and also on the line $\overline{y_{N+1} x_N}$ extended. This is shown in Figure 5.4. For the case shown, $x_{\Delta a}$ and $x_{\Delta b}$ are negative, and $x_{\Delta c}$ is positive and greater than 1.0. This indicates that the net flows of the solute and major raffinate component are in a direction opposite to the total net flow, but the net flow of solvent is in the same direction as the total net flow and is larger than the total net flow. Even for the fictitious stream Δ, $x_a + x_b + x_c$ must equal 1.0. Therefore, if x_a and x_b are negative, x_c must be greater than 1.0. The delta point located by using the concentrations at the ends of the cascade will be used to calculate from stage to stage, as shown in the following discussion.

straight line $\overline{x_\Delta x_1}$. It must also be located on the phase envelope since V_2 is flowing from an equilibrium stage. The determination of the compositions of the phases in successive stages may now proceed by alternate use of the equilibrium data and the delta point. The location of x_1 from y_1 by tie-line data and of y_2 from x_1 by the delta point is shown in Figure 5.4. When y_2 has been located, x_2 can be found, and so forth by alternate application of equilibrium data and the delta point, until the composition x_n of the raffinate leaving a stage equals or surpasses the final desired composition, x_N. The stage-to-stage calculation can also be made in the opposite direction, from x_N to y_1. The choice of direction depends upon the original information given. In some cases the percentage recovery in the extract of the solute originally supplied is specified. It is possible to calculate the exit concentrations from this information.

Combination of Equations 5.6 and 5.8 gives the enthalpy coordinate of the delta point,

$$h_\Delta = \frac{L_0 h_0 - V_1 H_1}{L_0 - V_1} = \frac{L_n h_n - V_{n+1} H_{n+1}}{L_n - V_{n+1}}$$

$$= \frac{L_N h_N - V_{N+1} H_{N+1}}{L_N - V_{N+1}} \quad (5.10)$$

which may be used on enthalpy-composition diagrams together with Equation 5.9 to locate the delta point. Calculation of the over-all material and enthalpy balances and calculation from stage to stage on an enthalpy-composition diagram follows the procedures outlined previously for ternary composition diagrams. The enthalpy may be considered to replace the solvent.

Illustration 5.1. Acetic acid is to be extracted from 2000 lb/hr of a 40 per cent aqueous solution by countercurrent extraction with 3000 lb/hr of pure isopropyl ether.

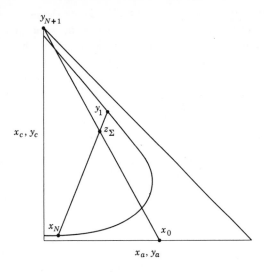

Figure 5.3. Graphical determination of the over-all material balance.

The concentration of acetic acid is to be reduced to 3 per cent in the exit raffinate. How many equilibrium stages are required to reduce the acetic acid concentration to 3 per cent?

SOLUTION. The complete solution is given on Figure 5.5. In this case there are three components and no conserved property to consider. Therefore, a three-component (ternary) diagram will be used. y_1 is located by an over-all material balance: First z_Σ is located on line $\overline{x_0 y_{N+1}}$ by the inverse lever-arm rule.

$$\frac{L_0}{V_{N+1}} = \frac{2000}{3000} = \frac{y_{N+1} - z_\Sigma}{z_\Sigma - x_0} = \frac{2}{3}.$$

Then y_1 must lie on the phase envelope on the extension of line $\overline{x_N z_\Sigma}$, since $L_N + V_1 = \Sigma$.

The delta point is now located on the extensions of the straight lines $\overline{x_0 y_1}$ and $\overline{x_N y_{N+1}}$. The calculation of stages may start at either y_1 or x_N. Starting with the final extract (y_1), x_1 must be in equilibrium with y_1, and the value of x_1 is determined from the equilibrium curve as shown by the construction. Since $L_1 - V_2 = \Delta$, y_2 lies at the intersection of the straight line $\overline{x_\Delta x_1}$ and the phase envelope. x_2 is now determined as before from equilibrium data, and the construction proceeds until x_n equals or surpasses x_N. This occurs between x_{14} and x_{15}. Therefore, approximately 14.5 equilibrium stages are required. The dashed lines of Figure 5.4 are equilibrium tie lines for the stages of this illustration. The construction lines to x_Δ have been omitted after the first two to avoid obscuring the stage tie lines.

The coordinates of the delta point can be read from Figure 5.5 or calculated from Equation 5.9 as $x_a = -0.012$, $x_b = -0.45$, $x_c = 1.46$. The negative values for x_a and x_b indicate that the net flows of acetic acid and water are in a direction opposite to the total net flow. In this case the total net flow is to the left. Therefore, the net flow of acetic acid and water is to the right. Even for this fictitious composition, $x_a + x_b + x_c = -0.012 + (-0.45) + 1.46 \doteq 1$.

OPERATING VARIABLES

If the flow rates of the L and V phases have been specified, the number of equilibrium stages required for a specified separation may be determined. On the other hand it may be necessary to evaluate the operation of existing equipment of a known number of equilibrium stages for a new separation. In this case usually either the L-phase or the V-phase may be specified, and the other phase may be adjusted to give the desired separation. If both L and V and the number of stages are specified, the separation is determined.

In a typical extraction problem, as the solvent flow (V) is decreased, more equilibrium stages are required to give the specified recovery of solute. Finally a value of solvent flow is reached where an infinite number of stages is required to give the recovery. This is referred to as the minimum V-phase flow or as the minimum V/L. Although the minimum V/L has no direct practical use, it is useful as a limiting value for the actual flow. Often the actual V/L will be taken as some arbitrary factor times the minimum.

A reduction in V/L will move the delta point farther from the triangular diagram; an increase shifts it closer. The minimum V/L may be determined by extending the tie lines in the direction of the delta point. The extended tie line on which the delta point first falls as it moves away from the diagram determines the minimum V/L. It is difficult to predict generally exactly which tie line will determine the minimum V/L. It may be a tie line through the feed composition, through the exit raffinate composition, or at any intermediate point, as demonstrated in the following illustration.

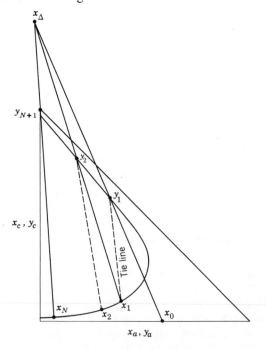

Figure 5.4. Graphical determination of delta point.

The minimum V/L occurs at the highest value of V which gives a line through the delta point that coincides with any extended tie line. Such a coincidence results in no change of composition from stage to stage and requires an infinite number of stages to give a finite change in composition. Because the lines representing the stages are crowded together, the point at which the infinite number of stages occurs is referred to as the "pinch." At values of V/L lower than the minimum, the desired separation cannot be made, even with an infinite number of stages. Since many of the subminimum values of V/L give delta points which lie on an extended tie line, it should be emphasized that the *largest* value of V/L which gives a delta point on any extended tie line is the correct minimum V/L.

Illustration 5.2. Determine the minimum solvent rate for the feed and recovery of Illustration 5.1.

SOLUTION. The location of the delta point in Figure 5.5 is for a solvent rate of 3000 lb/hr and results in 14.5 stages.

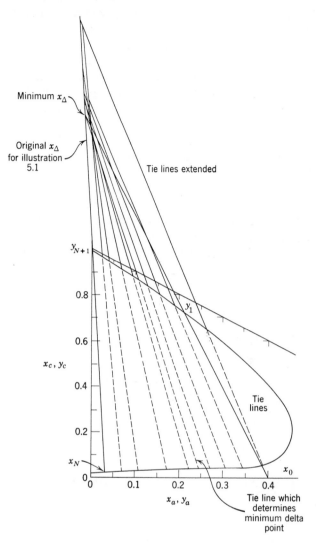

Figure 5.6. Solution to Illustration 5.2.

The minimum solvent rate will give a delta point farther away. Therefore, several tie lines are extended toward the delta point in Figure 5.6. The line $\overline{x_N y_{N+1}}$ is the same as in Illustration 5.1 since the compositions in Illustration 5.1 also apply to this illustration. The tie line which, when extended, intersects the extended line $\overline{x_N y_{N+1}}$ nearest the diagram gives the minimum V/L. All other extended tie lines intersect the extended line $\overline{x_N y_{N+1}}$ farther from the diagram. The value of y_1 may be determined by drawing the line $\overline{x_\Delta x_0}$, as shown. The quantity of solvent (V_{N+1}) is now determined by an over-all material balance. z_Σ is located at the intersection of $\overline{x_0 y_{N+1}}$ and $\overline{x_N y_1}$, and the inverse lever-arm rule is applied.

$$\frac{V_{N+1}}{L_0} = \frac{\overline{x_0 z_\Sigma}}{z_\Sigma y_{N+1}} = \frac{10.35 \text{ length units}}{7.1 \text{ length units}} = 1.46$$

$$V_{N+1} = 1.46 L_0 = (1.46)(2000) = 2920 \text{ lb}$$

Therefore, the minimum solvent flow rate is 2920 lb/hr.

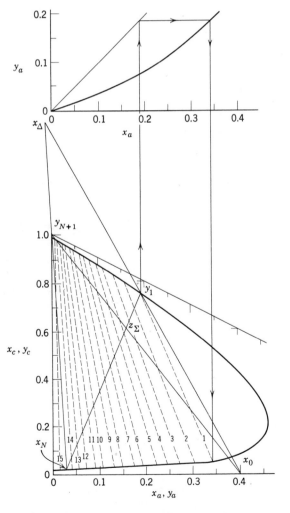

Figure 5.5. Solution to Illustration 5.1.

A minimum V/L can be defined in a separation process where the component of interest is to be transferred from the L-phase to the V-phase and the recovery of the component is specified, such as in the usual case of liquid extraction. On the other hand, in many operations, such as gas absorption, the component of interest is transferred in the opposite direction, from the V-phase to the L-phase. In this case it is possible to define a minimum L-phase flow or a minimum L/V ratio which requires an infinite number of stages to give the desired recovery. The methods of determining the minimum L/V are similar to those presented for minimum V/L. However, some of the procedures must be modified when L equals V and beyond, since in this case Δ equals zero and the delta point goes to infinity.

The maximum separation which is theoretically possible in simple countercurrent flow is obtained when an infinite number of stages is used. As the number of stages is increased for given feed rates, the degree of separation of components will increase until a pinch occurs in the cascade. The pinch will result in an infinite number of stages which gives a maximum separation of components. If, for example, the pinch occurs at the last stage at the left end of the cascade (Figure 5.2) the V-phase product leaving will be in equilibrium with the L-phase feed. Similarly, if the pinch occurs at the right end, the L-phase leaving will be in equilibrium with the V-phase entering. A pinch may also occur at an intermediate point.

THE SOLVENT-FREE BASIS AND ENTHALPY-CONCENTRATION DIAGRAMS

The equations derived in this chapter apply also to calculations on a solvent-free basis and on enthalpy-concentration diagrams. New symbols for extraction are defined and used in the derived equations.

L' = mass of two components (a and b) disregarding the solvent (c), in phase L

V' = mass of two components (a and b) disregarding the solvent (c), in phase V

X = mass of any component (a, b, or c) per unit mass of L'

Y = mass of any component (a, b, or c) per unit mass of V'

With this notation:

$X_a + X_b = 1$, X_c can have any value.

$Y_a + Y_b = 1$, Y_c can have any value.

Since Y_c is now the mass of solvent per unit mass of solvent-free material (a plus b), it may have any value between 0 and ∞. When pure solvent is used, $Y_{c(N+1)} = \infty$, and the inverse lever-arm expressions involving

$Y_{c(N+1)}$ are indeterminate. Material balances involving the solvent stream are, for this reason, more easily calculated analytically than graphically. For example, Equations 5.1 and 5.2 may be used to calculate the overall material balance.

Calculation of two-component distillation on an enthalpy-concentration diagram closely resembles calculations on the solvent-free basis. Since only two material components are present in distillation, the material balances of this chapter hold without redefining the symbols. The enthalpy coordinate of the delta point is defined by Equation 5.10. The solvent-free coordinates emphasize the analogy between enthalpy in distillation and solvent in extraction. Enthalpy H is analogous to the solvent concentration in the extract (Y_c) since they are both expressions of a property of the system per unit mass of two components of the system. A typical problem is shown on Figure 5.7. The diagram could apply to a hypothetical case in either extraction or distillation. Therefore, the diagram is labeled both for extraction and for distillation (in parentheses) to show the analogy. The solvent feed is shown to contain a small quantity of the major raffinate component, so

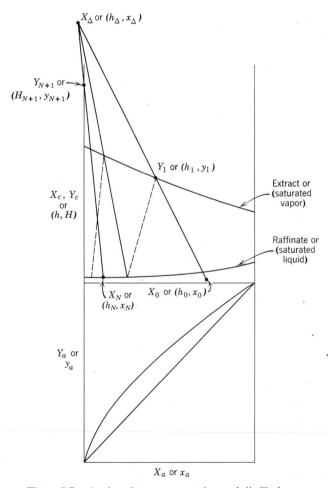

Figure 5.7. Analogy between extraction and distillation.

that Y_{N+1} does not lie at infinity. The vapor feed in distillation which corresponds to this point would be a superheated vapor composed of pure less-volatile component. The raffinate feed is shown to contain a small quantity of solvent (X_0). Analogously, the liquid feed in distillation contains a small quantity of enthalpy (h_0). Distillation calculations utilizing the enthalpy-composition diagram are usually referred to as the *Ponchon–Savarit* method. (1, 2).

The "adsorbent-free" diagram given in Figure 3.12 may also be used in calculation in the manner outlined above.

REFERENCES

1. Ponchon, M., *Tech. moderne* **13**, 20 (1921).
2. Savarit, R., *Arts et métiers*, 65ff. (1922).

PROBLEMS

5.1. Copper sulfate in a roasted ore is to be leached out with water in a continuous countercurrent extraction cascade. 1000 lb/hr of ore containing 10 per cent CuSO$_4$, 85 per cent insoluble gangue, and 5 per cent moisture will be extracted with water. The final underflow sludge will contain only 0.15 per cent CuSO$_4$. The underflow sludge retains 0.8 lb of solution per pound of gangue. Calculate the number of equilibrium stages required for an entering solvent flow of (a) 2000 lb/hr, (b) 4000 lb/hr. Compare the results with Illustration 4.4.

5.2. Copper sulfate in roasted ore is to be leached out with water in a continuous countercurrent extraction cascade. 100 tons/day of the ore containing 9 per cent CuSO$_4$, 86 per cent insoluble gangue, and 5 per cent moisture will be extracted with water. The strong extract solution will contain 10 per cent CuSO$_4$ and 90 per cent water. Since CuSO$_4$ is the valuable constituent, 95 per cent of it must be recovered in the extract. The gangue retains 1.5 tons of water per ton of gangue.

(a) Calculate the number of equilibrium stages required, without using graphical methods.

(b) Calculate the number of equilibrium stages required, using graphical methods.

(c) What is the flow rate of fresh water to the cascade?

(d) Calculate the total net flow at each stage.

(e) What are the coordinates of the delta point?

(f) In what direction is the net flow of water? Of inerts? Of CuSO$_4$?

5.3. From the equations defining the delta point derive

$$\frac{\Delta}{V_1} = \frac{\overline{y_1 x_0}}{\overline{x_0 x_\Delta}}$$

5.4. 700 lb/hr of halibut liver is to be extracted in a countercurrent cascade with ether to recover the oil. The ether, which has been partially purified, contains 2 per cent oil. The fresh livers contain 20 per cent oil and are to be extracted to a composition of 1 per cent oil (on a solvent-free basis). 500 lb of solvent is to be used.

(a) What percentage of the oil entering in the livers is recovered in the extract?

(b) How many equilibrium stages are required?

(c) Calculate the mass and direction of the total and component net flows.

(d) Calculate the coordinates of the delta point. Underflow data are given in Figure 3.14.

5.5. It is proposed to separate liquid *a* from its solution with liquid *b* by contacting the solution with solvent *c*. Liquids *b* and *c* are completely immiscible with each other at all concentrations of *a*. Liquid *a* is soluble in *b* and *c* at equilibrium such that the concentration of *a* in liquid *c* is always equal to that of *a* in *b*.

An existing mixer-settler unit is available for the separation. The unit is equivalent to five equilibrium stages. The feed solution is 20 per cent *a* and 80 per cent *b* and must be processed at the rate of 1000 lb/hr. The fresh solvent rate is limited to 1500 lb/hr by the design of the equipment.

(a) Plot the ternary diagram for the system *a–b–c*. Label the extract and raffinate loci and show typical tie lines.

(b) What percentage of the component *a* in the feed can be recovered in the leaving extract, using the maximum permitted solvent flow rate?

(c) What solvent flow rate would give 80 per cent recovery of *a*?

(d) Determine the coordinates of the delta point in parts (b) and (c). In which direction are the total and component net flows for each case?

5.6. A gaseous mixture of 50 mole percent propane, 50 mole percent propylene is to be separated by passing it countercurrent to silica gel in a rotating-plate column. The gas leaving the top of the unit is to be 95 mole percent propane.

(a) How many equilibrium stages would be required if 350 lb of silica gel are supplied for each pound mole of gas?

(b) What is the minimum mass of silica gel per pound mole of gas which will give a product with 95 mole percent propane?

5.7. A countercurrent extraction cascade is processing an aqueous solution of acetic acid with pure isopropyl ether. The following data are available:

Entering raffinate: 1000 lb/hr
    35 per cent acetic acid, 65 per cent water
Final Extract: 2500 lb/hr
    12 per cent acetic acid

How many equilibrium stages is the cascade equivalent to?

5.8. The following information is available on a multistage cod-liver oil extraction unit:

Entering Livers: 1000 lb/hr
    32.6 mass percent oil
    67.4 mass percent inerts
Entering Solvent: 2000 lb/hr
    1 mass percent oil
    99 mass percent ether
Exit Underflow: 1180 lb/hr
    1.14 mass percent oil
    41.0 mass percent ether
    57.9 mass percent inerts
Exit Extract: 1820 lb/hr
    18.3 mass percent oil
    81.7 mass percent ether

Calculate the three coordinates of the delta point.

5.9. Isopropyl ether is to be used to extract acetic acid from an aqueous solution. An existing countercurrent extraction column which is equivalent to four theoretical stages is to be used.

(a) Determine the flow rate of pure isopropyl ether solvent required for the following conditions:

Feed flow rate: 1000 lb/hr
Feed composition: 35 per cent acetic acid, 65 per cent H$_2$O
Final extract composition: 10 per cent acetic acid

(b) What percentage recovery of acetic acid is achieved?

5.10. Tung meal containing 55 mass percent oil is to be extracted at a rate of 3000 lb/hr, using 12,000 lb/hr of solvent. The solvent contains 98 per cent n-hexane and 2 per cent tung oil. The solution adhering to the insoluble meal in the underflow was determined experimentally, as tabulated below:

Composition of solution adhering, Mass Fraction Oil	lb Solution/lb Inerts
0.0	2.0
0.2	2.5
0.4	3.0
0.6	3.5

The tung meal was so finely divided that some of it goes out suspended in the overflow solution. This amounts to 0.05 lb of solids per pound of solution. The underflow must have a concentration no more than 2 mass percent oil.

(*a*) How many equilibrium stages are required?

(*b*) What is the percentage recovery of tung oil extracted from the meal?

5.11. Acetic acid is to be extracted from aqueous solution by countercurrent extraction with isopropyl ether. The feed is 10,000 lb/hr of 28 per cent acetic acid.

(*a*) Determine the number of equilibrium stages required for a solvent feed rate of 30,000 lb/hr and an extract composition of 8 per cent acetic acid.

(*b*) Determine the minimum solvent/feed ratio which will give the same raffinate composition as in part (*a*).

5.12. A liquid containing 40 per cent styrene and 60 per cent ethylbenzene is to be extracted in a countercurrent cascade by diethylene glycol solvent. The raffinate is to contain 10 per cent styrene on a solvent-free basis.

(*a*) Determine the minimum solvent to feed ratio.

(*b*) Using 1.3 times the minimum solvent to feed ratio, calculate the number of equilibrium stages required on ternary coordinates.

(*c*) Repeat parts (*a*) and (*b*) using coordinates plotted on a solvent-free basis.

5.13. A mixture of 50 mole percent ethanol, 50 mole percent water contains a very small quantity of nonvolatile impurity which discolors the solution. It has been proposed to concentrate and purify the ethanol by stripping the ethanol from solution with an available supply of superheated steam. The solution will flow downward in a bubble-cap column countercurrent to the upward-flowing steam. The solution is at 70°F and the steam at 350°F and 1 atm.

(*a*) What is the maximum ethanol vapor concentration that can be obtained with an infinite number of equilibrium stages?

(*b*) What ethanol concentration and recovery in the vapor is obtained with a column equivalent to five equilibrium stages with a steam to ethanol feed ratio of 1 : 2 mole/mole?

5.14. A new solvent extraction unit has just been installed, and it is not giving the desired separation. The unit was designed to extract 10,000 lb/hr of a 30 per cent diphenylhexane–70 per cent docosane mixture with 20,000 lb/hr of solvent. It was desired to produce an extracted docosane phase with only a 1 per cent concentration of diphenylhexane, but the actual concentration during the test run was much higher. Since a diphenylhexane concentration in the extracted docosane above 1 per cent is unacceptable, you must make a recommendation to the operating engineer to enable him to make the desired separation. Of course, it is necessary to maintain the feed rate of 10,000 lb/hr, and no more than 20,000 lb/hr of solvent can be used, since the unit can handle no more. The solvent used in the test was 98 per cent furfural and 2 per cent diphenylhexane, recycled from the solvent recovery system. The unit has fifteen actual stages and the over-all efficiency is about 30 per cent.

(*a*) Utilizing the principles and calculations of liquid-liquid extraction, determine why the desired separation was not accomplished in the test run.

(*b*) Suggest a change in operating conditions which will give the desired separation with no modification in the extraction equipment. Tell how you would calculate to determine that your suggestion is valid before it is tried.

Your answer should include any required calculations and a careful account of the reasoning used in arriving at your conclusions.

SOLUBILITY DATA FOR THE SYSTEM DOCOSANE–DIPHENYLHEXANE–FURFURAL AT 45°C.

Briggs, S. W., and E. W. Comings, *Ind. Eng. Chem.* Vol. 35, No. 4, pp. 411–417 (1943).

	Phase Envelope	
Mass Fraction Furfural	Mass Fraction Diphenylhexane	Mass Fraction Docosane
0.040	0.0	0.960
0.050	0.110	0.840
0.070	0.260	0.670
0.100	0.375	0.525
0.200	0.474	0.326
0.300	0.487	0.213
0.400	0.468	0.132
0.500	0.423	0.077
0.600	0.356	0.044
0.700	0.274	0.026
0.800	0.185	0.015
0.900	0.090	0.010
0.993	0.0	0.007

Tie Lines

Furfural phase	0.891	0.098	0.011	in equi-
Docosane phase	0.048	0.100	0.852	librium
Furfural phase	0.736	0.242	0.022	in equi-
Docosane phase	0.065	0.245	0.690	librium
Furfural phase	0.523	0.409	0.068	in equi-
Docosane phase	0.133	0.426	0.439	librium

chapter 6

Countercurrent Multistage Operations with Reflux

In many cases of mass transfer a more complete separation of components than that possible in simple counterflow is required. The best possible separation in simple countercurrent flow is that obtained when an infinite number of stages is required. A more complete separation may be obtained by the introduction of *reflux* at one or both ends of the cascade. In such a case part of one or both product streams is changed in phase and returned to the cascade as *reflux*. An industrial example of a stage operation with reflux is shown in Figure 6.1. Although reflux might be used in any mass-transfer operation, it is of value only in certain cases. Important applications of reflux will be considered in the following discussion. A general schematic representation of a countercurrent cascade with reflux is given in Figure 6.2. A refluxer unit is added at each end of the simple countercurrent cascade. The feed (F) is now introduced into the central part of the cascade. The V-phase product (D) is withdrawn after the L-phase refluxer (C). The stream L_0 is the L-phase reflux. At the right end of the cascade, the L-phase product is withdrawn before the V-phase refluxer (S). The V-phase reflux is V_S.

In liquid-liquid extraction, extract reflux (L_0) can be employed to obtain a greater recovery of solute in the extract. In this case solvent is removed from V_1 in the solvent separator (C) to produce a liquid (L_C) of low solvent concentration, part of which is extract product (D) and part extract reflux (L_0). The solvent separator usually is a distillation column. Extract reflux is particularly beneficial for ternary liquid systems with two pairs of partially miscible components, such as the diethylene glycol–styrene–ethyl benzene system (Figure 3.9). In such systems it is possible to obtain separation into an extract containing a negligible amount of unextracted raffinate component and a raffinate containing a negligible quantity of solute. On the other hand, a system with only one partially miscible pair, such as isopropyl ether–acetic acid–water (Figure 3.8), always yields one product phase with considerable quantities of both solute and unextracted raffinate component. In this case reflux is not so beneficial since separation becomes increasing difficult as the plait point is approached.

In liquid extraction (Figure 6.2) solvent (V_{S+1}) is added at the solvent mixer (S) where it dissolves with part (L_{S-1}) of the raffinate flow (L_N) from the last stage. The solvent added passes through the stages, coming to equilibrium with the raffinate at each stage. Ultimately, the extract (V_1) from the first stage is sent to the solvent separator, where the solvent is removed. Part (D) of the remaining solute-rich liquid is the extract product, and the remainder (L_0) is the extract reflux. With this arrangement, the concentration of solute in the V-phase is increased over that in equilibrium with the feed in stages F-1 to 1. Therefore, the resultant extract product may have a much higher concentration of solute than is possible in simple counterflow (Figure 5.2, feed $= L_0$). Since stages F-1 to 1 effectively enrich the V-phase in the solute, the section is called the "enriching section." Similarly, stages F to N strip solute from the raffinate and are called the "stripping section."

In distillation reflux at both ends of a cascade is very frequently used to obtain distillate (D) and bottoms (B) of high purity. The V-phase reflux (V_S) is produced by adding heat (q_S) in the reboiler (S) to part of the liquid from the cascade (L_N) and vaporizing it. Similarly, L-phase reflux is produced by removing heat ($-q_C$) in the condenser (C) from part of the vapor product (V_1) and condensing it to give liquid reflux (L_0). Use of reflux in distillation permits recovery of very high purity products, providing no azeotrope occurs.

Figure 6.1. Ethylene purification by distillation. The six columns pictured separate ethylene from other light hydrocarbons. The pure ethylene product is used in the manufacture of polyethylene plastic. The first column on the right (No. 1) is an absorption tower with reflux at the lower end which absorbs ethylene and less volatile materials with a heavier hydrocarbon solvent, separating them from methane and hydrogen. The sixth column from the right (No. 6) is a distillation column with reflux which removes propane and lighter components from the solvent. The second column separates ethane and ethylene from propane and propylene by distillation. The fourth and fifth columns separate ethylene from ethane. The two columns are in reality one distillation cascade, split into two sections to reduce the height of the unit. The third column separates the ethylene product from any methane which was not removed in the absorber. Physical data for the columns are tabulated below.

Column	Diameter, ft	Height, ft	No. of Actual Stages
1 Absorber	$3\frac{1}{2}$	75	30
2 De-ethanizer	$3, 4\frac{1}{2}$	77	30
3 Demethanizer	3	73	28
4 Ethylene fractionator, lower section	4	196	80
5 Ethylene fractionator, upper section	4	83	40
6 Depropanizer	4	67	26

Columns of this type may employ either bubble-cap or perforated plates. The operating pressures in this application vary between 280 and 500 psig. The temperatures in the columns range between $-45°F$ at the top of the demethanizer and $300°F$ at the base of the depropanizer. The dark-painted columns operate above atmospheric temperature, and the light painted columns operate below. The unit requires a substantial quantity of refrigeration for the reboilers and condensers on the columns operating below atmospheric temperature. Some of the large quantity of auxiliary equipment necessary to operate the columns is also pictured. (Designed for E. I. duPont de Nemours and Co. by The Lummus Company.)

In distillation (Figure 6.2), heat (q_S) is added at the still (S) to vaporize part (L_{S-1}) of the liquid flow (L_N) from the bottom stage. The vapor reflux (V_S) formed in the still rises through the stages, coming to equilibrium with the liquid downflow at each stage. Ultimately the vapor from the top stage (V_1) is condensed in the

ratios. The graphical calculations with reflux will be illustrated first for liquid extraction and then for distillation.

Liquid-Liquid Extraction. The number of equilibrium stages required for a given separation can be calculated using the concept of net flow. For extraction the net

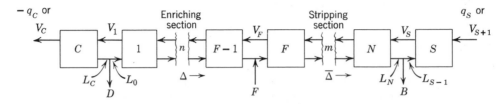

Figure 6.2. Countercurrent stage operation with reflux.

total condenser (C); part is removed as distillate product (D) and the remainder (L_0) is the liquid reflux. In binary distillation the succession of stages above the feed enrich the vapor in the more volatile component, and the stages below the feed strip the more volatile component from the liquid downflow.

The reflux rates of streams L_0 and V_S, as well as the number of equilibrium stages, determine the degree of separation of the two components present in a given feed. In extraction there is a net flow of solvent carrying the extracted solute from stage N to stage 1. Similarly, in distillation there is a net flow of heat which might be thought of as "carrying" the more volatile component from stage N to stage 1. Just as the solvent distribution between extract and raffinate must be considered in extraction, so must the heat contents (enthalpies) of the liquid and vapor be considered in distillation.

The reflux may constitute a large fraction of the material reaching the ends of the cascade. The products would then be but a small fraction of the phases circulating internally within the cascade.

Reflux may be employed in other stage operations where it proves beneficial. For example, certain separations in gas absorption and adsorption are improved by using reflux.

STAGE CALCULATIONS WITH REFLUX

In a typical design calculation the quantity and composition of the feed and products may be specified. If the reflux ratio (L_0/D) has been chosen, the required number of equilibrium stages for the separation may be calculated. Another situation often encountered involves prediction of the operation of existing equipment in the separation of a new system. Usually the construction of the cascade allows the variation of reflux ratio over a limited range. Then the possible product compositions for a given feed may be calculated at the allowable reflux

flow in the enriching section in Figure 6.2 is

$$\Delta = L_0 - V_1 = L_n - V_{n+1} = -(D + V_C) \quad (6.1)$$

where n is any stage in the enriching section. Similarly, the net flow in the stripping section is

$$\bar{\Delta} = L_N - V_S = L_m - V_{m+1} = (B - V_{S+1}) \quad (6.2)$$

where m is any stage in the stripping section. The introduction of feed changes the net flow between stages F-1 and F. An over-all material balance around the cascade in Figure 6.2 gives

$$F = V_C + D + B - V_{S+1} \quad (6.3)$$

Combining Equations 6.1 and 6.2 with 6.3 gives

$$F = -\Delta + \bar{\Delta} \quad \text{or} \quad \bar{\Delta} = \Delta + F \quad (6.4)$$

Equation 6.4 relates the net flows in the two sections of the cascade. The addition of the feed increases the net flow by an amount equal to F. The feed may consist of a mixture of L-phase and V-phase, such as liquid and vapor in distillation. In this situation the feed will split between the L-phase and V-phase when it is introduced into the cascade, but Equations 6.3 and 6.4 are still correct.

It should be emphasized that the magnitude of the internal circulation (L and V) bears no direct relation to the quantities of F, D, and B. It is therefore impossible to predict the values of L and V solely from the values of feed and products, in contrast to the case for a simple countercurrent cascade. It is necessary to know the ratio of reflux to product before the internal circulation can be evaluated. On the other hand, the net flows can be determined from the compositions and quantities of the external streams.

The change of net flow that occurs at the feed requires the use of a new delta point. In the enriching section the delta point implied by Equation 6.1 must be used to calculate from stage to stage. For the stripping section the delta point implied by Equation 6.2 should be used. Equation 6.4 indicates that the compositions of Δ, $\bar{\Delta}$, and F lie on a straight line. Calculations may be carried out on ternary or solvent-free coordinates. The procedure is shown on ternary coordinates in Figure 6.3

The two delta points may be located on the ternary diagram using Equations 6.1, 6.2, and 6.4 if the reflux ratio is known, but first it is necessary to locate the compositions of various streams on the diagram. The entering solvent V_{S+1} may contain a small quantity of solute (a) if the solvent is being reused after removal from the solvent separator. Similarly, the solvent (V_C) removed from the extract in the solvent separator may contain small quantities of a and b as impurities. The compositions y_{S+1} and y_C shown on Figure 6.3 indicate that V_C and V_{S+1} are impure solvents. The feed F may contain solvent, but it is shown by z_F on Figure 6.3 as containing no solvent ($z_{F_c} = 0$). In general, the feed may be L-phase or V-phase or a mixture. Therefore, its composition is denoted by z rather than x or y. Examination of Figure 6.2 shows that x_N, x_B, and x_{S-1} are identical. Since x_N is the composition of a stream flowing from an equilibrium stage, it must lie on the phase envelope, as shown in Figure 6.3. x_D, x_C, and x_0 are identical, since L_C is simply split into D and L_0. If the solvent separation is not complete, L_C will contain a small quantity of solvent, as indicated by x_C on Figure 6.3. By a material balance around the solvent separator, $V_1 = L_C + V_C$; and y_1 may be located at the intersection of the straight line $\overline{y_C x_C}$ and the phase envelope, since V_1 is flowing from an equilibrium stage.

The delta point (x_Δ) is shown by Equation 6.1 to be on line $\overline{x_0 y_1}$ or line $\overline{x_D y_C}$, which are coincident. The exact

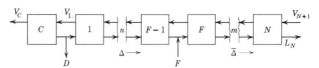

Figure 6.4. Extraction cascade with extract reflux.

location is given by

$$\frac{L_0}{D} = \frac{y_1 - x_\Delta}{x_0 - y_1} \cdot \left(\frac{-\Delta}{D} \right) \qquad (6.5)$$

$$= \frac{y_1 - x_\Delta}{x_0 - y_1} \cdot \frac{x_D}{x_\Delta} \text{ (when } y_{C_a} = 0) \qquad (6.5a)$$

which can be derived from Equation 6.1. The co-ordinates of the delta point (x_Δ) are determined with Equation 6.5 by substituting the known values of reflux ratio (L_0/D), y_1, x_0, Δ, and D and solving for x_Δ. If $y_{Ca} = 0$, Equation 6.5a is used with (L_0/D), y_{1a}, x_{0a}, and x_{Da}. The other delta point ($x_{\bar\Delta}$) is located by use of Equations 6.2 and 6.4, which show that $x_{\bar\Delta}$ lies at the intersection of line $\overline{x_B y_{S+1}}$ and line $\overline{x_{\bar\Delta} z_F}$. Since the values of x_B, x_Δ, y_{S+1}, and z_F are known, the two lines may be drawn and $x_{\bar\Delta}$ located at their intersection. After the two delta points are located as shown on Figure 6.3, they may be used along with equilibrium data to calculate the number of stages required. x_Δ is used to calculate from x_D to z_F, and $x_{\bar\Delta}$ from z_F to x_B.

The method outlined above also holds for calculations on solvent-free coordinates. The appropriate equations may be rewritten in terms of solvent-free masses and mass ratios. The primary difference involves a pure solvent stream which has a composition of $Y_c = \infty$.

Illustration 6.1. A mixture of styrene and ethylbenzene is to be separated by extraction with pure diethylene glycol in a countercurrent cascade with extract reflux. The feed contains 40 per cent styrene and the two products are to contain 90 per cent and 5 per cent styrene (on a solvent-free basis). The ratio of extract reflux to product is to be 10/1. The solvent from the solvent separator contains a negligible quantity of styrene and ethylbenzene. How many equilibrium stages are required? What is the solvent feed rate per 1000 lb of fresh feed?

SOLUTION. The extraction cascade modified for extract reflux alone is shown in Figure 6.4. The solution to the problem is given on Figure 6.5. The solvent-free coordinates are more convenient than the triangular coordinates.

From the data given in the statement of the problem the following points are located on the diagram:

$$Z_{Fa} = 0.40 \qquad Z_{Fc} = 0$$
$$X_{Da} = 0.90 \qquad X_{Dc} = 0$$
$$X_{Na} = 0.05 \qquad X_{Nc} = \text{equilibrium value}$$
$$= 0.008 \text{ (from Figure 3.8)}$$

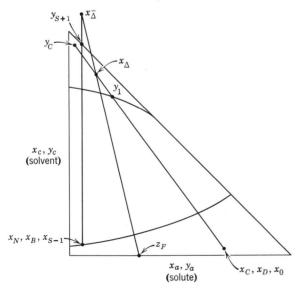

Figure 6.3. Graphical calculation of liquid extraction with reflux.

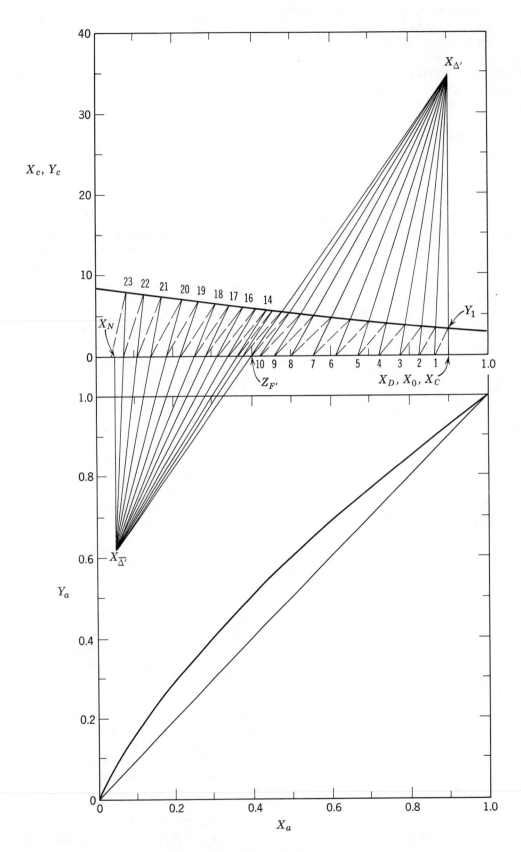

Figure 6.5. Solution to Illustration 6.1.

$Y_{N+1c} = \infty$ and $Y_{Cc} = \infty$, since both are pure solvent streams.

Equation 6.5 may be rewritten for solvent-free coordinates.

$$\frac{L_0'}{D'} = \frac{Y_1 - X_{\Delta'}}{X_D - Y_1} \qquad (6.5b)$$

Since $L_0'/D' = 10$, $X_{\Delta'}$ can be located if Y_1 can be evaluated. $V_1' = V_C' + (L_0' + D')$ by a material balance around the solvent separator (C). Y_1 is determined by the intersection of line $\overline{X_D Y_C}$ with the extract curve, since V_1 is flowing from an equilibrium stage. The line $\overline{X_D Y_C}$ is vertical through X_D because $Y_{Cc} = \infty$. After Y_1 is located Equation 6.5b can be applied.

$$10 = \frac{Y_1 - X_{\Delta'}}{X_D - Y_1}$$

Then $X_{\Delta'} - Y_1 = 10(Y_1 - X_D)$ and $X_{\Delta'}$ is located as shown on Figure 6.5. This delta point is valid for calculating stages from 1 to $F - 1$, but the feed changes the net flow for stages F to N.

Since
$$\bar{\Delta}' = \Delta' + F' \qquad (6.4a)$$

the points $X_{\Delta'}$, Z_F, and $X_{\bar{\Delta}'}$ lie on a straight line. In addition,

$$\bar{\Delta}' = L_N' - V_{N+1}' \qquad (6.2a)$$

But $V_{N+1}' = 0$, and so $X_{\bar{\Delta}a} = X_{Na}$, and $X_{\bar{\Delta}}$ is located at the intersection of the line $X_{\Delta'}Z_F$ and a vertical line through X_N, as shown. Stages are now stepped off using $X_{\Delta'}$ from stage 1 to the feed and $X_{\bar{\Delta}'}$ from the feed to the end of the cascade. Figure 6.5 shows eleven stages in the enriching section and about twelve stages in the stripping section.

The required solvent flow rate can be determined.

Basis: 1000 lb of fresh feed

Since $F' = L_N' + D'$,

$$\frac{L_N'}{F'} = \frac{\overline{X_D'Z_{F'}}}{\overline{X_D'X_N}} = \frac{0.9 - 0.4}{0.9 - 0.05} = 0.59$$

Therefore, $L_N' = 590$ lb

With Equation 6.2a

$$\bar{\Delta}' = L_N' = 590 \text{ lb}$$

From Figure 6.5,

$$X_{\bar{\Delta}'c} = -23.8$$

Therefore, the net flow of solvent $= (-23.8)(590) = -14,030$ lb. The solvent in the raffinate is $(590/0.992)0.008 = 4.7$ lb. Therefore the solvent feed rate is

$$V_{N+1} = 4.7 - (-14,030) = 14,035 \text{ lb per 1000 lb feed}$$

At the solvent separator, since $\bar{\Delta}' = \Delta' + F$

$$\Delta' = -410 \text{ lb}$$

$$X_{\Delta'c} = 34.2$$

Net flow of solvent $= (34.2)(-410) = -14,030$ lb. Since there is no solvent in L_0 or D,

$$V_C = 14,030 \text{ lb per 1000 lb feed}$$

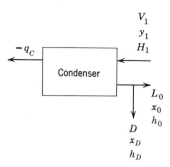

Figure 6.6. Enthalpy and material balances around the condenser.

Distillation. Calculations for distillation are made on the enthalpy-concentration diagram. The heat terms $-q_C$ and q_S replace solvent streams V_C and V_{S+1}, respectively. q is defined as the heat *added* to a system, according to the usual convention in thermodynamics. Therefore, the heat added at the reboiler, or still, is the reboiler duty (q_S), and the heat removed at the condenser is the condenser duty ($-q_C$).

Since $-q_C$ and q_S have no mass, the net flows of mass defined by Equations 6.1, 6.2, and 6.3 become

Enriching: $\quad \Delta = L_0 - V_1 = L_n - V_{n+1} = -D \quad (6.6)$

Stripping: $\quad \bar{\Delta} = L_N - V_S = L_m - V_{m+1} = B \quad (6.7)$

and $\quad\quad F = D + B = \bar{\Delta} - \Delta \quad\quad (6.8)$

The net flow of enthalpy may be defined for the enriching section.

$$\Delta h_\Delta = L_0 h_0 - V_1 H_1 \qquad (6.9)$$

where h_Δ is equal to the net flow of enthalpy per unit of total mass net flow. For use in graphical calculations h_Δ may be related to q_C by an enthalpy balance around the condenser (Figure 6.6). The enthalpies of the streams entering must equal the enthalpies of the streams leaving. The condenser is a special type of heat exchanger, where the heat removed from the vapor in condensation is transferred to a coolant liquid through a metal wall. An enthalpy balance for the coolant gives

$$-q_C = \text{enthalpy of coolant leaving}$$
$$- \text{enthalpy of coolant entering}$$

An enthalpy balance around the condenser then gives

$$V_1 H_1 = L_0 h_0 + D h_D + (-q_C)$$

The quantity of heat added at the condenser per unit mass of distillate, Q_{CD}, can be defined

$$Q_{CD} = \frac{q_C}{D} \qquad (6.10)$$

The enthalpy balance becomes

$$V_1 H_1 = L_0 h_0 + D(h_D - Q_{CD})$$

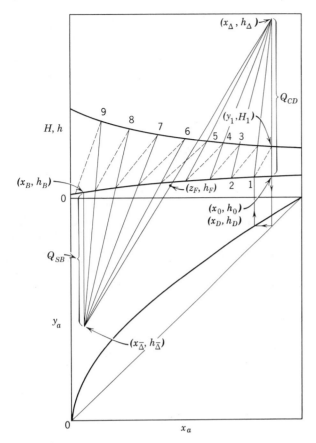

Figure 6.7. Distillation calculations with vapor and liquid reflux.

Combining this equation with Equation 6.9 gives

$$\Delta h_\Delta = L_0 h_0 - V_1 H_1 = -D(h_D - Q_{CD})$$

Solving for the enthalpy coordinate of the delta point gives

$$h_\Delta = \frac{-D(h_D - Q_{CD})}{\Delta} = h_D - Q_{CD} \quad (6.11)$$

$$= \frac{\text{net flow of heat}}{\text{total net flow of mass}}$$

The x-coordinate of the delta point is obtained from

$$\Delta x_\Delta = L_0 x_0 - V_1 y_1 = -D x_D$$

$$x_\Delta = \frac{-D}{\Delta} x_D = x_D \quad (6.12)$$

The two coordinates (h_Δ, x_Δ) of the delta point, as defined by Equations 6.11 and 6.12 can be used to locate the delta point when distillate quantity and composition and condenser duty are known. If the reflux ratio is known, h_Δ may be determined from it, analogous to the derivation of Equation 6.5 for extraction. A material balance around the condenser gives

$$L_0 + D = V_1$$

and an enthalpy balance around the condenser gives

$$L_0 h_0 + D(h_D - Q_{CD}) = V_1 H_1$$

Therefore,

$$L_0 h_0 + D h_\Delta = V_1 H_1$$

$$\frac{L_0}{D} = \frac{h_\Delta - H_1}{H_1 - h_0} \quad (6.13)$$

Equation 6.13 is useful in determining h_Δ from the reflux ratio. The values of L_0/D, H_1, and h_0 may be substituted into Equation 6.13 to evaluate h_Δ, or h_Δ may be located graphically using the equation as a guide. For a total condenser, $h_0 = h_D$ and $x_D = y_1$.

A similar analysis of the stripping section will yield the following relationships:

$$\bar{\Delta} = L_N - V_S = L_m - V_{m+1} = B \quad (6.14)$$

$$h_{\bar{\Delta}} = h_B - Q_{SB} = \frac{\text{net flow of heat}}{\text{total net flow of mass}} \quad (6.15)$$

where

$$Q_{SB} = \frac{q_S}{B} \quad (6.16)$$

$$x_{\bar{\Delta}} = x_B \quad (6.17)$$

$$\frac{V_S}{B} = \frac{h_N - h_{\bar{\Delta}}}{H_S - h_N} \quad (6.18)$$

The direction of the net flows in the distillation column should be considered. Distillation columns are built vertically with the condenser at the top and the reboiler at the bottom. Therefore, net flow defined as positive to the right in Figure 6.2 will be positive in a downward direction in a vertical distillation column. Equation 6.6 shows that the total net flow of mass in the enriching section is upwards, but Equation 6.7 shows that the total net flow in the stripping section is downward. Since heat is added at the reboiler and removed at the condenser, the net flow of heat is upward through both the stripping and enriching sections. Therefore, h_Δ is always positive and $h_{\bar{\Delta}}$ is always negative.

The use of the equations derived previously for graphical calculations is shown in Figure 6.7. For the illustration, the composition and thermal condition of the feed, bottoms, and distillate are assumed to be known. In addition, either the reboiler or condenser duty is known. The feed is shown as a subcooled liquid (z_F, h_F). The distillate is a liquid at its bubble point (x_D, h_D), and the bottoms is a liquid at its bubble point (x_B, h_B). The reflux is also a liquid at its bubble point (x_0, h_0). Since $h_\Delta = h_D - Q_{CD}$ and $x_\Delta = x_0 = x_D$, the delta point for the enriching section can be determined from the condenser duty. The delta point for the stripping section is located by the line $\overline{x_\Delta z_F}$ extended and the vertical line $x_B = x_{\bar{\Delta}}$. Stages may be stepped off from either end with use of the appropriate delta point. In Figure 6.7 the stages are stepped off from the top of the column downward, by alternate use of the equilibrium data and the delta point. Since the column has a total condenser $y_1 = x_0 = x_D$ and the point (y_1, H_1) can be located. This point represents the vapor leaving the first stage. The liquid leaving the first stage is in equilibrium with this vapor. Therefore, its composition (x_1) is determined from the equilibrium curve, and its enthalpy is determined from its composition and the saturated liquid enthalpy curve. To locate (y_2, H_2) a line is drawn from (x_1, h_1) to (x_Δ, h_Δ). From (y_2, H_2) the equilibrium relation is applied to find (x_2, h_2). This stepping-off of stages is continued using (x_Δ, h_Δ) until z_F is reached. Then the construction switches to the delta point for the stripping section $(x_{\bar\Delta}, h_{\bar\Delta})$. The calculation is completed when the liquid composition equals or exceeds the bottoms composition x_B. The calculation of stages across the feed stage as outlined is not rigorous, considering the enthalpy and phase characteristics of the feed. However, it is a sufficiently accurate approximation and is generally used. About 4.6 stages are required in the enriching section and about 4.2 are required in the stripping section. The steps in this calculation are identical to those out-

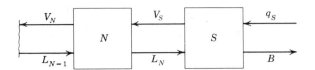

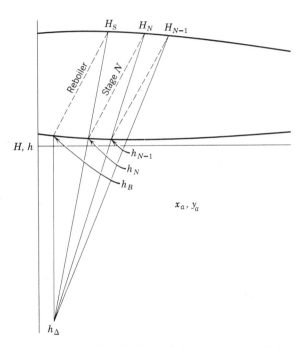

Figure 6.9. Partial reboiler equivalent to one equilibrium stage.

lined in Illustration 6.1 for extraction. The analogy between the two operations should be carefully studied.

Figures 6.2 and 6.6 show a *total condenser* in which all the vapor (V_1) is condensed and then split into distillate (D) and reflux (L_0). In a *partial condenser*, on the other hand, only the reflux is condensed and the distillate is drawn off as a vapor. Since the liquid reflux and vapor distillate are in contact, they may approach equilibrium and give as much as one additional equilibrium stage, as shown in Figure 6.8. In actual practice, the liquid and vapor may not have sufficient contact to equilibrate, so that a partial condenser functions as less than one equilibrium stage.

Figures 6.2 and 6.7 show the bottoms withdrawn before the reboiler, so that the stream entering the reboiler is totally vaporized. In most distillation towers the bottoms are withdrawn from the still, as shown in Figure 6.9. This results in a reboiler that may be equivalent to as much as one equilibrium stage. The reboiler is frequently a steam-heated heat exchanger. Many types of reboiler construction are used in industry. The design of reboilers is based upon heat-transfer considerations covered in later sections of this book.

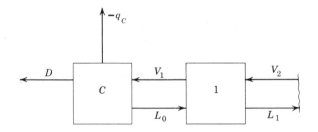

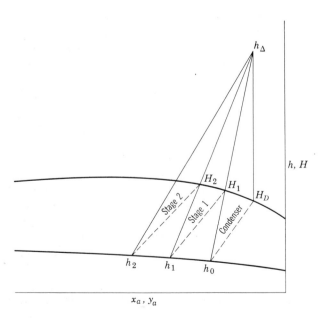

Figure 6.8. Partial condenser equivalent to one equilibrium stage.

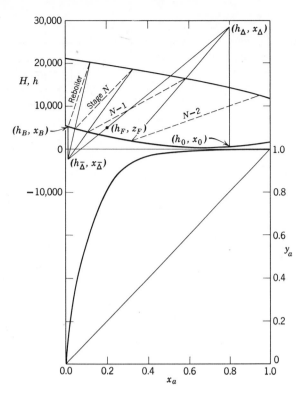

Figure 6.10. Solution to Illustration 6.2.

Illustration 6.2. One hundred pound moles per hour of a 20 mole percent ammonia–water solution is to be distilled at 100 psia in a countercurrent column with reflux. The distillate is to have a composition of 80 mole percent ammonia, and it is to contain 95 per cent of the ammonia charged. There is a total condenser, and the reflux ratio (L_0/D) is to be 1. The bottom product is withdrawn through the reboiler. The feed enters at a temperature of 250°F, and the reflux is at its bubble point. How many equilibrium stages are required? Where is the feed stage?

SOLUTION. $F = 100$ lb moles/hr, $z_F = 0.2$, $x_D = 0.8$ Ammonia in distillate is 95 per cent of that charged: $(0.95)(0.2)(100) = 19$ moles.

$$D = \frac{19}{0.8} = 23.8, \quad B = 76.2, \quad x_B = \frac{1.0}{76.2} = 0.0131$$

x_B, $x_D = x_0$, and z_F are located on Figure 6.10. Since the feed is at 250°F at $z_F = 0.2$, it is a mixture of liquid and vapor lying on the 250°F tie line. h_Δ is located from Equation 6.13.

$$\frac{L_0}{D} = 1 = \frac{h_\Delta - H_1}{H_1 - h_0}$$

and so $\qquad h_\Delta - H_1 = H_1 - h_0$

From Equation 6.8, $\bar{\Delta} = \Delta + F$, and $h_{\bar{\Delta}}$ is located at the intersection of $x_{\bar{\Delta}} = x_B$ and the line $\overline{h_\Delta h_F}$. The stages are now stepped off beginning with the still. $h_{\bar{\Delta}}$ is used from x_B to z_F and h_Δ from z_F to x_D. About 2.7 equilibrium stages plus the still as one stage are required. The feed is at the 2.3 stage from the bottom.

OPERATING VARIABLES

For a feed of given composition and flow rate, the separation at a given pressure into a distillate product and a bottoms product is dependent upon the variables h_F, x_B, x_D, $-q_C$, q_S, L_0/D, and N. When any four of these values are specified, the other three are thereby fixed and may be evaluated. Of the many possible combinations of four specified variables, a few represent most of the common engineering problems.

In the design of new equipment, the most common case involves knowledge of h_F, x_B, and x_D. L_0/D is then chosen to give the most economical combination of equipment costs and energy costs. Equipment costs are proportional to the number of stages N, and energy costs are directly related to the reboiler duty (q_S), since it represents steam supplied to the reboiler. As the number of stages increases the reboiler duty decreases, so that a minimum total cost of equipment plus energy will exist.

In evaluating the operation of existing equipment or in planning a new use for it, N is usually fixed, and L_0/D, q_S, and $-q_C$ may not exceed certain values determined by the size of the equipment. Usually the enthalpy and composition of the feed are known, and the possible compositions of the products are to be evaluated. Some flexibility exists, since L_0/D, q_S, and $-q_C$ may be adjusted over a range below their maximum values.

Reflux. As the ratio of reflux (L_0) to V-phase product (D) is decreased, more stages will be required for a specified separation. Examination of Equation 6.5 shows that, as L_0/D is reduced for fixed values of x_0, x_D, and y_1, x_Δ approaches y_1. This results in a smaller change in composition between stages and requires a greater number of stages. Similarly, Equation 6.13 shows that, as L_0/D is reduced, h_Δ approaches H_1, and a greater number of stages is required.

If L_0/D is reduced to a point where an extended tie line passes through either delta point, a pinch occurs, and an infinite number of stages is required to accomplish the specified separation. This value of L_0/D is referred to as the *minimum reflux ratio*. An L_0/D below the minimum value cannot give the specified separation even with an infinite number of stages. A reflux ratio higher than the minimum requires a finite number of stages for the specified separation.

As the reflux is increased, the number of stages required for a given separation decreases. At *total reflux* no products are withdrawn ($D = 0$, $B = 0$), no feed is supplied ($F = 0$), and a minimum number of stages is required for a specified separation. In such a case, $L_0/D = \infty$ and $V_S/B = \infty$. Under this condition both delta points lie at infinity, and construction lines to the delta point are vertical; or, in other words, $y_{n+1} = x_n$ and $y_{m+1} = x_m$.

The concepts of minimum reflux and minimum stages

at total reflux are useful in estimating the difficulty of separation, in setting the actual operating reflux ratio, and in analyzing column performance.

A higher reflux requires fewer stages but greater reboiler and condenser duties in distillation. Since the heat supplied to the reboiler costs money, an economic balance between operating costs and initial equipment costs will determine the best reflux ratio. Similarly, in extraction a higher reflux ratio gives fewer stages but necessitates a higher solvent rate and a greater load on the solvent separator, thereby increasing operating costs. An economic evaluation can be made to determine the optimum combination of equipment and operating costs for a specified separation. A typical evaluation is shown in Figure 6.11. The optimum design occurs at minimum total cost.

Illustration 6.3. Determine the minimum reflux ratio and the minimum number of stages for the separation of Illustration 6.2.

SOLUTION. From Illustration 6.2, $z_F = 0.2$, $x_D = 0.8$, and $x_B = 0.0131$. These points may be plotted on Figure 6.12. As the delta points are moved closer to H_1 and h_B along the vertical lines at x_D and x_B, respectively, more stages are required. Finally, a point is reached where an infinite number of stages is required. Points still closer to the enthalpy curves will also result in infinite stages, but they will not give the desired separation. The delta points shown on Figure 6.12 give the minimum reflux ratio for the desired separation. That these delta points give an infinite number of stages can be verified by attempting to step off stages from either end of the cascade. A pinch occurs because the line $\overline{x_\Delta x_{\bar\Delta}}$ coincides with a tie line. By Equation 6.13 the minimum reflux ratio is

$$\frac{L_0}{D} = \frac{h_\Delta - H_1}{H_1 - h_0} = \frac{1225 - 825}{825 - 25} = 0.5$$

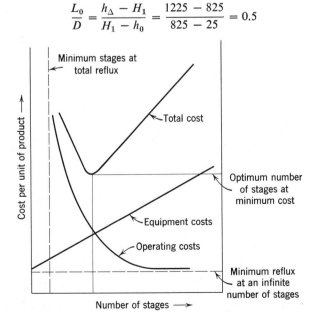

Minimum stages at total reflux

Total cost

Optimum number of stages at minimum cost

Equipment costs

Operating costs

Minimum reflux at an infinite number of stages

Cost per unit of product →

Number of stages →

Figure 6.11. Equipment costs and operating costs in multistage equipment with reflux.

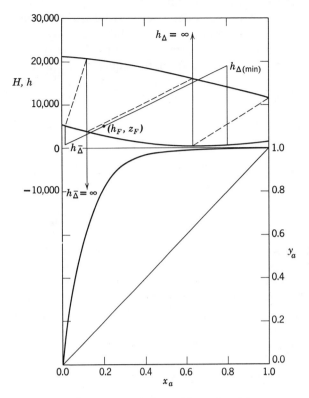

Figure 6.12. Solution to Illustration 6.3.

At total reflux, $L_0/D = \infty$, and both delta points are at infinity. Construction lines $\overline{x_n y_{n+1} x_\Delta}$ are vertical, as shown on Figure 6.12. The minimum number of stages at total reflux is about 2.5. Reference to Illustration 6.2 shows that at a reflux ratio of twice the minimum 3.7 stages are required.

The occurrence of the pinch at the feed point, as in Illustration 6.3, is characteristic of ideal solutions. For many real systems the equilibrium relationships produce a pinch at other compositions. This will be considered again in Chapter 7.

Subcooled Reflux. A total condenser may not only condense but also cool the liquid reflux below its bubble point. Such a reflux is referred to as a *subcooled reflux*. When a subcooled reflux is fed into the distillation column, it causes some vapor to condense and thereby produces a liquid downflow greater than the reflux supplied. The ratio L/V is often referred to as the *internal* reflux ratio. For the same value of the external reflux ratio (L_0/D) a subcooled reflux will produce a larger internal reflux ratio than will a reflux at its bubble point. All the derived equations hold for subcooled reflux, providing h_0 is properly plotted below the saturated-liquid enthalpy curve. It should be noted that subcooled reflux does not change the net flow. A subcooled reflux in distillation is analogous to an extract reflux in extraction which contains a quantity of solvent less than that which would give an over-all composition lying on the phase envelope.

Entrainment. Entrainment results when the heavier L-phase is not completely separated from the lighter V-phase before each leaves the equilibrium stage. The carry-over of L-phase results in an effective V-phase composition different from the equilibrium value. If the ratio of entrained L-phase to V-phase is known, a new effective V-phase locus can be determined and used

When the feed (F) is added, the net flow is

$$\bar{\bar{\Delta}} = \bar{\Delta} + F = \Delta + I + F \tag{6.21}$$

The stream I may have either a positive value (a feed) or a negative value (a product). Additional intermediate streams may be calculated by defining additional net flows.

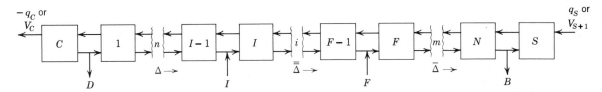

Figure 6.13. Intermediate stream.

to calculate the required number of equilibrium stages. Examples of entrainment of the L-phase in the V-phase include carry-over of liquid with the vapor in a bubble-cap column in distillation and the carry-over of insoluble solids with the extract in leaching. In liquid-liquid extraction both the extract and the raffinate may be entrained in the other phase.

INTERMEDIATE STREAMS

In any mass-transfer operation it is possible to have more than one feed or more than two product streams. For example, in crude-petroleum distillation, light gas, propane-butane, gasoline, naphtha, kerosene, gas oil, lubricating-oil stock, fuel oil, and asphalt may all be withdrawn from the column at different points. A typical industrial crude-petroleum distillation unit is shown in Figure 7.13. Calculation of the distillation behavior of such multicomponent mixtures is extremely complex. A simple case of an intermediate stream for binary distillation or extraction may be handled by the methods already covered. An example of this, with two feeds, is shown in Figure 6.13. The introduction or removal of an intermediate stream (I) changes the net flow. Therefore, a new delta point must be established. In Figure 6.13 there are three different net flows and, therefore, three delta points: Δ for stages 1 through I-1, $\bar{\Delta}$ for stages I through F-1, and $\bar{\bar{\Delta}}$ from stages F through N.

The equations for the total net flows are

$$\Delta = L_0 - V_1 = L_n - V_{n+1} = -(D + V_C) \tag{6.1}$$
$$\bar{\bar{\Delta}} = L_N - V_S = L_m - V_{m+1} = (B - V_{S+1}) \tag{6.2}$$
$$\bar{\Delta} = L_i - V_{i+1} \tag{6.19}$$

The subscript i refers to any stage in the intermediate section of the column between stage I and stage F-1.

The net flow at the left end of the cascade is Δ. After the intermediate stream (I) is added, the net flow becomes

$$\bar{\Delta} = \Delta + I \tag{6.20}$$

Since there are three different net flows, there will be three delta points. The determination of the three delta points is shown in Figure 6.14, for a typical case. If the reflux ratio is known, x_Δ may be located on the line $\overline{x_0 y_1}$ by use of Equation 6.5a. Once x_Δ has been located, $x_{\bar{\Delta}}$ may be determined by reference to Equations 6.2 and 6.21. Equation 6.2 implies that $x_{\bar{\Delta}}$ will be on the line $\overline{x_B y_{S+1}}$, so this line is drawn on the diagram. Equation 6.21 can be utilized to construct another line which intersects $\overline{x_B y_{S+1}}$ at $x_{\bar{\Delta}}$. Since there are three terms on the right-hand side of Equation 6.21, the addition must be carried out in two steps. First I and F are graphically added to give the sum Σ. The Σ and Δ may be

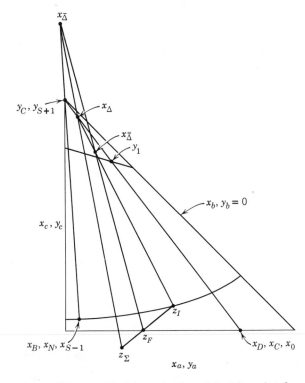

Figure 6.14. Graphical determination of the delta points for liquid extraction with an intermediate product stream.

added to give $\bar{\Delta}$; i.e., $x_{\bar{\Delta}}$ will lie on the straight line $\overline{x_\Delta z_\Sigma}$. The intersection of this line with $\overline{x_B y_{S+1}}$ is x_Δ.

The sum Σ may be considered as a "fictitious feed." The sections of the cascade between 1 and I and between F and N behave as if there were a single feed $\Sigma = I + F$ at some point between I and F. The net flows in these two sections may be determined by combining I and F into the fictitious feed (Σ) and treating the cascade as the usual case of one feed. Since the stream I was a product in the illustration of Figure 6.13, the numerical value of I is negative and z_Σ lies beyond z_F on the line $\overline{z_I z_F}$, as in the usual case of subtraction. The concept of the "fictitious feed" is abandoned in order to determine $x_{\bar{\Delta}}$, which is an expression of the net flow from I to F. The third delta point is located by inference from Equations 6.20 and 6.21 at the intersection of the lines $\overline{x_{\bar\Delta} z_I}$ and $\overline{x_{\bar\Delta} z_F}$.

The stages may now be stepped off using the appropriate delta point in each section of the cascade.

Intermediate streams in distillation and other operations are calculated in a similar manner. An "intermediate stream" in distillation may consist of only the addition or removal of heat.

PROBLEMS

6.1. Evaluate the total and component net flows in Illustration 6.1.

6.2. Using the same rates and compositions of solvent and feed as in Illustration 6.1, determine the maximum possible concentration of styrene in the extract product of a simple countercurrent cascade.

6.3. Locate the delta points of Illustration 6.1 on a ternary diagram. Attempt to step off the required number of stages on this diagram.

6.4. Prove that $L_N'/F' = \overline{X_D Z_F}/\overline{X_D X_N}$

6.5. Derive Equation 6.5b.

6.6. Derive Equations 6.14 through 6.18.

6.7. A flow of 100 lb moles/hr of a vapor containing 0.40 mole fraction NH_3 and 0.6 mole fraction H_2O is to be enriched in ammonia in a distillation tower which consists of an enriching section and a total condenser. The feed into the bottom of the column is a saturated vapor (at its dew point) at 100 psia, the operating pressure. The distillate is withdrawn from a total condenser and has a composition of 0.90 mole fraction NH_3. Part of the liquid condensed in the total condenser is withdrawn as the distillate product and part is returned to the column as reflux. 85 per cent of the ammonia charged must be recovered in the distillate.

(a) Calculate the number of equilibrium stages required and the reflux ratio (L_0/D).

(b) What is the condenser duty (heat removed per hour from the condenser)?

(c) After the column has been built and is operating, a change in the process changes the feed to 0.2 mole fraction NH_3 at its dew point. Can the existing column of part (a) be used? The reflux ratio can be adjusted for this new feed, but the number of theoretical stages is fixed. The distillate composition and percentage recovery of NH_3 must be unchanged.

6.8. A liquid ethanol–water feed is to be stripped of ethanol in distillation column consisting of a stripping section and a reboiler.

The liquid will be fed to the top plate and vapor reflux will be furnished by a reboiler which takes liquid from the bottom plate and totally vaporizes it. The bottoms product is also withdrawn from the bottom plate.

(a) Calculate the number of equilibrium stages required for the following conditions:

Feed: 100 lb moles/hr, 0.2 mole fraction ethanol, 0.8 mole fraction water at 100°F and 1 atm

Vapor product: 0.5 mole fraction ethanol

Bottom product: 0.03 mole fraction ethanol

(b) What is the reboiler duty (Btu/hr added at reboiler)?

(c) Can a vapor product of 0.7 mole fraction ethanol be obtained with a stripping unit like this? (You may use more stages and higher reflux ratio but use the same feed and bottoms composition.) Explain.

6.9. 72 lb moles/hr of an ammonia–water mixture at 100 psia and 70°F, containing 25 mole percent ammonia, is to be fractionated in a distillation tower into an overhead product of 95 per cent ammonia and a bottoms of 4 per cent ammonia. The overhead product is withdrawn as a vapor in equilibrium with the reflux from a partial condenser; and the bottoms are withdrawn from the reboiler. The reboiler duty is 700,000 Btu/hr.

(a) What is the condenser duty (Btu/hr)?

(b) What is the reflux ratio (L_0/D)?

(c) How many equilibrium stages are required?

(d) What is the net flow of heat in the column?

(e) What is the minimum reflux ratio for this separation?

(f) What is the minimum number of stages required at total reflux?

6.10. 100 lb moles/hr of an aqueous solution containing 20 mole percent ethanol is to be fractionated at 1 atm to produce a distillate of 80 mole percent ethanol and a bottoms of 2 mole percent ethanol. The total condenser has a duty of 1,000,000 Btu/hr. The bottoms are withdrawn from the bottom plate. The feed is a liquid at 140°F.

(a) How many equilibrium stages are required?

(b) What is the reboiler duty?

(c) Plot the number of stages required for this separation as a function of reflux ratio. The plot should extend from the minimum reflux to total reflux.

(d) How many stages would be required if the feed were half liquid and half vapor?

6.11. 2000 lb/hr of a cottonseed oil–oleic acid solution containing 30 per cent acid is to be extracted with propane in a continuous countercurrent cascade at 98.5°C using extract reflux. The extract product is to contain 85 per cent acid, and it is to have a negligible propane content. The raffinate product should contain 3 per cent acid. The extract reflux ratio is to be 3.5. The fresh solvent and the recovered solvent are pure propane.

(a) Calculate the number of equilibrium stages required on the triangular diagram.

(b) Calculate the number of equilibrium stages required on solvent-free coordinates.

(c) Determine the minimum reflux ratio and the minimum number of stages.

6.12. A gaseous mixture of propane and propylene can be separated by fractional adsorption using silica gel at 1 atm pressure. The silica gel will be fed into the top of the column and will pass countercurrent to the gas. At the bottom of the column all the adsorbed gas will be stripped from the silica gel. Part of the gas will be withdrawn as product, and part will be fed back into the column as reflux. The feed contains 60 mole percent propane, and the product streams are to contain 90 per cent and 5 per cent propane.

(a) What is the minimum reflux ratio?

(b) How many equilibrium stages are required at twice the minimum reflux ratio?

(c) How much silica gel per pound mole of feed is required at twice the minimum reflux ratio?

6.13. 50 lb moles/hr of an aqueous ethanol solution containing 23 mole percent ethanol is to be fractionated at 1 atm in a distillation column equipped with a total condenser and a reboiler from which the bottoms are withdrawn. The distillate is to have a composition of 82 mole percent ethanol and the bottoms 3 mole percent ethanol. The feed is at its bubble point.

(a) At a reflux ratio (L_0/D) of 3, how many equilibrium stages are required?

(b) Plot the number of stages required for this separation as a function of the reflux ratio from minimum reflux to total reflux.

(c) Plot the reboiler duty as a function of reflux ratio from minimum reflux to total reflux.

(d) How many stages would be required if the reflux ratio (L_0/D) is 3 but if the reflux is subcooled 30°F below its bubble point?

6.14. A mixture is to be fractionated into a distillate, a bottoms, and an intermediate stream between the feed and the bottoms.

(a) Draw a flow sheet for the fractionating column, and label streams.

(b) Sketch a typical enthalpy-concentration diagram and show how the location of the required delta points is determined. Assume that L_0/D is set and that the quantities, compositions, and enthalpies of F, D, B, and I are known.

6.15. An aqueous solution of ammonia containing 27 mole percent ammonia is to be fractionated at 100 psia in a tower equivalent to five equilibrium stages. It is desired to recover 95 per cent of the ammonia in a distillate which has a concentration of 98 mole percent ammonia. The feed solution is 20°F below its bubble point. What reflux ratio (L_0/D) should be used?

6.16. Diphenylhexane is to be separated from docosane by extraction with furfural in a simple countercurrent cascade at 45°C. There are two sources of furfural solvent, one pure and one containing a small amount of diphenylhexane and docosane. Since it is desired to process a maximum quantity of feed, both solvents must be used. It will be desirable to feed the impure solvent separately from the pure solvent.

Stream	Mass Flow Rate, lb/hr	Mass Fraction Furfural	Mass Fraction Docosane	Mass Fraction Diphenyl-hexane
Entering raffinate		0.00	0.70	0.30
Entering solvent 1	5000	1.00	0	0
Entering solvent 2	3000	0.92	0.05	0.03
Exit extract				0.15
Exit raffinate				0.01

(a) How much entering raffinate can be processed with the available solvent?

(b) Determine the mass of the exit extract and exit raffinate.

(c) Determine the number of equilibrium stages required.

(d) At what stage should the impure solvent be fed? (Data for this system are in Chapter 5, Problem 5.13.)

6.17. In Problem 6.10 how many equilibrium stages are required if the liquid entrainment amounts to 10 mole percent of the vapor leaving each stage?

6.18. Derive Equations 6.20 and 6.21.

6.19. The oil from halibut livers is to be extracted using ethyl ether as a solvent in a simple countercurrent cascade. There are available two sources of halibut livers: 200 lb/hr of livers of 10 per cent oil, and 90 per cent insoluble; 300 lb/hr of livers of 35 per cent oil and 65 per cent insoluble. The entering ethyl ether solvent contains 2 per cent halibut-liver oil. The final extract is to contain 75 per cent oil and 25 per cent ether, and 95 per cent of the oil charged in the livers is to be recovered in the extract.

(a) Calculate the number of equilibrium stages required when each liver source is fed separately into the cascade at the best location. Underflow data are in Fig. 3.14.

(b) Calculate the number of equilibrium stages required when the two sources of livers are mixed and fed together into the cascade. Use the same solvent rate and recovery as in part (a).

6.20. A mixture of 100 lb moles/hr of 40 mole percent a and 60 mole percent b is to be separated into a distillate of composition 90 mole percent a and a bottoms of 5 mole percent a. Instead of using a conventional reboiler or condenser, the following proposal has been made.

Each theoretical plate in the enriching section will have a cooling coil which will remove 100,000 Btu/hr from the liquid on the plate.

Each theoretical plate in the stripping section will have a heating coil which will supply 100,000 Btu/hr to the liquid on the plate.

(a) Calculate the number of equilibrium stages required to give the desired separation.

Relative volatility: $\alpha_{a-b} = 3.0$

Enthalpy: $H = 9000 - 5000x$; $h = 1000x$

where x = mole fraction a

(b) Is there any advantage to this arrangement over the conventional condenser and reboiler?

6.21. Is reflux useful in leaching? Explain.

6.22. Although Figure 6.2 implies a raffinate reflux when applied to liquid extraction, raffinate reflux is never used in practice because it does not improve the separation. Using an appropriate extraction diagram, show that raffinate reflux does not reduce the number of stages required for a given separation, either with or without extract reflux.

6.23. Compare the effect of using a total reboiler (Fig. 6.2) in place of a partial reboiler (Fig. 6.9) on the number of equilibrium stages required for a given separation. What conclusion can be drawn from this comparison?

chapter 7

Special Cases in Stage Operations: Simplified Calculation Methods

The methods presented in the previous chapters can be applied to the calculation of multistage equipment when the data for the system of interest are available. In many cases more simplified graphical and analytical methods may be used. A number of such simplified methods and their limitations will be considered in this chapter. The methods are in many cases more rapid than those considered previously, and often they require less physical data. However, the methods are based upon certain simplifying assumptions, which the system under consideration must follow. Simplified graphical procedures will be developed for simple countercurrent flow and for countercurrent flow with reflux. All the concepts covered in the preceding chapters will be redeveloped using the simplified graphical procedures. In addition, certain analytical procedures will be considered.

GRAPHICAL CALCULATIONS ON THE EQUILIBRIUM DIAGRAM

In certain cases of stage operations the equilibrium x–y diagram may be used alone for a simplified calculation of the number of equilibrium stages required for a given separation.

A material balance around any stage n and the left end of the simple countercurrent cascade in Figure 5.2 gives

Total material balance:

$$L_0 + V_{n+1} = L_n + V_1 \qquad (7.1)$$

Component balance:

$$L_0 x_0 + V_{n+1} y_{n+1} = L_n x_n + V_1 y_1 \qquad (7.2)$$

Solving Equation 7.2 for y_{n+1} gives

$$y_{n+1} = \frac{L_n}{V_{n+1}} x_n + \frac{V_1 y_1 - L_0 x_0}{V_{n+1}} \qquad (7.3)$$

Equation 7.3 relates the compositions of a V-phase (y_{n+1}) and an L-phase (x_n) flowing past each other between stages. This was precisely the purpose of the delta point defined in an earlier chapter. If the line represented by Equation 7.3 is plotted on x–y coordinates it will be the locus of all possible values of (x_n, y_{n+1}). It is usually referred to as the "operating line." If the composition (x_n) of the L-phase is known for any point in the cascade, the composition (y_{n+1}) of the V-phase flowing in the opposite direction at the same point can be determined from the plot of Equation 7.3. If in addition the equilibrium curve is plotted on the same x–y diagram, it is possible to calculate from stage to stage graphically.

Equation 7.3 will yield a straight operating line if L and V are constant through the cascade. The terms L_0, V_1, x_0, and y_1 have unique values, and therefore, if V is constant, the second term of Equation 7.3 is a constant. If in addition L is constant, the equation has the form of the standard slope-intercept equation of a straight line. The subscripts on L and V may be dropped, since the terms are constant.

$$y_{n+1} = \frac{L}{V} x_n + \frac{V y_1 - L x_0}{V} \qquad (7.3a)$$

The slope of the operating line is L/V, and the y-intercept is the last term of Equation 7.3a. The line representing 7.3a is easily plotted if one point and the slope are known. If L and V are not constant, the line will not be straight and more detailed calculations are required to plot it.

For most purposes the simplified method is more convenient only if Equation 7.3 describes a straight line. For this reason, it is of interest to examine several stage operations to determine under what circumstances L and V are constant.

In *distillation* if the heat effects in the column do not change the molar flow rate of liquid or vapor from stage to stage, Equation 7.3 will describe a straight line. The liquid and vapor flows in a binary distillation column are influenced by four factors:

1. The molar heat of vaporization of mixtures.
2. The heat of mixing in the vapor and liquid.
3. The increase in sensible heat with increase in temperature through the column.
4. Heat losses from the column walls.

If the molar heats of vaporization of all mixtures are constant and if the other factors are negligible in comparison, for every mole of any liquid vaporized one mole of vapor will be condensed, and L and V will be constant through the cascade. Constant molar heats of vaporization give parallel vapor and liquid lines on the enthalpy-molar composition diagram. Under this condition it can be shown geometrically that L and V do not vary. It is possible to have a straight operating line even if all four factors vary and are influential, as long as the net effect gives constant liquid and vapor flow. Many binary systems give sufficiently constant molar flow rates to permit use of the simplified calculation methods. When applied to distillation, the simplified procedures are often referred to as the *McCabe-Thiele method* (2).

When the two solvent components (b and c) in *liquid-liquid extraction* are completely immiscible at the concentrations of solute (a) under consideration, data can be reported as concentration (y_a) of a in phase V in equilibrium with concentration (x_a) of a in phase L. Since components b and c are mutually insoluble, a ternary diagram is unnecessary. Equilibrium data may be presented on x-y diagrams if the two solvent phases are mutually insoluble. However, the total mass flow rate of the raffinate (L) will decrease as the solute is transferred to the extract phase (V), and the extract phase will increase in total mass. This will not give a straight line with Equation 7.3. The mass of each solvent will not vary from stage to stage, so that Equation 7.3 will be a straight line if the symbols are redefined in terms of mass ratios instead of mass fractions, as follows:

$$Y_{n+1} = \frac{L'}{V'} X_n + \frac{V'Y_1 - L'X_0}{V'} \qquad (7.3b)$$

where $L' =$ mass of unextracted raffinate component (b), not including solute (a)

$V' =$ mass of extract solvent (c), not including solute (a)

$Y_{n+1} =$ mass of solute (a) per unit mass solvent (c)

$X_n =$ mass of solute (a) per unit mass of unextracted raffinate component (b)

The extraction equilibrium data must be replotted on this basis on X-Y coordinates. The equations derived subsequently in the dimensions of Equation 7.3 may also be written in the dimensions of Equation 7.3b.

In many cases of *gas absorption*, a single gas is absorbed from another nonabsorbed gas into a relatively nonvolatile liquid. An example is the absorption of ammonia from air-ammonia mixtures by water. Equation 7.3b can be used, where now

$L' =$ moles of liquid absorbent (b)

$V' =$ moles of the nonabsorbed constituent of the gas (c)

$Y_{n+1} =$ mole ratio: moles of solute (a) in gas per mole of nonabsorbed gas (c)

$X_n =$ mole ratio: moles of solute (a) in liquid per mole of liquid absorbent (b)

Where a gas is adsorbed by a solid from a nonadsorbed gas, Equation 7.3b may be used. An example of this would be the adsorption of water vapor by silica gel from humid air. Equilibrium data must be expressed in the appropriate units.

GRAPHICAL CALCULATIONS FOR SIMPLE COUNTERCURRENT OPERATIONS

The graphical determination of the number of equilibrium stages involves alternate use of the operating line and the equilibrium curve. An example of stage calculations for simple countercurrent operation is shown in Figure 7.1. The operating line may be plotted either by knowing all four of the compositions at both ends of

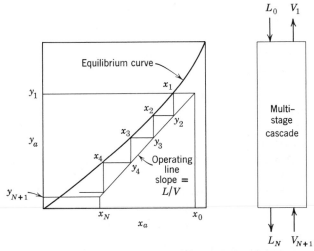

Figure 7.1. Calculations on the equilibrium diagram for simple countercurrent flow with transfer from the L-phase to the V-phase.

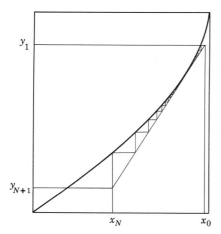

(a) Transfer from L-phase to V-phase.
Recovery specified; y_1 to be determined.

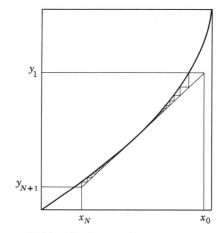

(b) Transfer from L-phase to V-phase.
y_1 specified; x_N to be determined.

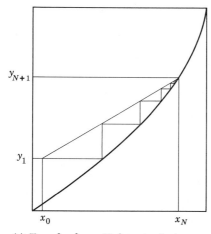

(c) Transfer from V-phase to L-phase.
Recovery specified; x_N to be determined.

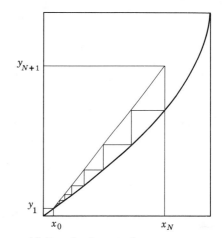

(d) Transfer from V-phase to L-phase.
x_N specified; y_1 to be determined.

Figure 7.2. Limiting values of the ratio of phase flow rates.

the cascade or by knowing three compositions and the slope (L/V) of the operating line. Equation 7.3a can be rearranged to give

$$L(x_n - x_0) = V(y_{n+1} - y_1) \qquad (7.3c)$$

or

$$\frac{L}{V} = \frac{y_{n+1} - y_1}{x_n - x_0} \qquad (7.3d)$$

Even though the mathematical expression for the operating line is valid for any values of y_{n+1} and x_n, it has physical meaning only for compositions which actually occur in the cascade. That is, the actual operating line extends from the point (x_0, y_1) at one end of the cascade to the point (x_N, y_{N+1}) at the other end, as shown on Figure 7.1.

The equilibrium curve can be plotted in the same range of x and y, as shown in Figure 7.1. The procedure for

stepping off stages may begin at either end of the cascade. For example, if the calculation is begun with the composition (y_1) of the V-phase leaving the cascade, the composition (x_1) of the L-phase leaving the first stage is determined by drawing a horizontal at y_1, on the equilibrium diagram. The horizontal intersects the equilibrium curve at x_1, as shown. Next, the composition y_2 must be determined from x_1 by use of the equation for the operating line. This equation written specifically for the flow between stages 1 and 2 is

$$y_2 = \frac{L}{V} x_1 + \frac{V y_1 - L x_0}{V} \qquad (7.3a)$$

Since the operating line is a plot of the general form of this equation, the value of y_2 is determined by the intersection of a vertical through x_1 with the operating line, as shown. Now x_2 is determined by the intersection of a horizontal at y_2 with the equilibrium curve, and this

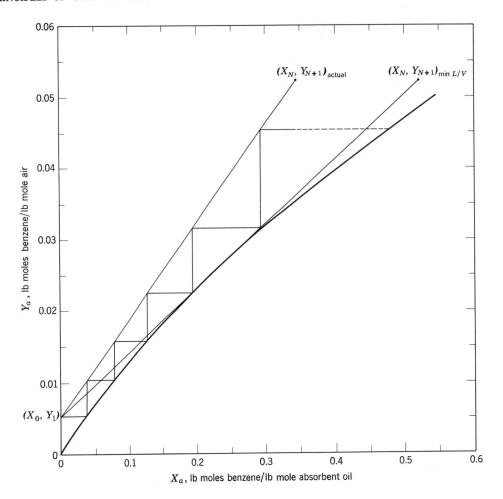

Figure 7.3. Solution to Illustration 7.1.

stepwise calculation is continued until (x_N, y_{N+1}) is reached. As shown in Figure 7.1 slightly over four stages are required. The location of the operating line below the equilibrium curve indicates that mass transfer is from the L-phase to the V-phase. On the other hand, an operating line above the equilibrium curve shows that mass transfer is from the V-phase to the L-phase.

Operating Variables. Limiting values of the L/V ratio may be determined for several cases. For transfer from the L-phase to the V-phase, the minimum V/L ratio, which gives the desired separation with an infinite number of stages, can be determined as shown in Figure 7.2. For a specified recovery from the L-phase, the point (x_N, y_{N+1}) is fixed. As the quantity of V-phase is reduced, the slope (L/V) of the operating line through (x_N, y_{N+1}) increases until the operating line first touches the equilibrium curve, as shown in Figure 7.2a. An attempt to step off stages shows that an infinite number of stages is required to change the composition across the point at which the operating line and equilibrium curve intersect. The term "pinch" applies here to the

pinched-in stages at the point of intersection. If the concentration of the V-phase leaving the cascade is specified, the point (x_0, y_1) is fixed, and the pinch occurs as shown in Figure 7.2b. However, the recovery is not specified; that is, x_N is not fixed. In this case the operating line through (x_0, y_1) which first intersects the equilibrium curve gives the limiting value of V/L, which is a maximum value.

Limiting ratios of L/V can be determined in a similar manner for the case where mass transfer is from the V-phase to the L-phase, such as in gas absorption. For transfer from the V-phase to the L-phase, the operating line will be above the equilibrium curve. When the recovery of component a is specified, the value of y_1 is set since y_{N+1} is known. Since the entering L-phase composition (x_0) is usually known, the point (x_0, y_1) is fixed and the limiting L/V is located as shown in Figure 7.2c. In this case the limiting L/V is a minimum. On the other hand, when x_0 and y_{N+1} are known instead of the recovery, x_N may be specified. Then the point (x_N, y_{N+1}) is fixed and the limiting L/V is determined as shown in Figure 7.2d. This limiting ratio is a maximum.

Illustration 7.1. A benzene–air mixture is to be scrubbed in a simple countercurrent absorption tower using a non-volatile hydrocarbon oil as solvent. The inlet gas contains 5 per cent benzene and the entering gas flow is 600 lb moles/hr. Solubility of benzene in oil follows Raoult's law. The tower operates isothermally at 80°F. The average molecular weight of the oil is 200 and the tower pressure is 1 atm.

(*a*) What is the minimum oil rate (lb/hr) needed to recover 90 per cent of the entering benzene?

(*b*) How many theoretical stages are required if the oil rate is 1.5 times the minimum?

SOLUTION. Since the recovery is specified and the mass transfer is from the *V*-phase to the *L*-phase, the limiting L/V ratio will be determined as shown in Figure 7.2*c*. To assure a straight operating line, mole ratios will be used. This necessitates calculating the equilibrium curve for mole-ratio coordinates.

$$y_a = \frac{P_a}{P} x_a \quad \text{for benzene}$$

At 80°F, $P_a = 103$ mm Hg; $P = 760$ mm Hg.

$$y_a = \frac{103}{760} x_a = 0.136 x_a$$

Since $y_a = \dfrac{Y_a}{1 + Y_a}$ and $x_a = \dfrac{X_a}{1 + X_a}$

$$\frac{Y_a}{1 + Y_a} = 0.136 \frac{X_a}{1 + X_a}$$

Y_a	0	0.005	0.01	0.02	0.03	0.04	0.05
X_a	0	0.038	0.0785	0.168	0.272	0.395	0.539

These values are plotted on Figure 7.3.

$$V = 600 \text{ lb moles/hr}$$

$$X_0 = 0.0, \qquad Y_{N+1} = \frac{0.05}{0.95} = 0.0526$$

For 90 per cent recovery, the benzene leaving in the gas will be $(0.10)(0.05)(600) = 3.0$ lb moles. Therefore, $Y_1 = 3.0/570 = 0.00526$ lb mole benzene/lb mole air, since $V_1' = V_{N+1}' = (0.95)(600) = 570$ lb moles air. The conditions at the upper end of the tower are set and can be plotted: $X_0 = 0$, $Y_1 = 0.00526$. The minimum liquid rate occurs when the operating line through (X_0, Y_1) first touches the equilibrium curve, as shown. The slope of this line is

$$\left(\frac{L'}{V'}\right)_{\min} = \frac{0.0526 - 0.00526}{0.52 - 0} = 0.091$$

Therefore, the minimum oil rate is

$$(0.091)(570)(200) = 10{,}390 \text{ lb/hr}$$

At $(L'/V')_{\text{actual}} = 1.5(L'/V')_{\min} = (1.5)(0.091) = 0.137$,

$$L' = (0.137)(570)(200) = 15{,}600 \text{ lb/hr}$$

The actual operating line has a slope of 0.137, and it is determined that $X_N = 0.345$.

The stages may be stepped off from either end. About 5.3 stages are required for a liquid rate of 15,600 lb/hr, as shown on Figure 7.3. Use of mole fraction coordinates with the assumption of constant L/V in the problem would result in a substantial error. Although the gas phase is dilute, the liquid phase is not. The equilibrium curve shows appreciable curvature.

COUNTERCURRENT FLOW WITH REFLUX

The introduction of reflux (Figure 6.2) results in two net flows and two operating lines. A material balance from the solute-rich end to include any stage *n* in the enriching section gives

$$y_{n+1} = \frac{L_n}{V_{n+1}} x_n + \frac{Dx_D + V_C y_C}{V_{n+1}} \tag{7.4}$$

When *L* and *V* are constant, the subscripts may be dropped.

$$y_{n+1} = \frac{L}{V} x_n + \frac{Dx_D + V_C y_C}{V} \tag{7.4a}$$

A material balance between any stage *m* in the stripping section and the solute-poor end of the cascade gives

$$y_m = \frac{L_{m-1}}{V_m} x_{m-1} - \frac{Bx_B - V_{S+1} y_{S+1}}{V_m} \tag{7.5}$$

The values of *L* and *V* may change at the feed. Therefore,

$$y_m = \frac{\bar{L}}{\bar{V}} x_{m-1} - \frac{Bx_B - V_{S+1} y_{S+1}}{\bar{V}} \tag{7.5a}$$

where $\bar{L}$ and $\bar{V}$ are the constant values of *L*-phase and *V*-phase flow in the stripping section.

In distillation calculations $V_{S+1} = 0$ and $V_C = 0$; and Equations 7.4*a* and 7.5*a* become

$$y_{n+1} = \frac{L}{V} x_n + \frac{D}{V} x_D \tag{7.4b}$$

or

$$= \frac{L}{L + D} x_n + \frac{D}{L + D} x_D \tag{7.4c}$$

$$y_m = \frac{\bar{L}}{\bar{V}} x_{m-1} - \frac{B}{\bar{V}} x_B \tag{7.5b}$$

or

$$= \frac{\bar{V} + B}{\bar{V}} x_{m-1} - \frac{B}{\bar{V}} x_B \tag{7.5c}$$

Inspection of the equation for the enriching-section operating line (Equation 7.4*b*) shows that it is a straight line when *L* and *V* are constant. The slope of the line is L/V and its intercept at $x = 0$ is $y = Dx_D/V$. Furthermore, the operating line intersects the diagonal ($x = y$) at $x_n = y_{n+1} = x_D$. Similarly, for the stripping section (Equation 7.5*b*), the operating line has a slope $\bar{L}/\bar{V}$ and a *y*-intercept at $-Bx_B/\bar{V}$. It intersects the diagonal at

$x_{m-1} = y_m = x_B$. The method of plotting the two operating lines depends upon the known information. The intersections with the diagonal and the slopes or y-intercepts are commonly used.

The two operating lines can be plotted on the equilibrium diagram if sufficient information is available. Equation 7.4 or 7.4b will give a straight operating line for the enriching section of the cascade as long as the L-phase and the V-phase flows are constant. However, the addition of the feed may change both of the phase flow rates. This results in a new operating line of a different slope for the stripping section, as given by Equation 7.5 or 7.5b. The changes in the L-phase and V-phase flows depend upon the quantity and properties of the feed. The following derivation will interrelate the two operating lines with the feed characteristics. The intersection of the two operating lines properly occurs at the feed. The equations for the two operating lines will be solved simultaneously to develop an equation which will give, in a general form, the locus of all possible values of the compositions at the intersection. This equation will be of great use in locating the operating lines on the equilibrium diagram. The following derivation shows that the locus of the intersections can be expressed in terms of properties of the feed alone.

At the intersection of the two operating lines a point on one line must be identical to a point on the other, so that $x_m = x_i = x_n$ and $y_m = y_i = y_{n+1}$ where the subscript i refers to the intersection value. Equations 7.4a and 7.5a then become, at their intersections,

$$Vy_i = Lx_i + Dx_D + V_Cy_C \qquad (7.6)$$

$$\bar{V}y_i = \bar{L}x_i - Bx_B + V_{S+1}y_{S+1} \qquad (7.7)$$

Subtracting Equation 7.6 from Equation 7.7 gives

$$(\bar{V} - V)y_i = (\bar{L} - L)x_i - Bx_B - Dx_D$$
$$+ V_{S+1}y_{S+1} - V_Cy_C \qquad (7.8)$$

This equation may be simplified by combining it with a material balance around the entire cascade (Figure 6.2), which is

$$Fz_F = Bx_B + Dx_D + V_Cy_C - V_{S+1}y_{S+1} \qquad (7.9)$$

and $\qquad F = B + D + V_C - \bar{V}_{S+1} \qquad (7.10)$

Combination of Equation 7.8 with Equation 7.9 gives

$$(\bar{L} - L)x_i - Fz_F = (\bar{V} - V)y_i \qquad (7.11)$$

From net flow considerations

$$\bar{V} = \bar{L} - \bar{\Delta} = \bar{L} - (B - V_{S+1}) \qquad (7.12)$$

$$V = L - \Delta = L + (D + V_C) \qquad (7.13)$$

Substitution in Equation 7.11 for $\bar{V}$ and V from Equations 7.12 and 7.13 gives

$$(\bar{L} - L)x_i - Fz_F = (\bar{L} - L - B + V_{S+1} - D - V_C)y_i$$
$$= (\bar{L} - L - F)y_i \quad \text{from Equation 7.10}$$

and

$$\frac{\bar{L} - L}{F}x_i - z_F = \left[\frac{\bar{L} - L}{F} - 1\right]y_i \qquad (7.14)$$

The ratio i is now defined as the increase at the point of feed introduction in total flow of the L-phase per unit of feed, or

$$i = \frac{\bar{L} - L}{F} \qquad (7.15)$$

Equation 7.14 becomes

$$y_i = \frac{i}{i - 1}x_i - \frac{1}{i - 1}z_F \qquad (7.16)$$

which is the equation for the locus of all possible intersections of the two operating lines. The i-line defined by Equation 7.16 can be plotted if i and the feed composition are known. The line has a slope of $i/(i - 1)$ and intersects the diagonal ($x = y$) at z_F.

In extraction using immiscible solvents the feed will very likely contain none of the extract solvent. Converting to the mass-ratio coordinates gives

$$i = \frac{\bar{L}' - L'}{F'} \qquad (7.15a)$$

where i is the change in mass of the unextracted raffinate component per unit of feed, not counting the solute in the feed. If the feed contains no extract solvent, F' is equal to the unextracted raffinate component in the feed and $F' = \bar{L}' - L'$. Therefore $i = 1$ and the i-line is a vertical line from the diagonal at z_F. In the unusual case of extract solvent in the feed, F' will be greater than $\bar{L}' - L'$, and i will be less than one.

Equations 7.15 and 7.16 also hold for distillation. In distillation, the enthalpy of the feed determines the value of i. By use of Equation 7.15 and an enthalpy balance around the feed stage, the following definition of i can be derived:

$$i = \frac{\begin{array}{c}\text{heat required to convert one mole}\\\text{of feed to a saturated vapor}\end{array}}{\begin{array}{c}\text{latent heat of vaporization of one mole}\\\text{of the feed composition}\end{array}} \qquad (7.15b)$$

Equation 7.15b is an expression equivalent to the original definition of i (Equation 7.15). The variation in i and in the slope of the i-line with the thermal condition of the feed is outlined in Figure 7.4.

The simplified graphical method for distillation does not usually make use of enthalpy data, since the method assumes that the latent heat of vaporization of all mixtures is constant. However, the thermal condition of the feed must be considered. For example, if the feed is a subcooled liquid, it must pick up heat as it enters the cascade, since all liquids in the cascade must be saturated liquids. It picks up heat at the expense of the condensation of some vapor, which increases the liquid downflow.

Therefore, $(L - L) > F$ and $i > 1$, as shown on Figure 7.4. Since the slope of the i-line is $i/(i - 1)$, the slope for a subcooled liquid will be positive. The other cases shown in Figure 7.4 can be treated by an analysis similar to that above. Consideration must also be given to a subcooled liquid reflux from the condenser, and a super-heated vapor reflux from the reboiler must be allowed for in calculations on the equilibrium diagram. A subcooled reflux yields a liquid downflow in the column (L) which exceeds the external reflux (L_0).

A partial condenser or reboiler may contribute up to one equilibrium stage, as discussed in Chapter 6. The equations for the operating lines can be derived for these cases from enthalpy and material balances.

The reboiler may be eliminated in favor of direct addition of heat as steam in distillations where water is one of the components. The direct feeding of steam into the bottom of the distillation column is called "open steam." The operating line in the stripping section for open steam can be derived from material balances around the lower part of the column.

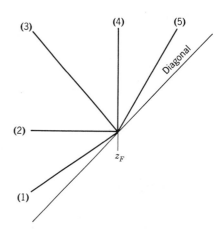

Figure 7.4. Location of i-lines in distillation.

	i	Slope of i-Line
(1) Superheated vapor	<0	$+$
(2) Saturated vapor	0	0
(3) Liquid and vapor	0 to 1	$-$
(4) Liquid at bubble point	1	∞
(5) Liquid below bubble point	>1	$+$

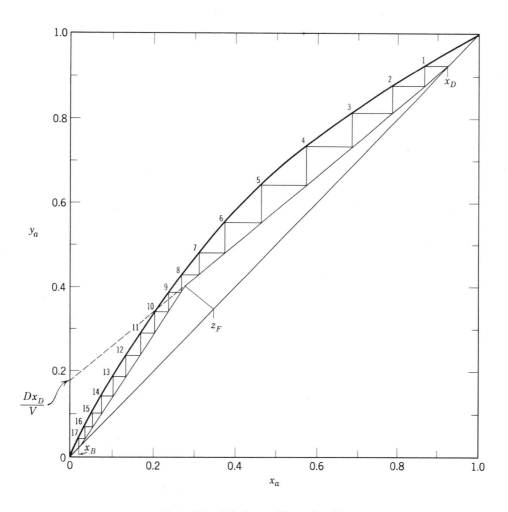

Figure 7.5. Solution to Illustration 7.2.

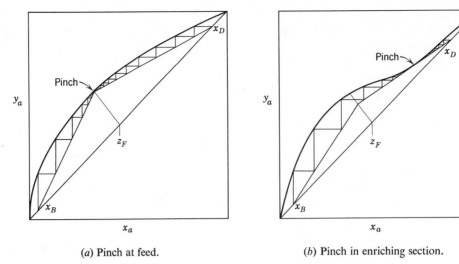

(a) Pinch at feed. (b) Pinch in enriching section.

Figure 7.6. Determination of minimum reflux on the x–y diagram.

Illustration 7.2. A mixture of 35 mole percent a and 65 mole percent b is to be separated in a distillation tower. The concentration of a in the distillate is 93 mole percent and 96 per cent of all product a is in the distillate. The feed is half vapor and the reflux ratio (L_0/D) is to be 4. The relative volatility of a to b is $\alpha_{a-b} = 2$. How many equilibrium stages are required in each section of the column?

SOLUTION. An x–y diagram (Figure 7.5) is drawn for $\alpha_{a-b} = 2$.

x_a	0	0.20	0.40	0.50	0.60	0.80	1.0
y_a	0	0.333	0.571	0.667	0.75	0.889	1.0

$$z_F = 0.35 \qquad i = \frac{\frac{1}{2}}{1} = \frac{1}{2}$$

$$\text{Slope of the } i\text{-line} = \frac{\frac{1}{2}}{\frac{1}{2} - 1} = -1$$

$$x_D = 0.93 \qquad \frac{L_0}{D} = 4.0$$

Slope of the enriching operating line

$$= \frac{L}{L + D} = \frac{\frac{L_0}{D}}{\frac{L_0}{D} + 1} = \frac{4}{5} = 0.8$$

To find x_B: Basis: 100 lb moles feed

Moles a in feed = 35

Moles a in distillate = (35)(0.96) = 33.6

Moles a in bottoms = 1.4

Total moles in distillate = 33.6/0.93 = 36.1

Total moles in bottoms = 100 − 36.1 = 63.9

Therefore, $$x_B = \frac{1.4}{63.9} = 0.0219$$

The enriching-section operating line is located by the point $x_D = 0.93 = y_1$ and the slope = 0.8. Alternatively, the enriching-section operating line can be plotted by the point $x_0 = 0.93 = y_1$, and the intercept $x = 0$ at

$$y = \frac{Dx_D}{V} = \frac{x_D}{\left(\frac{L_0}{D}\right) + 1} = \frac{0.93}{4 + 1} = 0.186$$

The stripping-section operating line is a straight line from the intersection of the enriching line and the i-line to the point $x_B = 0.0219 = y_S$. The stages may be stepped off from x_B or x_D. Since no information on the reboiler and condenser is available, neither is assumed to contribute to the separation. There are 9.5 stages in the stripping section and 7.5 stages in the enriching section.

Operating Variables. As the reflux ratio is reduced, the slope of the enriching-section operating line decreases until one of the operating lines first intersects the equilibrium curve. For ideal mixtures this occurs at the feed, but in other cases it may occur at any point between x_B and x_D, as shown in Figure 7.6. At total reflux, the slope of the operating lines becomes 1.0, and they coincide with the diagonal. The minimum number of stages required can be determined by stepping off stages from x_D to x_B between the equilibrium curve and the diagonal, as shown in Figure 7.7.

Calculation of Intermediate Streams. The reasoning used in calculating intermediate streams on the equilibrium x–y diagram is the same as that used on the three-variable diagrams considered in Chapter 6. The addition or removal of a stream changes the L- and V-phase flow rates and results in a new operating line. Two methods are available for locating the new operating line in the intermediate section of the cascade shown in Figure 6.13.

The i-line for the intermediate stream may be plotted regardless of whether the stream is a product or a feed. If the top reflux ratio is given, the upper operating line may be drawn from x_D to the intersection with the i-line for the intermediate stream. The operating line is continued beyond the intersection but now at a new slope corrected for the addition or removal of the intermediate stream. The intermediate operating line intersects the i-line for the feed, changes slope, and continues to x_B. In this case only one intermediate stream was present. In general, any number may be calculated as outlined.

The second method of calculation is by defining a fictitious feed (Σ), which is the algebraic sum of the feed (F) and the intermediate stream (I). This is shown in Figure 7.8. First the three i-lines are plotted. That for the fictitious feed is calculated from the properties of the two component streams. The slope of the enriching-section operating line is usually set by fixing the top reflux ratio. This operating line is drawn in and continued until it intersects the i-line for the fictitious feed (Σ). The stripping-section operating line is then located by drawing a straight line from this intersection to $x_B = y$. However, the enriching operating line has physical significance only until I is added and the stripping operating line is valid only from x_B to the i-line for the feed. The operating line between the F and I i-lines is easily located by the straight line connecting the intersections of the i-lines with the stripping and enriching operating lines, as shown. This method may be extended to multiple intermediate streams, but it then is less convenient than the first method outlined.

Interrelation between the Concepts of an Operating Line and a Delta Point. Although calculation on the equilibrium x–y diagram is most advantageous when straight operating lines occur, the method may be applied to cases of curved operating lines. Curved operating lines result when the assumptions made earlier in this chapter do not hold. For example, a variation in the molar overflow in distillation caused by nonconstant latent heats of vaporization or significant sensible-heat

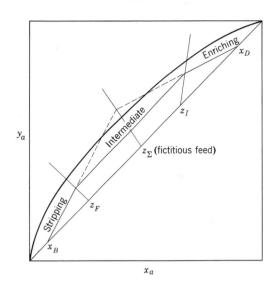

Figure 7.8. Fictitious feed for an intermediate stream.

changes would give a variable slope to the operating lines. Similarly, in liquid extraction a variable partial solubility of the two solvents would result in curved operating lines.

The delta point and the operating line are both used to relate the compositions of the streams flowing *between* two stages. For any stage n (Figure 7.9a) either the delta point or the operating line may be used to determine y_n from x_{n-1}, or vice versa, or to determine y_{n+1} from x_n, or vice versa. Figure 7.9b shows the construction required to locate the point (x_n, y_{n+1}) on the operating line by use of the delta point. The compositions x_n and y_n of the streams flowing from stage n are known and are plotted on the upper diagram. A straight line from x_Δ through x_n locates y_{n+1} at its intersection with the V-phase locus, as shown. The values of x_n and y_{n+1} are now known. For calculations on the equilibrium diagram, Equation 7.3 shows that the point (x_n, y_{n+1}) must be on the operating line. The values for x_n and y_{n+1} found on the upper diagram may now be transposed to the equilibrium diagram as shown by the construction lines, thereby establishing a point on the operating line. Additional points on the operating line may be determined in the same manner, and the curved operating line may then be drawn in, as shown. It is not necessary to use streams actually flowing between equilibrium stages to determine the operating line. Any straight line from the delta point cuts the L-phase and V-phase curves at compositions which represent a point on the operating line. The complete determination of curved operating lines is shown in Figure 7.10 for a case in distillation where the latent heat varies.

Applications of Simplified Graphical Methods. In many cases in stage operations insufficient physical data are available to permit rigorous calculations utilizing the methods of the previous chapters. In such cases, the methods outlined in this chapter may be used with a

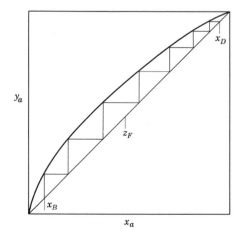

Figure 7.7. Determination of minimum stages at total reflux.

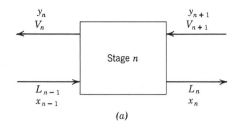

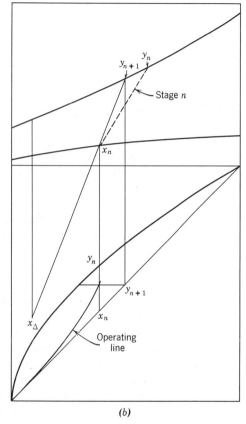

Figure 7.9. Interrelation between the delta point and the operating line.

minimum of physical data, even though the physical system does not fall within the simplifying assumptions made at the beginning of the chapter. Often the error introduced is small, and a sufficient factor of safety may be included in the design. For example, distillation of ammonia–water systems may be evaluated by the methods of this chapter, even though the latent heats, and therefore the phase flow rates, are not constant.

An intermediate degree of accuracy may be achieved where partial data are available for the phases over the composition range of interest. For example, in distillation if only the heats of vaporization of the pure components are known, they can be plotted on an enthalpy-composition diagram, and a straight line may be drawn between the two values of the saturated-vapor

enthalpy to give an approximate vapor enthalpy curve over the entire composition range. This method accounts for the latent heat of vaporization but neglects other heat effects, which are often of a much smaller magnitude. A similar improvement for the simplified procedure can be made by recomputing the equilibrium diagram using for one of the components a fictitious molecular weight, chosen so that the molar flow rates remain constant.

Stage Efficiencies. In actual countercurrent multi-stage equipment, the two phases leaving a stage are not in equilibrium, because of insufficient time of contact or inadequate dispersion of the two phases in the stage. As a result the concentration change for each phase in an actual stage is usually less than that possible in an equilibrium stage.

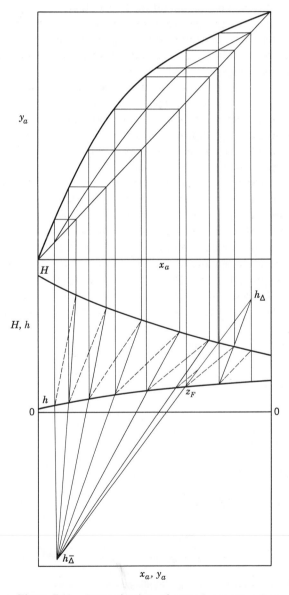

Figure 7.10. Determination of curved operating lines.

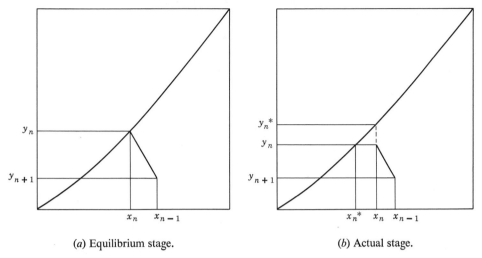

(*a*) Equilibrium stage. (*b*) Actual stage.

Figure 7.11. Operating lines for a single stage.

A stage efficiency may be defined to describe the lack of equilibrium. The *over-all stage efficiency* is defined as the ratio of the number of equilibrium stages required for a given separation to the number of actual stages required. Although it is permissible to report a fractional number of equilibrium stages, it is obvious that only an integral number of actual stages can be built. Although the over-all efficiency is simple to use in calculations, it does not allow for the variations in efficiency which may occur from stage to stage.

The *Murphree stage efficiencies* (4) are used for individual stages. The *Murphree V-phase efficiency* is defined as the actual change in average *V*-phase composition divided by the change that would occur if the total *V*-phase were in equilibrium with the *L*-phase actually leaving the stage (Figure 7.9*a*), or

$$E_V = \frac{y_n - y_{n+1}}{y_n^* - y_{n+1}} \tag{7.17}$$

A *Murphree L-phase efficiency* can also be defined

$$E_L = \frac{x_{n-1} - x_n}{x_{n-1} - x_n^*} \tag{7.18}$$

The composition y_n^* is that in equilibrium with the *L*-phase *leaving* the actual stage. Because the actual *L*-phase composition may vary (across the width of a bubble-cap plate, for example), the Murphree stage efficiency can exceed 100 per cent.

Murphree point efficiencies may be defined by Equations 7.17 and 7.18. A point efficiency refers to a single point on a bubble-cap plate at the liquid-vapor interface. For this definition y_n^* refers to a vapor-phase composition in equilibrium with the actual liquid at the point being considered. Point efficiencies cannot exceed 100 per cent.

The Murphree stage efficiency can be used with the methods outlined in this chapter to calculate the number

of actual stages required for a given separation. Consider the operating lines for a single stage shown in Figure 7.11. The operating line for the equilibrium stage runs from the entering conditions (x_{n-1}, y_{n+1}) to the leaving conditions (x_n, y_n). The latter point lies on the equilibrium curve since x_n and y_n are in equilibrium. In an actual stage (Figure 7.11*b*) the entering conditions are the same, but the compositions $(x_n$ and $y_n)$ of the phases leaving do not lie on the equilibrium curve, since the phases are not in equilibrium. The composition y_n^* is in equilibrium with the liquid composition x_n leaving the *actual* stage, as shown. The definition of Murphree *V*-phase efficiency can be more fully understood by reference to Figure 7.11*b*, where it is seen to be the ratio of the two distances $(y_n - y_{n+1})$ and $(y_n^* - y_{n+1})$.

When actual stages are stepped off between the equilibrium curve and operating line, the full vertical distance $(y_n^* - y_{n+1})$ is not used. Rather, some fraction of the vertical distance as determined by the efficiency of the stage is stepped off. Figure 7.12 illustrates a typical problem in distillation with a Murphree vapor efficiency of 50 per cent. It is convenient to draw an "effective separation curve" between the equilibrium curve and operating line at a distance determined by the efficiency. In this case the curve is located at half the vertical distance between the equilibrium curve and operating line, since the vapor efficiency is 50 per cent. The stages are then stepped off between the effective separation curve and the operating line.

Stage efficiencies depend on many factors, such as the time of contact and degree of dispersion of the phases, the geometry of the stage, the rate of mass transfer, and the physical properties and flow rates of the fluids. A great deal of research has been directed toward the prediction of stage efficiencies from theoretical or empirical relationships. Correlations for a number of

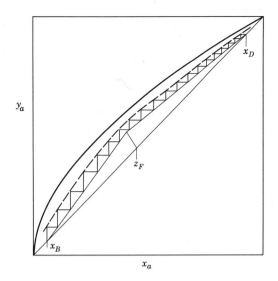

Figure 7.12. Stepping off actual stages for a Murphree vapor efficiency of 50 per cent.

systems are available. In many cases it is still necessary to use values determined experimentally. The numerical value of the stage efficiency often involves the greatest uncertainty in the design of multistage equipment. The relationship between the stage efficiency and the *rate* of mass transfer will be considered in Chapter 16.

Multicomponent Distillation and Absorption. Calculation of multicomponent systems is much more complex than that of binary systems in distillation. There are more variables and a correspondingly greater number of degrees of freedom. It is possible to have a pinch in several components and still effect a separation. That is, the concentration of certain components may not change over several stages, but the concentration of other components may vary greatly over the same stages. It is possible for the concentration of a given component to reach a maximum value at some stage within the column and then to decrease before the end of the column is reached. A rigorous calculation of multicomponent distillation involves laborious trial-and-error procedures. Many of the calculations in the petroleum industry are now being performed automatically on digital computers. For example, distillation calculations for crude petroleum, a multicomponent mixture, may be carried out on a computer. A multistage industrial crude-distillation unit is pictured in Figure 7.13.

Azeotropic and Extractive Distillation. Many mixtures are difficult to separate because their relative volatility is close to 1.0. Such mixtures may be separated by the addition of another component which will increase the relative volatility of the original constituents.

In azeotropic distillation the added component is relatively volatile and goes overhead in the distillate. In many cases it forms a low-boiling-point azeotrope with one of the original components. The low-boiling azeotrope goes overhead in the distillate. It is then necessary to separate the azeotropic agent from the other component. Absolute (100 per cent) ethyl alcohol can be produced from the 89.4 mole percent ethyl alcohol–water azeotrope by introducing benzene into the distillation column. The benzene goes overhead as a ternary azeotrope with nearly all the water and some alcohol.

Extractive distillation involves the addition of a

Figure 7.13. Crude-petroleum distillation battery. The three towers separate the crude petroleum feed into eleven products ranging from asphalt to light hydrocarbon gases. The product streams are sent to other processing equipment in the refinery to produce a wide range of fuels, lubricants, and petrochemicals. The crude petroleum is fed to the first tower on the left, which is the prefractionator. Here the most volatile components of the crude are removed. Next, the flashed crude is sent to the atmospheric fractionating column, on the right in the photograph. In this column the crude is fractionated into a number of streams of intermediate volatility and a reduced crude, which is sent to the vacuum fractionating tower. The vacuum fractionating column (the center tower in the photograph) operates under a vacuum, so that the distillation may proceed at low enough temperatures to minimize thermal decomposition of the petroleum. In this tower the least volatile components are separated.

The prefractionator varies in diameter between 10 ft and 12 ft, as can be seen in the photograph. It is $65\frac{1}{2}$ ft high and contains twenty-four bubble-cap plates. The atmospheric column is $15\frac{1}{2}$ ft in diameter, $83\frac{1}{2}$ ft high, and contains thirty-four plates. The vacuum tower has three diameters: 9 ft at the bottom, 25 ft in the middle, and 18 ft at the top. The tower is approximately 80 ft high and contains about twenty-eight plates. (Designed and constructed by the Foster-Wheeler Corporation for the Atlantic Refining Company.)

relatively nonvolatile solvent which increases the relative volatility of the original components. The "solvent" which is added goes out in the bottoms, since it is non-volatile, along with the original component whose volatility is lower in the ternary system, although not necessarily lower in the original binary system.

ANALYTICAL CALCULATION OF STAGES

For simple countercurrent flow in cases where both the operating line and equilibrium curve are straight, the number of equilibrium stages may be calculated analytically. This situation occasionally arises in operations such as gas absorption.

A material balance around the first stage in Figure 5.2 gives for dilute gases (where L and V are nearly constant)

$$Lx_0 + Vy_2 = Lx_1 + Vy_1$$

or

$$L(x_0 - x_1) = V(y_1 - y_2) \qquad (7.19)$$

The equilibrium curve will be a straight line if Henry's law applies

$$y_1 = Kx_1$$

Substituting for x_1 in Equation 7.19 gives

$$L\left(x_0 - \frac{y_1}{K}\right) = V(y_1 - y_2) \qquad (7.20)$$

and

$$y_1 = \frac{y_2 + \dfrac{L}{V} x_0}{\dfrac{L}{KV} + 1} \qquad (7.21)$$

The group L/KV is called the absorption factor and is designated by A. Equation 7.21 becomes

$$y_1 = \frac{y_2 + KAx_0}{A + 1} \qquad (7.22)$$

A similar analysis for stage 2 gives

$$y_2 = \frac{y_3 + KAx_1}{A + 1} = \frac{y_3 + Ay_1}{A + 1} \qquad (7.23)$$

If the value of y_1 from Equation 7.22 is substituted in Equation 7.23

$$y_2 = \frac{(A + 1)y_3 + KA^2x_0}{A^2 + A + 1} \qquad (7.24)$$

If Equation 7.24 is multiplied by $(A - 1)/(A - 1)$,

$$y_2 = \frac{(A^2 - 1)y_3 + A^2(A - 1)Kx_0}{(A^3 - 1)} \qquad (7.25)$$

An expression for y_3 may be derived in a similar manner. If this process is continued to stage N,

$$y_N = \frac{(A^N - 1)y_{N+1} + A^N(A - 1)Kx_0}{(A^{N+1} - 1)} \qquad (7.26)$$

A material balance over the entire cascade gives

$$L(x_0 - x_N) = V(y_1 - y_{N+1}) \qquad (7.27)$$

Since $x_N = y_N/K$, Equation 7.27 becomes

$$L\left(x_0 - \frac{y_N}{K}\right) = V(y_1 - y_{N+1}) \qquad (7.28)$$

Eliminating y_N between Equations 7.26 and 7.28 gives

$$\frac{y_{N+1} - y_1}{y_{N+1} - Kx_0} = \frac{A^{N+1} - A}{A^{N+1} - 1} \qquad (7.29)$$

Equation 7.29 is useful in calculating the separation if A and N are known. It can be derived in a more direct manner using the calculus of finite differences (3). Equation 7.29 can be solved for the total number of stages. Cross multiplication gives

$$A^{N+1}y_{N+1} - A^{N+1}y_1 - y_{N+1} + y_1$$
$$= A^{N+1}y_{N+1} - A^{N+1}Kx_0$$
$$- Ay_{N+1} + AKx_0$$

$$A^{N+1}(y_1 - Kx_0) = A(y_{N+1} - Kx_0) - y_{N+1} + y_1$$
$$= (A - 1)(y_{N+1} - Kx_0) + y_1 - Kx_0$$

$$A^{N+1} = (A - 1)\left(\frac{y_{N+1} - Kx_0}{y_1 - Kx_0}\right) + 1$$

$$A^N = \left(\frac{A - 1}{A}\right)\left(\frac{y_{N+1} - Kx_0}{y_1 - Kx_0}\right) + \frac{1}{A}$$

Taking the logarithm of both sides gives

$$N = \frac{\log\left[\left(\dfrac{A - 1}{A}\right)\left(\dfrac{y_{N+1} - Kx_0}{y_1 - Kx_0}\right) + \dfrac{1}{A}\right]}{\log A} \qquad (7.30)$$

Equation 7.30 can be used to calculate the number of stages required for a given change in composition of the gas phase in absorption. The equation may also be written in mole-ratio dimensions, providing the equilibrium curve and operating line are straight on mole-ratio coordinates. Use of Equation 7.30 with mole ratios requires the assumption of dilute solutions to ensure a straight equilibrium curve.

For stripping columns where mass is transferred from the liquid to the vapor phase,

$$\frac{x_0 - x_N}{x_0 - \dfrac{y_{N+1}}{K}} = \frac{\left(\dfrac{1}{A}\right)^{N+1} - \dfrac{1}{A}}{\left(\dfrac{1}{A}\right)^{N+1} - 1} \qquad (7.31)$$

The term $(1/A)$ is referred to as the stripping factor.

Analytical Calculation of Minimum Stages at Total Reflux. An expression first developed by Fenske (1) is useful in determining analytically the minimum number of stages required at total reflux. Use of this equation is limited to systems with constant relative volatility.

By the definition of relative volatility,

$$\frac{y_N}{1 - y_N} = \alpha \frac{x_N}{1 - x_N} \qquad (7.32)$$

for the last stage. At total reflux the operating line coincides with the diagonal and $y_N = x_{N-1}$, so that

$$\frac{x_{N-1}}{1 - x_{N-1}} = \alpha \frac{x_B}{1 - x_B} \qquad (7.33)$$

For stage $N - 1$,

$$\frac{y_{N-1}}{1 - y_{N-1}} = \alpha \frac{x_{N-1}}{1 - x_{N-1}} = \alpha^2 \frac{x_N}{1 - x_N} \qquad (7.34)$$

If this procedure is continued to stage 1,

$$\frac{y_1}{1 - y_1} = \alpha^N \frac{x_N}{1 - x_N} \qquad (7.35)$$

For a total condenser $x_D = y_1$, and for the case of the bottoms withdrawn from the bottom plate $x_B = x_N$. Therefore,

$$\frac{x_D}{1 - x_D} = \alpha^N \frac{x_B}{1 - x_B} \qquad (7.36)$$

and

$$N = \frac{\log \dfrac{x_D(1 - x_B)}{x_B(1 - x_D)}}{\log \alpha} \qquad (7.37)$$

REFERENCES

1. Fenske, M. R., *Ind. Eng. Chem.*, **24**, 482 (1932).
2. McCabe, W. L., and E. W. Thiele, *Ind. Eng. Chem.*, **17**, 605 (1925).
3. Mickley, H. S., T. K. Sherwood, and C. E. Reed, *Applied Mathematics in Chemical Engineering*, McGraw-Hill Book Co. New York, 1957.
4. Murphree, E. V., *Ind. Eng. Chem.*, **17**, 747 (1925).

PROBLEMS

7.1. Solve Problem 5.5*b* and *c* on an equilibrium diagram.

7.2. An absorption column equivalent to five equilibrium stages is to be used to remove ethane from a hydrogen–ethane mixture at 300 psia. The entering absorbent liquid contains 99 per cent hexane and 1 per cent ethane. The entering gas contains 35 per cent ethane and 65 per cent hydrogen. The column is at a constant temperature of 75°F. Because of the fluid flow characteristics of the column, the mole ratio of entering absorbent to entering gas cannot exceed 3 to 1 at the desired flow rate of gas. What is the maximum possible recovery of ethane if this column is used?

7.3. Propane is to be stripped from a nonvolatile oil by steam in a countercurrent tower. 4 moles of steam will be supplied at the bottom of the tower for every 100 moles of oil–propane feed at the top. The oil originally contains 2.5 mole percent propane, and this concentration must be reduced to 0.25 mole percent. The tower is maintained at 280°F and 35 psia. The molecular weight

of the heavy oil is 300, and the molecular weight of propane is 44. The equilibrium relationship is $y = 33.4x$ where $y =$ mole fraction propane in vapor and $x =$ mole fraction propane in liquid.

(*a*) How many equilibrium stages are required?

(*b*) If the pressure is increased to 70 psia how many equilibrium stages would be required?

(*c*) What would be the effect of lowering the temperature of stripping?

(*d*) What would be the minimum steam flow rate necessary to give the same recovery of propane?

7.4. Acetone is to be removed from an acetone–air mixture at 30°C and 1 atm by adsorption on charcoal in a simple counter-current process. The following equilibrium data were taken on the charcoal to be used at 30°C:

gm acetone adsorbed/gm charcoal	0	0.1	0.2	0.3	0.35
partial pressure acetone, mm Hg	0	2.0	12	42	92

The flow rate of fresh gas is 4000 standard cu ft/hr and the flow rate of charcoal is 200 lb/hr.

(*a*) Calculate the number of theoretical stages required to reduce a 10 per cent acetone–air mixture to 1 per cent acetone (by volume) at 30°C and 1 atm total pressure.

(*b*) What effect would increasing the total pressure have?

(*c*) What effect would increasing the temperature have?

(*d*) What is the minimum flow rate of charcoal for the conditions of part (*a*)?

7.5. Aqueous triethanolamine solutions are used to absorb carbon dioxide from a gaseous mixture as a purification step in the manufacture of dry ice. The carbon dioxide is absorbed by the amine solution, but the other components of the gas are not. The triethanolamine solution is subsequently heated to recover the pure carbon dioxide.

A gaseous mixture containing 25 per cent CO_2 is to be scrubbed with a 1 molar solution of triethanolamine at 25°C and 1 atm.

(*a*) What is minimum flow rate of triethanolamine solution necessary to reduce 10 lb moles/hr of entering gas to an exit concentration of 1 per cent?

(*b*) How many equilibrium stages are required at 1.5 times the minimum amine flow rate? (The density of amine solution may be assumed to be 1.0 gm/cu cm.)

(*c*) In many cases the scrubbing solution contains a small amount of CO_2 as it is recycled from the CO_2 stripper. What would be the minimum solution rate if it enters with a concentration of 0.10 lb moles CO_2 per pound mole amine?

(*d*) How many stages would be required for 1.5 times the minimum solution rate with the entering solution of part (*c*)? (*Note:* The mole ratio of CO_2 to solution may be used, but since the ratio of water to triethanolamine is constant, the ratio $X =$ lb moles CO_2 per pound mole amine can be used directly.)

Equilibrium Data for 1 Molar Triethanolamine at 25°C
[Mason and Dodge, *Trans. Am. Inst. Chem. Eng.*, **32**, 27 (1936).]

Partial pressure of CO_2 in gas, mm Hg	1.4	10.8	43.4	96.7	259	723
Liquid concentration, lb mole CO_2/lb mole amine	0.0587	0.161	0.294	0.424	0.612	0.825

7.6. Using an enthalpy-concentration diagram, show geometrically that constant molar overflow results when the latent heats of vaporization are equal for all compositions. Neglect heat effects other than the latent heat, but do not assume that the enthalpy of either phase is constant.

7.7. Derive an equation for the enriching-section operating line in distillation with a partial condenser. Draw a sketch to show how this line is plotted, and label important points.

7.8. Derive an equation for the stripping-section operating line where the bottoms are withdrawn through the reboiler. Draw a sketch showing a plot of the equation, and label points.

7.9. Solve Problems 6.7 and 6.8 on an equilibrium diagram, assuming constant molar overflow. For which case is the assumption of constant molar overflow more valid?

7.10. A mixture containing 30 mole percent benzene and 70 mole percent toluene is to be fractionated at 1 atm in a distillation column which has a total condenser and a still from which the bottoms are withdrawn. The distillate is to contain 95 mole percent benzene and the bottoms 4 mole percent benzene. The feed is at its dew point.

(a) What is the minimum reflux ratio (L_0/D)?

(b) What is the minimum number of equilibrium stages in the column required at total reflux?

(c) How many equilibrium stages are required at a reflux ratio of 8?

(d) How many equilibrium stages would be required at a reflux ratio of 8 if the feed were a liquid at its bubble point?

7.11. Solve Problem 6.9 on an equilibrium diagram, assuming constant molar overflow.

7.12. For Problem 6.11 determine the number of equilibrium stages required using the solvent-free equilibrium diagram. Assume that the mutual solubility of cottonseed oil and propane does not vary with concentration of oleic acid. Is this assumption reasonable?

7.13. An aqueous solution containing 20 mole percent ethanol is to be fractionated at 1 atm into a distillate of composition 80 mole percent ethanol and a waste of composition 3 mole percent ethanol. The feed is at 60°F. A distillation column equivalent to seven theoretical plates is available. What reflux ratio (L_0/D) should be used?

7.14. Solve Problem 6.13a and b on an equilibrium x–y diagram.

7.15. Solve Problem 6.15 on an equilibrium x–y diagram.

7.16. A mixture of 30 mole percent ethanol, 70 mole percent water is to be fractionated at 1 atm into a distillate of 82 mole percent ethanol and a bottoms of 4 mole percent ethanol. There will be a total condenser at the top of the tower, but, instead of a reboiler, saturated steam will be fed directly into the bottom of the tower to supply necessary heat. The feed is 40 mole percent vapor, 60 mole percent liquid.

(a) Draw a flow diagram for the tower, and label entering and leaving streams.

(b) Derive an equation for the operating line in the stripping section.

(c) Determine the number of equilibrium stages required.

(d) Determine the reflux ratio (L_0/D).

(e) Determine the minimum steam rate (per mole of bottoms) which would permit this separation.

7.17. 1000 lb moles/hr of a 25 mole percent ethanol–water mixture is to be fractionated at 1 atm into two ethanol-rich streams of compositions 60 per cent and 80 per cent ethanol. 98 per cent of the ethanol in the feed is to be recovered in these two products. That is, only 2 per cent of the feed ethanol may go out in the bottoms, which are withdrawn from the reboiler. Equal quantities (in moles) of 60 percent and 80 percent products are to be produced. The feed, the bottoms, and the 60 percent product are liquids at the bubble point. The 80 percent product is a saturated vapor withdrawn from a partial condenser. The reflux ratio (L_0/D) at the top of the column is to be 3.0. The over-all plate efficiency is 25 per cent. Assuming constant molar overflow, and, using the x–y diagram, calculate:

(a) The number of actual plates required for the desired separation.

(b) The actual plate on which to put the feed.

(c) The actual plate from which the 60 percent product is withdrawn.

(d) The reboiler duty (Btu/hr).

(e) The condenser duty (Btu/hr).

(f) The minimum reflux ratio at the top of the column.

(g) The minimum number of stages at total reflux.

7.18. Two mixtures of a and b are to be fractionated at constant pressure in a distillation column equipped with a total condenser and a total reboiler.

	Quantity	Composition x_a	Thermal Condition
Feed 1	50 lb moles	0.50	saturated liquid
Feed 2	100 lb moles	0.35	saturated vapor
Distillate		0.9	saturated liquid
Bottoms		0.05	saturated liquid

$\alpha_{a-b} = 4.0$ Reflux Ratio $(L_0/D) = 1.5$

The Murphree Vapor Efficiency varies with the liquid composition:

x_a	0.05	0.2	0.4	0.6	0.8	0.9
E_v	0.67	0.67	0.67	0.50	0.50	0.50

(a) How many actual stages are required?

(b) At what stage is each feed introduced?

(c) What is the over-all stage efficiency?

7.19. An aqueous solution containing 35 mole percent ammonia at 100 psia and 150°F is to be fractionated into a distillate of 95 mole percent ammonia and a bottoms of 3 mole percent ammonia. There is a total condenser, and the bottoms product is withdrawn from the bottom plate. The reflux ratio L_0/D is 1.

(a) Construct curved operating lines on an equilibrium diagram for this separation.

(b) Determine the number of actual stages required at a Murphree vapor efficiency of 25 per cent.

(c) Determine the over-all stage efficiency.

7.20. In Chapter 2 the stage efficiency was defined as the ratio of the composition change in an actual stage to the composition change which would result in an equilibrium stage. Using a diagram similar to Figure 7.11, show how this definition differs from the Murphree stage efficiency.

7.21. Determine the number of actual stages required in Problem 7.10c for:

(a) An over-all stage efficiency of 60 per cent.

(b) A Murphree vapor efficiency of 60 per cent.

(c) A Murphree liquid efficiency of 60 per cent.

7.22. Under what circumstances is the over-all efficiency of a column numerically equal to the Murphree vapor efficiency?

7.23. An existing distillation column is being used to fractionate a feed into the desired distillate and bottom products. Predict what the effect each of the following changes will have on the purity of the distillate and bottoms, other independent variables in the column remaining unchanged. Assume that the stage efficiency never changes and that the feed can be relocated over several plates on either side of the original feed nozzle, if this is necessary to maintain high purity distillate and bottoms.

(a) The relative volatility of the feed is lowered. (Feed composition and enthalpy characteristics unchanged.)

(b) The reflux ratio (L_0/D) is increased.

(c) The concentration of the more volatile component in the feed is reduced.

(d) The reboiler becomes fouled and the heat it can supply is substantially decreased.

(e) The feed (formerly a liquid) is now fed in as a vapor.

7.24. Derive an expression for $(L/V)_{min}$ from Equation 7.29. Can A for the minimum (L/V) be greater than 1.0?

7.25. The ammonia concentration of an air–ammonia mixture is to be reduced from 3 per cent to 0.05 per cent by volume by scrubbing the gas with water in a countercurrent tower at 90°F and 1 atm pressure. The flow rate of entering gas is 5000 cu ft/hr (90°F, 1 atm). For ammonia in aqueous solution in the concentration range the equilibrium is expressed $y = 0.185x$.

(*a*) Determine graphically the minimum water rate and the number of stages required at twice the minimum water rate.

(*b*) Determine analytically the minimum water rate and the number of stages required at twice the minimum water rate.

7.26. Derive Equation 7.31.

7.27. In what way should Equation 7.37 be modified to be valid for a partial condenser and for bottoms withdrawn through the reboiler?

7.28. Solve Problem 7.10*b* analytically.

chapter 8

Unsteady-State Stage Operations

The calculation methods developed in the previous chapters have applied to steady-state operation in which the compositions and flow rates did not vary with time. An unsteady-state or transient process involves changes of conditions with time. There are many transient operations in chemical processing. All batch mass-transfer operations involve a change in composition with time. Although the continuous stage operations are assumed to be at steady state for calculation, the question of starting the equipment naturally arises.

Start-up of Continuous Equipment. In the start-up of multistage equipment compositions and flow rates may vary with time. The length of time after start-up required to reach a steady state is an important consideration in process operation. The time depends upon the time of passage of the two phases through the equipment and upon the rate of mass transfer. The time of passage depends upon the volume of material held up in the equipment and upon the flow rates.

In the start-up of a distillation column, feed is supplied to the column, where it runs down to the still and is vaporized. The vapor rises to the condenser, where it is condensed and returned to the column. Frequently a distillation column is started up at total reflux. It is run at total reflux for a time sufficient to allow the distillate composition to approach that desired for steady-state operation. The time for this period can be predicted approximately. Once the distillate composition is achieved at total reflux, the distillate and bottom products may be withdrawn and feed added to complete the transition to steady-state operation.

There will also be a transient period at the start-up of a simple countercurrent cascade, such as in extraction or absorption. The length of this period will depend upon the flow rates of the two phases, the hold-up

volume of the equipment, and the rate of approach to equilibrium. The transient period may range from minutes in small equipment to several hours in large-scale installations, to several months in some special applications.

Batch Operations. When the quantity of material to be processed is small, batch methods are often used. Although the trend in the chemical process industry is toward continuous processing wherever possible, batch operations are still very widely used.

Many batch operations can be calculated using methods developed in earlier chapters. For example, a single-stage batch operation, such as the extraction of coffee from the grounds into a hot-water phase, can be calculated assuming equilibrium between the two phases before they are separated. The two phases (initially pure water and ground coffee) are mixed and undergo composition changes with time. The time required to reach equilibrium is an important process variable. It will depend upon the rate of mass transfer and upon the degree of contact between the phases. The composition of the two product phases in any batch single-stage process can be calculated by methods developed in earlier chapters if the two phases are allowed to remain in contact until all of one phase is in equilibrium with all of the other phase. On the other hand, in some cases one phase may be added or withdrawn during the process, as in differential distillation.

Differential Distillation. In batch distillation from a single stage a liquid is vaporized, and the vapor is removed from contact with the liquid as it is formed. Each differential mass of vapor is in equilibrium with the remaining liquid. The composition of the liquid will change with time, since the vapor formed is always richer in the more volatile component than the liquid

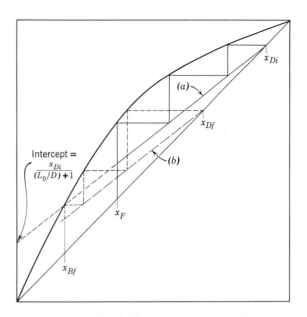

Figure 8.1. Batch distillation at constant reflux ratio. (a) Initial operating line. (b) Final operating line.

from which it is formed. This results in a continuing impoverishment of the liquid in the more volatile component.

Since the liquid composition varies, so also must the equilibrium vapor composition. The original composition of the total liquid phase (L) is taken as x and of the vapor phase formed, y. By an over-all material balance for the vaporization of a differential quantity of liquid,

$$dL = -dV \tag{8.1}$$

If dV moles of average composition (y) are formed by vaporization, a material balance for the more volatile component gives

$$d(Lx) = -y\, dV \tag{8.2}$$

and

$$x\, dL + L\, dx = -y\, dV \tag{8.3}$$

Combination of Equations 8.1 and 8.3 gives

$$x\, dL + L\, dx = y\, dL \tag{8.4}$$

and

$$\int \frac{dL}{L} = \int \frac{dx}{y - x} \tag{8.5}$$

which is called the Rayleigh equation. Integrating over a finite change from state 1 to state 2 gives

$$\ln \frac{L_2}{L_1} = \int_{x_1}^{x_2} \frac{dx}{y - x} \tag{8.6}$$

If the equilibrium relationship between y and x is known, the right-hand side of Equation 8.6 may be integrated. If Henry's law holds, $y = Kx$, and

$$\ln \frac{L_2}{L_1} = \frac{1}{K - 1} \ln \frac{x_2}{x_1} \tag{8.7}$$

For a constant relative volatility,

$$\ln \frac{L_2}{L_1} = \frac{1}{\alpha - 1} \left(\ln \frac{x_2}{x_1} - \alpha \ln \frac{1 - x_2}{1 - x_1} \right) \tag{8.8}$$

Batch Distillation. Batch distillation is often used for separating small quantities of liquids. Often the batch still is used for a large variety of separations, and therefore it must be versatile. Since a batch distillation is usually carried out in an existing column equivalent to a known number of equilibrium stages, it is necessary to determine the reflux ratio required to give the desired distillate purity.

In the typical batch distillation the liquid to be processed is charged to a heated kettle, above which is mounted the distillation column equipped with a condenser. Once the initial liquid is charged no more feed is added. The liquid in the kettle is boiled, and the vapors pass upward through the column. Part of the liquid from the condenser is refluxed, and the remainder is withdrawn as distillate product. Nothing is withdrawn from the still pot until the run is completed.

Because the distillate which is withdrawn is richer in the more volatile component than the residue in the still pot, the residue will become increasingly depleted in the more volatile component as the distillation progresses. Since the number of equilibrium stages in the column is constant, the concentration of more volatile component in the distillate will decrease as the still-pot concentration decreases, if the reflux ratio is held constant. This is shown in Figure 8.1. F is the quantity of original charge of composition x_F, D the quantity of distillate of composition x_D, and B the residue in the still pot of composition x_B. At any time during the distillation,

$$F = D + B \tag{8.9}$$

$$Fx_F = Dx_D + Bx_B \tag{8.10}$$

Since the entire column is an enriching section, there will be only an enriching operating line. It extends initially between x_F and an initial distillate composition (x_{Di}) which is determined by the reflux ratio. The equation of the initial operating line is

$$y_{n+1} = \frac{L}{V} x_n + \frac{D}{V} x_{Di} \tag{8.11}$$

The column represented in Figure 8.1 is equivalent to three equilibrium stages. The initial distillate composition is determined by adjusting the operating line of a given slope until exactly three stages fit between x_F and x_{Di}. As the distillation continues the concentration of the more volatile component in the still-pot residue decreases to the final value x_B. The composition of the final distillate (x_{Df}) is determined by an operating line of the same slope as earlier and by the requirement of exactly three equilibrium stages.

At any time during the distillation there are B moles of composition x_B in the still pot. If dB moles of composition x_D are removed from the column, a material balance for the more volatile component gives

$$
\begin{array}{ccc}
\text{More volatile} & \text{more volatile} & \text{more volatile} \\
\text{component in} & \text{component in} & \text{component in} \\
\text{the original} & = \text{the } dB \text{ moles} & + \text{the } (B - dB) \\
B \text{ moles} & \text{removed} & \text{moles remaining}
\end{array}
$$

or $\qquad Bx_B = (dB)x_D + (B - dB)(x_B - dx_B)$ (8.12)

Neglecting second-order differentials yields

$$\frac{dB}{B} = \frac{dx_B}{x_D - x_B} \qquad (8.13)$$

Integrating from the initial to final still-pot conditions (F to B) gives

$$\int_B^F \frac{dB}{B} = \ln \frac{F}{B} = \int_{x_{Bf}}^{x_F} \frac{dx_B}{x_D - x_B} \qquad (8.14)$$

This equation is of the same form as the Rayleigh equation for differential distillation (Equation 8.6). However, a number of equilibrium stages are available so that the relationship between x_D and x_B must be determined graphically and the right-hand side of Equation 8.14 is integrated graphically, after several corresponding values of x_D and x_B have been determined. It should be noted that when only one stage is available, Equation 8.14 reduces to Equation 8.6.

Illustration 8.1. One hundred pound moles of a mixture of 20 mole percent ethanol, 80 mole percent water is charged to the still pot of a batch distillation column equivalent to three equilibrium stages. The distillation at a reflux ratio of 3 and 1 atm is continued until the residue in the still pot reaches a composition of 0.03 mole fraction ethanol. What is the quantity and average composition of the distillate?

SOLUTION. (Figure 8.2a.) The slope of the operating line is 3/4. The three stages in the column plus the still pot give a total of four equilibrium stages. The initial x_D is determined by adjusting the operating line of slope 3/4 until exactly four stages fit between x_F and x_D; $x_{Di} = 0.75$. Now various values of x_D are chosen arbitrarily and x_B is determined graphically for each x_D. Then $1/(x_D - x_B)$ is plotted as a function of x_B and a graphical integration (Figure 8.2b) is made.

x_D	x_B	$\dfrac{1}{x_D - x_B}$
0.76	0.22	1.85
0.75	0.20	1.80
0.74	0.10	1.6
0.72	0.04	1.5
0.70	0.03	1.5
0.68	0.02	1.5

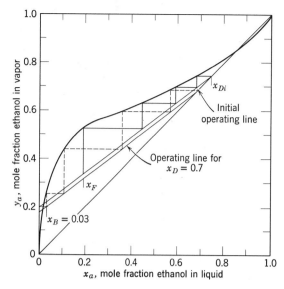

(a) Determination of x_B from x_D.

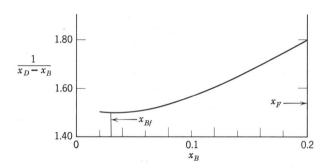

(b) Graphical integration of Equation 8.14.

Figure 8.2. Solution to Illustration 8.1.

$$\int_{x_{Bf}}^{x_F} \frac{dx_B}{x_D - x_B} = 0.275$$

$$\ln \frac{100}{B} = 0.275$$

$$B = 76.2, \qquad D = 23.8$$

$$\text{Average distillate composition} = \frac{Dx_D}{D} = \frac{Fx_F - Bx_B}{D}$$

$$= \frac{(100)(0.2) - (76.2)(0.03)}{23.8}$$

$$= 0.745$$

In this case the average composition of the distillate is close to its initial value since the change during the distillation was small. In other cases the average distillate composition may be considerably less than the initial value.

It is possible to maintain a constant distillate composition in batch distillation if the reflux ratio is continuously increased during the run. Calculations in this case are

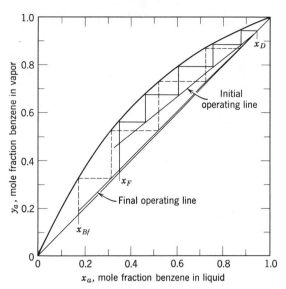

Figure 8.3. Solution to Illustration 8.2.

identical with those illustrated previously, except that x_D remains constant and the slope of the operating line (L/V) must change to maintain a constant number of equilibrium stages between x_D and x_B.

Illustration 8.2. A mixture of 35 mole percent benzene and 65 mole percent toluene is to be fractionated at 1 atm pressure in a batch column to recover 60 per cent of the benzene in a distillate of composition 0.95 mole percent benzene. The column is equivalent to four equilibrium stages.

(*a*) What is the initial reflux ratio?

(*b*) What is the final reflux ratio?

SOLUTION. (*a*) The initial reflux ratio is determined by adjusting the slope of the operating line through $x_D = 0.95 = y$ until exactly five stages fit between x_F and x_D. From Figure 8.3, the slope is 0.8, so that the initial reflux ratio is $L_0/D = 0.8/(1 - 0.8) = 4$.

(*b*) One hundred pound moles charged: 35 lb moles benzene, 65 lb moles toluene.

Benzene recovered in distillate $= (0.60)(35) = 21$

Toluene in distillate $= \dfrac{0.05}{0.95}(21.0) = 1.1$

Final still-pot composition: Benzene $= 14.0$

Toluene $= 63.9$

$$x_{Bf} = \frac{14.0}{63.9 + 14.0} = 0.18$$

The operating line is pivoted around $x_D = 0.95 = y$ until

exactly five stages fit between $x_{Bf} = 0.18$ and $x_D = 0.95$. The slope is 76/77, and the reflux ratio is $(76/77)/(1 - 76/77) = 76$. This value is close to total reflux. Operation at such a high reflux ratio would require a large quantity of heat per unit of product and would be economically unjustifiable. More benzene could be recovered, but the purity of the distillate product would have to be reduced.

It is frequently found most feasible to maintain a constant distillate composition at the beginning of batch distillation. When the reflux ratio has increased to an economic or physical limit, it is held constant for the remainder of the run, and the distillate composition is allowed to decrease until it reaches a predetermined limit, at which time distillation is stopped.

In many batch distillation columns the liquid held up on each plate of the column is appreciable. This must be taken into account when predicting the yield and composition of distillate.

PROBLEMS

8.1. Derive Equations 8.7 and 8.8.

8.2. A liquid mixture of 65 mole percent benzene and 35 mole percent toluene is distilled with the vapor continuously withdrawn as it is formed.

(*a*) What is the composition of the liquid after 25 mole percent of the original liquid has been vaporized?

(*b*) How much of the original liquid will have vaporized when the equilibrium vapor has a composition of 65 mole percent benzene?

8.3. A vapor mixture of 40 mole percent hexane and 60 mole percent heptane is slowly condensed at 1 atm and the equilibrium liquid is withdrawn as it is formed.

(*a*) What is the composition of the vapor after 50 mole percent of it has condensed?

(*b*) If all the liquid formed were allowed to remain in contact with the vapor what would be the composition of the vapor after 50 mole percent had condensed?

8.4. An equimolar mixture of benzene and toluene is charged to the still pot of a batch distillation tower which is equivalent to six equilibrium stages. The distillation is carried out at 1 atm at a constant reflux ratio of 3 until the concentration of benzene in the still pot is reduced to 5 mole percent.

(*a*) What is the average composition of the distillate?

(*b*) What is the over-all percentage recovery of benzene in the distillate?

(*c*) What is the temperature in the still pot at the beginning and at the end of the run?

8.5. 95 per cent of the ethylene glycol in a 20 mole percent ethylene glycol–80 mole percent water mixture must be recovered in the residue from a batch distillation tower equivalent to three equilibrium stages (including the still pot) operating at 228 mm Hg.

EQUILIBRIUM DATA:* ETHYLENE GLYCOL–WATER AT 228 mm Hg

Mole percent ethylene glycol

Liquid	0	10	20	30	40	50	60	70	80	90	92	95	97	99	100
Vapor	0	0.2	0.4	0.7	1.1	1.8	2.8	5.0	10.0	21.4	25.5	35.0	46.5	69.0	100
Temperature, °C	69.5	72.8	75.6	78.8	82.9	87.7	93.1	100.5	111.2	127.5	132.0	139.5	145.1	152.4	160.6

* Trimble and Potts, *Ind. Eng. Chem.*, **27**, 66 (1935).

An over-all distillate composition of 5 mole percent ethylene glycol is desired.

(*a*) If the distillate composition is held constant, what must be the initial and final reflux ratios?

(*b*) If the reflux ratio is held constant, what must be its value?

8.6. A mixture of 36 mole percent chloroform and 64 per-cent benzene is to be separated using a batch distillation tower equivalent to ten equilibrium stages operating 760 mm Hg. The distillate composition will be held constant at 96 mole percent chloroform during the run.

(*a*) Plot the reflux ratio as a function of the percentage of original charge distilled.

(*b*) What is the theoretical maximum recovery of chloroform in the 96 per cent distillate?

EQUILIBRIUM DATA:* CHLOROFORM–BENZENE

Mole percent chloroform

Liquid	0	8	15	22	29	36	44	54	66	79	100
Vapor	0	10	20	30	40	50	60	70	80	90	100
Temperature, °C	80.6	79.8	79.0	78.2	77.3	76.4	75.3	74.0	71.9	68.9	61.4

* *International Critical Tables*, McGraw-Hill Book Company, New York (1926).

PART I: NOTATION AND NOMENCLATURE FOR STAGE OPERATIONS

	General Designation	Typical Units	Specific Designation				
			Distillation	Liquid–Liquid Extraction	Solid–Liquid Extraction	Gas Absorption	Gas or Liquid Adsorption
a	Component	—	more volatile component	solute	solute	solute gas	solute, adsorbate
B	L-phase product	lb or lb moles	bottoms or waste	raffinate product		rich oil	adsorbent product
b	Component	—	less volatile component	major raffinate component	insoluble inerts	solvent, absorbent liquid	adsorbent
C	V-phase refluxer	—	condenser	solvent separator			
c	Component	—	(not defined for binary distillation)	extract solvent	solvent	inert gas, carrier gas	nonadsorbed fluid, carrier fluid
D	V-phase product	lb or lb moles	distillate or tops	extract product			
E_L	L-phase Murphree stage efficiency	per cent	Murphree liquid efficiency	Murphree raffinate efficiency		Murphree liquid efficiency	
E_V	V-phase Murphree stage efficiency	per cent	Murphree vapor efficiency	Murphree extract efficiency		Murphree vapor efficiency	
F	Feed	lb or lb moles	*	*		*	*
H	Vapor-phase enthalpy concentration	Btu/lb or lb mole	*				
h	Liquid-phase enthalpy concentration	Btu/lb or lb mole	*				
I	Intermediate stream	lb or lb moles	*	*	*	*	*
i	Change in L-phase flow per unit of feed at the feed (at the intersection of the operating lines)	—	*	*		*	*
K	Distribution coefficient	various	Henry's law constant	*	*	Henry's law constant	adsorption isotherm
L	A phase	lb or lb moles	liquid	raffinate phase	underflow, sludge	liquid	adsorbent phase
m	Any stage in stripping section	—	*	*			
N	Last stage in cascade	—	*	*	*	*	*
n	Any stage in enriching section	—	*	*	*	*	*
Q_{SB}	Heat added at still per lb or lb mole of bottoms	Btu/lb or lb mole	*				
Q_{CD}	Heat added at condenser per lb or lb mole of distillate	Btu/lb or lb mole	*				
q_s	Heat added at still	Btu/hr	*				
q_c	Heat added at condenser	Btu/hr	*				
S	L-phase refluxer	—	still or reboiler	Mixer			*
V	A phase	lb or lb moles	vapor	extract phase	overflow, extract phase	gas	liquid or gas
X	Ratio of components in phase L	mass or mole ratio		*	*	*	*
x	Fraction of any component in phase L	mass or mole fraction	*	*	*	*	*
Y	Ratio of components in phase V	mass or mole ratio		*	*	*	*
y	Fraction of any component in phase V	mass or mole fraction	*	*	*	*	*
Z	Ratio of components in streams other than L or V (such as I, F, Σ)	mass or mole ratio		*	*	*	*
z	Fraction of any component in streams other than L or V	mass or mole fraction	*	*	*	*	*
α	Separation factor	—	relative volatility	*	*	*	*
Δ	Net flow, difference	lb or lb moles	*	*	*	*	*
Σ	Sum	lb or lb moles	*	*	*	*	*
0, 1, 2	Stage number	—	*	*	*	*	*

* Indicates that specific designation is identical with general designation.
Blank space indicates that the term is not defined for that operation.
— Indicates that the term has no dimensions.

Superscripts—$\bar{L}, \bar{V}, \bar{\Delta}$: same as L, V, Δ, but in the stripping section (L and V in enriching section when a distinction is made). $\bar{\bar{L}}, \bar{\bar{V}}, \bar{\bar{\Delta}}$: same but between feeds. L', V': partial masses of phases L, V (not counting one or more components).

Subscripts—Many of the symbols defined above will appear as subscripts. For example, x_{3a}, refers to the concentration of a in the L-phase flowing from stage 3. Subscripts of stage numbers always denote the stage *from* which the stream is flowing. Thus, L_0 is flowing *from* stage 0 (which is nonexistent) to stage 1.

Molecular and Turbulent Transport

Nearly all unit operations involve the transport of mass, heat, or momentum. Such transport may occur within one phase or between phases. In many cases it is necessary to know the rate of transport of mass, heat, or momentum in order to design or to analyze industrial equipment for unit operations. For example, in the stage operations covered in the preceding chapters, the *rate* of mass transfer between phases was assumed to be great enough to permit the rapid establishment of equilibrium. In reality, however, the rate of mass transfer is an important factor in determining the stage efficiency, which must be known to determine the actual number of stages necessary for a given separation.

A knowledge of the principles of momentum transport is a prerequisite to the proper design of the pumps and piping systems so essential to the chemical process industry. Similarly, the design of industrial heat exchangers rests on an understanding of the fundamental transport mechanism for thermal energy.

The fundamental mechanisms of mass, heat, and momentum transport are closely related. The rate equations for the three transport systems are of the same form, and in certain simple physical situations the mechanisms of transport are identical. *Molecular* transport depends on the motion of individual molecules. *Turbulent* transport results from the motion of large groups or clusters of molecules.

The following six chapters will consider the fundamental mechanisms of molecular and turbulent heat, mass, and momentum transfer. Since the three transport processes are closely related, they will be developed in a general way wherever possible.

The terms *transport* and *transfer* are used interchangeably in many cases. If a distinction is made, *transport* usually is used when referring to the fundamental mechanism within a *single* phase, whereas *transfer* refers to the over-all process.

After the transport mechanisms have been developed, a number of fundamental applications will be considered. Methods of evaluating and correlating transport properties will be discussed. The application of the transport principles to interphase transfer will be developed. The integration of the rate equation will be demonstrated where concentration varies along the length of the contacting path. Finally, after the fundamentals of transport systems have been thoroughly developed, applications to the design of industrial equipment will be considered in Part III of this book. If the reader is not familiar with the dimensions and units used in chemical engineering, he should read Appendix A before proceeding. A table of the notation used in Part II, together with dimensions and typical units, is given after Chapter 14.

chapter 9

Molecular-Transport Mechanism

Molecular transport of mass, heat, and momentum may occur in a solid, liquid, or gas. A simple example of molecular transport is the conduction of heat in a metal bar. Molecular transport, as the term implies, depends upon the motion of individual molecules for the transport of mass, heat, or momentum. The transport mechanism may be developed from the kinetic theory of gases and liquids or from a consideration of the physics of the solid state. Neither of these fundamental subjects will be considered in detail here. Instead, an extremely simplified physical model of a gas will be used to derive expressions for the rate of transport. The resulting equations will be extended to include real gases, liquids, and solids. Finally, the evaluation and prediction of transport properties will be considered.

MASS, HEAT, AND MOMENTUM TRANSPORT

Each molecule of a system has a certain quantity of mass, thermal energy, or momentum associated with it. Mass transport occurs when different kinds of molecules are present in the same gas phase. If the concentration of one kind of molecule is greater in one region of the gas than in another, mass will be transferred from the region of higher concentration to a region of lower concentration. *Thermal energy* may be generally defined as that part of the internal energy of a molecule which may be transferred under the influence of the available temperature gradient. *Heat* is defined as that portion of the thermal energy which is actually being transferred along a temperature gradient. If a molecule possesses greater internal energy by virtue of having a higher temperature than its neighbors, it can transfer the excess energy to its less energetic neighbors. Momentum transport in a fluid depends upon the transfer of the macroscopic momentum of molecules of the system. If a fluid is in motion, the molecules will possess a macroscopic momentum in the direction of flow. If there is a variation in flow velocity, the faster moving molecules possess a greater momentum in the direction of flow and can transfer the excess momentum to their slower moving neighbors.

The practical applications of mass-, heat-, and momentum-transport principles developed independently in the earlier days of industry. The engineers who applied the principles of the three types of transport were not primarily interested in the similarities among the mechanisms of transport. As a result, three distinct systems of notation and nomenclature developed. Although it might be desirable from a fundamental point of view to use one set of notation in discussing the three systems, the traditional three sets of notation are so widely used that the engineer using a unified notation would not be understood by the majority of engineers in industry. For this reason, after the rate equation is derived using a completely general notation to show the interrelation, it will then be rewritten in the traditional terminology of mass, heat, and momentum transfer for application to industrial problems.

The transport of mass by individual molecular motion is usually referred to as *molecular diffusion*. Molecular transport of heat is called *conduction*. Molecular momentum transport occurs in *laminar flow*. These terms are widely used in discussing molecular transport.

The General Molecular-Transport Equation. The general rate equation for molecular transport may be derived using a simple physical model of a gas. Although the resultant equation is strictly applicable only to the model gas, it may be extended to real gases, liquids, and solids.

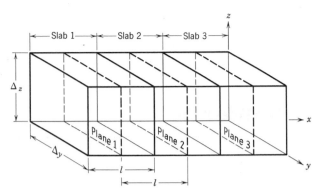

Figure 9.1. Volume element of the model gas.

A simplified kinetic theory of gases postulates the following model:

1. The gas is made up of molecules each of which is a perfect sphere of diameter σ.

2. No attractive or repulsive forces exist between gas molecules.

3. The actual volume of the molecules is negligible compared to the volume between molecules.

4. All collisions between molecules are perfectly elastic.

5. Each molecule is in random motion at a mean speed $(\bar{c})$ in the random direction.

6. Each molecule will move a distance l between collisions with other molecules. The distance l is called the *mean free path*.

7. The time required for a molecule to travel a mean free path traveling at the mean speed is the mean time between collisions $\bar{\theta}$. That is, $\bar{\theta} = l/\bar{c}$.

8. The number of molecules is large enough that statistically average values of properties can be used to describe all the molecules.

This is a highly idealized molecular picture of a gas. Molecules of real gases are not spherical, and there may be strong attractive or repulsive forces among molecules. Furthermore, the molecules will move at various speeds for various distances between collisions. A more rigorous treatment of a real gas involves complex physical and mathematical concepts which are beyond the scope of this book. Such treatments may be found in any book on kinetic theory, such as reference 1.

The derivation will consider the molecules of gas occupying the volume shown in Figure 9.1. Since the molecules are in random motion, they will move in all possible directions. To simplify the situation, the derivation will consider that the molecules move in directions parallel with the coordinate axes x, y, and z. Then, one-sixth of the total number of molecules will move in the $+x$-direction at any instant, one-sixth in the $-x$-direction, one-sixth in the $+y$-direction, and so

forth. Attention will be focused only on the molecules moving in the $+x$- or $-x$-direction. Three planes are spaced a distance l apart, and each has an area $\Delta y \, \Delta z$. Each plane can be considered the midpoint of a slab of gas of thickness l, as shown in Figure 9.1. All properties of the gas are uniform within each slab.

Small groups of molecules will have certain properties associated with them. For example, in mass transfer, some molecules will be gas a and others gas b. In heat transfer the thermal energies of the molecules are different, and in momentum transfer the momentum of molecules will vary. Each molecule can be considered to possess a certain *concentration* of one of the three transferent properties. Rather than for a single molecule, the concentration will be defined for a given volume of molecules. That is,

Γ = concentration of the property to be transferred, in dimensions of quantity of transferent property per unit volume of gas

Each slab of gas will have a different concentration of transferent property during transport. Since the concentration within each slab is considered uniform, the average uniform concentration of a slab is equal to the concentration at the plane at the center of each slab. Therefore, it is possible to consider the transport from slab to slab as equivalent to the transport from plane to plane. The concentration in slab 2 can be related to the concentration in slab 1 by

$$\Gamma_1 = \Gamma_2 + \frac{d\Gamma}{dx}(-l) \qquad (9.1)$$

where $d\Gamma/dx$ is the increase in concentration with distance in the $+x$-direction and $-l$ is the distance *from* plane 2 *to* plane 1.

Similarly,

$$\Gamma_3 = \Gamma_2 + \frac{d\Gamma}{dx}(l) \qquad (9.2)$$

The concentration gradient is shown schematically in Figure 9.2a. The gradient is assumed to be constant in the volume element.

In the time $\bar{\theta}$, one-sixth of the molecules in the slab centered around plane 3 will move a distance l to occupy exactly the volume of the slab centered around plane 2. Similarly, one-sixth of the molecules in slab 1 will move to the right into slab 2, one-sixth of the molecules in slab 2 will move to the right into slab 3, and one-sixth will move to the left into slab 1. If the molecules move, the property associated with the molecules will also move. The quantity of property in slab 1 is equal to the concentration times the volume of the slab, i.e., $\Gamma_1 \, \Delta y \, \Delta z \, l$. Similarly the quantity of property in slab 2 is $\Gamma_2 \, \Delta y \, \Delta z \, l$, and in slab 3, $\Gamma_3 \, \Delta y \, \Delta z \, l$. The *flux* ψ of the transferent property can be defined as the rate of transport of

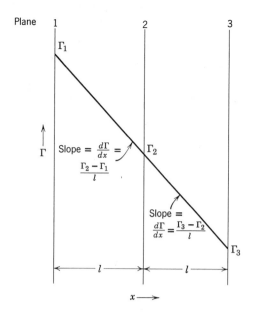

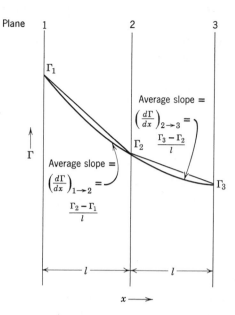

(a) Concentration gradient at steady state. (b) Concentration gradient at unsteady state.

Figure 9.2. Concentration gradients in molecular transport.

property per unit of transport area. Since $\bar{\theta}$ is the time taken to transfer one-sixth of the quantity of property $\Gamma_1 \, \Delta y \, \Delta z \, l$ from plane 1 to plane 2, the flux is

$$\psi_{(1 \to 2)} = \frac{1}{6} \frac{\Gamma_1 \, \Delta y \, \Delta z \, l}{\bar{\theta} \, \Delta y \, \Delta z} = \frac{\Gamma_1 l}{6\bar{\theta}} \qquad (9.3)$$

where $\psi_{(1 \to 2)}$ is the flux of transferent property from plane 1 to plane 2. It has dimensions of quantity of transferent property per unit time and unit transport area.

Similarly, the flux from plane 3 to plane 2 is

$$\psi_{(3 \to 2)} = \frac{-\Gamma_3 l}{6\bar{\theta}} \qquad (9.4)$$

The minus sign arises because the flux is always defined as positive if it is moving in the $+x$-direction. Thus, since the flux from plane 3 to plane 2 is in the $-x$-direction, the numerical value of this flux must be negative. The two fluxes leaving plane 2 are

$$\psi_{(2 \to 1)} = \frac{-\Gamma_2 l}{6\bar{\theta}} \qquad (9.5)$$

$$\psi_{(2 \to 3)} = \frac{\Gamma_2 l}{6\bar{\theta}} \qquad (9.6)$$

The *net flux* in the $+x$-direction between planes 1 and 2 is defined in the same way as net flow was for stage operations. It is the difference between the flow in the $-x$-direction and the $+x$-direction. Since flux is always positive with $+x$,

$$\psi_{(\text{net } 1 \to 2)} = \psi_{(1 \to 2)} + \psi_{(2 \to 1)} = \frac{1}{6} \frac{l}{\bar{\theta}} [\Gamma_1 - \Gamma_2] \quad (9.7)$$

Substitution of the concentration Γ_1 given in Equation 9.1 into Equation 9.7 yields

$$\psi_{(\text{net } 1 \to 2)} = \frac{1}{6} \frac{l}{\bar{\theta}} \left[\Gamma_2 + \frac{d\Gamma}{dx}(-l) - \Gamma_2 \right]$$

$$= -\frac{1}{6} \frac{l^2}{\bar{\theta}} \frac{d\Gamma}{dx} \qquad (9.8)$$

A similar treatment gives for the net flow between planes 2 and 3

$$\psi_{(\text{net } 2 \to 3)} = -\frac{1}{6} \frac{l^2}{\bar{\theta}} \frac{d\Gamma}{dx} \qquad (9.9)$$

Thus, the flux into slab 2 is the same as the flux out of the slab, and a steady state exists. The concentration gradient $(d\Gamma/dx)$ was taken to be the same on both sides of plane 2 when Equations 9.1 and 9.2 were written. This is shown in Figure 9.2a.

A constant concentration gradient is a necessary condition for steady state. *Steady state* means that there is no accumulation of the property with time or that the flux into plane 2 is equal to the flux out, as shown by Equations 9.8 and 9.9. A property balance shows that the accumulation within any slab is

Rate of accumulation

$$= \text{rate of input} - \text{rate of output}$$

$$= \psi_{(\text{net } 1 \to 2)} - \psi_{(\text{net } 2 \to 3)}$$

$$= 0 \quad \text{for steady state} \qquad (9.10)$$

On the other hand, if the concentration gradient is not constant, as shown in Figure 9.2b, the accumulation is not zero, and an unsteady state exists. The unsteady state will be considered further in Chapter 11.

Since the flux is constant at steady state, Equations 9.8 and 9.9 apply at any value of x,

$$\psi_{\text{net}} = -\frac{1}{6}\frac{l^2}{\bar{\theta}}\frac{d\Gamma}{dx} \qquad (9.11)$$

Since $l = \bar{c}\bar{\theta}$, Equation 9.11 becomes

$$\psi_{\text{net}} = -\frac{1}{6}l\bar{c}\frac{d\Gamma}{dx} \qquad (9.12)*$$

Equation 9.12 is the final result of the derivation. It relates the *rate* of transport per unit area (i.e., the flux) to the concentration gradient ($d\Gamma/dx$). The proportionality constant is equal to one-sixth of the product of the mean velocity and the mean free path of the molecules. If the gradient is negative, as shown in Figure 9.2a, the flux will be positive in the $+x$-direction, since there is a minus sign in Equation 9.12. The result of this derivation is completely general for any transport depending upon random molecular motion. It will be applied to the specific cases of mass, heat, and momentum transport in the following sections.

Nomenclature for Mass, Heat, and Momentum Transport. As was mentioned earlier, the terminology for the three transport systems is traditionally different. The notation is given in Table 9.1. Dimensions and typical units are also included. The definition of the terms for the three systems is given at the end of Part II, and they will be considered in detail below. Equation 9.12 is the general transport equation. It may be written in the terms listed in Table 9.1 for any specific transport system. For the model gas *only*, $\mathscr{D} = \alpha = \nu = \frac{1}{6}l\bar{c}$. Generally, the transport diffusivities are not equal. In the following sections, the mechanism of each of the three transport systems will be considered separately, and the three specific transport equations will be derived independently. The form of each derivation is identical with that used for the general equation. However, special attention will be given to the physical significance of the specific fluxes and gradients.

Mass Transport. If the model gas consists of two

---

* A common derivation of the transport equation found in books on physics and physical chemistry yields

$$\psi_{\text{net}} = -\frac{1}{3}l\bar{c}\frac{d\Gamma}{dx}$$

The difference in the factors of $\frac{1}{6}$ and $\frac{1}{3}$ occurs because of the manner in which net flux is defined. The definition used in this book is consistent with the chemical-engineering concept of net flow. A typical derivation yielding the factor $\frac{1}{3}$ may be found in reference 1. Other factors result when more complex physical models of a gas are used. Since the mean free path is usually evaluated from experimental data using the transport equation based on a given model, the resulting value of mean free path or molecular diameter will depend upon the model.

Table 9.1. NOTATION FOR MASS, HEAT, AND MOMENTUM TRANSPORT

Transferent Property	General Notation for Model Gas	Specific Notation		
		Mass Transport	Thermal-Energy Transport	Momentum Transport
Flux of transferent property	ψ_{net}	$\dfrac{N_a}{A}$ $\dfrac{\text{lb moles } a}{\text{hr sq ft}}$	$\dfrac{q}{A}$ $\dfrac{\text{Btu}}{\text{hr sq ft}}$	$\tau_y g_c$ $\dfrac{\text{lb-ft/hr}}{\text{hr sq ft}}$
Concentration of transferent property	Γ	c_a $\dfrac{\text{lb moles } a}{\text{cu ft}}$	$\rho c_P T$ $\dfrac{\text{Btu}}{\text{cu ft}}$	$v\rho$ $\dfrac{\text{lb-ft/hr}}{\text{cu ft}}$
Proportionality constant, the transport diffusivity	$\delta = \frac{1}{6}l\bar{c}$	$\mathscr{D}$ $\dfrac{\text{sq ft}}{\text{hr}}$	$\alpha = \dfrac{k}{\rho c_P}$ $\dfrac{\text{sq ft}}{\text{hr}}$	$\nu = \dfrac{\mu}{\rho}$ $\dfrac{\text{sq ft}}{\text{hr}}$

N_a = rate of mass transport $(\mathbf{M}/\mathbf{\theta})$, lb moles/hr
q = rate of transport of thermal energy $(\mathbf{H}/\mathbf{\theta})$, Btu/hr
τ_y = shear stress on the fluid $(\mathbf{F}_y/\mathbf{L}_y\mathbf{L}_z)$, $\text{lb}_f/\text{sq ft}$
g_c = dimensional constant $\mathbf{ML}_j/\mathbf{\theta}^2\mathbf{F}_j$, 4.18×10^8 lb-ft/hr²-lb$_f$
A = transport area $(\mathbf{L}_y\mathbf{L}_z)$ sq ft
ρ = density $(\mathbf{M}/\mathbf{L}_x\mathbf{L}_y\mathbf{L}_z)$ lb/cu ft
c_P = heat capacity $(\mathbf{H}/\mathbf{MT})$ Btu/lb °F
T = temperature $(\mathbf{T})$ °F
v = fluid velocity $(\mathbf{L}_y/\mathbf{\theta})$ ft/hr
k = thermal conductivity $[\mathbf{H}/\mathbf{\theta L}_y\mathbf{L}_z(\mathbf{T}/\mathbf{L}_x)]$ Btu/hr sq ft (°F/ft)
μ = absolute viscosity $(\mathbf{ML}_x/\mathbf{L}_y\mathbf{L}_z\mathbf{\theta})$ lb/ft hr
δ = generalized transport diffusivity $(\mathbf{L}_x{}^2/\mathbf{\theta})$, sq ft/hr
$\mathscr{D}$ = mass diffusivity $(\mathbf{L}_x{}^2/\mathbf{\theta})$, sq ft/hr
α = thermal diffusivity $(\mathbf{L}_x{}^2/\mathbf{\theta})$, sq ft/hr
ν = momentum diffusivity $(\mathbf{L}_x{}^2/\mathbf{\theta})$, sq ft/hr

different kinds of molecules, mass transport may occur. It will be assumed that the volume element pictured in Figure 9.1 contains two different species of gas, designated by a and b. Mass transport will occur only if the concentration of the gas phase is nonuniform. According to the original model, all molecules are equal in size and move at equal speeds in a random direction. If more molecules of gas a are present in a given volume, they will tend to migrate by random molecular motion to a neighboring region of lower concentration of gas a. If the difference in concentration is maintained, a steady flux of gas a will move from the region of higher concentration to the region of lower concentration.

It would be possible to derive the mass-transport equation by considering individual molecules and concentrations in terms of molecules per unit volume. However, the engineer is interested in average properties of groups of molecules, and it is more convenient to express concentration in terms of moles per unit volume, where

$$\text{Moles of gas } a = \frac{\text{molecules of gas } a}{\text{Avogadro's number}}$$

In the metric system Avogadro's number is 6.02×10^{23} molecules per gram mole of gas. The engineering notation for concentration in mass transport is

$$c_a = \text{concentration of gas } a, \frac{\text{lb moles gas } a}{\text{cu ft of total gas}}$$

The concentration of gas a in the volume element pictured in Figure 9.1 will vary from plane to plane. It will be assumed that the concentration varies linearly from plane 1 to plane 3. That is, the concentration gradient is constant, and a steady state exists. The concentration at plane 1 (or in slab 1) may be related to the concentration at plane 2 (or in slab 2) by

$$c_{a1} = c_{a2} + \frac{dc_a}{dx}(-l) \qquad (9.13)$$

This equation is equivalent to Equation 9.1 for the general case. The arguments presented for the general case also apply here. The concentrations at planes 2 and 3 can be related by

$$c_{a3} = c_{a2} + \frac{dc_a}{dx}(l) \qquad (9.14)$$

which is equivalent to Equation 9.2.

The random motion of the molecules in the volume element is such that one-sixth of the molecules in any slab will move in the $+x$-direction and one-sixth will move in the $-x$-direction. The total quantity of gas a in any slab is $c_a \, \Delta y \, \Delta z \, l$. Therefore, by the same reasoning used to derive Equation 9.3, the flux of gas a from slab 1 to slab 2 is

$$\left(\frac{N_a}{A}\right)_{(1 \to 2)} = \frac{1}{6} \frac{c_{a1} \, \Delta y \, \Delta z \, l}{\bar{\theta} \, \Delta y \, \Delta z} = \frac{c_{a1} l}{6 \bar{\theta}} \qquad (9.15)$$

The flux (N_a/A) may be split into two terms:

N_a = the rate of mass transport, lb moles/hr
A = the area across which mass transport occurs, sq ft

Analogous to the general case, the other fluxes of gas a are

$$\left(\frac{N_a}{A}\right)_{(3 \to 2)} = -\frac{c_{a3} l}{6 \bar{\theta}} \qquad (9.16)$$

$$\left(\frac{N_a}{A}\right)_{(2 \to 1)} = -\frac{c_{a2} l}{6 \bar{\theta}} \qquad (9.17)$$

$$\left(\frac{N_a}{A}\right)_{(2 \to 3)} = \frac{c_{a2} l}{6 \bar{\theta}} \qquad (9.18)$$

The net flux in the $+x$-direction between planes 1 and 2 is equal to the sum of Equations 9.15 and 9.17.

$$\left(\frac{N_a}{A}\right)_{(\text{net } 1 \to 2)} = \frac{1}{6} \frac{l}{\bar{\theta}}(c_{a1} - c_{a2}) \qquad (9.19)$$

Substitution of Equation 9.13 into Equation 9.19 gives

$$\left(\frac{N_a}{A}\right)_{(\text{net } 1 \to 2)} = -\frac{1}{6} \frac{l^2}{\bar{\theta}} \frac{dc_a}{dx} = -\frac{1}{6} l\bar{c} \frac{dc_a}{dx} \qquad (9.20)$$

Since the gradient is constant in the volume element the net flux from plane 2 to plane 3 is

$$\left(\frac{N_a}{A}\right)_{(\text{net } 2 \to 3)} = -\frac{1}{6} l\bar{c} \frac{dc_a}{dx} \qquad (9.21)$$

and the net flux is constant throughout the volume element. In Table 9.1, the term $\frac{1}{6}l\bar{c}$ is designated as $\mathscr{D}$, the mass diffusivity, for the *model gas*.

For *real* gases and liquids the diffusivity is defined by

$$\frac{N_a}{A} = -\mathscr{D} \frac{dc_a}{dx} \qquad (9.22)$$

where N_a/A is the net flux at steady state. Generally, molecules of two different real gases or liquids have different sizes and move at different velocities, contrary to our model gas. For these and other reasons, $\mathscr{D}$ is generally not equal to $\frac{1}{6}l\bar{c}$ for real gases and liquids. It must be determined experimentally or calculated utilizing the kinetic theory of real gases and liquids. Prediction and correlation of mass diffusivities will be considered in the latter part of this chapter. Mass transport is of great importance in real gases and liquids. In solids, the relative immobility of the molecules suppresses mass transport.

The molecules of gas b have the same random motion as those of gas a. Therefore, if a difference in concentration of gas b exists in a given volume, the gas will diffuse from the region of higher concentration to that of lower concentration. Considering gas b, Equation 9.21 may be written

$$\frac{N_b}{A} = -\frac{1}{6} l\bar{c} \frac{dc_b}{dx} \qquad (9.23)$$

and Equation 9.22 may be derived for gas b,

$$\frac{N_b}{A} = -\mathscr{D} \frac{dc_b}{dx} \qquad (9.24)$$

For dilute real gases or for liquids the diffusion coefficients defined by Equations 9.22 and 9.24 are identical.

The diffusion rates of gases a and b are interrelated as shown in the following paragraph. The total concentration c_t for the gaseous mixture of a and b is given by

$$c_t = c_a + c_b \qquad (9.25)$$

If the temperature and total pressure are constant, the total number of moles per unit volume (c_t) is constant. Differentiation of Equation 9.25 with respect to the distance x gives

$$0 = \frac{dc_a}{dx} + \frac{dc_b}{dx}$$

or

$$\frac{dc_a}{dx} = -\frac{dc_b}{dx} \qquad (9.26)$$

Therefore, if a gradient exists in gas a, a gradient exists in gas b. The gradient of gas b is equal but opposite in sign to that of gas a. Since a gradient for b exists, there must be a mass flux of b, as stated by Equation 9.24. Combination of Equations 9.22, 9.24, and 9.26 gives

$$N_b = -N_a \qquad (9.27)$$

This shows that the rates of diffusion are equal but in opposite directions. This phenomenon is known as *equimolar counterdiffusion*. It occurs when the concentration gradients and transport area are constant in the steady state. In diffusion through a stationary gas, which will be considered in Chapter 10, the concentration gradients are not constant at steady state. Therefore, the gradients shown in Figure 9.2 apply only to equimolar counterdiffusion.

Counterdiffusion is a unique feature of mass transport. Since two distinct species are present and since no mechanism is available to convert one to the other, two transport processes occur simultaneously. In contrast, heat or momentum transport concerns a single species, and consequently only one transport process can occur. That is, there is only one kind of heat (or thermal energy) for transport and only one kind of momentum for transport in a given direction, in contrast to the minimum of two kinds of gas necessary for detectable mass transport.

It is of interest to consider a gas phase consisting of a single species a. The random molecular motion would be identical with that discussed earlier for the two-component model gas. However, there would be no net flux of gas a. Equations 9.15, 9.17, and 9.20 can be rewritten for this case where $c_{a1} = c_t = c_{a2}$.

$$\left(\frac{N_a}{A}\right)_{(1 \to 2)} = \frac{c_t l}{6\bar{\theta}} = \frac{c_t \bar{c}}{6} \qquad (9.28)$$

$$\left(\frac{N_a}{A}\right)_{(2 \to 1)} = -\frac{c_t l}{6\bar{\theta}} = -\frac{c_t \bar{c}}{6} \qquad (9.29)$$

and

$$\left(\frac{N_a}{A}\right)_{(\text{net } 1 \to 2)} = 0 \qquad (9.30)$$

which shows that there is no net flux of gas a.

Equation 9.22 may be written in terms of partial pressure of gas a. From the perfect-gas law, $p_a V = n_a RT$

$$c_a = \frac{n_a}{V} = \frac{p_a}{RT} \qquad (9.31)$$

Substitution of c_a from Equation 9.31 into Equation 9.22 gives

$$\frac{N_a}{A} = -\frac{\mathscr{D}}{RT}\frac{dp_a}{dx} \qquad (9.32)$$

where p_a = partial pressure of gas a, atm

T = absolute temperature, °R

R = gas constant, cu ft atm/lb mole °R

This equation is a common alternate form of Equation 9.22.

Illustration 9.1. A large tank filled with a mixture of gases a and b at 1 atm and 32°F is connected to a large tank filled with a different mixture of a and b at 1 atm and 32°F. The connection between the tanks is a tube 2 in. inside diameter and 6 in. long. Calculate the steady-state rate of transport of gas a through the tube, when the concentration of a in one tank is 90 mole percent and the concentration of a in the other tank is 5 mole percent. Assume that the gas in each tank is uniform in composition and that transfer between the tanks is by molecular diffusion.

SOLUTION. First the mass diffusivity must be evaluated. Since no experimental data are available for gases a and b, Equation 9.20 will be used. From the kinetic theory of gases as shown in a later section,

$$\bar{c} = 1.7 \times 10^5 \text{ cm/sec}$$

$$l = 1.5 \times 10^{-5} \text{ cm}$$

$$\mathscr{D} = \tfrac{1}{6} l \bar{c} = \tfrac{1}{6}(1.5 \times 10^{-5})(1.7 \times 10^5) = 0.425 \text{ sq cm/sec}$$

$$= \left(0.425 \frac{\text{sq cm}}{\text{sec}}\right)\left(\frac{1 \text{ ft}}{30.4 \text{ cm}}\right)^2\left(\frac{3600 \text{ sec}}{1 \text{ hr}}\right)$$

$$= 1.66 \text{ sq ft/hr}$$

Equimolar counterdiffusion of gases a and b will occur. The steady-state gradient is shown in Figure 9.3. It would be possible to calculate c_a but instead Equation 9.32 will be used.

$$\frac{N_a}{A} = -\frac{\mathscr{D}}{RT}\left(\frac{dp_a}{dx}\right) \qquad (9.32)$$

At $x = 0$, $c_a = 0.9c_t$, or $p_a = 0.9P$, where P is the total pressure. At $x = 0.5$ ft, $p_a = 0.05P$. Integrating Equation 9.32 between these limits gives

$$\frac{N_a}{A} = -\frac{\mathscr{D}}{RT}\frac{(0.05P - 0.9P)}{(0.5 - 0)}$$

Substituting $P = 1$ atm, $T = 492°$R; $R = 0.7302$ atm cu ft/lb mole °R gives

$$\frac{N_a}{A} = -\frac{(1.66)(0.05 - 0.9)}{(0.7302)(492)(0.5)} = 7.85 \times 10^{-3} \text{ lb moles/hr sq ft}$$

Since $$A = \frac{\pi(2)^2}{4 \times 144} = 0.0218 \text{ sq ft}$$

$$N_a = (7.85 \times 10^{-3})(0.0218) = 1.71 \times 10^{-4} \text{ lb moles/hr}$$

It should be pointed out that the quantities $\mathscr{D}$ and A were

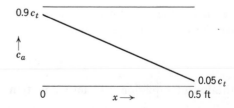

Figure 9.3. Concentration gradient for Illustration 9.1.

constant in this illustration. If they had varied with x or c_a, they would be included within the integrated value.

Thermal-Energy Transport. If the molecules in one region of a gas possess greater thermal energy than those in a neighboring region, part of the thermal energy will be transported by random molecular motion from the region of higher energy to that of lower energy. Thermal energy in the process of transfer is referred to as *heat*. The measure of the thermal energy of a single-phase system is its temperature. The higher the temperature of a system, the greater is the concentration of thermal energy. Therefore, heat will be transported from the region of higher temperature to the region of lower temperature.

A constant temperature gradient may be assumed to exist through the volume element of the model gas pictured in Figure 9.1. The concentration of thermal energy in each slab of gas is equal to $\rho c_P T$ Btu/cu ft, where ρ is the density, c_P is the specific heat, and T is the temperature of the gas. This expression is a *relative* measure of the thermal energy content of the gas. It is based on an arbitrary datum of zero thermal energy at $T = 0°R$ or $T = 0°F$. Since T generally appears as a differential or a difference, the arbitrary datum temperature cancels out. The *British thermal unit* (Btu) may be defined as the increase in thermal energy of one pound mass of water when its temperature is increased by one Fahrenheit degree.

The concentration of thermal energy at plane 1 (or in slab 1) is $(\rho c_P T)_1$, the concentration in slab 2 is $(\rho c_P T)_2$, and in slab 3, $(\rho c_P T)_3$. The mean free path and mean velocity of the molecules of the model gas vary with temperature. In this case an appropriate constant *average* value of l and $\bar{c}$ will be assumed. If the thermal concentration gradient is constant in the x-direction, as shown in Figure 9.2, the concentration of the slabs can be interrelated.

$$(\rho c_P T)_1 = (\rho c_P T)_2 + \frac{d(\rho c_P T)}{dx}(-l) \quad (9.33)$$

and

$$(\rho c_P T)_3 = (\rho c_P T)_2 + \frac{d(\rho c_P T)}{dx}(+l) \quad (9.34)$$

The heat flux is written as q/A, where q is the rate of thermal-energy transport (Btu/hr) and A is the transport area (sq ft). By the procedure used to derive Equation 9.3, the flux of heat from slab 1 to slab 2 is

$$\left(\frac{q}{A}\right)_{(1 \to 2)} = \frac{1}{6}\frac{(\rho c_P T)_1 \, \Delta y \, \Delta z \, l}{\bar{\theta} \, \Delta y \, \Delta z} = \frac{(\rho c_P T)_1 l}{6\bar{\theta}} \quad (9.35)$$

The heat flux from plane 3 to plane 2 is

$$\left(\frac{q}{A}\right)_{(3 \to 2)} = -\frac{(\rho c_P T)_3 l}{6\bar{\theta}} \quad (9.36)$$

and

$$\left(\frac{q}{A}\right)_{(2 \to 1)} = -\frac{(\rho c_P T)_2 l}{6\bar{\theta}} \quad (9.37)$$

$$\left(\frac{q}{A}\right)_{(2 \to 3)} = \frac{(\rho c_P T)_2 l}{6\bar{\theta}} \quad (9.38)$$

The net flux in the $+x$-direction between planes 1 and 2 is equal to the sum of Equations 9.35 and 9.37,

$$\left(\frac{q}{A}\right)_{(net \, 1 \to 2)} = \frac{1}{6}\frac{l}{\bar{\theta}}\left[(\rho c_P T)_1 - (\rho c_P T)_2\right] \quad (9.39)$$

Substitution of Equation 9.33 into Equation 9.39 gives

$$\left(\frac{q}{A}\right)_{(net \, 1 \to 2)} = -\frac{1}{6}\frac{l^2}{\bar{\theta}}\frac{d(\rho c_P T)}{dx}$$

$$= -\frac{1}{6}l\bar{c}\frac{d(\rho c_P T)}{dx} \quad (9.40)$$

By a similar derivation the net flux from plane 2 to plane 3 is

$$\left(\frac{q}{A}\right)_{(net \, 2 \to 3)} = -\frac{1}{6}l\bar{c}\frac{d(\rho c_P T)}{dx} \quad (9.41)$$

Comparison of Equations 9.40 and 9.41 shows that the net flux is constant if the thermal concentration gradient $(d(\rho c_P T)/dx)$ is constant. In this case, the net flux from plane 1 to plane 2 is equal to the net flux from plane 2 to plane 3. Therefore, by Equation 9.10 the accumulation is zero.

Reference to Table 9.1 shows that the thermal diffusivity (α) is equal to $\frac{1}{6}l\bar{c}$ for the model gas. For real gases, liquids and solids, the thermal diffusivity can be defined by writing Equation 9.41 as

$$\frac{q}{A} = -\alpha\frac{d(\rho c_P T)}{dx} \quad (9.42)$$

Thermal diffusivities must be evaluated experimentally or calculated using kinetic theory. Equation 9.42 may be written for constant c_P and ρ as

$$\frac{q}{A} = -\alpha\rho c_P\frac{dT}{dx} = -k\frac{dT}{dx} \quad (9.43)$$

where k = the thermal conductivity = $\alpha\rho c_P$, Btu/hr sq ft (°F/ft) and dT/dx is the temperature gradient, °F/ft. Values of thermal conductivities may be found in tables of physical data. If k and A are constant, integration of Equation 9.43 yields

$$\frac{q}{A} = -\frac{k(T_2 - T_1)}{(x_2 - x_1)} \quad (9.44)$$

The mechanism of transport of thermal energy may be considered to have two parts. First, as shown earlier, the molecules in a given volume element will migrate by random molecular motion. Second, if a molecule of higher thermal energy migrates to a region of lower energy, it must distribute its excess energy among the

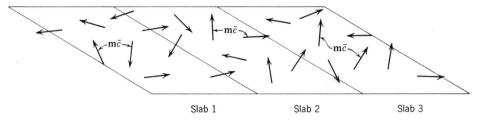

(*a*) Momentum of random molecular motion.

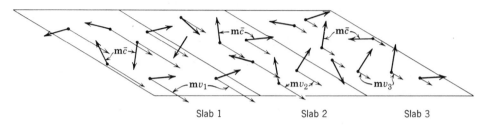

(*b*) Flow momentum superimposed on the random molecular momentum ($v_1 > v_2 > v_3$).

Figure 9.4. Vectorial representation of momentum in the model gas.

less energetic molecules, since the concentration of thermal energy was assumed to be constant in a given slab of gas. The distribution of energy may occur by collisions between molecules. The two-step thermal-transport mechanism differs from the mass-transport mechanism which is essentially one step, since the transport of mass is complete upon arrival of the molecule in the host region. The differing gas molecules retain their identity, and there is no second step.

In the simple model gas it is assumed that the second step for heat transport occurs instantaneously when the molecule arrives in the host region. At steady state the net flux of thermal energy entering a slab is equal to that leaving. Therefore, no *net* change in the thermal-energy concentration occurs in the second step, even though there is a continual interchange of energy.

In real gases, liquids, and solids, the two steps of the thermal-energy-transport mechanism occur simultaneously. In a real gas each collision of a migrant molecule may result in an exchange of thermal energy. In solids the free random motion of the molecules in the first step of transport is restricted by the amplitude of vibration of the individual molecules in the crystal structure of the solid. In this case, the molecules vibrate around fixed points and transfer energy only by collision with the nearest neighbors, equivalent to the second step of the model-gas mechanism. Since the first step of transport is restricted in solids and since mass cannot be transferred by collision alone, it is obvious that mass transport must be very small in solids. Experimental studies of mass diffusion in solids show that it occurs at a very low rate. In liquids, transfer of energy occurs by both steps of the mechanism to an extent intermediate

between that for gases and solids, because liquid molecules have less mobility than gases.

Illustration 9.2. If the tanks in Illustration 9.1 are at 0°F and 64°F, calculate the rate of heat transfer by molecular transport. The density of the gas is 0.08 lb/cu ft and the heat capacity is 0.25 Btu/lb °F.

SOLUTION. The thermal diffusivity will be evaluated at the average temperature, 32°F. From Illustration 9.1,

$$\alpha = \tfrac{1}{6} l \bar{c} = 1.66 \text{ sq ft/hr}$$

Integration of Equation 9.20 gives

$$\frac{q}{A} = -\alpha \frac{(\rho c_P T_2 - \rho c_P T_1)}{(x_2 - x_1)} = -\alpha \rho c_P \frac{(T_2 - T_1)}{(x_2 - x_1)}$$

The present limits are

$$x_1 = 0; \qquad T_1 = 64°F = 524°R$$
$$x_2 = 0.5 \text{ ft}; \qquad T_2 = 0°F = 460°R$$

Therefore

$$\frac{q}{A} = -\frac{(1.66)(0.08)(0.25)(460 - 524)}{(0.5 - 0)}$$

$$= 4.25 \text{ Btu/hr sq ft}$$

and $q = (4.25)(0.0218) = 0.0925$ Btu/hr. If any of the quantities c_P, ρ, or A varied, they would have to be integrated with T and x.

Momentum Transport. Since the concepts of momentum flux and momentum concentration are more difficult to visualize physically, the momentum-transport mechanism in the model gas will be considered in detail.

Each molecule in the model gas has a mass **m** and an

average speed $\bar{c}$. Therefore each molecule has a momentum $\mathbf{m}\bar{c}$ in a random direction. However, since the momentum of every molecule is in a random direction, the vectorial sum of all the momenta in a given volume is zero, and there is no excess momentum to be transferred. A gas will possess transferable momentum only if various regions of the gas possess different concentrations of momentum. Differences in momentum will occur if there are differences in velocity. The molecules of the model gas pictured in Figure 9.4a are in random motion at a speed $\bar{c}$. The length of the arrow on each molecule indicates the magnitude of its velocity and the direction of the arrow indicates the direction of the velocity. Each of the molecules in the three slabs shown is moving at the same speed and consequently possesses a momentum equal in magnitude to those of all other molecules. Since the direction of movement is random, the sum of all the momenta is zero.

On the other hand, the model gas in Figure 9.4b is shown to have in addition a *flow velocity v* in the +y-direction. Each molecule then may be considered to have two components of momentum, $\mathbf{m}\bar{c}$ in a random direction and $\mathbf{m}v$ in the +y-direction. If the flow velocity were constant in all three slabs, the momentum concentration of each slab would be the same, and there would be no net transport of momentum from slab to slab. However, in the gas pictured, the flow velocities in the slabs differ, as indicated by the length of the arrows representing the flow momentum. This is an idealized picture, since the flow velocities within any slab are assumed to be uniform. In reality, the velocities would vary uniformly in the x-direction. The velocity v in the +y-direction is a macroscopic or bulk velocity. It is equivalent to an actual *flow* of the gas in the +y-direction. If the molecules in slab 1 are flowing at a velocity v_1, each molecule has two components of momentum, $\mathbf{m}\bar{c}$ and $\mathbf{m}v_1$. Similarly, the momentum components in slab 2 are $\mathbf{m}\bar{c}$ and $\mathbf{m}v_2$, and in slab 3, $\mathbf{m}\bar{c}$ and $\mathbf{m}v_3$. In the model gas $\mathbf{m}\bar{c}$ is uniform throughout the volume element, and the vectorial sum of all $\mathbf{m}\bar{c}$ is zero. Consequently they do not contribute to the *excess* of momentum necessary for momentum transport. However, if the flow velocities of the slabs are not equal, an excess of momentum can exist in one slab compared to another. The momentum caused by the bulk flow of the gas *in the y-direction* will be transferred between slabs *in the x-direction*, as indicated in Figure 9.4b.

The model gas is considered to be flowing in a regular manner in the +y-direction. Average groups of molecules flow parallel with the xy and yz planes. Thus, there is no random motion so far as flow is concerned. Of course, individual molecules still move randomly in any direction, but the average movement is uniformly in the flow direction. This uniform macroscopic movement is called *laminar flow*. A more random macroscopic flow is referred to as *turbulent flow*; it will be discussed in Chapter 12.

The velocity gradient between the planes of gas shown in Figure 9.4b is assumed to be constant. Each slab of gas has a different momentum. *Individual* molecules traveling from one slab to another by random motion in the x-direction will carry with them their y-flow momentum. This effectively results in the transport of flow momentum from one slab to another.

The *concentration* of momentum for any volume of gas is equal to the total momentum possessed by the gas divided by the volume of the gas. If there are n molecules in slab 1, the momentum concentration is $nmv_1/\Delta y\,\Delta z\,l$. However, nm is equal to the total mass of gas in the volume $\Delta y\,\Delta z\,l$.

The *density* of the gas is therefore

$$\rho = \frac{m}{V} = \frac{nm}{\Delta y\,\Delta z\,l}$$

where m = the total mass = nm, and the momentum concentration in slab 1 becomes ρv_1. Similarly, the momentum concentrations in the other slabs are ρv_2 and ρv_3.

The *momentum flux* is equal to the rate of momentum transport divided by the transport area.

A general expression for momentum flux is

$$\text{Momentum flux} = \frac{1}{A}\frac{d(mv)}{d\theta} \qquad (9.45)$$

where A = transport area, sq ft
m = total mass being accelerated, lb
v = velocity, ft/hr
θ = time, hr

At steady state, the momentum flux is constant. Further, in the model gas v is constant for the molecules originating in any given slab. Therefore,

$$\text{Momentum flux} = \frac{1}{A}\frac{\Delta(mv)}{\Delta\theta} = \frac{v}{A}\frac{\Delta m}{\Delta\theta} \qquad (9.46)$$

For the model gas, $A = \Delta y\,\Delta z$, $\Delta\theta = \bar{\theta}$, and $\Delta m = \frac{1}{6}\rho\,\Delta y\,\Delta z\,l$, since only one-sixth of the molecules of a volume element move in the +x-direction. Equation 9.46 becomes

$$\frac{1}{A}\frac{\Delta(mv)}{\Delta\theta} = \frac{(\rho v)\,\Delta y\,\Delta z\,l}{6\bar{\theta}\,\Delta y\,\Delta z} = \frac{(\rho v)l}{6\bar{\theta}} \qquad (9.47)$$

The momentum concentrations of the slabs may be related by

$$(\rho v)_1 = (\rho v)_2 + \frac{d(\rho v)}{dx}(-l) \qquad (9.48)$$

$$(\rho v)_3 = (\rho v)_2 + \frac{d(\rho v)}{dx}(l) \qquad (9.49)$$

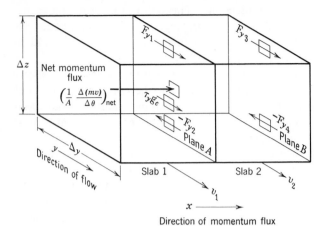

Figure 9.5. Accelerative forces in momentum transfer.

The momentum flux from plane 1 to plane 2 is

$$\left(\frac{1}{A}\frac{\Delta(mv)}{\Delta\theta}\right)_{(1\rightarrow2)} = \frac{(\rho v_1)\,\Delta y\,\Delta z\,l}{6\bar\theta\,\Delta y\,\Delta z} = \frac{(\rho v)_1 l}{6\bar\theta} \quad (9.50)$$

The other momentum fluxes are

$$\left(\frac{1}{A}\frac{\Delta(mv)}{\Delta\theta}\right)_{(3\rightarrow2)} = -\frac{(\rho v)_3 l}{6\bar\theta} \quad (9.51)$$

$$\left(\frac{1}{A}\frac{\Delta(mv)}{\Delta\theta}\right)_{(2\rightarrow1)} = -\frac{(\rho v)_2 l}{6\bar\theta} \quad (9.52)$$

$$\left(\frac{1}{A}\frac{\Delta(mv)}{\Delta\theta}\right)_{(2\rightarrow3)} = \frac{(\rho v)_2 l}{6\bar\theta} \quad (9.53)$$

The net flux between planes 1 and 2 is equal to the sum of Equations 9.50 and 9.52

$$\left(\frac{1}{A}\frac{\Delta(mv)}{\Delta\theta}\right)_{(net\ 1\rightarrow2)} = \frac{1}{6}\frac{l}{\bar\theta}[(\rho v)_1 - (\rho v)_2] \quad (9.54)$$

Substitution of Equation 9.48 into Equation 9.54 gives

$$\left(\frac{1}{A}\frac{\Delta(mv)}{\Delta\theta}\right)_{(net\ 1\rightarrow2)} = -\frac{1}{6}\frac{l^2}{\bar\theta}\frac{d(\rho v)}{dx} \quad (9.55)$$

$$= -\frac{1}{6}l\bar c\frac{d(\rho v)}{dx}$$

Similarly, the net flux from plane 2 to plane 3 is

$$\left(\frac{1}{A}\frac{\Delta(mv)}{\Delta\theta}\right)_{(net\ 2\rightarrow3)} = -\frac{1}{6}l\bar c\frac{d(\rho v)}{dx} \quad (9.56)$$

Equations 9.55 and 9.56 are of the same form as the other transport equations. At steady state the accumulation of momentum is zero.

It is now of interest to examine further the mechanics of momentum transport in order to define the momentum flux in conventional terms. The momentum flux cannot be directly measured, but it is related to the force or the shear stress acting on the fluid phase. (In Chapter 10 the force acting on the fluid phase will be evaluated using an over-all force balance.)

As was the case in thermal-energy transport, the arrival into a host region of molecules with flow momentum different from flow momentum of the host region requires the assumption that the new momentum contribution be instantaneously distributed throughout the host region. The migrant molecules have a velocity different from that of the molecules in the host region, and therefore the distribution of momentum requires an acceleration of the migrant molecules, with a resulting force exerted on the migrant molecules by the host region. The host region in turn suffers an equivalent reaction to the force. During the same time interval, the host region has sent an equivalent number of molecules to the original region in exchange for the arriving molecules, and these exchange molecules are subject to a similar force in the original region.

In the model gas the exchange of molecules and momentum between slabs 1 and 2 will result in accelerative forces between the slabs, as shown in Figure 9.5. It is assumed that v_1 is greater than v_2. The molecules leaving slab 1 that arrive in slab 2 must be decelerated so that their momentum may be distributed throughout slab 2. This deceleration exerts a force upon slab 2, and the origin of the force is slab 1. The force acts in the direction of flow, the y-direction. This force can be represented by the force vector F_{y1}, which is the force exerted on slab 2 by slab 1 at the plane between the slabs. Meanwhile the molecules arriving in slab 1 from slab 2 must be accelerated from v_2 to v_1. This creates a force $-F_{y2}$ acting in the $-y$-direction from slab 2 on slab 1 at the boundary plane. (Since the numbers of molecules moving between planes in the two directions are equal, $F_{y1} = -F_{y2}$.)

The negative sign appears with $-F_{y2}$ because the force acts in the $-y$-direction, and by convention all forces are defined as positive in the $+y$-direction. The two forces acting on the fluid at plane A are numerically equal, but opposite in sign. This is true at *any plane* in the fluid. For example, at plane B in Figure 9.5, there are two opposing forces F_{y3} and $-F_{y4}$. In Chapter 10 the force acting on a fluid plane will be evaluated using a *force balance* around a volume of gas. Since the forces acting on a plane are difficult to measure directly, the force balance permits their evaluation from other forces of the system. A force balance on a volume element must consistently utilize either the *external* or the *internal* forces acting on a volume element. For slab 2 the *external* forces are F_{y1} and $-F_{y4}$. The *internal* forces are $-F_{y2}$ and F_{y3}. The force balance states that the sum of the external forces is zero,

$$F_{y1} + (-F_{y4}) = 0$$

or that the sum of the internal forces is zero,

$$-F_{y2} + F_{y3} = 0$$

These equations assume no other forces acting upon the slab.

The force acting between two surfaces, such as that at plane A of Figure 9.5, is called a *shearing force*, since it tends to deform the fluid. The laws of motion state that for this case the shearing force is equal to the rate of change of momentum at the surface,

$$F_y g_c = \frac{d(mv)}{d\theta} \tag{9.57}$$

where m is the mass which is accelerated. The dimensional conversion factor (g_c) is included because customarily force is expressed in dimensions of *pounds force* ($\mathbf{F}_j$, lb$_f$) and mass in dimensions of *pounds mass* ($\mathbf{M}$, lb). These two systems of dimensions are related by g_c, as discussed in Appendix A.

The shearing force per unit of shear area is referred to as the *shear stress*, and designated by τ_y. At the plane between the slabs, the shear stress is

$$\tau_y = \frac{F_y}{A} = \frac{F_y}{\Delta y \, \Delta z} \tag{9.58}$$

Combining Equations 9.57 and 9.58 gives

$$\tau_y g_c = \frac{1}{A} \frac{d(mv)}{d\theta} \tag{9.59}$$

Comparison of Equations 9.45 and 9.59 shows that the shear stress acting on a plane is identical to the *net momentum flux* across that plane. Therefore, the shear stress in the $+y$-direction ($\tau_y g_c$) may be substituted for momentum flux in the $+x$-direction in any of the equations used in this derivation. For example, Equation 9.56 becomes

$$\tau_y g_c = -\frac{1}{6} l\bar{c} \frac{d(\rho v)}{dx} \tag{9.60}$$

The *momentum diffusivity* (ν) is equal to $\frac{1}{6}l\bar{c}$ for the model gas, listed in Table 9.1. For real gases and liquids the momentum diffusivity is defined by

$$\tau_y g_c = -\nu \frac{d(\rho v)}{dx} \tag{9.61}$$

The momentum diffusivity is more commonly referred to as the *kinematic viscosity*. The *absolute viscosity* is defined by

$$\tau_y g_c = -\mu \frac{dv}{dx} \tag{9.62}$$

where $\mu = \nu\rho$ = absolute viscosity, lb/ft hr. The absolute viscosity may be calculated from kinetic theory in a few cases and in general may be determined experimentally utilizing Equation 9.62. This equation is the conventional form of the momentum-transport equation.

Illustration 9.3. Two vertical parallel metal plates are spaced 1 in. apart. The left-hand plate is moving at a velocity of 5 ft/min, and the right-hand plate is stationary. The space

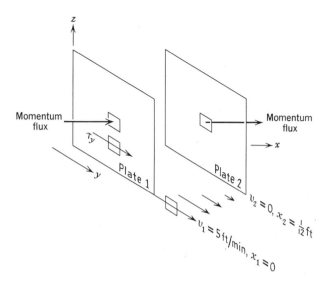

Figure 9.6. Momentum transport between two parallel plates in Illustration 9.3.

between the plates is filled with the model gas at 1 atm and 32°F, as described in Illustrations 9.1 and 9.2.

(a) Calculate the force necessary to maintain the movement of the left plate.

(b) Calculate the momentum flux at the surface of the left plate and at the surface of the right plate.

SOLUTION. At steady state, momentum will be transferred from the left plate through the gas to the right plate, and the velocity of the gas will vary linearly with distance between the two plates (Figure 9.6). The velocity of the gas at the surface of the moving plate is equal to the velocity of the plate. The velocity of the gas at the surface of the stationary plate is zero.

(a) Since the area of the plates is not given, the force per unit area of plate will be evaluated. This is the shear stress (τ_y) which can be evaluated from Equation 9.61. For the model gas (Illustration 9.1),

$$\nu = \tfrac{1}{6}l\bar{c} = 1.66 \text{ sq ft/hr}$$

From Illustration 9.2, the density of the gas is 0.08 lb/cu ft. Integration of Equation 9.61 with constant flux, viscosity, and density gives

$$\tau_y g_c = -\nu\rho \frac{(v_2 - v_1)}{(x_2 - x_1)}$$

Since $x_1 = 0$, $x_2 = 1/12$ ft, $v_1 = 5 \times 60$ ft/hr, $v_2 = 0$,

$$\tau_y g_c = -\left(1.66 \frac{\text{sq ft}}{\text{hr}}\right)\left(0.08 \frac{\text{lb}}{\text{cu ft}}\right)\left(\frac{0\text{–}300 \text{ ft/hr}}{1/12\text{–}0 \text{ ft}}\right)$$

$$= 478 \frac{\text{ft-lb/hr}}{\text{sq ft hr}}$$

Now $\quad g_c = 32.2 \dfrac{\text{ft-lb/sec}}{\text{lb}_f\text{-sec}} = 4.17 \times 10^8 \dfrac{\text{ft-lb/hr}}{\text{lb}_f\text{-hr}}$

Therefore $\quad \tau_y = \dfrac{478}{4.17 \times 10^8} = 1.15 \times 10^{-6} \text{ lb}_f/\text{sq ft}$

This is the force per unit area required to maintain the motion of the left plate.

(b) The momentum flux from the left plate to the gas is equal to the shear stress at the left plate.

$$\text{Momentum flux} = 478 \frac{\text{ft-lb/hr}}{\text{hr sq ft}}$$

Note that the dimensions are consistent with the definition of momentum flux. Since steady state exists, the momentum flux is uniform throughout the gas and the flux at the right plate is also $478 \dfrac{\text{ft-lb/hr}}{\text{hr sq ft}}$.

NON-NEWTONIAN FLUIDS

The momentum transport equations developed above are written for fluids with a viscosity that is constant at constant temperature and independent of rate of shear and time of application of shear. Fluids with this property are called "Newtonian" fluids. All gases and pure low-molecular-weight liquids are Newtonian. Miscible mixtures of low-molecular-weight liquids are also Newtonian.

Non-Newtonian fluids are characterized by viscosities that change with rate of shear or time of application of shear. Of the two classifications, shear-rate-dependent non-Newtonian fluids will be described first. Shear-rate-dependent non-Newtonian fluids are usually represented by an equation of the form

$$\tau_y g_c = \phi\left(\frac{dv}{dx}\right) \tag{9.63}$$

where
$\tau_y g_c =$ the shear stress at the point of consideration

$\phi\left(\dfrac{dv}{dx}\right) =$ some function of the velocity gradient at the point of consideration

At any point in the system the "apparent" viscosity is useful in characterizing local behavior. The "apparent" viscosity can be defined in the same manner as the Newtonian viscosity,

$$\tau_y g_c = -\mu_a \frac{dv}{dx} \tag{9.64}$$

Equating Equations 9.63 and 9.64 gives

$$\mu_a = -\frac{\phi(dv/dx)}{(dv/dx)} \tag{9.65}$$

Note that, if any point on the flow curve is joined to the origin with a straight line, the slope of the line is the apparent viscosity (Figure 9.7). Figures 9.7 and 9.8 and subsequent text describe various types of non-Newtonian flow behavior.

Bingham Plastic Fluids. (See Figure 9.7.) These substances require a threshold stress ($\tau_0 g_c$) which must be exceeded before flow can occur; usually called the

yield value. Examples of Bingham plastics are suspensions of rock and clay. For Bingham plastics,

$$\tau_y g_c - \tau_0 g_c = -\mu_B\left(\frac{dv}{dx}\right) \tag{9.66}$$

where
$\tau_0 g_c =$ the yield value, the initial value of the shear stress which must be exceeded
$\mu_B =$ defined by Equation 9.66

For Bingham plastics,

$$\mu_a = \frac{\mu_B \dfrac{dv}{dx} - \tau_0 g_c}{\left(\dfrac{dv}{dx}\right)} \tag{9.67}$$

From Equation 9.67 it is evident that μ_a decreases with rate of shear dv/dx.

Dilatant Fluids. The stress-shear rate diagram, Figure 9.7 shows that the equation for dilatant fluids is

$$\tau_y g_c = -K\left(\frac{dv}{dx}\right)^n; \quad n > 1 \tag{9.68}$$

and

$$\mu_a = -K\left(\frac{dv}{dx}\right)^{n-1}; \quad n > 1 \tag{9.69}$$

Thus, the apparent viscosity *increases* with increase in rate of shear. Starch suspensions, potassium silicate "solutions" and gum arabic "solutions" are examples of dilatant fluids.

Pseudoplastic Fluids. (See Figure 9.7.) This is probably the largest class of non-Newtonian fluids. The stress-rate of shear pattern for pseudoplastic fluids shows that the equation for pseudoplastic fluids is

$$\tau_y g_c = -K\left(\frac{dv}{dx}\right)^n; \quad n < 1 \tag{9.70}$$

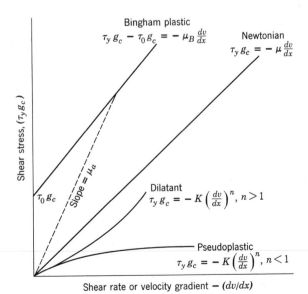

Figure 9.7. Shear behavior of non-Newtonian fluids.

and

$$\mu_a = -K\left(\frac{dv}{dx}\right)^{n-1}; \qquad n < 1 \qquad (9.71)$$

Note that the apparent viscosity of pseudoplastic fluids *decreases* with increase in rate of shear. Examples are solutions of high polymers, paper pulp, mayonnaise.

Newtonian Fluids. On the stress-shear rate diagram (Figure 9.7), Newtonian fluids are represented by a straight line through the origin. In this case, the apparent viscosity is equal to the absolute viscosity at all parts of the curve. Newtonian fluids can be represented by Equation 9.68, where $n = 1$, at which time $\mu = \mu_a = K$.

Time-Dependent Non-Newtonian Fluids. The time-dependent non-Newtonian fluids are very common. As might be expected, the additional variable of time complicates the analysis. One procedure of analysis is the so-called loop technique in which a substance is subjected to an increase in shear rate and then to a decrease in shear rate, returning to a shear rate of zero. If no time dependence exists, the two curves should be coincident. However, if the apparent viscosity changes with time, two separate curves will be traced (Figure 9.8).

Rheopectic Fluids. Rheopectic fluids show an increase in apparent viscosity with time. Figure 9.8a is the flow curve for a rheopectic fluid. The arrows indicate the path with time; in other words, the loop is traversed by increasing then decreasing the shear rate, and the arrows indicate the chronological order in which the data were taken. Some investigators refer to the area within the loop as an index of rheopexy (8). From this diagram, the change in apparent viscosity can be determined for a specific shear-rate–time history. The curve would not be the same for another time history. Points A and B are the shear-stress values at constant rate of shear but different duration of shear. Bentonite clay suspensions and some sols are rheopectic. After standing, rheopectic substances revert to the original condition.

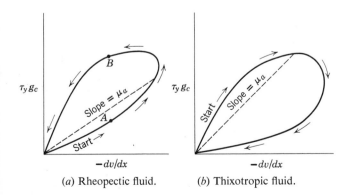

(a) Rheopectic fluid.   (b) Thixotropic fluid.

Figure 9.8. Flow curves for time-dependent non-Newtonians.

Thixotropic Fluids. The opposite of rheopexy is thixotropy. Thixotropic fluids show a decrease in apparent viscosity with time of application of shear. Data taken in the manner described for rheopectic fluids appear as a loop diagram but the path indicated by the arrows is reversed. (See Figure 9.8b.) Thixotropic behavior is found in paint, catsup, etc. After standing, thixotropic fluids revert to the original condition.

Non-Newtonian properties are sometimes desirable. For example, thixotropic behavior is desirable in paints. During brush working, certain paints flow readily to cover the surface, but upon standing the original highly viscous condition returns and the paint will not run.

Unfortunately, the study of non-Newtonian fluids has not progressed far enough to develop many useful generalizations. Useful engineering procedures for design are described in reference 8.

Non-Newtonian behavior is attributed to the combination of the properties of typical solids with the properties of typical liquids. Yield value, for example, is a solid property. The decrease of viscosity of solutions of polymers is attributed to the presence, at low rate of shear, of masses of solvent attracted to the solid to create a gel structure. As the rate of shear is increased, the structure breaks down, solvent is liberated, and apparent viscosity decreases.

OTHER TRANSPORT PHENOMENA

The three major types of transport considered in this chapter are not the only transport systems. For example, mass transport may occur by any of a number of mechanisms. Transport across a concentration gradient is the most significant mechanism, and it has been considered in detail earlier in this chapter.

Mass transport across a *temperature* gradient is called thermal diffusion. If a temperature gradient is applied to a uniform mixture of gases, the heavier molecules will tend to migrate to the lower temperatures, and the lighter molecules to the higher temperatures. The effect can be predicted from the rigorous kinetic theory of gases (1, 2), and it has been verified experimentally on many occasions. Thermal diffusion has been applied to the separation of gaseous mixtures. It is particularly useful in the separation of gaseous isotopes. For example, two isotopes of uranium were separated from each other by the thermal diffusion of $U^{235}F_6$ and $U^{238}F_6$ in the early stages of the nuclear-energy program in the Second World War.

Mass transport across a *pressure* gradient can be predicted from kinetic theory, but the effect is small and has not been investigated to any extent. Mass transport under the influence of an externally applied force is called *forced diffusion* (3). If the force acts uniformly upon all molecules of the system (as does gravity), there is no net transport. On the other hand, an electromagnetic field imposed upon a mixture of ionized gases may cause forced diffusion.

A binary mixture of gases may be separated by allowing part of the gas to flow through a barrier containing very small holes (diameter less than the mean free path of the molecules). The separation factor is equal to the square root of the ratio of the molecular weights of the gases. The process is generally called *gaseous diffusion* and is used to separate $U^{235}F_6$ and $U^{238}F_6$ in the United States nuclear energy program at plants built at a cost of over three billion dollars (4).

The thermal-energy flux is the sum of four fluxes (2). The flux caused by a temperature gradient has already been considered. Energy flux is also caused by the intrinsic energy associated with the molecules in mass transport. Thermal-energy transport also occurs across a *concentration* gradient. This is known as the "Dufour effect." Finally, heat may be transferred by *radiation*, a topic which is covered in a later chapter.

The transport of subatomic particles is important in a number of fields. For example, electron transport in solids, liquids, and gases is the basis for electrical and electronic developments. Neutron-transport considerations are essential to the design of nuclear reactors.

Few of the transport phenomena described in this section follow the simple flux-gradient relationship derived earlier in this chapter. For example, thermal diffusion depends upon the difference in the logarithms of the temperatures. On the other hand, electron and neutron diffusion follow equations of the flux-gradient form in certain elementary cases. For example, the law (electrical potential/resistance) = (current) is of the flux-gradient form.

APPLICATIONS OF HEAT, MASS, AND MOMENTUM TRANSFER

Applications to industrial problems of the transport equations derived in this chapter will be shown in subsequent chapters. The mass-transfer case has been examined, in part, in Part I. For example, on an actual plate of a distillation column, bubbles of vapor rise through a liquid. During the period when the two phases are in contact (the bubbles are submerged), mass is transferred from phase to phase. The rate of transfer of mass from the bubble to the liquid and from the liquid to the bubble may be expressed with a transfer-rate equation. The absorption tower, packed with shaped ceramic pieces, is used to contact liquid (which enters the top) and vapor or gas (which enters the bottom). When the two phases are in contact, the transfer of mass can be expressed by a rate equation.

In the heat-transfer application, a typical apparatus might be a straight pipe, exposed to condensing steam on the outside of the pipe and a cold fluid flowing inside the pipe. Heat from the steam is transferred to the pipe wall, then through the pipe wall to the inside surface of pipe, and then into the cold water. A rate equation can be used to define the rate of transfer through each of the three steps in the process. All applications of heat transfer with the exception of heat transfer by thermal radiation may be analyzed with the simple rate equations. Other examples of heat transfer are condensers, combustion furnaces, and reboilers.

Momentum transfer occurs in liquids, in gases, and in solids suspended in liquids or gases as long as the mixture exhibits an apparent or absolute viscosity and flows in contact with a solid barrier that has a different velocity relative to the average velocity of the fluid. A particle of fluid that flows within a pipe possesses some momentum based on the mass and velocity of the fluid particle. This momentum will be transferred to the wall of the pipe and will exert a stress on the pipe wall in the direction of flow of the fluid; and conversely the pipe will tend to reduce the velocity of the flowing fluid. This is a friction phenomenon. Thus, the power required to keep the fluid in motion at steady state is a measure of the energy necessary to overcome friction. In a pipe, momentum is transferred radially from one particle to another until the total transferable momentum is transferred to the pipe wall. Other applications where momentum is transferred are airfoils, flow of rivers, and flow of fluid through beds of solids as in filtration.

In some cases, two transport phenomena occur simultaneously. For example returning to the heat exchanger described earlier, as a fluid flows in a heat exchanger it is apparent that momentum is transferred from fluid to wall and heat is transferred from the wall to the fluid simultaneously. The wet bulb used in wet and dry bulb thermometry is analyzed as an example of simultaneous heat and mass transfer. The bulb is cooled by evaporation of water (transfer of mass from the liquid phase to the vapor phase), and heat is transferred from the warm gas to the cold bulb. The analysis of the wet bulb is not complete until it is realized that a stationary solid surface is in contact with a moving gas stream with consequent transfer of momentum. To analyze the complete system, at least one rate equation is necessary for each of heat, momentum, and mass transfer. A similar application on a large scale is a water cooling tower.

Applications of the molecular-transport equations will be considered in the following two chapters. Many of the applications mentioned in this section are based upon turbulent transport, as well as molecular transport. Turbulent transport will be discussed in Chapters 12 and 13 and will be applied in later chapters.

TRANSPORT PROPERTIES OF GASES

The general term *transport properties* includes the mass diffusivity, thermal conductivity, and absolute viscosity of gases, liquids, and solids. Much work has been done on the theoretical and experimental evaluation of transport properties. This section will summarize a few of the more useful theoretical and empirical methods for the evaluation of transport properties. Although many of the theoretical derivations involve very complex physical and mathematical concepts, their application usually depends upon the experimental evaluation of certain terms in the final equations. Consequently, even the theoretical equations require empirical verification before they may be used for the prediction of properties of real materials. Nevertheless, the theoretical expressions are invaluable for interrelating data for transport properties. For example, it is possible to predict mass diffusivities and thermal conductivities of certain real gases from viscosity data for the gases.

This section will discuss methods of evaluation of the transport properties of the model gas, real gases, liquids, and solids. Additional methods of evaluation of properties of real gases and liquids can be found in reference 9.

The Model Gas. For the simple model gas the transport diffusivities are equal to $\frac{1}{6}l\bar{c}$; that is,

$$\mathscr{D} = \alpha = \nu = \tfrac{1}{6}l\bar{c} \qquad (9.72)$$

With the assumption that the model gas follows the perfect gas law, it can be shown that (1)

$$\bar{c} = \left(\frac{8RT}{M\pi}\right)^{1/2} \qquad (9.73)$$

where
$\bar{c}$ = arithmetic mean speed, cm/sec
R = gas constant, 8.314×10^7 ergs/°K gm mole
T = absolute temperature, °K
M = molecular weight, gm/gm mole

The mean free path is

$$l = \frac{R'T}{(\frac{1}{3} + \frac{2}{3}\sqrt{2})\tilde{A}P\pi\sigma^2} = \frac{R'T}{7.67 \times 10^{23}P\pi\sigma^2} \qquad (9.74)*$$

---

* The molecular diameter σ appearing in this equation cannot be measured directly. It is usually evaluated from viscosity data using a transport equation such as Equation 9.12 or that given in the footnote to Equation 9.12. Equation 9.74 holds for either equation. Obviously, the numerical value of σ will depend upon the model chosen and hence upon the equation used. The molecular diameters listed in Appendix D-5 and D-6 do *not* apply to Equation 9.12 or its footnote. Therefore, the molecular diameter appearing in Equations 9.75, 9.76, 9.77, 9.78, and 9.79 must be evaluated from experimental viscosity data, as shown in Illustration 9.4.

where
$\tilde{A}$ = Avogadro's number, 6.02×10^{23} molecules/gm mole
P = pressure, atm
R' = gas constant, 82.06 atm cu cm/°K gm mole
T = absolute temperature, °K
σ = molecular diameter, cm
l = mean free path, cm

The different dimensions for the gas constants found in Equations 9.73 and 9.74 are necessary for dimensional consistency. However, since 1 atm cu cm = 1.013×10^6 ergs, Equation 9.74 may be written as

$$l = \frac{RT}{7.78 \times 10^{29}P\pi\sigma^2} \qquad (9.74a)$$

Combining Equations 9.72, 9.73, and 9.74a gives

$$\mathscr{D} = \alpha = \nu = \tfrac{1}{6}l\bar{c} = \frac{8.28 \times 10^{-20}T^{3/2}}{P\sigma^2 M^{1/2}} \qquad (9.75)$$

Equation 9.75 shows that the transport diffusivities vary with the 3/2 power of the temperature and inversely with the total pressure.

The transport diffusivities $\mathscr{D}$, α, and ν have the same dimensions and are equivalent for the simple model gas. However, they are not in the most convenient form for correlation and tabulation of data. It is more customary in chemical engineering to use the thermal conductivity (k) instead of the thermal diffusivity (α), and the absolute viscosity (μ) in place of the momentum diffusivity (ν). Since $\nu = \mu/\rho$ Equation 9.75 may be written

$$\mu = \frac{8.28 \times 10^{-20}T^{3/2}\rho}{P\sigma^2 M^{1/2}} \qquad (9.76)$$

From the perfect gas law $\rho = PM/RT$, and substitution in Equation 9.76 gives

$$\mu = \frac{8.28 \times 10^{-20}\sqrt{MT}}{R\sigma^2} \qquad (9.77)$$

The absolute viscosity is shown to be independent of the pressure of the system, and thus it is a more convenient quantity to tabulate. The absolute viscosity also is shown to be proportional to the square root of the absolute temperature. Since these conclusions are based upon the simple model gas, they hold only approximately for real gases at moderate pressures. Viscosity data are reported in the chemical-engineering literature in terms of μ, since it is more nearly pressure independent than ν.

The thermal conductivity is related to the diffusivity by $\alpha = k/\rho c_P$. Substitution of this into Equation 9.75 gives

$$k = \frac{8.28 \times 10^{-20}T^{3/2}\rho c_P}{P\sigma^2 M^{1/2}} \qquad (9.78)$$

For a monatomic perfect gas $c_P = 5R/2M$ and $\rho = PM/RT$. Equation 9.78 then becomes

$$k = \frac{20.7 \times 10^{-20}\sqrt{T/M}}{\sigma^2} \qquad (9.79)$$

which is correct only for a monatomic perfect model gas. Combination of Equations 9.77 and 9.79 gives

$$k = \frac{5}{2}\frac{R}{M}\mu \qquad (9.80)$$

which is a simple relationship between the two transport properties. It should be noted that the thermal conductivity is independent of the pressure and proportional to the square root of the temperature.

Traditional chemical-engineering practice reports the mass diffusivity ($\mathscr{D}$) as such, even though it varies with temperature and pressure, as shown by Equation 9.75. It may be written as

$$\mathscr{D} = \frac{8.28 \times 10^{-20}\sqrt{T^3/M}}{P\sigma^2} \qquad (9.75a)$$

Combination of Equations 9.77 and 9.75a gives

$$\mathscr{D} = \frac{RT}{PM}\mu = \frac{\mu}{\rho} \qquad (9.81)$$

which is not unexpected.

In order to utilize these equations to evaluate transport properties, it is necessary to know the molecular diameter σ. This must be determined experimentally for the substance of interest. Since the diameter cannot be measured directly, it must be determined from some measurable property, such as the viscosity. Even in the most rigorous kinetic theory, the viscosity is often used to determine molecular diameters of real gases.

Illustration 9.4. The viscosity of argon at 0°C and 1 atm has been measured as 2.096×10^{-4} gm/cm sec. Assuming that argon is a perfect model gas:

(a) Calculate the equivalent molecular diameter.
(b) Predict the viscosity of argon at 400°C and 1 atm.
(c) Predict the mass diffusivity of argon at 0°C and 1 atm.
(d) Predict the thermal conductivity of argon at 0°C and 1 atm.

SOLUTION. (a) The diameter (σ) may be evaluated from Equation 9.76.

$$\sigma = \sqrt{\frac{8.28 \times 10^{-20} T^{3/2}\rho}{\mu P M^{1/2}}}$$

where

$$\rho = \frac{39.94}{22400} = 1.78 \times 10^{-3} \text{ gm/cu cm}$$

$$\sigma = \sqrt{\frac{(8.28 \times 10^{-20})(273)^{3/2}(1.78 \times 10^{-3})}{(2.096 \times 10^{-4})(1)(39.94)^{1/2}}}$$

$$= 2.24 \times 10^{-8} \text{ cm}$$

(b) The absolute viscosity may be evaluated with Equation 9.77, noting that the absolute viscosity is proportional to $T^{1/2}$:

$$\frac{\mu_{673°}}{\mu_{273°}} = \left(\frac{673}{273}\right)^{1/2}$$

Therefore, $\mu_{673°} = (2.096 \times 10^{-4})\left(\frac{673}{273}\right)^{1/2}$

$$= 3.28 \times 10^{-4} \text{ gm/cm sec}$$

This value may be compared with the experimental value of 4.11×10^{-4} gm/cm sec.

(c) By Equation 9.73

$$\bar{c} = \left(\frac{(8)(8.314 \times 10^7)(273)}{(39.94)(3.14)}\right)^{1/2} = 3.8 \times 10^4 \text{ cm/sec}$$

By Equation 9.74a, with $\sigma = 2.24 \times 10^{-8}$ cm,

$$l = \frac{(8.314 \times 10^7)(273)}{(7.78 \times 10^{29})(1)(3.14)(2.24 \times 10^{-8})^2} = 1.85 \times 10^{-5} \text{ cm}$$

Therefore,

$$\mathscr{D} = \tfrac{1}{6}l\bar{c} = \tfrac{1}{6}(1.85 \times 10^{-5})(3.8 \times 10^4) = 0.117 \text{ sq cm/sec}$$

This value may be compared to an experimental value of 0.162. Use of Equation 9.81 gives

$$\mathscr{D} = \frac{RT}{PM}\mu = \frac{(82.06)(273)}{(1)(39.94)}(2.09 \times 10^{-4}) = 0.117 \text{ sq cm/sec}$$

(d) The thermal conductivity may be evaluated from Equation 9.79 or 9.80. Using Equation 9.79,

$$k = \frac{20.7 \times 10^{-20}\sqrt{273/39.94}}{(2.24 \times 10^{-8})^2}$$

$$= 1.08 \times 10^3 \frac{\text{ergs}}{\text{sec sq cm °C/cm}}$$

or $\qquad k = 2.6 \times 10^{-5} \dfrac{\text{cal}}{\text{sec sq cm °C/cm}}$

This value is lower than the experimental value of 3.89×10^{-5}.

The prediction of transport coefficients based on the simple model gas gives values within 35 per cent of the experimental values. This agreement is surprisingly good, considering how simple the model was, but it is not sufficiently accurate for most flux-gradient calculations.

A more rigorous theory for spherical molecules predicts that the diffusivities are not simply equal to $l\bar{c}/6$. With the more rigorous kinetic theory, the coefficients of Equations 9.75a, 9.77, and 9.79 change, and

$$\mu = \frac{2.6693 \times 10^{-21}\sqrt{MT}}{\sigma^2} \qquad (9.82)$$

where μ is in gm/cm sec,

$$k = \frac{1.989 \times 10^{-20}\sqrt{T/M}}{\sigma^2} \qquad (9.83)$$

where k is in cal/sec sq cm (°C/cm),

and
$$\mathscr{D} = \frac{2.628 \times 10^{-19}\sqrt{T^3/M}}{P\sigma^2} \qquad (9.84)$$

where $\mathscr{D}$ is in sq cm/sec. Values for the molecular diameter (σ) for this model have been determined experimentally from viscosity data. They are tabulated in Appendix D-5. Examination of the three equations above shows that $\mathscr{D} = \frac{6}{5}\nu$, and $\alpha = \frac{3}{2}\nu$, unlike the original simple model gas where all were equal. Furthermore,

$$k = \frac{15}{4}\frac{R}{M}\mu \qquad (9.85)$$

where R = gas constant, 1.987 cal/gm mole °C.

Illustration 9.5. Using the data of Appendix D-5 and the more rigorous theory for spherical molecules:
(a) Calculate the viscosity of argon at 0°C and 1 atm.
(b) Calculate the viscosity of argon at 400°C and 1 atm.
(c) Calculate the mass diffusivity of argon at 0°C and 1 atm.
(d) Calculate the thermal conductivity of argon at 0°C and 1 atm.

SOLUTION. From Appendix D-5 for argon

$$\sigma = 3.64 \times 10^{-8}\text{ cm}$$

(a) Using Equation 9.82

$$\mu = \frac{(2.6693 \times 10^{-21})\sqrt{(39.94)(273)}}{(3.64 \times 10^{-8})^2}$$

$$= 2.1 \times 10^{-4}\text{ gm/cm sec}$$

This value checks exactly with the experimental value, since the molecular diameter was determined from viscosity data.
(b) At 400°C, using Equation 9.82

$$\mu = \frac{(2.6693 \times 10^{-21})\sqrt{(39.94)(673)}}{(3.64 \times 10^{-8})^2} = 3.3 \times 10^{-4}\text{ gm/cm sec}$$

This result is no better than that in Illustration 9.4.
(c) Using Equation 9.84

$$\mathscr{D} = \frac{(2.628 \times 10^{-19})\sqrt{(273)^3/39.94}}{(1)(3.64 \times 10^{-8})^2}$$

$$= 0.142\text{ sq cm/sec}$$

This is closer to the experimental value than that calculated in the previous illustration.
(d) Using Equation 9.85

$$k = \frac{15}{4}\frac{(1.987)}{(39.94)}(2.1 \times 10^{-4})$$

$$= 3.91 \times 10^{-5}\text{ cal/sec sq cm (°C/cm)}$$

which compares well with the experimental value of 3.89×10^{-5}.

The simple model gas has been used to predict certain transport properties for argon. In general it is not an adequate model for many real gases and is totally useless for predicting properties of liquids and solids. Prediction of properties of real substances will be considered in the following sections.

Real Gases. Real gases differ from the simple model gas in many ways. The molecules of a real gas are not usually spherical, and they travel at a wide range of speeds. The volume of the molecules may be significant compared to the volume between molecules. Of great importance are the attractive and repulsive forces between the molecules, and these forces were neglected in the model gas. Molecules of real gases attract each other when they are far apart and repel each other when they are close together. Even in the most rigorous treatments of the kinetic theory of gases, it is necessary to assume the form of this molecular interaction. An expression for the potential energy of interaction which can be used to predict transport properties of certain gases with good accuracy is known as the Lennard–Jones 6-12 potential:

$$\phi_{(r)} = 4\epsilon\left[\left(\frac{\sigma}{r}\right)^{12} - \left(\frac{\sigma}{r}\right)^{6}\right] \qquad (9.86)$$

where $\phi_{(r)}$ = intermolecular potential energy function
ϵ = maximum energy of attraction of two molecules
σ = distance of closest approach of two molecules which collide with zero initial relative kinetic energy
r = distance between the molecules.

For large separations between molecules, r is much greater than σ, and the second term in Equation 9.86 predominates and causes an attractive force between molecules. When the molecules are close together, r is much less than σ, and the first term in Equation 9.86 is dominant, and the molecules repel each other. Many other potential functions have been suggested, but the Lennard–Jones 6-12 potential has proved most useful in predicting transport properties for many nonpolar gases. The constants ϵ and σ must be evaluated from experimental data. This potential is not useful for polar molecules such as water or for long molecules, excited molecules, free radicals, or ions. The engineer must therefore resort to more empirical methods of prediction of transport properties for these molecules.

A detailed treatment of the rigorous kinetic theory using the Lennard–Jones model is beyond the scope of this book. However, the resulting equations are useful in predicting transport properties. The equations will be presented and applied to the calculation of transport properties. For the viscosity of a pure gas,

$$\mu = 2.6693 \times 10^{-21}\frac{\sqrt{MT}}{\sigma^2\Omega_1} \qquad (9.87)$$

where μ = absolute viscosity, gm/cm sec

M = molecular weight, gm/gm mole

σ = collision diameter, cm (Appendix D-6)

Ω_1 = collision integral, a function of T^* (Appendix D-6)

T = absolute temperature

T^* = reduced temperature = kT/ϵ

ϵ/k = potential parameter, °K (Appendix D-6)

Note that this equation is identical to Equation 9.82, except for Ω_1. The function Ω_1 is a second-order correction taking into account attractive forces and other characteristics of real gases.

Illustration 9.6. Calculate the viscosity of argon at 0°C and 1 atm.

SOLUTION. For argon from Appendix D-6a, $\epsilon/k = 124$°K, $\sigma = 3.418 \times 10^{-8}$ cm. The molecular weight is 39.94. Then $T^* = 273/124 = 2.20$. From Appendix D-6b at $T^* = 2.20$, $\Omega_1 = 1.138$. Substitution of these values in Equation 9.87 gives

$$\mu = \frac{(2.6693 \times 10^{-21})\sqrt{39.94 \times 273}}{(3.418 \times 10^{-8})^2(1.138)}$$

$$= 2.1 \times 10^{-4} \text{ gm/cm sec}$$

This value is in exact agreement with the experimental value, as expected.

For the thermal conductivity of a pure monatomic gas

$$k = 1.989 \times 10^{-20} \frac{\sqrt{T/M}}{\sigma^2 \Omega_1} \qquad (9.88)$$

where k = thermal conductivity, cal/sec sq cm °C/cm and the other symbols are as defined earlier. This equation is identical to Equation 9.83 except for the collision function Ω_1.

Combination of Equations 9.82 and 9.83 yields

$$k = \frac{15}{4} \frac{R}{M} \mu \qquad (9.85)$$

which was also derived earlier for the simple model gas. Use of this equation to calculate k gives the same result as Illustration 9.5d.

For polyatomic molecules Equation 9.85 must be modified to take into account internal degrees of freedom in the molecule. For polyatomic molecules,

$$k = \frac{15}{4} \frac{R}{M} \left[\frac{4}{15} \frac{c_V}{R} + \frac{3}{5} \right] \mu \qquad (9.89)$$

where c_V is the molar heat capacity at constant volume.

Illustration 9.7. Calculate the thermal conductivity of carbon dioxide at 100°C and 1 atm.

SOLUTION. Equation 9.88 may be modified for use.

$$k = \frac{1.989 \times 10^{-20}\sqrt{T/M}}{\sigma^2 \Omega_1} \left[\frac{4}{15} \frac{c_V}{R} + \frac{3}{5} \right]$$

$T = 373$°K, $c_V = 7.15$ cal/gm mole °C (Appendix D-13)

$R = 1.987$ cal/gm mole °C, $M = 44.01$

From Appendix D-6, for CO_2, $\epsilon/k = 190$°K

$\sigma = 3.996 \times 10^{-8}$ cm, $T^* = \dfrac{373}{190} = 1.96$, $\Omega_1 = 1.184$

$$k = \frac{1.989 \times 10^{-20}\sqrt{373/44.01}}{(3.996 \times 10^{-8})^2(1.184)} \left[\left(\frac{4}{15} \right) \left(\frac{7.15}{1.987} \right) + \frac{3}{5} \right]$$

$$= 4.8 \times 10^{-5} \text{ cal/sec sq cm °C/cm}$$

This value agrees well with an experimental value of 5.06×10^{-5}.

For calculation of mass diffusivity the following equation has been derived for binary mixtures.

$$\mathscr{D}_{ab} = 2.628 \times 10^{-19} \frac{\sqrt{(T^3)\left(\dfrac{1}{2}\right)\left(\dfrac{1}{M_a} + \dfrac{1}{M_b}\right)}}{P\sigma_{ab}^2 \Omega_2} \qquad (9.90)$$

where $\mathscr{D}_{ab}$ = mass diffusivity, sq cm/sec

M_a = molecular weight of species a

M_b = molecular weight of species b

P = total pressure, atm

σ_{ab}, Ω_2, T_{ab}^* = Lennard–Jones constants

This equation has the same form as Equation 9.84, but the weighted average values of molecular weight and constants are used. The Lennard–Jones constants are evaluated from the following relationships:

$$\sigma_{ab} = \tfrac{1}{2}(\sigma_a + \sigma_b) \qquad (9.91)$$

where σ_a and σ_b are the collision diameters for each molecular species (Appendix D-6a).

$$\frac{\epsilon_{ab}}{k} = \sqrt{\frac{\epsilon_a}{k} \times \frac{\epsilon_b}{k}} \qquad (9.92)$$

where the individual potential parameters are found in Appendix D-6a. With $T_{ab}^* = kT/\epsilon_{ab}$, Ω_2 is evaluated using Appendix D-6b.

Illustration 9.9. Calculate the mass diffusivity for the mixture carbon dioxide–nitrogen at 25°C and 1 atm.

SOLUTION.

$$T = 298\text{°K}, \quad M_a = 44.01, \quad M_b = 28.02$$

For CO_2 $\dfrac{\epsilon_a}{k} = 190$, $\sigma_a = 3.996 \times 10^{-8}$

For N_2 $\dfrac{\epsilon_b}{k} = 79.8$, $\sigma_b = 3.749 \times 10^{-8}$

Therefore

$$\sigma_{ab} = \tfrac{1}{2}(3.996 + 3.749)(10^{-8}) = 3.872 \times 10^{-8} \text{ cm}$$

$$\frac{\epsilon_{ab}}{k} = \sqrt{190 \times 79.8} = 123.0°\text{K}$$

Then

$$T_{ab}^* = 298/123 = 2.42$$

Then from Appendix D-6b, $\Omega_2 = 1.010$. Using Equation 9.90 gives

$$\mathscr{D} = 2.628 \times 10^{-19} \frac{\sqrt{(298)^3 \left(\frac{1}{2}\right)\left(\frac{1}{44.01} + \frac{1}{28.02}\right)}}{(1)(3.872 \times 10^{-8})^2(1.010)}$$

$$= 0.154 \text{ sq cm/sec}$$

The experimental value of $\mathscr{D}$ is 0.165 sq cm/sec.

An additional higher-order correction factor may be applied to Equations 9.87, 9.88, and 9.90 for greater accuracy (1). However, since this correction is less than 1 per cent, it has been omitted here.

As was mentioned earlier, the Lennard–Jones (6-12) model does not hold for polar molecules, free radicals, or long molecules. Two important polar molecules are water and ammonia. Values for the transport properties of these anomolous molecules have in some cases been correlated by semi-empirical methods. For example, viscosities can be correlated with the Sutherland model (reference 1, p. 565); Gilliland has developed an empirical correlation for the mass diffusivities of many gases (5). Tables of transport properties are included in Appendix D.

The Gilliland equation is

$$\mathscr{D} = \frac{0.0043 \sqrt{T^3 \left(\frac{1}{M_a} + \frac{1}{M_b}\right)}}{P(V_a^{1/3} + V_b^{1/3})^2} \quad (9.93)$$

where V is the molar volume, as given in Appendix D-7. The term $V^{1/3}$ is a measure of the molecular diameter of each constituent and is analogous to σ. Thus the form of Gilliland's empirical equation is similar to the previous theoretical expressions for mass diffusivity. The constant was evaluated by measuring the diffusivities of many real gases, including water vapor.

Illustration 9.10. Calculate the mass diffusivity for the mixture carbon dioxide–nitrogen at 25°C and 1 atm using the Gilliland correlation.

SOLUTION. From Appendix D-7, the molar volume of N_2 is $V_b = 2 \times 15.6 = 31.2$ cu cm/gm mole, and for CO_2, $V_a = 14.8 + 7.4 + 12 = 34.2$ cu cm/gm mole. $M_a = 44.01$, and $M_b = 28.02$.

$$\mathscr{D} = \frac{0.0043 \sqrt{(298)^3 \left(\frac{1}{44.01} + \frac{1}{28.02}\right)}}{(1)(34.2^{1/3} + 31.2^{1/3})^2}$$

$$= 0.13 \text{ sq cm/sec}$$

The result is not as close to the experimental value as is that of Illustration 9.9. Generally, the Gilliland correlation should be used only when the Lennard-Jones (6–12) model cannot be used.

TRANSPORT PROPERTIES OF LIQUIDS

The kinetic theory of gases has developed to such a degree that considerable extension of existing experimental data is possible, and the available data are well organized so that useful generalizations are available, as described earlier in this chapter. Similar theory of liquids is not as well organized, but it is evident that the proposed theories are progressing rapidly and extensive developments can be expected in the future.

The model for gases described earlier is made up of continuous free space, throughout which are distributed moving molecules. The molecules make up a small fraction of the available volume. The model for liquids is made up of space filled with a continuous phase of molecules in close array. The space between molecules, similar to the mean free path, is very small, less than the actual molecular dimension. The molecules are maintained in this array by intermolecular forces. The molecules can move only to the extent of the free space available to them. Since the molecule is subject to restraining forces, it may be considered that the molecule vibrates within the limited space and that the vibration is restrained by the intermolecular forces. Thus each molecule maintains an average "equilibrium" position such that the intermolecular forces are balanced. This model is relatively inflexible, as described, but the model is not complete. Distributed throughout the continuous molecular array there are "holes," or elements of free space each of about molecular dimensions. In the liquid model, the molecular array is the continuous phase and the "holes" or free space are the dispersed phase. Contrast this to the gas model in which free space is the continuous phase and molecules are the dispersed phase. This relatively ordered model may be inadequate as a physical picture of *real* liquids, but it gives a qualitative picture upon which is based a theory of liquid transport properties.

Transport in a liquid is accomplished when a molecule in the array migrates into a "hole." In order for a molecule to leave the array, enough energy must be furnished to overcome the intermolecular forces that hold the molecule in the "equilibrium" position, after which the molecule can migrate into the "hole," leaving a new "hole" at the original site of the migrating molecule. Figure 9.9 is a two-dimensional diagram of the molecular array with the "holes" indicated within the array.

The number of "holes" in a liquid is related to the density of the liquid, with more holes present at lower density. The greater number of "holes," the greater is

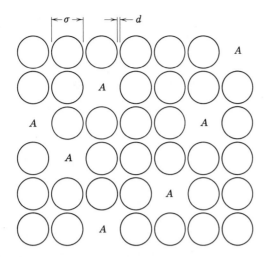

Figure 9.9. Simple model of a liquid. σ is the molecular diameter; d is the free distance between molecules; A is a "hole" between molecules.

the degree of molecular migration. In addition, the greater the ease with which a molecule is dislodged from the equilibrium position for migration, the greater is the degree of molecular migration. This second phenomenon is a function of the magnitude of the intermolecular forces. The dislodgement of a molecule from the continuous array is closely related to the vaporization of a liquid, in fact a simple proportion exists between the energy (or latent heat) of vaporization and the activation energy necessary for molecular migration.

The simple model for liquids has been used by Eyring (6) and others to explain parts of the transport behavior of liquids, but it has not been developed to the same extent as the kinetic theory of gases. The same technique that is used in the kinetic theory of gases, namely, the use of the easily determined viscosity data to estimate diffusivity, is used with liquids. For reasons to be mentioned later, thermal conductivities of liquids cannot be handled in the same manner at the present state of development of the theory.

With the gas model and liquid model in mind, consider an evacuated vessel into which a pure liquid is added such that vapor and liquid exist. The two phases are assumed to be in equilibrium. At moderate temperatures, the vapor will be described by the gas model and the liquid will be described by the liquid model. As the temperature is increased the pressure will increase, and the gas molecules will move closer together, so that the vapor will behave less like the gas model. The liquid on the other hand will behave substantially like the liquid model, except near the critical point. At the critical point, all physical properties including transport properties must be identical for the vapor and the liquid. Near the critical point, the gas model must begin to resemble the liquid model, and vice versa. At the

critical point, neither the molecular continuum of liquids nor the free-space continuum of gases is well defined. Either may be considered to be the continuum. Enough free space exists in the liquid model to serve as a continuum, or the molecules are at sufficient concentration so that the molecules may constitute the continuum. This is indicative of the behavior of gases and liquids at the critical point and should illustrate the significance of the critical point.

In the section on the kinetic theory of gases, the various transport properties were described in terms of the migration and distribution as two separate steps. In liquids the distance between molecules, indicated by d in Figure 9.9 is very short, consequently no migration can occur by the molecule as it vibrates in the confined space, if there is no "hole" adjacent to the molecule. Without migration, mass cannot be transported. However, the vibrating molecule can transmit thermal energy *without leaving the fixed array*. In momentum transport, as a molecule migrates into a hole in an adjacent moving region, it carries with it the momentum of its origin. In addition, however, those molecules in the original region that do *not* migrate are still subject to the attractive forces of neighboring molecules in adjacent regions of different velocity. The attractive forces create "drag" on the molecules, which constitutes the greater part of the mechanism of transport of momentum yet is not dependent upon migration of a molecule, as was the case for the model gas.

As molecules migrate from the array into "holes," mass, momentum, and thermal energy can all be transported. In addition, thermal energy and momentum can also be transported by collision of molecules. Thus, thermal energy and momentum can be transported by two separate mechanisms in liquids, but mass can be transported by only one mechanism. To show this, the transport properties of water at 0°C may be examined. $\mathscr{D} = 5.25 \times 10^{-5}$, $\alpha = 550 \times 10^{-5}$, and $\nu = 6950 \times 10^{-5}$ all in square feet per hour. The value of $\mathscr{D}$ is the diffusivity of water diffusing through water or the coefficient of self-diffusion. Comparison of these values of diffusivities shows that migration is of minor importance for transport in liquids compared to gases.

Binary mass diffusion is further complicated by the necessity for the presence of two species. The rate of diffusion is therefore a function of the properties of the two components.

According to the theory of absolute reaction rates (6), the equations for momentum and mass transport properties take the general forms

$$\frac{\mathscr{D}}{T} = Ae^{-B/RT} \qquad (9.94)$$

$$\mu = Ce^{B/RT} \qquad (9.95)$$

for the same substance, where

$\mathcal{D}$ = mass diffusivity, sq cm/sec

μ = absolute viscosity, gm/cm sec

A = a function of the density of the fluid specific for mass transfer

C = a function of the density of the fluid specific for momentum transfer

B = an energy function related to the latent heat of vaporization, assumed to be the same for diffusion and viscosity

R = gas constant in units consistent with B and T

Viscosity. From Equation 9.95, the two constants B and C can be determined from two or more experimental measurements of viscosity at different temperatures. If the logarithm of Equation 9.95 is taken,

$$\ln \frac{\mu}{C} = \frac{B}{RT} \qquad (9.96)$$

so that $\ln (\mu/C)$ plotted against $1/T$ results in a straight line of slope B/R. In the absence of sufficient viscosity data, B may be taken as $\Delta H_v/2.45$ for associated or nonassociated liquids but not for liquid metals. ΔH_v is the latent heat of vaporization of the liquid.

Illustration 9.11. The following experimental data are available for the viscosity of water:

Temperature, °C	Viscosity, gm/cm sec
0	0.01792
20	0.01005
40	0.00656
60	0.00469
80	0.00357

(a) Evaluate the constants of Equation 9.95 for water.
(b) Predict the viscosity of water at 100°C.

SOLUTION. (a) It would be possible to use any two of the data points given to evaluate B and C. However, since the data are subject to experimental error, it is preferable to use all the data. The data will be plotted as indicated by Equation 9.96

T, °K	$\dfrac{1}{T}$	μ
273	0.00366	0.01792
293	0.00342	0.01005
313	0.00320	0.00656
333	0.00300	0.00469
353	0.00283	0.00357

The data are plotted on semilogarithmic coordinates, as shown on Figure 9.10. It is noticed that the data do not fall

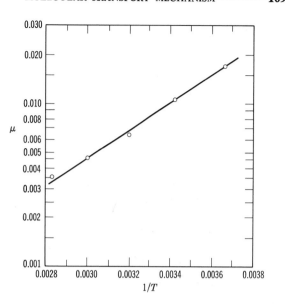

Figure 9.10. Solution to Illustration 9.11.

on a straight line. Nevertheless, the best straight line is drawn through the data to evaluate the constants. Equation 9.96 may be rewritten in the standard slope-intercept form

$$\ln \mu = \frac{B}{RT} + \ln C$$

It is more convenient to evaluate B and C by taking two points from the straight line through the data.

$$\text{At } \mu = 0.004 \qquad \frac{1}{T} = 0.00293$$

$$\text{At } \mu = 0.015 \qquad \frac{1}{T} = 0.00360$$

Then

$$\ln 0.004 = 0.00293 \frac{B}{R} + \ln C \quad (a)$$

$$\ln 0.015 = 0.00360 \frac{B}{R} + \ln C \quad (b)$$

Subtracting (a) from (b) gives

$$0.00067 \frac{B}{R} = \ln \frac{0.015}{0.004}$$

$$\frac{B}{R} = 1970$$

Then

$$C = \frac{\mu}{e^{B/RT}} = \frac{0.015}{e^{(1970)(0.00360)}}$$

$$C = 1.25 \times 10^{-5}$$

Therefore

$$\mu = 1.25 \times 10^{-5}(e^{1970/T})$$

(b) At $T = 100°C = 373°K$

$$\mu = 1.25 \times 10^{-5}(e^{1970/373}) = 0.00244 \text{ gm/cm sec}$$

This is lower than the experimental value of 0.00284 because the actual viscosity does not exactly follow Equation 9.95. This equation should be used only over short temperature ranges.

Mass Diffusivity. According to Wilke (7) if Equations 9.94 and 9.95 are combined, the function F may be defined

$$F = \frac{T}{\mathscr{D}\mu} = \frac{1}{AC} \qquad (9.97)$$

Collected experimental data for numerous systems of *dilute solutions of nonelectrolytes* were correlated with F plotted against the molar volume of the solute in Appendix D-8. The function ϕ in Appendix D-8 takes into account certain solvent properties that are not accounted for in the molar volume. The molar volume can be calculated by Kopp's law, based upon the addition of atomic volumes to produce the molar volume. Atomic volumes are given in Appendix D-7.

The value of ϕ that must be used depends upon properties of the solvent other than density and molecular size that can have an effect upon transport properties, such as hydrogen bonding. For water $\phi = 1.0$, for methanol, 0.82, and for benzene, 0.70. For other substances ϕ can be determined from a known value of viscosity and mass diffusivity, or, in the absence of any other information, ϕ may be assumed to be 0.90.

The usefulness of the plot lies in the large amount of viscosity data available and the ease of experimentally determining viscosity. Diffusivity is difficult to determine, but, with the procedure set forth, considerable extension of available data can be made with confidence.

Illustration 9.12. Calculate the mass diffusivity of ethanol in water at 10°C for a dilute solution.

SOLUTION. The molar volume of ethanol can be calculated from atomic volumes from the table in Appendix D-7.

$$
\begin{aligned}
2(C) &= 2(14.8) = 29.6 \\
6(H) &= 6(3.7) = 22.2 \\
(O) &= 7.4 = 7.4 \\
\hline
&= 59.2 \text{ cu cm/gm mole}
\end{aligned}
$$

From Appendix D-8 at $\phi = 1$, for water,

$$F = 2.4 \times 10^7, \quad \mu = 1.308 \text{ centipoises}$$

$$\mathscr{D} = \frac{T}{F\mu} = \frac{283}{(2.4 \times 10^7)(1.308)} = 0.90 \times 10^{-5} \text{ sq cm/sec}$$

The experimental diffusivity at 10°C is 0.83×10^{-5} sq cm/sec.

Illustration 9.13. Calculate the diffusivity of chloroform in ethanol at 20°C for dilute solutions.

SOLUTION. Values of ϕ are not known for ethanol, consequently any available data for diffusion through ethanol

may be used to evaluate ϕ. An experimental value of $\mathscr{D}$ for CO_2 through ethanol is 3.25×10^{-5} sq cm/sec at 17°C. The viscosity of ethanol is 1.29 centipoises at 17°C and 1.20 centipoises at 20°C.

The molar volume of CO_2 can be found from Appendix D-7 as follows:

$$
\begin{aligned}
\text{Carbon} &= 14.8 \\
\text{Oxygen (carbonyl)} &= 12.0 \\
\text{Oxygen} &= 7.4 \\
\hline
& 34.2 \text{ cu cm/gm mole}
\end{aligned}
$$

At 17°C $\quad \dfrac{T}{\mathscr{D}\mu} = \dfrac{290}{(3.2 \times 10^{-5})(1.3)} = 0.698 \times 10^7$

From Appendix D-8:

$$\phi = 0.43$$

The molar volume of chloroform can be calculated from Appendix D-7

$$
\begin{aligned}
\text{Carbon} &= 14.8 \\
\text{Hydrogen} &= 3.7 \\
\text{3 chlorine} &= 3(24.6) \\
\hline
& 92.3 \text{ cu cm/gm mole}
\end{aligned}
$$

From Appendix D-8 at $V = 92.3$, using $\phi = 0.43$ determined above gives

$$\frac{T}{\mathscr{D}\mu} = 1.4 \times 10^7$$

$$\mathscr{D} = \frac{293}{(1.4 \times 10^7)(1.20)} = 1.75 \times 10^{-5} \text{ sq cm/sec}$$

This may be compared to 1.25×10^{-5} sq cm/sec determined experimentally.

Thermal Conductivity. Thermal energy can be transmitted through liquids without migration of molecules. The liquid model developed earlier postulates migration as the means of transport, whether the migration is in the flow direction or in the transfer direction. Therefore the model is not particularly helpful in organizing thermal conductivity data.

Liquids generally show a decrease in thermal conductivity with increase in temperature but exceptions exist. For example, between 0° and 300°F, k for gasoline varies between 0.09 and 0.082, whereas water varies between 0.298 and 0.470 Btu/hr sq ft (°F/ft). No generalization of data exists, but fortunately thermal conductivity is relatively insensitive to temperature and almost independent of pressure for liquids. Thermal conductivity data for liquids can be found in Appendix D.

TRANSPORT PROPERTIES OF SOLIDS

The model for the solid state is similar to the model for the continuous phase in the liquid state except that essentially no "holes" are present for migration by molecules; consequently, except for extremely high-pressure and high-temperature operations such as forging, solids are not considered to flow.

Thermal Conductivity. The mechanism of thermal-energy transport in solids is independent of any molecular migration. The molecules in a solid crystal vibrate over a limited range, so that energy transport is due to transfer on collision. Alloys and crystalline materials show varying behavior of thermal conductivity with change in temperature, with all materials relatively insensitive. The thermal conductivity of solids is dependent upon the gross state of aggregation. For example, cellular and fibrous materials which may include entrapped gas, such as glass wool and asbestos, have low thermal conductivities and are used as insulating material.

Mass Diffusion. Although molecular migration is not considered to occur in solids, apparently a very limited degree of mass transport can occur. The diffusion can occur through crystal boundaries or through vacant lattice positions. Diffusion of a metal through a metal is extremely slow and has little practical importance except in specialized applications, such as transistors. Certain gases, especially hydrogen, tend to diffuse through metals. This process is also slow and is of importance in specialized cases only.

Flow of fluids through porous solids can occur in a manner similar to diffusion, except that the pore diameters are large compared to the molecular size. This process is not true diffusion but is a special case of momentum transfer and will be treated later.

DIMENSIONLESS RATIOS OF TRANSPORT DIFFUSIVITIES

Ratios of diffusivities are extremely useful in analysis of transport phenomena, as will be shown later. The two ratios that are most useful are the Prandtl number, $(N_{Pr} = \nu/\alpha)$ and the Schmidt number, $(N_{Sc} = \nu/\mathscr{D})$. For the simple model gas both of these ratios are equal to unity, and for simple real gases they are close to unity. For liquids the ratios vary widely. The ratios are useful for analysis when several transport phenomena occur simultaneously. Characteristic values of N_{Pr} and N_{Sc} are shown in Table 9.2.

Table 9.2. TYPICAL VALUES FOR DIFFUSIVITY RATIOS

	Prandtl No. = ν/α	Schmidt No. = $\nu/\mathscr{D}$
Simple model gas	1	1
Complex model gas*	0.67	0.83
Inorganic gases	0.7–1.1	0.6–1.1
Liquid water	7	1200
Organic liquids	10–1000 or more	300–2000
Liquid metals	0.001–0.1	—

* As described by Equations 9.82, 9.83, and 9.84.

REFERENCES

1. Hirschfelder, J. O., C. F. Curtiss, and R. B. Bird, *Molecular Theory of Gases and Liquids*, John Wiley and Sons, New York, 1954.
2. Bird, R. B., "Theory of Diffusion", *Advances in Chemical Engineering*, ed. T. B. Drew and J. W. Hoopes, Vol. 1, Academic Press, New York, 1956.
3. Chapman, S., and T. G. Cowling, *The Mathematical Theory of Non-Uniform Gases*, Cambridge, 1939.
4. Benedict, Manson, and T. H. Pigford, *Nuclear Chemical Engineering*, McGraw-Hill Book Co., New York, 1957.
5. Gilliland, E. R., *Ind. Eng. Chem.*, **26**, 681 (1934).
6. Glasstone, S., K. J. Laidler, H. Eyring, *The Theory of Rate Processes*, McGraw-Hill Book Co., New York, 1941.
7. Wilke, C. R., *Chem. Eng. Progr.*, **45**, 95 (1950).
8. Metzner, A. B., "Non-Newtonian Technology", *Advances in Chemical Engineering*, ed. T. B. Drew and J. W. Hoopes, Vol. 1, Academic Press, New York, 1956.
9. Reid, R. C., and T. K. Sherwood, *The Properties of Gases and Liquids*, McGraw-Hill Book Co., New York, 1958.

PROBLEMS

9.1. Derive Equation 9.9.

9.2. Derive Equation 9.23 using a positive gradient for gas *b*.

9.3. Show that for a perfect gas the total concentration (c_t) is constant if pressure and temperature are constant.

9.4. Derive Equation 9.41, justifying each step.

9.5. By adding vectors graphically show that vectorial sum of the momentum due to random molecular motion is zero. (Suggestion: Instead of using a large number of randomly oriented speeds, represent a general randomness with six vectors 60° apart.)

9.6. Calculate the mass flux of benzene through a slab of air 1 cm thick at 25°C and 2 atm. The partial pressure of benzene is 50 mm Hg at the left face of the slab and 10 mm Hg at the right face of the slab. The mass diffusivity at 25°C and 2 atm is 0.044 sq cm/sec.

9.7. Calculate the mass flux of methanol in water at 25°C. The concentrations of methanol at two points 0.1 in. apart are 0.05 lb moles/cu ft and 0.10 lb moles/cu ft. The mass diffusivity of methanol in liquid water is 1.28×10^{-5} sq cm/sec at 25°C.

9.8. A furnace wall consists of 2 ft of brick. The brick has a thermal conductivity of 0.6 Btu/hr sq ft (°F/ft), a specific heat of 0.2 Btu/lb °F, and a density of 110 lb/cu ft. The temperature at the inside surface of the wall is 1100°F, and at the outside surface, 200°F.

(*a*) Calculate the thermal diffusivity of the brick.

(*b*) Calculate the heat loss per hour through a wall 10 ft high and 10 ft long.

(*c*) Calculate the temperature of the brick $4\frac{1}{2}$ in. from the inside surface.

9.9. A copper rod 1 in. in diameter and 2 ft long is heated at one end and cooled at the other end. Heat is supplied at a rate of 40 Btu/hr. What is the difference in temperature between the two ends of the rod, assuming no loss of heat through the cylindrical surface of the rod? The thermal conductivity of copper is 215 Btu/hr sq ft (°F/ft).

9.10. Two horizontal plates are spaced 3 in. apart. The space between the plates is filled with mercury at 90°F. The upper plate is moving at 3 ft/sec, and the lower plate is moving in the same direction at 1 ft/sec.

(*a*) Why does the liquid velocity vary linearly between the plates?

(b) Calculate the shear stress at a plane 1 in. from the slower moving plate.

(c) Calculate the momentum flux at the surface of the faster moving plate.

9.11. Calculate the thermal conductivity, mass diffusivity for self-diffusion, and absolute viscosity for neon at 100°C and 3 atm.

(a) Assuming that neon follows the model of Equations 9.82, 9.83, and 9.84.

(b) Assuming that neon follows the Lennard–Jones (6.12) model.

9.12. Methane has a viscosity of 0.020 centipoise at 380°C and 0.0226 centipoise at 499°C at 1 atm pressure.

(a) Calculate the viscosity of methane at 100°C and 700°C, assuming it is a simple model gas.

(b) Calculate the viscosity of methane at 100°C and 700°C assuming it follows the Lennard–Jones (6-12) model.

(c) Calculate the thermal conductivity of methane at 500°C assuming it follows the Lennard–Jones model.

(d) Calculate the mass diffusivity of methane in methane at 500°C, assuming it follows the Lennard–Jones model.

9.13. Calculate accurately the mass diffusivity of carbon dioxide in benzene vapor at 100°C and 1.5 atm.

9.14. Calculate the mass diffusivity of water vapor in nitrogen at 200°C and 2 atm.

9.15. Calculate the mass diffusivity of hydrogen in water vapor at 25°C and 2 atm.

9.16. The viscosity of liquid benzene is given below. Develop an equation useful in predicting the viscosity of benzene in this temperature range

Temperature, °C	Viscosity, centipoise
10	0.758
20	0.652
30	0.564
40	0.503
50	0.442
60	0.392
70	0.358
80	0.329

9.17. The viscosity of liquid acetone is 2.148 centipoises at −92.5°C and 0.399 centipoise at 0°C. Predict the values of viscosity at −30°C and at 30°C and compare with experimental values of 0.575 centipoise and 0.295 centipoise, respectively.

9.18. Calculate the mass, thermal, and momentum diffusivities of air and of liquid water at 32°F and 1 atm. Compare these values. Calculate the Schmidt number for self-diffusion and the Prandtl number at 32°F and 1 atm for air and for liquid water. Discuss the meaning of these dimensionless ratios in terms of a unit concentration gradient. (Calculate the mass diffusivity for self-diffusion.)

9.19. Calculate the Prandtl number for pure ethyl alcohol, water, mercury, and glycerol at approximately 68°F. Compare the resulting values in terms of the individual properties.

9.20. Discuss the reasons for the differences in Prandtl number and Schmidt number for the various materials given in Table 9.2.

chapter 10

Applications of Molecular-Transport Theory to the Steady State

The molecular-transport equations developed in Chapter 9 may be applied in the analysis of many problems encountered in unit operations. The differential transport equations must be integrated, taking into account all pertinent physical variables. The resultant constants of integration may be evaluated, utilizing known values of the variables at a boundary or at any other point in the system where information is available.

A general procedure may be outlined for the analysis of molecular-transport problems. The individual steps are listed below.

1. A balance on the transferent property is written for a specified volume. The balance equation states that the rate of efflux of transferent property from the specified volume must equal the rate of influx plus the rate of production of the transferent property within the volume.

2. The appropriate differential transport equation is substituted into the balance equation. The differential transport equations for mass, heat, and momentum transport were developed in Chapter 9.

3. Variables are separated for integration. Often the physical properties or transport area vary with distance or with concentration of the transferent property. In such cases the properties must be expressed in terms of the distance or concentration. For example, the thermal conductivities of many materials vary with temperature and must be expressed as functions of temperature before integration.

4. The resulting equation is integrated and the constants of integration are evaluated from known boundary conditions. In some cases, it is more convenient to

evaluate a definite integral to eliminate the evaluation of the integration constants.

These steps will be followed in each of the applications discussed in this chapter. In some cases the resultant differential equation cannot be integrated by simple analytical procedures. It is then necessary to use more complex methods, such as numerical or series approximations or graphical procedures.

In practical application two classes of transfer occur. In *simple transfer* the transferent property enters through one boundary of the system, passes through by molecular transport, and leaves at another boundary. An example of simple transfer is heat conduction through a furnace wall. The second class of transfer results when a quantity of the transferent property is actually generated within the volume under consideration. This is called *transfer with internal generation.* Transfer may still occur into the volume, and the transfer out will be equal to the transfer in plus the internal generation. A common example of transfer with internal generation of heat is an electric heating element. In such an element heat is generated within the metal element by resistance to an electric current. The internally generated heat must be transferred from the element to the surroundings to maintain a steady state. A common but less obvious example of internal generation is a fluid flowing in a pipe. Momentum is generated at every point in the flowing fluid and is transferred to the wall of the pipe. An example of internal generation of mass is a nuclear reactor, where neutrons are produced upon the fission of uranium. After generation, the neutrons diffuse through the reactor core, obeying the simple mass diffusion equation.

The term *generation* should not be confused with *creation*. The law of conservation of mass and energy states that neither can be created nor destroyed, although it is possible to convert mass into energy and vice versa. Therefore, the mass, heat, or momentum generated within the volume must have been present in a different form before it was transformed into that form being generated. For example, in the electrically heated element, electrical energy is converted into thermal energy. In the flowing fluid, a portion of the pressure increase supplied by a pump to the fluid is transformed into momentum which is transferred to the wall. During this process, a portion of the mechanical energy supplied by the pump is dissipated as heat in the fluid. In a nuclear reactor the neutron was originally part of the mass of the uranium nucleus. The fission of the nucleus liberates not only neutrons but also energy which appears as heat. Thus, the nuclear reactor core is also an example of transfer with internal generation of heat.

The applications of molecular-transport theory usually will be limited to one-dimensional transfer and special cases of two- and three-dimensional transfer. Other examples of two- and three-dimensional transfer often occur, but the treatment usually involves complex mathematics and is in most cases beyond the scope of this book. The principles of transfer can be satisfactorily illustrated using one dimension. Two- or three-dimensional transfer will be studied for only a few simple geometries.

SIMPLE TRANSFER

In all applications of simple transfer at steady state the rate of input of transferent property into a volume must equal the rate of output. In the simple one-dimensional case, the rate of input at the entrance boundary must equal the rate of output at the exit boundary. As a result, the rate of transfer at any point in a transfer medium is a constant with distance in the transfer direction. Since the rate of transfer is the product of the flux ψ and transfer area A, the following equation may be written.

$\psi A =$ constant with distance in the transfer direction

$$(10.1)$$

Another expression for this relationship may be obtained by differentiating Equation 10.1 with respect to the direction of transfer (x).

$$\frac{d(\psi A)}{dx} = 0 \qquad (10.2)$$

These equations are expressions of the transferent-property balance discussed earlier. Equation 10.2 is a balance for a small volume element of thickness dx.

Equation 10.1 may be applied to elements of finite thickness and will be the starting point for the applications considered in the following sections.

Mean Transfer Area. In the simple examples of Chapter 9, the transfer area was constant in the transfer direction. In many applications more complex geometries are encountered. For example, in a cylindrical pipe the transfer area in the radial direction varies with radius.

In many cases it is convenient to introduce a *mean transfer area* to use as a constant in the transfer equations. In general, if y is a function of x, a mean value of y may be obtained by the following definition:

$$\bar{y} \, \Delta x = \int_{x_1}^{x_2} y \, dx \qquad (10.3)$$

where $\bar{y}$ is the mean value of y in the interval from x_1 to x_2. Before this equation can be integrated, it is necessary to express y as a function of x.

Illustration 10.1. Calculate the mean transfer area in the radial direction through the wall of a pipe of inside radius r_1 and outside radius r_2.

SOLUTION. The general transfer equation for the radial direction is

$$\psi = -\tfrac{1}{6} l \bar{c} \frac{d\Gamma}{dr} \qquad (9.12a)$$

If the left side of the equation is multiplied by A/A,

$$\frac{(\psi A)}{A} = -\tfrac{1}{6} l \bar{c} \frac{d\Gamma}{dr}$$

where ψA is the *constant* rate of transfer as shown in Equation 10.1. Separation of variables gives

$$(\psi A) \int_{r_1}^{r_2} \frac{dr}{A} = -\tfrac{1}{6} l \bar{c} \int_{\Gamma_1}^{\Gamma_2} d\Gamma \qquad (10.4)$$

The left side of this equation may be integrated if A can be expressed as a function of r. However, it is often convenient to use the mean area ($\bar{A}$). Therefore, the mean flux is $(\psi A)/\bar{A}$, and Equation 10.4 integrates to

$$(\psi A) \frac{r_2 - r_1}{\bar{A}} = -\tfrac{1}{6} l \bar{c} (\Gamma_2 - \Gamma_1) \qquad (10.5)$$

Inspection of Equations 10.4 and 10.5 shows that

$$\frac{r_2 - r_1}{\bar{A}} = \int_{r_1}^{r_2} \frac{dr}{A} \qquad (10.6)$$

which is of the form of Equation 10.3. Since the transfer area $A = 2\pi r L$, where L is the length of the pipe,

$$\frac{r_2 - r_1}{\bar{A}} = \int_{r_1}^{r_2} \frac{dr}{2\pi r L} = \frac{1}{2\pi L} \ln\left(\frac{r_2}{r_1}\right)$$

The mean area is then

$$\bar{A} = A_{lm} = 2\pi L \frac{(r_2 - r_1)}{\ln\left(\dfrac{r_2}{r_1}\right)} = 2\pi L r_{lm} \qquad (10.7)$$

where r_{lm} is logarithmic-mean radius as defined by Equation 10.7, and A_{lm} is the particular mean area for the case where A varies linearly with distance. An alternate form of Equation 10.7 is

$$A_{lm} = \frac{A_2 - A_1}{\ln (A_2/A_1)} \qquad (10.7a)$$

which relates the mean area (A_{lm}) to the inside and outside areas of the pipe. The mean area defined by Equation 10.7a is called the *logarithmic-mean area*. It always arises when the transfer area varies linearly with distance, such as in transfer in the radial direction in a cylindrical geometry. As A_1 approaches A_2, the logarithmic-mean area approaches the *arithmetic mean*

$$A_{am} = \frac{A_2 + A_1}{2} \qquad (10.7b)$$

For example, if $A_2 = 1.5\ A_1$, the arithmetic mean is only 1.4 per cent higher than the logarithmic mean.

Heat Transfer. The model for heat transfer consists of a specified volume with entry and exit areas A_1 and A_2, and with heat fluxes $(q/A)_1$ and $(q/A)_2$. At steady state, Equation 10.1 becomes

$$\left(\frac{q}{A}\right) A = q = \text{constant} \qquad (10.8)$$

or

$$q_1 = q_2$$

which is the heat balance for the specified volume. For the second step as outlined at the beginning of the chapter, it is necessary to substitute the transport equation derived in Chapter 9,

$$\frac{q}{A} = -k \frac{dT}{dx} \qquad (9.43)$$

into Equation 10.8. The alternate form of the transport equation using the thermal diffusivity (α) may also be used. Substituting Equation 9.43 into Equation 10.8 yields

$$q = \text{constant} = -kA \frac{dT}{dx} \qquad (10.9)$$

For the third step in the general procedure, the variables in Equation 10.9 are separated.

$$q \frac{dx}{A} = -k\ dT \qquad (10.10)$$

If A is not constant but a function of x, it is now necessary to substitute for A in terms of x. The thermal conductivity may also vary with temperature.

Finally, in the fourth step, Equation 10.10 is integrated and the integration constant is evaluated.

$$q \int \frac{dx}{A} = -\int k\ dT + C_1 \qquad (10.11)$$

or, as a definite integral

$$q \int_{x_1}^{x_2} \frac{dx}{A} = -\int_{T_1}^{T_2} k\ dT \qquad (10.12)$$

which is applicable for simple transfer.

The choice of integration procedure depends upon the variation of A with x and k with T. If these variations can be expressed as simple functions of x or T, it may be possible to integrate analytically. Otherwise, numerical or graphical methods may be used.

Illustration 10.2. A steam pipe 2 in. in outside diameter has an outside surface temperature of 350°F. The pipe is covered with a coating material 2 in. thick. The thermal conductivity of the coating varies with temperature such that $k = 0.5 + 5 \times 10^{-4}T$ where T is in degrees Fahrenheit and k in Btu/hr sq ft (°F/ft). The outside surface of the coating is at 100°F. Calculate the heat loss per foot of pipe length.

SOLUTION. Let 1 designate the inside of the insulation and 2 the outside. Then $r_1 = 1$ in., $r_2 = 3$ in.; $T_1 = 350°F$, $T_2 = 100°F$; and $q_1 = q_2 = q$ for simple transfer at steady state. Using Equation 10.12 for transfer in the radial direction,

$$q \int_{r_1}^{r_2} \frac{dr}{A} = -\int_{T_1}^{T_2} k\ dT \qquad (10.12a)$$

Since $A = 2\pi rL$ for L feet of pipe length, and $k = 0.5 + 5 \times 10^{-4}T$,

$$q \int_{r_1}^{r_2} \frac{dr}{2\pi rL} = -\int_{T_1}^{T_2} (0.5 + 5 \times 10^{-4}T)\ dT$$

$$\frac{q}{2\pi L} \left(\ln \frac{r_2}{r_1}\right) = -\left[(0.5)(T_2 - T_1) + \left(\frac{5 \times 10^{-4}}{2}\right)(T_2^2 - T_1^2)\right]$$

Then for 1 ft of pipe

$$q = -\frac{2\pi[(0.5)(350 - 100) + (5 \times 10^{-4}/2)(350^2 - 100^2)]}{\ln \left(\dfrac{3/12}{1/12}\right)}$$

$$= 877\ \text{Btu/hr}$$

Illustration 10.3. Solve Illustration 10.2 using the concept of mean area.

SOLUTION. Combining Equation 10.12a with Equation 10.6 gives

$$q \frac{r_2 - r_1}{\bar{A}} = -\int_{T_1}^{T_2} k\ dT \qquad (10.13)$$

Therefore

$$q = -\frac{\bar{A} \displaystyle\int_{T_1}^{T_2} k\ dT}{r_2 - r_1}$$

For transfer in the radial direction for cylinders, from Equation 10.7 the mean area is

$$\bar{A} = A_{lm} = 2\pi L \frac{(r_2 - r_1)}{\ln \left(\dfrac{r_2}{r_1}\right)} = \frac{(2)(3.14)(3/12 - 1/12)}{\ln \left(\dfrac{3/12}{1/12}\right)}$$

$$= 0.955\ \text{sq ft for 1 ft of pipe}$$

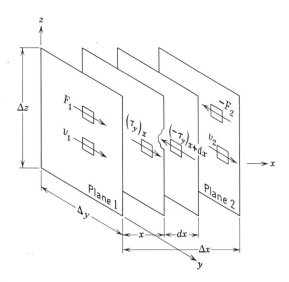

Figure 10.1. The external stress pattern in simple momentum transfer.

Then

$$q = -\frac{(0.955)[(0.5)(350 - 100) + (5 \times 10^{-4}/2)(350^2 - 100^2)]}{(3/12 - 1/12)}$$

$$= 877 \text{ Btu/hr}$$

which agrees with Illustration 10.2.

Momentum Transfer. The momentum-balance equation for a volume of fluid is, from Equation 10.2,

$$(\tau_y g_c A) = \text{constant} \tag{10.14}$$

The physical mechanism of momentum transfer will be considered to verify Equation 10.14. The external stress pattern in simple transfer is shown in Figure 10.1. Two solid planes, 1 and 2, are moving at velocities v_1 and v_2 as a result of forces F_1 and F_2 applied to the planes. The planes move at constant velocities, and constant forces are required to overcome the resistance to flow offered by the fluid between the planes. If the fluid and planes are at steady state, an *external* force balance on the fluid between the planes gives

$$\Sigma F = F_1 + (-F_2) = 0 \tag{10.15}$$

where the forces are defined as positive in the $+y$-direction. At any plane in the fluid, a $+y$-directed stress exists as a result of *entry* of momentum into the plane, and a $-y$-directed stress exists as a result of *departure* of momentum from a plane. An *external* force balance over the volume between plane 1 and the plane at $x + dx$ can be written, noting that momentum enters plane 1 and leaves at the plane $(x + dx)$, with appropriate effect upon the stress signs.

$$F_1 + (-\tau_y A)_{x+dx} = 0 \tag{10.16}$$

where $A = \Delta y\, \Delta z =$ area of each boundary plane.

A force balance over the volume between plane 2 and the plane at x gives

$$(-F_2) + (\tau_y A)_x = 0 \tag{10.17}$$

Combining Equations 10.15, 10.16, and 10.17 gives

$$(\tau_y A)_x + (-\tau_y A)_{x+dx} = 0 \tag{10.18}$$

or

$$(\tau_y A) = \text{constant} \tag{10.14a}$$

which is equivalent to Equation 10.14 after inserting the dimensional constant g_c.

The momentum-transport equation is

$$\tau_y g_c = -\mu \frac{dv}{dx} \tag{9.62}$$

Combining Equations 9.62 and 10.14 gives

$$(\tau_y g_c A) = \text{constant} = -\mu A \frac{dv}{dx} \tag{10.19}$$

Separation of variables and integration gives

$$\int_{v_1}^{v_2} \mu\, dv = -(\tau_y g_c A) \int_{x_1}^{x_2} \frac{dx}{A} \tag{10.20}$$

which is applicable for simple transfer.

Generally, the viscosity may vary with velocity (in non-Newtonian fluids) and area of transfer may vary with distance. The variations must be expressed in terms of velocity or distance before the integration can be completed.

Illustration 10.4. Two parallel flat plates are spaced 2 in. apart. One plate is moving at a velocity of 10 ft/min and the other is moving in the opposite direction at 35 ft/min. The viscosity of the fluid between the plates is constant at 150 centipoises.

(a) Calculate the stress on each plate.

(b) Calculate the fluid velocity at $\frac{1}{2}$ in. intervals from plate to plate.

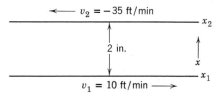

SOLUTION.

$$x_2 = 2 \text{ in.}, \quad v_2 = -35 \text{ ft/min}$$
$$x_1 = 0, \quad\quad v_1 = 10 \text{ ft/min}$$

(a) For a differential thickness of fluid (dx) Equation 10.20 applies.

$$\int_{v_1}^{v_2} \mu\, dv = -(\tau_y g_c A) \int_{x_1}^{x_2} \frac{dx}{A} \tag{10.20}$$

In this case μ and A are constant, so that Equation 10.20 gives on integration

$$\tau_y g_c = -\mu \frac{(v_2 - v_1)}{(x_2 - x_1)} \tag{10.21}$$

Since $\mu = 150$ centipoises $= 363$ lb/ft hr

$v_1 = 10$ ft/min $= 600$ ft/hr

$v_2 = -35$ ft/min $= -2100$ ft/hr

$x_2 = \frac{1}{6}$ ft

Then the momentum flux is

$$\tau_y g_c = \frac{-(363)(-2100 - 600)}{(1/6 - 0)} = 5{,}880{,}000 \; \frac{\text{ft-lb/hr}}{\text{sq ft hr}}$$

Since $g_c = 32.2 \, \dfrac{\text{lb-ft}}{\text{lb}_f \, \text{sec}^2}$, the stress on either plate is

$$\tau_y = \frac{5{,}880{,}000}{32.2 \times (3600)^2} = 0.0141 \; \frac{\text{lb}_f}{\text{sq ft}}$$

The stress is positive because momentum is transferred in the direction of decreasing velocity.

(b) The velocity at any value of x may be obtained by integrating between $(x = x_1, v = v_1)$ and $(x = x, v = v)$. Then Equation 10.20 gives

$$v - v_1 = -\frac{\tau_y g_c}{\mu}(x - x_1) \qquad (10.22)$$

Eliminating $\tau_y g_c / \mu$ between Equations 10.21 and 10.22 gives

$$v - v_1 = (v_2 - v_1)\left(\frac{x - x_1}{x_2 - x_1}\right) \qquad (10.23)$$

Therefore the velocity v at any value of x is

$$v = 10 + (-35 - 10)\left(\frac{x}{2}\right)$$

At $x = \frac{1}{2}$ in.,

$$v = 10 - (45)\left(\frac{1/2}{2}\right)$$

$$= -1.25 \text{ ft/min}$$

at $x = 1$ in., $\qquad v = -12.5$ ft/min

and at $x = 1\frac{1}{2}$ in., $\quad v = -22.75$ ft/min

Equation 10.23 shows that the velocity varies linearly with distance in the x-direction.

An example of simple momentum transfer with varying transfer area is the Couette viscometer. This device is used experimentally to determine the viscosity of liquids. It consists of a cylindrical bob set concentrically inside of a rotating cylindrical cup (Figure 10.2). The annular space is filled with the liquid whose viscosity is to be determined. As the cup is rotated by an external force, the liquid in the annular space transmits momentum radially from the cup to the bob, thereby imparting a moment or torque to the bob. The torque on the bob is measured by the angular displacement of the bob and the viscosity can be calculated using the equations derived below. It will be assumed that no liquid exists between the bottom of the bob and the cup, which actually is the case in precision Couette viscometers.

At steady state the moment on the cup must equal

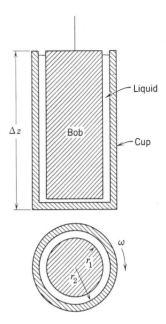

Figure 10.2. Schematic diagram of the Couette viscometer.

the moment on the bob, since no other moment acts on the fluid. Momentum transfer will occur in the radial direction from the cup to the bob. The moment-balance equation is

$$\Sigma \mathbf{M} = \mathbf{M}_{\text{cup}} + (-\mathbf{M}_{\text{bob}}) = 0$$

or $\qquad \mathbf{M} = $ constant with radius $\qquad (10.24)$

where $\mathbf{M}$ is the moment at any radial position. The moment is equal to the product of a force and the radius at which the force acts. The force is equal to the shear stress times the area over which the stress is exerted. In this case the transfer area is $2\pi r \, \Delta z$ and the moment is

$$\mathbf{M} g_c = F g_c r = \tau g_c (2\pi r \, \Delta z) r \qquad (10.25)$$

The momentum-transport equation may be written for an angular velocity (ω) and a radius (r). Since

$$v = r\omega$$

$$dv = r \, d\omega *$$

and $\qquad \dfrac{dv}{dr} = r \dfrac{d\omega}{dr} \qquad (10.26)$

The transport equation is then

$$\tau g_c = -\mu r \frac{d\omega}{dr} \qquad (10.27)$$

---

* One might expect that $dv = r \, d\omega + \omega \, dr$. However, it has been shown that $\omega \, dr = 0$ at a fixed boundary, but for solids $\omega \, dr$ has a finite value. (Loeb, L. B., *The Kinetic Theory of Gases*, 2nd ed., McGraw-Hill Book Co., New York, 1934, p. 232.)

The momentum flux τg_c is eliminated between Equations 10.25 and 10.27,

$$\frac{\mathbf{M} g_c}{2\pi r^2 \, \Delta z} = -\mu r \frac{d\omega}{dr} \qquad (10.28)$$

Integrating and assuming μ is constant gives

$$\int_{\omega_1}^{\omega_2} d\omega = -\frac{\mathbf{M} g_c}{2\pi \, \Delta z \, \mu} \int_{r_1}^{r_2} \frac{dr}{r^3}$$

$$(\omega_2 - \omega_1) = \frac{\mathbf{M} g_c}{4\pi \, \Delta z \, \mu} \left[\frac{1}{r_2{}^2} - \frac{1}{r_1{}^2} \right] \qquad (10.29)$$

The moment on the bob is measured experimentally when the bob is stationary in the rotating cup. From this, the viscosity can be calculated.

Illustration 10.5. Determine the viscosity of a liquid with the following data taken on a Couette viscometer: $r_1 = 0.9$ in., $r_2 = 1.0$ in., $\Delta z = 2$ in. The cup is rotating at 10 rpm and the moment on the bob is -0.00060 ft-lb$_f$.

SOLUTION. Since the bob is stationary, $\omega_1 = 0$.

$$\omega_2 = \left(2\pi \frac{\text{radians}}{\text{rev}} \right) \left(10 \frac{\text{rev}}{\text{min}} \right) = 20\pi \text{ radians/min}$$

Solving Equation 10.29 for μ gives

$$\mu = \frac{\mathbf{M} g_c}{4\pi \, \Delta z \, \omega_2} \left[\frac{1}{r_2{}^2} - \frac{1}{r_1{}^2} \right]$$

$$= \frac{-(0.00060)(32.2)}{4(3.14)(2/12)(20\pi/60)} \left[\frac{1}{(1/12)^2} - \frac{1}{(0.9/12)^2} \right]$$

$$= 0.30 \text{ lb/ft sec}$$

$$= 446 \text{ centipoises}$$

Mass Transfer. Two types of simple molecular mass transfer may be analyzed using the balance and transport equations. In *equimolar counterdiffusion*, component a diffuses through component b, which is diffusing at the same molar rate as a but in an opposite direction. The boundaries of the system are permeable to both components. Equimolar counterdiffusion occurs in distillation. Mass transfer occurs between the liquid on a distillation plate and the vapor bubble rising through it. If the molar latent heats of vaporization of the mixtures are constant and the system adiabatic, the moles of more volatile component vaporized must equal the moles of less volatile component condensed. In such a case the more volatile component diffuses in the direction opposite to that of the less volatile component but at the same rate. Equimolar counterdiffusion was discussed in Chapter 9, where it was shown that if the temperature and total pressure are constant, a gradient in component a will of necessity result in a gradient in component b. If the boundaries are permeable to both components (i.e., if a and b can leave or enter at

any boundary), both gases will diffuse because of their gradients and

$$N_b = -N_a \qquad (9.27)$$

For mass transfer the balance equations are

$$N_a = \text{constant} \qquad (10.30)$$

and

$$N_b = \text{constant} \qquad (10.31)$$

The second type of molecular mass transfer is called *diffusion through a stationary gas*,* which occurs when one boundary of the system is permeable to only one component. In this case it will be shown that there is no net movement of the other component which is said to be *stationary*. In gas absorption, diffusion through a stationary gas occurs. For example, consider the absorption of ammonia from an air–ammonia mixture by water. The boundary between the gas and the water is permeable only to ammonia, since air has a negligible solubility in water. Therefore, ammonia will diffuse from the bulk of the gas through stationary air to the water surface, where it will be absorbed. There is no net transfer of air.

The development of a mass-transfer equation for diffusion through a stationary gas requires the consideration of several transfer processes which occur simultaneously. Typical concentration gradients are shown in Figure 10.3, where boundary 2 is permeable only to component a. For example, boundary 2 might be the gas-water interface, component a the ammonia, and component b the air. The balance equations for this type of transfer are

$$(N_a)_{\text{total}} = \text{constant} \qquad (10.30a)$$

and

$$(N_b)_{\text{total}} = 0 \qquad (10.31a)$$

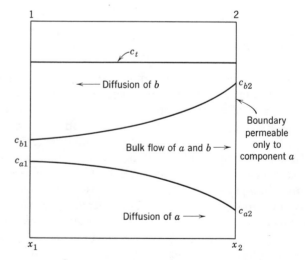

Figure 10.3. Concentration gradients in diffusion through a stationary gas.

---

* Most cases of interest involve gases, but the following development also applies to liquids.

since component b is stationary. Equation 10.30a states that the total rate of transfer of component a is constant at any position. This is a requirement of simple transfer at steady state. Equation 10.31a defines that stationary character of component b. There is no total (or net) transfer of b. It will be shown in the following paragraphs that both a and b are transferred not only by molecular transport but also by bulk flow. *Bulk flow* is the movement of a volume element of gas with respect to the stationary coordinate axes.

In considering the transfer of component a it will be helpful to look first at a volume element of the gas itself, without regard to the movement of the volume element with respect to a stationary coordinate system. If a gradient in component a exists within the volume element, molecular transport will occur according to the transport equation.

$$\frac{N_a}{A} = -\mathscr{D}\frac{dc_a}{dx} \qquad (9.22)$$

At a constant temperature and total pressure, the total concentration (or molar density) (c_t) is constant. Then, as shown in Chapter 9,

$$c_t = c_a + c_b \qquad (9.25)$$

and

$$\frac{dc_b}{dx} = -\frac{dc_a}{dx} \qquad (9.26)$$

Equation 9.26 shows that a gradient must exist in the concentration of component b if one exists in the concentration of a. The gradient will be equal in magnitude but opposite in sign to that of a, as shown in Figure 10.3.

Since a gradient exists in component b, it will diffuse to the left opposite to the diffusion of component a. The flux of b is

$$\frac{N_b}{A} = -\mathscr{D}\frac{dc_b}{dx} \qquad (9.24)$$

Combination of Equations 9.22, 9.24, and 9.26 shows that for molecular transport

$$N_b = -N_a \qquad (9.27)$$

for the volume element of gas. Therefore, equimolar counterdiffusion by molecular transport occurs at every point in the gas. However, because boundary 2 is impermeable to b another transfer mechanism occurs which just balances the transport of b, as shown below.

The gradient in b shown in Figure 10.3 will result in the transfer of b away from boundary 2. Since this boundary is permeable only to a, no b can be supplied across it. Therefore, another transfer mechanism must supply b at the same rate it is being removed by molecular transport. This new mechanism is *bulk flow*, i.e., the movement of a volume element of the gaseous mixture toward the impermeable boundary. Therefore, *any*

molecule in the gas is subject to two motions: the normal molecular transport due to concentration gradient and bulk flow in the x-direction due to depletion at the boundary. The volume element of gas is made up of the mixture of a and b, which varies with position. The rate of transfer of b *toward* boundary 2 must just balance the rate of transfer *away* from the boundary, to maintain a steady state. That is,

$$\begin{bmatrix} \text{Rate of} \\ \text{bulk flow of } b \end{bmatrix} = -\begin{bmatrix} \text{rate of} \\ \text{molecular transport of } b \end{bmatrix}$$
$$(10.32)$$

The minus sign indicates that the bulk flow is in a direction opposite to the molecular transport.

Since the rate of molecular transport of b is N_b, from Equation 10.32,

$$\text{Bulk flux of } b = \frac{\text{bulk flow of } b}{A} = -\frac{N_b}{A} \quad (10.33)$$

In Equation 10.33, the bulk flow of b was considered, however the gas in the volume element consists of both a and b. By Dalton's law,

$$\frac{\text{Total moles of gas}}{\text{Moles of gas } b} = \frac{c_t}{c_b} \qquad (10.34)$$

and

$$\frac{\text{Moles of gas } a}{\text{Moles of gas } b} = \frac{c_a}{c_b} \qquad (10.35)$$

Combination of Equations 10.33 and 10.34 gives

$$\text{Bulk flux of } a \text{ and } b = \left(\frac{c_t}{c_b}\right)\left(-\frac{N_b}{A}\right) \quad (10.36)$$

Combination of Equations 10.33 and 10.35 gives

$$\text{Bulk flux of } a = \left(\frac{c_a}{c_b}\right)\left(-\frac{N_b}{A}\right) \qquad (10.37)$$

The bulk flow of the gas in the x-direction aids in transfer of a, thus component a is transferred toward the boundary by bulk flow according to Equation 10.37 and by diffusion according to Equation 9.22. The total rate of transfer of a toward the boundary is the sum of the two,

$$\left(\frac{N_a}{A}\right)_t = \left(\frac{c_a}{c_b}\right)\left(-\frac{N_b}{A}\right) + \frac{N_a}{A} \qquad (10.38)$$

Since by Equation 9.27 the diffusional flux of b is numerically equal but opposite to that of a, Equation 10.38 becomes

$$\left(\frac{N_a}{A}\right)_t = \left(\frac{c_a}{c_b}\right)\left(\frac{N_a}{A}\right) + \frac{N_a}{A}$$

or

$$\left(\frac{N_a}{A}\right)_t = \left(1 + \frac{c_a}{c_b}\right)\left(\frac{N_a}{A}\right) \qquad (10.39)$$

It is important to note that (N_a/A) is the flux due to molecular transport only and that $(N_a/A)_t$ is due to both transport and flow. Substitution for N_a/A from Equation 9.22 into 10.39 gives

$$\left(\frac{N_a}{A}\right)_t = \left[1 + \frac{c_a}{c_b}\right]\left[-\mathscr{D}\frac{dc_a}{dx}\right]$$

or

$$\left(\frac{N_a}{A}\right)_t = -\mathscr{D}\left(\frac{c_t}{c_b}\right)\left(\frac{dc_a}{dx}\right) \qquad (10.40)$$

This is the general equation for transfer through a stationary gas. From a simple material balance it can be shown that $(N_a/A)_t$ is constant. However, c_b varies with c_a as shown by Equation 9.25. Equation 10.40 may be integrated if the variations in $\mathscr{D}$ and A are known.

The total bulk flow of a and b may be shown to be equal to the total transfer of a and b, as follows. Since $N_a = -N_b$, Equation 10.38 may be written as

$$\left(\frac{N_a}{A}\right)_t = \left(\frac{c_a}{c_b} + 1\right)\left(-\frac{N_b}{A}\right)$$

$$\left(\frac{N_a}{A}\right)_t = \left(\frac{c_t}{c_b}\right)\left(-\frac{N_b}{A}\right) = \left(\frac{c_t}{c_b}\right)\left(\frac{N_a}{A}\right) \qquad (10.41)$$

Comparison of Equations 10.36 and 10.41 shows that the bulk flow of a and b is equal to the total transfer of a and that transfer through a stationary gas is greater by a factor c_t/c_b than transfer by equimolar counterdiffusion.

Equation 10.31a states that the total transfer of b, $(N_b)_t$, is zero. Therefore

$(N_b)_t = $ bulk flow of b + molecular transport of $b = 0$

$$(10.42)$$

In this sense, the component b is stationary. There is no *net* transfer of b, since the bulk flow balances the molecular transport.

For the case where $\mathscr{D}$ and A are constants, Equation 10.40 can be integrated analytically. Since $dc_a/dx = -dc_b/dx$, Equation 10.40 becomes upon integration

$$\left(\frac{N_a}{A}\right)_t \int_{x_1}^{x_2} dx = \mathscr{D}c_t \int_{c_{b1}}^{c_{b2}} \frac{dc_b}{c_b}$$

$$\left(\frac{N_a}{A}\right)_t (x_2 - x_1) = \mathscr{D}c_t \ln\left(\frac{c_{b2}}{c_{b1}}\right) \qquad (10.43)$$

The logarithmic-mean concentration is defined as

$$c_{bm} = \frac{c_{b2} - c_{b1}}{\ln\left(\frac{c_{b2}}{c_{b1}}\right)} \qquad (10.44)$$

Combining Equations 10.43 and 10.44 gives

$$\left(\frac{N_a}{A}\right)_t = \mathscr{D}\frac{c_t(c_{b2} - c_{b1})}{c_{bm}(x_2 - x_1)} \qquad (10.45)$$

Since $c_{a1} + c_{b1} = c_t = c_{a2} + c_{b2}$, Equation 10.45 may be written as

$$\left(\frac{N_a}{A}\right)_t = -\mathscr{D}\frac{c_t(c_{a2} - c_{a1})}{c_{bm}(x_2 - x_1)} \qquad (10.46)$$

This equation applies only where $\mathscr{D}$ and A are constant. It should be noted that at steady state $(N_a)_t$ must be constant, but that N_a for transport alone varies with distance, as indicated by the curved concentration gradients in Figure 10.3.

Illustration 10.6. An open cylindrical tank is filled to within 2 ft of the top with pure methanol. The tank is tapered, as shown in Figure 10.4. The air within the tank is stationary, but circulation of air immediately above the tank is adequate to assure a negligible concentration of methanol at this point. The tank and air space are at 77°F and 1 atm. The diffusivity of methanol in air at 77°F and 1 atm is 0.62 sq ft/hr. Calculate the rate of loss of methanol from the tank at steady state.

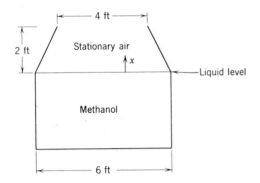

Figure 10.4. Cylindrical tank in Illustration 10.6.

SOLUTION. This is a case of methanol diffusing through 2 ft of stationary air from the surface of the liquid methanol to the top of the tank. The transfer area varies. The concentration of methanol in the air at the surface of the liquid is determined from the vapor pressure of methanol at 77°F, which is $p_a = 135$ mm Hg. Since by the perfect-gas law $c_a = p_a/RT$, Equation 10.40 may be written as

$$\left(\frac{N_a}{A}\right)_t = -\frac{\mathscr{D}P}{RTp_b}\frac{dp_a}{dx} \qquad (10.40a)$$

To integrate this equation A must be expressed in terms of x. Geometrical considerations show that

$$A = \frac{\pi}{4}(6 - x)^2$$

where $x = 0$ is the surface of the methanol and $x = 2$ ft is the top of the tank. Furthermore, $p_b = P - p_a$. Then Equation 10.40a may be written as

$$(N_a)_t \int_{x_1}^{x_2} \frac{dx}{(\pi/4)(6 - x)^2} = -\mathscr{D}\frac{P}{RT}\int_{p_{a1}}^{p_{a2}} \frac{dp_a}{P - p_a}$$

$$(N_a)_t \left[\frac{4}{\pi}\right]\left[\frac{1}{6 - x_2} - \frac{1}{6 - x_1}\right] = \frac{\mathscr{D}P}{RT}\ln\left(\frac{P - p_{a2}}{P - p_{a1}}\right)$$

Now $x_1 = 0$, $x_2 = 2$ ft; $p_{a2} = 0$, $p_{a1} = 135$ mm Hg; $P = 760$ mm, $R = 0.7302$ cu ft atm/lb mole °R, $T = 537$°R, and

$$(N_a)_t \left(\frac{4}{\pi}\right)\left(\frac{1}{4} - \frac{1}{6}\right) = \frac{(0.62)(1)}{(0.7302)(537)} \ln \left(\frac{760 - 0}{760 - 135}\right)$$

and

$$(N_a)_t = 0.00286 \text{ lb moles/hr}$$

TRANSFER WITH INTERNAL GENERATION

In transfer with internal generation some of the transferent property is generated within the medium and may appear at all points of the medium. The transferent property is continuously generated and must be continuously transferred to a boundary to maintain steady state. Obviously, the rate of transfer is not constant with distance but increases as the boundary is approached.

A system with internal generation is shown in Figure 10.5. The transferent property is generated at a rate of $\mathscr{G}$ units of property per unit time and unit volume. A property balance can be made around the element of volume of thickness dx. The rate of influx to the element plus the rate of generation must equal the rate of efflux, or

$$(\psi A)_x + \mathscr{G} \, dV = (\psi A)_{x+dx} \qquad (10.47)$$

where the volume $dV = \Delta y \, \Delta z \, dx$ and $A = \Delta y \, \Delta z$. Since $(\psi A)_{x+dx} - (\psi A)_x = d(\psi A)$, Equation 10.47 becomes

$$d(\psi A) = \mathscr{G} \, dV \qquad (10.48)$$

This is the general transferent-property balance equation for internal generation. It merely states that the increase in rate of transfer across a volume element is equal to the rate of generation within the element. If $\mathscr{G}$ is constant for the entire volume between planes 1 and 2, the equation may be integrated to

$$(\psi A)_2 - (\psi A)_1 = \mathscr{G}V \qquad (10.49)$$

where the volume $V = \Delta x \, \Delta y \, \Delta z$. The transferent property generated within the volume may leave through both planes, as determined by the property gradient. If one plane is impermeable to the transferent property, all the internal generation must leave the volume through the other plane.

Equation 10.48 may be integrated when the variation of A and V are known. It applies to mass, heat, and momentum transfer, as shown in the following sections.

Heat Transfer. Examples of internal generation of heat include nuclear-reactor fuel elements and electrically heated wires. The generation is not necessarily uniform within the volume. When the generation is not uniform, it must be expressed as a function of distance or another variable. For heat transfer, the balance equation, Equation 10.48, becomes

$$d\left[\left(\frac{q}{A}\right)A\right] = \mathscr{G}_q \, dV \qquad (10.50)$$

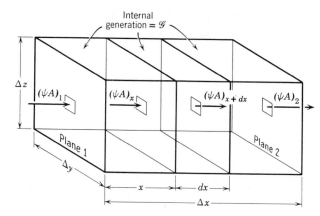

Figure 10.5. Transfer with internal generation.

where $\mathscr{G}_q$ is the internal generation rate for heat, Btu/hr cu ft. Substitution of q/A from the transport equation gives

$$d\left[-kA\frac{dT}{dx}\right] = \mathscr{G}_q \, dV \qquad (10.51)$$

This equation may be integrated twice,

$$-kA\frac{dT}{dx} = \int \mathscr{G}_q \, dV + C_1$$

$$-\int k \, dT = \int \frac{(\int \mathscr{G}_q \, dV)}{A} \, dx + \int \frac{C_1 \, dx}{A} + C_2 \qquad (10.52)$$

To complete the integration, the variations of k, A, V, and $\mathscr{G}_q$ must be known as functions of T or x.

One geometry of interest is the cylinder. This could be an electrical resistance wire or a nuclear fuel element. The heat generated uniformily within the element is transferred in a radial direction to the surface. For constant $\mathscr{G}_q$ and k Equation 10.51 may be integrated. For the cylinder, the following substitutions may be made into Equation 10.51.

$$V = \pi r^2 L$$
$$dV = 2\pi L r \, dr$$
$$A = 2\pi r L$$
$$\frac{dT}{dx} = \frac{dT}{dr}$$

where L is the length of the cylinder.

Equation 10.51 becomes

$$d\left(-2\pi k L r \frac{dT}{dr}\right) = 2\pi L \mathscr{G}_q r \, dr \qquad (10.53)$$

which integrates to

$$-2\pi k L r \frac{dT}{dr} = \pi L \mathscr{G}_q r^2 + C_1 \qquad (10.54)$$

The temperature at the center of the cylinder must be a maximum if heat is uniformly generated. Therefore, one boundary condition is at $r = 0$, $T = T_0$, or $dT/dr = 0$.

Substitution of this boundary condition into Equation 10.54 permits evaluation of C_1.

$$-2\pi kLr(0) = \pi L \mathscr{G}_q(0)^2 + C_1$$

Therefore,

$$C_1 = 0$$

Equation 10.54 may now be rewritten as

$$-k\left(\frac{dT}{dr}\right) = \frac{\mathscr{G}_q r}{2} \qquad (10.55)$$

which shows that the temperature gradient (dT/dr) varies linearly with distance in the radial direction. Substitution of the transport equation gives

$$\frac{q}{A} = \frac{\mathscr{G}_q r}{2} \qquad (10.56)$$

which shows that the flux also varies linearly in the radial direction. The flux at the boundary is

$$\left(\frac{q}{A}\right)_1 = \frac{\mathscr{G}_q r_1}{2} \qquad (10.57)$$

Combining Equations 10.56 and 10.57 yields

$$\frac{(q/A)}{(q/A)_1} = \frac{r}{r_1} \qquad (10.58)$$

Further integration of Equation 10.55 gives

$$-kT = \frac{\mathscr{G}_q r^2}{4} + C_2 \qquad (10.59)$$

The integration constant C_2 may be evaluated by applying either the boundary condition at the center ($T = T_0$ at $r = 0$) or the boundary condition at the outside surface ($T = T_1$ at $r = r_1$). Using the latter gives

$$-kT_1 = \frac{\mathscr{G}_q r_1^2}{4} + C_2$$

or

$$C_2 = -kT_1 - \frac{\mathscr{G}_q r_1^2}{4}$$

Substituting this expression for C_2 into Equation 10.59 gives

$$-k(T - T_1) = \frac{\mathscr{G}_q(r^2 - r_1^2)}{4} \qquad (10.60)$$

Use of the center boundary condition gives

$$-k(T_0 - T) = \frac{\mathscr{G}_q(-r^2)}{4} \qquad (10.61)$$

Either Equation 10.60 or Equation 10.61 may be used to calculate the temperature at any point in the cylinder, depending upon the available data. These equations show that the radial temperature distribution is parabolic in a cylinder with uniform internal generation.

Illustration 10.7. An electrical resistance wire has a melting point at 2500°F. The electrical input to a wire 10 ft long and $\frac{1}{4}$ in. diameter gives a uniform volumetric heat generation totaling 1,400,000 Btu/hr. The surface temperature of the wire is 1500°F and the thermal conductivity is 10 Btu/hr sq ft (°F/ft).

(a) Can the wire be safely used, or will the center reach its melting point?

(b) Calculate the radius of any molten core formed assuming that the properties of the molten metal are the same as those of the solid.

SOLUTION. (a) This is a case of transfer of heat with internal generation. The volumetric rate of generation is

$$\mathscr{G}_q = \frac{1,400,000}{(10)(\pi/4)(1/4 \times 1/12)^2} = 4.1 \times 10^8 \text{ Btu/hr cu ft}$$

For $T = T_0$ at $r = 0$, Equation 10.60 becomes

$$-k(T_0 - T_1) = \frac{\mathscr{G}_q(0 - r_1^2)}{4}$$

Since $k = 10$ Btu/hr sq ft (°F/ft), $T_1 = 1500$, and $r_1 = 1/96$ ft,

$$-(10)(T_0 - 1500) = \frac{(4.1 \times 10^8)(0 - 1/96)^2}{4}$$

and

$$T_0 = 2612°F$$

Therefore, the wire is molten at the center.

(b) The molten core will extend outward to the radius at which $T = 2500°F$. Division of Equation 10.60 by Equation 10.61 gives

$$\frac{T - T_1}{T_0 - T} = \frac{r^2 - r_1^2}{-r^2} = \left(\frac{r_1}{r}\right)^2 - 1$$

$$\frac{2500 - 1500}{2612 - 2500} = \left(\frac{1/8}{r}\right)^2 - 1$$

and

$$r = 0.04 \text{ in.}$$

Momentum Transfer. The transfer of internally generated momentum occurs in all fluids flowing in a stationary duct. In laminar flow, momentum is generated uniformly throughout the fluid and is transferred to the boundaries by molecular transport. For momentum transfer, Equation 10.48 becomes

$$d(\tau_y g_c A) = \mathscr{G}_\tau \, dV \qquad (10.62)$$

where $\mathscr{G}_\tau$ is the volumetric rate of generation of momentum. In many applications, the shear stress is not easily measured. It is therefore desirable to relate it and the generation rate to a more easily measured quantity, as shown in the following discussion. It will also be shown that the internal-generation rate is related directly to the pressure drop in the flow direction.

Consider the external forces acting upon the volume of fluid shown in Figure 10.6. The fluid at the plane at $x + dx$ is moving at a lower velocity than at plane x, and therefore a negative velocity gradient is present at all values of x within the element. The element is chosen so that there is no change in any property with z. According to Chapter 9, a stress in the $+y$-direction occurs on the face of a plane into which momentum is transferred. Momentum is transferred into the external face of plane x, and therefore, a $(+y)$-directed force $(F_y)_x$ is present over the area $A = \Delta z \, dy$.

$$(F_y)_x = (\tau_y A)_x \qquad (10.63)$$

At $(x + dx)$, momentum is leaving the external face of plane $(x + dx)$ and therefore a $(-y)$-directed force is present over the area A.

$$(-F_y)_{x+dx} = (-\tau_y A)_{x+dx} \qquad (10.64)$$

In simple transfer, it has been demonstrated that the momentum transfer in a system of this geometry is constant with x. In this example, the rate of transfer at $x + dx$ is greater than the transfer at x by an amount equal to the rate of internal generation. Therefore, the generation should be included in a momentum balance. It is necessary to obtain an expression for the internal momentum generation in terms of the forces acting on the volume element. This will be accomplished by writing a force balance.

The pressure of the fluid at planes y and at $y + dy$ produces forces acting on the volume element. Pressure is exerted in all directions at any point in the fluid; however, the pressure exerted upon the specific face of a plane may be considered directional. For example, the pressure force exerted on the external face of plane y over an area S is equal to the pressure times the area upon which the pressure is applied

$$(F_y)_y = (PS)_y \qquad (10.65)$$

and is $+y$-directed. The pressure force acting upon the external face of plane $y + dy$ is

$$(-F_y)_{y+dy} = (-PS)_{y+dy} \qquad (10.66)$$

and is $-y$-directed.
where P = pressure of the fluid.
$\quad S$ = area over which the pressure is applied
$\quad\quad = \Delta z \, dx$

No other forces are acting upon the volume element. Note that the transfer area (A) is perpendicular to the flow area (S). Pressure acts in all directions in a fluid. Here, Equations 10.65 and 10.66 define the external forces which are applied to the chosen volume element by the presence of a pressure. The external forces are balanced by equivalent internal pressure forces.

A total force balance around the element states that the sum of the external forces must equal zero at steady state. Therefore, a balance of the four external forces acting on the volume element gives

$$(F_y)_x + (-F_y)_{x+dx} + (F_y)_y + (-F_y)_{y+dy} = 0 \quad (10.67)$$

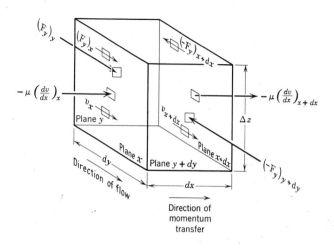

Figure 10.6. Internal momentum generation.

Substitution of Equations 10.63, 10.64, 10.65 and 10.66 into Equation 10.67 gives

$$(\tau_y A)_x - (\tau_y A)_{x+dx} + (PS)_y - (PS)_{y+dy} = 0$$

$$(\tau_y A)_{x+dx} - (\tau_y A)_x = -[(PS)_{y+dy} - (PS)_y] \quad (10.68)$$

Equation 10.68 reduces to

$$d(\tau_y A) = -d(PS) \qquad (10.69)$$

for the differential distances dx and dy. This equation states that the change in shear force (or rate of momentum transport) in the x-direction across a volume element is equal to the change in pressure force across the element in the y-direction.

If S is constant and the right side of Equation 10.69 is multiplied by dy/dy,

$$d(\tau_y A) = -\frac{dP}{dy} S \, dy = -\frac{dP}{dy} dV \qquad (10.70)$$

where $S \, dy = dV$
Combination of Equations 10.62 and 10.70 gives

$$\mathscr{G}_\tau = -g_c \left(\frac{dP}{dy} \right) \qquad (10.71)$$

This equation shows that the rate of generation of momentum is equal to the decrease of pressure with distance in the y-direction. The pressure gradient can be easily measured in most cases. In the common engineering application of flow in pipes, the pressure gradient is maintained with pumps or compressors. The loss in pressure with distance represents a loss in the mechanical energy of the fluid. This energy is lost overcoming fluid friction and is transformed into heat.

If the fluid is incompressible, the density and velocity will be independent of pressure and

$$\mathscr{G}_\tau = -g_c \frac{\Delta P}{\Delta y} \qquad (10.72)$$

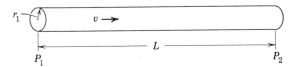

Figure 10.7. Momentum transfer in a fluid in laminar flow in a circular pipe.

Substitution of the momentum transport equation, 9.62, into Equation 10.62 gives

$$d\left(-\mu A \frac{dv}{dx}\right) = \mathscr{G}_\tau \, dV \qquad (10.73)$$

which may be integrated if the variations in A and V are known.

In the derivation discussed previously the external force on each plane of the volume element was considered. A corresponding balance of internal forces on each of the planes could have been written with the same result because each plane has associated with it two forces on opposite sides of planes, equal in magnitude but opposite in direction.

The most frequently encountered application of momentum transfer with internal generation is the flow of fluids through a circular pipe. This application has the same geometry as that considered in heat transfer with internal generation, and final equations of identical form result. Consider a pipe of inside radius r_1 and length L with an incompressible Newtonian fluid of viscosity μ (Figure 10.7). Pressure gages measure the difference in pressure of the fluid across the length L. Equation 10.73 may be used with the following substitutions

$$V = SL = \pi r^2 L \qquad \Delta P = P_2 - P_1$$

$$dV = 2\pi L r \, dr$$

$$A = 2\pi r L$$

$$\mathscr{G}_\tau = -g_c \frac{\Delta P}{\Delta y} \text{ (incompressible fluid)}$$

Then

$$d\left(-2\pi L \mu r \frac{dv}{dr}\right) = -\frac{\Delta P g_c}{\Delta y} 2\pi L r \, dr \qquad (10.74)$$

Integrating once gives

$$-2\pi L \mu r \frac{dv}{dr} = -\frac{\Delta P g_c}{\Delta y} \pi L r^2 + C_1 \qquad (10.75)$$

At the center of the pipe the velocity will be a maximum; therefore at $r = 0$ $dv/dr = 0$, and from Equation 10.75 $C_1 = 0$. Rearrangement of Equation 10.75 gives

$$\mu \frac{dv}{dr} = \frac{\Delta P g_c}{2 \Delta y} r \qquad (10.76)$$

which is of the form of Equation 10.56 and shows that the velocity gradient varies linearly with radial distance.

By substitution of the transport equation, Equation 10.76 becomes

$$(\tau_y g_c) = -\frac{\Delta P g_c}{2 \Delta y} r \qquad (10.77)$$

For the boundary at r_1

$$(\tau_y g_c)_1 = -\frac{\Delta P g_c}{2 \Delta y} r_1 \qquad (10.78)$$

Combining Equations 10.77 and 10.78 gives

$$\frac{\tau_y g_c}{(\tau_y g_c)_1} = \frac{r}{r_1} \qquad (10.79)$$

Therefore, the shear stress also varies linearly with radius. It is zero at the center of the pipe.

An integration of Equation 10.76 gives

$$v = \frac{\Delta P g_c r^2}{4 \Delta y \, \mu} + C_2 \qquad (10.80)$$

The integration constant C_2 may be evaluated from the boundary condition at $r = r_1$, $v = 0$. Since the pipe wall is stationary, the fluid particles immediately adjacent must have zero velocity. Therefore, with this boundary condition

$$C_2 = -\frac{\Delta P g_c r_1^2}{4 \Delta y \, \mu} \qquad (10.81)$$

and Equation 10.80 becomes

$$v = -\frac{\Delta P g_c}{4 \Delta y \, \mu} (r_1^2 - r^2) \qquad (10.82)$$

Examination of Equation 10.82 shows that the velocity distribution is parabolic. This equation is of the same form as Equation 10.60 for heat transfer.

It is usually more convenient to measure an average velocity of flow rather than a point velocity as defined in Equation 10.82. By definition, the average velocity of flow is equal to the total volumetric flow rate divided by the total flow area (S_1). Then, as defined by Equation 10.3,

$$\bar{v} S_1 = \int_0^{S_1} v \, dS \qquad (10.83)$$

Since $S = \pi r^2$, $dS = 2\pi r \, dr$, and $S_1 = \pi r_1^2$

$$\bar{v}(\pi r_1^2) = \int_0^{r_1} 2\pi v r \, dr \qquad (10.84)$$

Substituting v from Equation 10.82 into Equation 10.84 gives

$$\bar{v}(\pi r_1^2) = 2\pi \left(\frac{-\Delta P g_c}{4 \Delta y \, \mu}\right) \int_0^{r_1} (r_1^2 - r^2) r \, dr$$

$$= 2\pi \left(\frac{-\Delta P g_c}{4 \Delta y \, \mu}\right) \left(\frac{r_1^4}{2} - \frac{r_1^4}{4}\right)$$

and

$$\bar{v} = \frac{-\Delta P g_c r_1^2}{8 \Delta y \, \mu} \qquad (10.85)$$

Since $r_1 = D/2$ where D is the pipe diameter,

$$\bar{v} = \frac{-\Delta P g_c \, D^2}{32 \, \Delta y \, \mu} \qquad (10.86)$$

This equation expresses Poiseuille's law for laminar flow in circular ducts. It is useful in calculating the pressure loss in laminar flow and in determining the viscosity of a fluid using a flow-tube viscometer.

An equation which is useful in calculating the point velocity from the average velocity is obtained by combining Equations 10.82 and 10.85.

$$\frac{v}{\bar{v}} = 2\left(\frac{r_1^2 - r^2}{r_1^2}\right) = 2\left[1 - \left(\frac{r}{r_1}\right)^2\right] \qquad (10.87)$$

For non-Newtonian fluids where the viscosity varies with velocity gradient, an expression for the viscosity in terms of other variables must be substituted before the general Equation 10.73 may be integrated. Since the viscosities of many non-Newtonian fluids do not vary in a simple manner, many of the resultant equations cannot be integrated analytically.

Illustration 10.8. An oil is in laminar flow in a $\frac{1}{2}$-in. I.D. tube at 6 gal/min. The oil viscosity is 300 centipoises, and its density is 60 lb/cu ft. Calculate:

(a) The pressure drop per foot of pipe length (lb$_f$/sq in. ft).
(b) The wall stress (lb$_f$/sq ft).
(c) The velocity at the center of the tube.
(d) The radial position at which the point velocity is equal to the average velocity.

SOLUTION. The average velocity is calculated by dividing the volumetric flow rate by the flow area.

$$\bar{v} = \frac{\left(\dfrac{6 \text{ gal}}{\text{min}}\right)\left(\dfrac{1 \text{ cu ft}}{7.48 \text{ gal}}\right)\left(\dfrac{1 \text{ min}}{60 \text{ sec}}\right)}{\left[\dfrac{\pi}{4}(1/2 \times 1/12)^2 \text{ sq ft}\right]} = 9.8 \text{ ft/sec}$$

(a) From Equation 10.86

$$-\frac{\Delta P}{\Delta y} = \frac{32\bar{v}\mu}{D^2 g_c} = \frac{(32)(9.8)(300 \times 0.000672)}{(1/2 \times 1/12)^2(32.2)}$$

$$= 1130 \text{ lb}_f/\text{sq ft per foot of pipe}$$

or $\qquad = 7.84 \text{ lb}_f/\text{sq in per foot of pipe}$

(b) From Equation 10.78, the momentum flux at the wall is

$$(\tau_y g_c)_1 = \frac{-\Delta P g_c}{2 \, \Delta y} r_1$$

$$= \frac{(1130)(32.2)(1/48)}{2}$$

$$= 380 \frac{\text{ft-lb/sec}}{\text{sq ft sec}}$$

and the shear stress is

$$\tau_y = \frac{380}{32.2} = 1.18 \text{ lb}_f/\text{sq ft}$$

(c) From Equation 10.87 at $r = 0$

$$v = 2\bar{v}\left[1 - \left(\frac{r}{r_1}\right)^2\right]$$

$$= (2)(9.8)(1 - 0)$$

$$= 19.6 \text{ ft/sec}$$

(d) From Equation 10.87 when $v = \bar{v}$

$$\bar{v} = 2\bar{v}\left[1 - \left(\frac{r}{r_1}\right)^2\right]$$

$$1 - \left(\frac{r}{1/4}\right)^2 = \frac{1}{2}$$

$$r = 0.177 \text{ in.}$$

Mass Transfer. Examples of mass transfer with internal generation can be found in systems in which diffusion and a chemical or nuclear reaction occur simultaneously. As mentioned earlier, the generation and diffusion of neutrons in a nuclear reactor is an illustration of mass transfer with internal generation. Of more interest to the chemical engineer is diffusion with a chemical reaction. An example of this is the absorption of sulfur trioxide by water to form sulfuric acid. Both absorption and reaction occur simultaneously, and both phenomena progress at a given rate. The sulfur trioxide will diffuse a short distance in the liquid water before it reacts. When it reacts, it disappears. This is mass transfer with *negative* generation.

The general equation may be written for mass transfer with internal generation

$$d(N_a) = \mathscr{G}_N \, dV \qquad (10.88)$$

where $\mathscr{G}_N$ is the volumetric rate of internal generation of mass and N_a is the rate of transport. In the typical case $\mathscr{G}_N$ varies and must be written as a variable in terms of distance x or composition c_a.

Consider the case of the equimolar counterdiffusion of component a through a region in which the component undergoes a slow first-order irreversible decomposition $(a \rightarrow b)$. The rate of the decomposition will be determined by the first-order mechanism, which states that the rate of change of concentration of a is directly proportional to the concentration of a,

$$\frac{dc_a}{d\theta} = -kc_a \qquad (10.89)$$

where k is an experimentally determined constant.

The minus sign shows that the concentration is decreasing with time. The left side of the equation is equal to the change in moles of a per unit time and unit volume. This is exactly equal to the internal generation rate of mass ($\mathscr{G}_N$), that is,

$$\mathscr{G}_N = \frac{dc_a}{d\theta} \qquad (10.90)$$

Therefore,
$$\mathcal{G}_N = -kc_a \qquad (10.91)$$

The minus sign indicates *negative* generation, i.e., decomposition or disappearance.

Combining the transport equation with Equations 10.88 and 10.91 gives

$$d\left(-\mathcal{D}A\frac{dc_a}{dx}\right) = -kc_a \, dV \qquad (10.92)$$

Taking $\mathcal{D}$ and A constant and $V = Ax$ for a slab geometry gives

$$d\left(-\mathcal{D}A\frac{dc_a}{dx}\right) = -kc_a(A \, dx)$$

and
$$\frac{d}{dx}\left(\frac{dc_a}{dx}\right) - \frac{k}{\mathcal{D}}c_a = 0$$

or
$$\frac{d^2c_a}{dx^2} - \frac{k}{\mathcal{D}}c_a = 0 \qquad (10.93)$$

This is a linear second-order differential equation with constant coefficients. By routine analytical techniques this equation is integrated to

$$c_a = C_1 e^{\sqrt{k/\mathcal{D}}\,x} + C_2 e^{-\sqrt{k/\mathcal{D}}\,x} \qquad (10.94)$$

where C_1 and C_2 are integration constants which must be evaluated from boundary conditions. With the boundary conditions $c_a = c_{a1}$ at plane 1 ($x = 0$) and $c_a = c_{a2}$ at plane 2 ($x = x_2$), Equation 10.94 becomes

$$c_a = \frac{c_{a2}\sinh\left(\sqrt{\frac{k}{\mathcal{D}}}\,x\right) + c_{a1}\sinh\left(\sqrt{\frac{k}{\mathcal{D}}}\,(x_2 - x)\right)}{\sinh\left(\sqrt{\frac{k}{\mathcal{D}}}\,x_2\right)} \qquad (10.95)$$

where the hyperbolic sine,

$$\sinh\left(\sqrt{\frac{k}{\mathcal{D}}}\,x\right) = \frac{e^{\sqrt{k/\mathcal{D}}\,x} - e^{-\sqrt{k/\mathcal{D}}\,x}}{2}$$

Equation 10.95 may be used to calculate the composition c_a at any point x between planes 1 and 2.

Since component *a* decomposes to component *b*, component *b* undergoes internal generation and will also diffuse. This could be treated by the methods illustrated for component *a*.

If the rate of the chemical reaction cannot be expressed by a simple first-order equation such as Equation 10.89, the integration of the resulting differential equation may be difficult. Additional discussion of diffusion and chemical reaction may be found in reference 1.

REFERENCE

1. Sherwood, T. K., and R. L. Pigford, *Absorption* and *Extraction*, McGraw-Hill Book Co., New York, 1952.

PROBLEMS

10.1. Derive an expression for the mean area of transfer through the wall of a thick-walled hollow sphere of radii r_1 and r_2.

10.2. Ammonia is being absorbed from an air–ammonia mixture by a sulfuric acid solution. The concentration of ammonia in the air 1 in. from the surface of the acid is 40 volume percent. The concentration at the acid surface is 0 per cent, since the ammonia reacts with the acid. The total pressure of the system is 400 mm Hg and the temperature is 60°F.

(*a*) Calculate the rate of absorption of ammonia across 0.5 sq ft of acid surface.

(*b*) Calculate the concentration of ammonia $\frac{1}{2}$ in. from the acid surface.

(*c*) Calculate the rate of transfer of ammonia by molecular transport and the rate of transfer by bulk flow 1 in. from the acid surface.

(*d*) Repeat part (*c*) at the acid surface.

10.3. A well located in the desert is 35 ft deep to the water level and 3 ft in diameter. The stagnant air and the water in the well are at 90°F and 1 atm pressure. A slight breeze of dry air is blowing across the top of the well. Calculate the rate (lb/hr) of diffusion at steady state of water vapor from the surface of water in the well. (Assume the partial pressure of water vapor in the air at the surface of the water is equal to the vapor pressure of water at 90°F.)

10.4. Calculate the rate of diffusion of sodium chloride at 20°C through a stationary film of water 1 mm thick, where the concentrations are 20 and 10 weight percent, respectively, on either side of the film.

10.5. Sulfuric acid is diffusing through liquid water at a rate of 6.0×10^{-6} lb moles/hr sq ft. What will be the concentration of acid 1 in. from a point where the concentration is 5 mass percent acid? (Assume the density of the solution is that of water and that the diffusion is one dimensional. The temperature is 68°F.)

10.6. A chimney brick is shown in the sketch. Heat is transferred from the 3-in. end to the 5-in. end. No heat is lost through

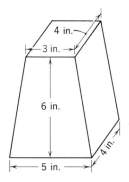

the other sides, since these sides adjoin other bricks. The 3-in. side is at 800°F since it is exposed to hot gases inside the chimney. The 5-in. side is exposed to the atmosphere and is at 200°F. The thermal conductivity of the brick is 0.4 Btu/hr sq ft (°F/ft). Calculate the heat loss per hour through the brick.

10.7. Steam at 500°F is flowing through a thick-walled nickel pipe. The temperature of the outside pipe surface is 200°F. The inside diameter of the pipe is 1 in. and the outside diameter is $1\frac{1}{4}$ in.

(*a*) How much heat is lost from 20 ft of this pipe in 24 hr?

(*b*) What is the temperature halfway through the pipe wall?

10.8. A spherical furnace has an inside radius of 3 ft and an outside radius of 4 ft. The thermal conductivity of the wall is 0.10 Btu/hr sq ft (°F/ft). The inside furnace temperature is 2000°F; the outside surface is at 175°F.

(a) Calculate the total heat loss for 24 hr of operation.

(b) What is the heat flux and temperature at a radius of $3\frac{1}{2}$ ft?

10.9. A Couette viscometer ($r_1 = 1$ in., $r_2 = 1.25$ in. and length = 4 in.) is filled with a fluid of viscosity 3000 centipoises.

(a) Calculate the moment on the bob if the cup rotates at 20 rpm.

(b) Calculate the moment on the bob, if the bob rotates at 20 rpm in the same fluid contained in a large tank ($r_2 = \infty$).

10.10. The viscosity of a liquid is to be determined with the following data taken on a Couette viscometer.

Viscometer: $r_1 = \frac{1}{2}$ in., $r_2 = \frac{3}{4}$ in.,

length = 5 in.

The measured moment on the bob is -0.005 ft-lb$_f$ when the cup is rotating at 30 rpm.

10.11. An electrically heated resistance wire has a diameter of 2 mm and a resistance of 0.10 ohm per foot of wire. The thermal conductivity of the wire is 10 Btu/hr sq ft (°F/ft). At a current of 100 amp, calculate the temperature difference between the center and surface of the wire at steady state.

10.12. Thin flat plates of uranium are used as fuel elements in nuclear reactors. Heat is generated uniformly within the uranium metal by fission. This heat flows to the surface of the metal and is removed by a liquid coolant. Consider a fuel element which is 3 mm thick whose surface temperatures are both 200°F. The volumetric heat generation rate ($\mathscr{G}_q$) is 2×10^8 Btu/hr cu ft. Calculate and plot the temperature profile across the 3-mm thickness, starting at one surface and ending at the other.

Uranium: $\rho = 1155$ lb/cu ft; $k = 17.5$ Btu/hr sq ft (°F/ft);

$c_p = 0.032$ Btu/lb°F

10.13. Heat is generated within a sphere at 2×10^8 Btu/hr cu ft. The sphere is 3 in. in diameter. The surface temperature is 200°F. The thermal conductivity is 200 Btu/hr sq ft (°F/ft).

(a) Calculate the temperature at the center of the sphere.

(b) Calculate the temperature at $r = \frac{3}{4}$ in.

10.14. For volumetric heat generation in a flat slab whose surfaces are at different temperatures as shown,

(a) Derive an expression for T in terms of x and the constants of the system.

(b) Derive an expression for the value of x at which T is a maximum.

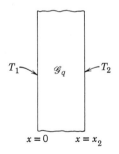

$x = 0 \qquad x = x_2$

10.15. A heavy oil is pumped through a pipe with a 2-in. inside diameter. The pressure drop over 10 ft of pipe is 10 lb/sq in. The viscosity of the oil is 200 centipoises and the density is 50 lb/cu ft.

(a) Calculate the volumetric flow rate of oil through the pipe (cu ft/min).

(b) Calculate and plot the momentum flux profile across the pipe.

10.16. A fluid of viscosity 200 centipoises is flowing between two flat plates $\frac{1}{2}$ in. apart. The plates are 1 ft by 1 ft in area. The average velocity is 4 ft/sec. Calculate the pressure drop. Plot the velocity and the stress as functions of distance across the space between the plates.

10.17. 20 gal/hr of benzene at 70°F is flowing through a pipe of 0.957 in. inside diameter.

(a) What is the pressure drop (psi) through 100 ft of pipe?

(b) What is the velocity (ft/sec) at the center of the pipe?

(c) What is the momentum flux at the pipe wall?

10.18. A common channel for fluid flow is the *annulus*, the space between two concentric circular pipes, as shown. The inside of the annulus has a radius r_1 and the outside, r_2.

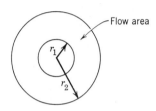

—Flow area

(a) Derive an expression for the velocity of flow v at any radius r in the annulus.

(b) Derive an expression for the radius at which the maximum velocity occurs, in terms of r_1 and r_2.

10.19. Derive an expression for the average velocity of flow between parallel flat plates of the form of Equation 10.85. Let the channel be $2x_1$ high, Δy long, and Δz wide. The channel is much wider than it is high, so edge effects may be neglected.

10.20. Equation 9.68 is a general equation for dilatant, Newtonian, and pseudoplastic liquids, depending upon the value of the exponent n. Consider a liquid flowing between two stationary flat plates spaced 2 in. apart.

(a) Plot the shear stress and velocity profiles for each of the three liquids.

(b) List typical dimensions for K for each liquid.

Data: Assume that the velocity at the center (v_0) is 1 ft/sec and that $K = 1$ in dimensions appropriate for each fluid.

Dilatant: $n = 2.0$
Newtonian: $n = 1.0$
Pseudoplastic: $n = 0.5$

10.21. Show that if there is no reaction ($k = 0$) Equation 10.95 reduces to an expression for c_a at any value of x for simple diffusion, which is

$$\frac{c_a - c_{a1}}{c_{a2} - c_{a1}} = \frac{x}{x_2}$$

chapter 11

Applications of Molecular-Transport Theory to the Unsteady State

Transport processes wherein the concentration of transferent property at a point varies with time are referred to as *unsteady-state* processes. The variation in concentration is accompanied by a variation in the flux and rate of transport. Chapter 10 considered *steady-state* molecular transfer of heat, mass, and momentum, where the concentrations were constant with time. This chapter will develop equations which express the *variation* with time of the concentration of transferent property. The unsteady-state differential equations are simple to derive; however, their solution in most cases involves more complex mathematics than the steady-state equations.

Many common examples of unsteady-state transfer may be cited. Consider an electric heater which is turned on after having been idle. Before it is turned on, the resistance element is at the temperature of the air in the room. When the current to the element is turned on, the resistance to the current generates heat within the element. During the heating period, the energy supplied to the element is distributed in two ways. Some of the energy accumulates within the element and results in an increase in temperature of the element. As the energy is accumulated, the higher temperature of the element will result in a temperature gradient between it and the surroundings, and heat will be transferred from the element to the surroundings. Ultimately the temperature of the element will reach a value where the rate of heat transfer to the surroundings will equal the rate of energy input, and steady state is attained. Such an unsteady-state start-up period precedes steady-state operations in many cases. As the element heats up, the concentration of thermal energy increases, and the heat flux increases until a steady state is reached. If the current to the element is shut off, its temperature will again approach that of the surroundings as it cools off by unsteady-state heat transfer. The time of approach to steady state depends upon the capacity of the system and upon the gradient available. For example, it would take a longer time for a heavier heating element to reach steady state at the same rate of energy input.

Unsteady-state transfer is also important in batch systems. For example, consider a thermometer suddenly placed in an ice bath. Initially, the surface temperature of the bulb is high compared to that of the ice bath, and the rate of heat transfer from the bulb to the ice is rapid. The thermal energy in the bulb decreases (a negative accumulation) as the heat is transferred. Ultimately, the temperature of the bulb reaches that of the bath and the flux becomes zero.

For unsteady-state transfer, the balance equation is

$$\text{Input} = \text{output} + \text{accumulation} \qquad (11.1)$$

When steady state is reached, the accumulation is zero, and input = output. The accumulation may have a positive or negative value. In the heating element, the current supplies the heat input, and the heat transfer to the surroundings is the output. The accumulation is positive, since the thermal-energy content increases as the temperature rises. For the thermometer bulb, the input is zero, and the output is the heat transferred to the ice bath. The accumulation is negative, since the bulb loses energy to the bath.

THE GENERAL UNSTEADY-STATE TRANSPORT EQUATION

The general equation for simple unsteady-state molecular transport can be derived using Equation 11.1. This equation may be written as

$$\text{Rate of input} - \text{rate of output} = \text{rate of accumulation} \qquad (11.1a)$$

Each of the rates may be expressed in general terms by considering the volume element pictured in Figure 11.1. The rate of input into the volume element is equal to the flux times the transport area (A). From Equation 9.12,

$$\psi A = -\frac{1}{6} l\bar{c}A \frac{d\Gamma}{dx} = -\delta A \frac{d\Gamma}{dx} \qquad (9.12b)$$

where δ is the generalized diffusivity. This expression is valid for steady or unsteady state. For simple transfer at steady state $d\Gamma/dx$ and ψA are constant, but at unsteady state they vary with time and position. The gradient at plane 1 is $(d\Gamma/dx)_1$ and the rate of input is

$$(\psi A)_1 = -\delta \, \Delta y \, \Delta z \left(\frac{d\Gamma}{dx}\right)_1 \qquad (11.2)$$

where $A = \Delta y \, \Delta z$. The gradient at plane 2 is $(d\Gamma/dx)_2$ and the rate of output is

$$(\psi A)_2 = -\delta \, \Delta y \, \Delta z \left(\frac{d\Gamma}{dx}\right)_2 \qquad (11.3)$$

The rate of accumulation will be equal to the increase in transferent property in the volume element per unit time. The total transferent property in the element is equal to the concentration times the volume, i.e., $\Gamma \, dx \, \Delta y \, \Delta z$. Then,

$$\text{Rate of accumulation} = d(\Gamma \, dx \, \Delta y \, \Delta z)/d\theta \qquad (11.4)$$

Substituting Equations 11.2, 11.3, and 11.4 into the over-all balance Equation 11.1 gives

$$(\psi A)_1 - (\psi A)_2 = d(\Gamma \, dx \, \Delta y \, \Delta z)/d\theta \qquad (11.5)$$

Substitution for the rates of transport in terms of the concentration gradients gives

$$-\left[\delta \, \Delta y \, \Delta z \left(\frac{d\Gamma}{dx}\right)\right]_1 + \left[\delta \, \Delta y \, \Delta z \left(\frac{d\Gamma}{dx}\right)\right]_2 = \frac{d\Gamma}{d\theta} \, dx \, \Delta y \, \Delta z$$

or

$$-\left[\delta\left(\frac{d\Gamma}{dx}\right)\right]_1 + \left[\delta\left(\frac{d\Gamma}{dx}\right)\right]_2 = \frac{d\Gamma}{d\theta} \, dx \qquad (11.6)$$

In the left-hand term of Equation 11.6, Γ is a function of x alone, but in the right-hand term Γ is a function of θ alone. Therefore, to indicate this, Equation 11.6 must be written in the partial-derivative form

$$-\left[\delta\left(\frac{\partial\Gamma}{\partial x}\right)\right]_1 + \left[\delta\left(\frac{\partial\Gamma}{\partial x}\right)\right]_2 = \left(\frac{\partial\Gamma}{\partial\theta}\right) dx \quad (11.7)$$

The fluxes at the two planes may be related by

$$\left[\delta\left(\frac{\partial\Gamma}{\partial x}\right)\right]_2 = \left[\delta\left(\frac{\partial\Gamma}{\partial x}\right)\right]_1 + \frac{\partial}{\partial x}\left[\delta\left(\frac{\partial\Gamma}{\partial x}\right)\right] dx \quad (11.8)$$

Combination of Equations 11.7 and 11.8 gives

$$\frac{\partial}{\partial x}\left[\delta\left(\frac{\partial\Gamma}{\partial x}\right)\right] = \frac{\partial\Gamma}{\partial\theta} \qquad (11.9)$$

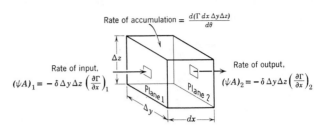

Figure 11.1. Unsteady-state molecular transport.

If δ is constant, then Equation 11.9 becomes

$$\delta \frac{\partial^2\Gamma}{\partial x^2} = \frac{\partial\Gamma}{\partial\theta} \qquad (11.10)$$

Equation 11.10 is the general unsteady-state molecular-transport equation. It states that the change of concentration with time is proportional to the change in concentration gradient with position.

The unsteady-state equation may be written in the specific terms of mass, heat, and momentum transport.

Mass:
$$\frac{\partial c_a}{\partial\theta} = \mathscr{D} \frac{\partial^2 c_a}{\partial x^2} \qquad (11.11)$$

Heat:
$$\frac{\partial(\rho c_P T)}{\partial\theta} = \alpha \frac{\partial^2(\rho c_P T)}{\partial x^2} \qquad (11.12)$$

or
$$\frac{\partial T}{\partial\theta} = \alpha \frac{\partial^2 T}{\partial x^2} \qquad (11.12a)$$

Momentum:
$$\frac{\partial(v\rho)}{\partial\theta} = \nu \frac{\partial^2(v\rho)}{\partial x^2} \qquad (11.13)$$

or
$$\frac{\partial v}{\partial\theta} = \nu \frac{\partial^2 v}{\partial x^2} \qquad (11.13a)$$

Equation 11.11 is generally referred to as Fick's law for unsteady-state mass transfer, and Equation 11.12a is Fourier's law. Equation 11.13a is a form of Euler's law for unsteady-state momentum transport.

The unsteady-state equations reduce to the familiar steady-state form if the accumulation is zero. For example, if the accumulation in molecular transport is zero, $d\Gamma/d\theta = 0$ and Equation 11.10 becomes

$$0 = \delta \frac{\partial^2\Gamma}{\partial x^2} \qquad (11.14)$$

Integration gives

$$\text{constant} = \delta \frac{d\Gamma}{dx} \qquad (11.15)$$

Thus, the gradient is a constant, and this expression is equivalent to Equation 9.12.

Equations 11.11, 11.12, and 11.13 were derived for transport in the x-direction only. In general, transport may be in all three directions and the following equations may be developed:

General:
$$\frac{\partial \Gamma}{\partial \theta} = \delta \left[\frac{\partial^2 \Gamma}{\partial x^2} + \frac{\partial^2 \Gamma}{\partial y^2} + \frac{\partial^2 \Gamma}{\partial z^2} \right] = \delta \nabla^2 \Gamma$$
(11.16)

Mass:
$$\frac{\partial c_a}{\partial \theta} = \mathscr{D} \left[\frac{\partial^2 c_a}{\partial x^2} + \frac{\partial^2 c_a}{\partial y^2} + \frac{\partial^2 c_a}{\partial z^2} \right] = \mathscr{D} \nabla^2 c_a$$
(11.17)

Heat:
$$\frac{\partial T}{\partial \theta} = \alpha \left[\frac{\partial^2 T}{\partial x^2} + \frac{\partial^2 T}{\partial y^2} + \frac{\partial^2 T}{\partial z^2} \right] = \alpha \nabla^2 T$$
(11.18)

Momentum:
$$\frac{\partial v}{\partial \theta} = \nu \left[\frac{\partial^2 v}{\partial x^2} + \frac{\partial^2 v}{\partial y^2} + \frac{\partial^2 v}{\partial z^2} \right] = \nu \nabla^2 v \quad (11.19)$$

where
$$\nabla^2 = \left[\frac{\partial^2}{\partial x^2} + \frac{\partial^2}{\partial y^2} + \frac{\partial^2}{\partial z^2} \right]$$

SOLUTION OF THE UNSTEADY-STATE EQUATIONS

Although the unsteady-state equations are relatively easy to set up, their solution is in many cases very difficult. The solution is dependent on the geometry and boundary conditions. The equations have been solved for a number of simple geometries and boundary conditions. A detailed discussion of the mathematics involved is inappropriate here. The reader is referred to books on applied mathematics, such as reference 1.

Analytical Solution. Working with Equation 11.12a, Fourier developed an analytical solution which utilizes an infinite trigonometric series, now known as the Fourier series. For a flat slab of thickness $2x_1$, where the slab is originally at temperature $_0T$ and is suddenly exposed to surroundings at time zero such that its surface temperature is T_1, the solution to Equation 11.12a is

$$\frac{T_1 - T}{T_1 - {_0}T} = \frac{4}{\pi} \left[e^{-\alpha \pi^2 \theta / 4x_1{}^2} \sin \frac{\pi x}{2x_1} \right.$$
$$+ \tfrac{1}{3} e^{-9\alpha \pi^2 \theta / 4x_1{}^2} \sin \frac{3\pi x}{2x_1}$$
$$+ \tfrac{1}{5} e^{-25\alpha \pi^2 \theta / 4x_1{}^2} \sin \frac{5\pi x}{2x_1}$$
$$\left. + \ldots + \text{etc.} \right]$$
(11.20)

where
T = temperature at point x at time θ
$_0T$ = original uniform temperature of the slab at all values of x
T_1 = new surface temperature of the slab at all values of θ
x = distance from center plane to any point in the slab
x_1 = distance from center plane to surface.

The expression may be evaluated for any value of x and θ.

Table 11.1. NOTATION FOR FIGURES 11.2, 11.3, AND 11.4

Parameter Symbol	General Notation	Specific Notations		
		Heat	Mass	Momentum
Unaccomplished change Y	$\dfrac{\Gamma_1 - \Gamma}{\Gamma_1 - {_0}\Gamma}$	$\dfrac{T_1 - T}{T_1 - {_0}T}$	$\dfrac{c_{a1} - c_a}{c_{a1} - {_0}c_a}$	$\dfrac{v_1\rho_1 - v\rho}{v_1\rho_1 - {_0}v_0\rho}$
Relative time X	$\dfrac{\delta\theta}{x_1{}^2}$	$\dfrac{\alpha\theta}{x_1{}^2}$	$\dfrac{\mathscr{D}\theta}{x_1{}^2}$	$\dfrac{\nu\theta}{x_1{}^2}$
Relative position n	$\dfrac{x}{x_1}$	$\dfrac{x}{x_1}$	$\dfrac{x}{x_1}$	$\dfrac{x}{x_1}$
Relative resistance m	$\dfrac{\delta}{\mathscr{E}x_1}$	$\dfrac{k}{hx_1}$	$\dfrac{\mathscr{D}}{k_c x_1}$	—

X, Y, m, n are dimensionless

T = temperature
c_a = concentration
v = velocity
x = distance from center to any point
θ = time
k = thermal conductivity

h, k_c = transfer coefficients*
$\mathscr{E}$ = generalized transfer coefficient
α = thermal diffusivity
$\mathscr{D}$ = mass diffusivity
ν = momentum diffusivity
δ = generalized diffusivity

The assumptions of the solution are seldom met in momentum transport

Postsubscripts:
  1 = boundary
  a = component a

Presubscript:
  0 = at time 0 (initial condition)

Generally, the presubscript will refer to time, the postsubscript to position. For momentum transport $v_1 = 0$.

---

* These terms will be defined in later chapters. For this chapter $m = 0$. The use of the charts for $m > 0$ will be demonstrated in Chapter 14.

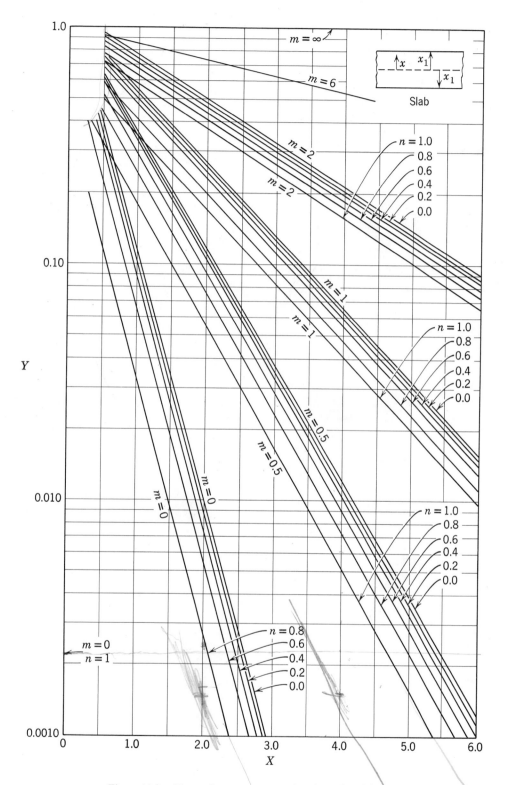

Figure 11.2. Unsteady-state transport in a large flat slab.

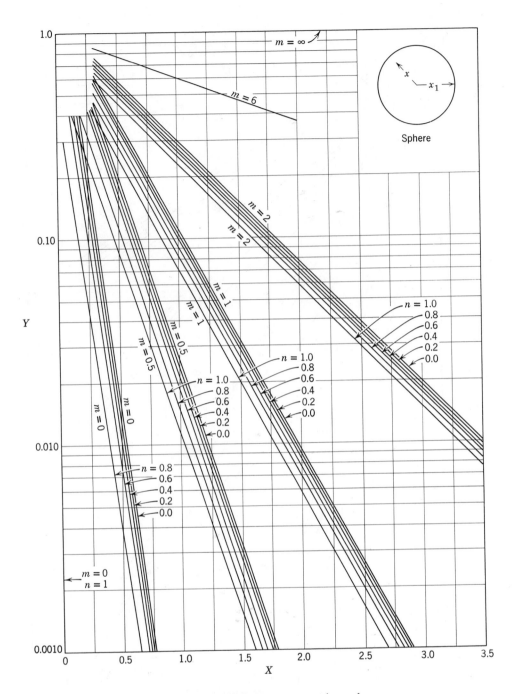

Figure 11.3. Unsteady-state transport in a sphere.

To facilitate calculation, solutions to the unsteady-state equations have been prepared in graphical form for a few simple geometries. Figure 11.2 presents the solution for the flat slab, Figure 11.3 for the sphere, and Figure 11.4 for the cylinder. The charts may be used only if the following conditions are met:

(a) The diffusivity is constant.

(b) The body initially has a uniform concentration of transferent property.

(c) The boundary is brought to a new condition which is constant for the entire time.

The notation for use of the charts is given in Table 11.1. The charts are only approximate since a limited number of terms of the infinite series was used. The charts were originally prepared by Gurney and Lurie (2). The use of the charts will be shown in the following illustration.

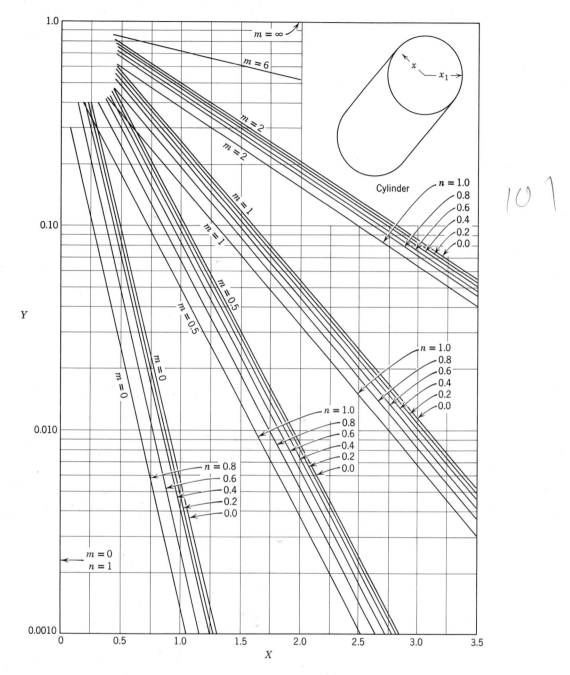

Figure 11.4. Unsteady-state transport in a long cylinder.

Illustration 11.1. A cylindrical steel shaft 4 in. in diameter and 8 ft long is heat treated to give it desired physical properties. It is heated to a uniform temperature of 1100°F and then plunged into an oil bath which maintains the surface temperature at 300°F. Calculate the radial temperature profile at 2 min and at 3 min after immersion.

SOLUTION. First it is necessary to check that the three conditions for use of the charts are met.

(*a*) Constant diffusivity. The rod will vary in temperature between 1100°F and 300°F. It is desirable to check α at these two temperatures.

	c_P, $\dfrac{\text{Btu}}{\text{lb°F}}$	ρ, $\dfrac{\text{lb}}{\text{cu ft}}$	k, Btu/hr sq ft (°F/ft)	α, $\dfrac{\text{sq ft}}{\text{hr}}$
300°F	0.122	487	26	0.44
1100°F	0.173	480	21	0.25

This is a substantial difference in thermal diffusivities. Since no other simple method is available for calculation, an average of these values will be used and assumed constant.

Let α = 0.35 sq ft/hr.

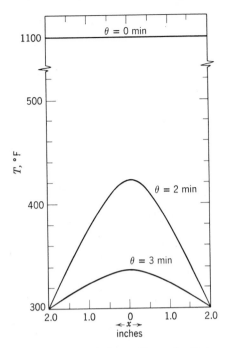

Figure 11.5. Solution to Illustration 11.1.

(b) The rod is initially at a uniform temperature of $_0T = 1100°F$.

(c) The boundary is maintained at the new temperature of $T_1 = 300°F$.

Figure 11.4 will be used and the temperature at various values of position will be evaluated.

For $\theta = 2$ min,

$$\mathbf{Y} = \frac{T_1 - T}{T_1 - {}_0T} = \frac{300 - T}{300 - 1100}$$

$$\mathbf{X} = \frac{\alpha\theta}{x_1^2} = \frac{(0.35)\left(\dfrac{2}{60}\right)}{\left(\dfrac{2}{12}\right)^2} = 0.42$$

$$\mathbf{n} = \frac{x}{2} \quad \text{where } x \text{ is in inches}$$

$$\mathbf{m} = 0$$

For $x = 1$ in., $n = \frac{1}{2}$ and from Figure 11.4

$$\mathbf{Y} = 0.08 = \frac{300 - T}{300 - 1100}$$

and $T = 364°F$

Additional points at 2 min and at 3 min may be calculated. The resulting profiles are shown in Figure 11.5. Theoretically it would require an infinite time to cool the rod uniformly to 300°F. In practice, after 6 min the center of the rod is at about 301°F. This solution is only approximate, since an average thermal diffusivity was used.

Numerical and Graphical Solutions. Differential equations may be solved approximately by use of the calculus of finite differences. In certain cases the finite-difference equations may be solved graphically. The general differential equation, Equation 11.10, may be rewritten in terms of finite differences.

$$\frac{\Delta\Gamma_x}{\Delta\theta} = \delta\frac{\Delta\left(\dfrac{\Delta_0\Gamma}{\Delta x}\right)}{\Delta x} \tag{11.21}$$

The finite-difference terms may be expanded

$$\frac{\Delta\Gamma_x}{\Delta\theta} = \delta\frac{\left(\dfrac{\Delta_0\Gamma}{\Delta x}\right)_B - \left(\dfrac{\Delta_0\Gamma}{\Delta x}\right)_A}{\Delta x} \tag{11.22}$$

where $(\Delta_0\Gamma/\Delta x)_B$ = value at position B and time θ (Figure 11.6)

$(\Delta_0\Gamma/\Delta x)_A$ = value at position A and time θ (Figure 11.6)

Examination of Figure 11.6 will show the meaning of these finite differences. Further expansion gives

$$\frac{_{\theta+\Delta\theta}\Gamma_x - {}_\theta\Gamma_x}{\Delta\theta} = \frac{\delta\left(\dfrac{_\theta\Gamma_{x+\Delta x} - {}_\theta\Gamma_x}{\Delta x} - \dfrac{_\theta\Gamma_x - {}_\theta\Gamma_{x-\Delta x}}{\Delta x}\right)}{\Delta x} \tag{11.23}$$

where $_\theta\Gamma_x$ = concentration at any position x, time θ.

$_{\theta+\Delta\theta}\Gamma_x$ = concentration at position x at time an interval $\Delta\theta$ later than θ.

Solution of Equation 11.23 for $_{\theta+\Delta\theta}\Gamma_x$ gives

$$_{\theta+\Delta\theta}\Gamma_x = \frac{\delta\,\Delta\theta}{\Delta x^2}\left[{}_\theta\Gamma_{x+\Delta x} - 2{}_\theta\Gamma_x + {}_\theta\Gamma_{x-\Delta x}\right] + {}_\theta\Gamma_x \tag{11.24}$$

If a modulus $\mathcal{M}$ is defined

$$\mathcal{M} = \frac{\Delta x^2}{\delta\,\Delta\theta} \tag{11.25}$$

Equation 11.24 becomes

$$_{\theta+\Delta\theta}\Gamma_x = \frac{{}_\theta\Gamma_{x+\Delta x} + (\mathcal{M} - 2){}_\theta\Gamma_x + {}_\theta\Gamma_{x-\Delta x}}{\mathcal{M}} \tag{11.26}$$

This equation permits the prediction of the value of concentration at a point from the values of concentration at that point and those adjacent to it at a time $\Delta\theta$ preceding. The left-hand term is at time $\theta + \Delta\theta$, the others at time θ. Therefore, given an initial distribution of concentration, it is possible to calculate the distribution as time passes by taking increments of time $\Delta\theta$.

Equation 11.26 may also be derived by considering a transferent-property balance around the volume between planes A and B in Figure 11.6. In a time interval $(\Delta\theta)$ the input is equal to the rate times the time interval.

$$-\delta A\left(\frac{\Delta\Gamma}{\Delta x}\right)_A \Delta\theta = -\delta A\frac{({}_\theta\Gamma_x - {}_\theta\Gamma_{x-\Delta x})}{\Delta x}\Delta\theta$$

The output is

$$-\delta A\left(\frac{\Delta \Gamma}{\Delta x}\right)_B \Delta\theta = -\delta A\left(\frac{{}_\theta\Gamma_{x+\Delta x} - {}_\theta\Gamma_x}{\Delta x}\right)\Delta\theta$$

The accumulation is equal to the volume times the change in concentration in the interval $\Delta\theta$.

$$(\Delta\Gamma_x)(A\ \Delta x) = ({}_{\theta+\Delta\theta}\Gamma_x - {}_\theta\Gamma_x)(A\ \Delta x)$$

Since the input = output + accumulation,

$$-\delta A\frac{({}_\theta\Gamma_x - {}_\theta\Gamma_{x-\Delta x})}{\Delta x}\Delta\theta = \delta A\frac{({}_\theta\Gamma_{x+\Delta x} - {}_\theta\Gamma_x)}{\Delta x}\Delta\theta$$

$$+ ({}_{\theta+\Delta\theta}\Gamma_x - {}_\theta\Gamma_x)(A\ \Delta x)\quad(11.27)$$

Solving for ${}_{\theta+\Delta\theta}\Gamma_x$ yields Equation 11.24.

If the modulus $\mathscr{M}$ is chosen to equal 2, Equation 11.24 becomes

$$_{\theta+\Delta\theta}\Gamma_x = \frac{{}_\theta\Gamma_{x+\Delta x} + {}_\theta\Gamma_{x-\Delta x}}{2}\quad(11.28)$$

Thus, the value of Γ_x is equal to the arithmetic average of the concentration on either side of it one time interval earlier. Setting $\mathscr{M} = 2$ gives $\Delta x^2/\delta\ \Delta\theta = 2$, so that if the distance interval Δx is chosen, the time interval $\Delta\theta$ is fixed. Any value of $\mathscr{M}$ greater than 2 may be chosen and used in a numerical solution. However, $\mathscr{M} = 2$ permits a simple *graphical* solution.

Equation 11.28 may be used with any initial-concentration distribution. This gives it an advantage over the analytical solution and charts presented earlier in the chapter, since they required a uniform initial concentration. However, the graphical method presented here is limited to flat slabs.

Equation 11.28 may be rewritten in specific terms for

Mass transfer: $\quad _{\theta+\Delta\theta}c_{a,x} = \dfrac{{}_\theta c_{a,x+\Delta x} + {}_\theta c_{a,x-\Delta x}}{2}$ (11.29)

Heat transfer: $\quad _{\theta+\Delta\theta}T_x = \dfrac{{}_\theta T_{x+\Delta x} + {}_\theta T_{x-\Delta x}}{2}$ (11.30)

Momentum transfer: $\quad _{\theta+\Delta\theta}v_x = \dfrac{{}_\theta v_{x+\Delta x} + {}_\theta v_{x-\Delta x}}{2}$

(11.31)

The graphical application of these equations is usually called the Schmidt method (3, 4). Further information on numerical methods may be found in reference 5.

Special Treatment for the Surface Concentration. This convenient graphical technique is based on proper choice of time and distance intervals such that the concentration at a point one time interval later is equal to the average of the concentrations at the two adjacent distance intervals at the start of the time interval. The development assumes that the whole volume is homogeneous in all respects except in concentration of the transferent property. In the derivation of Equation

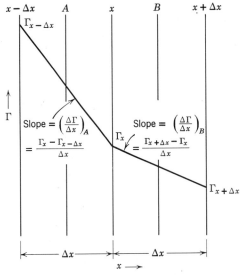

(a) Point within the slab.

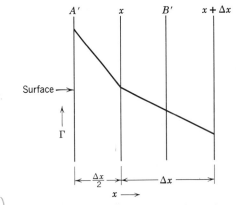

(b) The special case of a point on the surface.

Figure 11.6. Finite differences in transfer at time θ.

11.26, the property balances assume that the concentration at plane A may be taken as representing a volume element extending $\Delta x/2$ on either side of the plane. Obviously, the plane A' at the surface (Figure 11.6b) does not fit the model, and its concentration cannot represent the average concentration in the last half slab. Some kind of special treatment is necessary.

The special procedure for handling surface concentration described here is empirical, but solutions using this method have been tested and are fair approximations for more rigorous methods. The procedure is as follows: In the first time interval at the surface location, use the arithmetic-mean value of the surface concentrations before and after the initial change that takes place at time zero. This procedure is demonstrated in the illustration that follows.

The Schmidt method may be applied to any initial condition. For example, the final result of Illustration 11.2 may be taken as the initial condition for further drying.

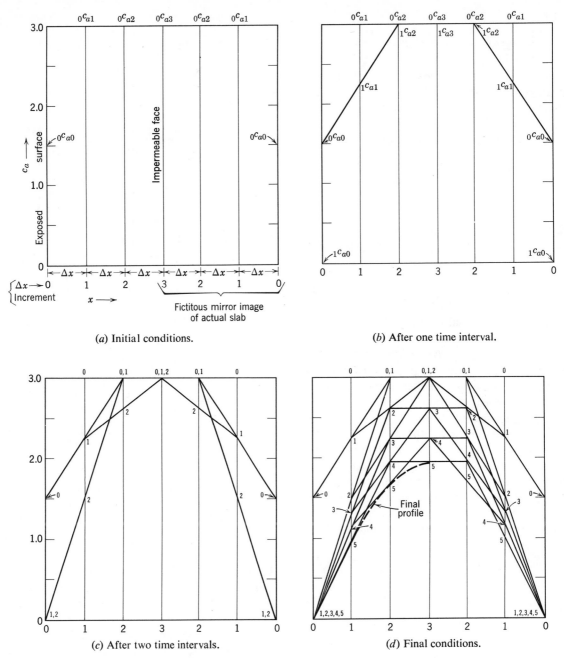

(a) Initial conditions.

(b) After one time interval.

(c) After two time intervals.

(d) Final conditions.

Figure 11.7. Solution to Illustration 11.2

Illustration 11.2. A colloidal gel wet with alcohol is to be dried. The material is packed into a pan 2 ft by 2 ft by 1 in. thick, and the top face is exposed to a dry-air stream. The alcohol will be transferred through the gel by molecular transport with a mass diffusivity of 0.00096 sq ft/hr. The initial gel concentration is uniform at 3 per cent alcohol by volume. Determine the concentration profile 2 hr after the surface of the gel is exposed to the air containing no alcohol.

SOLUTION. The slab has one face impermeable to alcohol (the bottom of the pan). This face would be equivalent to a 2-in. thick slab with both sides exposed to air. Then the alcohol would diffuse to both faces, but none would cross the center plane. Alternatively, it is evident that the concentration gradient at the gel-pan interface must be zero since there is no transfer. The 2-in. slab is shown in Figure 11.7. The position increment Δx is chosen at $\frac{1}{3}$ in. to give a reasonable number of increments over the total thickness for the purposes of this illustration. A smaller increment would give greater accuracy. Since $\mathcal{M} = 2$, $\Delta\theta$ may be evaluated

$$\mathcal{M} = \frac{\Delta x^2}{\mathcal{D}\,\Delta\theta} = 2$$

$$\Delta\theta = \frac{(\frac{1}{3} \times \frac{1}{12})^2}{(0.00096)(2)} = 0.40 \text{ hr}$$

Therefore $2/0.4 = 5.0$ time increments will be required. The initial condition of the slab is shown in Figure 11.7a. The notation $_0c_{a2}$ refers to the concentration at zero time at 2 distance increments from the surface. The presubscript will in general give the time increment, and the postsubscript, the distance increment from the surface of the slab. The initial concentration gradient is horizontal at 3 per cent alcohol. However, on exposure to dry air the surface concentration drops to zero. Just before exposure it is 3 per cent. Thus $_0c_{a0}$ may be either 0 or 3 per cent. *For the first time increment only* it is customary to use the arithmetic average of these values, i.e., $_0c_{a0} = 1.5$. After the first time interval, $_0c_{a0}$ is taken as zero. To find the concentrations after one time interval, Equation 11.29 is used. For $_1c_{a1}$ it becomes

$$_1c_{a1} = \frac{_0c_{a0} + _0c_{a2}}{2}$$

This can be accomplished graphically by connecting the points $_0c_{a0}$ and $_0c_{a2}$ by a straight line, as shown in Figure 11.7b. Thus, the concentration $\frac{1}{3}$ in. from the surface decreases to 2.25 per cent after a time interval of 0.4 hr. The other concentrations at the first time interval are determined graphically in the same manner.

$$_1c_{a2} = \frac{_0c_{a1} + _0c_{a3}}{2}$$

$$_1c_{a3} = \frac{_0c_{a2} + _0c_{a2}}{2}$$

This last equation shows the utility of the mirror image of the slab. $_1c_{a3}$ is evaluated using $_0c_{a2}$ in the slab and $_0c_{a2}$ in the mirror image. Since $_0c_{a2} = 3.0$, $_1c_{a3} = 3.0$. Also, by the previous equation, since $_0c_{a1}$ and $_0c_{a3} = 3.0$, $_1c_{a2} = 3.0$. The resulting gradient after 0.4 hr shows the slab next to the pan still at 3 per cent, but near the surface the concentration has decreased. The concentration profile is not a smooth curve because finite differences have been taken. The smaller the distance increment, the smoother will be the resulting profile.

Figure 11.7c shows the profile after two time increments. It is determined as before. For example,

$$_2c_{a1} = \frac{_1c_{a0} + _1c_{a2}}{2}$$

After the first interval $_1c_{a0} = 0$ per cent, as shown. Each point is labeled only with its time interval, for brevity. After two time intervals, the concentration at the pan is 3 per cent, but it drops rapidly nearer the surface.

The calculation is continued for five time intervals, as shown in Figure 11.7d. The points labeled 5 represent the final concentration profile, as shown by the dotted line. The final concentration at the pan has dropped to slightly less than 2 per cent.

UNSTEADY-STATE TRANSFER WITH INTERNAL GENERATION

Transfer with internal generation may also occur in the unsteady state. The balance equation is

$$\text{Input} + \text{generation} = \text{output} + \text{accumulation} \quad (11.32)$$

This may also be expressed in terms of the rate of each term. As before, $\mathscr{G}$ is the volumetric rate of generation of the transferent property. With the substitutions made earlier in the derivation of Equation 11.10 the rate-balance equation is

$$(\psi A)_1 + \mathscr{G}(A\,dx) = (\psi A)_2 + \frac{d\Gamma}{d\theta} A\,dx \quad (11.33)$$

Appropriate rearrangement and substitution yields

$$\frac{\partial \Gamma}{\partial \theta} = \delta \left[\frac{\partial^2 \Gamma}{\partial x^2} \right] + \mathscr{G} \quad (11.34)$$

which is the general equation for one-dimensional transfer with internal generation. If there is no internal generation, Equation 11.34 reduces to the simple transfer Equation 11.10.

For transfer in the x-, y- and z-directions, the following equation applies

$$\frac{\partial \Gamma}{\partial \theta} = \delta \left[\frac{\partial^2 \Gamma}{\partial x^2} + \frac{\partial^2 \Gamma}{\partial y^2} + \frac{\partial^2 \Gamma}{\partial z^2} \right] + \mathscr{G}$$

$$= \delta \nabla^2 \Gamma + \mathscr{G} \quad (11.35)$$

This equation may be written for mass, heat, and momentum transfer

Mass transfer: $\qquad \dfrac{\partial c_a}{\partial \theta} = \mathscr{D} \nabla^2 c_a + \mathscr{G}_N \quad (11.36)$

Heat transfer: $\qquad \dfrac{\partial (\rho c_P T)}{\partial \theta} = \alpha \nabla^2 (\rho c_P T) + \mathscr{G}_q \quad (11.37)$

or $\qquad \dfrac{\partial T}{\partial \theta} = \alpha \nabla^2 T + \dfrac{\mathscr{G}_q}{\rho c_P} \quad (11.38)$

Momentum transfer: $\qquad \dfrac{\partial (v\rho)}{\partial \theta} = \nu \nabla^2 (v\rho) + \mathscr{G}_\tau \quad (11.39)$

or $\qquad \dfrac{\partial v}{\partial \theta} = \nu \nabla^2 v + \dfrac{\mathscr{G}_\tau}{\rho} \quad (11.40)$

The solutions of the equations for unsteady-state transfer with internal generation are difficult even for simple geometries and are beyond the scope of this book.

REFERENCES

1. Mickley, H. S., T. K. Sherwood, and C. E. Reed, *Applied Mathematics in Chemical Engineering*, McGraw-Hill Book Co., New York, 1957.
2. Gurney, H. P., and J. Lurie, *Ind. Eng. Chem.*, **15**, 1170 (1923).
3. Binder, L., dissertation, Munich (1911).
4. Schmidt, E., *Foeppls Festschrift*, Springer, Berlin, 1924, pp. 179–198.
5. Dusinberre, C. M., *Numerical Analysis of Heat Flow*, McGraw-Hill Book Co., New York, 1949.

PROBLEMS

11.1. Derive Equation 11.12a starting with a heat balance and $q/A = -k\,dT/dx$.

11.2. Derive Equation 11.17.

11.3. A sphere of 6-in. radius at a temperature of 600°F is suddenly placed in a cool fluid so that its surface is at 100°F. The thermal diffusivity of the sphere is 0.045 sq ft/hr. Plot the radial temperature profile after 1 hr and 2 hr.

11.4. A platinum catalyst is made by immersing spherical alumina pellets in a chlorplatinic acid solution until an appropriate amount of acid diffuses into the pellet. The acid is then reduced to release a finely divided platinum on the alumina. $\frac{1}{2}$-in. diameter pellets are initially wet with pure water. They are immersed in acid solution such that the surface concentration is maintained at 50 per cent acid, 50 per cent water. The transfer of acid is by molecular transport and $\mathscr{D} = 5 \times 10^{-5}$ sq ft/hr.

(a) Calculate the acid concentration $\frac{1}{8}$-in. from the center after 3 hr immersion.

(b) Calculate the time required to reach a concentration of 40 mole percent acid at the center.

11.5. A large steel slab initially at 2000°F is quenched in oil which maintains its surface temperature at 300°F. The slab is 6 in. thick by 6 ft square. Calculate and plot the temperature profile across the thickness after 10 min and after 30 min. Let $\alpha = 0.30$ sq ft/hr.

11.6. A steel rod 3 ft long is heated until the rod has a linear gradient running between 1100°F at one end and 700°F at the other. The temperature at the 1100°F end is suddenly lowered to 100°F. The sides and the other end of the rod are insulated. Calculate the temperature profile after $7\frac{1}{2}$ hr.

(*Hint:* Since the sides and one end are insulated, the rod can be considered part of a flat slab with the 1100°F end at the surface of the slab.)

11.7. Calculate the concentration profile of the gel in Illustration 11.2 after 2 hr of drying, using Figure 11.2. Compare the result with that of the illustration.

11.8. Calculate the concentration profile of the gel in Illustration 11.2 four hours after drying begins.

11.9. A large brick wall 1 ft thick is heated on one side and cooled on the other so that at steady state a linear temperature gradient exists. The temperature varies between 1500°F at one face and 100°F at the other. Suddenly the heating and cooling are reversed, so that the two surface temperatures are effectively reversed. Determine the resulting temperature gradient 3 hr and 6 hr after the reversal. Let $\alpha = 0.02$ sq ft/hr.

11.10. Derive Equation 11.34.

11.11. The walls of a house are built of red brick, 13 in. thick. The walls are heated to 75°F during the day and the inside wall is allowed to fall to 60°F at night by adjusting the thermostat. For a day on which outside wall temperature is 0°F,

(a) Compute the steady-state heat loss per square foot of wall for daytime.

(b) For nighttime compute the time after change of thermostat required to accomplish 63 per cent of the change expected.

chapter 12

Turbulent-Transport Mechanism

Molecular transport, which was described in Chapter 9, depends on the random motion of individual molecules for transport of mass, heat, or momentum. *Turbulent transport*, which will be discussed in this chapter, is due to the random movement of large groups or clusters of molecules. These groups or clusters of molecules are called *eddies*, and in some cases they are large enough to be visible to the naked eye. Mass, heat, and momentum may be transferred by turbulent transport in any fluid.

The conditions under which turbulent transport occurs will be discussed in the following sections. Turbulent transport cannot be described completely by a single mathematical expression, because the mechanism is much more complex than that for molecular transport. As a result it is necessary to combine theory with experimental evidence to obtain an adequate picture of turbulence.

In order to understand turbulent transport it is first necessary to examine turbulent momentum transport in a flowing fluid. Once a picture of turbulent motion has been developed for momentum transport, it can be extended to heat and mass transport.

TURBULENT MOMENTUM TRANSPORT

Molecular momentum transport is variously named laminar flow, streamlined flow, or viscous flow. As was shown in Chapter 9, movement of the fluid as a whole is in the y-direction only, whereas momentum is transferred by random motion of individual molecules in the x-direction. Using this relatively simple model, expressions were developed in Chapter 9 relating shear stress, diffusivity, and concentration gradient. In practice, turbulent transport is of more importance to the chemical engineer than molecular transport. Therefore,

the concepts developed for laminar flow will now be applied to transport in turbulent flow.

Pressure Drop in Turbulent Flow. If the pressure drop $(-\Delta P)$ is measured in a pipe at a relatively high fluid velocity, it is found to be much higher than the value predicted by the laminar-flow equation (Equation 10.86). For example, consider that water at 70°F flows through a smooth circular pipe of 1 in. I.D. at a mean velocity of 10 ft/sec. If the pressure drop is calculated according to Equation 10.86, the loss in pressure per hundred feet of pipe would be 0.663 psi. However, if an apparatus is physically tested under the described conditions the measured pressure loss per hundred feet of pipe would be 15.4 psi. The experimental value would be twenty-three times the calculated value. On the other hand, if the water were flowing at 0.1 ft/sec, the experimental value and calculated value would agree exactly. This indicates that at low velocities the fluid behavior is identical to the model of momentum transport in laminar flow chosen for derivation of Equation 10.86. However, at some higher mean velocity, the fluid behavior is not the same as in the laminar regime. The purpose of this chapter then is to examine this new regime and to describe the regime as exactly as can be done at this time.

The Reynolds Experiment. A classical experiment reported by Osborne Reynolds (1) in 1883 is still used to demonstrate the qualitative difference between laminar and turbulent flow.

Consider a glass tube in which water flows at any desired mean velocity. At the center of the tube a fine jet of water-soluble dye is introduced through a capillary tube so that a thin filament of dye is injected into the stream of water; the velocity of the dye stream is equal to that of the water at the point of introduction. Reference to Figure 12.1a shows that at low water velocity the

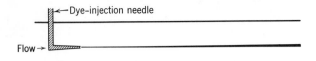

(a) Flow pattern at low mean velocity with dye injection.

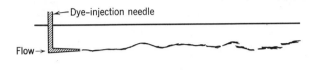

(b) Flow pattern at higher velocity with dye injection.

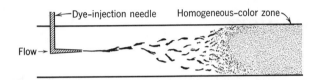

(c) Flow pattern at high velocity with dye injection.

Figure 12.1. The Reynolds experiment. (Vertical scale exaggerated.)

dye filament retains its identity in the water stream, tending to widen very slightly during the downstream passage because of molecular diffusion of dye. At a slightly higher mean velocity, as shown in Figure 12.1*b* the dye filament breaks up into *finite* large eddies. Farther downstream the eddies break up further, and the dye that has been introduced tends to become homogeneously dispersed. At much higher mean velocity (Figure 12.1*c*) the eddy activity becomes extremely violent, and the region of homogeneous dye color approaches the point of dye entry. From visual observation it is evident that the eddies in normal pipe flow are of the order of one-tenth the pipe diameter, and that the eddies move in completely random pattern. In some experimental studies the eddy formation has been suppressed by careful control so that high mean velocities were maintained with absence of eddies, but these conditions demand, for example, deaeration of the fluid, insulation of the apparatus from any external vibration, and supersmooth finish on the duct work. This evidence supports the notion that the incidence of eddies is brought about by external imperfections of the systems of the type that cannot be practically eliminated. In those experiments in which high velocity was maintained with absence of eddies, pressure-drop data were correlated with Equation 10.86. A system operating at abnormally high velocity in viscous flow could be made to revert to turbulent flow by simply physically jarring the pipe. This indicates that high-velocity viscous flow is an abnormal or metastable state.

It should also be noted that an upper limit of viscous flow as well as a lower limit of turbulent flow seems to exist and that the limits are separated by a transition region. The Reynolds apparatus and other means of examination (2) also indicate that the eddy formation begins at the center of the tube to form a central core of eddy activity. The diameter of the core increases with increase in mean velocity.

The conclusions that can be drawn from the Reynolds experiment are: (1) above a certain mean velocity *for a given system* relatively large eddies form that flow cross-stream in some random behavior; (2) these eddies are larger and more abundant at the center of the tube; and (3) an increase in mean velocity of the fluid widens the turbulent core until the tube is essentially filled with the core of eddy activity.

Momentum Transport by Eddy Activity. In Chapter 9, the transport of momentum by single molecules moving in natural random motion was described. The conclusion drawn there was that the fraction of the total molecules that possessed a cross-current component of velocity could move to regions in which the flow velocity was different from the origin velocity of the migrant molecule, with consequent transport of momentum. In certain respects, turbulent flow is similar to laminar flow, except that the scale of the migrant mass is much larger in turbulent flow and that additional variable dependencies exist. In the Reynolds experiment the eddies are distinguishable by eye, which is indicative of the magnitude of the eddies. Since the eddy has a random crossflow component superimposed upon the bulk flow, it is reasonable to suppose that momentum can be transferred by eddy crossflow motion as well as by molecular crossflow motion. The visible examinations are sufficient to indicate that the crossflow motion exists on a macroscopic scale.

The absence of agreement between fluids moving at higher velocities and equations that are derived based upon a molecular-transport model such as Equation 10.86 is sufficient to indicate that a heretofore unevaluated mechanism is operative *in addition to* molecular transport. This does not mean to imply the molecular transport is absent, because in any fluid sample in which a velocity gradient occurs, even within an eddy, random *molecular* motion will transport some momentum. A major difference between molecular behavior and eddy behavior is apparent from the Reynolds experiments. The eddy activity is violent at the center of the pipe and decreases in intensity away from the center. Furthermore, the eddy activity becomes more violent with increase in velocity. Therefore, the *eddy momentum diffusivity* is a function of both position and velocity in turbulent flow. In laminar flow on the other hand, the molecular momentum diffusivity is independent of tube position or velocity.

Subsequent developments will attempt to evaluate the

contribution to the momentum transport made by each of the mechanisms.

Velocity Distribution in Turbulent Flow. Nikuradse (3) measured the point velocity (in the y-direction) of turbulently flowing fluids and found that the velocity profile lost the parabolic character described by Equation 10.82 and tended to approach *plug flow*, almost as if the velocity of the fluid was independent of radial position. The tendency toward plug flow increased with increase in mean velocity for a given system, as shown in Figure 12.2. Note that with increase in velocity the "point" of the parabola is blunted, so that as $\bar{v} \to \infty$ the flat profile of plug flow appears to be the limit.

Von Kármán (4, 5) proposed that the Nikuradse velocity-distribution data be represented by three separate equations for all Newtonian fluids flowing in "smooth"† tubes regardless of density, velocity, wall stress, and viscosity for any radial position from the wall. The velocity profile would naturally include all the foregoing terms. The generalized velocity distribution is given in Equations 12.1, 12.2, and 12.3 and is presented graphically in Figure 12.3.

The parameter y^+ is a generalized position in a flowing fluid system at a particular radial position r. The parameter u^+ is a generalized velocity in a flowing system including the point velocity at y^+, where u^+ and y^+ are related in Equations 12.1, 12.2, and 12.3 below.

The first region of a turbulent-flow system is bounded by $y^+ = 0$, which is at the wall of a tube, and $y^+ = 5$, a short distance from the wall. This is called the *laminar sublayer*. In this region, the point velocity and position are related by

$$u^+ = y^+ \tag{12.1}$$

The region between the radial positions defined by $y^+ = 5$ and $y^+ = 30$ is called the *buffer layer* represented by the empirical equation

$$u^+ = -3.05 + 5.00 \ln y^+ \tag{12.2}$$

The region between the radial positions defined by $y^+ = 30$, and the tube center is called the *turbulent core*

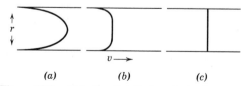

(a)   *(b)*   *(c)*

Figure 12.2. Velocity distribution in laminar and turbulent flow. (*a*) Low velocity laminar flow. (*b*) High velocity turbulent flow. (*c*) Very high velocity ($v \to \infty$) plug flow.

---

† "Smooth" means glass tubing or commercial drawn-metal tubing. Commercial pipe is somewhat rougher, which changes the velocity profile.

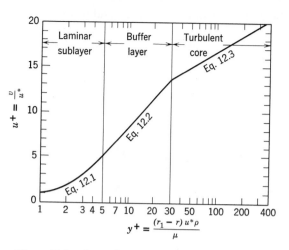

Figure 12.3. Generalized velocity profile for turbulent flow in tubes (5).

and is represented by the empirical equation

$$u^+ = 5.5 + 2.5 \ln y^+ \tag{12.3}$$

The definitions of the terms in these equations are:

$u^+ = v/u^*$, a point-velocity parameter with vectorial dimensions $\sqrt{\mathbf{L}_y/\mathbf{L}_r}$. Where nonvectorial dimensions are in use, this is dimensionless.

$u^* = \sqrt{(\tau_y g_c)_1/\rho}$ called the "friction velocity" with dimensions $\sqrt{\mathbf{L}_y \mathbf{L}_r/\mathbf{\theta}^2}$ and units in ft/sec

$v =$ the point velocity at a position r, $\mathbf{L}_y/\mathbf{\theta}$

$y^+ = [(r_1 - r)u^*\rho/\mu]$, a position parameter with vectorial dimensions $\sqrt{\mathbf{L}_y/\mathbf{L}_r}$. In nonvectorial dimensions, the parameter is dimensionless.

$r_1 =$ radius at the boundary, $\mathbf{L}_r$

$r =$ any radial position, $\mathbf{L}_r$

$\rho =$ density of the fluid, $\mathbf{M}/\mathbf{L}_r^2\mathbf{L}_y$

$\mu =$ absolute viscosity, $\mathbf{M}/\mathbf{L}_y\mathbf{\theta}$

$(\tau_y)_1 =$ fluid stress at the pipe wall, $\mathbf{F}_y/\mathbf{L}_y\mathbf{L}_r$.

Equation 12.1 will be examined more critically since the peculiar nature of the velocity and position parameters requires some elaboration. The derivation of Equation 12.1 follows. Consider Equation 10.82 which defines the position-velocity relationship in *laminar* flow through a circular duct.

$$v = \left(\frac{-\Delta P\, g_c}{\Delta y\, \mu}\right) \frac{(r_1{}^2 - r^2)}{4} \tag{10.82}$$

From Equation 10.78,

$$\left(\frac{-\Delta P\, g_c}{2\, \Delta y}\right) = \frac{(\tau_y g_c)_1}{r_1} $$

Substituting Equation 10.78 into Equation 10.82 gives

$$v = \left(\frac{(\tau_y g_c)_1}{\mu r_1}\right)\left(\frac{(r_1{}^2 - r^2)}{2}\right) \tag{12.4}$$

Equation 12.4 can be rearranged and factored.

$$v = \frac{(\tau_y g_c)_1}{\rho} \left(\frac{\rho(r_1 - r)}{\mu} \right) \left(\frac{(r_1 + r)}{2r_1} \right) \quad (12.5)$$

If r is very nearly equal to r_1 as is the case near the wall of the tube ($y^+ < 5$), then the term $(r_1 + r)/2r_1 \cong 1$ and can be therefore omitted.† With this modification Equation 12.5 can be rearranged to

$$\frac{v}{\sqrt{(\tau_y g_c)_1/\rho}} = \frac{(r_1 - r)\rho \sqrt{(\tau_y g_c)_1/\rho}}{\mu} \quad (12.6)$$

Combining Equation 12.6 with the definitions of u^+ and y^+ gives

$$u^+ = y^+ \quad (12.1)$$

Equation 10.82 was derived based upon $\tau_y g_c = -\mu(dv/dx)$ and consequently is a laminar-flow equation. Therefore, Equation 12.1 is also a laminar-flow equation. In turbulent flow the *laminar sublayer* is defined as the region extending between $y^+ = 0$ (at the boundary) and $y^+ = 5$ (a short distance from the boundary). Flow in the laminar sublayer is true laminar flow. For a given value of y^+ the actual thickness of the laminar sublayer ($r_1 - r$) depends upon u^*. As the wall stress τ_{y1} increases (as would be the case if the velocity were increased) the laminar-sublayer thickness decreases.

The actual existence of the laminar sublayer has been debated‡ for some time. The actual velocity in this region is very difficult to measure and has always been subject to doubt. There is no doubt, however, that the velocity distribution in this region can be *approximated* by Equation 12.1 and therefore in a practical sense the assumption of laminar flow in the small region would lead to an error in theory rather than to an error in prediction of rate of transport. The buffer layer extends from $y^+ = 5$ to $y^+ = 30$. The Reynolds experiment and others have shown that eddy activity is most pronounced at the center of a tube and is absent§ in the laminar sublayer, so that, as the name implies, the buffer layer is a region of transition from the laminar sublayer with no eddy activity to the violent eddy activity at the core of the tube. As in the case of the laminar sublayer, the thickness of the buffer region is a function of wall stress, so that at high wall stress the thickness is small. Since the buffer region is bounded on one side by a region of no eddy activity and on the other by a region of considerable eddy activity, it is reasonable to assume that there is some degree of eddy activity in the region. Velocity distribution in the buffer layer is experimentally

† In effect, this is the neglect of the second-order difference $(r_1 - r)^2$.

‡ See penetration theory, Chapter 13.

§ A few experimenters have visually observed some motions that appear to be eddy activity near the walls (6). This had led to another model for turbulence, but the model has not been confirmed. This topic will be dealt with later as "penetration theory."

determined with considerable precision, so that there is little doubt about the manner in which the data fit the curve. The closeness of fit does not imply description of mechanism.

The *turbulent core* is represented by Equation 12.3. It is the region of maximum eddy activity. Comparison of the constant 2.5 in Equation 12.3 with 5.0 in Equation 12.2 indicates a decrease in rate of change of the velocity gradient with increase in position parameter or approach to the center of the circular duct. The logarithmic function of position also inherently indicates decreasing gradient. It is in this region that the extreme "blunting" of the velocity profile takes place. The blunted profile is the first noticeable difference in velocity pattern between laminar and turbulent flow.

There is one point of inconsistency in Equation 12.3. The function $u^+ = 5.5 + 2.5 \ln y^+$ is continuous for all positive values of y^+ and will have a finite value for the first derivative for all positive values of y^+. At the center of the tube $r = 0$ and $y_0^+ = r_1 u^* \rho/\mu$. Consequently, according to Equation 12.3 dv/dr must be a real nonzero number at the center of the tube. All velocity measurements show that dv/dr is zero at the center of a tube; therefore, Equation 12.3 is incorrect for the center of a tube and probably in the region near the center. Away from the center the equation is a reasonable empirical representation of the velocity at various points. Actually, it is unreasonable to suppose that the velocity profile should follow three distinct equations; more likely one very complex function is correct, but this function has not been proposed as yet. In spite of obvious mathematical inconsistencies, the three equations still provide a basis for calculation that yields valid results. For example, even though a serious mathematical inconsistency exists when Equation 12.3 is used near the center, it should be remembered that very little momentum is transferred in that region, so that even a large percentage error in that region amounts to very little actual error when the entire tube is considered.

In summary, the available information for analysis is:

(1) The experiments of Reynolds that show formation of gross eddies moving in random manner, with the eddy activity starting in a central core with the core increasing in radial dimension with increase in mean velocity.

(2) The velocity-profile data of Nikuradse that was expressed as three equations by von Kármán. The three equations represent three separate regions of flow behavior: (a) laminar flow near the walls, (b) limited eddy activity through the buffer layer, and (c) the turbulent core. The point velocities are correlated in terms of the wall stress, geometry, density and viscosity.

Mathematical Analysis of Turbulent Flow. In simplest form the increase in fluid stress due to combined laminar and turbulent flow can be represented by the general equation

$$\tau_y g_c = -(v + E_\tau)\frac{d(\rho v)}{dx}$$

or   (12.7)

$$\tau_y g_c = -(\mu + \rho E_\tau)\frac{dv}{dx}$$

where E_τ is the eddy diffusivity of momentum. This equation is in the same form as the flux-gradient equation written in Chapter 9 with an additional coefficient, E_τ. This equation does *not* imply that E_τ is an increase in viscosity of the fluid at high velocity. It does imply that additional momentum is transferred beyond that predicted by the laminar flow mechanism alone, and the transfer may be represented by separate equations for the two mechanisms. For molecular transport,

$$(\tau_y g_c)_m = -v\frac{d(\rho v)}{dx}$$ (12.8)

For eddy transport,

$$(\tau_y g_c)_e = -E_\tau\frac{d(\rho v)}{dx}$$ (12.9)

and $$\tau_y g_c = (\tau_y g_c)_m + (\tau_y g_c)_e$$ (12.10)

E_τ varies from zero at the wall (laminar flow) to high values in the turbulent core region. It should be remembered that the force-balance equation derived in Chapter 10 is applicable regardless of the origin of the forces and therefore can be used here. For any section of flowing fluid, the net force applied to the ends of the tube because of pressure drop is $(-\Delta P)S$ and the resisting force due to fluid stress is $\tau_y A$. These terms may be equated in a force balance as

$$-(\Delta P)S = \tau_y A$$ (12.11)

where S and A are the areas of application of ΔP and τ, and, for a tube whose radius is r_1, $S = \pi r_1^2$ and $A = 2\pi r_1 L$; and

$$(-\Delta P)\pi r_1^2 = \tau_{y1}(2\pi r_1 L)$$ (12.12)

For a part of the tubular section bounded by any radius r,

$$(-\Delta P)\pi r^2 = \tau_y(2\pi r L)$$ (12.13)

Equations 12.11, 12.12 and 12.13 are similar to those derived in Chapter 10 for molecular transport with internal generation. If Equation 12.13 is divided by Equation 12.12,

$$\tau_y = \tau_{y1}\left(\frac{r}{r_1}\right)$$ (12.14)

This is equivalent to Equation 10.79. If Equation 12.14 is substituted into Equation 12.7 with modification of the derivative for cylindrical geometry,

$$(\tau_y g_c)_1\left(\frac{r}{r_1}\right) = -(v + E_\tau)\left(\frac{d(\rho v)}{dr}\right)$$ (12.15)

When the density can be taken as constant, the derivative (dv/dr) can be evaluated by differentiating Equations 12.1, 12.2, and 12.3 over the ranges of y^+ where these equations apply. If τ_y, v, ρ, and the geometry are known, the eddy diffusivity of momentum, E_τ can be determined at various radial positions.

Illustration 12.1 below demonstrates the calculation and illustrates several inconsistencies in Equations 12.2 and 12.3.

Illustration 12.1. Water is flowing through a 2 in. I.D. tube. The pressure drop is 10 lb$_f$/sq ft per foot of tube. Find the point velocity and eddy diffusivity at various positions between the wall and the center of the tube. ($\mu = 1.0$ centipoise, $\rho = 62.4$ lb/cu ft.)

SOLUTION. Using the pressure-drop data and the geometry, the wall stress τ_{y1} can be calculated from the force balance, Equation 12.11.

$$(-\Delta P)S_1 = (\tau_y A)_1$$

Rearranging and inserting data gives

$$\tau_{y1} = \frac{(-\Delta P)S_1}{A} = \frac{(10)\pi(1/12)^2}{2\pi(1/12)(1)}$$

$$\tau_{y1} = 0.417 \text{ lb}_f/\text{sq ft}$$

From the wall stress and density data the friction velocity u^* can be calculated.

$$u^* = \sqrt{\frac{\tau_{y1}g_c}{\rho}} = \left(\frac{(0.417)(32.2)}{(62.4)}\right)^{1/2}$$

$$u^* = 0.464 \text{ ft/sec}$$

Then $$u^+ = \frac{v}{u^*} = \frac{v}{0.464}$$

and $$y^+ = \frac{(r_1 - r)u^*\rho}{\mu}$$

Consider first a position at the wall, $r = r_1$, $y^+ = 0$. Equation 12.1 applies, so that $u^+ = 0$ and $v = 0$. This of course is a basic assumption of fluid flow: that the velocity of the fluid at a boundary is equal to the boundary velocity.

Next consider a point at $y^+ = 5$. This point is the upper limit of Equation 12.1 and the lower limit of Equation 12.2. Using Equation 12.1

$$u^+ = y^+ = 5$$

By the definition of u^+

$$v = 5(0.464) = 2.32 \text{ ft/sec}$$

If Equation 12.2 had been used,

$$u^+ = -3.05 + 5.00 \ln 5$$

$$u^+ = 5.00$$

$$v = 2.32 \text{ ft/sec}$$

The actual position represented by $y^+ = 5$ can be calculated from the definition of y^+ given above.

$$(r_1 - r) = \frac{y^+\mu}{u^*\rho} = \frac{(5)(1 \times 6.72 \times 10^{-4})}{(0.464)(62.4)}$$

$$(r_1 - r) = 0.0001162 \text{ ft}$$

This is the distance of the point $y^+ = 5$ from the wall for the conditions of this example. The same position measured from the center is

$$r = (\tfrac{1}{12} - 0.0001162) = 0.0831838 \text{ ft}$$

Notice the inconvenient number of significant figures required to define the limit of the laminar sublayer. Is the assumption that $(r_1 + r)/2r_1 \cong 1$ used in Equation 12.5 justified in this case? The eddy diffusivity E_τ is zero in this region because by definition the relationship $u^+ = y^+$ is a viscous-flow equation based in its original definition upon $\tau_y g_c = -\mu(dv/dr)$.

The next logical point to test is $y^+ = 30$, the upper limit of the buffer region and the lower limit of the turbulent core. The velocity at $y^+ = 30$ can be calculated from Equation 12.2 or 12.3. Notice that between $y^+ = 5$ and $y^+ = 30$ Equation 12.2 is the only equation that applies. If Equation 12.2 is used, then

$$u^+ = -3.05 + 5.00 \ln 30 = 13.96$$

and $\qquad v = 6.48 \text{ ft/sec}$

The position represented by $y^+ = 30$ can be calculated from the definition of y^+ and known values of μ, ρ, and u^+.

$$(r_1 - r) = 0.000697 \text{ ft}$$

$$r = 0.082603 \text{ ft}$$

The same result can be calculated from Equation 12.3 since $y^+ = 30$ is a valid limit for Equations 12.2 and 12.3.

$$u^+ = 5.5 + 2.5 \ln 30 = 14.0$$

$$v = 6.5 \text{ ft/sec}$$

The eddy diffusivity at this point can be calculated from Equation 12.15, provided that dv/dr is known at $y^+ = 30$. To evaluate dv/dr, Equation 12.2 is differentiated.

$$\frac{du^+}{dy^+} = \frac{5.0}{y^+}$$

Differentiating the definition of u^+ gives

$$du^+ = \frac{dv}{u^*}$$

and differentiating the definition of y^+ yields

$$dy^+ = \frac{-u^* \rho \, dr}{\mu}$$

Then, by combination of these three equations,

$$\frac{dv}{dr} = \frac{-5.00 u^{*2} \rho}{\mu y^+}$$

From the definition of u^*,

$$u^{*2} = \frac{(\tau_y g_c)_1}{\rho}$$

so that

$$\frac{dv}{dr} = \frac{-5.00(\tau_y g_c)_1}{\mu y^+}$$

From Equation 12.15

$$(\tau_y g_c)_1 \frac{r}{r_1} = (v + E_\tau)\rho \left(\frac{5.00(\tau_y g_c)_1}{\mu y^+}\right)$$

$(\tau_y g_c)_1$ cancels, so that by rearrangement

$$(v + E_\tau) = \frac{(r \mu y^+)}{(5 \rho r_1)}$$

At $y^+ = 30$, $r = 0.0826$ ft, and

$$(v + E_\tau) = \frac{(0.0826)(6.72 \times 10^{-4})(30)}{(5.00)(62.4)(0.0833)}$$

$$(v + E_\tau) = 6.41 \times 10^{-5} \text{ sq ft/sec}$$

$$v = \frac{\mu}{\rho} = \frac{6.72 \times 10^{-4}}{62.4} = 1.077 \times 10^{-5} \text{ sq ft/sec}$$

$$E_\tau = (6.42 - 1.077) \times 10^{-5}$$

$$E_\tau = 5.33 \times 10^{-5} \text{ sq ft/sec}$$

Note that E_τ is five times as large as v. This indicates that the momentum transferred by the turbulent mechanism is five times as great as that transferred by the molecular mechanism at a point $r = 0.0826$ ft defined by the position parameter $y^+ = 30$. If the velocity can be defined at $y^+ = 30$ by both Equations 12.2 and 12.3 it is reasonable to expect that E_τ might be the same regardless of the equation used. However, this is not the case. Using Equation 12.3 to find dv/dr gives

$$\frac{dv}{dr} = \frac{-2.5 u^{*2} \rho}{\mu y^+}$$

and

$$(v + E_\tau) = 12.84 \times 10^{-5}$$

$$E_\tau = 11.76 \times 10^{-5} \text{ sq ft/sec}$$

This value is slightly more than twice the value found using Equation 12.2. Thus the values of E_τ at $y^+ = 30$ differ because the derivatives of the two equations differ, even though the values of u^+ are the same at $y^+ = 30$. This discontinuity in function is a fault with this empirical generalized velocity distribution which utilizes three distinct equations.

The next radial position that will be considered is the midpoint between $r = 0$ and $r = r_1$. In this example the point lies well into the turbulent core, and therefore Equation 12.3 applies.

At $r = r_1/2$,

$$y^+ = \left(\frac{r_1 u^* \rho}{2 \mu}\right) = 1795$$

$$u^+ = 5.5 + 2.5 \ln 1795$$

$$v = 11.22 \text{ ft/sec}$$

E_τ can be determined using Equation 12.15 and the value of dv/dr determined by differentiating and rearranging Equation 12.3.

$$(\tau_y g_c)_1 \frac{(r_1)}{(2r_1)} = \frac{(v + E_\tau)(2.5\rho)(\tau_y g_c)_1}{(\mu y^+)}$$

$$(v + E_\tau) = \frac{(\mu y^+)}{(2)(2.5)(\rho)}$$

$$(v + E_\tau) = \frac{(1 \times 6.72 \times 10^{-4})(1795)}{(2)(2.5)(62.4)}$$

$$E_\tau = 386 \times 10^{-5} - 1.077 \times 10^{-5}$$

$$E_\tau = 385 \times 10^{-5} \text{ sq ft/sec}$$

It is evident that the eddy diffusivity is nearly 400 times the molecular diffusivity of momentum at the point $r = r_1/2$.

At the center of the tube, $r = 0$ and $y^+ = r_1 u^* \rho/\mu = 3590$.

$$u^+ = 5.5 + 2.5 \ln 3590$$

$$v = 12.02 \text{ ft/sec}$$

Using Equation 12.15 at $r = 0$ gives

$$(v + E_\tau) = 0$$

$$E_\tau = -1.077 \times 10^{-5}$$

This result shows another serious inconsistency. The Reynolds experiments indicate visually that eddy activity is a maximum at the center of the pipe, whereas the calculations above show a negative value. The negative value implies that a negative transfer of momentum is taking place. Differentiation of Equation 12.3 results in a finite value of dv/dr at the center of the tube; however, all experimental evidence points to $dv/dr = 0$ at the center of the tube. If $dv/dr = 0$ is substituted into Equation 12.15 at $r = 0$,

$$\frac{(\tau_y g_c)_1(0)}{r_1} = -(v + E_\tau)\rho(0)$$

This equation is indeterminate, so that no light can be shed on the eddy diffusivity at the center of the tube by this method.

All the foregoing indicates that the velocity-distribution equations are inconsistent in some respects. The velocity data derived from these equations are useful empirical values, but the equation forms are apparently incorrect. If the equations were correct, the derivatives of the equations would be valid at all points, and dv/dr would be zero at the center of the tube.

Illustration 12.1 offers some indication of the difficulties encountered when working in a flow regime that is not rigorously described mathematically. The velocity-profile equations result from the correlation of experimental data without having available an adequate theory as a foundation. Consequently, the data are reported in a convenient orderly manner (as u^+ and y^+) which is not rigorous. The general aim of the experimenters was to describe the velocity profile with special attention focused upon the condition near the tube wall. No particular effort was made to describe conditions at the center of the tube and therefore a term in y^+ at the center was omitted. All the above emphasizes that the present empirical mathematical formulations of turbulence are certainly not theoretically adequate. In order to complete the picture of turbulence, a rigorous mathematical representation is required.

Mechanism of Turbulent Flow. The eddy-diffusivity concept is a useful mathematical expression for turbulence but sheds no light on the mechanism. As the eddy diffusivity has been defined, it is a counterpart in the turbulent regime of the molecular diffusivity. In Chapter 9 it was reported that for gases the molecular diffusivity is proportional to the product of molecular velocity and mean free path of the molecules. Since

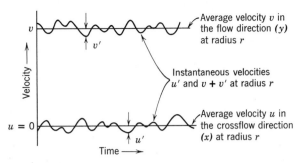

Figure 12.4. Instantaneous point velocities in the x- and y-directions.

turbulence is observed to be motion of macroscopic rather than microscopic portions of the fluid in the transfer direction, it might be assumed that the eddy diffusivity is a product of a crossflow velocity and some appropriate distance of travel in the transfer direction for an eddy. Without an exact mathematical representation of turbulence it becomes necessary to derive an expression in a less rigorous manner.

Consider an extremely sensitive and responsive device for measuring point velocity, so arranged that y- and x-components of velocity can be determined at any point in a turbulently flowing system. If the instrument response is plotted for a period of time, as shown in Figure 12.4, two curves result.

The response curve for velocity in the x-direction consists of a series of velocity pulsations, the average of which is zero which shows that the average velocity is zero in the x- or crossflow direction. The response curve for velocity in the y-direction indicates pulsations, but the average flow velocity is v, the average point velocity at the point of measurement. Therefore, although there is zero average velocity in the x-direction, momentary pulses occur in the $+x$- or $-x$-direction. Similar random pulsations are superimposed on the flow velocity in the y-direction. Thus, the steady-state point velocity is actually the average of many pulses, each of which is an unsteady-state case. The instantaneous y-component of velocity is actually $v \pm v'$, where $\pm v'$ is the contribution of the pulsation above or below the average v. The x-component of velocity is $\pm u'$, the pulsation above or below the zero average.

Consider a small region of fluid in a turbulent stream. If the region is subject to a pulsation of magnitude v' above the average v in the region, the increase must be compensated for by a pulse u'. This is necessary to satisfy the need for constant inventory of mass and momentum in any region. As shown in Figure 12.5, u' may be directed into the region in the $+x$-direction, such as u_1' or in the $-x$-direction, as $-u_2'$. The presence of v' presumes no knowledge of the direction of u', therefore,

$$|u'| = |v'|$$

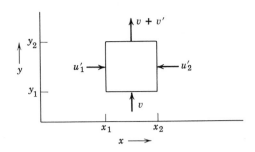

Figure 12.5. Momentum and mass balance around an element of fluid subject to a pulse.

By similar reasoning, a pulse w' can exist in the $\pm z$-direction also; therefore,

$$|u'| = |v' + w'| \qquad (12.16)$$

Using this model as a basis, any small volume in a fluid in turbulent flow can have associated with it randomly directed momentary pulses of magnitude $|u'| = |v' + w'|$. The equality is determined by the need for constant mass, and the random direction is associated with equal probability of each directional component. In this model, the random pulse velocity is analogous to the molecular velocity as described in Chapter 9. The component u' in the crossflow direction is a partial description of a mechanism for transfer of momentum.

The Prandtl Mixing Length. To complete the description of the model it is necessary to examine the distance that any pulse will travel before its identity is lost. Consider an element of fluid in a region A with mean velocity v_A. As a result of a pulse, the element accelerates to $v_A + v'$ and adopts a crossflow component u'. (See Figure 12.6.) With the new components of motion, the fluid element is now an eddy. It is *assumed* that this eddy will retain its identity in crossflow over a distance $x_B - x_A$ until a new region B is reached, wherein the mean velocity is $v_B = v_A + v'$, and there the eddy will decay or be absorbed. The velocities at A and B can be related by

$$v_B = v_A + \left(\frac{dv}{dx}\right)(x_B - x_A) = v_A + v' \quad (12.17)$$

If the distance $x_B - x_A$ is designated by λ, the *Prandtl mixing length*, (7) then Equation 12.17 can be written for any region as

$$v + v' = v + \left(\frac{dv}{dx}\right)\lambda \qquad (12.18)$$

and

$$v' = \lambda\left(\frac{dv}{dx}\right) \qquad (12.19)$$

where $\lambda =$ Prandtl mixing length ($\mathbf{L}_x$)

The Prandtl mixing length is the distance in the x-direction, measured in a field of gradient dv/dx, between the point of origin and point of decay of an eddy. No

meaningful counterpart of the Prandtl mixing length exists in the y- and z-directions for a fluid with momentum transfer in the x-direction and bulk motion in the y-direction. The Prandtl mixing length is the turbulent counterpart of the mean free path l described in Chapter 9.

By analogy, the eddy diffusivity may be considered as some simple multiple of the product $u'\lambda$. If the simple multiple is presumed unity in the absence of further information,

$$E_\tau = u'\lambda$$

If $|v'|$ is substituted for u',

$$E_\tau = |v'|\,\lambda \qquad (12.20)$$

The velocity v' is related to λ by Equation 12.19. Equation 12.19 can be substituted into Equation 12.20 to eliminate v'.

$$E_\tau = \lambda^2\left|\frac{dv}{dx}\right| \qquad (12.21)$$

By use of Equation 12.21, u', v', and λ can be determined if the eddy diffusivity and velocity profile are known. Equation 12.21 may also be substituted into Equation 12.7 to eliminate the eddy diffusivity,

$$\tau_y g_c = -\left[\nu + \lambda^2\left|\frac{dv}{dx}\right|\right]\left(\frac{d(\rho v)}{dx}\right)$$

or

$$\tau_y g_c = -\left[\mu + \rho\lambda^2\left|\frac{dv}{dx}\right|\right]\left(\frac{dv}{dx}\right) \qquad (12.22)$$

The Prandtl mixing length is defined in terms of a *constant-velocity-gradient field*. In practice, the velocity gradient frequently is not constant, and therefore the Prandtl mixing length is *not* the true length of an eddy. The *eddy length* is the actual distance measured in the *existing* velocity field between the origin and decay of an eddy. Or, conversely, the eddy length is the distance measured in the x-direction between the points at which the mean point velocities are v_A and $v_A + v'$, in the existing velocity field. From the definitions of Prandtl mixing length and eddy length, the two quantities are equal only when the existing velocity gradient is constant. The Prandtl mixing length and eddy length are compared in Illustration 12.2.

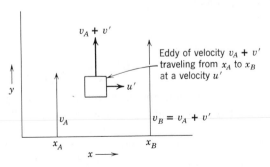

Figure 12.6. The crossflow movement of an eddy.

Illustration 12.2. For the conditions set forth in Illustration 12.1 calculate and compare the Prandtl mixing length and the eddy length at various positions.

SOLUTION. Between $y^+ = 0$ and $y^+ = 5$ viscous flow occurs. Consequently $E_\tau = 0$ and $\lambda = 0$. The point $y^+ = 30$ was tested in Illustration 12.1 using the derivatives of Equations 12.2 and 12.3 as a means of determining dv/dr with two different results. From Illustration 12.1

$$E_\tau = 5.34 \times 10^{-5} \text{ sq ft/sec}$$

Since from Equation 12.21, $E_\tau = \lambda^2 |dv/dx|$ the same value for dv/dr that was used to determine E_τ in Illustration 12.1 should be used in Equation 12.21.

From Illustration 12.1,

$$\frac{dv}{dr} = \frac{-5.00u^{*2}\rho}{\mu y^+}$$

$$= -\frac{(5.00)(0.464)^2(62.4)}{(1 \times 6.72 \times 10^{-4})(30)} = -0.334 \times 10^4 \text{ sec}^{-1}$$

$$\left|\frac{dv}{dr}\right| = 0.334 \times 10^4 \text{ sec}^{-1}$$

and also $\qquad E_\tau = 5.34 \times 10^{-5} \text{ sq ft/sec}$

From Equation 12.21,

$$\lambda = \sqrt{\frac{5.34 \times 10^{-5}}{0.334 \times 10^4}} = 1.264 \times 10^{-4} \text{ ft}$$

The eddy length can be determined from the difference in position between the point at which the average velocity is v and the point at which the average velocity equals the eddy velocity $v \pm v'$. The value $\pm v'$ can be evaluated from Equation 12.19.

$$v' = \pm\lambda\left|\frac{dv}{dr}\right| = (1.264 \times 10^{-4})(0.334 \times 10^4)$$

$$= \pm 0.422 \text{ ft/sec}$$

It is now necessary to find the point at which the average point velocity is equal to $v \pm 0.422$. From Illustration 12.1 the velocity at the eddy source ($y^+ = 30$) is 6.48 ft/sec. Assuming first that v' is negative, the eddy will move toward the wall until it reaches a point at which the velocity is $(6.48 - 0.422) = 6.06$ ft/sec. From Equation 12.2,

$$u^+ = \frac{v}{u^*} = \frac{6.06}{0.464} = 13.1$$

$$y^+ = 25.2$$

The position corresponding to this y^+ is

$$(r_1 - r) = \frac{\mu y^+}{u^*\rho} = \frac{(1 \times 6.72 \times 10^{-4})(25.2)}{(0.464)(62.4)} = 0.000581 \text{ ft}$$

At $y^+ = 30$, the source of the eddy, $(r_1 - r) = 0.000697$ ft. Therefore, the eddy length is

$$(0.000697 - 0.000581) = 1.16 \times 10^{-4} \text{ ft}$$

Thus, when the eddy moves toward the wall, the velocity gradient is increasing and the eddy length is less than the Prandtl mixing length. The rapid change in velocity gradient in the buffer region gives assurance that any eddy moving toward the tube wall will decay upon reaching the laminar sublayer.

If the eddy has a velocity $v + v'$, the eddy would have to move away from the wall to find a field with an average point velocity $v + v'$. In that case Equation 12.3 must be used to evaluate dv/dr. From Illustration 12.1, $E_\tau = 11.76 \times 10^{-5}$ at $y^+ = 30$. From Equation 12.3,

$$\frac{dv}{dr} = \frac{-2.5u^{*2}\rho}{\mu y^+}$$

$$= -\frac{(2.5)(0.464)^2(62.4)}{(1 \times 6.72 \times 10^{-4})(30)} = -0.167 \times 10^4 \text{ sec}^{-1}$$

$$\left|\frac{dv}{dr}\right| = 0.167 \times 10^4 \text{ sec}^{-1}$$

$$\lambda = \sqrt{\frac{11.76 \times 10^{-5}}{0.167 \times 10^4}} = 2.65 \times 10^{-4} \text{ ft}$$

It will be noted that this value of $|dv/dr|$ is just half that computed above from the equation for the buffer layer for the same point in the tube. The real value is probably between those computed from the two empirical equations. Since this particular eddy is being assumed to migrate into the turbulent core, the value of the gradient used will be that from the equation for the turbulent core.

In this case the eddy can be considered to decay at the point where the average point velocity is $v + v'$. Then,

$$v' = \lambda\left(\frac{dv}{dr}\right) = (2.65 \times 10^{-4})(1.67 \times 10^4) = 0.443 \text{ ft/sec}$$

The velocity at the source of the eddy is 6.48 ft/sec; therefore, the velocity at the point of decay is $6.48 + 0.443 = 6.92$ ft/sec. From Equation 12.3,

$$u^+ = \frac{6.92}{0.464} = 14.92$$

$$y^+ = 43$$

The position at $y^+ = 43$ is

$$(r_1 - r) = \frac{\mu y^+}{u^*\rho} = \frac{1 \times 6.72 \times 10^{-4} \times 43}{0.464 \times 62.4} = 0.000996 \text{ ft}$$

At $y^+ = 30$, the source of the eddy, $(r_1 - r) = 0.000697$; therefore, the eddy length is

$$0.000996 - 0.000697 = 3.0 \times 10^{-4} \text{ ft}$$

In this case, since the eddy moves toward the center of the tube, the eddy length is greater than the Prandtl mixing length. In the case of motion toward the center, dv/dr decreases, and therefore the eddy must move farther than the distance predicted from the velocity gradient at the source. This is consistent with observations that eddy activity increases toward the center of the tube.

At $r = r_1/2$, from Illustration 12.1 using Equation 12.3 for evaluation of dv/dr, $E_\tau = 385 \times 10^{-5}$ and $y^+ = 1795$; $v = 11.22$ ft/sec.

$$\frac{dv}{dr} = \frac{-2.5u^{*2}\rho}{\mu y^+}$$

$$= -\frac{(2.5)(0.464)^2(62.4)}{(1 \times 6.72 \times 10^{-4})(1795)} = -27.8 \text{ ft/sec ft}$$

$$\left|\frac{dv}{dr}\right| = 27.8 \text{ ft/sec ft}$$

$$\lambda = \sqrt{\frac{385 \times 10^{-5}}{27.8}} = 0.01177 \text{ ft}$$

Considering the eddy to move toward the wall and to have a velocity component in the y-direction of $v - v'$, then, by Equation 12.19,

$$v' = 0.01177 \times 27.8 = 0.327 \text{ ft/sec}$$

The eddy length is the distance between the source and the point at which the average point velocity is $(11.22 - 0.327) = 10.89$ ft/sec.

$$u^+ = 5.5 + 2.5 \ln y^+$$

$$y^+ = 1320$$

The position corresponding to $y^+ = 1320$ is $(r_1 - r) = 0.0301$. The position corresponding to $r_1/2 = 0.0417$. Therefore, the eddy length is $(0.0417 - 0.0301) = 0.0116$ ft.

Note that, as the center is approached, the eddy length and Prandtl mixing length are nearly equal. In this region, the velocity gradient changes very slowly with position; consequently, the Prandtl mixing length increases rapidly, and the eddy length and mixing length are nearly equal.

From the foregoing material, a model has been described that results in a mathematical definition of turbulent flow. There are intuitive objections to a model that considers gross particles of fluid that move in small sporadic motions. Although the model is probably oversimplified, it can be visualized and analyzed, and it is consistent with observation.

A more rigorous model may be derived if each pulsation in velocity is recognized and treated mathematically as an unsteady-state case. An attempt in this direction is presented in Chapter 13 under "Penetration Theory." Other models of turbulence have been described (8, 9, 10, 11).

From Equation 12.7 it is evident that at any point at which turbulence occurs, molecular transport occurs simultaneously and by a parallel mechanism. In Illustration 12.1 the value of E_τ was usually much greater than v. Consequently, although the molecular mechanism existed, the magnitude was almost negligible. In systems with high momentum diffusivities, v may be of such magnitude that a significant part of the momentum transport is by the molecular mechanism.

The initial production of eddies has not been discussed. It is generally supposed and supported experimentally that eddy activity is caused by minute imperfections in the system (12). The experiments of Reynolds and others confirm this. If eddy activity begins, secondary eddies are formed by the primary eddies so that unless some mechanism exists to cause decay, the eddy activity will multiply indefinitely. The viscous shear of the fluid and especially the high shear rate near the walls tend to cause eddy decay so that the system of eddies comes to steady state as the rate of eddy generation equals the rate of eddy decay.

Extensive investigations have been made of the size of eddies and their energy content in relation to that of the over-all stream. It is to be hoped that the elaborate studies of turbulence, its origin, and other factors will give complete understanding sometime in the future. The oversimplified developments of the counterpart quantities in molecular and turbulent transport should present an adequate picture of the radical differences between them. Simultaneously the similarity of the mean free path of the molecule and the distance traveled by an eddy before it loses its identity and of the molecular speed and crossflow velocity should be appreciated.

It is necessary to appreciate the existence of a laminar-flowing layer and a turbulent core, both vaguely bounded and overlapping in a buffer or transition layer. When the flow pattern is known, extension of the same concepts to transport of heat and mass follows simply and logically.

HEAT AND MASS TRANSFER IN TURBULENT FLOW

The eddy activity in turbulence serves as a transfer mechanism for momentum. In addition, since gross particles of fluid move in crossflow, this eddy activity also serves as a medium for physical mixing. The physical mixing of fluids is important in mass and heat transfer. Consider a wall boundary that is capable of delivering heat or mass to a flowing stream. If the concentration of transferent property at the wall is significantly higher than in the main stream of fluid, a gradient will be set up at the wall, and transfer will occur into the fluid. After steady-state operation prevails, the concentration gradients in the various sections of the

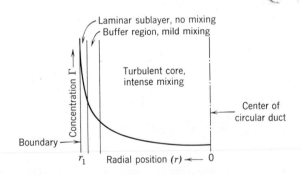

Figure 12.7. Heat and mass transfer in turbulent flow. Variation of concentration of transferent property with radial position. (Not to scale.)

fluid streams will be as shown in Figure 12.7 and as described below:

(1) The gradient across the laminar sublayer is large, with most of the difference in concentration between the wall and center of the fluid occurring across the very thin laminar sublayer. Molecular transfer occurs across this layer, with the rate dependent upon the gradient and molecular diffusivity.

(2) The gradient across the buffer region is considerably less than that in the laminar sublayer with less difference in temperature or concentration across the buffer region than across the laminar sublayer. Eddy motion tends to carry (by the mass-mixing process) some of the transferent property from the buffer region into the turbulent core and to return fluid containing less of the property or solute from the turbulent core. In addition, since a gradient is present, some of the property will transfer by the molecular-transport mechanism.

(3) The gradient through the turbulent core is very small compared to the buffer region. Some transfer will occur because of the molecular-transport mechanism, but most of the transport will occur by physical mass mixing. Any heat or solute that passes through the buffer region will circulate rapidly throughout the turbulent core. The rapid circulation tends to offset the establishment of a gradient, and the small gradient reduces the amount of transfer that can occur by molecular transport.

A general equation for transfer in the turbulent regime may be written in terms of the eddy diffusivity of transferent property.

$$\psi = -(\delta + E)\frac{d\Gamma}{dx} \qquad (12.23)$$

where E = eddy diffusivity of transferent property. This equation can be applied to cylindrical geometry by modification of the concentration gradient.

$$\psi = -(\delta + E)\frac{d\Gamma}{dr} \qquad (12.24)$$

Equation 12.24 can be used in a transferent-property balance to develop a relationship for the flux at any position in terms of the flux at the boundary. Consider a test section extending from y to $y + \Delta y$ of radius r_1 (see Figure 12.8.) In Figure 12.7 the concentration at

all radial positions was described as being very nearly constant except in the small region of the laminar sublayer and buffer region. Assume that, after passing through the test section, the displacement of concentration $\Delta\Gamma = \Gamma_{y+\Delta y} - \Gamma_y$ is constant at all radial positions.* As shown in Figure 12.8, the transferent property entering the flow section at y is $\bar{v}S\Gamma_y$, and the transferent property leaving the flow section at $y + \Delta y$ is $\bar{v}S\Gamma_{y+\Delta y}$. The transferent property leaving the test section through the wall (note that the transfer is taken as positive with increasing r) is ψA. S is the flow-tube cross-sectional area, and A is the transfer area. The balance equation may be written in terms of cylindrical geometry for a section extending from the center to radius r as

$$\bar{v}\pi r^2 \Gamma_y = \bar{v}\pi r^2 \Gamma_{y+\Delta y} + \psi 2\pi r\, \Delta y$$

or, rearranging,

$$\psi = \frac{\bar{v}r}{\Delta y}(\Gamma_y - \Gamma_{y+\Delta y}) \qquad (12.25)$$

Equation 12.25 may be written for a test section extending from the center to the boundary at r_1.

$$\psi_1 = \frac{\bar{v}r_1}{\Delta y}(\Gamma_y - \Gamma_{y+\Delta y}) \qquad (12.26)$$

If Equation 12.25 is divided by Equation 12.26,

$$\psi = \psi_1\left(\frac{r}{r_1}\right) \qquad (12.27)$$

In Equation 12.27 the flux at any radial position is shown to be linear with radius, zero at $r = 0$ and a maximum at $r = r_1$. This is a condition that is unique to "transfer with internal generation" as described in Chapter 10 and shown in Equation 10.58. The similarity of these equations suggests that transfer through the wall is similar in behavior to the internal-generation application described earlier. This is actually the case. Even though the initial mechanism of transfer through the laminar sublayer is simple transfer, the intense physical mixing in the buffer layer and turbulent core transfers the property rapidly enough so that it appears *almost* instantaneously in all parts of the turbulent core. In this sense, the rapid transfer is similar in behavior to the internal-generation applications described in Chapter 10. The transfer is not *absolutely* instantaneous as can be seen from the existence of a small gradient $(d\Gamma/dr)$ in the turbulent core, as shown in Figure 12.7. The small gradient indicates that there is some transfer

---

* For this condition to hold, the surface concentration would have to change by $\Delta\Gamma$ over the range Δy. Γ is usually dependent upon conditions outside the flow tube, and the assumption would probably not be rigorous. However, the small thickness of fluid involved in that region and the low velocity of fluid in that region would have little effect upon the entire behavior of the system.

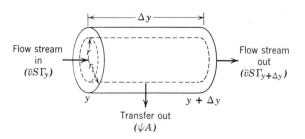

Flow stream in $(\bar{v}S\Gamma_y)$

Flow stream out $(\bar{v}S\Gamma_{y+\Delta y})$

y $y + \Delta y$

Transfer out (ψA)

Figure 12.8. Transfer in turbulent flow.

by the molecular mechanism. Thus two parallel transfer mechanisms are operating, with transport due to turbulence or physical mixing, usually predominating. An exception is the case of heat transfer in liquid metals, in which the thermal diffusivities are extremely high; therefore, a significant amount of transfer can occur with a small gradient.

Equation 12.27 can be substituted into Equation 12.24

$$\psi_1 \left(\frac{r}{r_1}\right) = -(\delta + E)\frac{d\Gamma}{dr} \qquad (12.28)$$

Specific expressions for mass and heat transfer can be written by analogy.

For mass transfer

$$\left(\frac{N_a}{A}\right)_1 \left(\frac{r}{r_1}\right) = -(\mathscr{D} + E_N)\frac{dc_a}{dr} \qquad (12.29)$$

For heat transfer

$$\left(\frac{q}{A}\right)_1 \left(\frac{r}{r_1}\right) = -(\alpha + E_q)\frac{d(\rho c_p T)}{dr}$$

or

$$\left(\frac{q}{A}\right)_1 \left(\frac{r}{r_1}\right) = -(k + \rho c_p E_q)\frac{dT}{dr} \qquad (12.30)$$

Equations 12.29 and 12.30 are working equations if the diffusivity and gradient can be evaluated. In Chapter 9, the *simple* molecular-transport theory shows equal diffusivities for mass, heat, and momentum, equal to $\frac{1}{6}\bar{c}l$. Using the same reasoning in turbulent transfer, it is reasonable to assume that $E_N \cong E_q \cong E_\tau \cong \lambda^2 |dv/dx|$; thus, E_q and E_N can be evaluated from knowledge of wall stress and velocity profile. Evidence substantiating this assumption will be included in the next chapter when the work of Martinelli is introduced.

Illustration 12.3. From the data given in Illustration 12.1, calculate dT/dr in the laminar sublayer and also at $r = r_1/2$ when the heat flux at the wall and directed toward the wall is 10,000 Btu/hr sq ft. Assume that the average properties of the fluid are those given in Illustration 12.1.

SOLUTION. In the laminar sublayer, all heat passing through the wall must pass through the laminar sublayer, and the mechanism is simple transfer. From Illustration 12.1, at $y^+ = 5$, $(r_1 - r) = 0.0001162$ ft. The laminar sublayer is thin with respect to the pipe diameter so that the change in transfer area through the laminar sublayer is negligible; therefore, from Equation 9.43 in which dT/dr is taken equal to dT/dx, the thermal conductivity of water = 0.354 Btu/hr sq ft (°F/ft). Then,

$$\frac{dT}{dr} = -\left(\frac{q}{Ak}\right) = -\frac{10,000}{(0.354)} = -28,300°\text{F/ft}$$

At $r = r_1/2$, from Illustration 12.1, $E_\tau = 385 \times 10^{-5}$ sq ft/sec, which is the assumed value of E_q. Other data are, $r_1 = 1$ in., $\rho = 62.4$ lb/cu ft, $c_P = 1$ Btu/lb°F. These values are substituted into Equation 12.30 in consistent hour units.

$$(10,000)\left(\frac{\frac{0.5}{12}}{\frac{1}{12}}\right)$$

$$= -[0.354 + (62.4)(1)(3600 \times 3.85 \times 10^{-3})]\frac{dT}{dr}$$

Therefore,

$$\frac{dT}{dr} = -5.78 \ °\text{F/ft}$$

This chapter emphasizes the similarity between heat, mass, and momentum transfer for a point between $r = 0$ and $r = r_1$ and between $y = y$ and $y = y + \Delta y$. In the following chapter the functions described here will be integrated with respect to r to form transfer coefficients that are valid at any axial position (y). Along with the integration, the analogous behavior will be examined further.

Von Kármán's equations are adequate empirical relationships for predicting the velocity profile in turbulent flow in circular conduits. In this chapter the equations were differentiated in order to evaluate the eddy diffusivity and Prandtl mixing length. Although such a differentiation is theoretically correct, it proved to lead to impossible values of the eddy diffusivity at several positions in the flow channel. This is an example of the weakness of empirical equations which have inadequate theoretical basis. Such empirical equations often cannot be extended beyond their intended purpose. It is hoped that adequate simple expressions for the velocity profile in turbulent flow may ultimately be developed. Such adequate expressions would lead to correct values of eddy diffusivity after differentiation by the methods outlined in this chapter.

REFERENCES

1. Reynolds, O., *Phil. Trans. Roy. Soc. London, Ser. A*, **174**, 935 (1883).
2. Prengle, R. S., and R. R. Rothfus, *Ind. Eng. Chem.*, **47**, 379 (1955).
3. Nikuradse, J., *VDI-Forschungsheft* **356** (1932).
4. Von Kármán, T., *Engineering*, **148**, 210–213 (1939).
5. Von Kármán. T., *Trans. ASME*, **61**, 705–710 (1939).
6. Fage, A., and H. C. H. Townend, *Proc. Roy. Soc. London, A*, **135**, 656 (1932).
7. Prandtl, L., *Essentials of Fluid Dynamics*, Hafner, New York, 1952, p. 118.
8. Taylor, G. I., *Proc. London Math. Soc.*, **20**, 196 (1921).
9. Taylor, G. I., *Proc. Roy. Soc. London, A*, **151**, 421 (I–IV) (1935).
10. Von Kármán, T., *Natl. Advisory Comm. Aeronaut. Tech. Mem. No. 611* (1921).
11. Von Kármán, T., *J. Aeronaut. Sci.*, **1**, 1 (1934).
12. Knudsen, J. G., and D. L. Katz, *Fluid Dynamics and Heat Transfer*, McGraw-Hill Book Co., New York (1958) p. 107.

PROBLEMS

12.1. Combine Equations 12.3 and 12.7 and find the radial position in the turbulent core at which the eddy diffusivity is a maximum.

12.2. Air is in turbulent flow in a smooth tube of 3 in. I.D. The air is at 70°F and 1 atm, and the pressure drop $(-\Delta P)$ is 0.025 $\mathrm{lb}_f/\mathrm{sq}$ ft per foot of tube.

(a) Calculate and plot the velocity as a function of radial position from the wall to the center of the tube on linear coordinates.

(b) Calculate and plot the eddy diffusivity (E_τ) as a function of radial position from the wall to the center of the tube. Discuss the resulting plot, with particular regard to the eddy diffusivity at $y^+ = 5$, at $y^+ = 30$, and at the center of the tube.

12.3. The pressure drop for molten sodium at 300°F flowing through a tube 1 in. I.D. is 230 $\mathrm{lb}_f/\mathrm{sq}$ ft over 10 ft of tube. The flow is turbulent.

(a) Calculate the thickness of the laminar sublayer and of the buffer layer.

(b) Calculate and plot the velocity as a function of radial position from the wall.

(c) Plot the velocity profile for laminar flow which has the same maximum velocity as the case for turbulent flow in part (b).

(d) Would the two cases of parts (b) and (c) have the same average velocity?

(e) What is the pressure drop in part (c)?

12.4. A valve on a 2 in. line carrying benzene at 70°F can be used to adjust the flow rate. Pressure gages measure the pressure drop across the pipe for various flow rates. For pressure drops of 0.01 and 1.0 $\mathrm{lb}_f/\mathrm{sq}$ ft per foot of tube:

(a) Calculate and compare the thicknesses of the laminar sublayer for turbulent flow.

(b) Calculate and plot the velocity as a function of radial position for turbulent flow.

(c) The pressure drop of 0.01 $\mathrm{lb}_f/\mathrm{sq}$ ft indicates sufficiently low velocity to give laminar flow. Calculate and plot the laminar-flow velocity profile for the pressure drop on the same plot as in part (b).

12.5. The velocity at the center of a 6 in. duct was found to be 40 ft/sec for air at 70°F and 1 atm. At this velocity turbulent flow exists.

(a) Calculate the shear stress at the wall.

(b) Calculate the thickness of the laminar sublayer.

(c) Calculate the eddy diffusivity at $y^+ = 5$ and $y^+ = 30$. Discuss.

12.6. Calculate the Prandtl mixing length, eddy length, and cross-flow velocity for the conditions stated in Problem 12.2:

(a) At $y^+ = 30$.

(b) At $r = r_1/2$.

Discuss the results.

12.7. Discuss the applicability of the eddy length and Prandtl mixing length near the center of the turbulent core.

12.8. Show that with the assumption $u^+ = y^+$ in the laminar sublayer, the eddy diffusivity is *not* zero from $y^+ = 0$ to $y^+ = 5$. Explain the reason and suggest a correction so that $E_\tau = 0$ at all values of y^+ between 0 and 5.

12.9. Calculate the eddy diffusivity, Prandtl mixing length, eddy length, and crossflow velocity for the conditions of Problem 12.3:

(a) At $y^+ = 2$.   (c) At $y^+ = 20$.

(b) At $y^+ = 5$.   (d) At $y^+ = 40$.

Discuss the results.

12.10. For heat transfer in turbulent flow, calculate the temperature gradient (dT/dr). The heat flux through the wall is 500 Btu/hr sq ft.

(a) In the laminar sublayer.

(b) At $y^+ = 25$.

(c) At $r = r_1/2$ for the conditions of Problem 12.2.

chapter 13

Fundamentals of Turbulent Transfer

In Chapter 12 turbulence was studied in detail in an effort to describe the individual mechanisms that are operating at each radial position at steady state. The mathematical manipulation of velocities, gradients, and diffusivities at points in a system is cumbersome. It would be advantageous for design purposes if the same information were available in the form of simpler functions. The simple functions should respond readily to mathematical handling. In addition, the data required to evaluate the functions should not be difficult to determine and, if possible, should be in terms of gross dimensions, mean velocities, and the physical properties of the system. Equations expressed in gross dimensions and mean velocities are the result of integration between boundaries of equations describing point conditions such as Equation 12.24. The integrated equations describe over-all conditions at some position along a flow duct and define *transfer coefficients*. Certain of the integrations cannot be performed by ordinary analytical procedures, in which case the integrated form may be indicated or defined and then evaluated or correlated by empirical methods based upon experimental data. The correlation of transfer coefficients is a complex procedure because the transfer coefficients are usually dependent upon several simultaneous parallel mechanisms, such as momentum transfer by turbulent transport and heat transfer by molecular transport. The general form of the simplest function for this purpose can be predicted by *mechanism-ratio analysis*, a procedure described in this chapter. An alternative procedure, *dimensional analysis*, which is based upon measurable variables, such as velocity, diffusivity, and length, is described in Appendix A.

In addition to the definition and evaluation of transfer coefficients, the analogous behavior of mass, heat, and momentum transfer is examined in this chapter. There are other applications of mass, heat, and momentum transport that are dependent upon conditions other than those heretofore described. For example, in certain momentum-transfer phenomena, the velocity distributions vary with axial position as well as with radial position as in the entry length to a pipe or around bodies that offer large projected areas normal to flow. These applications are examined in a section devoted to skin friction in the boundary layer on a smooth surface and form drag due to accelerative forces in fluids flowing over and around bulky shapes. In addition, heat and mass transfer in laminar flow, heat transfer associated with motion due to natural rather than forced convection, and heat transfer associated with condensing vapors and boiling liquids are considered in this chapter.

MECHANISM-RATIO ANALYSIS

Most of the practical applications of the transfer phenomena are more complex than the systems heretofore described. For example in a heat exchanger, a moving fluid may be heated. In this simple application both heat and momentum are transferred. If a moderate velocity is maintained, heat and momentum are transferred by parallel molecular and turbulent mechanisms. As has been shown in Chapters 9, 10, and 12, the mechanisms are interrelated. However, the mathematical expression of the interrelations is cumbersome. In addition, certain of the information is empirical, particularly in those applications involving turbulent flow. Empirical information is not based upon theoretical foundations derived from first principles but is obtained from prior experience. Therefore, a procedure will be developed by which the operations involving complex multiple mechanisms can be expressed in relatively simple mathematical form for the correlation of experimental data.

In Appendix A, the ordering of data by dimensional analysis is described. In this chapter, the more direct mechanism-ratio analysis is used to predict the form of the function required to describe the interrelation of mechanisms in a complex system. According to dimensional analysis, the appropriate *variables* must be assumed or predicted before the analysis can be completed. According to mechanism-ratio analysis, the over-all operating *mechanisms* must be predicted. Typical mechanisms are momentum transfer by molecular interaction, heat transfer due to eddy activity, and so forth. Mechanisms are usually easier to predict than variables. Each of the mechanisms can usually be described in terms of a combination of several variables. Consider for example a system in which three mechanisms, A, B, and C, are predicted by qualitative examination, experience, or intuition. Mechanisms A, B, and C are represented by variables, such as τ, ρ, $\bar{v}$, k, μ, and so forth, arranged in groups so that the dimensions of mechanisms A, B, and C are the same. Therefore the ratios (A/B), (B/C), and (A/C) are dimensionless. The empirical relationship between A, B, and C may be written as an exponential series of terms.

$$A = KB^b C^c + K'B^{b'}C^{c'} + \ldots \quad (13.1)$$

where K, K' = dimensionless constant terms
b, b', c, c' = dimensionless constant exponents

In an equation of this type, mechanism A is completely described in terms of mechanisms B and C. This is an exponential series of similar terms, with the terms differing only in the numerical value of the constants and exponents. With this equation, only the first term need be examined for form, and conclusions drawn from this term may be applied to all terms. The first term is

$$A = KB^b C^c \quad (13.2)$$

Equation 13.2 may be divided by B.

$$\left(\frac{A}{B}\right) = K\frac{B^b}{B}C^c = KB^{b-1}C^c \quad (13.3)$$

The left-hand term (A/B) is dimensionless, and K is defined as dimensionless. Therefore, if the equation is dimensionally consistent, $B^{b-1}C^c$ must be dimensionless, or $(-b+1) = c$. Then Equation 13.3 may be written

$$\frac{A}{B} = K\left(\frac{C}{B}\right)^c \quad (13.4)$$

The complete series may be written

$$\left(\frac{A}{B}\right) = K\left(\frac{C}{B}\right)^c + K'\left(\frac{C}{B}\right)^{c'} + \ldots \quad (13.5)$$

The same reasoning may be applied to an equation including mechanisms totaling n.

$$\left(\frac{A}{B}\right) = K\left(\frac{C}{B}\right)^c \left(\frac{D}{B}\right)^d \ldots \left(\frac{N}{B}\right)^n + \ldots$$
$$+ K'\left(\frac{C}{B}\right)^{c'} \left(\frac{D}{B}\right)^{d'} + \ldots + \left(\frac{N}{B}\right)^{n'} + \ldots \quad (13.5)$$

From this development, the following general properties of mechanism-ratio analysis may be stated:

1. If n mechanisms are operating, an exponential-series equation written in $(n-1)$ dimensionless ratios in each term is required to define the system.

2. For each term of an exponential series, one constant and $(n-2)$ exponents must be evaluated. Similar reasoning for dimensionally consistent equations was used in Appendix A in the description of dimensional analysis.

The mechanisms that depend upon turbulence usually cannot be described exactly, but usually a group of variables can be arranged that results in a number proportional to, rather than equal to, the effect of the mechanism. If a mechanism-ratio analysis is used, the proportionate descriptions are applicable when constants can be determined empirically. Examples of mechanism-ratio analysis follow throughout this chapter. At the end of the chapter the subject will be summarized.

THE STRESS-MEMBRANE MODEL

A momentum-transfer equation especially suited for design can be written based upon a mechanism-ratio analysis. In order to do so, a model must be described that includes all the mechanisms. From the discussion in Chapter 12, a qualitative picture of a turbulent stream can be drawn, made up of a laminar portion of the fluid near the boundary and the turbulent core characterized by eddy activity. At a low mean velocity of fluid, the turbulent core does not exist; therefore, for simplicity, laminar flow will be examined first. In *laminar flow*, the fluid moves in the flow direction, with no component of velocity in any other direction. Stress is set up in the fluid according to the expression $\tau_y g_c = -\mu \, dv/dx$, as derived in Chapter 9. Thus, the linear-flow path and the stress are interrelated, so that the stress results in maintaining the flow linear, and the linear flow produces the stress. For the case of laminar flow through tubes, the tube may be considered to be filled with an infinite number of concentric tubular stress membranes, each of which tends to confine the flow within the annuli formed between the membranes. The *strength* of the membranes (or the stress in the fluid) increases linearly from zero at the center to a maximum at the wall (see Equation 10.79).

If some eddy-forming event occurs, a portion of the fluid may be subject to a force tending to give a component of velocity in a direction other than y. Opposing this force, the stress membrane will tend to confine the flow in the orderly manner. If the kinetic energy of the distorted fluid element is high enough, it can penetrate the stress membrane and form a true eddy. In this model, the region of zero stress at the center of the tube is the most likely portion of the tube to exhibit eddy activity, which is in agreement with observation. In addition, if an eddy forms and penetrates the stress membrane, the penetration continues until a region of high stress is encountered, high enough to overcome the kinetic energy of the eddy. At this point the eddy is absorbed and the orderly flow resumes. The region of highest stress is in the vicinity of the wall, so that little, if any, eddy activity occurs near the wall. This hypothesis is also in agreement with observation. At the wall, the stress membrane is supported by the wall, since the wall is strong enough that no eddy can penetrate it. The support by the wall extends into the fluid to form the laminar sublayer, in which no eddy activity occurs. This hypothesis is also in agreement with observation.

REYNOLDS NUMBER–FRICTION FACTOR CORRELATION

In an effort to set up mechanism-ratio relationships, it is necessary to examine the stress–kinetic-energy relationships that apply in fluid systems as expressed in the stress-membrane model. It is desirable to use mean velocities and gross dimensions for ease in calculation and analysis of experimental data. Expressions for the laminar regime will be developed first, for circular conduits. In the *laminar* regime a relationship between point velocity and mean velocity in a circular duct is given by Equation 10.87

$$v = 2\bar{v}\left(1 - \frac{r^2}{r_1^2}\right) \qquad (10.87)$$

where $\bar{v}$ is the average velocity and r_1 is the radius at the boundary. This equation can be differentiated to

$$\frac{dv}{dr} = \frac{-4\bar{v}r}{r_1^2} \qquad (13.6)$$

This equation can be substituted into the stress-gradient equation, Equation 9.62, for laminar flow to give

$$\tau_y g_c = \frac{4\mu\bar{v}r}{r_1^2} \qquad (13.7)$$

At the wall, $(\tau_y g_c) = (\tau_y g_c)_1$ and $r = r_1$, so that

$$(\tau_y g_c)_1 = \frac{4\mu\bar{v}}{r_1} \qquad (13.8)$$

Substituting the duct diameter (D) for the radius gives

$$(\tau_y g_c)_1 = \frac{8\mu\bar{v}}{D} \qquad (13.9)$$

Thus the stress at the wall is written in terms of the diameter and mean velocity. Equation 13.9 states that the total transfer of momentum to the wall $(\tau_y g_c)_1$ is equal to the rate of transfer of momentum through the fluid by molecular transport $(8\mu\bar{v}/D)$. Thus a relationship is written for two mechanisms, total transfer to the wall, and transfer by molecular diffusion. The mechanisms are equal in *laminar* flow because no other mechanisms are operative in the system described.

The determination of wall stress in a fluid system is usually indirect and is based upon the pressure drop per unit length of duct. The wall stress and pressure drop are related by a force balance.

$$(\tau_y A)_1 = (-\Delta P)S_1 \qquad (13.10)$$

In terms of momentum, the force balance states that the rate of transfer of momentum *to* the wall $(\tau_y A)_1$ is equal to the rate of transfer of momentum *from* the fluid $(-\Delta P)S_1$. The force balance represents total forces and therefore total momentum transfer irrespective of mechanism. For cylindrical geometry $S_1 = \pi D^2/4$ and $A_1 = \pi DL$ where L is the length of the tube; then, Equation 13.10 can be used to replace $(\tau_y)_1$ in Equation 13.9.

$$(\tau_y g_c)_1 = \frac{-(\Delta P)g_c D}{4L} = \frac{8\mu\bar{v}}{D} \qquad (13.11)$$

or

$$4(\tau_y g_c)_1 = \frac{-(\Delta P)g_c D}{L} = \frac{32\mu\bar{v}}{D} \qquad (13.12)$$

In a system in which turbulent flow is also present, the turbulent mechanism must be combined with the two presented in Equation 13.9. The stress-membrane model considers an eddy with crossflow-directed kinetic energy sufficient to pierce the stress membranes. The mechanism-ratio analysis requires that a group of variables functionally related to this kinetic energy be incorporated with the other mechanism expressions. The kinetic energy of a particle of mass (**m**) moving in the y-direction at velocity v is represented by $\mathbf{m}v^2/2g_c$. In fluids it is more convenient to use density rather than mass; therefore, the kinetic energy per unit volume of a fluid element of volume dV is $d(\rho v^2/2g_c)$. If the differential expression is integrated over all positions in the flow duct based upon average velocity (2), the kinetic energy per unit volume of the fluid is

$$\frac{\text{K.E.}}{V} = \frac{\rho\bar{v}^2}{2\alpha g_c}$$

In tubular ducts α is a number ranging between 0.5 for the usual velocity profile of laminar flow and unity for the plug-flow condition postulated at $\bar{v} = \infty$, in which turbulence is a maximum and the velocity is invariant across the duct. A complete mathematical statement of turbulence includes recognition of the intensity and direction for each eddy in the system. As stated in Chapter 12, this information is not presently known. Since eddy activity exhibits a random behavior, the mean y-directed kinetic energy of all eddies in a system may be assumed proportional to the kinetic energy of the fluid, as represented by $\rho\bar{v}^2/2\alpha g_c$. Then $\rho\bar{v}^2/2\alpha g_c$, or multiples thereof such as $\rho\bar{v}^2/2*$, are sufficient mathematical description of the turbulence mechanism to include in a mechanism-ratio analysis.

Consider the laminar-flow mechanism described in Equation 13.12. As the mean velocity in the system is increased, at some velocity, momentum transfer by the turbulent mechanism will begin. If turbulence occurs, the appropriate mechanism expression should be present in the equation. The term $\rho\bar{v}^2/2$ has the same dimensions as the terms in Equation 13.12. If Equation 13.12 is divided by $\rho\bar{v}^2/2$,

$$\left[\frac{8(\tau_y g_c)_1}{\rho\bar{v}^2}\right] = \left[\frac{2(-\Delta P)g_c D}{\bar{v}^2 \rho L}\right] = 64\left[\frac{\mu}{D\bar{v}\rho}\right] \quad (13.13)$$

Now each term in Equation 13.13 is a dimensionless ratio of two mechanisms. The two bracketed left-hand terms represent the ratio

$$\frac{\text{Total momentum transfer}}{\text{Momentum transfer by the turbulence mechanism}}$$

The mathematical expressions for this ratio define the *friction factor* (f) where

$$f = \frac{8(\tau_y g_c)_1}{\rho\bar{v}^2} = \frac{2(-\Delta P)g_c D}{\bar{v}^2 \rho L} \quad (13.14)$$

The bracketed right-hand in Equation 13.13 term represents the ratio

$$\frac{\text{Momentum transfer by molecular transport}}{\text{Momentum transfer by the turbulence mechanism}}$$

Traditionally, the reciprocal of this ratio is defined as the *Reynolds number*, N_{Re}.

$$N_{\text{Re}} = \frac{D\bar{v}\rho}{\mu} \quad (13.15)$$

Equation 13.13 may be written

$$f = \frac{64}{N_{\text{Re}}} \quad (13.16)$$

which holds for laminar flow only. Since f and N_{Re} each represent dimensionless ratios of two mechanisms, Equation 13.16 may be written more generally for turbulent flow as

$$f = \phi(N_{\text{Re}}) = K(N_{\text{Re}})^a \quad (13.17)$$

---

* In most instances of turbulent flow α varies between 0.90 at low turbulent velocities and 1.00 at high turbulent velocities. See Figure 20.2.

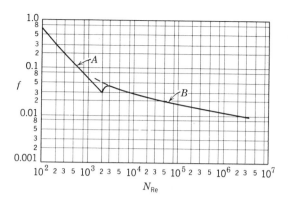

Figure 13.1. Variation of friction factor with Reynolds number. (A more complete chart is presented in Appendix C-3.)

Equation 13.17 has the same form as Equation 13.4 which was used to describe a three-mechanism analysis. A counterpart derivation of Equation 13.17 may be performed by dimensional analysis (see Appendix A).

Equation 13.13 is a laminar-flow equation. It is equivalent to the Poiseuille equation for laminar flow through smooth tubes.

$$-\Delta P = \frac{32\mu L\bar{v}}{g_c D^2} \quad (13.13a)$$

In laminar flow or in turbulent flow, the term $\rho\bar{v}^2$ is finite. Recall that Equation 13.12 was divided by $\rho\bar{v}^2$ to form Equation 13.13. If no turbulence is present, Equation 13.12 is valid as written, and $\rho\bar{v}^2$ is extraneous and cancels from Equation 13.13. However, if turbulence is present, the finite turbulence mechanism can be expected to change the nature of the exponents and constant of Equation 13.17. The validity of the assumptions used in this statement of mechanism-ratio relationships can be tested with experimental data in the laminar and turbulent regime.

Figure 13.1 is a logarithmic plot of friction factor as a function of Reynolds number over a range of Reynolds numbers between 100 and 10,000,000 for flow in smooth tubes.

The data for the plot were taken over a wide range of each of the variables (velocity, density, tube diameter, etc., using liquids and gases) in smooth tubes. The system was operating in *well-developed flow*. Well-developed flow means that the velocity pattern in the tube is the same at all points along the test length. For the tests all pressures were measured far enough downstream of the tube entrance so that entrance effects do not contribute any error. The subject of entrance effects is covered later in this chapter in the section on boundary layers. An important feature to notice is that the data fall into two distinct curves separated by a sharp discontinuity at $2100 < N_{\text{Re}} < 3500$.

Curve A in Figure 13.1 is continuous from low Reynolds numbers and ends at $N_{Re} = 2100$. This curve is in exact agreement with Equation 13.16. It is in this range of Reynolds number that all flow is laminar. According to the stress-membrane model, it is in this region that no incipient eddies have sufficient energy to break through the stress membrane. Above $N_{Re} = 3500$ in smooth tubes the fluid is normally in turbulent flow, as represented by curve B in Figure 13.1. It is in this region that eddy activity is violent enough to break through the stress membrane and that momentum is transferred by eddy activity as well as by molecular transport.

In the range between $N_{Re} = 2100$ and $N_{Re} = 3500$, unstable *transition* flow occurs, and either laminar flow, turbulent flow, or a combination of both may exist. Behavior in the transition region is a function of fluid properties, system geometry, system kinematics, and system history. These data are consistent with the stress-membrane model and with the observations described in the Reynolds experiment in Chapter 12. The simplicity of the plotted data for wide variation of all variables confirms the usefulness of the analysis.

The Reynolds Number. The Reynolds number was originally proposed by Sir Osborne Reynolds about the middle of the last century as a criterion to delineate the nature of the flow in ducts and pipe. Many additional forms of the Reynolds number have been proposed and used for systems other than circular pipe. For example, data on structures subject to force from wind, such as buildings, bridges, ship hulls in water, airfoils on aircraft, or any shape of a solid boundary subject to a moving fluid, are correlated using the Reynolds number in conjunction with other dimensionless groups. In each of the examples listed above, the ratio of momentum transferred by the eddy mechanism to momentum transferred by molecular transport is of major significance. The terms used to express the ratio may differ in general form for different systems, but the significance of the ratio is the same. The Reynolds number appears in equations for boundary layers, form drag, agitation, classification of solid particles, fluidization, and many other unit operations.

The Reynolds number may be derived in several different ways. For example, consider the equation for momentum transfer in cylindrical geometry

$$\tau_y g_c = -(\nu + E_\tau)\rho \frac{dv}{dr} \qquad (12.7a)$$

The right-hand term of this equation expresses the rate of transfer of momentum by two mechanisms, and the two mechanisms may be separated into

Rate of transfer by molecular transport, $-\nu\rho(dv/dr)$

and

Rate of transfer by turbulent transport, $-E_\tau\rho(dv/dr)$

Combining the two separate terms in the appropriate mechanism ratio gives

Momentum transfer by the turbulence mechanism
————————————————————————————
Momentum transfer by the molecular mechanism

$$= \frac{-E_\tau\rho(dv/dr)}{-\nu\rho(dv/dr)} \qquad (13.18)$$

The right-hand term of Equation 13.18 is made up of the *point values* of the two transfer mechanisms. At any point, (dv/dr) is constant irrespective of mechanism; therefore, Equation 13.18 may be written to define a *point value* of the Reynolds number, $(N_{Re})_r$

$$(N_{Re})_r = \frac{(E_\tau)_r}{\nu} \qquad (13.18a)$$

where the subscript r represents any radial position.

In Chapter 12, the eddy diffusivity of momentum is defined as the product of the crossflow pulsation velocity and the Prandtl mixing length $(E_\tau = \lambda|u'|)$, and the crossflow pulsation velocity is equal to the axial-flow pulsation velocity, or $|u'| = |v'|$; therefore, the point value of the Reynolds number may be written

$$(N_{Re})_r = \frac{(\lambda|v'|)_r}{\nu} \qquad (13.19)$$

The quantities λ and v' may be integrated over the turbulent core to give the mean values $\bar{\lambda}$ and $\bar{v}'$ in which case $\bar{\lambda} \sim D$ and $|\bar{v}'| \sim \bar{v}\sqrt{f/8}$. If the small variation in $\sqrt{f/8}$ is neglected, then $(\bar{\lambda}\bar{v}') \sim (D\bar{v})$ and the Reynolds number may be written

$$N_{Re} = \frac{D\bar{v}}{\nu} = \frac{D\bar{v}\rho}{\mu} \qquad (13.15)$$

In Appendix A dynamic similarity is described as an unique condition of two systems in which all counterpart forces bear a constant ratio. The forces dependent upon momentum transfer in cylindrical geometry can be extracted from Equation 12.7.

$$(\tau_y g_c) = \frac{Fg_c}{A} = -(\nu + E_\tau)\rho\left(\frac{dv}{dr}\right)$$

or

$$F = F_m + F_e = -(\nu + E_\tau)\frac{A\rho}{g_c}\left(\frac{dv}{dr}\right) \qquad (13.20)$$

where F = total forces resulting from momentum transfer at any position r

F_m, F_e = forces associated with molecular transport and eddy activity, respectively at any position r

A = transfer area at any position r

The ratio of forces associated with eddy activity and molecular transport can be extracted from Equation 13.20 and written as a ratio,

$$\left(\frac{F_e}{F_m}\right) = \frac{\dfrac{-E_\tau A\rho}{g_c}\left(\dfrac{dv}{dr}\right)}{\dfrac{-\nu A\rho}{g_c}\left(\dfrac{dv}{dr}\right)} = \frac{(E_\tau)_r}{\nu} \qquad (13.21)$$

E_τ can be eliminated by use of Equation 13.19 and the Reynolds number for the duct may be written

$$N_{\text{Re}} = \frac{D\bar{v}\rho}{\mu} \qquad (13.15)$$

From the considerations leading to Equation 13.21 any two systems that operate at the same Reynolds number are dynamically similar with respect to forces associated with momentum transfer.

A useful relationship can be developed by writing a simple mass-balance equation for flow through ducts.

$$w = \bar{v}\rho S \qquad (13.22)$$

where w = mass rate of flow
$\bar{v}$ = mean velocity
ρ = density
S = duct cross-sectional area

This equation is called the *continuity* equation. In ducts of constant cross section operating at steady state, w and S are constant, and

$$\frac{w}{S} = \bar{v}\rho = \text{constant}$$

The product $\bar{v}\rho$ is constant for any fluid irrespective of change in density due to temperature or pressure in compressible fluids. The velocity adjusts to changes in density. This useful product is called the *mass velocity* and given the symbol $G = \bar{v}\rho$. Using this symbol, the Reynolds number is frequently written as

$$N_{\text{Re}} = \frac{D\bar{v}\rho}{\mu} = \frac{DG}{\mu} = \frac{D\bar{v}}{\nu} \qquad (13.23)$$

The Friction Factor. As discussed above, the friction factor is proportional to the ratio of the momentum loss of the fluid and the momentum loss by eddy activity. The friction-factor–Reynolds-number correlation chart, Figure 13.1, is based upon *smooth* tubes and consequently is dependent upon *skin friction*. *Skin friction* is that portion of the fluid friction that is associated with a *tangential* force on a smooth surface that is oriented *parallel* to the direction of flow. Reference will be made to *form friction* or *form drag*, for nontangential forces later in this chapter. In contrast to smooth tubing, the inside surface of commercial *pipe* is not

"smooth" in the sense that the word is used here. The friction for new commercial pipe in the turbulent regime is some 20 to 30 per cent higher than for smooth tubing. This behavior of pipe of varying roughness will be discussed in Chapter 20.

The data for Figure 13.1 were taken after flow was *fully developed*. *Fully developed flow* exists in a tube in which the velocity distribution pattern does not change with length along the flow duct. Consider, for example, a large filled tank with a tube connected near the bottom through which flow occurs. As fluid enters the tube from the tank, it will be flowing at a uniform velocity and will develop a velocity distribution pattern because of friction on the walls. As the moving fluid adjusts itself to new surroundings, the point velocities change until some predictable laminar or turbulent pattern results. The adjustment requires some twenty-five to seventy-five tube diameters of travel in the tubing. This portion is called the entry length. Beyond the entry length, the flow is fully developed.

A number of expressions of friction factor appear in present-day literature. None of these expressions differ in the over-all concept. The difference lies in the value and placement of a constant. The friction factor used in *this book* is written as

$$f = \frac{\left(\dfrac{-\Delta P}{\rho}\right)\left(\dfrac{D}{L}\right)}{\left(\dfrac{\bar{v}^2}{2g_c}\right)} \qquad (13.24)$$

The groups $(-\Delta P/\rho)$ and $(\bar{v}^2/2g_c)$ will be encountered in the mechanical energy balance for fluids, in Chapter 20. Another definition of the friction factor that may be encountered elsewhere (1) is

$$f' = \frac{(-\Delta P)g_c D}{2\bar{v}^2 L\rho} \qquad (13.25)$$

Then

$$f = 4f'$$

The form described in Equation 13.25 is *not* used in this book. When using friction-factor–Reynolds-number plots, or the equivalent equations, note carefully the definition of f associated with that particular correlation.

Fluid-Flow Equations. The data reported in Figure 13.1 may be expressed by several equations. Curve A, in the laminar-flow region is represented by Equation 13.16,

$$f = \frac{64}{N_{\text{Re}}} \qquad (13.16)$$

Equation 13.16 is also plotted in Appendix C-3.

Several equations, all empirical are offered for the turbulent-flow region, the first is valid over the limited range for N_{Re} between 5000 and 200,000.

$$f = \frac{0.184}{(N_{Re})^{0.2}} \qquad (13.26)$$

A plot of $f/8$ as a function of N_{Re} is also plotted in Appendix C-5. A more exact equation that is valid over a wider range, for N_{Re}, between 3000 and 3,000,000.

$$f = 0.00560 + \frac{0.5}{(N_{Re})^{0.32}} \qquad (13.27)$$

Forms of other equations may be postulated from the mechanism-ratio analysis (Equation 13.5). The curvature of curve B indicates that the first term of the infinite series is not an adequate description of this phenomenon. The choice of a graphical representation or a more complex equation must be faced.

In Equations 13.26 and 13.27 the velocity appears in both the Reynolds number and the friction factor. When velocity is the unknown variable, these equations are inconvenient, since a trial-and-error solution is required. Equation 13.28 is another equation form for the same data so arranged that velocity appears in the left term only.

$$\frac{1}{\sqrt{f}} = 2 \log_{10} (N_{Re} \sqrt{f}) - 0.80 \qquad (13.28)$$

Note the similarity in form between $1/\sqrt{f}$ and u^+ of Chapter 12 and also $(N_{Re}\sqrt{f})$ and y^+.

The foregoing equations are the results of analyses of experimental data and are therefore empirical. However, the means of presentation has at least some basis in theory.

Illustration. 13.1 Calculate the pressure drop through 100 ft of smooth tubing for an oil flowing at a mean velocity of 8 ft/sec. The tubing diameter is 3 in., $\mu = 5$ centipoises, $\rho = 60$ lb/cu ft.

SOLUTION. Since no knowledge of the flow regime is available, the first step in the calculation is to determine the Reynolds number.

$$N_{Re} = \frac{D\bar{v}\rho}{\mu} = \frac{(3/12)(8)(60)}{(5 \times 6.72 \times 10^{-4})} = 35,700$$

This is in the turbulent range and therefore Figure 13.1, Equation 13.26, or Equation 13.27 may be used. From Figure 13.1, at $N_{Re} = 35,700$

$$f = 0.023 = \frac{2(-\Delta P)g_c D}{\bar{v}^2 L \rho}$$

Therefore,

$$(-\Delta P) = \frac{f\bar{v}^2 L \rho}{2g_c D} = \frac{(0.023)(64)(100)(60)}{(2)(32.2)(3/12)} = 548 \text{ lb}_f/\text{sq ft}$$

$$(-\Delta P) = \frac{548}{144} = 3.81 \text{ psi}$$

Illustration 13.2. Calculate the velocity of an oil flowing through a 3-in. tube. The pressure drop through the tube is 548 lb$_f$/sq ft per 100 ft of tube. Oil properties are $\mu = 5$ centipoises, $\rho = 60$ lb/cu ft.

SOLUTION. In this problem, velocity is unknown. Since velocity appears in both N_{Re} and f, neither of these groups can be evaluated, so that for turbulent flow a direct solution is not possible with Figure 13.1 nor with Equations 13.26 and 13.27. In the viscous-flow regime a direct solution is possible because Equation 13.16 can be rearranged to Equation 13.12 for direct solution.

Assuming turbulent flow, Equation 13.28 can be used.

$$\frac{1}{\sqrt{f}} = 2 \log_{10} (N_{Re} \sqrt{f}) - 0.80$$

Expanding f and N_{Re} gives

$$\sqrt{\frac{\bar{v}^2 L \rho}{2(-\Delta P)g_c D}} = 2 \log_{10} \left(\frac{D\bar{v}\rho}{\mu} \sqrt{\frac{2(-\Delta P)g_c D}{\bar{v}^2 L \rho}} \right) - 0.80$$

The velocity may be canceled in the right-hand term to give

$$\sqrt{\frac{\bar{v}^2 L \rho}{2(-\Delta P)g_c D}} = 2 \log_{10} \left(\frac{D\rho}{\mu} \sqrt{\frac{2(-\Delta P)g_c D}{L\rho}} \right) - 0.80$$

The right-hand term can be evaluated with the data at hand.

$$\sqrt{\frac{\bar{v}^2 L \rho}{2(-\Delta P)g_c D}} = 2 \log_{10} \left(\frac{(3/12)(60)}{(5)(0.000672)} \right.$$

$$\left. \times \sqrt{\frac{2(548)(32.2)(3/12)}{(100)(60)}} \right) - 0.80$$

$$\sqrt{\frac{\bar{v}^2 L \rho}{2(-\Delta P)g_c D}} = 6.67$$

Then

$$\bar{v}^2 = \frac{(6.67)^2(2)(548)(32.2)(3/12)}{(100)(60)} = 64$$

and $\quad \bar{v} = 8.0$ ft/sec

Shapes Other Than Cylindrical. For duct shapes other than cylindrical, the above equations may be used with appropriate modification for the new geometry. The flow-duct geometry in the friction factor and Reynolds number is defined by the diameter of the tube. For shapes other than circular the diameter must be replaced by an appropriately chosen variable or group of variables that describe the system with a single linear dimension which is equivalent in behavior to D. If a variable were chosen that was not equivalent, curves A and B in Figure 13.1 would be shifted, and the curves as shown would be useless.

The geometrical factor is introduced into the flow equations in the force balance.

$$(-\Delta P)S_1 = (\tau_y A)_1$$

In cylindrical geometry after substituting $S_1 = \pi D^2/4$ and $A_1 = \pi DL = bL$,

$$D = \frac{4(\tau_y)_1 L}{(-\Delta P)} = \frac{4S_1}{b} \qquad (13.29)$$

where b is the wetted perimeter of the duct $= \pi D$.

For any other shape a more general form of Equation 13.29 must be developed. If more than one surface is present, as in the annulus, the various surfaces probably exert different stresses on the fluid so that the force balance is

$$(-\Delta P)S_1 = \tau_1 A_1 + \tau_2 A_2 + \ldots \qquad (13.30)$$

If the stresses are not equal, Equation 13.30 does not reduce to any simple function. If as an *approximation* the mean stress is used to replace the different stresses on the various surfaces, Equation 13.30 reduces to Equation 13.29. For the shapes normally encountered, Equation 13.29 is a reasonable approximation except in extremes, such as annuli made up of small wires running through the center of large tubes.

For shapes other than those circular in cross section, *assuming* that the mean stress is a satisfactory replacement for the actual wall stresses, a replacement term D_{eq} for the geometrical factor D may be written,

$$D_{eq} = \frac{4S_1}{b} \qquad (13.31)$$

where D_{eq} = the equivalent diameter.

In other words, the equivalent diameter is equal to four times the cross-sectional area of the duct divided by the wetted perimeter. By substitution of appropriate terms for cylindrical geometry, it can be shown that for cylinders $D_{eq} = D$. The friction factor measured for the special case of round pipe cannot be expected to fit *rigorously* for noncircular conduits of various configurations.

Illustration 13.3. Calculate the equivalent diameter for a rectangular duct 3 ft high and 5 ft wide.

SOLUTION. For a rectangular duct the stress at every point on the wall is not constant. Since the dimensions of the duct are of about the same order, in the absence of more exact information Equation 13.31 may be used as an *approximation* for the equivalent diameter.

$$D_{eq} = \frac{4S_1}{b} = \frac{(4)(5 \times 3)}{2(3 + 5)} = 3.75 \text{ ft}$$

THE BOUNDARY LAYER

The term *fully developed turbulence* mentioned earlier refers to the condition which occurs after a fluid has progressed through a duct far enough so that no further change will take place in the velocity pattern with further progress through the duct. Fully developed turbulence is assumed in all work previous to this topic. This section will examine the behavior of fluids prior to the full development of turbulence. Examples of this area of study are (1) entry section of long tube and (2) large-diameter short ducts. Outside the field of chemical

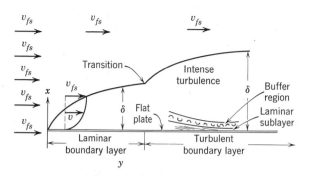

Figure 13.2. Boundary-layer build-up on a flat plate. (Vertical scale magnified.)

engineering, those involved in the design of ship hulls and airfoils are more concerned with the study of this flow behavior.

Consider a flat plate with an edge facing the direction of flow. The flat plate can be visualized as suspended in an infinite duct so that the volume of the plate has no effect upon the fluid velocity in the duct; i.e., the volume of the plate is negligible with respect to the volume of the duct. The velocity of the bulk of the fluid (free stream) is designated as v_{fs}. At the instant a differential element of moving fluid contacts the leading edge of the flat plate, the velocity of that fluid element immediately decreases to zero, which is consistent with the concept of no slip at a boundary. Any acceleration or deceleration is associated with a force; therefore, the deceleration of the fluid element generates a stress in the flow direction on the surface. At the leading edge of the plate, an element of fluid of differential volume is decelerated instantaneously by a differential area of the surface. Instantaneous deceleration requires an infinite force. Mathematically, $(\tau_{y1})_{y=0} = \infty$. The next element of fluid (measured in the x-direction) would still be moving at $v = v_{fs}$. Successive layers from the wall will be retarded as y increases and the stress distributes itself over a layer of some thickness.

At a distance y along the surface, a finite velocity gradient will be set up so that at some small distance $x = \delta$ from the surface, the fluid velocity would approach (say within 99 per cent) the free-stream velocity. The region in which the fluid velocity is less than 99 per cent of the free-stream velocity is defined as the *boundary layer*. The boundary layer extends from the leading edge ($y = 0$) to the end of the plate ($y = y_t$) and extends from the surface ($x = 0$) to the boundary-layer limit ($x = \delta$). At $y = 0$, $\delta = 0$ and δ increases to a maximum at $y = y_t$. The boundary-layer geometry is shown in Figure 13.2. The *point stress* (or stress at a specific position y) on the plate surface also varies so that the stress is a maximum at the leading edge and diminishes as y increases. It may be noted that a plate as described can operate at *steady state*, in that at any position y there is no change in any property with time, but it can never operate in *fully developed flow*, because the boundary layer continues to change regardless of how long the plate is made.

It is evident that near the leading edge δ is small; consequently, the distance between $v = 0$ and $v = v_{fs}$ is small. The velocity gradient must be high, and therefore the stress on the fluid ($\tau_y g_c = -\mu \, dv/dx$) is large. Regions of high fluid stress inhibit the formation of eddies, and consequently laminar flow can be expected. Moving in the y-direction, thickness (δ) increases but v_{fs} remains constant so that there is a subsequent decrease in gradient and fluid stress. Under these conditions of reduced stress the bulk of the boundary layer can exist in the turbulent regime. In the turbulent regime a laminar sublayer as well as a buffer region exists near the wall. The high stress near the wall and the impenetrable nature of the fluid adjacent to the wall assure this.

Excellent derivations of the boundary-layer equations are offered in most fluid-mechanics textbooks (22). No complete derivations will be offered here. However, certain significant relationships will be cited for comparison with equations already known. For example, the ratio of momentum transfer by turbulence to momentum transfer by molecular mechanism may be used to write an appropriate Reynolds number for the boundary layer. Intuitively, this may be written $(N_{Re})_\delta = (\delta v_{fs} \rho / \mu)$. The boundary-layer thickness (δ) is difficult to measure; consequently, since δ is a function of y, for convenience the Reynolds number may be written $(N_{Re})_y = (y v_{fs} \rho)/\mu$. Both forms [$(N_{Re})_\delta$ and $(N_{Re})_y$] of the Reynolds number have specific uses.

A boundary-layer counterpart of the friction factor in ducts can also be written. First note that in an infinite duct pressure drop, or more exactly the momentum lost by the fluid, cannot be measured. In effect, this measurement would result from subtraction of a finite number, the momentum gained by the plate, from infinity, the momentum content in an infinite duct. This difference is indeterminate. The momentum gained by the plate, represented by the stress on the plate, is a finite and measurable quantity. Therefore, any counterpart of the friction factor for ducts can only be expressed in terms of surface stress. In a finite-sized duct through which a finite amount of fluid is flowing the pressure-drop term is no longer indeterminate. The ratio of total momentum transferred to the plate to the momentum transferred by the turbulence mechanism may be written as $(\tau_y g_c)_1/v_{fs}^2 \rho$, which is identical in form to the friction factor in ducts as shown in Equation 13.14.

A number of equations for use in boundary layers will be written here (3). There are three equations for laminar flow and three equations for turbulent flow. The equations are each limited in application. The greater number of equations is required in this application

because δ varies with y and $(\tau_y)_1$ varies with y. The counterpart variables in fully developed flow in cylindrical ducts, D and $(\tau_y)_1$, are constant with L. The additional variables call for either a more complex equation or a greater number of simple equations.

In the *laminar* regime, the stress at any point y may be determined in terms of the Reynolds number based upon boundary-layer thickness at the same point y.

$$\frac{(\tau_y g_c)_{1y}}{v_{fs}^2 \rho} = \frac{1.5}{(N_{Re})_\delta} \qquad (13.32)*$$

In the *laminar* regime, the boundary-layer thickness (δ) may be determined at a point y in terms of the Reynolds number based upon the same point y.

$$\frac{\delta}{y} = \frac{4.64}{(N_{Re})_y^{0.5}} \qquad (13.33)$$

In the *laminar* regime, the mean stress on the surface between $y = 0$ and $y = y$ is related to the Reynolds number based upon y,

$$\frac{\overline{(\tau_y g_c)_1}}{v_{fs}^2 \rho} = \frac{0.65}{(N_{Re})_y^{0.5}} \qquad (13.34)$$

As stated earlier, near the leading edge the boundary layer is laminar and is therefore described by Equations 13.32 to 13.34. On a long plate as y increases, $(N_{Re})_y$ increases. At the position y corresponding to $(N_{Re})_y = 10^5$ to 10^6, transition occurs and a turbulent boundary layer forms. Thus any turbulent boundary layer is preceded by a laminar boundary layer near the leading edge. Equations for the turbulent boundary layer are written below. The turbulent-boundary-layer equations presented here are written *as if the entire plate were in turbulent flow*.

In the *turbulent* regime, the stress at any point y is related to the Reynolds number based upon the boundary layer thickness (δ) at the point y.

$$\frac{(\tau_y g_c)_{1y}}{v_{fs}^2 \rho} = \frac{0.0228}{(N_{Re})_\delta^{0.25}} \qquad (13.35)$$

In the *turbulent* regime the boundary layer thickness (δ) at any point y is related to the Reynolds number based upon the same point y.

$$\frac{\delta}{y} = \frac{0.376}{(N_{Re})_y^{0.2}} \qquad (13.36)$$

In the *turbulent regime*, the mean stress on the surface of the plate between $y = 0$ and $y = y$ is related to the Reynolds number based upon the same point y.

$$\frac{\overline{(\tau_y g_c)_1}}{v_{fs}^2 \rho} = \frac{0.037}{(N_{Re})_y^{0.2}} \qquad (13.37)$$

---

* The nomenclature for Equations 13.32 through 13.37 is presented immediately after Equation 13.37.

The nomenclature for Equations 13.32 through 13.37 is:

δ = The boundary layer thickness; ($\mathbf{L}_x$)

y = a position measured from the leading edge of a plate ($y = 0$) in the direction of flow. ($\mathbf{L}_y$)

v_{fs} = free-steam velocity ($\mathbf{L}_y/\theta$)

$(N_{\text{Re}})_y = \left(\dfrac{yv_{fs}\rho}{\mu}\right)$ = the Reynolds number based upon the position y, dimensionless.

$(N_{\text{Re}})_\delta = \left(\dfrac{\delta v_{fs}\rho}{\mu}\right)$ = the Reynolds number based upon the boundary-layer thickness (δ), dimensionless

$(\tau_y g_c)_{1y}$ = the stress on the plate at a point y downstream from the leading edge

$\overline{(\tau_y g_c)_1}$ = the mean value of the stress between $y = 0$ and $y = y$

$\dfrac{(\tau_y g_c)_{1y}}{v_{fs}^2\rho}$, $\dfrac{\overline{(\tau_y g_c)_1}}{v_{fs}^2\rho}$ = friction factors for the boundary layer

Equations 13.34 and 13.37 both presume that the flow regime is the same between $y = 0$ and $y = y$. In practice, the laminar boundary layer exists from the leading edge to some critical point y_c where $(N_{\text{Re}})_y$ is between 10^5 and 10^6. Therefore, to determine the mean stress on a plate in turbulent flow, Equations 13.34 and 13.37 must be used over the ranges in which they apply. Consider a plate of total length y_t and width z. Transition from laminar to turbulent flow is considered to occur at y_c. The transition point y_c can be calculated from $(N_{\text{Re}})_y = 10^5$ as shown in Equation 13.38.

$$y_c = \frac{10^5\mu}{v_{fs}\rho} \tag{13.38}$$

The mean stress for the laminar portion of the boundary layer which lies between $y = 0$ and $y = y_c$ can be calculated from Equation 13.34.

$$\overline{(\tau_y g_c)_1} = \frac{0.65v_{fs}^2\rho}{(10^5)^{0.5}} \tag{13.39}$$

The force on *one* side of the plate between $y = 0$ and $y = y_c$ is equal to the stress-area product.

$$F_{\text{lam}}g_c = \overline{(\tau_y g_c)_1}y_c z = \frac{0.65v_{fs}^2\rho z y_c}{(10)^{2.5}} \tag{13.40}$$

Equation 13.37 is written for a plate in *turbulent* flow between $y = 0$ and $y = y_t$. Therefore the force between y_c and y_t must be computed and added to the force due to laminar flow between $y = 0$ and $y = y_c$. The mean stress between $y = 0$ and $y = y_t$ is calculated from Equation 13.37.

$$\overline{(\tau_y g_c)_1} = \frac{(0.037)v_{fs}^2\rho}{\left(\dfrac{y_t v_{fs}\rho}{\mu}\right)^{0.2}} \tag{13.41}$$

and the force on *one* side of a plate between $y = 0$ and $y = y_t$ if the entire plate were in turbulent flow is

$$F_{\text{turb}}g_c = \overline{(\tau_y g_c)_1}z y_t = \frac{(0.037)v_{fs}^2\rho z y_t}{\left(\dfrac{y_t v_{fs}\rho}{\mu}\right)^{0.2}} \tag{13.42}$$

The mean stress for the region between $y = 0$ and $y = y_c$ for turbulent flow can be calculated as

$$\overline{(\tau_y g_c)_1} = \frac{(0.037)v_{fs}^2\rho}{(10^5)^{0.2}} = (0.0037)v_{fs}^2\rho \tag{13.43}$$

The force applied to one side of a plate in turbulent flow between $y = 0$ and $y = y_c$ is

$$F_{\text{turb}}g_c = \overline{(\tau_y g_c)_1}z y_c = 0.0037v_{fs}^2\rho y_c z \tag{13.44}$$

The net tangential force applied to one side of the plate is

$$F_{\text{net}} = F_{\text{turb}} - F_{\text{turb}} + F_{\text{lam}} \tag{13.45}$$
$$\phantom{F_{\text{net}} =}{\scriptstyle 0\to y_t} \quad {\scriptstyle 0\to y_c} \quad {\scriptstyle 0\to y_c}$$

The net mean stress $\overline{(\tau_y g_c)}_{\text{net}}$ is equal to the net force per unit area of plate

$$\overline{(\tau_y g_c)}_{\text{net}} = \frac{F_{\text{net}}g_c}{y_t z} = \frac{F_{\text{turb}}g_c - F_{\text{turb}}g_c + F_{\text{lam}}g_c}{y_t z} \tag{13.46}$$

Equations 13.40, 13.42, and 13.44 are substituted into Equation 13.46, and the equation is rearranged.

$$\frac{\overline{(\tau_y g_c)}_{\text{net}}}{v_{fs}^2\rho} = \frac{\left(\dfrac{0.037y_t z}{(N_{\text{Re}})_y^{0.2}} - 0.0037y_c z + \dfrac{0.65y_c z}{10^{2.5}}\right)}{y_t z} \tag{13.47}$$

$$\frac{\overline{(\tau_y g_c)}_{\text{net}}}{v_{fs}^2\rho} = \frac{0.037}{(N_{\text{Re}})_y^{0.2}} - 0.0017\frac{y_c}{y_t} \tag{13.48}$$

In the foregoing derivation, the transition point was arbitrarily taken as $N_{\text{Re}} = 10^5$ although the actual transition value lies between 10^5 and 10^6. The only justification for this arbitrary choice lies in the fact that at the low limit of transition (10^5) the calculated stress on a plate is safely high.

Illustration 13.4. A flat plate 5 ft long has an edge facing the flow direction. Air is flowing over the plate at 40 ft/sec. The air properties are $\rho = 0.075$ lb/cu ft and $\mu = 0.018$ centipoise.

(a) Calculate the boundary layer thickness 5 ft from the leading edge.

(b) Calculate the point stress 5 ft from the leading edge.

(c) Calculate the average stress over the plate from the leading edge to a point 5 ft from the leading edge.

SOLUTION. A test for flow regime is performed.

$$(N_{\text{Re}})_y = \left(\frac{yv_{fs}\rho}{\mu}\right) = \frac{(5)(40)(0.075)}{0.018 \times 6.72 \times 10^{-4}} = 1{,}240{,}000$$

The plate in this problem has a turbulent-flow portion beyond

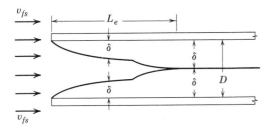

Figure 13.3. Boundary-layer build-up in the entry length.

some value y_c where $N_{\text{Re}} = 10^5$. The transition point can be calculated from Equation 13.38.

$$y_c = \frac{10^5 \mu}{v_{fs}\rho} = \frac{(10^5)(0.018)(6.72 \times 10^{-4})}{(40)(0.075)} = 0.403 \text{ ft}$$

(a) The boundary-layer thickness at $y = 5$ is calculated from Equation 13.36 for turbulent flow.

$$\delta = \frac{0.376y}{(N_{\text{Re}})_y^{0.2}} = \frac{(0.376)(5)}{(1{,}240{,}000)^{0.2}} = \frac{(0.376)(5)}{16.54} = 0.1133 \text{ ft}$$

(b) The point stress in the turbulent regime can be calculated using Equation 13.35.

$$(N_{\text{Re}})_\delta = \left(\frac{\delta v_{fs}\rho}{\mu}\right) = \frac{(0.1133)(40)(0.075)}{(0.018)(6.72 \times 10^{-4})} = 2.82 \times 10^4$$

$$(\tau_y)_{1y} = \frac{v_{fs}^2 \rho \, (0.0228)}{g_c (N_{\text{Re}})_\delta^{0.25}} = \frac{(40)^2(0.075)(0.0228)}{(32.2)(2.82 \times 10^4)^{0.25}}$$

$$= 0.00655 \text{ lb}_f/\text{sq ft}$$

(c) The net mean stress over the plate can be calculated using Equation 13.48.

$$\overline{(\tau_y)_{\text{net}}} = \frac{v_{fs}^2 \rho}{g_c} \left(\frac{0.037}{(N_{\text{Re}})_{yt}^{0.2}} - 0.0017 \frac{y_c}{y_t}\right)$$

$$= \frac{(40)^2(0.075)}{g_c} \left(\frac{0.037}{(12.4 \times 10^5)^{0.2}} - 0.0017 \frac{(0.403)}{5}\right)$$

$$= \frac{120}{32.2}(2.23 \times 10^{-3} - 0.137 \times 10^{-3})$$

$$= 0.0078 \text{ lb}_f/\text{sq ft}$$

For this particular example, the laminar portion of the boundary layer corresponds to the first 8 per cent of the plate. If the laminar correction is omitted and the entire plate is considered in the turbulent regime, the mean stress is 0.00833 lb$_f$/sq ft, which is about 6 per cent higher than without the correction.

The Entry Length. The entry length of a tube is an example of the build-up of a boundary layer. Consider a pipe suspended in a region of flow of uniform velocity (see Figure 13.3). At the immediate entrance to the pipe, a boundary layer is set up at the inside tube surface. Some length of tube is required as the boundary layer grows in thickness until all boundary layers meet

at the center of the pipe. This length is known as the entry length (L_e). Adjustments in the free-stream velocity in the entry length must be made because the free-stream velocity at the entry must equal the *average* velocity of the fluid in the tube. In this sense, boundary-layer flow within a finite duct differs from the boundary layer in an infinite duct described above. Thus, any flow within a tube is essentially all boundary-layer flow. In the entry length, a boundary layer and free stream exist within the tube. The entry length ends where the boundary layers meet at the center after which a constant velocity profile is maintained regardless of position in the y-direction. After the boundary layers meet at the center of the duct, the flow is "fully developed." All equations for tubing mentioned earlier apply to fully developed flow. If the boundary layer is laminar throughout the entire entry length, flow in the "fully developed flow" section of the tube is laminar. If turbulence develops in the boundary layer within the entry length, the remainder of the tube will operate in turbulent flow.

As might be expected from consideration of the boundary layer, the wall stress per unit length of entry is greater than that for fully developed flow, with a maximum at the point of entry, and the value diminishes with progress through the tube to a minimum at the end of the entry length. The minimum value at the end of the entry length is equal to the value in fully developed flow. In a bounded duct of finite dimensions as is described here, a pressure drop can also be expected. In the infinite duct described in the previous section, the pressure drop was unmeasureable. In the bounded duct the usual force balance may be written relating the tangential force at the wall and the force due to pressure on the fluid.

Equations for the entry length can be derived from boundary-layer considerations, but the acceleration of the free-stream fluid within the finite duct must be accounted for. The derivations are beyond the scope of this book, but several equations are given.

In the *laminar* regime, the entry length can be predicted from the following equation (4):

$$\frac{L_e}{D} = 0.0575(N_{\text{Re}}) \qquad (13.48a)$$

where L_e = entry length ($\mathbf{L}_y$)

D = tube diameter ($\mathbf{L}_r$)

(N_{Re}) = Reynolds number based upon the usual mean velocity and tube diameter; dimensionless

The entry length for a *turbulently flowing fluid* can be predicted (6) from the equation

$$\frac{L_e}{D} = 0.693 \, (N_{\text{Re}})^{0.25} \qquad (13.49)$$

Qualitatively the pressure drop through the entry length is considerably greater than the pressure drop after fully developed flow is established. The increase is the result of two effects. Boundary-layer equations reveal that the local wall stress near the leading edge is greater than the local wall stress downstream from the leading edge. In addition, the fluid in the boundary layer is moving more slowly than the fluid in the free stream. If, as in the entry to a tube, boundary-layer growth occurs, then to maintain a constant mass-flow rate in a bounded duct the free-stream fluid portion in the tube must accelerate. The twofold force effects of acceleration of the free stream and high stress near a leading edge increase the pressure drop over the entry length. As a general rule, the pressure drop through the entry length may be taken approximately as twice to three times the value for fully developed flow at the same Reynolds number.

Illustration 13.5. Calculate the entry length for water at 70°F flowing through a $\frac{1}{2}$-in. tube at 0.1 ft/sec.

SOLUTION. Since no flow regime is specified, the Reynolds number should be evaluated.

$$N_{\text{Re}} = \frac{D\bar{v}\rho}{\mu} = \frac{(0.5/12)(0.1)(62.4)}{(1)(0.000672)} = 387$$

This is laminar flow; therefore, Equation 13.48a may be used.

$$\frac{L_e}{D} = 0.0575 \, (N_{\text{Re}}) = 0.0575(387) = 22.3$$

and

$$L_e = \left(\frac{0.5}{12}\right)(22.3) = 0.927 \text{ ft}$$

FORM DRAG

In the examples of fluid friction that have been considered the transfer of momentum resulted in a tangential stress or drag on a smooth surface that was oriented parallel to the flow direction, This phenomenon is traditionally called *skin friction* or *skin drag*. From the foregoing, it is evident that, if *any* surface is in contact with a fluid and a relative motion exists between the fluid and the surface, skin friction will exist between the surface and the fluid. In addition to skin friction, significant frictional losses occur because of acceleration and deceleration of the fluid. The accelerative effects occur when the fluid changes path to pass around a solid body set in the flow path. This phenomenon is called *form drag*.

Consider the body shown in Figure 13.4a. The body is suspended in an infinite duct and is subject to a free-stream velocity v_{fs}. *Streamlines* are drawn to represent the path of fluid elements around the body. The boundary layer is shown as a dotted line close to the body.

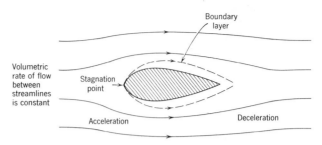

(a) Streamlined shape—no separation.

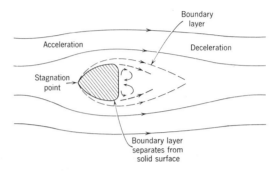

(b) Nonstreamlined shape—separation of the boundary layer.

Figure 13.4. Flow around submerged bodies.

The fluid approaching the center of the front face of a symmetrical body will impinge upon the body and be split into two portions, one half of which moves to each side of the body. At the exact center of the body the fluid will have zero velocity. This is known as a *stagnation point*. Boundary-layer growth begins at the stagnation point and continues over the entire surface. Beyond the trailing edge, the boundary layers revert into the free stream. The tangential stress on the body arising from transfer of momentum originating in the slowing down of the boundary layer is the *skin friction*. The boundary layer will have the same characteristics as described earlier; it will be laminar unless the Reynolds number exceeds a critical value, after which the boundary layer becomes turbulent.

However, the fluid outside the boundary layer is subject to acceleration due in part to change in path and in part to change in linear speed. As the fluid is diverted in path to pass around the body, a force is exerted upon the body by the fluid. Note that this force on the body is in addition to the skin friction associated with the boundary layer. This is not a tangential force, but it is directed according to the geometries of the body and streamlines. In Figure 13.4, the space between any two streamlines represents the duct cross section occupied by an equal volumetric flow rate. The position of the streamlines around the widest part of the body shows that the fluid is moving at a velocity greater than v_{fs}, so that the linear speed of the fluid has increased in that region. Beyond the widest part of the body, the fluid is subject to a directional acceleration opposite to that at the leading face of the body and to a deceleration of linear speed as the fluid returns to the normal free stream pattern downstream of the body. The summation of all forces on the body due to acceleration and deceleration constitutes the *form drag* of the body.

In Figure 13.4a, a shape was chosen such that no sharp discontinuities exist along the length. The boundary layer is continuous over the length of the body but obviously must adopt a shape and character subject to the changes in velocity and direction of the fluid beyond its bounds. Another shape is shown in Figure 13.4b, with a sharp discontinuity in the trailing portion. In the new shape, the boundary layer is shown, along with the stagnation point. The acceleration of fluid in the vicinity of the leading face is the same as in the body described in Figure 13.4a. The upper and lower boundary layers begin at the stagnation point and follow along the surface of the body. In the trailing portion of the body, beyond the widest point, the behavior of the boundary layer is significantly different. If the fluid streamlines were to conform exactly to the surface of the body, deceleration must necessarily be very rapid. The boundary layer must conform to the surface and the free stream. In this case, the free stream is undergoing rapid deceleration, and therefore the boundary layer must decelerate also. But the boundary layer is moving very slowly before deceleration; therefore, deceleration of the boundary layer brings about *reversal of direction of the boundary layer* and consequently *separation* of the boundary layer from the surface. As can be seen in Figure 13.4b, the boundary-layer reversal relieves the necessity of the extreme deceleration of the streamline by setting up a region of eddy activity after the body. This intense eddy activity results in a sizable force exerted on the body, much more than in the body described in Figure 13.4a in which the separation eddy is absent.

From the foregoing, it is evident that the geometry of a system is a determining factor in the amount of force exerted on the body. The "teardrop" shape shown in Figure 13.4a exerts less drag than the same body with the teardrop tail removed. On the other hand, exaggeration of the trailing portion can lead to a large amount of surface with consequent increase in skin friction. Form drag appears in many chemical-engineering applications. Metering devices are designed to maximize or minimize form drag, according to requirements. Roughness in pipe is an example of form drag. The finite constant rate at which particles settle in fluids is another. Other engineering applications are the stress on structures in wind and flowing water.

Correlations of flow characteristics and geometry for bodies in a free stream utilize dimensionless groups that are analogous in concept to the friction factor and

Reynolds number described in the sections of this chapter covering flow through pipe and momentum transfer through the boundary layer. The *drag coefficient* (C_D) is defined as

$$C_D = \frac{2F_D g_c}{S v_{fs}^2 \rho} \tag{13.50}$$

and the Reynolds number is

$$N_{Re} = \frac{L_D v_{fs} \rho}{\mu} \tag{13.51}$$

where F_D = total force exerted on the body (**F**)
  S = maximum projected area normal to flow (**$L_x L_z$**)
  L_D = length characteristic of the geometry (**L_j**)
  v_{fs} = free-stream velocity (**L_y/θ**)

Correlations are usually presented graphically on logarithmic plots of C_D as a function of N_{Re}. Several such *drag diagrams* are presented in Figure 13.5.

The diagrams are similar to the f–N_{Re} plot used for skin friction in pipe, but several significant differences should be noted. Below $N_{Re} = 0.1$ all shapes are represented by a single line which may be written in equation form as

$$C_D = \frac{24}{N_{Re}} \tag{13.52}$$

This is the region in which the boundary layer is laminar and accelerative effects are small enough to be negligible. Equation 13.52 is analogous to Equation 13.16. In the range $0.1 < N_{Re} < 10^5$ several different mechanisms contribute to the transfer of momentum. The accelerative forces become significant enough to affect the shape of the curve, and in addition the boundary layer becomes turbulent. In most instances the accelerative forces begin to operate at Reynolds numbers well below the onset of significant turbulence in the boundary layers. For example, the discontinuity in the drag diagram for spheres at $N_{Re} \cong 10^5$ is the result of change in the boundary layer from partially laminar and partially turbulent to completely turbulent. Above $N_{Re} = 10^5$ the boundary layer is considered turbulent for all shapes and the accelerative effects predominate. It is in this

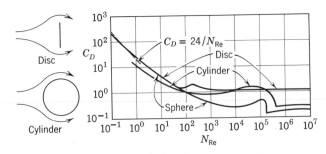

Figure 13.5. Form-drag diagram for various shapes[21].

region that the drag diagram can be represented by the equation

$$C_D = \text{constant} \tag{13.53}$$

In this respect the drag diagram differs from the f–N_{Re} correlation. At high values of the Reynolds number, the f–N_{Re} diagram still has a slight negative slope characteristic of skin friction. This same characteristic can be seen in the turbulent-boundary-layer equations, namely Equation 13.37. Therefore, it might be concluded that, where accelerative effects predominate, the drag diagram will approach Equation 13.53 very rapidly. This leveling-off will be seen in the f–N_{Re} plots for pipe with rough interior surfaces (Appendix C-3). The limiting constant values of C_D are listed in Table 13.1 for various shapes, along with the dimension used for L_D and the lower limit of applicability of Equation 13.53. For the transition range, the diagram must be used.

Table 13.1. VALUES OF C_D THAT MAY BE USED IN EQUATION 13.53

Shape	L_D	Lower Limit of N_{Re}	C_D
Circular disk, flat side perpendicular to flow	Diameter, D	10^3	1.12
Cylinder, infinite height, axis perpendicular to flow	Diameter, D	5×10^5	0.33
Sphere	Diameter, D	3×10^5	0.20

Drag diagrams like those of Figure 13.5 can be developed for use with shapes different from the spheres or cylinders specified in Figure 13.5. The correlation of data on the drag performance of irregular shapes has received particular attention since solids involved in chemical processes are so often irregular in size and shape. These correlations have required a more complicated specification of size and shape than is required by the shapes listed in Table 13.1. In addition to a size parameter, such as particle "diameter," a shape parameter is needed. Several such parameters have been developed, the most common of which is the *sphericity*. The sphericity (ψ) is defined as

$$\psi = \frac{\substack{\text{surface area of a sphere of the} \\ \text{same volume as the particle}}}{\text{surface area of the particle}} \tag{13.54}$$

Problems in the characterization of irregular particles are treated in Appendix B.

Illustration 13.6. A chimney 100 ft high and 5 ft in diameter is subject to a maximum wind of 100 mph. Calculate the force exerted on the chimney by the wind.

$$(\mu = 0.018 \text{ centipoise}, \rho = 0.075 \text{ lb/cu ft.})$$

SOLUTION. From Table 13.1, the characteristic dimension

for a cylinder is the diameter. The regime of flow is determined by the Reynolds number.

$$N_{Re} = \frac{L_D v_{fs}\rho}{\mu} = \frac{(5)[(100 \times 88)/(60)](0.075)}{(0.018 \times 6.72 \times 10^{-4})}$$
$$= 4,530,000$$

This value is above the lower limit of 5×10^5 given in Table 13.1, so that $C_D = \text{constant} = 0.33$. Then by Equation 13.50

$$\frac{2F_D g_c}{S v_{fs}^2 \rho} = 0.33$$

$$F_D = \frac{0.33 \, S v_{fs}^2 \rho}{2 g_c}$$

The maximum projected area normal to flow is the area of a rectangle 5 ft by 100 ft; therefore $S = 500$ sq ft.

$$F_D = \frac{(0.33)(500)(100 \times 88/60)^2(0.075)}{(2)(32.2)}$$

$$F_D = 4130 \text{ lb}_f$$

TRANSFER COEFFICIENTS FOR MASS AND HEAT TRANSFER IN THE TURBULENT REGIME

In Chapter 12, heat and mass transfer in the turbulent regime were introduced. Calculations were based upon concentrations of transferent property at various radial positions. The turbulent-flow properties at the various positions were defined in terms of eddy diffusivity, mixing lengths, and so forth. Although the point-condition behavior is useful in understanding the mechanism of turbulence and its effect upon heat and mass transfer, the solutions are tedious and cumbersome. Therefore, it is desirable to approach the evaluation of transfer in terms of mean fluid properties and gross dimensions rather than point conditions as used in Chapter 12. The mean properties and gross dimensions have an additional desirable feature; the experimental data may be determined accurately and reproducibly with the simplest of analytical tools. Useful equations that have the properties mentioned result from the integration of the point-condition equations introduced in Chapter 12.

The friction factor defined earlier in this chapter is, by its definition, an integrated form of the expressions for point conditions for turbulent flow. It would seem reasonable that similar concepts and equation forms would have been written for heat and mass transfer. Unfortunately, this is not the case. Early in the investigations of momentum-, heat-, and mass-transfer behavior, the similarities that exist among the three transfer phenomena were not recognized. The various equation *forms* that were developed independently are not exactly similar, but the content of the equations can be related.

The general transport equation that must be integrated is

$$\psi = -(\delta + E)\frac{d\Gamma}{dx} \qquad (12.23)$$

where ψ = flux of a property at any value of x
   δ = molecular diffusivity
   E = eddy diffusivity
   Γ = volume concentration of transferent property

In cylindrical geometry, with appropriate modification of the gradient, Equation 12.23 may be written

$$\psi = -(\delta + E)\frac{d\Gamma}{dr} \qquad (12.24)$$

In Equation 12.27 the flux is shown to be linear with radius, and therefore Equation 12.24 may be written

$$\psi_1\left(\frac{r}{r_1}\right) = -(\delta + E)\frac{d\Gamma}{dr} \qquad (12.28)$$

where ψ_1 = the flux at the boundary
   r_1 = boundary radius
   r = any radial position

Equation 12.28 may also be written as Equations 12.29 and 12.30 for heat and mass transfer respectively. At steady state Equation 12.28 may be integrated,

$$\psi_1\int_0^{r_1}\frac{r\,dr}{(\delta + E)} = -r_1\int_{\Gamma_0}^{\Gamma_1}d\Gamma \qquad (13.55)$$

where Γ_0 and $r = 0$ are the conditions at the center of the cylinder and Γ_1 and r_1 are the conditions at the wall. The right-hand term of Equation 13.55 may be integrated directly to $-r_1(\Gamma_1 - \Gamma_0)$. This form is written in terms of Γ_0, the concentration of transferent property at the center of the tube, which is difficult to determine by analytical means. Since the analysis is easier to perform, it is more desirable to use the mean value of transferent property that results from integrating over the entire duct. The appropriate mean value may be defined as

$$\bar{\Gamma} = \frac{1}{V_1}\int_0^{V_1}\Gamma\,dV \qquad (13.56)$$

where V is the volumetric flow rate at a point within the duct and V_1 is the total volumetric flow rate in the duct.

The mean value described is the result of integrating over all the fluid flowing in the duct. Note that $dV = v\,dS$, where v is the point velocity and S is the applicable area of flow, and for cylindrical geometry, $dV = v2\pi r\,dr$.

Physically, the mean value can be determined if the entire flow stream is collected and averaged. An appropriate mean *temperature* will result if the entire flow stream is passed through a mixing chamber so that every element of fluid is at the same temperature; the resulting temperature will be the mean value in question.

The mean *velocity* will be determined if the entire flow stream passes through a volume measuring meter and the volumetric flow rate considered to pass through the duct at uniform velocity across the duct.

The ratio γ may be defined as

$$\gamma = \frac{\Gamma_1 - \bar{\Gamma}}{\Gamma_1 - \Gamma_0} \qquad (13.57)$$

The ratio γ is the ratio of the difference in concentration of transferent property between the wall and the mean value of the fluid to the maximum difference between the wall and the center. The ratio γ is a function of the distribution of Γ with position.* Then Equation 13.55 may be written

$$\psi_1\int_0^{r_1}\frac{r\,dr}{(\delta + E)} = -r_1(\Gamma_1 - \Gamma_0) = -\frac{r_1}{\gamma}(\Gamma_1 - \bar{\Gamma}) \qquad (13.58)$$

The integration of the left-hand term of Equation 13.58 will not be performed,† but the *mean eddy diffusivity* $(\bar{E})$ that results from the indicated integration will be defined.

$$\int_0^{r_1}\frac{r\,dr}{(\delta + E)} = \frac{1}{(\delta + \bar{E})}\int_0^{r_1}r\,dr = \frac{r_1{}^2}{2(\delta + \bar{E})} \qquad (13.59)$$

Equation 13.59 can be substituted into Equation 13.58.

$$\frac{\psi_1 r_1 \gamma}{2(\delta + \bar{E})} = -(\Gamma_1 - \bar{\Gamma}) \qquad (13.60)$$

The tube radius $r_1 = D/2$. This value can be substituted into Equation 13.60, and, after rearrangement, the equation becomes

$$\psi_1 = -\frac{4(\delta + \bar{E})}{\gamma D}(\Gamma_1 - \bar{\Gamma}) \qquad (13.61)$$

Equation 13.61 can be written as a rate equation by multiplying by A_1,

$$(\psi A)_1 = -\frac{4(\delta + \bar{E})}{\gamma D}(\Gamma_1 - \bar{\Gamma})A_1 \qquad (13.62)$$

The *transfer coefficient* $(\mathscr{E})$ may be defined as

$$\mathscr{E} = \frac{4(\delta + \bar{E})}{\gamma D} \qquad (13.63)$$

Equation 13.62 becomes

$$(\psi A)_1 = -\mathscr{E}(\Gamma_1 - \bar{\Gamma})A_1$$

or

$$(\psi A)_1 = -\frac{(\Gamma_1 - \bar{\Gamma})}{\dfrac{1}{(\mathscr{E}A_1)}} \qquad (13.64)$$

---

* Figure 20.2 is a special case of this ratio for momentum transfer. In the figure $\gamma_\tau = \bar{v}/v_{\max}$, is the ratio of mean velocity to point velocity at the center of the tube. For a stationary tube $v_1 = 0$ at the wall.

† For momentum transfer the integration may be performed by writing E_τ as a function of r and dv/dr as in Equation 12.15. The derivative dv/dr can be written as a function of r by use of Equations 12.1, 12.2, and 12.3.

Equation 13.64 is an integrated form of the differential equation for transfer (Equation 12.24). The variables that make up the equation can be evaluated conveniently, and therefore the equation is convenient to use. The equation form warrants some attention. The integration of the gradient in the differential equation results in the term $(\Gamma_1 - \bar{\Gamma})$, and the rate of transfer is proportional to this term, which is called the *driving force* (or less frequently the *transfer potential*). The denominator term $(1/\mathscr{E}A)$ is called the *resistance to transfer*. Equation 13.64 may be written

$$\text{Rate of transfer} = \frac{\text{driving force}}{\text{resistance}} \qquad (13.65)$$

Note the exact analogy between transfer of heat, mass, and momentum as described in Equation 13.65 and the transfer of electrical current as described by Ohm's law

$$I_\epsilon = \frac{E_\epsilon}{R_\epsilon} \qquad (13.66)$$

where I_ϵ = current, or rate of transfer of electrons
E_ϵ = electrical potential
R_ϵ = resistance

$$\text{Rate of transfer of electrons} = \frac{\overset{\text{electrical potential}}{\text{(driving force)}}}{\text{resistance}} \qquad (13.67)$$

Since this analogous behavior exists, all the principles used for electrical transfer apply to the transfer of mass, heat, and momentum. This fact is the basis of analogue-computing equipment. Electrical analogues of transfer phenomena are built, and from electrical-transfer behavior the transfer systems can be analyzed. Equation 13.65 is extremely useful in chemical-engineering calculation, as will be shown in Chapter 14.

The definition of the transfer coefficient given in Equation 13.64 shows it to be a complex function of transfer by the molecular and turbulent mechanism, of the distribution of concentration of the transferent property as indicated by γ, and of the geometry of the system. It is also a function of the flow characteristics of the system, because the mean eddy diffusivity $(\bar{E})$ is dependent upon flow behavior. In spite of the complexity, it is reasonable to believe that the transfer coefficient may be related to pertinent variables by use of the mechanism-ratio-analysis procedure described in the earlier part of this chapter. This procedure will be described for the specific transfer coefficients for heat and mass.

Heat-Transfer Coefficients in Turbulent Flow in Tubes. Equation 13.64 is a general equation for transfer in integrated form and may be written for heat transfer

by substitution of the appropriate variables. The *heat-transfer coefficient* (h) may be defined as

$$\frac{h}{\rho c_P} = \mathscr{E}_q = \frac{4(\alpha + \bar{E}_q)}{D\gamma_q} \qquad (13.68)$$

The other variables for substitution are: $\delta = \alpha$, $\Gamma = \rho c_P T$, $\psi A = q$ and $\bar{E} = \bar{E}_q$ so that the heat-transfer equation for turbulent flow is

$$q_1 = -\left[\frac{(\rho c_P T)_1 - (\overline{\rho c_P T})}{\frac{\rho c_P}{h A_1}}\right] \qquad (13.69)$$

or

$$q_1 = \frac{-(T_1 - \bar{T})}{\left(\dfrac{1}{hA_1}\right)} \qquad (13.70)$$

where h = heat-transfer coefficient, with dimensions $\mathbf{H/L_y L_z \theta T}$, and typical units Btu/sq ft hr °F

The driving force for heat transfer is the difference in temperature between the wall and the bulk fluid, $(T_1 - \bar{T})$, and the resistance to transfer is $(1/hA)$.

By mechanism-ratio analysis, an equation form will be deduced for the heat-transfer coefficient (h). The individual transfer mechanisms are:
1. Heat transfer by molecular transport.
2. Heat transfer by turbulent transport.
From considerations in Chapter 12, the heat-transfer mechanisms are dependent upon flow conditions. The flow conditions of a system in turn are dependent upon momentum transfer, and therefore two further mechanisms can be stated:
3. Momentum transfer by molecular transport.
4. Momentum transfer by turbulent transport.
Each of the four individual mechanisms must be a part of a mechanism ratio. From the description of the mechanism-ratio analysis presented earlier in this chapter, three ratios are required to describe the system, but no particular order or arrangement need be followed. For convenience, consider first the ratio

$$\frac{\text{Total heat transfer, by both molecular and turbulent transport}}{\text{Heat transfer by molecular transport}} \qquad (13.71)$$

The total heat transfer may be represented by $(\alpha + \bar{E}_q)/\gamma_q$, and, in the same units, the heat transferred by molecular transport may be represented by α; therefore, the ratio described in Equation 13.71 is represented symbolically by

$$\frac{(\alpha + \bar{E}_q)}{\gamma_q \alpha} = \frac{(\alpha + \bar{E}_q)\rho c_P}{\gamma_q k} \qquad (13.72)$$

If the definition of the heat-transfer coefficient (Equation

13.68) is rearranged and divided by k, it is identical to the ratios of Equation 13.72.

$$\frac{(\alpha + \bar{E}_q)\rho c_P}{\gamma_q k} = \frac{hD}{4k} \qquad (13.68)$$

Accordingly, the ratio of Equation 13.71 can be written

$$\frac{hD}{4k} \qquad (13.73)$$

On subsequent use of this ratio in a mechanism-ratio correlation, the numerical constant can appear in the general numerical constant and hence may be dropped. The *Nusselt number* may be defined to represent this ratio

$$N_{\text{Nu}} = \frac{hD}{k} \qquad (13.74)$$

Two other ratios are required to complete the mechanism-ratio analysis. The ratio of momentum transfer by turbulent transport to momentum transfer by molecular transport has already been described as the Reynolds number (N_{Re}) in Equation 13.15.

$$N_{\text{Re}} = \frac{D\bar{v}\rho}{\mu} \qquad (13.15)$$

The Nusselt number is a ratio of mechanisms involving heat transfer, and the Reynolds number is a ratio of mechanisms involving momentum transfer. The third ratio must necessarily include one heat-transfer mechanism and one momentum-transfer mechanism. The ratio of momentum transfer by molecular transport to heat transfer by molecular transport has been described in Chapter 9 as the *Prandtl number* (N_{Pr}) and is defined

$$N_{\text{Pr}} = \frac{\nu}{\alpha} = \frac{c_P \mu}{k} \qquad (13.75)$$

The Prandtl number is a function of fluid properties only and is independent of flow characteristics. This fact facilitates the quantitative evaluation of this ratio.

Three ratios have been written for four mechanisms; therefore, an equation form can be written

$$N_{\text{Nu}} = \text{constant} \, (N_{\text{Re}})^a (N_{\text{Pr}})^b$$
$$+ \text{constant}' \, (N_{\text{Re}})^{a'} (N_{\text{Pr}})^{b'} + \dots \quad (13.76)$$

The constant and exponents must be evaluated from experimental data.

The early correlations for heat transfer in turbulent flow in tubes were restricted to fluids of moderate Prandtl number. These correlations (5) were written in terms of two equations. The first equation applies when the fluid is *heated*

$$N_{\text{Nu}} = 0.023 \, (N_{\text{Re}})^{0.8} (N_{\text{Pr}})^{0.4} \qquad (13.77)$$

and the second equation applies when the fluid is *cooled*

$$N_{\text{Nu}} = 0.023 \, (N_{\text{Re}})^{0.8} (N_{\text{Pr}})^{0.3} \qquad (13.78)$$

In Equations 13.77 and 13.78 the fluid properties are determined at the mean temperature of the fluid ($\bar{T}$). The equations are of the form of Equation 13.76, but the inconsistency of the Prandtl number exponents of the two equations had to be resolved.

In Chapter 12, the temperature profile of the fluid was examined, and the greater part of the gradient appears in the laminar sublayer and buffer region. The majority of the resistance to heat transfer occurs in this small resistance layer of fluid. Since the wall is at T_1 and the inner edge of the resistance layer of fluid is *approximately* at $\bar{T}$, the transferring medium in the resistance layer operates at some temperature other than $\bar{T}$. Since the properties of the fluid in Equations 13.77 and 13.78 were evaluated at $\bar{T}$, the discrepancy between the actual temperature of the actual resistance layer and the temperature used for evaluation serves as an explanation for the difference in the equations for heating and cooling. The presence of the temperature gradient in the resistance layer sets up density gradients due the density change with temperature, so that below $N_{\text{Re}} = 10,000$ some natural convection currents alter the transfer mechanism. This subject will be discussed in detail later in the chapter. Furthermore, the thickness of the resistance layer depends upon viscosity and density (from the definition of y^+ in Chapter 12), and the viscosity of liquids is extremely sensitive to temperature change. A new equation was proposed (7) that included a term to correct for the actual condition in the resistance layer. With a modification (8) in the constant this equation is

$$N_{\text{Nu}} = 0.023 \, (N_{\text{Re}})^{0.8} (N_{\text{Pr}})^{1/3} \left(\frac{\mu}{\mu_1}\right)^{0.14}$$

which expands to

$$\frac{hD}{k} = 0.023 \left(\frac{D\bar{v}\rho}{\mu}\right)^{0.8} \left(\frac{c_P\mu}{k}\right)^{1/3} \left(\frac{\mu}{\mu_1}\right)^{0.14} \quad (13.79)$$

where μ_1 is the viscosity of the fluid at the wall temperature T_1. Equation 13.79 is valid for heating and cooling gases and liquids with moderate or low viscosity, but it does *not* apply for liquid metals. The fluid properties, with the exception of μ_1, are evaluated at the mean temperature of the fluid $\bar{T}$. The equation is valid above $N_{\text{Re}} = 10,000$ for all fluids with Prandtl number in the range between 0.5 and 100.

Illustration 13.7. An SAE 10 oil flows through a heat exchanger tube 2 in. I.D. at 400 gal/min. At the point of examination the mean oil temperature ($\bar{T}$) is 120°F. Calculate the heat-transfer coefficient and the heat flux if the tube wall temperature is (*a*) 50°F and (*b*) 170°F.

Data:

$T,°F$	μ, centipoises	Data in this column at 120°F.
50	100	$\rho = 56.2$ lb/cu ft
120	20	$c_P = 0.465$ Btu/lb °F
170	6	$k = 0.070$ Btu/hr sq ft (°F/ft)

SOLUTION. The Reynolds number is calculated to determine the regime of flow, but first the flow rate may be converted from gal/min to ft/sec.

$$\bar{v} = \frac{400/(7.48 \times 60)}{[\pi(2/12)^2/4]} = 41.4 \text{ ft/sec}$$

$$\frac{Dv\rho}{\mu} = \frac{(2/12)(41.4)(56.2)}{(20 \times 6.72 \times 10^{-4})} = 28,600$$

At $N_{Re} = 28,600$ the flow is turbulent.

(a) If the oil is cooling, Equation 13.78 may be used, but, because of greater precision over a wide range of viscosity variations, Equation 13.79 is preferred. It is not difficult to apply because the wall temperature T_1 is known.

From Equation 13.79

$$h = 0.023 \frac{k}{D} (N_{Re})^{0.8} \left(\frac{c_P \mu}{k}\right)^{0.33} \left(\frac{\mu}{\mu_1}\right)^{0.14}$$

$$h = 0.023 \times \frac{0.070}{(2/12)} (28,600)^{0.8}$$

$$\times \left[\frac{(0.465)(20 \times 2.42)}{(0.070)}\right]^{1/3} \left[\frac{20}{100}\right]^{0.14}$$

$$h = (0.023)\left(\frac{0.070}{2/12}\right)(3640)(6.88)(0.798)$$
$$= 193 \text{ Btu/hr °F sq ft}$$

The heat flux can be determined from Equation 13.70

$$\frac{q}{A} = -h(T_1 - \bar{T})$$

$$\frac{q}{A} = -(50 - 120)(193) = 13,530 \text{ Btu/hr sq ft}$$

The positive sign indicates heat transfer is in the outward radial direction (cooling). In N_{Re} the velocity was calculated in feet per second; consequently, μ was converted to pounds per foot second. In N_{Pr} the thermal conductivity k is expressed in hour units, and consequently μ is converted to pounds per foot hour.

(b) Equation 13.79 applies for heating as well as cooling. All properties are evaluated at $\bar{T}$ except μ_1 which is evaluated at T_1.

$$h = 0.023 \frac{k}{D} (N_{Re})^{0.8} (N_{Pr})^{0.33} \left(\frac{\mu}{\mu_1}\right)^{0.14}$$

$$h = (0.023)\frac{(0.070)}{(2/12)}(3640)(6.88)\left(\frac{20}{6}\right)^{0.14}$$

$$h = 286 \text{ Btu/hr °F sq ft}$$

$$\frac{q}{A} = -(170 - 120)(286) = -14,300 \text{ Btu/hr sq ft}$$

The negative sign indicates heat transfer inward in the radial direction (heating).

Illustration 13.7 shows the difference in transfer coefficient if material that is at a given condition is subject to heating or to cooling. The viscosity of oils is particularly sensitive to temperature change. Water under the same conditions would respond in a similar manner but to a lesser degree. The viscosities of gases increase with increase in temperature; therefore, gases would respond in opposite fashion.

Mass Transfer Coefficients in Turbulent Flow in Tubes. Equation 13.63 the general equation for transfer in tubes, may be written specifically for mass transfer by substitution of the appropriate variables. For mass transfer $\delta = \mathscr{D}$, $\bar{E} = \bar{E}_N$, $(\psi A) = N_a$, and $\Gamma = c_a$. The *mass-transfer coefficient* for equimolar counter diffusion may be defined

$$k_c' = \mathscr{E}_N = \frac{4(\mathscr{D} + \bar{E}_N)}{D\gamma_N} \tag{13.80}$$

where k_c' is the mass-transfer coefficient,

$[M/\theta L_y L_z(M/L_x L_y L_z)]$, lb moles/hr sq ft (lb mole/cu ft).

After substitution of appropriate specific variables, the mass transfer equation is written

$$\frac{N_{a1}}{A_1} = -k_c'(c_{a1} - \bar{c}_a) \tag{13.81}$$

or upon rearrangement

$$N_{a1} = \frac{-(c_{a1} - \bar{c}_a)}{\dfrac{1}{A_1 k_c'}} \tag{13.82}$$

The driving force for mass transfer is the difference in concentration between the boundary and the bulk of the fluid, and the resistance to mass transfer is $(1/A_1 k_c')$.

For the evaluation of mass-transfer coefficients in gases and liquids in the turbulent regime, a correlation* similar to that for heat transfer has been developed. The data for this correlation were taken on a wetted-wall column, consisting of a vertical tube arranged so that liquid flows down the wall in a thin layer while a gas, in contact with the thin layer of liquid, flows upward through the core. If the liquid layer moves in laminar flow, the area of transfer is known. The wetted-wall column is useful for examination of mass-transfer data, but it has little commercial value when compared with the higher transfer rates per unit of volume in packed columns. However, the transfer area of the packed column is not known. Liquid mass-transfer rates were determined experimentally by permitting a liquid to flow in a pipe made of a soluble material. The equation

---

* The mechanism ratio analysis may be written based upon the four mechanisms for mass and momentum transfer by molecular and turbulent transport.

for the correlation (9) of all liquid and gas data takes the form*

$$\frac{k_c' D}{\mathscr{D}} = 0.023 \, (N_{\text{Re}})^{0.83} (N_{\text{Sc}})^{0.33} \qquad (13.83)$$

where $(N_{\text{Sc}}) = \nu/\mathscr{D}$, the ratio of momentum to mass transfer, both by the molecular mechanism. Mass-transfer data are not as common as heat-transfer data; consequently, the exact conditions of applicability of this equation cannot be stated with as much confidence as for heat transfer. Suffice it to say that this equation applies when the fluid descends the column in laminar flow and the gas ascends the column in well-developed turbulent flow.

The mechanical features of the wetted-wall column are such that the column for mass transfer and the tube for heat transfer are very nearly exact analogues. In addition, the equations resulting from correlation of heat- and mass-transfer data are very nearly identical in form differing only slightly in the exponent on the Reynolds number. The greater part of the concentration gradient exists in the resistance layer, that is, the fluid in the approximate region of the laminar sublayer and buffer region. Since a concentration gradient for two gases of differing molecular weight results in a density gradient, some degree of natural convection can occur in the resistance layer. Any natural convection in this layer will result in increase in the surface coefficient, and the probability exists that the small differences in the correlating equations can be attributed in part to this convection phenomenon.

In the integrated expression for molecular-mass transfer through a stationary film (Equation 10.45), the ratio c_t/c_{blm} was inserted as a correction to the equimolar transport equation. This correction for conversion of the coefficient† to the case of diffusion of

---

* Gas data alone are correlated (9) with

$$\frac{k_c' D}{\mathscr{D}} = 0.023 \, (N_{\text{Re}})^{0.83} (N_{\text{Sc}})^{0.44} \qquad (13.83a)$$

For these gas data N_{Sc} varied between 0.6 and 2.5. For liquids N_{Sc} approximates 1000. If the exponent 0.44 or 0.33 is used for common gases, the numerical difference in k_c' is very small and may be considered negligible compared to the experimental error.

† The mass-transfer coefficient may be written according to Equation 13.80 and then multiplied by c_t/c_{blm}.

$$k_c = \frac{4(\mathscr{D} + E_N)}{D \gamma_N} \cdot \frac{c_t}{c_{blm}}$$

Note that this expression implies turbulent transfer of gas (*a*) through a *stationary* gas (*b*). The assumption of existence of stationary gas (*b*) in the turbulent portion of the tube is obviously incorrect, but since none of the gas (*b*) passes through the boundary, traditional correlations of experimental data have been written on this basis, and until new information is forthcoming the practice will be continued here.

component *a* through stationary component *b* is

$$\frac{k_c' D}{\mathscr{D}} \cdot \frac{c_t}{c_{blm}} = \frac{k_c D}{\mathscr{D}}$$

Accordingly the equation for this case becomes

$$\frac{k_c D}{\mathscr{D}} \cdot \frac{c_{blm}}{c_t} = 0.023 \left(\frac{D \bar{v} \rho}{\mu}\right)^{0.83} \left(\frac{\mu}{\rho \mathscr{D}}\right)^{0.33} \qquad (13.84)$$

where k_c is the mass-transfer coefficient for gas *a* diffusing through a stationary gas *b*.

Equations 13.83 and 13.84 are reliable over the range between $N_{\text{Re}} = 2000$ and $N_{\text{Re}} = 35{,}000$ and between $N_{\text{Sc}} = 0.6$ and $N_{\text{Sc}} = 1000$. The gas velocity is measured relative to the container and not to the moving liquid. The data from which the equations are written are limited, and consequently some future improvement in the equations may be forthcoming.

Concentration data for gases and liquids may be reported in dimensions other than moles per unit volume. These other expressions of concentration lead to the writing of several different mass-transfer coefficients that are commonly used in the literature. For example, for gas concentrations expressed as partial pressures, the form k_G' is defined by the equation

$$\frac{N_a}{A_1} = -k_G' \, (p_{a1} - \bar{p}_a) \qquad (13.85)$$

where k_G' = the mass-transfer coefficient for equimolar counterdiffusion, for use with gas concentrations expressed as partial pressure, with dimensions $\mathbf{M}/\theta \, \mathbf{L}_y \mathbf{L}_z \, (\mathbf{F}/\mathbf{L}^2)$ and typical units (lb moles/hr sq ft atm).

p_{a1} = partial pressure of component *a* at the boundary $\mathbf{F}/\mathbf{L}^2$ (atm)

$\bar{p}_a$ = the mean value of partial pressure of component *a* in the flowing stream $\mathbf{F}/\mathbf{L}^2$ (atm)

The coefficient k_G' can be related to k_c' as follows. The rate of transfer of component *a* can also be represented by Equation 13.82.

$$\frac{N_a}{A_1} = -k_c' (c_{a1} - \bar{c}_a)$$

For perfect gases $c_a = p_a/RT$; therefore,

$$\frac{N_a}{A_1} = \frac{-k_c'}{RT} (p_{a1} - \bar{p}_a) \qquad (13.86)$$

By examination of Equations 13.85 and 13.86,

$$k_G' = \frac{k_c'}{RT}$$

Several other forms of mass-transfer coefficient are shown in Table 13.2. These forms can be related to k_c and k_c'.

Table 13.2. MASS-TRANSFER COEFFICIENTS FOR VARIOUS EXPRESSIONS FOR DRIVING FORCES

General equation: $\dfrac{N_a}{A_1} = -\dfrac{\text{(driving force)}}{\left(\dfrac{1}{\text{coefficient}}\right)}$

where N_a/A_1 = mass flux, lb moles/hr sq ft

Expression for Driving Force	Units of Driving Force	Coefficients		Units of Coefficient
		Equimolar Counterdiffusion	Transfer of (a) through Stationary (b)	
FOR GASES				
$c_{a1} - \bar{c}_a$	$\dfrac{\text{lb moles}}{\text{cu ft}}$	k_c'	k_c	$\dfrac{\text{lb moles of (a) transferred}}{\text{hr sq ft (lb moles/cu ft)}}$
$p_{a1} - \bar{p}_a$	atm	$k_G' = \dfrac{k_c'}{RT}$	$k_G = \dfrac{k_c}{RT}$	$\dfrac{\text{lb moles of (a) transferred}}{\text{hr sq ft atm}}$
$y_{a1} - \bar{y}_a$	mole fraction	$k_y' = \dfrac{k_c'P}{RT}$	$k_y = \dfrac{k_cP}{RT}$	$\dfrac{\text{lb moles of (a) transferred}}{\text{hr sq ft (mole fraction)}}$
$Y_{a1} - \bar{Y}_a$	mole ratio	$k_Y' = \dfrac{k_c'P(y_{b1})(\bar{y}_b)}{RT}$	$k_Y = \dfrac{k_cP(y_{b1})(\bar{y}_b)}{RT}$	$\dfrac{\text{moles of (a) transferred}}{\text{hr sq ft (mole ratio)}}$
FOR LIQUIDS				
$c_{a1} - \bar{c}_a$	$\dfrac{\text{lb moles}}{\text{cu ft}}$	$k_L' = k_c'$	$k_L = k_c$	$\dfrac{\text{lb moles of (a) transferred}}{\text{hr sq ft (lb mole volume)}}$
$x_{a1} - \bar{x}_a$	mole fraction	$k_x' = \dfrac{k_c'\bar{\rho}}{\bar{M}}$	$k_x = \dfrac{k_c\bar{\rho}}{\bar{M}}$	$\dfrac{\text{lb moles of (a) transferred}}{\text{hr sq ft (mole fraction)}}$
$X_{a1} - \bar{X}_a$	mole ratio	$k_X' = \dfrac{k_c'\bar{\rho}(x_{b1})(\bar{x}_b)}{\bar{M}}$	$k_X = \dfrac{k_c\bar{\rho}(x_{b1})(\bar{x}_b)}{\bar{M}}$	$\dfrac{\text{lb moles of (a) transferred}}{\text{hr sq ft (mole ratio)}}$

where x, y = mole fraction
X, Y = mole ratio, moles of (a)/moles of (b)
$\bar{\rho}$ = average density of a mixture of (a) and (b)
$\bar{M}$ = average molecular weight of a mixture of (a) and (b)

Illustration 13.8. A wetted-wall column 2 in. I.D. contains air and CO_2 flowing at 3 ft/sec. At one point in the column, the CO_2 concentration in the air is 0.1 mole fraction. At the same point in the column, the concentration of CO_2 in the water at the air–water interface is 0.005 mole fraction. The column operates at 10 atm and 25°C. Calculate the mass-transfer coefficient and mass flux at the point of consideration. Specify the direction of diffusion.

SOLUTION. From Appendix D-11, for CO_2 in air at 25°C and 1 atm $\mathscr{D} = 0.164$ sq cm/sec and $N_{Sc} = 0.94$. As shown in Chapter 9, the mass diffusivity of gases varies inversely with the pressure; therefore,

$$\mathscr{D}_2 = \mathscr{D}_1\left(\frac{P_1}{P_2}\right) = 0.164\frac{1}{10} = 0.0164 \text{ sq cm/sec}$$

This answer can be converted to English units of sq ft/hr.

$$\frac{(0.0164)(3600)}{(30.5)^2} = 0.0635 \text{ sq ft/hr}$$

In gases the Schmidt number $\nu/\mathscr{D}$ is independent of pressure because ν and $\mathscr{D}$ each vary in the same manner with pressure. The partial pressure of CO_2 in the gas phase is equal to the product of mole fraction of CO_2 and the total pressure.

$$\bar{p}_a = \bar{x}_a P = (0.1)(10) = 1 \text{ atm}$$

The value of p_{a1} must be calculated. The problem statement reports the concentration of CO_2 *in the liquid* at the interface as 0.005 mole fraction. The partial pressure of CO_2 in the gas phase at the interface is equal to the equilibrium vapor pressure of CO_2 over the liquid. For the carbon dioxide–water system, the Henry's law constant (H) can be used to calculate p_{a1}. In Appendix D-3 at 25°C, the Henry's law constant $H = 1640$ atm/unit mole fraction CO_2 in the liquid. Then

$$p_{a1} = Hx_{a1}.$$
$$p_{a1} = (1640)(0.005) = 8.2 \text{ atm}$$

Then, assuming that the properties of air are a satisfactory approximation for the air–carbon dioxide mixture, the properties of air at 25°C and 10 atm are

$$\mu_{\text{air}} = 0.018 \text{ centipoise}$$

$$\rho_{\text{air}} = 0.0808 \text{ lb/cu ft at standard conditions}$$

$$\rho_{\text{air}} \text{ at 10 atm, 25°C} = 0.080 \,(273/298)(10/1) = 0.74 \text{ lb/cu ft}$$

Testing for turbulence shows

$$N_{\text{Re}} = \frac{(D\bar{v}\rho)}{\mu} = \frac{\left(\dfrac{2}{12}\right)(3)(0.74)}{(0.018 \times 6.72 \times 10^{-4})} = 30,600$$

At this Reynolds number the air is in turbulent flow.

Air is essentially insoluble in water, so that this is an example of diffusion through a nondiffusing gas. In this case Equation 13.84 applies

$$\frac{c_{blm}}{c_t} = \frac{p_{blm}}{p_t}$$

where subscript b refers to the stationary component, air in this case.

By definition, $p_{blm} = \dfrac{(p_{b1} - \bar{p}_b)}{\ln\left(\dfrac{p_{b1}}{\bar{p}_b}\right)}$

$$p_{b1} = (p_t - p_{a1}) = (10 - 8.2) = 1.8 \text{ atm}$$

$$\bar{p}_b = (p_t - \bar{p}_a) = (10 - 1) = 9 \text{ atm}$$

$$p_{blm} = \frac{(1.8 - 9)}{\ln\left(\dfrac{1.8}{9}\right)} = \frac{-7.2}{-\ln\left(\dfrac{9}{1.8}\right)} = 4.47 \text{ atm}$$

$$\frac{p_{blm}}{p_t} = \frac{4.47}{10} = 0.447$$

Equation 13.84 may be solved for k_c

$$k_c = 0.023 \left(\frac{\mathscr{D}}{D}\right)\left(\frac{p_t}{p_{blm}}\right)(N_{\text{Re}})^{0.83}\,(N_{\text{Sc}})^{0.33}$$

$$= (0.023)\frac{(0.0635)}{\left(\dfrac{2}{12}\right)}\left(\frac{1}{0.447}\right)(30,600)^{0.83}\,(0.94)^{0.33}$$

$$= (0.023)\frac{(0.0635)}{\left(\dfrac{2}{12}\right)}\left(\frac{1}{0.447}\right)(5240)(0.98)$$

$$= 100 \text{ lb moles/hr sq ft (lb mole/cu ft)}$$

Equation 13.82 is used to calculate the mass flux.

$$N_a/A_1 = -k_c\,(c_{a1} - \bar{c}_a)$$

$$c_a = p_a/RT$$

$$R = 0.729 \text{ atm cu ft/lb mole °R}$$

$$T = (298)(1.8) \text{ °R}$$

$$N_a/A_1 = -\frac{k_c}{RT}(p_{a1} - \bar{p}_a)$$

$$= -\frac{100}{(0.729)(298)(1.8)}(8.2 - 1)$$

$$= -1.84 \text{ lb moles/hr sq ft}$$

The negative sign indicates transfer from the wall toward the center of the tube, since flux is positive in the direction of increasing radius. This is also evident from the high value of 8.2 atm for p_{a1}, the equilibrium vapor pressure of CO_2 over a liquid with concentration 0.005 mole fraction CO_2 in water.

ANALOGIES AMONG MASS, HEAT, AND MOMENTUM TRANSFER

Heretofore, mass, heat, and momentum transfer were examined almost simultaneously, with the general similarity of behavior always made evident. The similarity among molecular-transport phenomena was examined in Chapters 9 and 10. In Chapter 12, the similar dependence of mass, heat, and momentum transfer upon eddy activity was presented. Up to the present, a close-knit relationship seems to exist between the three transfer phenomena in the laminar and turbulent regime. Other evidence exists to show more quantitative similarity in the turbulent regime. In the following section several of many analogies between the transfer phenomena will be examined. Analogies will be drawn between the transfer-rate correlations presented previously in this chapter. In addition, some further information will be presented in the Martinelli analogy.

The analogies are useful tools to the student as an aid to rapid understanding of transfer phenomena and to the professional as a sound means to predict behavior of systems for which limited quantitative data are available. In this section, the analogies will be used to elucidate the mechanism of transfer.

The Transfer Coefficient for Momentum. The development of transfer-rate correlations for turbulent flow shows one inconsistency. The correlation of the momentum-transfer rate utilizes a friction factor–Reynolds number relationship, whereas transfer coefficients are used to correlate heat- and mass-transfer rates. These correlations are traditional practice in engineering.* If a surface coefficient for momentum transport is written, perhaps further similarities can be discovered. Equation 13.64 may be written for momentum transfer by substituting ρv for Γ and $(\tau_y g_c)_1$ for $(\psi)_1$.

$$(\tau_y g_c)_1 = -\mathscr{E}_\tau[\rho v_1 - \rho\bar{v}] \tag{13.87}$$

For the fluid adjacent to a stationary duct $v_1 = 0$, so that

$$(\tau_y g_c)_1 = \mathscr{E}_\tau \rho\bar{v}$$

---

* The early momentum-transfer work was done by civil engineers, who had no professional interest in heat or mass transfer. The early heat-transfer work was done by mechanical engineers, who had a small professional interest in momentum transfer. Mass transfer lies in the province of chemical engineering. It is this diverse background that probably hindered immediate recognition of the similar behavior of the transfer phenomena.

and

$$\mathscr{E}_\tau = \frac{(\tau_y g_c)_1}{\rho \bar{v}} \tag{13.88}$$

If Equation 13.88 is divided by $\bar{v}$, the result can be substituted into Equation 13.14.

$$\frac{f}{8} = \frac{(\tau_y g_c)_1}{\bar{v}^2 \rho} = \frac{\mathscr{E}_\tau}{\bar{v}}$$

or

$$\mathscr{E}_\tau = \frac{f\bar{v}}{8} \tag{13.89}$$

The Reynolds Analogy. The Reynolds analogy (10) is of historical importance as the first recognition of analogous behavior of momentum- and heat-transfer rates. As such it undoubtedly was the guidepost for much of the later work in the same areas. Although it is limited in application, it is useful in any work involving common gases, and in fact it has been one of the more powerful tools in analyzing the complex boundary-layer phenomena of aerodynamics.

Reynolds postulated that the mechanisms for transfer of momentum and heat are identical. The postulate may be written as

$$\frac{4(\alpha + \bar{E}_q)}{\gamma_q D} = \frac{4(\nu + \bar{E}_\tau)}{\gamma_\tau D} \tag{13.90}$$

or, if coefficients are substituted,

$$\frac{h}{c_P \rho} = \mathscr{E}_\tau \tag{13.91}$$

Equation 13.91 may be divided by $\bar{v}$, and the friction factor substituted for $\mathscr{E}_\tau/\bar{v}$ as in Equation 13.89.

$$\frac{h}{c_P \rho \bar{v}} = \frac{f}{8} \tag{13.92}$$

Equation 13.92 is a mathematical statement of the Reynolds analogy. The group $h/c_P \rho \bar{v}$ is the Stanton number (N_{St}) and represents the dimensionless ratio.

$$\frac{\text{Total (molecular and turbulent) heat transfer}}{\text{Turbulent momentum transfer}}$$

The Stanton number is related to the Nusselt, Reynolds, and Prandtl numbers as follows:

$$N_{St} = \frac{N_{Nu}}{N_{Re} N_{Pr}} \tag{13.93}$$

Therefore, Equation 13.92 may be written

$$\frac{N_{Nu}}{N_{Re} N_{Pr}} = N_{St} = \frac{f}{8} \tag{13.94}$$

When Equation 13.94 was tested experimentally, it was found to correlate data *approximately* for gases in turbulent flow. It did not correlate experimental data for liquids in turbulent flow nor for any fluids in laminar flow. All gases are characterized (see Chapter 9) by values of the Prandtl number in the general range between 0.6 and 2.5. Upon more critical examination, it was concluded (14) that the Reynolds analogy is valid *only* at $(N_{Pr}) = 1$. By similar reasoning, an expression may be written based upon the postulate that mass and momentum are transferred by an identical mechanism. This postulate may be stated as

$$\frac{4(\mathscr{D} + \bar{E}_N)}{\gamma_N D} = \frac{4(\nu + \bar{E}_\tau)}{\gamma_\tau D} \tag{13.95}$$

If the appropriate transfer coefficients are substituted and if Equation 13.95 is divided by $\bar{v}$, an analogy equation can be written

$$\frac{k_c'}{\bar{v}} = \frac{f}{8} \tag{13.96}$$

When Equation 13.96 was tested experimentally, it was found to correlate data *approximately* for gases in turbulent flow but did *not* correlate data for liquids in turbulent flow nor for any fluids in laminar flow. Gases are characterized by values of the Schmidt number near unity. More critical examination revealed that Equation 13.96 is valid (9) only at $N_{Sc} = 1$.

Although the Reynolds analogy is of limited utility, one significant conclusion may be drawn, and that is that, at $N_{Pr} = N_{Sc} = 1$, the mechanisms for mass, heat, and momentum are identical. A second conclusion may also be drawn, namely that, for other fluids, transfer processes differ in some manner functionally related to the Prandtl or Schmidt number. Application of the Reynolds analogy appears in Illustration 13.9, on page 177.

The Colburn Analogy. Colburn (11) examined the empirical equations for heat and momentum transfer in the turbulent regime and presented a more general analogy. If Equation 13.79 for heat transfer in the turbulent regime is divided by the product $(N_{Re})(N_{Pr})$, then

$$\frac{N_{Nu}}{N_{Re} N_{Pr}} = 0.023 \frac{(N_{Re})^{0.8} (N_{Pr})^{0.33} \left(\dfrac{\mu}{\mu_1}\right)^{0.14}}{(N_{Re})(N_{Pr})}$$

This expression can be rearranged to

$$(N_{St})(N_{Pr})^{2/3} \left(\frac{\mu_1}{\mu}\right)^{0.14} = 0.023 (N_{Re})^{-0.2} \tag{13.97}$$

Equation 13.97 is also plotted in Appendix C-5. Equation 13.26 for momentum transfer may be divided by 8 and written as

$$\frac{f}{8} = 0.023 (N_{Re})^{-0.2} \tag{13.98}$$

Equation 13.98 is also plotted in Appendix C-5. The term $0.023 \, (N_{Re})^{-0.2}$ can be eliminated from Equations 13.97 and 13.98

$$(N_{St})(N_{Pr})^{2/3} \left(\frac{\mu_1}{\mu}\right)^{0.14} = \frac{f}{8} \qquad (13.99)$$

The term j_q is defined as $(N_{St})(N_{Pr})^{2/3} \, (\mu_1/\mu)^{0.14}$ and is called the "j" factor for heat transfer. Equation 13.99 can be written

$$j_q = \frac{f}{8} \qquad (13.100)$$

This equation is a statement of the Colburn analogy between heat and momentum transfer. Equation 13.100 is derived from two equations that correlate experimental data over the range $10,000 < N_{Re} < 300,000$ and $0.6 < N_{Pr} < 100$. Equation 13.99 shows a functional relationship between heat and momentum transfer and the Prandtl number as predicted in the Reynolds analogy. Note that in Equation 13.99, if $N_{Pr} = 1$ and $(\mu_1/\mu) = 1$,

$$N_{St} = \frac{f}{8} \qquad (13.94)$$

which is the mathematical statement of the Reynolds analogy.

The empirical equations for mass transfer and momentum transfer in the turbulent regime may be compared in the same manner. Equation 13.83 for mass transfer can be divided by the product $(N_{Re}) \, (N_{Sc})$ and rearranged to

$$\frac{k_c'}{\bar{v}} (N_{Sc})^{2/3} (N_{Re})^{0.03} = 0.023 \, (N_{Re})^{-0.2} \qquad (13.101)$$

If $(N_{Re})^{0.03}$ is taken as unity,* Equation 13.101 may be written

$$\frac{k_c'}{\bar{v}} (N_{Sc})^{2/3} = 0.023 \, (N_{Re})^{-0.2} \qquad (13.102)$$

The factor $0.023 \, (N_{Re})^{-0.2}$ can be eliminated between Equations 13.102 and 13.98 to give

$$\frac{k_c'}{\bar{v}} (N_{Sc})^{2/3} = \frac{f}{8} \qquad (13.103)$$

The term j_N can be defined as $(k_c'/\bar{v})(N_{Sc})^{2/3}$ and called the "j" factor for mass transfer. Therefore

$$j_N = \frac{f}{8} \qquad (13.104)$$

---

* The error in this assumption may be considerable. Over the range of Reynolds number for which this assumption is applied, $(N_{Re})^{0.03}$ varies between 1.32 and 1.46. The surface of the fluid in a wetted-wall column may be rippled, which can introduce geometrical disturbances and increase the rate of mass transfer in the gas phase. The higher exponent on the Reynolds number may be a result of the rippling of the fluid surface.

Equations 13.100 and 13.104 can be combined to give

$$j_q = j_N = \frac{f}{8} \qquad (13.105)$$

This equation applies over the following ranges: For heat transfer $10,000 < N_{Re} < 300,000$ and $0.6 < N_{Pr} < 100$. For mass transfer,* $2000 < N_{Re} < 300,000$ and $0.6 < N_{Sc} < 2500$.

Some interesting conclusions may be drawn concerning Equation 13.105 if transfer coefficients are substituted in each of the terms. If this substitution is made, then

$$\frac{h}{c_P \rho \bar{v}} (N_{Pr})^{2/3} \left(\frac{\mu_1}{\mu}\right)^{0.14} = \frac{k_c'}{\bar{v}} (N_{Sc})^{2/3} = \frac{\mathscr{E}_\tau}{\bar{v}} \qquad (13.106)$$

The velocity $\bar{v}$ can be canceled from each term. In addition consider the equation to be under test at a negligible temperature driving force, in which time $(\mu_1/\mu)^{0.14}$ = unity, then

$$\frac{h}{c_P \rho} (N_{Pr})^{2/3} = k_c'(N_{Sc})^{2/3} = \mathscr{E}_\tau \qquad (13.107)$$

Equation 13.107 may be divided by $(\nu)^{2/3}$

$$\frac{h}{c_P \rho(\alpha)^{2/3}} = \frac{k_c'}{(\mathscr{D})^{2/3}} = \frac{\mathscr{E}_\tau}{(\nu)^{2/3}} \qquad (13.107a)$$

Equation 13.107 is understood to apply to a constant Reynolds number for each of the transfer phenomena. Equation 13.107a, demonstrates that the ratio of the transfer coefficient to the molecular diffusivity to the 2/3 power is a constant. This simple relationship is valid for the empirical equations used for the Colburn analogy and for the mass-transfer analogue of the Colburn analogy, subject to the correction mentioned above. However, if the mechanism definition of the coefficients is substituted into Equation 13.107 by use of Equations 13.90, 13.80, and 13.68,

$$\frac{4}{D} \frac{(\alpha + \bar{E}_q)}{\gamma_q(\alpha)^{2/3}} = \frac{4}{D} \frac{(\mathscr{D} + \bar{E}_N)}{\gamma_N(\mathscr{D})^{2/3}} = \frac{4}{D} \frac{(\nu + \bar{E}_q)}{\gamma_\tau(\nu)^{2/3}} \qquad (13.108)$$

The term $4/D$ cancels and Equation 13.108 may be written

$$\frac{(\alpha + \bar{E}_q)}{\gamma_q(\alpha)^{2/3}} = \frac{(\mathscr{D} + \bar{E}_N)}{\gamma_N(\mathscr{D})^{2/3}} = \frac{(\nu + \bar{E}_\tau)}{\gamma_\tau(\nu)^{2/3}} \qquad (13.109)$$

This equation is a restatement of Equation 13.107 and applies as long as the Reynolds number is the same for each transfer phenomenon. The Reynolds number may have any value within the range of validity of the equations used in the derivation. The concentration-distribution functions γ_q, γ_N, and γ_τ, by definition, (Equation 13.60) have a maximum value of unity and a probable minimum value not below 0.7. Within the range of validity of the correlation equations, transfer coefficients vary over about a fifteenfold range. If the concentration-distribution functions (γ_q, γ_τ, and γ_N) are taken as approximately equal whereas the molecular diffusivities (α, v, and $\mathscr{D}$) vary over the range of validity of the empirical equations, then at a given Reynolds number the *mean* eddy diffusivities of mass, heat, and momentum will not be equal. In Chapter 12 the *point* value of the eddy diffusivity was assumed equal at a given flow rate irrespective of the transferent property. There is no inconsistency in these observations. The definition of mean eddy diffusivity as given by Equation 13.59 shows the integration of the point value over the entire duct. The integration includes the use of the gradients $d(\rho v)/dr$, $d(\rho c_P T)/dr$ and dc_a/dr. If at a particular flow condition the *point* eddy diffusivities are equal but the gradients differ with transferent property, the *mean* values that result from integration will be different. In the Reynolds analogy, in which $\alpha = v = \mathscr{D}$, the values of the mean eddy diffusivities were equal, implying that the gradients were equal; but for the Colburn analogy, in which $\alpha \neq v \neq \mathscr{D}$, the *mean* eddy diffusivities are not equal, implying differences in gradient for different transferent properties. This concept will be amplified further in the next section on the Martinelli analogy.

The empirical equations in this section are valid for gases and liquids within the range $0.6 < N_{Pr} < 100$ and $0.6 < N_{Sc} < 2500$. This range includes most fluids of practical importance except liquid metals. Little data are available for liquids of extremely high Prandtl number because such liquids must have extremely high viscosity, high enough that pumping machinery to maintain the high velocity of turbulent flow does not exist. The liquid metals are a major class of real fluids whose heat-transfer behavior cannot be predicted with the empirical equations cited above. These fluids are characterized by low viscosity (approximately that of water) but high thermal conductivities (approximately thirty to sixty times higher than water). Other properties are such that a typical value for liquid metals is $N_{Pr} = 0.01$. This value is well outside the range of validity of the empirical correlations cited above. If the empirical equations were faithful statements of mechanism, they would undoubtedly apply over the entire range of Prandtl numbers. The criticism here is much the same as the criticism of Equations 12.1, 12.2, and 12.3, the velocity-distribution equations in the turbulent regime. The equations are valid for predicting behavior but do not faithfully follow the mechanism of the behavior. Once again, it might be concluded that these problems will disappear as soon as a complete mathematical picture of turbulence is written. Application of the Colburn analogy appears in Illustration 13.9, on page 177.

The Martinelli Analogy. The Reynolds analogy is a significant contribution because it demonstrates similarity of mechanism among transfer phenomena, but it is limited to unit values of the Prandtl and Schmidt numbers. The Reynolds analogy does nothing to elucidate the mechanism. The Colburn analogy demonstrates numerical similarity among transfer phenomena over a wider but still limited range of Prandtl and Schmidt numbers. The Colburn analogy implies that the correlation equations are *not* faithful statements of the mechanism, however reliable they may be for predicting numerical values of coefficients.

Other analogies have been written that may be applied with varying success over the entire range of Prandtl number. These analogies are based upon the model for turbulent flow described in Chapter 12, each with its own set of assumptions. The work of Martinelli (17) is a good example of an analogy between heat and momentum transfer that can be derived from a more detailed model. Martinelli made the following assumptions.

1. The temperature driving force between the wall and the fluid is small enough so that $\mu/\mu_1 \cong 1$.
2. Well-developed turbulent flow exists within the test section.
3. Heat flux across the tube wall is constant along the test section.
4. Both stress and heat flux are zero at the center of the tube and increase linearly with radius to a maximum at the wall. (See Equations 10.58 and 12.27.)
5. At any point $E_q = E_\tau$.
6. The velocity-distribution equations, Equations 12.1, 12.2, and 12.3, are valid. The equations

$$\tau_y g_c = -(v + E_\tau)\frac{d(\rho v)}{dx} \qquad (12.7)$$

and

$$\frac{q}{A} = -(\alpha + E_q)\frac{d(\rho c_P T)}{dx}$$

are modified according to assumption 4 and written for cylindrical geometry

$$(\tau_y g_c)_1 \left(\frac{r}{r_1}\right) = -(v + E_\tau)\frac{d(\rho v)}{dr} \qquad (12.15)$$

and

$$\left(\frac{q}{A}\right)_1 \left(\frac{r}{r_1}\right) = -(\alpha + E_q)\frac{d(\rho c_P T)}{dr} \qquad (12.30)$$

The velocity-distribution equations are used to evaluate E_τ in Equation 12.15, and these values may be substituted into Equation 12.30 according to assumption 5. Equation 12.30 is integrated so that the point value of temperature is expressed as a function of position, and this equation is converted into the form

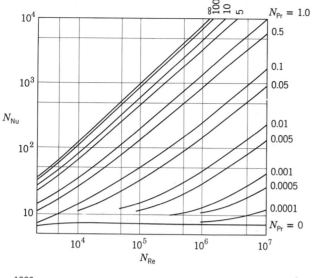

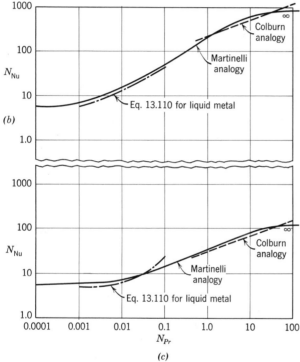

Figure 13.6. The Martinelli analogy. (a) N_{Nu} as a function of N_{Re} and N_{Pr}. Comparison of the Martinelli analogy and empirical equations at: (b) $N_{Re} = 100,000$, $\mu/\mu_1 \cong 1$. (c) $N_{Re} = 10,000$, $\mu/\mu_1 \cong 1$. (After McAdams, *Heat Transmission*, 3rd ed.)

$N_{Nu} = \phi(N_{Re}, N_{Pr}, f)$. The equations are complex and are not presented here. Instead the results are presented in the form of a plot of N_{Nu} as a function of N_{Re} at values of N_{Pr} varying between 0 and ∞, in Figure 13.6a.

Since the Martinelli analogy can be applied over the entire range of N_{Pr}, it can be tested with experimental data. Lyon (18) predicted the form of a correlation equation that is valid over the Prandtl number range of

liquid metals by a method similar to the work of Martinelli and tested the equation with experimental data using sodium and potassium alloys. The equation resulting from this correlation is

$$N_{Nu} = 7 + 0.025(N_{Pe})^{0.8}* \qquad (13.110)$$

where N_{Pe} = the dimensionless group called the Peclet number. This group is equal to the product $(N_{Re})(N_{Pr})$ and represents the mechanism ratio:

$$\frac{\text{Total (molecular + turbulent) momentum transfer}}{\text{Heat transfer by molecular mechanism}}$$

For convenience, Equation 13.110 is plotted in Appendix C-5.

Data for other liquid metals such as mercury, lead, and bismuth show a somewhat lower Nusselt number than that predicted by Equation 13.110 at comparable values of the Peclet number. The lower values of the Nusselt number are presently attributed to the non-wetting characteristics of these liquid metals with respect to the heat-transfer tubing, although the hypothesis is not clearly established (51). For mercury, lead, and bismuth, Equation 13.110a is recommended (52).

$$N_{Nu} = 0.625 (N_{Pe})^{0.4} \qquad (13.110a)$$

Equation 13.110a is plotted in Appendix C-5 for convenience.

The Martinelli analogy predicts the values of N_{Nu} for liquid metals, with fair precision and also predicts heat-transfer coefficients for those fluids that fall within the valid range of the Colburn analogy. The agreement between this analogy and experimental data can be seen in Figures 13.6b and 13.6c. Figure 13.6b is a plot of N_{Nu} as a function of N_{Pr} as predicted by Martinelli at $N_{Re} = 100,000$. Data for the same conditions taken from Equation 13.110 and Equation 13.79 are plotted on the same figure. Figure 13.6c is a similar plot taken at $N_{Re} = 10,000$. Examination of these plots shows that the Martinelli analogy is in fair agreement with experimental data over the total available range of Prandtl number. Thus this work is a significant contribution to the understanding of the *mechanism* of heat and momentum transfer in the turbulent regime.

Analogies have been written for mass transfer using techniques similar to those of Martinelli and others (19, 20). The experimental data and analogies agree over a wide range, provided the assumption $E_N = 1.6E_r$ is used instead of the equality of eddy diffusivities assumed by Martinelli for heat transfer. A reason

---

* This equation does not contain a term (μ/μ_1) and attaches greater importance to the Prandtl number. In contrast to gases, water, and organic liquids, the molten metals have viscosities which are relatively insensitive to temperature, as well as exhibiting greater molecular conductivity.

is offered for this difference. For example, the surface of fluid in a wetted-wall column may be rippled and therefore tend to transfer mass in the gas phase because of an increase in turbulence due to the surface irregularities. Heat transfer in this same kind of apparatus invariably is greater (20) than under similar conditions in dry tubes.

The foregoing material is impressive evidence of the value of the analogies to define a mechanism for turbulent transfer of mass, heat, and momentum. In summary, transfer depends upon the product of the molecular diffusivity and the gradient for a part of the transfer and upon the product of the eddy diffusivity and the gradient for the remainder of the transfer. The eddy diffusivities for mass, heat, and momentum are probably identical in physically identical apparatus. These observations support the assumption that $E_\tau = E_N = E_q$ made in Chapter 12. The molecular diffusivities are certainly not identical, and the effect of this difference is shown when heat-transfer characteristics of substances such as water and mercury are compared. The high thermal conductivity of mercury results in much higher heat-transfer coefficients at constant Reynolds numbers. For water, the contribution of molecular transfer to the total transfer is almost negligible, whereas, for mercury, the contribution of molecular transfer to the total transfer is of the same order of magnitude as the contribution due to turbulence.

The Martinelli analogy is a valuable starting point for the prediction of transfer behavior of substances whose properties are outside the range previously investigated. For example, for any fluid an equation of the form

$$N_{Nu} = \text{constant } (N_{Re})^a (N_{Pr})^b \qquad (13.111)$$

might be predicted.

The behavior of fluids of very high Prandtl number may be predicted from Figure 13.6a in which the curves for $N_{Pr} = 100$ and $N_{Pr} = \infty$ each have a slope of approximately 0.8. The two curves are approximately coincident, and therefore $a = 0.8$ and $b = 0$ in Equation 13.111. No prediction can be made as to the effect of the viscosity ratio (μ/μ_1).

The analogies have also been used to predict behavior for transfer to flat plates, as in the boundary layer. These correlations are shown in a later part of this chapter. Although the analogies have definite usefulness in predicting behavior beyond the range of known data, the tested empirical equation is still the most valuable source for design data.

Effect of Prandtl Number on the Temperature Profile. The ratio $\gamma = (\Gamma_1 - \bar{\Gamma})/(\Gamma_1 - \Gamma_0)$ defined in Equation 13.57 can be examined qualitatively for heat transfer. Temperature–radial-position data are plotted in Figure 13.7 using a coordinate system that is applicable to

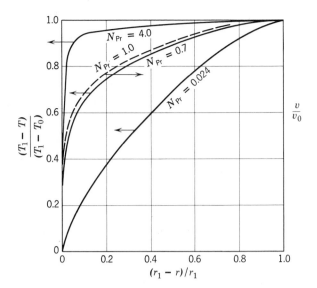

Figure 13.7. Temperature and velocity profiles at $N_{Re} = 40,000$. (After McAdams, *Heat Transmission*, 3rd ed.)

cylindrical geometry irrespective of the size of the cylinder and irrespective of the actual values of temperatures. The temperature function is $(T_1 - T)/(T_1 - T_0)$, and the position function is $(r_1 - r)/r_1$. The subscripts are indicative of location, where T_1 and r_1 represent the condition at the wall, T_0 represents the condition at the center of the tube, and T and r represent any radial position in the duct. The temperature profiles for fluids of Prandtl numbers 4, 0.7, and 0.024 are plotted. All data are taken for systems at $N_{Re} \cong 40,000$. In addition, the dotted line represents the velocity profile (v/v_c), plotted against $(r_1 - r)/r_1$ for all fluids. The data of the three temperature curves was cross-plotted to find the temperature profile curve at $N_{Pr} = 1$. This curve is congruent with the dotted velocity profile.

The plot may be used to predict qualitatively that γ_q, which is based upon the mixing-cup mean value of Γ (see Equation 13.59), will increase with increase in Prandtl number at constant N_{Re}. An examination of Figure 20.2 shows that γ_τ (or in this case $\bar{v}/v_{\max}$) increases with increase in Reynolds number. It may be assumed then that γ_q is a function of both the Reynolds and Prandtl number.

Comparable data for mass transfer are not available, but in the light of analogous behavior it may be assumed that γ_N increases with increase in Reynolds number and Schmidt number.

Illustration 13.9. Compare the value of the Nusselt number, given by the appropriate empirical equation, to that predicted by the Reynolds, Colburn, and Martinelli analogies for each of the following substances at $N_{Re} = 100,000$. (a) SAE 10 lube oil, (b) water, (c) air (1 atm), and (d) liquid mercury. Consider all substances at 100°F, subject to heating with the tube wall at 150°F.

Data: Properties at 1 atm, in Btu, hrs, ft, and lbs

	μ_{100} (centi-poises)	k_{100}	ρ_{100}	c_{P100}	$(N_{Pr})_{100}$	μ_{150} (centi-poises)
Air	0.018	0.0157	–	0.2401	0.71	0.02
Water	0.69	0.365	61.9	1.	4.57	0.42
SAE 10	31.	0.071	56.6	0.45	47.5	12.
Mercury	1.45	5.1	84.3	0.33	0.0234	1.40

At $N_{Re} = 100{,}000$, $f = 0.0184$ (from Equation 13.26).

SOLUTION FOR AIR. For air, *the appropriate empirical equation* is Equation 13.79.

$$N_{Nu} = 0.023(N_{Re})^{0.8}(N_{Pr})^{1/3}\left(\frac{\mu}{\mu_1}\right)^{0.14}$$

$$N_{Nu} = 0.023(100{,}000)^{0.8}(0.71)^{1/3}\left(\frac{0.018}{0.02}\right)^{0.14}$$

$$= 0.023(10^4)(0.893)(0.9914)$$

$$N_{Nu} = 204, \text{ most accurate value}$$

By the Reynolds analogy, use Equation 13.94.

$$N_{St} = \frac{f}{8}$$

According to Equation 13.93, $N_{St} = N_{Nu}/N_{Re}N_{Pr}$. If the term $f/8$ is used to replace the Stanton number in the Reynolds analogy, then

$$N_{Nu} = (N_{Re})(N_{Pr})\left(\frac{f}{8}\right)$$

$$N_{Nu} = (10^5)(0.71)\left(\frac{0.0184}{8}\right)$$

$$N_{Nu} = 164.2, \text{ Reynolds analogy}$$

By the Colburn analogy, use Equation 13.99.

$$(N_{St})(N_{Pr})^{2/3}\left(\frac{\mu_1}{\mu}\right)^{0.14} = \frac{f}{8}$$

By substituting $N_{Nu}/(N_{Re}N_{Pr})$ for N_{St} and rearranging, this equation becomes

$$N_{Nu} = (N_{Re})(N_{Pr})^{1/3}\left(\frac{f}{8}\right)\left(\frac{\mu}{\mu_1}\right)^{0.14}$$

$$= (10^5)(0.71)^{1/3}\left(\frac{0.0184}{8}\right)\left(\frac{0.018}{0.020}\right)^{0.14}$$

$$= (10^5)(0.893)\left(\frac{0.0184}{8}\right)(0.9914)$$

$$N_{Nu} = 204, \text{ Colburn analogy}$$

By the Martinelli analogy, the data are available in Figure 13.6a.

$$N_{Nu} = 170$$

SOLUTION FOR MERCURY. Equation 13.110a is the appropriate empirical equation that can be used to predict the Nusselt number.

$$N_{Nu} = 0.625(N_{Pe})^{0.4}$$

$(N_{Pe}) = (N_{Re})(N_{Pr})$; therefore

$$N_{Nu} = 0.625(10^2)(0.0234)^{0.4}$$

$$= 0.625(22.1)$$

$$N_{Nu} = 13.9, \text{ most accurate value}$$

By Reynolds analogy,

$$N_{St} = \frac{f}{8}$$

$$N_{Nu} = N_{Re}N_{Pr}\frac{f}{8}$$

$$N_{Nu} = (10^5)(0.0234)\left(\frac{0.0184}{8}\right)$$

$$= 5.38, \text{ Reynolds analogy}$$

By Colburn analogy,

$$N_{St}(N_{Pr})^{2/3}\left(\frac{\mu_1}{\mu}\right)^{0.14} = \frac{f}{8}$$

Since $N_{St} = N_{Nu}/(N_{Re}N_{Pr})$ the Colburn analogy equation can be written

$$N_{Nu} = N_{Re}(N_{Pr})^{1/3}\left(\frac{f}{8}\right)\left(\frac{\mu}{\mu_1}\right)^{0.14}$$

$$N_{Nu} = (10^5)(0.0234)^{1/3}\left(\frac{0.0184}{8}\right)\left(\frac{1.45}{1.40}\right)^{0.14}$$

$$= (10^5)(0.285)(0.0023)(1.01)$$

$$= 66.1, \text{ Colburn analogy}$$

By the Martinelli analogy, from Figure 13.6a,

$$N_{Nu} = 20$$

SOLUTION FOR WATER. Equation 13.79 can be used to predict the heat-transfer coefficient for water.

$$N_{Nu} = 0.023(N_{Re})^{0.8}(N_{Pr})^{0.33}\left(\frac{\mu}{\mu_1}\right)^{0.14}$$

$$= 0.023(10^5)^{0.8}(4.57)^{0.33}\left(\frac{0.69}{0.42}\right)^{0.14}$$

$$= 0.023(10^4)(1.658)(1.066)$$

$$N_{Nu} = 406, \text{ most accurate value}$$

By Reynolds analogy,

$$N_{St} = \frac{f}{8}$$

Since $N_{St} = N_{Nu}/N_{Re}N_{Pr}$

$$N_{Nu} = N_{Re}N_{Pr}\left(\frac{f}{8}\right)$$

$$N_{Nu} = (10^5)(4.57)\left(\frac{0.0184}{8}\right)$$

$$N_{Nu} = 1050, \text{ Reynolds analogy}$$

By the Colburn analogy,

$$N_{St}(N_{Pr})^{2/3}\left(\frac{\mu_1}{\mu}\right)^{0.14} = \frac{f}{8}$$

Since $N_{St} = N_{Nu}/N_{Re}N_{Pr}$, the equation may be written

$$N_{Nu} = (N_{Re})(N_{Pr})^{1/3}\left(\frac{\mu}{\mu_1}\right)^{0.14}\left(\frac{f}{8}\right)$$

$$= (10^5)(4.57)^{1/3}\left(\frac{1.66}{1.05}\right)^{0.14}\left(\frac{0.0184}{8}\right)$$

$$= 406$$

By the Martinelli analogy, from Figure 13.6a

$$N_{Nu} = 400$$

SOLUTION FOR SAE 10 OIL. The heat-transfer coefficient for a viscous oil can be predicted by Equation 13.79.

$$N_{Nu} = 0.023(N_{Re})^{0.8}(N_{Pr})^{1/3}\left(\frac{\mu}{\mu_1}\right)^{0.14}$$

$$= 0.023(10^5)^{0.8}(47.5)^{1/3}\left(\frac{31}{12}\right)^{0.14}$$

$$= 0.023(10^4)(3.56)(1.142)$$

$$= 935, \text{ most accurate value}$$

By the Reynolds analogy,

$$N_{St} = \frac{f}{8}$$

Since $N_{St} = N_{Nu}/(N_{Re}N_{Pr})$ the equation may be written

$$N_{Nu} = N_{Re}N_{Pr}\left(\frac{f}{8}\right)$$

$$= 10^5(47.5)\left(\frac{0.0184}{8}\right)$$

$$N_{Nu} = 10,500, \text{ Reynolds analogy}$$

By the Colburn analogy,

$$(N_{St})(N_{Pr})^{2/3}\left(\frac{\mu_1}{\mu}\right)^{0.14} = \frac{f}{8}$$

Since $N_{St} = N_{Nu}/N_{Re}N_{Pr}$, the equation may be written

$$N_{Nu} = (N_{Re})(N_{Pr})^{0.33}\left(\frac{\mu}{\mu_1}\right)^{0.14}\left(\frac{f}{8}\right)$$

$$= 10^5(47.5)^{0.33}\left(\frac{31}{12}\right)^{0.14}\left(\frac{0.0184}{8}\right)$$

$$= 10^5(3.56)(1.142)\left(\frac{0.0184}{8}\right)$$

$$= 935, \text{ Colburn analogy}$$

By the Martinelli analogy, from Figure 13.6a,

$$N_{Nu} = 800$$

The values of the Nusselt numbers are tabulated below for easy comparison. The substances are written in terms of increasing Prandtl number.

	N_{Pr}	Empirical Equation Number	Nusselt Numbers			
			Empirical Equation (Most probable)	Rey-nolds	Col-burn	Mar-tinelli
Mercury	0.0234	(13.110a)	13.9	5.38	66.1	20
Air	0.71	(13.79)	204	164	204	170
Water	4.57	(13.79)	406	1050	406	400
SAE 10 oil	47.5	(13.79)	935	10,500	935	800

Note that over the entire range, the Martinelli analogy is in good agreement with the empirical correlation of experimental data. The Colburn analogy agrees perfectly, as it should, for air, water, and oil, because the Colburn analogy for these substances is the same as the empirical equation. The discrepancy for oil between the Martinelli analogy and the empirical equation is due to the absence of a viscosity ratio term in the Martinelli analogy. If this factor is included, the value becomes $(800)(1.142) = 913$, where $(\mu/\mu_1)^{0.14} = (31/12)^{0.14} = 1.142$.

Summary. This section on analogous behavior of transfer phenomena has described three analogies. The Reynolds analogy is based upon the assumption that the mechanisms for heat, and momentum are identical; subsequent extensions include mass transport. Test of the analogy shows this assumption to be true at $N_{Pr} = 1$ and approximately true at $N_{Sc} = 1$. The Colburn analogy is based entirely upon similarity of empirical equations, and therefore, over the range of application, it is in good agreement with experiment. Critical examination of the analogy shows that it does not represent a mathematical statement of the mechanism. The Colburn analogy when applied to mass transfer results in an error of 1.32 to 1.46 as mentioned in an earlier footnote. The reason for the discrepancy is attributed at least in part to surface effects on the descending film of fluid. The discrepancy then is geometrical, and the implication may be that the systems are not exactly analogous in a geometrical sense. The Martinelli analogy is based upon equality of eddy diffusivities of heat and momentum as a statement of mechanism. The Martinelli analogy shows remarkable agreement over the entire range of Prandtl number, indicating that the analogy is a reasonable statement of the mechanism. A similar analogy for mass transfer agrees with mass-transfer data if $E_N = 1.6\,E_\tau$ is used. This discrepancy is attributed to the difference in geometry between a cylindrical tube of dry metal and a wetted-wall column. The mechanism for *geometrically* *similar systems* may be considered dependent upon the general equation

$$\psi = -(\delta + E)\frac{d\Gamma}{dx} \qquad (13.55)$$

where in the light of present knowledge

$$E_\tau = E_q = E_N$$

HEAT AND MASS TRANSFER IN THE TURBULENT BOUNDARY LAYER

It was shown earlier that a flat plate in a stream produces a boundary layer of fluid of velocity lower than the free-stream velocity and that this layer lies close to and is affected by the surface. If the same shape at some temperature T_1 is placed in a free stream at temperature T_{fs}, a temperature boundary layer will also form, with a similar (but not necessarily identical) shape, as shown in Figure 13.8. Just as the boundary layer is a region of velocity gradient in the gas phase adjacent to the boundary through which momentum is transferred from the fluid free stream to the boundary, the thermal boundary layer is a region of temperature gradient through which heat is transferred to or from the free stream to the boundary. In the event that the plate surface temperature is T_{fs} between $y = 0$ and $y = y_1$, and then at y_1 the temperature changes from T_{fs} to T_1, the temperature boundary layer begins to form at y_1. Meanwhile the velocity boundary layer would have begun to form at $y = 0$. The two layers develop according to the appropriate fluid and system properties. An analogous mass-transfer boundary layer can also occur. Consider a pan of water subject to a gas flow parallel to the water surface. If the surface stress of the wind does not cause ripples on the surface, the smooth surface is geometrically similar to the flat plate, and, if the partial pressure of the water in the air is different from the vapor pressure of water at the liquid surface, a boundary layer of water-vapor concentration will form along the water surface. The entrance to a heat exchanger and the entrance to a wetted-wall tower are other examples of thermal or mass boundary layers combined with a momentum boundary layer. As in momentum transfer, when the edges of the thermal or mass boundary layer meet at the center of the tube, the flow is fully developed, and no further growth of boundary layer takes place with progress through the tube. The temperature or concentration can continue to respond to transfer.

As shown in Figure 13.8, the momentum boundary layer begins to form at some point $y = 0$, and the thermal or mass boundary layer begins at a different point y_1. For example, a heat exchanger tube may extend through a tube sheet in such a manner that the length of tube between y_0 and y_1 is at the same temperature as the free stream, in which case no thermal boundary layer forms upstream of y_1. Beyond the tube sheet, beginning

at y_1, the tube may be heated. In this case the momentum boundary layer begins at $y = 0$ and the thermal boundary layer begins at $y = y_1$.

Consideration of the above suggests that the analogous behavior of fluids in similar geometries can be analyzed using the Reynolds or the Colburn analogy.

The Reynolds Analogy Applied to the Boundary Layer. The Reynolds analogy was applied to turbulent flow through smooth tubes and found to be valid if the fluid Prandtl number was unity. The same reasoning might be applied to a flat plate in a free stream. The Reynolds analogy for tubes is given by Equation 13.92,

$$\frac{h}{c_P \bar{v} \rho} = \frac{f}{8} \qquad (13.92)$$

Equation 13.14 states that $f/8 = (\tau_y g_c)_1/\bar{v}^2\rho$ for smooth tubes, so that Equation 13.92 may be written

$$\frac{h}{c_P \bar{v} \rho} = \frac{(\tau_y g_c)_1}{\bar{v}^2 \rho} \qquad (13.112)$$

If a flat plate with a turbulent boundary layer in a free stream behaves in an analogous manner to fully developed turbulent flow in a tube, the following equations may be *assumed* to apply for boundary layers:

$$\frac{\bar{h}}{c_P v_{fs} \rho} = \frac{\overline{(\tau_y g_c)_1}}{v_{fs}^2 \rho} \qquad (13.113)$$

and

$$\frac{h_y}{c_P v_{fs} \rho} = \frac{(\tau_y g_c)_{1y}}{v_{fs}^2 \rho} \qquad (13.114)$$

where h_y = the heat-transfer coefficient at a specific point y measured from the leading edge. The entire plate is heated; that is, the momentum and thermal boundary layers begin at the same point

$(\tau_y g_c)_{1y}$ = the wall stress at a point y measured from the leading edge

$\bar{h}$ = the average value of the heat-transfer coefficient for the section between the leading edge and some position y

$\overline{(\tau_y g_c)_1}$ = the average wall stress for the section between the leading edge and some point y

v_{fs} = the free-stream velocity

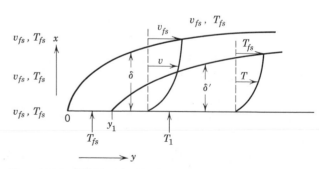

Figure 13.8. Simultaneous formation of momentum boundary layer and mass or thermal boundary layer.

The friction factors have been related to Reynolds numbers in equations written earlier. If Equation 13.37 is combined with Equation 13.113, then

$$\frac{\bar{h}}{c_P v_{fs} \rho} = \frac{0.037}{(N_{Re})_y^{0.2}} \qquad (13.113a)$$

and, if Equation 13.35 is combined with Equation 13.114, then

$$\frac{h_y}{c_P v_{fs} \rho} = \frac{0.0228}{(N_{Re})_\delta^{0.25}} \qquad (13.114a)$$

where $(N_{Re})_y = (y v_{fs} \rho / \mu)$
   $y =$ any position on the plate downstream of the leading edge
   $(N_{Re})_\delta = (\delta v_{fs} \rho / \mu)$
   $\delta =$ the boundary-layer thickness at some position y

The boundary-layer thickness (δ) and position (y) are related by Equation 13.36

$$\frac{\delta}{y} = \frac{0.376}{(N_{Re})_y^{0.2}} \qquad (13.36)$$

In a similar manner, equations can be derived for mass transfer through the boundary layer. The conditions where the Reynolds analogy holds are that the fluid be characterized by a Schmidt number of unity, that the boundary layer for momentum and mass start at the same point, and that the flat plate is smooth. For smooth tubes, the mass-transfer counterpart equation to the Reynolds analogy is

$$\frac{k_c'}{\bar{v}} = \frac{\tau_y g_c}{\bar{v}^2 \rho} \qquad (13.96)$$

where $k_c' =$ the mass-transfer coefficient for equimolar counterdiffusion

If a flat plate with a turbulent boundary layer in a free stream behaves in analogous manner to fully developed flow in a tube, the following equations may be *assumed*:

$$\frac{\overline{(k_c')}}{v_{fs}} = \frac{(\tau_y g_c)_1}{v_{fs}^2 \rho} \qquad (13.116)$$

and

$$\frac{(k_c')_y}{v_{fs}} = \frac{(\tau_y g_c)_{1y}}{v_{fs}^2 \rho} \qquad (13.116a)$$

Equations 13.116 and 13.116a can be combined with Equations 13.37 and 13.35 respectively to give

$$\frac{\overline{(k_c')}}{v_{fs}} = \frac{0.037}{(N_{Re})_y^{0.2}} \qquad (13.116b)$$

and

$$\frac{(k_c')_y}{v_{fs}} = \frac{0.0228}{(N_{Re})_\delta^{0.25}} \qquad (13.116c)$$

where $\overline{(k_c')} =$ the average mass-transfer coefficient from the leading edge to a position y
   $(k_c')_y =$ the point value of the mass-transfer coefficient at a position y

The mass-transfer coefficient for diffusion of gas a through a stationary gas b may be written by noting that

$$k_c \frac{c_{blm}}{c_t} = k_c' \qquad (13.117)$$

where $c_{blm} =$ logarithmic mean concentration of gas b
   $c_t =$ total concentration of gases a and b
   $k_c =$ mass-transfer coefficient for diffusion of gas a through stationary gas b

Equations 13.116b and 13.116c have been tested experimentally and agree with experiment within the limits of the assumptions used in derivation.

The Colburn Analogy Applied to the Boundary Layer. If the fluid properties in the system are such that the Prandtl number is greater than 0.6 and less than 100, then in a similar manner the Colburn analogy can be used to deduce equations for heat and mass transfer in the boundary layer. For turbulent flow in which all boundary layers begin at $y = 0$, and $(\mu / \mu_1) \cong 1$, the Colburn analogy may be stated

$$\frac{h_y}{c_P v_{fs} \rho} (N_{Pr})^{2/3} = \frac{k_{cy}'}{v_{fs}} (N_{Sc})^{2/3} = \frac{(\tau_y g_c)_{1y}}{v_{fs}^2 \rho} = \frac{0.0228}{(N_{Re})_\delta^{0.25}} \qquad (13.114b)$$

and

$$\frac{\bar{h}}{c_P v_{fs} \rho} (N_{Pr})^{2/3} = \frac{\overline{k_c'}}{v_{fs}} (N_{Sc})^{2/3} = \frac{\overline{(\tau_y g_c)_1}}{v_{fs}^2 \rho} = \frac{0.037}{(N_{Re})_y^{0.2}} \qquad (13.113b)$$

From the foregoing material, it has been established that the analogous behavior of mass, heat, and momentum in smooth tubes can be extended to other examples of transfer to a fluid stream from a smooth surface in which the flow is parallel to the surface. The boundary-layer equations for heat and mass transfer find occasional application in specialized-equipment components that do not fall within the conditions of fully developed flow in ducts.

Illustration 13.10. Stack gases leave a boiler, pass through a square steel duct 3 ft by 3 ft and 16 ft long, and enter a masonry chimney. Estimate the heat-transfer coefficient in the duct if the gases are at 900°F and 1 atm. Consider the stack gases to have the same properties as air. The gas velocity is 50 ft/sec.

SOLUTION. In a duct of this size and length, the boundary-layer equations may be used, provided that the boundary layer is small with respect to the duct size and the effect of the interactions of the boundary layers at the corners of the duct can be neglected. For this example the boundary-layer thickness can be calculated using Equation 13.36

$$\frac{\delta}{y} = \frac{0.376}{(N_{Re})_y^{0.2}} \qquad (13.36)$$

The physical properties of air at 900°F and 1 atm can be found in Appendix D.

$$\mu_{air\ 900°F} = 0.0353 \text{ centipoise}$$

$$\rho_{air\ 900°F} = \left[\frac{28.9 \text{ lb}}{359 \text{ cu ft}}\right]_{STP} \cdot \frac{492}{460 + 900} = 0.0291 \text{ lb/cu ft}$$

$$c_{P,\ air\ 900°F} = 0.260 \text{ Btu/lb °F}$$

$$(N_{Re})_y = \frac{y v_{fs} \rho}{\mu} = \frac{(16)(50)(0.0291)}{(0.0353 \times 6.72 \times 10^{-4})} = 985000$$

$$= 9.85 \times 10^5$$

This answer is in excess of 10^5, and therefore most of the boundary layer is turbulent. Equation 13.36 may be applied to determine the thickness of the boundary layer.

$$\delta = \frac{y(0.376)}{(N_{Re})_y^{0.2}} = \frac{16(0.376)}{(9.85 \times 10^5)^{0.2}} = \frac{16(0.376)}{15.8} = 0.38 \text{ ft}$$

The boundary-layer thickness is in the order of 10 per cent of the duct dimension. The remainder of the duct constitutes the free stream. Since the problem statement requests an estimation rather than a rigorous calculation, the boundary-layer equations may be used for estimation of heat-transfer coefficient.

Using the Reynolds analogy, Equation 13.113a can be used to determine the average heat-transfer coefficient over the entire length of duct.

$$\frac{h}{c_P v_{fs} \rho} = \frac{0.037}{(N_{Re})_y^{0.2}}$$

$$h = \frac{c_P v_{fs} \rho (0.037)}{(N_{Re})_y^{0.2}}$$

$$h = \frac{(0.260)(50 \times 3600)(0.0291)(0.037)}{(9.85 \times 10^5)^{0.2}}$$

$$h = 3.2 \text{ Btu/hr sq ft °F}$$

The same calculation can be made with Equation 13.113b which was derived using the Colburn analogy.

$$\frac{h}{c_P v_{fs} \rho}(N_{Pr})^{2/3} = \frac{0.037}{(N_{Re})_y^{0.2}}$$

$$h = \frac{c_P v_{fs} \rho (0.037)}{(N_{Re})_y^{0.2}}\left(\frac{1}{N_{Pr}}\right)^{2/3}$$

$$h = 3.2\left(\frac{1}{N_{Pr}}\right)^{2/3}$$

For air at 900°F and 1 atm, $N_{Pr} = 0.69$.

$$h = \frac{3.2}{(0.69)^{2/3}} = \frac{3.2}{0.78}$$

$$h = 4.11 \text{ Btu/hr sq ft °F}$$

This latter value is probably more nearly correct than that obtained using the Reynolds analogy.

More rigorous analogies have been written for the boundary layers in a manner similar to the work of Martinelli described above. The differences between the analogies are dependent upon the choice of assumptions and the velocity-distribution function chosen for the derivation. The reader may refer to textbooks in fluid mechanics for the development of these equations (22).

MASS AND HEAT TRANSFER IN SYSTEMS EXHIBITING FORM DRAG

All the material on heat and mass transfer that has been presented in this chapter was written specifically for smooth surfaces over which fluids moved tangentially. Many of the chemical-engineering applications of transfer occur in systems that do not meet these conditions exactly. For example, commercial pipe of the commonest variety has a sufficient degree of surface roughness to increase the friction factor by 20 to 30 per cent. Although for smooth surfaces heat, mass, and momentum transfer followed a remarkably analogous pattern, when a system is subject to form drag, the previous analogous behavior is altered. In one study (23) the internal surfaces of tubes were artifically roughened so that the friction was increased. At maximum roughness, the pressure drop was increased sixfold, but the heat-transfer coefficient was increased only twofold. It has been universally observed that, for systems exhibiting form drag, heat or mass transfer is not increased to as great an extent as momentum transfer as a consequence of the form drag.

To explain this phenomenon, an element of roughness as shown in Figure 13.9 will be examined. In the same manner as the shape shown in Figure 13.4, the boundary layer separates and an eddy forms behind the roughness element. The eddy circulates fluid within a very small finite part of the free volume of the tube as a result of the position of the roughness element. The local turbulence intensity in the eddy is greater than in the region nearby. The eddy transfers momentum from the flowing stream with an increase in measured stress.

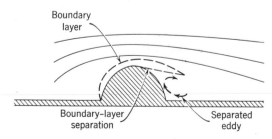

Figure 13.9. Eddy behavior behind a roughness element.

The intense circulation *within the eddy* does not contribute proportionately to the bulk mixing. It should be remembered that the local eddy is "fed" by the surrounding streams and in turn feeds the surrounding streams. Therefore, there is a small increase in circulation throughout the stream because of the interchange between the eddy and the surroundings. This small interchange is a partial contribution to the mixing phenomenon as understood in turbulent flow and can explain the small increase in heat- and mass-transfer coefficient. That is, the local circulating eddy transfers more momentum without proportional increase in bulk mixing in the main stream.

Several geometric shapes have been studied experimentally and a partial compilation of the work is presented graphically in Figure 13.10. Figure 13.10a reports experimental work on a single cylinder with flow normal to the axis of the cylinder. The *j* factors for mass transfer and heat transfer are plotted as a function of the Reynolds number as defined in the figure. The heat- and mass-transfer data are in excellent agreement. In a system that offers a projected area normal to flow, both form drag and skin friction are present. Goldstein (37) reports a procedure by which the measured forces normal to a body can be subtracted from the measured total forces on a body. The remaining force is the tangential force or skin friction. For the single cylinder these tangential forces are combined with appropriate variables to form the friction factor, and this factor is plotted as $f/8$, consistent with previous considerations in the Colburn analogy. This plot is in excellent agreement with the data for j_N and j_q. Finally, to describe the magnitude of the total force on the body, $C_D/8$ is plotted and shown to be very large compared to j_q, j_N, and $f/8$. The foregoing indicates that for systems in which form drag occurs, if the tangential force or skin friction can be isolated, it will correlate with mass and heat transfer, but when total force, as in $C_D/8$, is used no correlation can be written. This analogous behavior is consistent with all previous analogies in which skin friction was the *only* means of momentum transfer.

Similar data for the single sphere are plotted in Figure 13.10b, but the experimental data for $f/8$ are not available. In the case of the sphere for the data presented, the agreement between j_q and j_N is as good as it was for single cylinders.

Illustration 13.11. The "Lister" bag is a common means of storage of potable water by army field forces. The bag is made of porous canvas. A small amount of water diffuses through the canvas and evaporates from the surface. The evaporation of the water cools the surface of the bag and a temperature driving force is established. Estimate the heat- and mass-transfer coefficients if the bag is a 2.5-ft diameter sphere hanging in a 0.5 mph wind at 90°F.

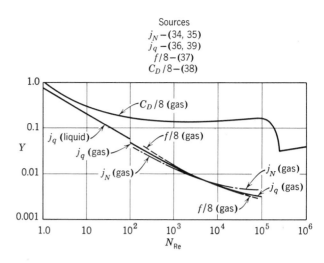

Sources
j_N—(34, 35)
j_q—(36, 39)
$f/8$—(37)
$C_D/8$—(38)

(a) Heat, mass, and momentum transfer to single cylinders.

$Y = f/8 = \tau_y g_c / v_{fs}^2 \rho$ where D_2 = diameter of cylinder

$Y = C_D/8 = F_T g_c / 4 S v_{fs}^2 \rho$ $\tau_y g_c$ = tangential stress

$Y = j_q = \left(\dfrac{h}{\rho c_P v_{fs}}\right)(N_{Pr})^{2/3}$ F_T = Sum of normal and tangential force

$Y = j_N = \left(\dfrac{k_c'}{v_{fs}}\right)(N_{Sc})^{2/3}$ S = projected area

$N_{Re} = (D_2 v_{fs} \rho / \mu)$ v_{fs} = free-stream velocity

h = heat-transfer coefficient

k_c' = mass-transfer coefficient

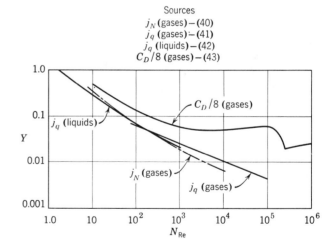

Sources
j_N (gases)—(40)
j_q (gases)—(41)
j_q (liquids)—(42)
$C_D/8$ (gases)—(43)

(b) Heat, mass, and momentum transfer to single spheres.

$Y = C_D/8 = F_T g_c / 4 S v_{fs}^2 \rho$ where D = diameter of sphere

$Y = j_q = \left(\dfrac{h}{\rho c_P v_{fs}}\right)(N_{Pr})^{2/3}$ v_{fs} = free-stream velocity

$Y = j_N = \left(\dfrac{k_c'}{v_{fs}}\right)(N_{Sc})^{2/3}$ F_T = total force on the sphere

S = projected area to flow

$N_{Re} = (D v_{fs} \rho / \mu)$ h = heat transfer coefficient

k_c' = mass-transfer coefficient

Figure 13.10. Heat, mass, and momentum transfer with form drag.

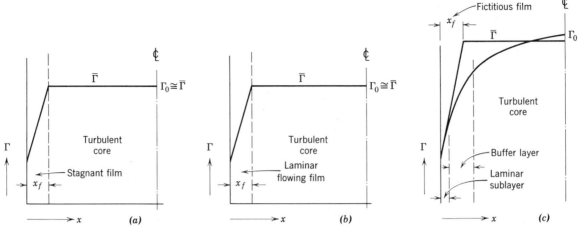

Figure 13.11. Concentration gradients in film theory.

Data (Appendix D): Air at 90°F

$$\mu = 0.0183 \text{ centipoise}$$

$$k = 0.0140 \text{ Btu/hr sq ft (°F/ft)}$$

$$c_P = 0.25 \text{ Btu/lb °F}$$

$$\rho = \frac{(28.9)}{(359)}\left[\frac{(460 + 32)}{(460 + 90)}\right] = 0.0721 \text{ lb/cu ft}$$

$$\mathscr{D}_{H_2O\text{-air}} = 1 \text{ sq ft/hr}$$

$$N_{Pr} = \frac{c_P \mu}{k} = \frac{(0.25)(0.0183 \times 2.42)}{(0.0140)} = 0.79$$

$$N_{Sc} = \frac{\mu}{\rho \mathscr{D}} = \frac{(0.0183 \times 2.42)}{(0.0721)(1)} = 0.613$$

SOLUTION. The Reynolds number can be calculated

$$N_{Re} = \frac{Dv_{fs}\rho}{\mu} = \frac{(2.5)(0.5 \times 5280)(0.0721)}{(0.0183 \times 2.42)}$$

$$N_{Re} = 10,750$$

Figure 13.10b is entered at $N_{Re} = 10,750$ and j_N and j_q can be read

$$j_N = 0.007$$

$$j_q = 0.010$$

From the definition of j_N

$$j_N = \frac{k_c{}'}{v_{fs}}(N_{Sc})^{2/3} = 0.007$$

$$k_c{}' = \frac{(0.007)(0.5 \times 5280)}{(0.613)^{2/3}}$$

$$k_c{}' = 25.6 \text{ lb moles/hr sq ft (lb mole/cu ft)}$$

From the definition of j_q

$$j_q = \frac{h}{\rho c_P v_{fs}}(N_{Pr})^{2/3} = 0.01$$

$$h = \frac{(0.01)(0.0721)(0.25)(0.5 \times 5280)}{(N_{Pr})^{2/3}}$$

$$h = 0.557 \text{ Btu/hr sq ft °F}$$

OTHER MODELS FOR TURBULENT TRANSFER COEFFICIENTS

Transfer coefficients have been in use for many years in the design of heat-transfer and mass-transfer equipment. In most cases they have been considered to be empirical coefficients which could be determined from experimental data. In certain cases the experimental data could be correlated to give empirical equations useful in predicting transfer coefficients, as discussed in earlier sections. Parallel with the increased use of such coefficients, attempts to explain them physically have been made. Understanding of the physical mechanism of heat and mass transfer in turbulent flow rests on an understanding of the mechanism of turbulent flow itself. In the earlier part of this chapter, transfer coefficients were defined using the *eddy-diffusivity model*. Two other physical models for transfer will be considered here. These are the widely applied *film theory* and the more recently developed *penetration theory*.

Film Theories. The most rudimentary film theory states that all the resistance to transfer in a turbulently flowing fluid is concentrated in a *stagnant* film next to the wall or stationary boundary of the fluid. According to this model *all* the driving force (or concentration gradient) acts across the stagnant film, and also the concentration in the bulk of the fluid is constant, since it is in highly turbulent motion. This profile is shown in Figure 13.11a. The transfer in the stagnant film would be by molecular transport, and the general equation for transfer would be

$$\psi = -\frac{\delta}{x_f}(\Gamma_1 - \bar{\Gamma}) \tag{13.118}$$

where x_f is the thickness of the stagnant film, ψ is the flux, and δ is the diffusivity. Since the film thickness cannot be measured directly, it is combined with the diffusivity to give a *transfer coefficient* ($\mathscr{E}$), where

$$\mathscr{E} = \frac{\delta}{x_f} \qquad (13.118a)$$

and Equation 13.118 becomes

$$\psi = -\mathscr{E}(\Gamma_1 - \bar{\Gamma}) \qquad (13.118b)$$

A more satisfactory concept of the film states that it is *not* stationary but is in *laminar flow*. Transfer in laminar flow would still be by molecular transport, and the gradient would still be only across the film, as shown in Figure 13.11b. The resulting equations would be identical to Equations 13.118 and 13.118b.

Neither the stagnant film nor the laminar-flowing film is consistent with what is known about turbulent flow, as outlined in Chapter 12. The most accurate observations show no stagnant film near the wall. Although evidence supports the existence of a *laminar sublayer* next to the wall, this is not equivalent to a laminar-flowing film, since only *part* of concentration gradient is in the laminar sublayer and *all* of it is postulated to occur across the laminar film. These inconsistencies with observed data have led to a redefinition of the film as a fictitious thickness of laminar-flowing fluid next to the boundary which offers the same resistance to transfer as actually exists in the entire turbulently flowing fluid. That is, all resistance to transfer is concentrated in a *fictitious* film in which transfer is by molecular transport only. The thickness of this fictitious film is shown with a typical concentration gradient for turbulent flow in Figure 13.11c. The fictitious film extends beyond the laminar sublayer to allow for the change in concentration in the buffer layer and turbulent core. The center-line composition Γ_0 is somewhat greater than the average composition $\bar{\Gamma}$, as expected. Equations 13.118, 13.118a, and 13.118b may be used to define transfer across the fictitious film, where x_f is now the thickness of the fictitious film. Obviously, x_f can never be measured, since it does not exist. For this and other reasons, this most sophisticated version of film theory is of little use in understanding the mechanism of turbulent transfer. Equations 13.118a and 13.118b may be written specifically for heat transfer

$$\frac{h}{c_P \rho} = \frac{\alpha}{x_f{}'} \qquad (13.118c)$$

$$\frac{q}{A} = -h(T_1 - \bar{T}) \qquad (13.118d)$$

and for mass transfer

$$k_c{}' = \frac{\mathscr{D}}{x_f{}''} \qquad (13.118e)$$

$$\frac{N_a}{A} = -k_c{}'(c_{a1} - \bar{c}_a) \qquad (13.118f)$$

The fictitious-film thicknesses for heat, mass, and momentum at equal flow rates are not equal except under the limiting conditions of the Reynolds analogy.

Because of its apparent lack of adequate physical basis, the film theory is losing favor, even though it has proved very useful in interpreting and correlating turbulent transfer data for equipment design. The term *film coefficient* for heat or mass transfer is still widely used, but it is less desirable than the terms *transfer coefficient* or *surface coefficient*.

Penetration Theory. The penetration theory was originally applied by Higbie (30) to analyze the liquid phase in gas absorption where the liquid could be assumed to be in laminar flow or stationary. It was more recently applied to turbulent flow by Danckwerts (31) and Hanratty (32).

Higbie considered mass to be transferred in the liquid by unsteady-state molecular transport. This concept resulted in an equation for the mass flux at a point on the surface of a liquid exposed to an absorbent gas

$$\frac{N_a}{A} = -\sqrt{\frac{\mathscr{D}}{\pi\theta}} \cdot (c_{a1} - \bar{c}_a) \qquad (13.119)$$

where θ = time of contact of the liquid with the gas
c_{a1} = concentration in the liquid at the gas interface
$\bar{c}_a$ = average bulk concentration in the liquid

This relationship has proved useful in analyzing absorption data from wetted-wall columns.

Danckwerts (31) applied this unsteady-state concept to absorption in a turbulent liquid. He postulated *random* surface renewal in the liquid at the gas-liquid boundary. In this model there is no laminar sublayer at the boundary. The liquid eddies that originate in the turbulent core migrate to the gas-liquid boundary where they are exposed to the gas for a short period before they are displaced by other eddies arriving at the surface. During the short exposure period (θ) of an eddy, it absorbs gas in unsteady-state according to Equation 13.119. It then is displaced from the surface and returns to the bulk of the liquid, where its absorbed gas is distributed by turbulence. Fresh surface is continually created by the random motion of the eddies. The mean rate of production of fresh surface is constant for a given degree of turbulence and equal to s. The surface-renewal factor (s) is defined statistically and cannot be determined directly at present. The rate of absorption for the turbulent liquid with random surface renewal is

$$\frac{N_a}{A} = -\sqrt{\mathscr{D}s}(c_{a1} - \bar{c}_a) \qquad (13.120)$$

where s is the surface renewal factor, $(1/\theta)$, sec^{-1}. Hanratty (32) extended this model and showed that it was consistent with experimental data for mass and momentum transfer. To do this it is necessary to assume the form of the statistical probability function for the distribution of eddy residence time at the boundary. Microscopic examination of the motion of dust particles in a turbulent-flowing liquid very near the boundary (33) indicates a random motion which *may* be interpreted as large eddies. This observation tends to cast doubt on the existence of a laminar sublayer and to support a turbulent exchange at the boundary. More evidence is needed before any definite conclusions can be drawn.

It is possible to generalize the model-of-penetration theory to apply to mass, heat, and momentum transfer. The general equations for the three models considered in this chapter are

Film theory:
$$\psi = -\frac{\delta}{x_f}(\Gamma_1 - \bar{\Gamma}) \quad (13.118)$$

Eddy-diffusivity model:
$$\psi = -\frac{4(\delta + \bar{E})}{D\gamma}(\Gamma_1 - \bar{\Gamma}) \quad (13.121)$$

Penetration theory
(random surface renewal):
$$\psi = -\sqrt{\delta s}(\Gamma_1 - \bar{\Gamma}) \quad (13.122)$$

In each case it is possible to define a transfer coefficient.

Film theory:
$$\mathscr{E} = \frac{\delta}{x_f} \quad (13.118a)$$

Eddy-diffusivity model:
$$\mathscr{E} = \frac{4(\delta + \bar{E})}{D\gamma} \quad (13.123)$$

Penetration theory:
$$\mathscr{E} = \sqrt{\delta s} \quad (13.124)$$

In the film theory the fictitious-film thickness (x_f) cannot be measured or predicted directly. Although the film theory has proved useful, it contributes little to the understanding of turbulent transport. The eddy-diffusivity model considers the random motion of eddies and attempts to define the effect of this gross eddy motion upon the transport in the system. At the present time the direct evaluation of either point or average eddy diffusivities is difficult, and no general predictive equations are available. Much of the recent advances in theory of turbulence is based upon statistical consideration (13). In penetration theory with random surface renewal, rudimentary statistical methods have been applied to develop expressions for predicting transfer caused by random eddy motion at a boundary. In this case the behavior of a single eddy is examined and generalized by statistical means. The mathematical treatment is based upon unsteady-state transfer, in contrast to the steady state assumed in the eddy-diffusivity model. Penetration theory appears to be a more powerful tool in the analysis of turbulent transport, although it is not fully developed at the present time.

OTHER TRANSFER COEFFICIENTS*

The foregoing sections described the transfer phenomena associated with turbulent flow of fluids. There are several other applications of transfer that involve relative motion of solids and fluids in which the nature of the motion is not well defined. The specific applications are (1) heat transfer in laminar flow, (2) heat transfer in which the motion of the fluid is caused by natural convection, and (3) the transfer processes associated with a phase change, as in condensation or boiling.

Heat Transfer in the Laminar Regime. In isothermal laminar flow in tubes as shown in Chapters 9 and 10, the velocity distribution can be described exactly. It might be expected that heat transfer to a fluid flowing in the laminar regime would respond to a rigorous analysis.

Numerous rigorous solutions have been developed for various boundary conditions. Analyses have been made for the parabolic velocity distribution and constant wall temperature (25), constant wall heat flux (26) and with the wall temperature varying linearly with position in the flow direction (26). Other analyses were written for heat transfer during the development of the boundary layer within a tube at constant wall temperature (27), constant heat input (27), and constant temperature difference (27). All the analyses are of the general form

$$N_{\text{Nu}} = \Phi\left(N_{\text{Pe}}, \frac{L}{D}\right) \quad (13.125)$$

where N_{Pe} = the Peclet number, $(D\bar{v}\rho c_P/k)$
  L/D = length-diameter ratio
  Φ = unspecified function

With simplifying assumptions, several of the analyses reduce to

$$N_{\text{Nu}} = \text{constant}\,(N_{\text{Pe}})^a \left(\frac{L}{D}\right)^b \quad (13.126)$$

The rigorous analyses contribute some knowledge of the form of the equation, but the boundary conditions are restricted to conditions that are not usually encountered in practical heat transfer. In addition, the assumption of a parabolic velocity profile is not correct if any temperature driving force is operative, since the viscosity of the fluid is affected by the thermal gradient.

---

* The material covered on pp. 186–199 represents applications of the principles to various heat-transfer coefficients, in each of which application only one or two factors are different from cases previously treated. These cases can be adequately covered by a high-speed reading on first assignment, particularly because subsequent applications of these coefficients in connection with later chapters of the book will force reference back to these discussions. In order to assure that the reference back is successful, each is presented in rather considerable detail.

In gases the viscosity increases with increase in temperature, whereas in liquids the viscosity decreases with increase in temperature. The shear stress in a tube is linear with radius with the maximum at the walls; and this stress pattern is independent of the transfer mechanism. At any radial position the stress is constant. If a change in viscosity occurs at that position the velocity gradient must change to maintain constant stress. In a heated or cooled tube with laminar flow, a temperature gradient can be expected between the wall and the center of the tube. The thermal gradient tends to distort the velocity profile in a manner similar to that shown in Figure 13.12, for a hot wall in contact with a cold fluid.

The rigorous analysis of heat transfer in laminar flow is complicated further by the establishment of density gradients in the fluid as a result of thermal expansion. The density gradients establish natural convection currents in the fluid, with consequent cross-current flow in horizontal ducts, or locally accelerated flow in vertical ducts. These cross currents contribute to heat and momentum transfer in the same manner as the cross currents in turbulence. At low values of the Reynolds number (that is, below $N_{\text{Re}} = 2100$) the effect can be considerable, and, in fact, the natural convection effect is detected until $N_{\text{Re}} \geqslant 8000$; at which time, natural convection currents are overshadowed by the cross currents of turbulent flow.

The distortion of the velocity profile and the effect of natural convection currents can be expected to affect the rate of heat transfer. Two empirical equations are presented here with a statement of range of usefulness. The first equation (7) is limited to horizontal tubes one inch or less in diameter and temperature driving forces less than $100°F$.

$$(N_{\text{Nu}})_{am} = 1.86(N_{\text{Pe}})^{1/3} \left(\frac{D_1}{L}\right)^{1/3} \left(\frac{\mu}{\mu_1}\right)^{0.14} \quad (13.127)$$

where $(N_{\text{Nu}})_{am}$ = the Nusselt number $(h_{am} D_1/k)$
 h_{am} = the heat-transfer coefficient for a tube of length L. This heat-transfer coefficient is based upon the equation $q/A = h_{am}(\Delta T)_{am}$.
 $(\Delta T)_{am}$ = the arithmetic-mean-temperature driving force. The general subject of mean driving force will be discussed in Chapter 15.
 μ_1 = the viscosity of the fluid at the *wall* temperature (T_1)
 k, c_P, ρ, μ = the fluid properties evaluated at the mean bulk temperature $(\bar{T})$
 N_{Pe} = the Peclet number $(D_1 \bar{v} \rho c_P/k)$

The fluid variables in Equation 13.127 are evaluated at the arithmetic-mean bulk temperature $(\bar{T})$ with the

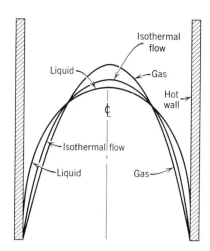

Figure 13.12. Velocity profiles with a thermal gradient.

exception of μ_1, which is evaluated at the temperature of the tube wall. However, Equation 13.127 does not acknowledge natural convection, and the equation is limited to small tubes $(D_1 < 1 \text{ in.})$ and small temperature driving force $((T_1 - \bar{T}) < 100°F)$. At a small driving force in small tubes the natural convection phenomenon is limited in magnitude. Although the natural-convection phenomenon is not recognized with appropriate variables, the constant, 1.86 is somewhat higher than the constants found in the theoretical derivations that are written without accounting for natural convection such as Equation 13.126. The data used to write Equation 13.127 scatter considerably, which is further confirmation of the presence of natural convection. The distortion of the velocity profile is recognized by use of the viscosity ratio (μ/μ_1).

For convenience, Equation 13.127 may be written in terms of the j factor for heat transfer by first substituting $N_{\text{Re}}N_{\text{Pr}}$ for N_{Pe} so that Equation 13.127 becomes

$$(N_{\text{Nu}})_{am} = 1.86(N_{\text{Re}})^{1/3}(N_{\text{Pr}})^{1/3} \left(\frac{D_1}{L}\right)^{1/3} \left(\frac{\mu}{\mu_1}\right)^{0.14} \quad (13.128)$$

The substitution introduces the viscosity into two dimensionless groups. This does not necessarily imply that the Nusselt number is dependent upon viscosity. Note that viscosity cancels from N_{Re} and N_{Pr} when both have the same exponent. Equation 13.128 may be divided by $N_{\text{Re}}N_{\text{Pr}}$ to give

$$\frac{(N_{\text{Nu}})_{am}}{(N_{\text{Re}}N_{\text{Pr}})} = 1.86(N_{\text{Re}})^{-2/3}(N_{\text{Pr}})^{-2/3} \left(\frac{D_1}{L}\right)^{1/3} \left(\frac{\mu}{\mu_1}\right)^{0.14} \quad (13.129)$$

Since $N_{\text{St}} = N_{\text{Nu}}/(N_{\text{Re}}N_{\text{Pr}})$, Equation 13.129 may be rearranged to

$$(N_{\text{St}})_{am}(N_{\text{Pr}})^{2/3} \left(\frac{\mu_1}{\mu}\right)^{0.14} = 1.86(N_{\text{Re}})^{-2/3} \left(\frac{D_1}{L}\right)^{1/3} \quad (13.130)$$

and since $(N_{St})_{am}(N_{Pr})^{2/3}(\mu_1/\mu)^{0.14} = j_q$,

$$j_q = 1.86(N_{Re})^{-2/3}\left(\frac{D}{L}\right)^{1/3} \qquad (13.131)$$

For convenience in use, Equation 13.131 is plotted in Appendix C-5, in which j_q is plotted as a function of N_{Re} and the parameter L/D. Although the actual heat transfer is not dependent upon a Reynolds number, the equation form of Equation 13.131 is desirable because the Reynolds number is the criterion for designation of laminar and turbulent flow. The chart in Appendix C-5 is recommended for all heat-transfer calculations in cylindrical geometry because the chart applies to both laminar and turbulent flow.

A correlation for laminar flow in horizontal tubes which recognizes the variables of natural convection has been offered (28).

$$(N_{Nu})_{am}\left(\frac{\mu_1}{\mu}\right)^{0.14} =$$

$$1.75\sqrt[3]{\frac{\pi(N_{Pe})D_1}{4L} + 0.04\left(\frac{(N_{Gr})(N_{Pr})}{\dfrac{L}{D_1}}\right)^{0.75}} \qquad (13.132)$$

where N_{Gr} = the Grashof number $(D_1{}^3\rho^2 g\beta\,\Delta T/\mu^2)$, dimensionless

g = the gravitational acceleration = 32.2 ft/sec^2

β = volumetric coefficient of thermal expansion of the fluid, (°F)$^{-1}$

ΔT = the temperature difference between the wall and the bulk of the fluid $(T_1 - \bar{T})$, °F

The general significance of the Grashof number will be discussed in a later section.

Although Equation 13.132 includes the variables of natural convection in the form of the Grashof number, the results are so close to those given by Equation 13.131 that the extra labor of using Equation 13.132 is usually not warranted.

Heat Transfer in the Transition Region. In the range between $N_{Re} = 2100$ and $N_{Re} = 10,000$ the flow regime is uncertain, and the effects of convection and entry length may both partially override the incompletely developed turbulence. The transition region has not been thoroughly examined experimentally but the scattered data have been plotted and smoothed as shown in Appendix C-5. It is suggested that, for *any* heat-transfer application in smooth tubes, Appendix C-5 be used in preference to an equation. This will minimize errors due to choice of regime of flow.

Mass Transfer in Laminar Flow. In a manner similar to heat transfer in laminar flow, mass transfer in laminar flow is characterized by entry length phenomena and by convection currents induced by a difference in concentration where the two components have different densities.

The small amount of data that is available can be correlated when a distorted velocity profile is assumed. The conventional parabolic velocity profile does not offer a means of correlating data. The distortion that is assumed for correlation was considered to result from natural convection. Too little data have been accumulated to write any general equation.

Illustration 13.12. Calculate the heat-transfer coefficient for a $\frac{3}{8}$-in. I.D. horizontal tube in which water flows at 0.3 ft/sec. The mean bulk temperature of the water is 125°F. The tube wall is maintained at 225°F. The tube is 3.21 ft long.

SOLUTION. First, the Reynolds number should be calculated. The data for this problem can be found in Appendix D.

$$\mu = 0.58 \text{ centipoise}$$
$$\rho = 61.5 \text{ lb/cu ft}$$
$$N_{Re} = D\bar{v}\rho/\mu = \frac{(3/8 \times 1/12)(0.3)(61.5)}{(0.58 \times 0.000672)}$$
$$= 1490$$

At $N_{Re} = 1490$, the flow is laminar and Equation 13.131 or Appendix C-5 applies. $D_1/L = (3/8 \times 1/12)/3.21 = 0.01$, $L/D_1 = 100$. If Equation 13.131 is used

$$j_q = 1.86(N_{Re})^{-2/3}\left(\frac{D_1}{L}\right)^{1/3}$$

$$j_q = 1.86\left(\frac{1}{1490}\right)^{2/3}(0.01)^{1/3}$$

$$j_q = 1.86\left(\frac{1}{130}\right)(0.216)$$

$$j_q = 0.00309$$

This value is in agreement with the chart of Appendix C-5 for $N_{Re} = 1490$ and $L/D_1 = 100$. Then

$$j_q = (N_{St})_{am}(N_{Pr})^{2/3}\left(\frac{\mu_1}{\mu}\right)^{0.14} = 0.00309$$

Numerical values can be substituted into N_{St}, N_{Pr}, μ_1/μ. $\mu_1 = 0.23$ centipoise, $c_P = 1.0$ Btu/lb °F, $k = 0.370$ Btu/hr sq ft (°F/ft)

$$(N_{St})_{am} = \frac{h_{am}}{c_P\bar{v}\rho} = \frac{h_{am}}{(1)(0.3 \times 3600)(61.5)}$$

$$N_{Pr} = \frac{c_P\mu}{k} = \frac{(1)(0.58 \times 2.42)}{0.370} = 3.8$$

$$\left(\frac{\mu_1}{\mu}\right) = \frac{0.23}{0.58} = 0.397$$

Then

$$\frac{h_{am}}{(1)(0.3 \times 3600)(61.5)}(3.8)^{2/3}(0.397)^{0.14} = 0.00309$$

$$\frac{h_{am}}{(1)(0.3 \times 3600)(61.5)}(2.44)(0.88) = 0.00309$$

$$h_{am} = \frac{0.00309(1)(0.3 \times 3600)(61.5)}{(2.44)(0.88)}$$

$$h_{am} = 95.7 \text{ Btu/hr sq ft °F}$$

Heat-Transfer Coefficients for Natural Convection.
When a fluid is heated in an unstirred vessel or when heat is lost from a pipe or furnace wall to the room or outside air, motion occurs in the fluid due to natural convection.

To begin the analysis of natural convection, consider a vertical plane wall at a constant temperature in contact with a large mass of air at a lower temperature. The air near the plane will be heated; its density will decrease with respect to the unheated air nearby, and the lighter heated air will rise along the wall to be replaced by the heavier colder air from the region more remote from the wall. This simple model is a starting point for an analysis, but a more minute examination of the mechanisms is in order. Figure 13.13 (8, 29) is a plot of local velocity and local temperature at points 1 cm from the bottom and 24 cm from the bottom of the vertical plate. The temperature gradient is steepest near the bottom and decreases with progress along the plate. The velocity is a minimum near the bottom and increases with progress along the plate.

If the system described above is operating at steady state, the heat added to the air at the hot wall must be removed at some other remote part of the cycle, or the entire air mass would eventually reach the wall temperature and heat transfer would cease. The history of an element of fluid will be traced during passage along the hot wall. The particular element starts at the bottom of the hot wall. At that point, the motion is slow, consisting of the motion imparted to the element by the entire moving mass. The temperature of the air element is low compared to the wall, since no heat has been transferred to the element during this cycle. Con-

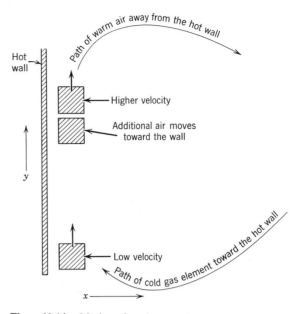

Figure 13.14. Motion of an element of gas along a hot plate.

sequently, a large temperature driving force exists. Heat is transferred to the element of fluid under the influence of the large driving force, and in turn the fluid expands, with consequent decrease in density. A natural convection current is created by the gas of reduced density, with the gas motion upward along the flat plate. An increase in velocity for a given element of fluid will mean that the element sweeps out a greater volume per unit time, which will tend to bring new air into the element in diagonal path from the outside cold-air supply. The process is shown qualitatively in Figure 13.14. All phenomena described here take place within a momentum and thermal boundary layer which is established near the wall, with consequent velocity and thermal gradients developing in the process. At the start, the velocity pattern of the boundary layer is characteristic of laminar flow, but with progress along the flat plate turbulent flow may occur. After some progress along the plate, the original element which is diluted somewhat with additional "outside" air reaches a temperature closer to the surface of the wall, and the thermal driving force is diminished. Further qualitative analysis of the boundary-layer behavior can be made by examining the effect of each phenomenon on the rate of acceleration of the fluid. The acceleration is always positive, and the velocity of the fluid is always increasing, but the rate of increase of acceleration may be positive or negative. The effect of the phenomena, *each taken alone*, is tabulated.

1. The decrease in thermal driving force tends to decrease the rate of heat transfer, rate of expansion, and consequently the rate of increase of acceleration.

2. The dilution of the boundary layer with new air

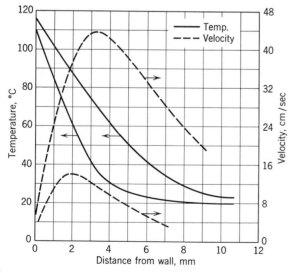

Figure 13.13. Local temperature and local velocity in the natural-convection boundary layer at a heated vertical plate (8, 29). Lower curve of each pair, 1 cm from bottom; upper, 24 cm. (After McAdams, *Heat Transmission*, 3rd ed.)

from the outer region tends to dilute the element of air with colder air and thus maintain the thermal driving force, with consequent increase in acceleration of the fluid.

3. Acceleration of the fluid tends to increase the flow of air from the outer regions.

4. The increase in velocity tends to increase the rate of heat transfer, rate of increase of volume, and consequently the rate of increase of acceleration. This is especially true if the boundary layer is turbulent.

5. The increase in velocity tends to increase the wall stress, which tends to decrease the rate of increase of acceleration of the fluid. Again, this is especially true if the boundary layer is turbulent.

6. The increase in temperature of a gas results in increased viscosity of gases, or decreased viscosity of liquids, with the usual effects on wall stress and acceleration of the fluid, especially if the boundary layer is laminar.

7. At all times, the acceleration of the gas results from the gravitational effect upon the expanded gas. The acceleration may be due to change in speed or change in direction.

The foregoing list shows that common natural convection is an extremely complex process for mathematical analysis. However, all seven phenomena described above are interrelated in terms of the mechanisms of flow described earlier in this chapter.

If experimental data are available, the complex system may yield to a mechanism-ratio analysis. As mentioned earlier, the mechanism-ratio analysis should be written in terms of "easily available data." The mechanisms that are postulated for this transfer are:

1. Momentum transfer by molecular transport.
2. Heat transfer by molecular transport.
3. Momentum transfer by the turbulent transport.
4. Heat transfer by the turbulent transport.
5. The gravity-dependent force on an element of fluid due to differences in density.

Several ratios have already been written, for example, the ratio of (4) to (2) is the traditional Nusselt number, $(N_{Nu}) = (hL/k)$.* The ratio of (3) to (1) is the traditional Reynolds number, $(N_{Re}) = (Lv\rho/\mu)$.* The ratio (1) to (2) is the traditional Prandtl number, $(N_{Pr}) = c_P\mu/k$. Item 5 must be combined in a ratio since, as stated earlier, with five mechanisms, four ratios are required. The friction factor discussed earlier is a ratio of force applied to the kinetic energy resulting from the applied force. If the force applied to a gas element can be written in terms of readily determinable system variables, a similar ratio can be written. For the gas element

---

* The Reynolds and Nusselt numbers may be written with any significant length dimension.

adjacent to the wall the fluid stress on the wall $(\tau_y)_1$ is balanced by a gravity-related buoyant force (F_B) because of a difference in density between the fluid element and its surroundings. For an element of gas of mass m and volume $\Delta x\, \Delta y\, \Delta z$,

$$F_B g_c = \tau_y g_c\, \Delta y\, \Delta z \qquad (13.133)$$

The mass of the gas element may be written as $\rho\, \Delta x\, \Delta y\, \Delta z$. The origin of force lies in the difference in density between the gas near the hot wall and the gas remote from the hot wall $(\Delta\rho)$. Therefore, the buoyant force F_B is equal to the volume of the gas element times the density difference, and Equation 13.133 becomes

$$\Delta x\, \Delta y\, \Delta z\, \Delta\rho\, g = (\tau_y g_c)_1\, \Delta y\, \Delta z \qquad (13.134)$$

where $g =$ the acceleration due to gravity $= 32.2$ ft/sec² or, canceling Δy and Δz

$$(\tau_y g_c)_1 = \Delta x\, \Delta\rho\, g \qquad (13.135)$$

where $\Delta x =$ the x dimension of the gas element
$\Delta\rho =$ the difference in density between the gas element and the gas remote from the wall

The stress term $(\tau_y g_c)_1$ can be combined with $v^2\rho$, the velocity and density of the gas element, to form a special friction factor f'', so that

$$f'' = \frac{(\tau_y g_c)_1}{v^2\rho} \qquad (13.136)$$

The stress term can be replaced by $\Delta x\, \Delta\rho g$ from Equation 13.135

$$f'' = \frac{\Delta x\, \Delta\rho\, g}{v^2\rho} \qquad (13.137)$$

The term $\Delta\rho$ is the difference between the density of the gas in at a distance x and the density near the wall. The volumetric coefficient of thermal expansion (β) of the fluid is an appropriate expression to evaluate $\Delta\rho$. Note that β represents the fractional volume increase with increase in temperature. For a given mass of gas, β also represents the fractional decrease of density with increase in temperature; therefore

$$\frac{\Delta\rho}{\rho} = \beta\, \Delta T \qquad (13.138)$$

The ratio $\Delta\rho/\rho$ in Equation 13.137 can be replaced by $\beta\, \Delta T$; therefore

$$f'' = \frac{\Delta x\, \beta\, \Delta T\, g}{v^2} \qquad (13.139)$$

For geometrically similar systems, Δx is proportional to L, where L is the length of the plate measured in the y-direction.

With this definition of f'', the mechanism ratio equation can be written

$$N_{Nu} = \phi((N_{Re}), (N_{Pr}), (f'')) \qquad (13.140)$$

The variable v appears in N_{Re} and f'' but does not fall in the category of "easily determined data," because v varies with x and y. Nothing further can be done with this equation to eliminate v until the experimental data are examined. Experimental data from many investigators (8) revealed an equation of the form

$$N_{Nu} = \text{constant} \, [(N_{Re})^2(f'')(N_{Pr})]^a \qquad (13.141)$$

correlated the data satisfactorily. Note that when the product $(N_{Re})^2(f'')$ is written in terms of group variables, v^2 cancels so that

$$(N_{Re})^2(f'') = \left[\frac{L^3 \rho^2 g \beta(-\Delta T)}{\mu^2}\right]$$

The resulting group of terms is traditionally called the Grashof number and assigned the symbol N_{Gr}. Therefore, for natural convection, the equations take the general form

$$(N_{Nu}) = \text{const} \, [N_{Gr} \cdot N_{Pr}]^a \qquad (13.142)$$

The cancellation of v, which is certainly fortuitous, probably occurs because the only source of velocity is the buoyancy force, so that buoyancy-force variables can exactly replace velocities. Note that Equation 13.142 is made up of three ratios, whereas four were predicted. This indicates that one of the five proposed mechanisms was redundant. The redundant mechanism is number 5. The gravitational force does not contribute in itself to transfer but merely acts as the driving force for velocity, and velocity is included in mechanisms 3 and 4.

The assembled data for many investigators are plotted in Figure 13.15 (49) in which $(N_{Nu})_f = (\bar{h}L/k_f)$ is plotted as a function of

$$(N_{Gr})_f(N_{Pr})_f = (L^3 \rho_f^{\,2} g \beta_f(-\Delta T)/\mu_f^{\,2})\left(\frac{c_P \mu}{k}\right)_f$$

in which the subscript f signifies that the fluid properties are evaluated at a special film temperature $T_f = (T_1 + \bar{T})/2$, the arithmetic mean temperature between the plate and the ambient fluid. The plotted data can be represented by two curves, one for the turbulent range and one for the laminar range.

For the turbulent range, $(N_{Gr})(N_{Pr}) = 10^9$ to 10^{12}, and

$$\frac{\bar{h}L}{k_f} = 0.13\left[\frac{L^3 \rho_f^{\,2} g \beta_f(-\Delta T)}{\mu_f^{\,2}} \cdot \left(\frac{c_P \mu}{k}\right)_f\right]^{1/3} \qquad (13.143)$$

For the laminar range, $(N_{Gr})(N_{Pr}) = 10^4$ to 10^9, and

$$\frac{\bar{h}L}{k_f} = 0.59\left[\frac{L^3 \rho_f^{\,2} g \beta_f(-\Delta T)}{\mu_f^{\,2}} \cdot \left(\frac{c_P \mu}{k}\right)_f\right]^{1/4} \qquad (13.144)$$

where $\bar{h}$ = the average heat-transfer coefficient for a flat plate of vertical length L

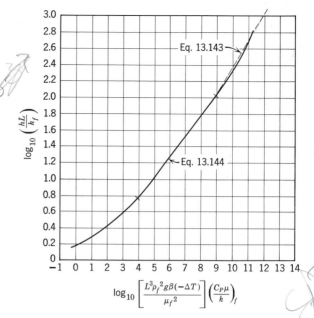

Figure 13.15. Natural convection from short vertical plates (49). (After McAdams, *Heat Transmission*, 3rd ed.)

$c_{Pf}, k_f, \rho_f, \mu_f$ = the fluid properties evaluated at the arithmetic mean temperature $(\bar{T} + T_1)/2$

$\bar{T}$ = the ambient temperature of the main body of the fluid

T_1 = the temperature of the plate

$-\Delta T = T_1 - \bar{T}$

β_f = the volumetric coefficient of expansion of the fluid, $(1/°F)$

Several simple dimensional equations (8) can be used for vertical planes exposed to air at atmospheric pressure and temperatures near room temperature.

For $(N_{Gr})(N_{Pr}) = 10^9$ to 10^{12},

$$\bar{h} = 0.19(-\Delta T)^{1/3} \qquad (13.145)$$

and for $(N_{Gr})(N_{Pr}) = 10^4$ to 10^9

$$\bar{h} = 0.29(-\Delta T/L)^{1/4} \qquad (13.146)$$

where $\bar{h}$ is in Btu/hr sq ft °F, ΔT in °F, and L in ft.

A similar correlation of experimental data for heat transfer from horizontal cylinders of outside diameter D_2 is

$$\frac{\bar{h}D_2}{k_f} = 0.53\left[\frac{D_2^{\,3} \rho_f^{\,2} g \beta_f(-\Delta T)}{\mu_f^{\,2}} \cdot \left(\frac{c_P \mu}{k}\right)_f\right]^{0.25} \qquad (13.147)$$

This equation applies over a range of

$$\frac{D_2^{\,3} \rho_f^{\,2} g \beta_f(-\Delta T)}{\mu_f^{\,2}} \cdot \left(\frac{c_P \mu}{k}\right)_f$$

between 10^3 and 10^9. The maximum length of travel for a fluid passing one side of a horizontal cylinder is $\pi D_2/2$, or one-half the perimeter. If $\pi D_2/2$ is substituted for L in Equation 13.144, Equation 13.147 would result.

Illustration 13.13. Calculate the heat-transfer coefficient for a vertical flat plate 3 ft high at 200°F in a room in which the air temperature is 80°F.

SOLUTION. This problem can be solved by use of Figure 13.15 or an equivalent equation. In any case the fluid properties must be determined at T_f. Since $T_f = (T_1 + \bar{T})/2$,

$$T_f = \frac{200 + 80}{2} = 140°F$$

At 140°F $\mu_f = 0.0195$ centipoise

$\quad\quad k_f = 0.0162$ Btu/hr sq ft (°F/ft)

$\quad\quad c_{Pf} = 0.255$ Btu/lb °F

The density of air can be computed from the density at standard conditions using as a basis the perfect-gas law.

$$\rho_f = \frac{28.9 \text{ lb}}{359 \text{ cu ft}}\left[\frac{492}{460 + 140}\right] = 0.066 \text{ lb/cu ft}$$

The volumetric coefficient of thermal expansion can be computed from the perfect-gas law. Since β is the fractional volume increase per degree of temperature rise, or fractional density decrease per degree of temperature rise, then

$$\beta = -\frac{1}{\rho}\frac{d\rho}{dT} = \frac{1}{V}\frac{dV}{dT}$$

Since, for perfect gases

$$PV = P/\rho = RT$$

or

$$\rho = \frac{P}{RT}$$

then, differentiating with respect to T

$$\frac{d\rho}{dT} = -\frac{P}{R}\frac{1}{T^2}$$

$\beta = -\dfrac{1}{\rho}\dfrac{d\rho}{dT} = \dfrac{1}{T}$ where T is absolute temperature. At

$T_f = 140°F = 600°R$, $\beta_f = 1/600$.

The data can be substituted into the product $(N_{Gr})(N_{Pr})$

$$(N_{Gr})(N_{Pr}) = \left(\frac{L^3\rho_f^2 g\beta_f(-\Delta T)}{\mu_f^2}\right)\left(\frac{c_P\mu}{k}\right)_f$$

$$= \left[\frac{(3)^3(0.066)^2(32.2)(1/600)(200-80)}{(0.0195 \times 6.72 \times 10^{-4})^2}\right.$$

$$\left. \cdot \frac{(0.255)(0.0195 \times 2.42)}{0.0162}\right]$$

$$= (4.40 \times 10^9)(0.743)$$

$(N_{Gr})(N_{Pr}) = 3.29 \times 10^9$

Therefore the flow is turbulent, and Equation 13.143 applies.

$$\frac{hL}{k_f} = 0.13(N_{Gr}N_{Pr})^{1/3}$$

$$= (0.13)(1.48 \times 10^3)$$

$$\frac{hL}{k_f} = 1.93 \times 10^2$$

$$h = \frac{0.0162 \times 1.93 \times 10^2}{3} = 1.04 \text{ Btu/hr sq ft °F}$$

Heat-Transfer Coefficients for Condensing Vapors. The transport of vapor from bulk to cooling surface offers little resistance. The accumulation of even small quantities of condensate on the surface introduces significant thermal resistance, and the calculation of the coefficient is based on the accumulation of liquid. The removal by condensation of an element of pure vapor from the region of a cold surface leaves a void that is immediately filled by flow of new vapor under a pressure gradient. The replacement is an extremely rapid process, unlike the diffusional processes described earlier. Generally, in a well-designed condensation chamber, the resistance to the bulk flow is negligible compared to other resistances that exist in the system. In the rare case in which the bulk-flow resistance is high enough to be meaningful, the principles outlined in earlier parts of this chapter can be adapted to the particular geometry and system. This case will be restricted to vapor consisting of one or more components, *all of which can condense at the existing conditions.*

With the appearance of condensate on the surface, the system becomes somewhat more complex. At this point the properties of the condensate and the geometry and physical character of the surface require separate consideration. Two types of surfaces are likely to be encountered. In the most frequent case, the condensate "wets" the surface and forms a continuous film of condensate over the cool surface; this is called *film-type* condensation. In the less common case, the condensate does not wet the surface but forms in discrete drops that barely adhere to the cool surface. This is called *dropwise* condensation.

Whatever the nature of condensate formation, the condensate leaves the surface by gravity flow except in specialized equipment. Therefore, the orientation and shape of the condensing surface is of considerable importance.

Film-Type Condensation. The usual design for condensers employs tubes, for simplicity of fabrication, with the condensate formed inside or outside of the tubes. A cold fluid on the opposite side of the tube wall serves to remove the latent heat of condensation. In film-type condensation, the condensate that forms at a point will flow downward over the remaining length of film-covered surface until a discontinuity is reached so that a pendant drop can form and break off. Since condensate forms over the entire surface, the thickness of the film will increase and will be a maximum at the point of exit. (See Figure 13.16.) As might be expected, the film may flow in the laminar regime or in the turbulent regime, depending upon the rate of condensation, length of path of the growing condensate film, fluid properties and geometry. The film velocity will be zero at the solid surface and a maximum at the condensate-vapor interface. It is evident that the mechanism of heat transfer will be "simple transfer," with heat entering the vapor-liquid boundary and leaving through the liquid-solid boundary. For simple transfer to occur, a temperature gradient must exist through the flowing film

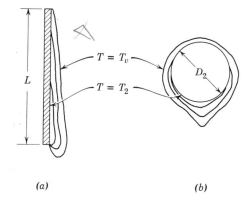

Figure 13.16. Condensate growth over vertical surfaces and horizontal tubes. (a) Vertical wall or tube. (b) Horizontal tube. T_v = saturation temperature of condensing vapor. T_2 = temperature of the solid surface.

of condensate, so that the vapor-liquid surface of the condensate is usually considered to be at the vapor temperature with the liquid surface in thermodynamic equilibrium with the vapor. The condensate in contact with the solid surface is considered to be at the temperature of that surface. The driving force for the downward flow of condensate is the gravitational field of the earth.

An over-all mechanism might be proposed that is made up of the individual mechanisms listed below. These proposed mechanisms include factors for laminar or turbulent flow *of the liquid condensate.* Flow characteristics of the vapor are *not* considered.

1. Total heat transfer (sum of turbulent and molecular transport).
2. Heat transfer by molecular transport.
3. Driving force for flow, a gravitational force.
4. Momentum transfer by turbulent transport.
5. Momentum transfer by molecular transport.

None of the condensate velocities, point or mean, is measurable in systems of this kind. A measurable over-all factor is the mass rate of flow for an entire unit element (for example, one tube in a tube bundle). Specifically, the linear mass flow rate (Λ) may be defined as $\Lambda = w/b$, where w is the mass rate of flow, and b is the perimeter measured normal to the flow direction. From the tabulated mechanisms above, an over-all mechanism equation based upon ratios can be written.

$$N_{\mathrm{Nu}} = \frac{hD}{k} = \text{the ratio of mechanism 1 to 2 above}$$

$$N_{\mathrm{Re}} = \frac{\Lambda}{\mu} = \text{the ratio of 4 to 5}$$

$$N_{\mathrm{Pr}} = \frac{c_P \mu}{k} = \text{the ratio of 5 to 2}$$

One additional ratio is required, and it must include the gravitational driving force for flow. This immediately suggests a counterpart friction factor. In the previous section, a "friction factor" was written based upon the buoyancy effect of natural convection. In this application, the difference in density between liquid and vapor is such that the buoyancy effect is usually negligible. A force balance in which fluid stress is balanced against the gravitational force on an element of fluid gives

$$(F_\tau)_1 g_c = m_\epsilon g \qquad (13.148)$$

where $(F_\tau)_1$ = the force on the surface associated with the stress in the fluid element
m_ϵ = the mass of the fluid element
g = the gravitational acceleration

$(F_\tau)_1$ may be written $(\tau_y)_1 \Delta y \Delta z$, and the mass m_ϵ may be written in terms of the density-volume product of the element as $(\rho_l - \rho_v) \Delta x \Delta y \Delta z$. The dimensions Δx, Δy, and Δz are the dimensions of the fluid element with x and y taken as the directions of transfer and flow respectively. The vapor density (ρ_v) is included to correct for buoyancy and ρ_l is the density of the liquid. These values are substituted into Equation 13.148 to give

$$(\tau_y g_c)_1 \Delta y \Delta z = (\rho_l - \rho_v) \Delta x \Delta y \Delta z\, g$$

which, after cancellation of Δy and Δz, becomes

$$(\tau_y g_c)_1 = (\rho_l - \rho_v) \Delta x\, g \qquad (13.149)$$

A friction factor may be defined

$$f''' = \frac{(\tau_y g_c)_1}{v^2 \rho_l} \qquad (13.150)$$

and Equation 13.149 can be used to replace $(\tau_y g_c)_1$ in Equation 13.150 so that

$$f''' = \frac{(\rho_l - \rho_v) \Delta x\, g}{v^2 \rho_l} \qquad (13.151)$$

If the length Δx is taken as the total thickness of the descending film, then

$$\Lambda = \Delta x\, v \rho_l$$

and

$$v^2 \rho_l = \frac{\Lambda^2}{\Delta x^2 \rho_l} \qquad (13.152)$$

Equation 13.152 can be used to replace $v^2 \rho_l$ in Equation 13.151.

$$f''' = \frac{(\Delta x)^3 \rho_l (\rho_l - \rho_v) g}{\Lambda^2}$$

For a geometrically similar system all linear dimensions bear a constant proportional relationship, and therefore $\Delta x \sim D_2 \sim L$, where D_2 and L are the diameter and length of the tube respectively. The commonest heat exchangers used for condenser service consist of banks of tubes mounted vertically or horizontally. The two geometries are not similar, and consequently two separate correlations are to be expected. For vertical tubes, the dimension L represents the full path of condensate flow and consequently is the significant length dimension. For horizontal tubes, the diameter (D_2) is certainly related to the full path of condensate flow. Therefore, the special friction factor (f''') should take two forms.

$$f'''_\uparrow = \frac{L^3 \rho_l (\rho_l - \rho_v) g}{\Lambda^2} \qquad (13.153)$$

and

$$f'''_\rightarrow = \frac{D_2{}^3 \rho_l (\rho_l - \rho_v) g}{\Lambda^2} \qquad (13.154)$$

where $f'''_\rightarrow$ and $f'''_\uparrow$ are the friction factors specific for horizontal and vertical tubes, respectively. The buoyancy effect of the vapor is usually negligible except at very high pressures approaching the critical pressure so that in subsequent writing ρ_v will be neglected. The friction factors then become

$$f'''_\uparrow = \frac{L^3 \rho_l{}^2 g}{\Lambda^2} \qquad (13.153a)$$

and

$$f'''_\rightarrow = \frac{D_2{}^3 \rho_l{}^2 g}{\Lambda^2} \qquad (13.155)$$

Sufficient ratios have been written in terms of simple system variables, and therefore the following mechanism ratio can be proposed:

$$N_{\text{Nu}} = \Phi(N_{\text{Re}}, f''', N_{\text{Pr}}) \qquad (13.156)$$

where Φ is an unknown function. The mechanism equation can now be tested with experimental data.

Vertical Tubes.* Experimental data were correlated (45) for *laminar flow* of condensate down a vertical tube as

$$\frac{\bar{h}L}{k} = 1.18 \left(\frac{L^3 \rho_l{}^2 g}{\Lambda^2} \right)^{1/3} \left(\frac{\Lambda}{\mu} \right)^{1/3} \qquad (13.157)$$

where $\bar{h}$ = the heat-transfer coefficient for a tube of length L
L = the length of the tube
$\Lambda = w/b$
w = mass rate of flow of condensate
b = the perimeter of the tubes, measured normal to flow = πD_2
D_2 = the outside diameter of the tube

---

* Nusselt (44) derived a similar equation

$$\frac{\bar{h}D}{k} = 0.926 \left(\frac{L^3 \rho_l{}^2 g}{\Lambda^2} \right)^{1/3} \left(\frac{\Lambda}{\mu} \right)^{1/3} \qquad (13.157b)$$

The difference in the equations rests only in the smaller coefficient in Equation 13.157b. The difference is attributed to rippling of the condensate surface. Nusselt did not allow for the rippling in his derivation.

The fluid properties, k, μ, and ρ_l, are evaluated at the temperature

$$T_f{}' = [T_v - (\tfrac{3}{4})(T_v - T_1)] \qquad (13.157a)$$

where $T_f{}'$ = a special film temperature
T_1 = wall temperature
T_v = vapor temperature

For *laminar* flow, mechanisms 1 and 4 tabulated above are not operative. With three mechanisms, only two mechanism ratios are required. Since both dimensionless groups have the same exponent, they can be combined to form a single group to give

$$\frac{\bar{h}L}{k} = 1.18 \left(\frac{L^3 \rho_l{}^2 g}{\mu \Lambda} \right)^{1/3} \qquad (13.158)$$

Equation 13.158 includes an extraneous variable. Note that L cancels to give, after rearrangement, a *dimensional* equation.

$$\bar{h} = 1.18 \left(\frac{k^3 \rho_l{}^2 g}{\mu \Lambda} \right)^{1/3} \qquad (13.159)$$

This equation must be used with special care paid to units. The units chosen must be consistent with the desired units of $\bar{h}$, usually Btu/hr sq ft °F. Equation 13.158 may be written with Λ eliminated and replaced by L as follows: The heat-transfer coefficient $\bar{h}$ is defined as

$$\bar{h} = - \frac{q}{A(T_1 - T_v)} \qquad (13.160)$$

and

$$A = \pi DL$$

$$q = w(\Delta H_v)$$

where (ΔH_v) = the latent heat of vaporization

Equation 13.160 may be written

$$\bar{h} = - \frac{w(\Delta H_v)}{\pi DL(T_1 - T_v)} \qquad (13.161)$$

Since $w/\pi D = \Lambda$, then Λ can be substituted into Equation 13.161, and the equation can be rearranged to

$$\Lambda = - \frac{\bar{h}L(T_1 - T_v)}{(\Delta H_v)} = \frac{\bar{h}L(T_v - T_1)}{(\Delta H_v)} \qquad (13.162)$$

The equation can be introduced into Equation 13.158 to eliminate Λ.

$$\frac{\bar{h}L}{k} = 1.18 \left(\frac{L^3 \rho_l{}^2 g(\Delta H_v)}{\mu \bar{h}(T_v - T_1)L} \right)^{1/3} \qquad (13.163)$$

Equation 13.163 may be solved for $\bar{h}$.

$$\bar{h}^{4/3} = 1.18 \left(\frac{k^3 \rho_l{}^2 g(\Delta H_v)}{L(T_v - T_1)\mu} \right)^{1/3} \qquad (13.164)$$

If each side of Equation 13.164 is raised to the 3/4 power,

$$\bar{h} = 1.13 \left(\frac{k^3 \rho_l{}^2 g(\Delta H_v)}{L(T_v - T_1)\mu} \right)^{1/4}$$

and each side may be multiplied by L/k to give the dimensionless equation

$$\frac{\bar{h}L}{k} = 1.13 \left(\frac{L^3 \rho_l^2 g (\Delta H_v)}{k(T_v - T_1)\mu} \right)^{1/4} \quad (13.165)$$

Equation 13.165 is useful if L is known but Λ is not known.

At high rates of condensate flow down a vertical surface, the film flows in the turbulent regime. The criterion for turbulence is $N_{\mathrm{Re}} = (\Lambda/\mu) \doteq 525$. If this condition is satisfied by a tube at its base, Equation 13.166 is recommended (46).

$$\frac{\bar{h}D_2}{k} = 0.0134 \left(\frac{\Lambda}{\mu} \right)^{1.07} \left(\frac{D_2^3 \rho_l^2 g}{\Lambda^2} \right)^{1/3} \quad (13.166)$$

where D_2 is the outside diameter of the tube. If Equation 13.162 is used to eliminate Λ in Equation 13.166,

$$\frac{\bar{h}D_2}{k} = 0.00071 \left(\frac{kL(T_v - T_1)}{D_2(\Delta H_v)\mu} \right)^{0.67} \left(\frac{D_2^3 \rho_l g}{\mu^2} \right)^{0.56} \quad (13.167)$$

Equations 13.157 through 13.167 neglect the effect of vapor velocity. High vapor velocities over parts of a tube bundle will tend to decrease the thickness of the layer of flowing condensate, with consequent increase in the coefficient. The fluid properties in these equations must be evaluated at T_f' described by Equation 13.157a.

Banks of Horizontal Tubes. The construction of horizontal condensers is such that condensate from tubes above drains onto the tubes below, thus increasing the condensate load. Figure 13.17 shows a bank of

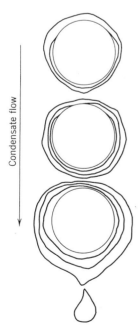

Figure 13.17. Condensate flowing down $N = 3$ rows of tubes.

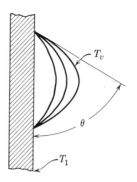

Figure 13.18. Discrete drop as found in dropwise condensation.

θ = receding contact angle.
T_1 = surface temperature.
T_v = vapor temperature.

$N = 3$ tubes. For tubes in horizontal rows with *laminar* flow of condensate down over the N tubes

$$\frac{\bar{h}D_2}{k} = 0.725 \left[\frac{\rho_l^2 g (\Delta H_v) D_2^3}{N\mu(T_v - T_1)k} \right]^{1/4} \quad (13.168)$$

The fluid properties ρ_l, k, and μ must be determined at T_f' as described in Equation 13.157a.

Dropwise Condensation. In the event that the condensate does not wet the surface and therefore collects in droplets on the surface, the configuration will be similar to the droplet shown in Figure 13.18.

The entire surface does not become covered with fluid. Instead, as condensate forms, it quickly coalesces to form discrete larger drops. The adhesion of the larger drops to the surface is weak, and they soon fall leaving behind a dry path. Since the total surface is at best partially covered with drops, the uncovered surface at T_1 serves as a film-free condenser with a large driving force maintained.

The obvious advantages of dropwise condensation would lead to the opinion that this condition should be used wherever possible. However, the physical preparation of surfaces that exhibit permanent dropwise condensation is impractical. Special temporary coatings on surfaces that produce dropwise condensation were tested experimentally with resultant increases of five to ten times the value for film-type condensation coefficients.

Noncondensable Gases. If noncondensable gases are present, the condensation coefficient is materially decreased. The noncondensable gas collects in the vicinity of the surface, and the condensing component must diffuse through the film. The introduction of the diffusion resistance into the path decreases the rate of condensation far below that for a pure material. For example, as little as 1 or 2 per cent of noncondensable gases in a steam system may reduce the rate of transfer of heat by 75 or 80 per cent. In practice, a steam condenser is operated with appropriate auxiliaries (steam traps or air vents) to eliminate air in the system. The calculation of coefficients for systems in which noncondensables are present requires consideration of simultaneous mass and heat transfer (53).

Illustration 13.14. A condenser is fitted with 1-in. O.D. tubes 3 ft long. If the condenser is mounted horizontally, the condensate will contact five tubes in its descent. If the tube surface temperature is 150°F and steam is condensing at 1 atm, recommend whether to mount the condenser (a) vertically or (b) horizontally.

Data: The fluid property data must be computed at T_f' according to Equation 13.157a.

$$T_f' = T_v - (\tfrac{3}{4})(T_v - T_1)$$

$$T_f' = 212 - 0.75(62) = 165.5°F$$

$$\mu_{165.5} = 0.37 \text{ centipoise}$$

$$k_{165.5} = 0.383 \text{ Btu/hr °F sq ft/ft}$$

$$\rho_{l165.5} = 61.0 \text{ lb/cu ft}$$

$$\Delta H_v = 970 \text{ Btu/lb (from steam tables)}$$

SOLUTION. (a) The calculation for vertical mounting can be tested with Equation 13.165 provided that the condensate moves in laminar flow.

$$\frac{hL}{k} = 1.13\left[\frac{L^3 \rho_l^2 g(\Delta H_v)}{k(T_v - T_1)\mu}\right]^{1/4}$$

$$h = \frac{1.13(0.383)}{3}\left[\frac{(3)^3(61.0^2)(32.2 \times 3600^2)(970)}{0.383(212 - 150)(0.37 \times 2.42)}\right]^{1/4}$$

$$h = \frac{1.13(0.383)}{3}[1.95 \times 10^{15}]^{1/4}$$

$$h = \frac{1.13(0.383)}{3}(6640)$$

$$h = 955 \text{ Btu/hr °F sq ft}$$

These data can be tested for turbulence. Equation 13.162 can be used to find Λ.

$$\Lambda = \frac{hL(T_v - T_1)}{\Delta H_v}$$

$$\Lambda = \frac{955(3)(62)}{970}$$

$$\Lambda = 178 \text{ lb/hr}$$

$$N_{Re} = \frac{\Lambda}{\mu} = \frac{178}{0.37(2.42)} = 200$$

This number is below the turbulent range; therefore

$$h = 955 \text{ Btu/hr °F sq ft}$$

(b) For the horizontally mounted condenser, Equation 13.168 applies.

$$\frac{hD_2}{k} = 0.725\left[\frac{\rho_l^2 g(\Delta H_v)D_2^3}{N\mu(T_v - T_1)k}\right]^{1/4}$$

$$h = \frac{(0.725)(0.383)}{(1/12)}\left[\frac{(61.0)^2(32.2 \times 3600^2)(970)(1/1728)}{5(0.37 \times 2.42)(62)0.383}\right]^{1/4}$$

$$= \frac{(0.725)(0.383)}{(1/12)}[82.0 \times 10^8]^{1/4}$$

$$= \frac{(0.725)(0.383)(313)}{(1/12)}$$

$$= 1045 \text{ Btu/hr sq ft °F}$$

For vertical mounting $h = 955$ Btu/hr °F sq ft, and for horizontal mounting $h = 1045$. There is a small advantage to horizontal mounting. If the tubes were shorter, then, as can be seen in Equation 13.165, there would be some possibility of increasing the coefficient to swing the advantage toward vertical mounting.

Boiling-Liquid Heat-Transfer Coefficients. Boiling, the opposite of condensation, is also of primary importance. Apparatus in which boiling may occur are many and varied in application; for that matter, it is almost safe to say that every condensing apparatus must have associated with it some boiling apparatus. In spite of the many necessary applications, boiling of liquids remains a phenomenon about which little is understood. The apparatus under discussion consists of a submerged solid, usually in the form of tube or a nearly flat plate, as in a kettle jacket, whose temperature exceeds the boiling point of the liquid which it contacts. Thus the driving force for boiling is the temperature difference $T_1 - T_B$, the difference between the surface temperature and the boiling point of the liquid under the existing pressure. As boiling progresses, heat transferred to the liquid appears as latent energy of vaporization, and vapor bubbles are formed. In a microscopic sense, an element of fluid, however small, must be elevated in temperature above its boiling point to the superheated state so that it will contain energy in excess of that necessary to bring about boiling. When this element of liquid is in the described state, the change of state is *initiated* in the form of a bubble nucleus. The nucleus formation requires this abnormal superheated state for some short period of time to overcome what seems to be a "reluctance" on the part of the fluid to change phase, and a metastable high-energy state must be achieved before this change can occur. In some systems the amount of liquid superheat necessary is greater than in others. The conditions favorable to bubble-nucleus formation are at best only partially understood. For example, the "bumping" of a laboratory flask occurs when a high degree of liquid superheat is accumulated before formation of a bubble nucleus. With the formation of the nucleus, the super-heated liquid "flashes," that is, forms a volume of vapor with nearly explosive force. The addition of boiling chips of rough porous media increases the ease of nucleus formation and therefore reduces the accumulation of liquid superheat.

If a system were operated at constant boiling temperature (T_B) but increasing surface temperature (T_1), the response of the system can be traced. As T_1 increases, formation of bubble nuclei increases, and the increase in bubble activity produces turbulence near the surface with consequent increase in heat-transfer coefficient. The increase in heat-transfer coefficient is favored by the decrease in viscosity in the high-temperature region near the surface. In this range the bubbles originate at favored sites on the surface. These sites are surface imperfections at which the shape of the imperfection, and perhaps a residual fragment of vapor or dissolved or adsorbed gas result in conditions favorable for bubble formation and growth. As T_1 is increased, additional less-favorable sites become active until the entire surface is actively forming bubbles. This condition is called *nucleate boiling*. If the increase in driving force is continued by increasing T_1 at constant T_B, the heat-transfer coefficient decreases after passing through a maximum because the heating surface becomes "blanketed" by a film of vapor. This second regime of boiling is called *film boiling*. This decrease continues until T_1 reaches such a level that additional heat is transferred by radiation. The thermal conductivity of vapor is considerably less than that of the liquid (for water at 212°F, $k_{\text{vapor}} = 0.0115$ and $k_{\text{liquid}} = 0.350$). Therefore, the resistance to heat transfer in the vicinity of the wall increases sharply because of the vapor blanket. Figure 13.19 is a plot of h as a function of $T_1 - T_B$ and shows the maximum value.

Any alteration of the conditions of surface and fluid can affect the heat-transfer coefficient. For example, if the pressure of the system is increased, the bubble volume per mole of vapor is decreased. This decrease will tend to suppress the incidence of film boiling by decreasing vapor volume and permit increased heat-transfer coefficients and increased values of the maximum temperature difference. The vapor blanket in film boiling is present because the density difference between liquid and vapor in the gravitational field is not sufficient to remove the vapor as rapidly as it is formed. A recent work (47) in which the rate of migration of bubbles away from the surface was increased in a centrifugal-force field showed marked increase in heat flux and heat-transfer coefficient.

Surface tension is said (48) to have an effect upon film boiling. If the liquid "wets" the surface, the liquid will spread under vapor bubbles tending to "snip" them off and away from the solid surface, whereas, if

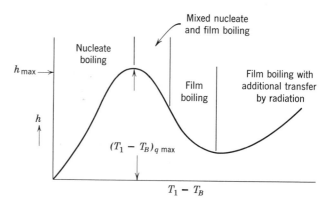

Figure 13.19. Plot of h as a function of $T_1 - T_B$.

the liquid does not wet the surface, there will be less tendency for the vapor bubbles to disengage. (Contrast this with dropwise condensation.) Thus a low surface tension or tendency to wet a solid is favorable to heat transfer. The wettability of liquid-solid systems is related to adsorption phenomena and is not well understood at this time.

The design of apparatus for boiling liquids introduces an unusual problem. If for some reason, the demand for vapor from a boiling apparatus should increase, the usual practice of increasing the temperature of the boiling surface will satisfy the demand *if the system normally operates at a temperature difference below that which gives maximum flux*. If the system is operating at or above the temperature difference for maximum flux, an increase in the heating-surface temperature will result in a decrease in vapor output.

The maximum heat flux $(q/A)_{\text{max}}$ and the temperature difference for maximum flux $(T_1 - T_B)_{q\,\text{max}}$ can be estimated in terms of the operating pressure and critical pressure of the system with two separate criteria existing for pure substances and binary mixtures. Figure 13.20a plots the ratio of maximum heat flux $(q/A)_{\text{max}}$ to the critical pressure P_c as a function of the reduced pressure P/P_c. Figure 13.20b is an expansion of the same plot in the low-pressure range, and the choice of the plot depends upon the reduced pressure. The temperature difference $(T_1 - T_B)_{q\,\text{max}}$, which occurs at $(q/A)_{\text{max}}$ can be estimated from Figure 13.20c in which $(T_1 - T_B)_{q\,\text{max}}$ is plotted as a function of the reduced pressure. Note that the *surface* temperature is significant. Estimation of surface temperatures in practical equipment is described in Chapter 15. The heat-transfer coefficient for boiling *at the maximum condition* can be calculated from the equation

$$\left(\frac{q}{A}\right)_{\text{max}} = h_{\text{max}}(T_1 - T_B)_{q\,\text{max}} \qquad (13.169)$$

where h_{max} = the heat-transfer coefficient at $(q/A)_{\text{max}}$. This heat-transfer coefficient can be used to compute the

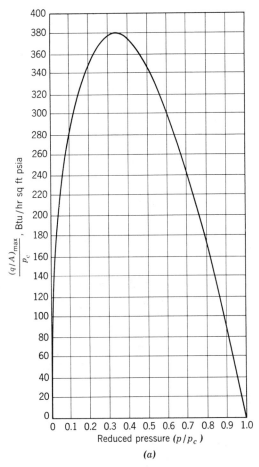

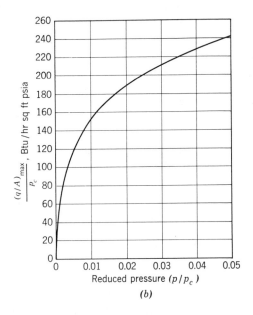

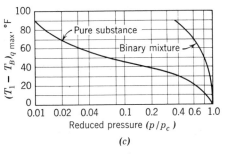

Figure 13.20. (a) Plot of maximum heat flux for boiling as a function of reduced pressure—high-pressure range (50). (b) Plot of maximum heat flux for boiling as a function of reduced pressure—low-pressure range (50). (c) Plot of maximum temperature difference for boiling as a function of the reduced pressure (50). (After Brown et al., *Unit Operations*, John Wiley and Sons.)

minimum area to satisfy the duty. Practical designs should always be made considerably larger, frequently corresponding to a coefficient 1–10 per cent of this maximum. Further discussion on this will be found in Chapter 15.

Illustration 13.15. For purposes of design, calculate the maximum heat flux and temperature difference for maximum flux for water boiling at 500°F.

Data: The critical pressure of water is 3206 psia and the operating pressure at 500°F is 681 psia.

SOLUTION. The reduced pressure for this system can be computed

$$\frac{P}{P_c} = \frac{681}{3206} = 0.222$$

Figure 13.20a is entered at $P/P_c = 0.222$, and $(q/A)_{max}/P_c$ is read as 355 Btu/hr sq ft psia.

$$\left(\frac{q}{A}\right)_{max} = (3206)(355)$$

$$\left(\frac{q}{A}\right)_{max} = 1,103,000 \text{ Btu/hr sq ft}$$

The temperature difference designated $(T_1 - T_B)_{max}$ associated with $(q/A)_{max}$ can be estimated from Figure 13.20c. The chart is entered at $P/P_c = 0.222$ and $(T_1 - T_B)_{q\ max}$ is read as 37°F.

The heat flux $(q/A)_{max}$ and temperature difference $(T_1 - T_B)_{q\ max}$ can be used to determine h_{max} according to Equation 13.169.

$$\left(\frac{q}{A}\right)_{max} = h_{max}(T_1 - T_B)_{q\ max}$$

$$h_{max} = \frac{1,103,000}{37}$$

$$h_{max} = 29,900 \text{ Btu/hr °F sq ft}$$

Summary of Mechanism Ratio Analysis. In the early part of this chapter, the mechanism-ratio-analysis technique was described. Throughout the chapter the method was illustrated in the prediction of equation forms for correlation of empirical data. At this point the various mechanisms will be presented in tabular form, arranged so that appropriate ratios can be written on sight. In Table 13.3 the mechanisms are stated,

Table 13.3. GROUPS OF SYSTEM VARIABLES THAT DESCRIBE VARIOUS MECHANISMS

	Diffusivity Units	Stress Units
(A) Molecular transport	δ	$\delta^2\rho/L'^2$
(1) Mass	$\mathscr{D}$	$\mathscr{D}^2\rho/L'^2$
(2) Heat	$\alpha, k/\rho c_P$	$\alpha^2\rho/L'^2$
(3) Momentum	$\nu, \mu/\rho$	$\nu^2\rho/L'^2$
(B) Turbulent transport	$\bar{E}$	$\bar{E}^2\rho/L'^2$
(1) Mass	$\bar{E}_N$	$\bar{E}_N^2\rho/L'^2$
(2) Heat	$\bar{E}_q$	$\bar{E}_q^2\rho/L'^2$
(3) Momentum	$\bar{E}_\tau, L'\bar{v}$	$\bar{E}_\tau^2\rho/L'^2, (D\bar{v})^2\rho/L'^2$
(C) Force-dependent mechanisms		
(1) Gravitational force	$\sqrt{L'^3 g}$	$L'g\rho$
(2) Centrifugal force	$\sqrt{L'^3 r\omega^2}$	$L'r\omega^2\rho$
(3) Surface forces	$\sqrt{\gamma g_c L'/\rho}$	$\gamma g_c/L'$
(D) Total transfer (by all mechanisms operating)	$(\mathscr{E}L')$	$\mathscr{E}^2\rho$
(1) Mass	$k_c'L'$	$k_c'^2\rho$
(2) Heat	$hL'/c_P\rho$	$(h^2/c_P^2\rho)$
(3) Momentum (loss by fluid)	$\dfrac{(-\Delta P)g_c L'^2}{vL'\rho}$	$\dfrac{(-\Delta P)g_c L'}{L''}$
(4) Momentum (gain by wall)	$\dfrac{(\tau_y g_c)_1}{v\rho}$	$[\tau_y g_c]_1$

Linear Dimensions may appear as ratios L'/L''
Definitions: L', L'' = characteristic lengths
γ = surface tension (F/L)
ω = angular velocity $(1/\theta)$
r = radius (L_r)
g = gravitational acceleration

and the groups of variables that describe the mechanisms will be presented in two forms. The diffusivity form, in which each group of variables has dimensions of L^2/θ, and the stress form, in which the group of variables has the dimensions of Fg_c/L^2 are both presented. Any pair of variable groups with the same dimensions constitutes a mechanism ratio. In addition, several other mechanisms, namely those mechanisms dependent upon surface forces, gravitational forces, and centrifugal forces, are also presented in such form that these mechanisms can be used in combination with the others.

In a geometrical system that requires n linear dimensions for complete description, $(n - 1)$ dimensionless ratios of these must be written.

Illustration 13.16. A long cold vertical tube that is used for condensing vapor on the outside rotates rapidly enough to throw condensate from the tube by centrifugal force. Predict the equation form that might be used to correlate experimental data.

SOLUTION. The mechanisms that may be assumed as operating are:

(1) Molecular transport of heat.
(2) Molecular transport of momentum.
(3) Total (molecular and turbulent) transfer of heat.
(4) Turbulent transport of momentum.
(5) A centrifugal force dependent mechanism.

There are five mechanisms considered as significant. The geometry, namely a long tube, can probably be characterized by a single linear dimension L'. The ratio can be set up in terms of the mechanisms tabulated above.

$$\frac{(3)}{(1)} = \Phi\left[\left(\frac{(5)}{(1)}\right)\left(\frac{(2)}{(4)}\right)\left(\frac{(2)}{(1)}\right)\right]$$

The mechanisms can be substituted from Table 13.3.

$$\frac{\left(\dfrac{hL'}{c_P\rho}\right)}{\alpha} = \Phi\left[\left(\frac{\sqrt{L'^3 r\omega^2}}{\alpha}\right)\cdot\left(\frac{\nu}{L'v}\right)\cdot\left(\frac{\nu}{\alpha}\right)\right]$$

This expression can be simplified

$$\frac{hL'}{k} = \Phi\left[\left(\frac{L'^3 r\omega^2}{\alpha^2}\right)\left(\frac{\nu}{L'v}\right)\left(\frac{\nu}{\alpha}\right)\right]$$

A system consisting of a long cylindrical tube can be characterized by the single dimension D, representing the outside diameter of the tube, therefore, $L' \sim r \sim D$. The velocity term $\bar{v}$ may be written as $\pi D\omega \sim D\omega$. This gives

$$\frac{hD}{k} = \Phi\left[\left(\frac{D^2\omega}{\alpha}\right)\left(\frac{\nu}{D^2\omega}\right)\left(\frac{\nu}{\alpha}\right)\right]$$

This expression can be tested with experimental data.

REFERENCES

1. Perry, J. H., *Chemical Engineers Handbook*, 3rd ed., McGraw-Hill Book Co., New York, 1950, p. 377.
2. Walker, W. H., W. K. Lewis, W. H. McAdams, and E. R. Gilliland, *Principles of Chemical Engineering*, 3rd ed., McGraw-Hill Book Co., New York, 1937, p. 43.
3. Coulson, J. M., and J. F. Richardson, *Chemical Engineering*, Vol. I, McGraw-Hill Book Co., New York, 1954.
4. Langhaar, H. L., *Trans. ASME*, **64**, A-55 (1942).
5. McAdams, W. H., *Heat Transmission*, 1st ed., McGraw-Hill Book Co., New York, 1933.
6. Latzko, H., *Z. angew. Math. u. Mech.*, **1**, 268 (1921).
7. Sieder, E. N., and G. E. Tate, *Ind. Eng. Chem.*, **28**, 1429–1436 (1936).
8. McAdams, W. H., *Heat Transmission*, 3rd ed., McGraw-Hill Book Co., New York, 1954.
9. Treybal, R. E., *Mass Transfer Operation*, McGraw-Hill Book Co., New York, 1955.
10. Reynolds, O., *Proc. Manchester Lit. Phil. Soc.*, **8** (1874).
11. Colburn, A. P., *Trans. Am. Inst. Chem. Engrs.* **29**, 174–210, (1933).
12. Prandtl, L., *Z. Physik*, **11**, 1072 (1910); *Z. Physik*, **29**, 487–489 (1928).
13. Taylor, G. I., *Brit. Advisory Comm. Aeronaut.*, Rept. Mem. 272, 31, pp. 423–429 (1916).
14. von Kármán, T., *Proc. Intern. Congr. Appl. Mech.*, 4th Congr., 1934, pp. 54–91; *Engineering*, **148**, 210–213 (1939); *Trans. ASME*, **61**, 705–710 (1939).
15. Hoffman, E., *Forsch. Gebeite Ingenieurw.*, **11**, 159–169 (1940).
16. Boelter, L. M. K., R. C. Martinelli, and F. Jonassen, *Trans. ASME*, **63**, 447–455 (1941).
17. Martinelli, R. C., *Trans. ASME*, **69**, 947–959 (1947).

18. Lyon, R. N., *Chem. Eng. Progr.*, **47**, 75–79 (1951).
19. Sherwood, T. K., *Trans. Am. Inst. Chem. Engrs.*, **36**, 817 (1940).
20. Perry, J. H., *Chemical Engineers Handbook*, 3rd ed., McGraw-Hill Book Co., New York, 1950, Sec. 8.
21. Rouse, H., and J. W. Howe., *Basic Fluid Mechanics*, John Wiley and Sons, New York, 1953, p. 181, Fig. 107; p. 188, Fig. 112.
22. Schlichting, H., *Boundary Layer Theory*, Pergamon Press, London, 1955.
23. Cope, W. F., *Proc. Inst. Mech. Engrs. London*, **145**, 99–105 (1941).
24. Bergelin, O. P., G. A. Brown, and S. C. Doberstein, *Trans. ASME*, **74**, 953–960 (1952).
25. Graetz, L., *Ann. Phys. u. Chem.*, **25**, 337 (1885).
26. Sellars, J. R., M. Tribus, and J. S. Klein, *Trans. ASME*, **78**, 441 (1956).
27. Kays, W. M. *Trans. ASME*, **77**, 1265 (1955).
28. Eubank, O. C., and W. S. Proctor, S. M. Thesis, Department of Chemical Engineering, Massachusetts Institute of Technology, 1951.
29. Schmidt, E., *Z. ges. Kälte-Ind.*, **35**, 213 (1928).
30. Higbie, Ralph, *Trans. Am. Inst. Chem. Engrs.*, **31**, 365–389 (1935).
31. Danckwerts, P. V., *Ind. Eng. Chem.*, **43**, 1460–1467 (1951).
32. Hanratty, T. J., *A.I.Ch.E. Journal*, **2**, 359–363 (1956).
33. Fage, A. and H. C. H. Townend, *Proc. Roy. Soc. London, Ser. A*, **135**, 656 (1932).
34. Powell, R. W., *Trans. Inst. Chem. Engrs. London*, **18**, 36 (1940); and Powell, R. W., and E. Griffiths, *Trans. Inst. Chem. Engrs. London*, **13**, 175 (1935).
35. Lohrisch, W., *Forsch. a.d. Geb. d. Ingen.* (*VDI*) **322**, 1 (1929).
36. McAdams, W. H., *Heat Transmission*, 2nd ed., McGraw-Hill Book Co., New York, 1942, p. 221, Fig. 111.
37. Goldstein, S. (ed.), *Modern Developments in Fluid Dynamics*, Oxford University Press, New York, 1938.
38. Rouse, H., and J. W. Howe, *Basic Mechanics of Fluids*, John Wiley and Sons, New York, 1953, p. 188, Fig. 112.
39. McAdams, W. H., *Heat Transmission*, 3rd ed., McGraw-Hill Book Co., New York, 1954, p. 267, Fig. 10–12.
40. Sherwood, T. K., and R. L. Pigford, *Absorption and Extraction*, McGraw-Hill Book Co., New York, 1952, p. 74, Fig. 22.
41. McAdams, W. H., *Heat Transmission*, 3rd ed., McGraw-Hill Book Co., New York, 1954, p. 266, Fig. 10–11.
42. McAdams, W. H., *Heat Transmission*, 3rd ed., McGraw-Hill Book Co., New York, 1954, p. 267, Fig. 10–13.
43. Rouse, H., and J. W. Howe, *Basic Mechanics of Fluids*, John Wiley and Sons, New York, 1953, p. 181, Fig. 107.
44. Nusselt, W., *Z. Ver. deut. Ing.*, **60**, 541 (1916).
45. Brown, G. G., et al., *Unit Operations*, John Wiley and Sons, New York, 1950, p. 449.
46. Brown, G. G., et al., *Unit Operations*, John Wiley and Sons, New York, 1950, p. 451.
47. Gambill, W. R., and N. D. Greene, *Chem. Eng. Progr.*, **54**, No. 10, 68–76 (1958).
48. Kreith, F., *Principles of Heat Transfer*, International Textbook Co., Scranton, pp. 406–407.
49. McAdams, W. H., *Heat Transmission*, 3rd ed., McGraw-Hill Book Co., New York, 1954, p. 173, Fig. 7–7.
50. Bonilla, C. F., and M. T. Cichelli, *Trans. Am. Inst. Chem. Engrs.*, **51**, 761 (1945).
51. Knudsen, J. G., and D. L. Katz, *Fluid Dynamics and Heat Transfer*, McGraw-Hill Book Co., New York, 1958, p. 467.
52. Lubarsky, B., and S. J. Kaufman, *Natl. Advisory Comm. Aeronaut. Tech. Notes*, No. 3336, 1955.
53. McAdams, W. H., *Heat Transmission*, 3rd ed., McGraw-Hill Book Co., New York, 1954, p. 355.

PROBLEMS

13.1. Calculate the pressure drop in 10 ft of smooth tubing of 1 in. I.D. for:

(*a*) Carbon dioxide at 100°F and 1 atm flowing at 100 cu ft/min.

(*b*) Benzene at 100°F and 1 atm flowing at 100 cu ft/min.

(*c*) Mercury at 100°F and 1 atm flowing at 100 cu ft/min.

(*d*) Carbon dioxide at 100°F and 1 atm flowing at 0.1 cu ft/min.

(*e*) Benzene at 100°F and 1 atm flowing at 0.1 cu ft/min.

(*f*) Mercury at 100°F and 1 atm flowing at 0.1 cu ft/min. What flow regime exists in each case above?

13.2. Estimate the flow rate at which the transition from laminar to turbulent flow will begin in a 2 in. I.D. smooth tube for:

(*a*) Propane at 70°F, 1 atm

(*b*) Acetone at 70°F

(*c*) SAE 10 lubricating oil at 70°F

13.3. The pressure drop is 10 psi over 100 ft of smooth $1\frac{1}{2}$ in. I.D. pipe with SAE 10 lubricating oil flowing at 100°F.

(*a*) What is the average velocity of flow?

(*b*) What is the flow regime?

13.4. Calculate the pressure drop per foot for methane at 1 atm and 400°F flowing through a smooth $\frac{1}{4}$ in. I.D. tube at 1 cu ft/min.

13.5. Air at 150°F and 1 atm flowing through a smooth circular duct of 6 in. diameter and 100 ft long gives a pressure drop of 10 in. of water. Calculate the flow rate of air (cu ft/min).

13.6. Liquid sodium at 600°F is to be supplied to a nuclear reactor through 100 ft of 3 in. I.D. stainless steel tubing. Calculate the pressure drop for a flow rate of 100 gal/min.

13.7. Estimate the pressure drop through a square duct 2 ft on a side and 40 ft long for an air flow of 10,000 cu ft/min at 70°F and 1 atm.

13.8. Estimate the pressure drop per foot for the flow of carbon tetrachloride at 100°F through an annular space whose dimensions are 2 in. I.D. and 3 in. O.D. The flow rate is 100 gal/min.

13.9. Methane at 70°F and 1 atm is flowing parallel to a flat plate 4 ft long. The methane velocity is uniform at 10 ft/sec as it passes the leading edge of the plate.

(*a*) At what point (if any) from the leading edge does the boundary layer become turbulent?

(*b*) What is the thickness of the boundary layer at 3 in. from the leading edge?

(*c*) What is the thickness of the boundary layer 3 ft from the leading edge?

(*d*) Calculate the point stress 3 ft from the leading edge.

(*e*) Calculate the average stress over the plate from the leading edge to a point 3 ft from the leading edge.

13.10. Air at 200°F and 1 atm flows over a long flat surface at 200 ft/sec free-stream velocity.

(*a*) Calculate the point at which the boundary layer becomes turbulent.

(*b*) Calculate the maximum thickness of the *laminar* boundary layer.

(*c*) Calculate the thickness of the boundary layer 10 ft from the leading edge.

13.11. Calculate the entry length for SAE 10 lubricating oil at 100°F flowing into a 1 in. I.D. tube for an average velocity in the tube of:

(*a*) 0.05 ft/sec.

(*b*) 1 ft/sec.

13.12. Determine the entry length for air at 100°F and 1 atm flowing into a 1 in. I.D. tube at 0.5 ft/sec. average velocity in the tube.

13.13. A spherical gas storage tank 40 ft in diameter is exposed to winds up to 75 mph. Calculate force exerted by the wind at 70°F on the tank.

13.14. Water at 60°F is flowing past a cylindrical bridge pier 1 ft in diameter submerged to a depth of 10 ft. Calculate the force on the pier when the water velocity is:

(*a*) 0.1 ft/sec.

(*b*) 10 ft/sec.

13.15. Integrate Equation 13.58 for momentum transfer according to the directions given in the footnote on p. 166.

13.16. Water is heated in a copper tube of 1 in. I.D. At a point along the tube the average water temperature is 100°F and the tube wall temperature is 210°F. Calculate the heat-transfer coefficient and the heat flux for a water velocity of 5 ft/sec.

13.17. Air at atmospheric pressure and an average temperature of 100°F is flowing at 10,000 cu ft/min through a 3 in. circular duct. The wall temperature is 300°F. Calculate the heat flux.

13.18. SAE 10 lubricating oil flows at 40 gal/min through a smooth tube of 1 in. I.D. At a point in the tube the average oil temperature is 300°F and the tube wall temperature is 60°F. Determine the heat flux at this point.

13.19. Calculate the heat-transfer coefficient for the cooling of hydrogen from the following data at 1 atm:

Average hydrogen temperature = 300°F
Tube-wall temperature   = 60°F
Tube diameter   = 0.5 in.
Hydrogen velocity   = 200 ft/sec

13.20. A wetted-wall column of $1\frac{1}{2}$ in. I.D. is being used to humidify air. Air at 70°F and 1 atm flows upward in the column at 12 ft/sec, and water at 70°F flows downward along the wall. At a point where air of negligible humidity is in contact with water at 70°F:

(*a*) Calculate k_c.   (*c*) Calculate k_y.

(*b*) Calculate k_G.   (*d*) Calculate k_Y.

13.21. Ethyl alcohol is flowing downward countercurrent to a stream of air in a wetted-wall column which is 2 in. I.D. Both the alcohol and air are at 77°F. The total pressure is 700 mm Hg. At a point in the column the average partial pressure of ethanol vapor in the air is 25 mm Hg. The gas velocity is 2 ft/sec. Calculate the mass flux at this point.

13.22. Show that (*a*) $k_y = k_c P/RT$ and (*b*) $k_G = k_c/RT$

13.23. Show that $k_x = k_c \bar{p}/M$

13.24. Show that $k_Y = k_c P(y_{b2})(\bar{y}_b)$

13.25. Air at 212°F flows over a streamlined naphthalene body. Naphthalene sublimes into air, and its vapor pressure at 212°F is 20 mm Hg. The heat-transfer coefficient for the same shape and air velocity was previously found to be 4 Btu/hr sq ft °F. The concentration of naphthalene in the bulk air stream is negligibly small. The mass diffusivity of naphthalene in air at 212°F is 0.32 sq ft/hr. Calculate the mass-transfer coefficient and mass flux for the system.

13.26. Air at 100°F and 1 atm flowing through a smooth tube 1 in. I.D. 20 ft long yields a pressure drop of 0.1 psi. Estimate the heat-transfer coefficient for this system using:

(*a*) The recommended empirical equation.

(*b*) The Reynolds analogy.

(*c*) The Colburn analogy.

(*d*) The Martinelli analogy.

13.27. Calculate the heat-transfer coefficient for liquid sodium at 210°F flowing at 5 ft/sec in a 2 in. I.D. stainless-steel tube using:

(*a*) The recommended empirical equation.

(*b*) The Reynolds analogy.

(*c*) The Colburn analogy.

(*d*) The Martinelli analogy.

13.28. Hot air at 600°F and 1 atm flows from a large plenum chamber into a smooth circular duct 2 ft in diameter. The average velocity of air in the duct is 10 ft/sec.

(*a*) Determine the average heat-transfer coefficient over the first 3 ft of the duct.

(*b*) Calculate the entry length.

(*c*) Calculate the heat-transfer coefficient after the flow is fully developed. Assume the average velocity is still 10 ft/sec.

13.29. Air at 60°F and 1 atm flows parallel to a heated plate which is at 200°F.

(*a*) Calculate and plot the point value of the heat-transfer coefficient starting at the leading edge of the plate and continuing to 2 ft from the leading edge. The free-stream air velocity is 25 ft/sec. Evaluate the physical properties of the air at 130°F.

(*b*) Calculate the rate of heat transfer per foot of width over the first 2 ft of the plate.

13.30. A cylindrical steel chimney 3 ft in diameter and 50 ft high is exposed to a wind of 50 mph. The surface of the chimney is at 500°F and the average air temperature is 70°F.

(*a*) Calculate the force on the chimney.

(*b*) Calculate the heat loss from the chimney.

13.31. Calculate the fictitious-film thickness for heat transfer (Equation 13.117*a*) for the conditions of:

(*a*) Problem 13.16.

(*b*) Problem 13.19.

13.32. Calculate the fictitious-film thickness for mass transfer (Equation 13.117*b*) for the conditions of:

(*a*) Problem 13.20.

(*b*) Problem 13.21.

13.33. Estimate the surface-renewal factor *s* for the conditions of:

(*a*) Problem 13.16.

(*b*) Problem 13.19.

13.34. Calculate the heat-transfer coefficient for SAE 10 lubricating oil in laminar flow at an average velocity of 0.1 ft/sec in a 1 in. I.D. tube. The oil is at 70°F and the tube wall is at 120°F. The tube is 10 ft long. Calculate the heat flux at this point.

13.35. A heated vertical plate 2 ft high and 1 ft wide is exposed to air at 70°F and 1 atm. The plate temperature is 250°F. Evaluate the heat-transfer coefficient and the rate of heat transfer over the plate.

13.36. A large vertical tank has a flat side 8 ft high and 20 ft long. The tank is filled with benzene at 120°F and the tank wall is at 60°F. β for benzene is $6.8 \times 10^{-4} (°F)^{-1}$. Calculate the heat loss in 24 hr from the tank wall.

13.37. A horizontal steam pipe of 2 in. O.D. is exposed to air at 40°F and 1 atm. The surface of the steam pipe is at 200°F. Calculate the heat loss over 100 ft of this pipe.

13.38. Condensing equipment may be mounted horizontally or vertically. In Illustration 13.14 horizontal mounting was indicated based upon the size of the coefficients. *For laminar flow of condensate*, derive an expression in *D*, *L*, and *N* that defines the limiting tube length beyond which condensers should be mounted vertically for maximum value of the heat-transfer coefficient.

13.39. Pure saturated benzene vapor at 2-atm total pressure is to be condensed on vertical tubes. The tube surface temperature will be maintained at 80°F by cooling water. The tubes are 2 in O.D. and 5 ft long. Calculate the heat-transfer rate and the condensation rate. ΔH_v for benzene = 150 Btu/lb.

13.40. Estimate the heat-transfer rate for a bundle of 30 horizontal tubes placed on a square pitch 5 tubes wide and 6 tubes high. Acetone vapor at 20 psi and 148°F is condensing on the tubes whose surface is at 50°F. The tubes are 1 in. O.D. and 8 ft long.

13.41. Water vapor is to be condensed on a single tube of 2 in. O.D. at 1-atm pressure. The tube surface is 80°F. What length tube will give the same average heat-transfer coefficient whether mounted vertically or horizontally?

13.42. Acetone is to be condensed on a single 1-in. tube at 1 atm. The tube surface is at 60°F. At what length will the tube operate with equal effectiveness whether mounted vertically or horizontally?

13.43. Evaluate the temperature difference at maximum flux and the maximum heat-transfer coefficient for ethanol boiling at 3-atm pressure.

13.44. Estimate the maximum rate of heat transfer to water boiling at 1 atm on a horizontal plate of 2 sq ft area.

chapter 14

Interphase Transfer

Many industrial operations involve the transfer of heat, mass, and momentum from one phase to another. In a heat exchanger, for example, heat is being transferred from a hot fluid through a tube wall to a cold fluid. Gas absorption involves the transfer of mass from a solute-rich gas phase to a solute-poor liquid phase. In the case of a fluid flowing through a pipe, momentum is transferred through the fluid to the pipe wall.

In Chapter 13, the rate equation was integrated in the transfer direction at a given position in an apparatus to define a transfer coefficient. It is the purpose of this chapter to show the interrelation of resistances and driving forces in an interphase transfer process at a particular point in a piece of equipment and to integrate the relationships over the entire transfer area. In addition, a design equation relating the rate of transfer to the total required transfer area will be developed.

MULTIPLE PHASE RESISTANCES

As before, the rate of transfer may be expressed as the driving force divided by the resistance. In operational equipment where the rate of transfer dictates the transfer-area requirements, a new situation arises that has not been encountered before. Multiple resistances are encountered, and the rate of transfer varies from position to position within the apparatus. For example, at a particular point in a heat exchanger, an analysis of the transfer path reveals that several resistances are encountered in series. First, there is the resistance to transfer between the hot bulk phase and the tube wall. Then there is the resistance of the tube wall itself, and finally the resistance between the tube wall and the cold bulk phase. Within each of these resistances is found a temperature gradient.

The transfer process is more complex if the transfer path is made up of resistances in parallel. As an example, situations arise in heat transfer where radiation, conduction, and convection occur simultaneously. The conduction part of the transfer process has one resistance while parallel to the conduction resistance, there is a different resistance along the radiation path.

A further complication occurs if the heat-transfer process is considered at another point in the exchanger where the temperatures of both phases have changed. Although the same qualitative arrangement of resistances is present, the changes in temperature have resulted in different temperature gradients and different resistances at this point.

Series Resistances at a Point: Steady State. It has already been established that the rate of transfer, be it heat, mass, or momentum, is equal to the driving force divided by a resistance. This may be expressed mathematically as

$$\text{Rate} = \psi A = -\frac{\Delta \Gamma}{\mathbf{R}} = \frac{\text{driving force}}{\text{resistance}} \quad (14.1)$$

The above equation in its earlier application was limited to a single phase with the driving force and resistance indigenous to that phase. The same equation, however, may be applied to more than one phase. Imagine simple transfer through two resistances in series as illustrated in Figure 14.1. Assume steady-state operation exists, in which case the rate of transfer (ψA) is constant with respect to time.

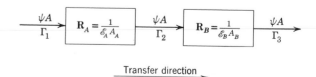

Figure 14.1. Series resistances.

In Figure 14.1, $\mathbf{R}_A$ and $\mathbf{R}_B$ schematically depict resistances associated with phases A and B. It should be recalled that these resistances are comprised of the transfer distance and the diffusivities. Solving Equation 14.1 for the driving force across each phase gives

for phase A:

$$(\Delta\Gamma)_A = -(\Gamma_2 - \Gamma_1) = (\psi A)\mathbf{R}_A \qquad (14.2)$$

For phase B:

$$(\Delta\Gamma)_B = -(\Gamma_3 - \Gamma_2) = (\psi A)\mathbf{R}_B \qquad (14.3)$$

Addition of Equations 14.2 and 14.3 gives

$$-(\Gamma_3 - \Gamma_1) = (\psi A)\mathbf{R}_A + (\psi A)\mathbf{R}_B \qquad (14.4)$$

For simple transfer at steady state, Equation 14.4 becomes

$$-(\Gamma_3 - \Gamma_1) = (\psi A)[(\mathbf{R}_A + \mathbf{R}_B)] \qquad (14.5)$$

or

$$-(\Gamma_3 - \Gamma_1) = (\psi A)\left[\frac{1}{\mathscr{E}_A A_A} + \frac{1}{\mathscr{E}_B A_B}\right] \quad (14.5a)$$

Solving Equation 14.5 for the rate of transfer gives

$$(\psi A) = -\frac{(\Gamma_3 - \Gamma_1)}{\mathbf{R}_A + \mathbf{R}_B} = -\frac{(\Gamma_3 - \Gamma_1)}{\dfrac{1}{\mathscr{E}_A A_A} + \dfrac{1}{\mathscr{E}_B A_B}} \qquad (14.6)$$

Note that Equation 14.6 is written in terms of the driving force $(\Gamma_3 - \Gamma_1)$ across both resistance A and resistance B with the intermediate concentration Γ_2 not appearing. This driving force is called the total driving force, and $\mathbf{R}_A + \mathbf{R}_B$ is the total resistance. The same reasoning as used in the development of Equation 14.6 can be applied to any number of resistances in series. In general, then, for steady state

$$\psi A = -\frac{\sum\limits_{j=1}^{j=n} (\Delta\Gamma)_j}{\sum\limits_{j=1}^{j=n} \mathbf{R}_j} \qquad (14.7)$$

For steady state, a useful relationship showing the driving-force–resistance relationship for the total system or any part may be written by restating Equations 14.2, 14.3, and 14.6.

$$(\psi A) = -\frac{\Sigma\Delta\Gamma}{\Sigma\mathbf{R}} = -\frac{\Delta\Gamma_A}{\mathbf{R}_A} = -\frac{\Delta\Gamma_B}{\mathbf{R}_B} = \ldots -\frac{\Delta\Gamma_n}{\mathbf{R}_n}$$
$$(14.8)$$

where $\quad \Delta\Gamma_A$ = the driving force across $\mathbf{R}_A$
$\qquad\quad \Delta\Gamma_B$ = the driving force across $\mathbf{R}_B$
$\qquad\quad \Delta\Gamma_n$ = the driving force across $\mathbf{R}_n$

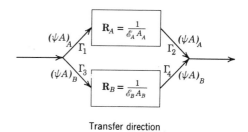

Figure 14.2. Parallel resistances.

Parallel Resistances at a Point: Steady State. Consider Figure 14.2 where the transfer process occurs through two parallel resistances $\mathbf{R}_A$ and $\mathbf{R}_B$ under conditions of steady-state operation. Let

$$\Gamma_2 - \Gamma_1 = \Delta\Gamma_A$$

and

$$\Gamma_4 - \Gamma_3 = \Delta\Gamma_B$$

For resistance $(\mathbf{R}_A)$ in accordance with Equation 14.7

$$(\psi A)_A = -\frac{\Delta\Gamma_A}{\mathbf{R}_A} \qquad (14.9)$$

and for $\mathbf{R}_B$

$$(\psi A)_B = -\frac{\Delta\Gamma_B}{\mathbf{R}_B} \qquad (14.10)$$

but

$$(\psi A)_A + (\psi A)_B = (\psi A)_{\text{total}} \qquad (14.11)$$

Therefore combining Equations 14.9 and 14.10 with Equation 14.11

$$(\psi A)_{\text{total}} = -\left[\frac{\Delta\Gamma_A}{\mathbf{R}_A} + \frac{\Delta\Gamma_B}{\mathbf{R}_B}\right] \qquad (14.12)$$

The nature of the driving force is such that, if the driving force in each branch is not equal, it will adjust itself until $\Delta\Gamma_A$ and $\Delta\Gamma_B$ do become equal, or

$$-\Delta\Gamma_A = -\Delta\Gamma_B \qquad (14.13)$$

As a result, Equation 14.12 may be written

$$(\psi A)_{\text{total}} = (-\Delta\Gamma_{A\,\text{or}\,B})\left(\frac{1}{\mathbf{R}_A} + \frac{1}{\mathbf{R}_B}\right) \qquad (14.14)$$

Generalizing Equation 14.14 for any number of branches gives

$$(\psi A)_{\text{total}} = -\Delta\Gamma_j \sum_{j=1}^{j=n} \frac{1}{\mathbf{R}_j} \qquad (14.15)$$

Equations 14.7 and 14.15 may now be applied to steady-state heat and mass transfer. Series or parallel resistances for momentum transfer have no significance since momentum transfer usually stops at a boundary.

HEAT TRANSFER

Heat transfer as carried out industrially involves the contacting of a hot phase and a cold phase, separated by a well-defined boundary. Both series and parallel resistances are frequently encountered. For example a typical heat exchanger involves at least three resistances in series, e.g., the fluid resistance of the hot phase, the resistance of the tube wall that keeps the two phases separate, and the fluid resistance of the cold phase. A similar physical situation exists in a furnace wall constructed of several different types of building material where each material constitutes a separate resistance in series in the direction of heat flow. This same furnace wall, on the other hand, with a steel door or a peep hole, will now present the more-complicated case of parallel resistance paths, each containing several resistances in series.

Series Resistances in Heat Exchangers. Consider a point in a heat exchanger where the hot phase has a temperature T_A and the cold phase has a temperature T_B, with both phases being in motion. For the time being, it is assumed that the two phases exchanging heat are kept separated by a boundary of negligible thickness and, hence, negligible resistance. The actual case of a tube wall serving as a boundary will be considered later in the chapter. The apparatus at the point in question will have two resistances in series in the transfer direction. However, as previously established in Chapter 13, because of boundary-layer phenomena only a portion of the phases on either side of the boundary will offer significant resistance to transfer. Figure 14.3 schematically depicts this physical situation.

Line *I–I* represents the imaginary interface common to both phases, and the dotted lines on either side of *I–I*, represent the extent of the region that comprises most of the resistance to transfer. It will be in these regions that the steep or large temperature gradients will occur, and the solid line illustrates the temperature profile as one proceeds from the hot phase to the cold. Since both phases have the same temperature at the interface, the interface is in equilibrium.

For a series of resistances, Equation 14.7 is applicable. Because the nature of the resistance layers is such that their thickness cannot be predicted, it will be necessary to use the surface-coefficient concept mentioned earlier. For heat transfer, the rate ψA becomes q and the resistance $1/\mathscr{E} A$ is designated as $1/hA$.

The rate of heat transfer from the bulk of the hot phase to the interface is

$$q_A = -\frac{(T_i - T_A)}{\dfrac{1}{h_A A_A}} \qquad (14.16)$$

and the rate of transfer from the interface to the bulk cold fluid is

$$q_B = -\frac{(T_B - T_i)}{\dfrac{1}{h_B A_B}} \qquad (14.17)$$

Each of the above equations is made up of a single resistance and a single driving force.

Rearranging Equations 14.16 and 14.17 yields

$$-(T_i - T_A) = q_A\left(\frac{1}{h_A A_A}\right) \qquad (14.16a)$$

and

$$-(T_B - T_i) = q_B\left(\frac{1}{h_B A_B}\right) \qquad (14.17a)$$

Addition and rearrangement of Equations 14.16a and 14.17a, with the condition that for steady state $q_A = q_B$, results in

$$q = \frac{-(T_B - T_A)}{\left[\dfrac{1}{h_A A_A} + \dfrac{1}{h_B A_B}\right]} \qquad (14.18)$$

Equation 14.18 is identical in form and meaning with Equation 14.7. The use of Equations 14.16 and 14.17 is perfectly valid, except that difficulties arise when measurement of T_i is attempted. However a relationship between bulk and interfacial temperatures is easily obtained by dividing Equation 14.16 by Equation 14.17. For steady state

$$\frac{T_A - T_i}{T_B - T_i} = -\left(\frac{h_B A_B}{h_A A_A}\right) \qquad (14.19)$$

Both h_B and h_A can be calculated by suitable correlations found in Chapter 13, and T_i can be determined by

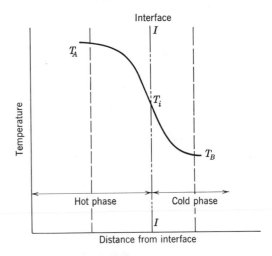

Figure 14.3. Temperature conditions in the heat exchange.

solving Equation 14.19 with $A_A = A_B$, since for well-defined boundaries the interfacial areas are common for both phases. The interfacial temperature, T_i, can also be determined graphically by plotting the usual equilibrium diagram. In this case the hot-phase temperature will be plotted against the cold-phase temperature with equilibrium being represented by a 45-degree line. Figure 14.4 shows this plot.

In Figure 14.4, point P gives the temperatures corresponding to the bulk conditions of both phases at the point of interest. This point would correspond to a point on the operating line relating the temperature history of the two phases throughout the exchanger. Equation 14.19 is plotted through point P; it is an equation of a straight line of slope $-(h_B A_B/h_A A_A)$. The intersection of this line with the equilibrium curve determines the interfacial temperature T_i. Figure 14.4 serves to demonstrate a concept that will be useful when considering series resistances in mass transfer. Equation 14.18 or a more complete form including *all* resistances is appropriate for studying heat transfer.

A practical case of heat exchange might be a furnace wall consisting of several layers of brick of various resistances. On one side of the wall will be hot gases, while on the other will be ambient air. In this situation both conduction and convection occur. There is the turbulent transfer by convection to and from both sides of the wall, and there is a molecular transfer, or conduction, through the wall. Both mechanisms occur simultaneously. Figure 14.5 illustrates this physical situation.

In this case there are three different types of brick making up the furnace wall. The brick layers are Δx_2, Δx_3, Δx_4 feet thick, and each layer will have a thermal conductivity dependent upon the material of construction and its temperature. The solid line represents the temperature profile through the various resistances, assuming constant thermal conductivity within each kind of brick.

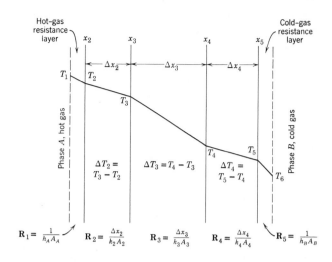

Figure 14.5. Series resistances in a furnace wall.

Applying the development used in Equation 14.7 for heat transfer, (ψA) is equal to q, and

$$-\Delta T_1 = q_A \left(\frac{1}{h_A A_A} \right) \qquad (14.20)$$

The heat-transfer coefficient is used in Equation 14.20 as an expression for the resistance to heat transfer offered by the fluid. Continuing across the wall, assuming intimate contacting at the brick interfaces,

$$-\Delta T_2 = q_2 \left(\frac{\Delta x_2}{k_2 A_2} \right) \qquad (14.21)$$

$$-\Delta T_3 = q_3 \left(\frac{\Delta x_3}{k_3 A_3} \right) \qquad (14.22)$$

$$-\Delta T_4 = q_4 \left(\frac{\Delta x_4}{k_4 A_4} \right) \qquad (14.23)$$

In these three equations, the transfer process is only molecular and the transfer path length is known. Finally across the outer resistance layer

$$-\Delta T_5 = q_B \left(\frac{1}{h_B A_B} \right) \qquad (14.24)$$

In these equations the symbols are:

ΔT = temperature driving force, °F across each resistance
q = heat-transfer rate, Btu/hr
k = thermal conductivity, Btu/hr sq ft (°F/ft)
A = heat-transfer area, sq ft
h = heat-transfer coefficient, Btu/hr sq ft °F

Adding Equation 14.20 through 14.24 gives

$$-(\Delta T_1 + \Delta T_2 + \Delta T_3 + \Delta T_4 + \Delta T_5) = q_A \frac{1}{h_A A_A}$$

$$+ q_2 \frac{\Delta x_2}{k_2 A_2} + q_3 \frac{\Delta x_3}{k_3 A_3} + q_4 \frac{\Delta x_4}{k_4 A_4} + q_B \frac{1}{h_B A_B} \qquad (14.25)$$

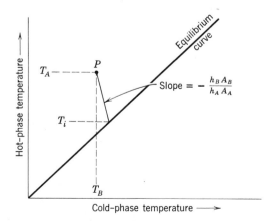

Figure 14.4. Equilibrium diagram for heat exchange.

Now, at steady state, all the heat passing through R_1 must pass through R_5 so $q_A = q_2 = \ldots = q$. Therefore Equation 14.25 may be written as

$$-\Sigma\Delta T = q\left[\frac{1}{h_A A_A} + \frac{\Delta x_2}{k_2 A_2} + \frac{\Delta x_3}{k_3 A_3} + \frac{\Delta x_4}{k_4 A_4} + \frac{1}{h_B A_B}\right]$$
$$(14.26)$$

or

$$-\Sigma\Delta T = q[R_1 + R_2 + R_3 + R_4 + R_5] = q\Sigma R \quad (14.27)$$

and the heat-transfer rate is

$$q = \frac{-\Sigma\Delta T}{\Sigma R} \quad (14.28)$$

Equation 14.28 is, of course, identical with Equation 14.7. Re-examination of the development of Equation 14.28 shows that

$$q = \frac{-\Sigma\Delta T}{\Sigma R} = \frac{-\Delta T_1}{R_1} = \frac{-\Delta T_2}{R_2} = \frac{-\Delta T_3}{R_3}$$
$$= \frac{-\Delta T_4}{R_4} = \frac{-\Delta T_5}{R_5} \quad (14.29)$$

Illustration 14.1. A flat furnace wall consists of a 6-in. layer of fire brick [$k = 0.95$ Btu/hr sq ft ($°$F/ft)], 4 in. of insulating brick ($k = 0.14$), and 4 in. of common brick ($k = 0.8$).

The fire-wall temperature is 1800°F, and the outer surface temperature is 120°F. Determine the heat loss through the wall and the temperature at the junctions between the different types of brick.

SOLUTION. Equation 14.28 clearly applies here. Since surface temperatures are given, there is no need to calculate the thermal resistance of the layer of gas on either side of the wall.

Basis: 1 sq ft of transfer area

$$\text{Resistance of fire brick} = \frac{\Delta x_2}{k_2 A} = \frac{(6/12)}{(0.95)(1.0)}$$
$$= 0.527 \text{ hr } °\text{F/Btu}$$

$$\text{Resistance of insulating brick} = \frac{\Delta x_3}{k_3 A} = \frac{(4/12)}{(0.14)(1.0)}$$
$$= 2.38 \text{ hr } °\text{F/Btu}$$

$$\text{Resistance of common brick} = \frac{\Delta x_4}{k_4 A} = \frac{(4/12)}{(0.8)(1.0)}$$
$$= 0.417 \text{ hr } °\text{F/Btu}$$

Therefore $\Sigma R = 0.527 + 2.38 + 0.417 = 3.32$ hr °F/Btu

and $\quad -\Sigma\Delta T = -(120 - 1800) = 1680°$F

Thus $\quad q = \frac{-\Sigma\Delta T}{\Sigma R} = \frac{1680}{3.32} = 506$ Btu/hr sq ft

Equation 14.29 is used to evaluate junction temperatures.

$$q = \frac{-\Sigma\Delta T}{\Sigma R} = \frac{-\Delta T_2}{R_2} = \frac{-\Delta T_3}{R_3} = \frac{-\Delta T_4}{R_4}$$

The temperature at the junction of the fire brick and insulating brick is determined.

$$q = -\frac{\Delta T_2}{R_2}$$
$$506 = -\frac{\Delta T_2}{0.527}$$

or

$$-\Delta T_2 = 0.527(506) = 267°\text{F}$$

which gives a temperature of $1800 - 267 = 1533°$F.

For the interface between the insulating brick and common brick

$$-\frac{\Delta T_2}{R_2} = -\frac{\Delta T_3}{R_3}$$

or

$$\frac{267}{0.527} = \frac{\Delta T_3}{2.38}$$

and

$$\Delta T_3 = 2.38(506) = 1205°\text{F}$$

which gives a junction temperature of $1533 - 1205 = 328°$F. This temperature could also have been calculated using the ΔT across both the fire brick and insulating brick and the sum of the resistances associated with both these materials.

This illustration was for a case of constant heat flow through a series of resistances of constant area. If the transfer area varies, the appropriate area must be used. (See Chapter 10.)

Over-all Heat-Transfer Coefficient. A glance at Figure 14.5 and Equation 14.29 shows that the heat-transfer rate for a steady-state process can be calculated if the temperatures at the two boundaries of any one resistance are known. Normally the bulk stream temperatures are the easiest to measure. The temperature difference between the two bulk phases will represent the temperature driving force across *all* resistances encountered between the two phases. This ΔT is referred to as the *total* driving force. Accordingly, an over-all coefficient may be defined as a function of the total resistance and transfer area as $1/UA = \Sigma R$, where U is the over-all heat-transfer coefficient, Btu/hr sq ft °F, and A is the transfer area consistent with the definition of U.

For example, the area term associated with the over-all heat-transfer coefficient may be the transfer area of either the hot or the cold phase. In a heat exchanger in which the two phases are separated by a tube wall, the outside surface area of the tube would be the transfer area of, say, the hot phase. On the other hand, the inside surface area of the tube would be the transfer area of the cold phase.

Since the total resistance is fixed in a given system, each area would have an unique numerical value of the over-all coefficient as

$$\Sigma R = \frac{1}{UA} = \frac{1}{U_A A_A} = \frac{1}{U_B A_B} = \frac{1}{h_B A_B}$$
$$+ \sum_{j=1}^{j=n} \frac{\Delta x_j}{k_j A_j} + \frac{1}{h_A A_A} \quad (14.30)$$

where U = over-all heat-transfer coefficient, Btu/hr sq ft °F

A_A or A_B = transfer area of hot or cold phase, sq ft

h_B = heat-transfer coefficient for cold phase, Btu/hr °F sq ft transfer area of cold phase

h_A = heat-transfer coefficient for hot phase, Btu/hr °F sq ft transfer area of hot phase

$\sum_{j=1}^{j=n} \frac{\Delta x_j}{k_j A_j}$ = sum of all resistances associated with the molecular transfer of heat, hr ft² °F/Btu

The heat-transfer equation may now be written as

$$q = -\frac{\Delta T}{\dfrac{1}{U_A A_A}} = -\frac{\Delta T}{\dfrac{1}{U_B A_B}} = -\frac{\Delta T}{\dfrac{1}{UA}} = \frac{\text{total driving force}}{\text{total resistance}}$$

$$(14.31)$$

where ΔT is the *total* temperature drop between the two phases.

By examination of Equation 14.30, some qualitative conclusions may be drawn regarding the relative effect of the various resistances encountered along the transfer path. Consider a heat-transfer apparatus where two fluids, separated by a tube wall, are exchanging heat. In this case Equation 14.30 reduces to

$$\frac{1}{UA} = \frac{1}{U_A A_A} = \frac{1}{U_B A_B} = \frac{1}{h_A A_A} + \frac{\Delta x_w}{k_w A_m} + \frac{1}{h_B A_B}$$
$$(14.32)$$

where A_m is the appropriately averaged mean of A_A and A_B. First, consider the pipe wall to be very thin; then, Δx_w approaches zero and the tube resistance is negligible, and with negligible wall resistance

$$\frac{1}{U_A A_A} = \frac{1}{U_B A_B} = \frac{1}{h_A A_A} + \frac{1}{h_B A_B} \quad (14.33)$$

but for a thin wall $A_A \cong A_B = A$, thus

$$\frac{1}{U} \cong \frac{1}{h_A} + \frac{1}{h_B} \quad (14.34)$$

Depending upon the magnitudes of h_A and h_B the resistances at both surfaces may be significant. The value of h may vary between about 2.0 Btu/hr sq ft °F and 40,000 or more, depending upon the nature of the heat-transfer process. For example, in a heat exchanger

using condensing steam to heat water, the magnitude of the steam coefficient is about 2000, whereas that of the flowing water might be about 400. From this one can deduce, using Equation 14.34, that the water-side resistance is a major portion of the total resistance. It is usual practice to base the over-all coefficient on the area associated with the greatest resistance. In general, however, before any simplification of Equation 14.30 can be made, each resistance should be analyzed individually and in comparison to the over-all resistance before it may be said to be negligible. Correlations found in Chapter 13 will permit the evaluation of the individual heat-transfer coefficients.

Illustration 14.2. A double pipe heat exchanger consists of a 1 in., 18 BWG copper tube inside a 2 in. Sch. 40 steel pipe.* Water† is flowing through the inner tube, while in the annular space lube oil flows countercurrent to the water. At a particular point in the exchanger the bulk temperature of the oil is 350°F, while the bulk water temperature at the same point is 95°F. The heat-transfer coefficients at this point are 100 Btu/hr sq ft °F for the oil and 400 for the water. Calculate the heat-transfer rate at this point. What is the temperature at the outer surface of the copper tube?

SOLUTION. The heat-transfer rate is found by Equation 14.28 and since bulk fluid temperatures are given the total driving forces and the total resistances must be employed.

$$\text{Total driving force} = 95 - 350 = -255°F$$

$$\text{Total resistance} = \frac{1}{UA}$$

From Equation 14.32

$$\frac{1}{U_A A_A} = \frac{1}{U_B A_B} = \frac{1}{h_A A_A} + \frac{\Delta x}{k A_m} + \frac{1}{h_B A_B}$$

Basis: 1 ft of tube length

The transfer area may be calculated as follows:
For the cold phase:

$$A_B = \pi D_B L = (3.14)(0.902/12)(1) = 0.236 \text{ sq ft}$$

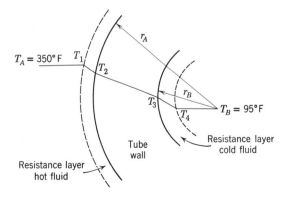

Figure 14.6. Sketch for Illustration 14.2.

---

* See Appendix C for dimensions of pipes and tubes.

† See Appendix D for physical properties of fluids and metals.

For the hot phase:

$$A_A = \pi D_A L = (3.14)(1/12)(1) = 0.262 \text{ sq ft}$$

For the tube wall:

$$A_{am} = \frac{0.236 + 0.262}{2} = 0.249 \text{ sq ft}$$

The arithmetic mean A_{am} area is justified as opposed to the use of A_{lm} in this case since the tube is thin-walled and the inside and outside areas are nearly equal.

$$\text{Tube wall thickness} = \frac{0.049}{12} = 0.00408 \text{ ft}$$

Thermal conductivity of copper = 215 Btu/hr sq ft (°F/ft)

Therefore $\dfrac{1}{UA} = \dfrac{1}{(400)(0.236)} + \dfrac{0.00408}{(215)(0.249)} + \dfrac{1}{(100)(0.262)}$

It should be evident from the above equation that the resistance of the copper wall is negligible for two reasons: first, it is thin, and, second, copper has a high thermal conductivity. Also by examination, both oil phase and water phase resistances are of the same order of magnitude.

Solving for the total resistance

$$\frac{1}{UA} = 0.0106 + 0.0000762 + 0.0382 = 0.0488 \text{ hr } °F \text{ sq ft/Btu}$$

Thus $\qquad q = \dfrac{-\Delta T}{\dfrac{1}{UA}} = \dfrac{255}{0.0488} = 5230 \text{ Btu/hr}$

The temperature of the outer surface of the copper tube may be calculated by using the single resistance associated with the oil film as in Equation 14.29.

$$5230 = \frac{-(T_2 - T_1)}{0.0382}$$

$$-\Delta T = (5230)(0.0382) = 200°F$$

or

$$T_2 = 350 - 200 = 150°F$$

This problem is solved in a manner identical with the previous illustration.

MASS TRANSFER

Industrial mass-transfer operations involve the contacting of two phases in various apparatus. Bubble- and sieve-tray columns, and cascade, packed and spray towers, are frequently used. Wetted-wall columns have the serious disadvantage of small transfer area; the equipment employing fine bubbles or sprays is much more practical. Whatever the mode of contacting, two dissimilar phases are brought together for transferring mass across the interface between the two phases.

One phase usually flows countercurrent to the other phase, with the phases in contact such that an interface exists between them. The solutes being transferred from phase to phase must accordingly pass through this interface. Because of the nature of the contacting apparatus the length of the transfer path or time of contact cannot be accurately established. Indeed, although a well-defined interface exists it is far from explicit in its geometry. However, it can be postulated that this interface is analogous to the tube wall in heat transfer; both phases are kept separated and on either side of their boundary the transfer-resistance layers are established. It is in these layers that the largest portion of the resistance to mass transfer is found.

To establish useful concepts in regard to resistances and driving forces in a mass-transfer operation, consider the process of gas absorption. Here a gas stream containing one or more transferable solutes is contacted with a solute-free liquid or a liquid containing little solute. Mass is transferred from the gas to the liquid at every point along the gas–liquid interface at a rate depending upon the driving force and resistance at each point. For this operation, assume steady state, simple transfer, and that true thermodynamic equilibrium exists at the interface and the interface itself offers no resistance to mass transfer. The equilibrium condition is in some manner common to both phases.

It should be pointed out that noticeable inconsistencies in some of the assumptions exist. For example, in a contacting device packed with ceramic rings, the liquid phase usually flows down each piece of packing in laminar flow, so that the entire liquid phase should offer resistance to transfer. The liquid phase is throughly mixed as it passes to the next piece of packing. In a stage contactor, the gas phase is usually in laminar flow, and, because of a relatively short flow path, well-developed turbulent velocity profiles do not exist. Furthermore, there is some evidence (1) that equilibrium does not exist at the interface. Despite these limitations the model is useful, and data analyzed according to the above assumptions are extendible to situations which may not necessarily fit the working model.

The problem is to relate the series of resistances encountered between the two bulk phases with the driving force across these resistances for the stated absorption process. At a particular point in the absorption tower the solid lines of Figure 14.7 illustrate the concentration variation from the bulk gas phase to the bulk liquid phase. Line *I–I* represents the interface between the two phases; and the dotted lines indicate the extent of the effective resistances. As shown in Figure 14.7 the liquid- and gas-phase solute concentrations have unrelated units. Obviously the concentration must have a continuous profile, and, if the solute concentration were expressed in terms of fugacity for the liquid and gas phases, the potential curve would be a continuous function in Figure 14.7 since the fugacity has the same significance and units for both phases. However, useful relationships can be developed without the concept of fugacity.

It is evident from Figure 14.7 that the transfer path

is through two resistances in series so that Equation 14.7 applies. Because the transfer mechanism is a combination of molecular and turbulent processes whose relative effects cannot be predicted, it is necessary to employ the transfer-coefficient concept.

The rate of solute transfer from the bulk gas phase to the interface is

$$N_a = -\frac{(p_{ai} - p_{aG})}{\dfrac{1}{Ak_G}} \qquad (14.35)$$

and the rate of transfer from the interface to the bulk liquid is

$$N_a = -\frac{(c_{aL} - c_{ai})}{\dfrac{1}{Ak_L}} \qquad (14.36)$$

where N_a = rate of solute transfer, lb moles/hr
$\quad\ p_a$ = solute partial pressure in gas phase, atm
$\quad\ c_a$ = solute concentration in liquid phase, lb moles/cu ft of liquid
$\quad\ A$ = transfer area, sq ft
$\quad\ k_G$ = mass-transfer coefficient for gas phase, lb moles/hr sq ft atm
$\quad\ k_L$ = mass-transfer coefficient for liquid phase, lb moles/hr sq ft (lb mole/cu ft)
subscripts G = gas phase
$\qquad\quad\ i$ = interface
$\qquad\quad\ L$ = liquid phase

It is important to re-emphasize the units of the mass-transfer coefficients. Note once again that the gas-phase coefficient is in terms of *gas-phase driving forces* and the liquid-phase coefficient is in terms of *liquid-phase driving forces*.

For steady-state transfer the two rates given by Equations 14.35 and 14.36 must be equal, or

$$N_a = -\frac{(p_{ai} - p_{aG})}{\dfrac{1}{Ak_G}} = -\frac{(c_{aL} - c_{ai})}{\dfrac{1}{Ak_L}} \qquad (14.37)$$

Equation 14.37 is of the form: rate is equal to the driving force divided by the resistance. Each side of the equation is made up of a single resistance and a single driving force. To use Equation 14.37, therefore, requires the interfacial compositions which are exceedingly difficult to obtain experimentally. However, bulk-phase solute concentrations are easily and reliably measured. Here then is an excellent case for the use of over-all driving forces and resistances.

Equations 14.35 and 14.36 may be rearranged as

$$-(p_{ai} - p_{aG}) = N_a \left[\frac{1}{Ak_G}\right] \qquad (14.35a)$$

and

$$-(c_{aL} - c_{ai}) = N_a \left[\frac{1}{Ak_L}\right] \qquad (14.36a)$$

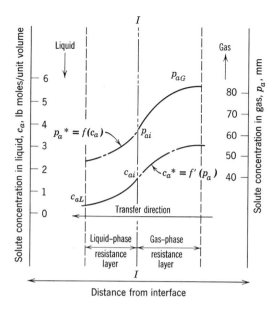

Figure 14.7. Concentration conditions in mass transfer.

Equation 14.35a determines the rate of transfer of solute from the bulk gas phase to the interface, and consequently the driving force is expressed in terms of *gas-phase* concentration units. Partial pressure is a useful gas-phase concentration unit and is solely a a gas-phase phenomenon, relating the concentration of the solute in the gas phase to the total gas-phase pressure. Equation 14.36a also determines the rate of transfer, but here the transfer is from the interface to the bulk liquid. Consequently driving forces are in terms of *liquid-phase* concentration units. Concentration units in the liquid phase employ the two components that make up the liquid, i.e., moles solute per unit volume of liquid.

Since p_a is the concentration of the solute in the gas phase and c_a is the concentration of the solute in the liquid, Equations 14.35a and 14.36a cannot be directly combined because they are expressed in dissimilar units. A relationship for solute concentration at the interface, where the concentrations of both phases are common, is needed before combination of these equations is possible. This relationship is the general phase-equilibrium equation which relates the concentration of the solute in the liquid phase to the concentration of the solute in the gas phase.

The composition of solute in the gas phase *in equilibrium with* liquid of solute composition c_a is

$$p_a^* = f(c_a) \qquad (14.38)$$

where c_a = liquid-phase concentration, moles solute/ unit volume of liquid
$\quad\ p_a^*$ = partial pressure of the solute in the gas phase that is in equilibrium with c_a

For dilute systems this function is frequently linear and becomes

$$p_a{}^* = mc_a \qquad (14.38a)$$

where m is a distribution factor relating $p_a{}^*$ and c_a; it might be pure component vapor pressure as in Raoult's law, or a Henry's law constant, or the slope of the equilibrium curve at c_a.

In the stage operations, two phases were brought together in a stage, mixed, and separated. The two phases that resulted from the separation were said to be in equilibrium. If one phase composition is known, the composition of the other phase in equilibrium could be calculated from an equation such as Equation 14.38a. For the case under study the two phases are not in equilibrium except at the interface. Nevertheless, *an equilibrium equivalent* for each phase may be conceived. For instance, through the use of Equation 14.38 or 14.38a a vapor composition that *would be in equilibrium* with a given liquid composition could be calculated, and this vapor sample could conceivably physically exist if conditions were appropriate. This vapor composition is referred to as the gas-phase equilibrium equivalent, for a stated liquid composition. In continuous-contacting equipment the equilibrium equivalents do not exist in the physical sense, but the compositions may be calculated and plotted on Figure 14.7 as indicated by the dotted lines. The equilibrium-equivalent composition profile now serves to make continuous the composition profile from bulk phase to bulk phase. Although the equilibrium equivalent does not exist, it can be used to relate the solute concentrations anywhere in the gas and liquid phases for particular calculations.

At the interface, by assumption of equilibrium

$$p_{ai} = f(c_{ai}) \qquad (14.39)$$

or, for simplicity, Equation 14.39 may be expressed in terms of the distribution factor as in Equation 14.38a.

$$p_{ai} = mc_{ai} \qquad (14.39a)$$

Multiplying Equation 14.36 by m yields upon rearrangement

$$-(mc_{aL} - mc_{ai}) = N_a \frac{m}{Ak_L} \qquad (14.40)$$

Combining Equations 14.38a and 14.39a with Equation 14.40 results in

$$-(p_a{}^* - p_{ai}) = N_a \frac{m}{Ak_L} \qquad (14.41)$$

Equation 14.41 is now written in terms of equilibrium-equivalent driving forces for the liquid phase. Equation 14.41 may now be added to Equation 14.35a since the driving force in the gas phase is expressed in pressure units and the driving force in the liquid phase is expressed as the pressure equivalent of the liquid-phase composition. Driving force units are now on the same basis. The addition of Equations 14.41 and 14.35a for a constant (N_a) gives

$$-(p_a{}^* - p_{aG}) = N_a \left[\frac{1}{Ak_G} + \frac{m}{Ak_L} \right] \qquad (14.42)$$

or

$$N_a = \frac{-(p_a{}^* - p_{aG})}{\left[\dfrac{1}{Ak_G} + \dfrac{m}{Ak_L} \right]} \qquad (14.43)$$

Equation 14.43 is now expressed in terms of total resistance and total driving force, in gas-phase units.

A similar expression may be written using liquid-phase driving forces,

$$N_a = \frac{-(c_{aL} - c_a{}^*)}{\left[\dfrac{1}{mAk_G} + \dfrac{1}{Ak_L} \right]} \qquad (14.44)$$

Either Equation 14.43 or 14.44 is appropriate for calculating mass-transfer rates. The choice depends upon the information at hand.

Over-all Mass-Transfer Coefficient. In the same manner as the total resistance for heat transfer, a total mass-transfer resistance may be defined with common areas A, as

$$\frac{1}{AK_G} = \frac{1}{Ak_G} + \frac{m}{Ak_L} \qquad (14.45)$$

and

$$\frac{1}{AK_L} = \frac{1}{mAk_G} + \frac{1}{Ak_L} \qquad (14.46)$$

where K_G = over-all gas-phase mass-transfer coefficient, lb moles/hr sq ft atm.

K_L = over-all liquid-phase mass-transfer coefficient, lb moles/hr sq ft (lb mole/cu ft)

Thus for steady-state transfer,

$$N_a = \frac{-(p_a{}^* - p_{aG})}{\dfrac{1}{AK_G}} = \frac{-(c_{aL} - c_a{}^*)}{\dfrac{1}{AK_L}}$$
$$= \frac{-(p_{ai} - p_{aG})}{\dfrac{1}{Ak_G}} = \frac{-(c_{aL} - c_{ai})}{\dfrac{1}{Ak_L}} \qquad (14.47)$$

The driving forces and resistances of Equation (14.47) may be visualized by referring to Figure 14.8. This diagram is analogous to that for heat transfer given in Figure 14.4.

In this diagram point M represents the concentration of solute in the bulk phases at the particular point under study. The gas-phase composition is p_{aG} and the liquid-phase composition is c_{AL}. The relationship between bulk compositions and interfacial compositions is obtained in a manner identical to that for the heat-transfer case. Rearranging the interfacial portions of Equation 14.47 yields

$$\frac{p_{aG} - p_{ai}}{c_{aL} - c_{ai}} = -\frac{Ak_L}{Ak_G} = -\frac{k_L}{k_G} \quad (14.48)$$

The reader is referred to Equation 14.19 for comparison. In the usual mass-transfer equipment both phases are in intimate contact, so that the transfer areas for each phase may be assumed equal. Equation 14.48 plotted through point M locates the interfacial compositions p_{ai} and c_{ai} at point D. It must be emphasized that point D corresponds only to bulk conditions (M), since concentrations will vary throughout the apparatus. Using Figure 14.8 and Equation 14.47, the following can be observed:

In terms of gas phase units:
1. The driving force from bulk gas to interface = $(p_{ai} - p_{aG})$.
2. The driving force from interface to bulk liquid = $(p_a^* - p_{ai})$.
3. The total driving force from bulk gas to bulk liquid = $(p_a^* - p_{aG})$.

In terms of liquid phase units:
1. The driving force from bulk gas to interface = $(c_a^* - c_{ai})$.
2. The driving force from bulk liquid to interface = $(c_{aL} - c_{ai})$.
3. The total driving force from bulk gas to bulk liquid = $(c_a^* - c_{aL})$.

Any driving force along with the appropriate resistance may be used to determine the transfer rate.

From Equation 14.47

$$\frac{p_{ai} - p_{aG}}{p_a^* - p_{aG}} = \frac{\dfrac{1}{k_G}}{\dfrac{1}{K_G}} \quad (14.49)$$

or

$$\frac{\text{Partial-pressure driving force in gas phase}}{\text{Over-all driving force in partial-pressure units}}$$
$$= \frac{\text{resistance in gas phase}}{\text{total resistance}}$$

and

$$\frac{c_{aL} - c_{ai}}{c_{aL} - c_a^*} = \frac{\dfrac{1}{k_L}}{\dfrac{1}{K_L}} \quad (14.50)$$

or

$$\frac{\text{Concentration driving force in liquid phase}}{\text{Over-all driving force in concentration units}}$$
$$= \frac{\text{resistance in liquid phase}}{\text{total resistance}}$$

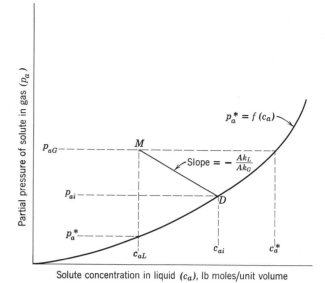

Figure 14.8. Phase relationships for mass transfer.

Equations 14.49 and 14.50 are analogous to the equations for heat transfer through series resistances. The ratio of the terms of Equations 14.49 and 14.50 will be constant only if the equilibrium curve is a straight line. Consequently the ratio of individual phase resistance to total resistance may vary from point to point throughout the unit. It is possible for the transfer resistance to change from the gas phase being the major resistance to the liquid phase being dominant between the terminals of the apparatus.

Some qualitative conclusions regarding the relative magnitude of the phase resistances and their influence on the total resistance may be made by using Equations 14.49 and 14.50 and a constant value of k_L/k_G.

Consider a case where the solute in the gas phase is very soluble in the liquid phase such that small changes of solute in the gas phase will produce large changes of solute concentration in the liquid phase. For this example, the equilibrium curve and bulk composition are illustrated in Figure 14.9.

Again, let M represent the concentration of the bulk phases at the particular point under study, with the interfacial and equilibrium-equivalent concentrations also shown. For this case, if Equation 14.39a is valid, m is relatively small, and from observation of Figure 14.9 as m becomes smaller and smaller or, in the limit, as $m \to 0$,

$$(p_{ai} - p_{aG}) \to (p_a^* - p_{aG}) \quad (14.51)$$

or from Equation 14.49

$$\frac{1}{k_G} \sim \frac{1}{K_G} \quad (14.52)$$

From Equation 14.52 it may be concluded that the

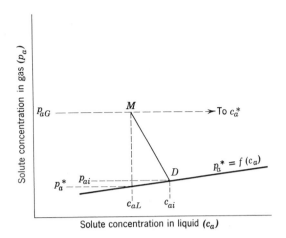

Figure 14.9. Phase relationships for soluble solute.

ratio of the gas phase resistance to the total resistance is approximately unity and the gas phase may be said to be the major resistance. The same conclusion may be reached by examination of the liquid-phase driving forces. From Figure 14.9

$$(c_{aL} - c_{ai}) \ll (c_{aL} - c_a{}^*) \qquad (14.53)$$

or from Equation 14.50

$$\frac{1}{k_L} \ll \frac{1}{K_L} \qquad (14.54)$$

and $(1/k_L)/(1/K_L)$ is very small.

Now consider the case where the solute in the gas phase is relatively insoluble in the liquid phase. Figure 14.10 shows the phase behavior of such a system. Once again, the same point conditions of bulk-, inter-facial-, and equilibrium-equivalent concentration are plotted.

Using the liquid-phase driving forces,

$$(c_{aL} - c_{ai}) \rightarrow (c_{aL} - c_a{}^*) \qquad (14.55)$$

or by Equation 14.50

$$\frac{1}{k_L} \sim \frac{1}{K_L} \qquad (14.56)$$

The conclusion reached from Equation 14.56 is that the liquid-phase resistance approximates the total resistance. In this case, the liquid phase is said to be the dominant resistance, or, traditionally, the "controlling" resistance.

To illustrate an earlier statement that the "controlling" resistance may change from liquid to gas phase, consider Figure 14.11. This curve is a combination of the phase diagrams of Figure 14.9 and Figure 14.10.

At the bottom of an absorption tower, for example, the gas and liquid streams passing will be rich in solute. This condition is shown at point M. From the previous

discussion, at this point in the unit the liquid-phase resistance is controlling. On the other hand, at the top of the column both streams are lean in solute concentration, as illustrated at point E. This is the exact situation as depicted in Figure 14.9 and the same conclusions derived from it can be made.

To summarize the conclusions regarding the over-all mass-transfer coefficient, it may be stated that just as in heat transfer, the over-all coefficient is based on the sum of the resistances, and, if the gas-phase transfer resistance is dominant, K_G may be used. If the liquid-phase resistance is dominant, K_L is used. The emphasis for deciding which phase is controlling appears to have been placed upon the value of the characteristic of the equilibrium relations. This is not the whole story however. In Figures 14.9, 14.10, and 14.11, k_L/k_G was held constant for illustration. The relative values of the individual transfer coefficients k_L and k_G are likewise important. The ratio k_L/k_G represents the ratio of the gas-phase resistance to the liquid-phase resistance. In Chapter 13 it was shown that mass-transfer coefficients are functions of flow rates of the liquid and gas phases and physical characteristics of the system, including diffusivity. Consequently, operating conditions can also influence the relative magnitudes of the two phase resistances. More will be said about the use of over-all mass-transfer coefficients in Chapter 16.

The development of the previous concepts pertinent to mass transfer has been accomplished using the operation of gas absorption as the instruction vehicle. The conclusions reached may be extended to other mass-transfer operations using suitable transfer coefficients. Distillation and extraction carried out in continuous-contacting devices are both analyzed using individual and over-all transfer coefficients. It is at times more convenient to express solute concentration in terms other than partial pressure in the gas and moles per unit volume of liquid; again the same principles apply.

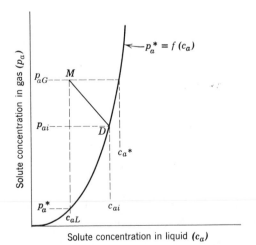

Figure 14.10. Phase relationships for insoluble solute.

Illustration 14.3. Air is sometimes dried by a sulfuric acid absorption process. The humid air flows counter-current to the H_2SO_4 through a column packed with suitable packing material. A test on a unit as described indicates that air enters the bottom of the tower with a relative humidity of 50 per cent and leaves with a relative humidity of 10 per cent. The entering acid at the top of the column is 67 per cent by weight H_2SO_4 and the acid leaving is 53 per cent by weight acid. Experimental evidence indicates that $k_G = 2.09$ lb moles/hr sq ft atm driving force and $k_L = 0.068$ lb moles H_2O/hr sq ft (lb moles/cu ft of liquid). For unity transfer area, a temperature of 25°C, and assuming k_G and k_L to be essentially constant, determine for the terminals of the tower:

a. The instantaneous rates of transfer.

b. The per cent of total diffusional resistance encountered in each phase. Use both gas- and liquid-phase driving forces.

c. The interfacial compositions.

d. The numerical values of K_G and K_L.

SOLUTION. The equilibrium data at 25°C for the H_2SO_4–water system is reported by Wilson (2) and tabulated below.

Per cent rel. humidity	0.8	2.3	5.2	9.8	17.2	26.8	36.8	46.8	56.8
Weight per-cent H_2SO_4	80.0	74.9	70.0	64.9	60.0	55.1	50.0	45.0	40.0

The equilibrium data as tabulated are not in as convenient a form as could be desired. This is frequently the case. For illustrative purposes these data will be converted to partial pressure of water in the gas phase and moles of water per unit volume of liquid. To convert the liquid phase (the calculation of only one point will be demonstrated):

For 60 per cent H_2SO_4: 40 lb H_2O, 60 lb H_2SO_4

Pound moles of water: $40/18 = 2.22$

Specific gravity of a 60 per cent acid = 1.498

Density of a 60 per cent acid = (1.498)(62.3) = 93.5 lb/cu ft

$$c_a = (2.22)\left(\frac{93.5}{100}\right) = 2.08 \text{ lb moles } H_2O/\text{cu ft solution}$$

The equilibrium relative humidity for a 60 per cent acid is 17.2 per cent.

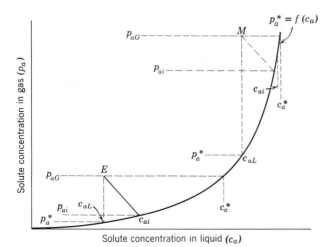

Figure 14.11. Typical phase diagram.

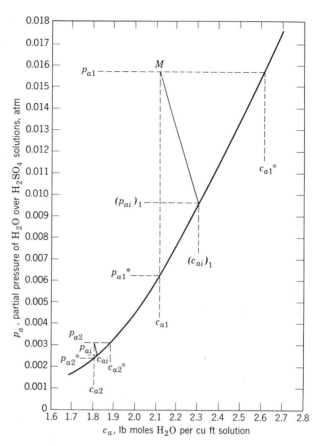

Figure 14.12. Graphical solution for Illustration 14.3.

Vapor pressure of water at 25°C = 23.75 mm

$$\text{Per cent relative humidity} = 100\frac{p_a}{P_a}$$

where p_a = partial pressure of water in air

P_a = vapor pressure of water at bulk air temperature

Thus for this particular acid concentration,

$$p_a = \left(\frac{17.2}{100}\right)\left(\frac{23.75}{760}\right) = 0.00538 \text{ atm}$$

The above calculations are repeated using the tabulated equilibrium data. The recalculated data are as follows and are plotted on Figure 14.12.

p_a^*, atm	0.000248	0.000721	0.001626	0.00305	0.00853	0.0115	0.01464	0.01775
c_a, lb moles H_2O/cu ft soln	1.1780	1.451	1.670	1.88	2.25	2.41	2.57	2.71

Entering acid, $c_{a2} = \left(\frac{33}{18}\right)\left(\frac{1}{100}\right)(1.576)(62.3)$

$$= 1.802 \text{ lb moles } H_2O/\text{cu ft solution}$$

Exit acid, $c_{a1} = \left(\frac{47}{18}\right)\left(\frac{1}{100}\right)(1.302)(62.3)$

$$= 2.12 \text{ lb moles } H_2O/\text{cu ft solution}$$

Entering air, $p_{a1} = \left(\frac{50}{100}\right)\left(\frac{23.75}{760}\right) = 0.01565 \text{ atm}$

Exit air, $p_{a2} = \left(\frac{10}{100}\right)\left(\frac{23.75}{760}\right) = 0.00313 \text{ atm}$

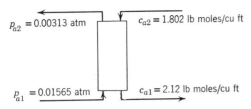

Figure 14.13. Sketch for Illustration 14.3.

For the bottom of the absorption tower

$$p_{a1} = 0.01565 \text{ atm}$$

$$p_{a1}* = 0.0062 \text{ atm} \quad \text{(From Figure 14.12)}$$

Thus the total driving force (gas units) $= 0.01565 - 0.0062 = 0.00945$ atm. From the statement of the problem and the use of Equation 14.48, $-k_L/k_G = 0.068/2.09 = 0.0325$. This is the slope of the line relating bulk phase conditions to those at the interface.

(*a*) The instantaneous rate of mass transfer is calculated using Equation 14.35 and obtaining p_{ai} from Figure 14.12.

$$N_a = \frac{-(p_{ai} - p_{aG})}{1/Ak_G} = \frac{-(0.00965 - 0.01565)}{1/(2.09)(1)}$$

$$= 0.0126 \text{ lb moles/hr sq ft}$$

Using Equation 14.36 should give the identical rate, or

$$N_a = -\frac{(c_{aL} - c_{ai})}{1/Ak_L} = -\frac{(2.12 - 2.315)}{1/(0.068)(1)}$$

$$= 0.0126 \text{ lb moles/hr sq ft}$$

(*b*) The fractional resistance across each phase is calculated from Equations 14.49 and 14.50 with the aid of Figure 14.12.

In gas-phase units

$$\text{Gas-phase resistance} = p_{aG} - p_{ai}$$
$$= 0.01565 - 0.00965$$
$$= 0.006 \text{ atm}$$

$$\text{Liquid-phase resistance} = p_{ai} - p_a*$$
$$= 0.00965 - 0.0062$$
$$= 0.00345 \text{ atm}$$

$$\text{Total resistance, both phases} = p_{aG} - p_a*$$
$$= 0.00945 \text{ atm}$$

Thus,

$$\frac{\text{Gas-phase resistance}}{\text{Total resistance}} = \frac{0.006}{0.00945} = 0.635$$

and

$$\frac{\text{Liquid-phase resistance}}{\text{Total resistance}} = 0.365$$

In liquid-phase units

$$\text{Liquid-phase resistance} = c_{ai} - c_a$$
$$= 2.305 - 2.12$$
$$= 0.185 \text{ lb moles/cu ft}$$

$$\text{Gas-phase resistance} = c_a* - c_{ai}$$
$$= 2.62 - 2.305$$
$$= 0.315 \text{ lb moles/cu ft}$$

$$\text{Total resistance, both phases} = (c_a* - c_a)$$
$$= 0.500 \text{ lb moles/cu ft}$$

Thus,

$$\frac{\text{Gas-phase resistance}}{\text{Total resistance}} = \frac{0.315}{0.500} = 0.630$$

and

$$\frac{\text{Liquid-phase resistance}}{\text{Total resistance}} = 0.370$$

One would expect these fractional resistances to be the same regardless of which set of driving forces is employed. The slight difference in this instance is due to errors in reading values from Figure 14.12 and to the curvature of the equilibrium line.

(*c*) The interfacial compositions have already been used and are (by Figure 14.12)

$$p_{ai} = 0.00965 \text{ atm}$$

$$c_{ai} = 2.305 \text{ lb moles H}_2\text{O/cu ft solution}$$

(*d*) The over-all coefficients K_G and K_L can be evaluated by several methods.

(1) From Equation 14.47

$$K_G = -\frac{N_a}{A(p_a* - p_{aG})} = \frac{0.0126}{1 \times 0.00945}$$

$$= 1.335 \text{ lb moles H}_2\text{O/hr sq ft }(\Delta p)$$

and

$$K_L = -\frac{N_a}{A(c_a - c_a*)} = \frac{0.0126}{1 \times 0.5}$$

$$= 0.0252 \text{ lb moles H}_2\text{O/hr sq ft }(\Delta c)$$

(2) An attempt to use Equation 14.45 and 14.46 for obtaining K_G and K_L will be made. Since a point slope at (c_{ai}, p_{ai}) must be used, good agreement is not expected.

$$m = 0.01955 \text{ (slope at } c_{ai} \text{ and } p_{ai})$$

From Equation 14.45

$$\frac{1}{K_G} = \frac{1}{k_G} + \frac{m}{k_L} = \frac{1}{2.09} + \frac{0.01955}{0.068} = 0.766$$

and $K_G = 1.308$ lb moles H_2O/hr sq ft (Δp).
From Equation 14.46

$$\frac{1}{K_L} = \frac{1}{mk_G} + \frac{1}{k_L} = \frac{1.0}{(0.01955)(2.09)} + \frac{1}{0.068} = 39.1$$

and $K_L = 0.0256$ lb moles/hr sq ft (Δc).

The agreement between, $d(1)$ and $d(2)$ is better than one can usually expect.

Exactly the same procedures as above are used to calculate the conditions at the top of the tower. In this case, only the results will be written.

For the top of the tower

$p_{a2} = 0.00314$ atm	$c_{a2} = 1.802$ lb moles H_2O/cu ft
$p_{ai} = 0.00255$ atm	$c_{ai} = 1.818$ lb moles H_2O/cu ft
$p_a* = 0.00239$ atm	$c_a* = 1.885$ lb moles H_2O/cu ft

(*a*) Instantaneous rate

$$N_a = -\frac{(p_{ai} - p_{aG})}{1/Ak_G} = -\frac{(0.00255 - 0.00314)}{1/(1)(2.09)}$$

$$= 0.001235 \text{ lb moles/hr}$$

(b) Using gas units

$$\frac{\text{Gas-phase resistance}}{\text{Total resistance}} = \frac{0.00314 - 0.00255}{0.00314 - 0.00239} = 0.788$$

Using liquid units

$$\frac{\text{Gas-phase resistance}}{\text{Total resistance}} = \frac{1.885 - 1.818}{1.885 - 1.802} = 0.832$$

(c) $p_{ai} = 0.00255$ atm $c_{ai} = 1.818$ moles/cu ft solution

(d) $K_G = 1.54$ lb moles/hr sq ft (Δp)

$\quad K_L = 0.18$ lb moles/hr sq ft (Δc)

Note how different the rates of absorption are at the tower ends. Examination of Figure 14.12 and a comparison of $(p_a{}^* - p_{aG})$ at the terminals show that the driving force at the bottom of the tower is considerably less than that at the top of the tower. The gas-phase resistance is also higher at the top of the tower than at the bottom.

If conditions with regard to driving forces, resistances, and rates vary from position to position within the tower, some means for accounting for these variations must be found, so that they may be related to the size of the equipment needed to do a specified transfer. This will be considered later in this chapter.

UNSTEADY-STATE TRANSFER BETWEEN PHASES

The general unsteady-state transport equation has been developed in Chapter 11.

$$\frac{\partial \Gamma}{\partial \theta} = \delta \left[\frac{\partial^2 \Gamma}{\partial x^2} \right] \qquad (11.16)$$

Equation 11.16 has been solved for several simple geometries and boundary conditions in terms of dimensionless parameters (see Table 11.1). In Chapter 11 transfer was limited to a single phase where the condition at the phase interface was known. In this case the relative-resistance term (**m**) was equal to zero. For **m** to be zero requires a negligible surface resistance to transfer, corresponding to an infinite value of the surface coefficient. In the practical case of either heat or mass transfer the values of the surface coefficients are finite, and the surface temperature or concentration is different from that of the surroundings.

Knowledge of surface resistances now makes it possible to solve Equation 11.16 in terms of more practical boundary conditions. For geometries discussed in Chapter 11 having a finite surface resistance, corresponding to a definite value of ε, the boundary conditions become: $\Gamma = {}_0\Gamma$ at $\theta = 0$; $\Gamma = \Gamma_2$ at $\theta = \infty$; at steady state

$$\delta \left(\frac{\partial \Gamma}{\partial x} \right)_{x=x_1} = \mathscr{E}(\Gamma_1 - \Gamma_2)$$

Γ_2 is the average concentration of transferent property

in the surroundings.

This interface condition simply states that the rate of transfer from within the solid to the surface must be equal to the rate of transfer from the surface to the surroundings. For heat transfer the interface condition is $k(\partial T/\partial x) = h(T_1 - T_2)$, or the rate of heat transfer by conduction to the surface equals the rate of heat transfer by convection away from the surface. Similarly for mass transfer, the interface condition is $\mathscr{D}(\partial c/\partial x) = k_c(c_{ai} - c_{a2})$.

Illustration 14.2. A cylindrical steel shaft 4 in. in diameter and 8 ft long is heat treated by quenching it in an oil bath. The shaft is heated to a uniform temperature of $1100°F$ and plunged into an oil bath maintained at $300°F$. The surface coefficient of heat transfer is 150 Btu/hr sq ft °F. Calculate the temperature of a point 1 in. from the center after 2 min of quenching time. Use a value of $\alpha = 0.35$ sq ft/hr and $k = 26$ Btu/hr sq ft (°F/ft).

SOLUTION. This illustration is similar to Illustration 11.1 with the exception that the oil bath is maintained at a constant temperature of $300°F$ rather than the shaft surface being held at a constant temperature. Because of the resistance to heat transfer between the oil and steel, the temperature at the surface of the steel will be somewhat higher than the bath temperature.

Figure 11.4 will be used for this solution. For $\theta = 2$ min

$$Y = \frac{T_2 - T}{T_2 - {}_0T} = \frac{300 - T}{300 - 1100}$$

$$X = \frac{\alpha \theta}{x_1{}^2} = \frac{(0.35)(2/60)}{(2/12)^2} = 0.42$$

$$n = \frac{x}{x_1} = \frac{1}{2}$$

$$m = \frac{k}{hx_1} = \frac{26}{(150)(2/12)} = 1.04$$

From Figure 11.4,

$$Y = 0.58 = \frac{300 - T}{300 - 1100}$$

and $\qquad\qquad T = 764°F$

This temperature may be compared with the value of $364°F$ at the same position for the solution to Illustrate 11.1.

THE DESIGN EQUATION

Thus far the transfer of heat or mass has been focused upon point conditions within a transfer apparatus. An understanding of driving forces and resistances at a point permits the evaluation of the rate of transfer only at that point. A need now exists for relating the rate of transfer at all points within the apparatus to the total transfer area required for process equipment. Specifically, it is desired to know how much heating surface is required for a specified heating assignment or how tall a contactor should be for a given separation in mass

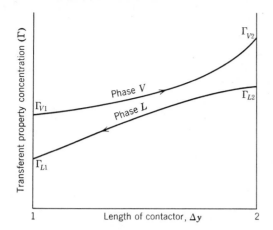

Figure 14.14. Concentration-length profile in process equipment.

transfer. The answer to these questions is best formulated as a design equation which will relate the rate of transfer across the total transfer area to the total change in transferent property. Further details of design will be considered in a later chapter.

Consider two phases in countercurrent flow undergoing an interchange of a particular transferent concentration (Γ). The phases are kept separated either by solubility characteristics as in mass transfer or by solid boundaries as in a heat exchanger. As these phases pass each other between the terminals of the contacting equipment, one phase will become depleted in the quantity of transferent property while the other will become enriched. This is illustrated in Figure 14.14.

In Figure 14.14 the terminals of the contactor are designated as 1 and 2. The *over-all* driving force has been established as being the difference in the concentration of transferent property between the two bulk phases. Note that in Figure 14.14, the over-all driving force is continuously varying from terminal to terminal. It is quite likely that the rate of transfer will also vary along the length of the equipment. Consequently, to calculate a rate of transfer for the entire apparatus will require an integration of the general rate equation. Although a knowledge of the rate of transfer is important, an equally important relationship is needed between the rate and length of contactor, or total transfer area.

Figure 14.15 represents an apparatus which is suitable for the general contacting of phases for exchange of heat or mass. For this apparatus,

V = volumetric flow rate of phase V, cu ft/hr
L = volumetric flow rate of phase L, cu ft/hr
Γ_V = transferent-property concentration in phase V,
    quantity of transferent property per cu ft of V
Γ_L = transferent-property concentration in phase L,
    quantity of transferent property per cu ft of L
subscripts: (1) = tower bottom
            (2) = tower top

Consider a differential section of this unit through which the two phases pass countercurrent to each other. This section will include a transfer area dA. Across this section, phase L will be changed in amount by $d\mathbf{L}$ and in composition by $d\Gamma_L$, and phase V will be changed by quantity $d\mathbf{V}$ and by composition $d\Gamma_V$.

For steady-state operation, a material balance over the differential section yields

$$d\mathbf{V} = d\mathbf{L} \qquad (14.57)$$

and a component balance over the same section

$$d(\Gamma_V \mathbf{V}) = d(\Gamma_L \mathbf{L}) \qquad (14.58)$$

Equation 14.58 is referred to as the operating-line equation, which relates the compositions and flow rates of the two phases passing each other at a particular point in the apparatus. This is the same operating line that appeared in Stage Operations, Part I.

Examination of Equation 14.58 shows that each term is a rate of addition or depletion of transferent property and is really a statement of the rate of transfer of transferent property across the total transfer area. Accordingly, Equation 14.58 must therefore be equal to the total driving force divided by the total resistance. Or, by analogy with Equation 14.1,

$$d(\Gamma_V \mathbf{V}) = d(\Gamma_L \mathbf{L}) = d(\psi A) = \frac{-(\Gamma_V - \Gamma_L)}{1/\mathscr{L} \, dA} \quad (14.59)$$

where $-(\Gamma_V - \Gamma_L)$ = total driving force between
            bulk phases V and L at the
            differential section
         $\mathscr{L}$ = over-all transfer coefficient

The area term in Equation 14.59 is of particular interest since it is a direct measure of the size of the equipment that is needed. Solving Equation 14.59 for dA gives

$$dA = -\frac{d(\Gamma_V \mathbf{V})}{\mathscr{L}(\Gamma_V - \Gamma_L)} = -\frac{d(\Gamma_L \mathbf{L})}{\mathscr{L}(\Gamma_V - \Gamma_L)} = -\frac{d(\psi A)}{\mathscr{L}(\Gamma_V - \Gamma_L)}$$
$$(14.60)$$

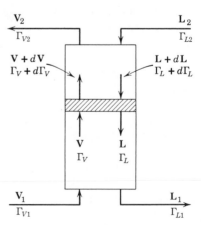

Figure 14.15. Continuous contactor.

Equation 14.60 defines the design equation for operational equipment where the rate of transfer, rather than the amount of enrichment attained at equilibrium dictates the size of equipment. Since the area must be a positive number, Equation 14.60 will adjust itself to account for the transfer proceeding towards a low potential. Integration of Equation 14.60 between the terminals of the apparatus will give the total transfer area required. Thus

$$\int_1^2 dA = -\int_1^2 \frac{d(\Gamma_V \mathbf{V})}{\mathscr{L}(\Gamma_V - \Gamma_L)} = -\int_1^2 \frac{d(\Gamma_L \mathbf{L})}{\mathscr{L}(\Gamma_V - \Gamma_L)}$$
$$= -\int_1^2 \frac{d(\psi A)}{\mathscr{L}(\Gamma_V - \Gamma_L)} \qquad (14.61)$$

The corresponding design equation for heat transfer is

$$\int_1^2 dA = -\int_1^2 \frac{d(\mathbf{V}c_P\rho T)_A}{U(T_A - T_B)} = -\int_1^2 \frac{d(\mathbf{V}c_P\rho T)_B}{U(T_A - T_B)}$$
$$= -\int_1^2 \frac{dq}{U(T_A - T_B)} \qquad (14.62)$$

where $\mathbf{V}_A, \mathbf{V}_B$ = hot and cold phase volumetric flow rates, cu ft/hr

c_{PA}, c_{PB} = hot and cold phase specific heats, Btu/lb °F

ρ_A, ρ_B = hot and cold phase densities, lb/cu ft

T_A, T_B = hot and cold phase temperatures, °R

For mass transfer, Equation 14.61 may be written as

$$\int_1^2 dA = -\int_1^2 \frac{d(\mathbf{V}c_a)}{K(c_a - c_a{}^*)} = -\int_1^2 \frac{d(\mathbf{L}c_a)}{K(c_a - c_a{}^*)}$$
$$\qquad (14.63)$$

where $\mathbf{V}$ = volumetric flow rate of phase V, cu ft/hr

$\mathbf{L}$ = volumetric flow rate of phase L, cu ft/hr

c_a = solute concentration (must be consistent with $\mathbf{V}$ and $\mathbf{L}$), lb moles/cu ft

K = over-all mass-transfer coefficient (must be consistent with $\mathbf{V}$ and $\mathbf{L}$), lb moles/hr sq ft (lb moles/cu ft)

The design equation for momentum transfer takes a different form from Equation 14.61 since momentum transfer is not an interphase phenomenon. Momentum is simply transferred through a fluid to the boundary where a stress is established. The design equation simply becomes a force balance, or

$$d(\Delta P g_c S) = d(\tau_y g_c A) = -d(\mathscr{E}_\tau v \rho A) \qquad (14.64)$$

where S = the cross-sectional area, sq ft

$\mathscr{E}_\tau$ = the momentum-transfer coefficient, ft/hr

A = the stress area, sq ft

ΔP = the pressure applied to S, lb$_f$/sq ft

τ_y = shear stress at the boundary, lb$_f$/sq ft

It is apparent that the solution of the heat- and mass-transfer design equations is quite complex and depends upon evaluation of resistances and driving forces at various points throughout the apparatus. In the general case, integration of Equation 14.62 or 14.63 is likely to be done graphically, since the probability that an analytical relationship exists among all the variables involved is small.

The next few chapters will devote considerable detail to the solution of these design equations along with descriptive material pertinent to an understanding of the process equipment being used.

REFERENCES

1. Schrage, R. W., *A Theoretical Study of Interphase Mass Transfer*, Columbia University Press, New York, 1953.
2. Wilson, R. E, *Ind. Eng. Chem.* **13**, 326 (1921).

PROBLEMS

14.1. The addition of insulation to a pipe increases the resistance of transfer path and hence decreases the heat loss from the hot fluid within the pipe. Under certain conditions the addition of too much insulation may cause an increase in the heat loss since the effect of added heat-transfer surface offsets the advantages of insulation. Show that the optimum thickness of insulation occurs when $r_c = k/h$, where r_c is outside radius of insulation where heat loss increases, h is heat-transfer surface coefficient, and k is thermal conductivity of the insulation. List the simplifying assumptions made.

14.2. Heat is flowing from steam on one side of a vertical steel sheet 0.375 in. thick to air on the other side. The steam heat-transfer coefficient is 1700 Btu/hr sq ft °F and that of the air is 2.0. The total temperature difference is 120°F. How would the rate of heat transfer be affected by making the wall of copper rather than steel? By increasing the steam coefficient to 2500? By increasing the air coefficient to 12.0? Express results in percentages of the rate for present operation.

14.3. If a scale layer 0.003 in. thick is deposited on the steam side of the steel sheet of Problem 14.2, what would be the rate of heat transfer. Thermal conductivity of the scale 0.5 Btu/hr sq ft (°F/ft.)

14.4. A furnace wall exposed to hot gases at 1400°F is constructed of 4 in. of fire brick and 10 in. of common brick. For safety purposes it is desirable to keep the outer surface wall temperature at 160°F in the surroundings at 80°F. Can an additional layer of magnesia insulation be used if it cannot withstand temperatures greater than 650°F? How thick a magnesia layer would you recommend? Assume flame-side heat-transfer coefficient is 12.0 Btu/hr sq ft °F and the room-side coefficient is 4.0.

14.5. Molten bismuth is flowing through a 2 in. I.D. type 304 stainless-steel pipe that is insulated with a 2-in. layer of molded diatomaceous earth ($k = 0.15$). An additional layer of laminated asbestos felt 1 in. thick is to be placed around the pipe and diatomaceous earth insulation to reduce the surface temperature of the complete assembly to 200°F. If asbestos felt is not recommended for use above 500°F, is this addition feasible? The inside surface temperature of the tube is 650°F, and the tube wall is 0.15 in. thick. Use a thermal conductivity for the asbestos felt equal to 0.044 Btu/hr sq ft (°F/ft.).

14.6. Water at an average temperature of 185°F is flowing inside a horizontal 1-in. sch. 40 steel pipe at a velocity of 10 ft/sec. Outside the pipe is saturated steam at 5 psig, condensing in such a manner that the steam heat-transfer coefficient is 2000 Btu/hr sq ft °F. Calculate:

(a) The over-all coefficient based on the water-side surface.

(b) The over-all coefficient based on the steam-side surface.

(c) The per cent of the total temperature drop that takes place across each phase and across the pipe wall.

14.7. In Illustration 14.2, what would be the maximum temperature the oil could reach without having the water boil? Assume the surface coefficients as given are constant.

14.8. A small electric furnace is 6 by 6 by 12 in. inside dimensions and has fire-brick walls ($k = 0.65$) 2 in. thick. The front of the furnace is a movable wall which permits entry into the furnace. In this section is a 2 by 2 by $\frac{1}{4}$ in. quartz observation window ($k = 0.040$). The inner surface temperature for all sides is 1200°F, and the outer surface temperature is 250°F. Assuming all joints perfectly made and neglecting the influence of the corners on the temperature distribution, what is the heat loss from this furnace?

14.9. One wall of a furnace includes a 2 ft by 2 ft by $\frac{1}{4}$ in. steel door. The wall is 8 ft high, 10 ft long and 1 ft thick (except at the door). The wall is constructed of three layers of brick: a 5 in. layer of refractory brick ($k = 2.7$ Btu/hr sq ft °F/ft, a 3-in. layer of insulating brick ($k = 0.15$), and a 4-in. layer of common brick ($k = 0.6$). The heat-transfer coefficient on the fire wall is 4.0 Btu/hr sq ft °F and the outside coefficient is 2.0. The flame temperature is 2200°F, and the room temperature is 80°F. Calculate the total heat loss from the wall and the temperature in the middle of the insulating brick.

14.10. Derive an expression for the over-all heat-transfer coefficient for the following case of a flat furnace wall in the x direction.

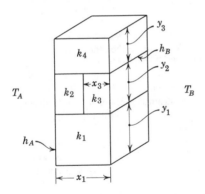

14.11. A vapor mixture of ethyl alcohol and air is separated from nitrogen by means of a membrane permeable only to the alcohol. The alcohol vapor is diffusing at a constant rate through the membrane into the nitrogen area. The air–alcohol mixture analyzes 1 per cent by volume ethyl alcohol, and the nitrogen–alcohol phase analyzes 0.01 per cent alcohol. It is estimated that on both sides of the membrane a 1 mm thickness of gas exists and that the membrane offers no resistance to transfer. A constant temperature of 25°C and a pressure of 1 atm prevails. Calculate (a) the rate of diffusion of ethyl alcohol and (b) the concentration of ethyl alcohol at the membrane.

14.12. Whitney and Vivian [*Chem. Eng. Prog.*, **45**, 323 (1949)] studied the absorption of SO_2 in water in a column packed with 1-in. ceramic rings. Their results show that

$$k_c a = 0.044 G_L^{0.82}$$
$$k_G a = 0.028 G_V^{0.7} G_L^{0.25}$$

where G_L = liquid rate, lb/hr sq ft of tower cross section

 G_V = gas rate, lb/hr sq ft of tower cross section

 a = sq ft of interfacial area/cu ft of packing

 $k_c a$ = liquid-phase surface coefficient, lb mole/hr cu ft (lb moles/cu ft)

 $k_G a$ = gas-phase surface coefficient, lb mole/hr cu ft atm

$$\left[\text{Note in this case, } \frac{1}{K_G a} = \frac{1}{k_G a} + \frac{m}{k_L a} \right]$$

An air–SO_2 mixture is to be contacted countercurrently with water at 20°C and 760 mm total pressure under such conditions that the findings of Whitney and Vivian are valid. The gas flow rate is to be 200 lb/hr sq ft and the water rate is 6000 lb/hr sq ft. At a particular point in the absorber the gas phase analyzes 0.12 mole fraction SO_2 while the liquid phase is 0.002 mole fraction SO_2.

(a) Determine the local mass-transfer rate.

(b) What per cent of the resistance to mass transfer is in the gas phase?

(c) Repeat parts *a* and *b* for the case where the gas mass velocity is doubled.

(d) Repeat parts *a* and *b* for the case where the liquid mass velocity is doubled.

14.13. Benzene and toluene are being separated by distillation in a packed column. At a point in the distilling column, the gas phase is 63 mole percent benzene, while the liquid is 50 mole percent benzene. The local mass-transfer rate is 0.05 lb moles/hr sq ft. If 85 per cent of the resistance to mass transfer is in the gas phase, determine the interfacial compositions and the local mass-transfer coefficients.

14.14. Solve Problem 11.3 such that the sphere is placed in a surroundings of 100°F. A thermal conductivity of $k = 26$ Btu/hr sq ft (°F/ft) is valid for this material. Plot the radial temperature profiles if:

(a) The sphere is placed in air at 100°F. Use $h = 2.0$ Btu/hr sq ft °F

(b) The sphere is placed in water at 100°F. Use $h = 50$ Btu/hr sq ft °F.

14.15. A reflecting mirror for a telescope is to be relieved of stresses by annealing in an oven in preparation for grinding. The mirror is a disk of plate glass 12 in. in diameter by 1 in. thick and is at a uniform temperature of 75°F before being placed in the annealing oven. How long will the disk have to remain in the oven so that every part of the disk reaches a temperature of 750°F? The air temperature in the oven is 790°F, and the heat-transfer coefficient is 3.0 Btu/hr sq ft °F. For the particular glass being used, $k = 0.45$ Btu/hr sq ft (°F/ft) and $\alpha = 0.15$ sq ft/hr.

14.16. A steel sphere 4 in. in diameter is at a uniform temperature of 900°F. This sphere is to be quenched in a water bath which is at 100°F. A thermocouple is embedded in the center of the sphere and after a period of 10 min the temperature at the center of the sphere is 325°F. Estimate the value of the heat-transfer coefficient between the sphere and the water bath.

PART II. NOTATION AND NOMENCLATURE

Wherever a variable in the table below has physical dimension, the dimensions as well as typical units will be stated.

Dimensions

$\mathbf{F}_j$ Force, with subscript indicating direction of application

$\mathbf{H}$ Thermal energy

$\mathbf{L}_j$ Length, with subscript indicating direction

$\mathbf{M}$ Mass

$\mathbf{T}$ Temperature

$\boldsymbol{\theta}$ Time

Symbols

A transfer area $(\mathbf{L}_y\mathbf{L}_z)$, sq ft

b wetted perimeter $(\mathbf{L})$, ft

c concentration $(\mathbf{M}/\mathbf{L}_x\mathbf{L}_y\mathbf{L}_z)$, lb moles/cu ft

$\bar{c}$ mean speed $(\mathbf{L}_j/\theta)$, cm/sec

c_P heat capacity $(\mathbf{H}/\mathbf{MT})$, Btu/lb °F

C_D drag coefficient

D diameter $(\mathbf{L}_r)$, ft

$\mathscr{D}$ mass diffusivity $(\mathbf{L}_x{}^2/\theta)$, sq ft/hr

E eddy diffusivity $(\mathbf{L}_x{}^2/\theta)$, sq ft/hr

$\mathscr{E}$ transfer coefficient

f friction factor, dimensionless

f' friction factor $(f = 4f')$

F force $(\mathbf{F}_j)$, lb$_f$

g acceleration of gravity $(\mathbf{L}/\theta^2)$, 32.2 ft/sec²

g_c dimensional const $(\mathbf{ML}_j/\theta^2\mathbf{F}_j)$, 32.174 lb-ft/sec² lb$_f$

G mass velocity $(\mathbf{M}/\mathbf{L}_r{}^2\theta)$, lb/sq ft, hr

$\mathscr{G}$ volumetric generation rate, quantity of transferent property$/\mathbf{L}_x\mathbf{L}_y\mathbf{L}_z\theta$

h heat-transfer coefficient $(\mathbf{H}/\theta\mathbf{L}_y\mathbf{L}_z\mathbf{T})$, Btu/hr sq ft °F

H_a Henry's law constant (Appendix D)

ΔH_v latent heat of vaporization $(\mathbf{H}/\mathbf{M})$, Btu/lb mole

j_N Colburn mass-transfer factor (Equation 13.104)

j_q Colburn heat-transfer factor (Equation 13.100)

k thermal conductivity $[\mathbf{H}/\theta\mathbf{L}_y\mathbf{L}_z(\mathbf{T}/\mathbf{L}_x)]$, Btu/hr sq ft (°F/ft)

k_c, k_c' mass-transfer coefficient $[\mathbf{M}/\theta\mathbf{L}_y\mathbf{L}_z(\mathbf{M}/\mathbf{L}_x\mathbf{L}_y\mathbf{L}_z)]$, lb moles/hr sq ft (lb mole/cu ft)

k_G, k_G' gas-phase mass-transfer coefficient $[\mathbf{M}/\theta\mathbf{L}_y\mathbf{L}_z(\mathbf{F}/\mathbf{L}^2)]$ (Table 13.2)

k_L, k_L' liquid-phase mass-transfer coefficient $[\mathbf{M}/\theta\mathbf{L}_y\mathbf{L}_z(\mathbf{M}/\mathbf{L}_x\mathbf{L}_y\mathbf{L}_z)]$ (Table 13.2)

k_x, k_x' liquid-phase mass-transfer coefficient $[\mathbf{M}/\theta\mathbf{L}_y\mathbf{L}_z(\mathbf{M}/\mathbf{M})]$ (Table 13.2)

k_X, k_X' liquid-phase mass-transfer coefficient $[\mathbf{M}/\theta\mathbf{L}_y\mathbf{L}_z(\mathbf{M}/\mathbf{M})]$ (Table 13.2)

k_y, k_y' gas-phase mass-transfer coefficient $[\mathbf{M}/\theta\mathbf{L}_y\mathbf{L}_z(\mathbf{M}/\mathbf{M})]$, (Table 13.2)

k_Y, k_Y' gas-phase mass-transfer coefficient $[\mathbf{M}/\theta\mathbf{L}_y\mathbf{L}_z(\mathbf{M}/\mathbf{M})]$

K constant

K_G over-all gas-phase mass-transfer coefficient $[\mathbf{M}/\theta\mathbf{L}_y\mathbf{L}_z(\mathbf{F}/\mathbf{L}^2)]$, lb moles/hr sq ft atm

K_L over-all liquid-phase mass-transfer coefficient $[\mathbf{M}/\theta\mathbf{L}_y\mathbf{L}_z(\mathbf{M}/\mathbf{L}_x\mathbf{L}_y\mathbf{L}_z)]$, lb moles/hr sq ft (lb moles/cu ft)

l mean tree path $(\mathbf{L}_j)$, cm

L length $(\mathbf{L}_j)$, ft

$\mathbf{L}$ volumetric flow rate of phase L, $(\mathbf{L}_x\mathbf{L}_y\mathbf{L}_z/\theta)$, cu ft/hr

L_e entry length $(\mathbf{L}_y)$, ft

m total mass $(\mathbf{M})$, lb

m Slope of equilibrium curve (Equation 14.39a)

$\mathbf{m}$ parameter (Table 11.1)

$\mathbf{m}$ molecular mass $(\mathbf{M})$, gm

M molecular weight $(\mathbf{M}/\mathbf{M})$, lb/lb mole

$\mathbf{M}$ moment $(\mathbf{FL}_r)$, ft-lb$_f$ (Chapter 10)

$\mathscr{M}$ modulus (Chapter 11)

n number of moles

$\mathbf{n}$ parameter (Table 11.1)

N rate of mass transfer $(\mathbf{M}/\theta)$, lb moles/hr

$\mathbf{N}$ number of rows of tubes

N_{Gr} Grashof number $(L^3\rho^2 g\beta\,\Delta T/\mu^2)$, dimensionless

N_{Nu} Nusselt number (hD/k), dimensionless

N_{Pe} Peclet number $(D\bar{v}\rho/c_P k)$, dimensionless

N_{Pr} Prandtl number $(c_P\mu/k)$, dimensionless

N_{Sc} Schmidt number $(\mu/\rho\mathscr{D})$, dimensionless

N_{St} Stanton number $(h/c_P\rho\bar{v})$, dimensionless

N_{Re} Reynolds number $(D\bar{v}\rho/\mu)$, dimensionless

p partial pressure $(\mathbf{F}/\mathbf{L}^2)$, psi

p^* equilibrium partial pressure $(\mathbf{F}/\mathbf{L}^2)$, atm

P total pressure $(\mathbf{F}/\mathbf{L}^2)$, psi

q rate of heat transfer $(\mathbf{H}/\theta)$, Btu/hr

r radius $(\mathbf{L}_r)$, ft

$\mathbf{R}$ resistance to transfer

R gas constant (Appendix A)

s surface renewal factor (Equation 13.122)

S area in flow direction $(\mathbf{L}_x\mathbf{L}_z)$, sq ft

T temperature $(\mathbf{T})$, °F or °R

T^* reduced temperature $(T^* = kT/\epsilon)$

T_B normal boiling point $(\mathbf{T})$, °F

u velocity in the x-direction $(\mathbf{L}_x/\theta)$, ft/sec

u^+ generalized velocity $(u^+ = v/u^*)$, $(\sqrt{\mathbf{L}_y/\mathbf{L}_r})$

u^* friction velocity $(u^* = \sqrt{(\tau_y g_c)_1/\rho}\,)$, $\sqrt{\mathbf{L}_y\mathbf{L}_r/\theta^2}$ ft/sec

u' fluctuating velocity component in the x-direction $(\mathbf{L}_x/\theta)$, ft/sec

U over-all heat-transfer coefficient $(\mathbf{H}/\theta\mathbf{L}_y\mathbf{L}_z\mathbf{T})$

v velocity in the y-direction $(\mathbf{L}_y/\theta)$, ft/sec

v' fluctuating velocity component in the y-direction $(\mathbf{L}_y/\theta)$, ft/sec

V volume $(\mathbf{L}_x\mathbf{L}_y\mathbf{L}_z)$, cu ft

$\mathbf{V}$ volumetric flow rate of phase V $(\mathbf{L}_x\mathbf{L}_y\mathbf{L}_z/\theta)$, cu ft/hr

w velocity in the z-direction $(\mathbf{L}_z/\theta)$, ft/sec

w mass rate of flow $(\mathbf{M}/\theta)$, lb/hr

w' fluctuating velocity component in the z-direction $(\mathbf{L}_z/\theta)$, ft/sec

x direction of molecular transport

x mole fraction, liquid phase $(\mathbf{M}/\mathbf{M})$

x_f film thickness $(\mathbf{L}_x)$, ft

X mole ratio $(\mathbf{M}/\mathbf{M})$, dimensionless

$\mathbf{X}$ parameter

y mole fraction, gas phase $(\mathbf{M}/\mathbf{M})$, dimensionless

PART II. NOTATION AND NOMENCLATURE Contd.

y direction of flow

y^+ generalized position $[y^+ = (r_1 - r)\rho u^*/\mu], (\sqrt{\mathbf{L}_y/\mathbf{L}_r})$

Y mole ratio $(\mathbf{M}/\mathbf{M})$

$\mathbf{Y}$ parameter

z direction of flow

$\mathscr{Z}$ generalized over-all transfer coefficient $(\mathbf{L}_x/\mathbf{\theta})$, ft/sec

α thermal diffusivity $(\alpha = k/c_P\rho)$ $(\mathbf{L}_x{}^2/\mathbf{\theta})$ sq ft/hr

β volumetric coefficient of thermal expansion $1/\mathbf{T}$, $1/°\text{F}$

γ ratio (Equation 13.57)

γ surface tension $(\mathbf{F}/\mathbf{L})$

Γ generalized concentration of transferent property, quantity per unit volume

δ generalized transport diffusivity $(\delta = \frac{1}{6}l\bar{c})$, $(\mathbf{L}_x{}^2/\mathbf{\theta})$, sq ft/hr

δ boundary-layer thickness $(\mathbf{L}_x)$, ft

Δ increment

ϵ/k potential parameter, °K

$\bar{\theta}$ mean travel time $(\bar{\theta} = l/\bar{c})$, $(\mathbf{\theta})$, sec

λ Prandtl mixing length $(\mathbf{L}_x)$, ft

Λ flow rate based upon circumference $(\mathbf{M}/\mathbf{L}\mathbf{\theta})$, lb/hr ft (Equation 13.157)

μ absolute viscosity $(\mathbf{M}\mathbf{L}_x/\mathbf{L}_y\mathbf{L}_z\mathbf{\theta})$, lb/ft-hr

μ_a apparent viscosity of a non-Newtonian fluid $(\mathbf{M}\mathbf{L}_x/\mathbf{L}_y\mathbf{L}_z\mathbf{\theta})$, lb/ft-hr

ν momentum diffusivity $(\nu = \mu/\rho)$, $(\mathbf{L}_x{}^2/\mathbf{\theta})$, sq ft/hr

ρ density $(\mathbf{M}/\mathbf{L}_x\mathbf{L}_y\mathbf{L}_z)$, lb/cu ft

σ molecular diameter $(\mathbf{L}_r)$, cm

τ stress $(\mathbf{F}/\mathbf{L}^2)$, lb$_f$/sq ft

τ_y shear stress on a fluid $(\mathbf{F}_y/\mathbf{L}_y\mathbf{L}_z)$, lb$_f$/sq ft

ψ generalized flux of transferent property, quantity per unit time and unit area

ω angular velocity $(1/\mathbf{\theta})$, 1/sec

Ω Lennard-Jones collision integral

Subscripts

a component

am arithmetic mean

A phase A

b component

B phase B

B boiling

c component

c critical, transition

D drag

e eddy, turbulent

f film

fs free stream

i interface

j general component, phase, position, etc.

l liquid

lm logarithmic mean

m molecular, laminar

N mass transport

q heat transport

r any radial position

t total

v vapor

x position in x-direction

y position in the y-direction

ϵ electrical

θ time

τ momentum transport

0 center position

1 position

2 position

3 position

Specialized subscripts are included only in the general notation.

Applications to Equipment Design

In the next several chapters, the fundamentals of heat, mass, and momentum transfer will be applied to the design of industrial equipment. As it is used here, design in most cases will be limited to the evaluation of the transfer area required. This evaluation is only the first step in a complete design. The detailed process and plant design which must follow the determination of transfer area will not be considered here. Such a design must include complete specification of mechanical details, materials of construction, electrical components, process control and operation, and many other factors. In addition, a detailed economic analysis must be made on the many process alternatives to determine which is the optimum, i.e., which will produce the product at the lowest unit cost.

The engineer may be asked to design new equipment or to evaluate and suggest improvements in the performance of existing equipment. The calculation methods considered in the succeeding chapters may be applied to either the synthesis of new processes or the analysis of existing processes. Although the fundamentals covered in Part II are completely appropriate in describing the transfer phenomena under study, it is in their practical application that many compromises must be made. Often the theoretical development underlying a particular application is incomplete. This lack of complete understanding leads to the use of empirical equations most commonly based upon experimental data taken in the laboratory or pilot plant. If the engineer is asked to design a plant, he must do it with the data available, or he must recommend what additional data are needed. He cannot expect a complete and precise theoretical basis for the proposed process. Therefore, he must be willing and able to use empirical methods and approximations based on experience whenever the theoretical background is inadequate to do the job.

There is no such thing as the permanently perfect design, because of the possibility of new and cheaper materials of construction or because of a shift in either the technical or economic specifications under which the equipment was originally designed. It is necessary that the design be technically sound, economically close to the optimum, and cognizant of safety and working conditions for operating personnel. In many plants, the value of continuing production justifies reasonable expenditures to avoid interruptions.

The design and operation of any chemical plant requires the close cooperation of specialists in many technical and nontechnical fields. The engineer must understand and appreciate the problems of his coworkers whenever they influence his own area of specialization.

This section will consider the unit operations based on heat transfer, mass transfer, simultaneous heat-and-mass transfer, and momentum transfer.

chapter 15

Heat Transfer

The chemical process industries all utilize to a very large extent the transfer of energy as heat. It is the purpose of this chapter to apply the design equation as it has been developed to the practical solution of industrial heat-transfer problems. Two mechanisms of heat transfer have already been established:

Molecular—the transfer of heat by molecular action; this is referred to as conduction.

Turbulent—the transfer of heat by a mixing process; this is usually referred to as convection.

It has been established that these two mechanisms can occur simultaneously or individually.

A third mechanism of heat transfer which is common but which has not yet been discussed is *radiation*, which is the transfer of heat by emission and absorption of energy without physical contact. Unlike conduction or convection, which depend upon physical contact for thermal-energy transfer, radiation depends upon electromagnetic waves as a means for transferring thermal energy from a hot source to a low-temperature sink. Radiation may occur simultaneously with, or independent of, the other two mechanisms of transfer.

INDUSTRIAL HEAT-EXCHANGE EQUIPMENT

The heat exchangers used by chemical engineers cannot be characterized by any one design; indeed, the varieties of such equipment are endless. However, the one characteristic common to most heat exchangers is the transfer of heat from a hot phase to a cold phase with the two phases being separated by a solid boundary.

Double-Pipe Heat Exchangers. The simplest type of heat exchanger is the double-pipe heat exchanger as shown in Figure 15.1. The double-pipe heat exchanger is essentially two concentric pipes with one fluid flowing through the center pipe while the other fluid moves cocurrently or countercurrently in the annular space. The length of each section is usually limited to standard pipe lengths, so that, if an appreciable heat-transfer surface is required, banks of sections are frequently used. If the required area is too large, a double-pipe exchanger is not recommended. The use of a double-pipe heat exchanger is not limited to liquid-liquid heat exchange but may also be used for gas-liquid exchange and for gas-to-gas exchange. Materials of construction may vary, depending upon the fluids being handled. Either fluid may be moved through the tube or annulus at relatively high velocities, thereby aiding in the heat-transfer process.

Shell-and-Tube Exchangers. When the required heat-transfer surface is large, the recommended type of exchanger is the shell-and-tube variety. In this type of heater or cooler, large heat-transfer surface can be achieved economically and practically by placing tubes in a bundle; the ends of the tubes are mounted in a tube sheet. This is very commonly accomplished by expanding the end of the tube into a close-fitting hole in the tube sheet by a process called "rolling." The resultant tube bundle is then enclosed by a cylindrical casing (the shell), through which the second fluid flows around and through the tube handle.

The simplest form of the shell-and-tube exchanger is shown in Figure 15.2. Here, a single-pass tubular heat exchanger is illustrated. The fluid flowing through the tubes enters a header or channel where it is distributed through the tubes in parallel flow and leaves the unit through another header. Either the hot or the cold fluid may flow in the shell of the exchanger surrounding the tubes.

Parallel flow through all tubes at a low velocity gives a low heat-transfer coefficient and low pressure drop. For higher rates of heat transfer, multipass operation

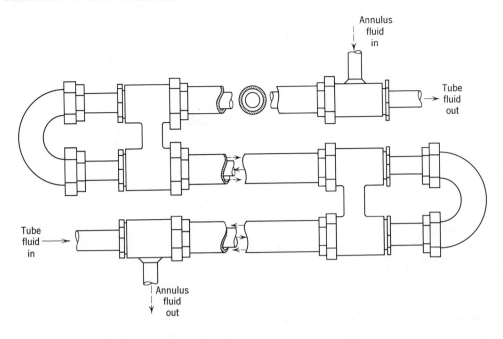

Figure 15.1. Double-pipe exchanger.

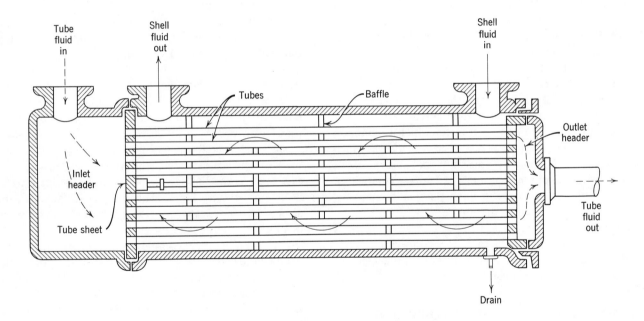

Figure 15.2. Single-pass shell-and-tube heat exchanger.

may be used. Such a heat exchanger is illustrated in Figure 15.3. In this type of construction, the fluid in the tubes is diverted by baffles within the distribution header. The liquid passes back and forth through some fraction of the tubes at high velocity which gives good heat-transfer coefficients. The number of tube passes employed depends upon the economics of the design and operation and upon the space available. Complexity in design sometimes results in expense in fabrication which must be balanced against improved performance. Another disadvantage of the multipass exchanger is the added frictional loss due to higher linear velocities and the entrance and exit losses in the headers. Only an economic balance can indicate the most judicious design.

Note that in Figure 15.3 baffles are placed within the shell in order to divert the flow of the shell fluid into a path predominantly *across* the tubes of the exchanger.

The constantly changing velocity of the shell fluid tends to impart turbulence to the stream, thereby improving heat transfer. Baffles in their simplest form consist of semicircular disks of sheet metal, pierced to accommodate the tubes. The baffles direct as much of the flow as possible normal to the tubes in the shell side as well

factors as cost, cleanability, temperatures, corrosion, operating pressure, pressure drop, and hazards. Large heat exchangers are designed specifically for one particular application, but small exchangers may be designed for multipurpose use. A fixed tube sheet is the cheapest because of its ease of fabrication. Unfortunately, high

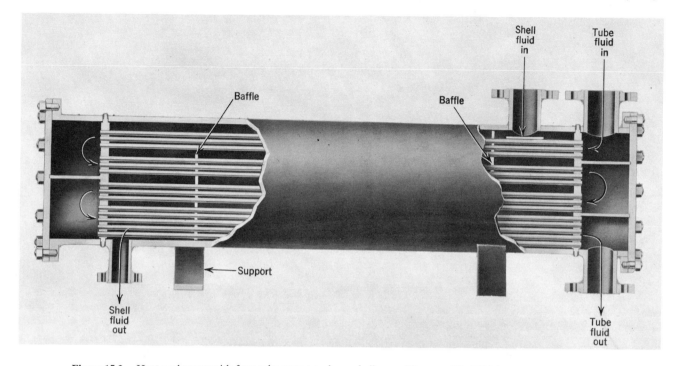

Figure 15.3. Heat exchanger with four tube passes and one shell pass. (Courtesy The Whitlock Manufacturing Co.)

as serve to keep the tubes from sagging. It may be desirable to have several shell-side passes, but this may cause complexities in construction and higher friction losses. Multiple passes on the shell side usually appear only in larger installations. Normally, close baffle spacing is employed. If a condensing vapor serves as the heating medium and is in the shell, baffles are usually not needed.

The two heat exchangers illustrated indicate that cleaning of both shell and tube bundle is difficult, at best. In addition, since large temperature differences between the two fluids exchanging heat are likely to exist, thermal expansion can be expected, for which no provision is indicated. Therefore, to allow provisions for easy removal of the tube bundle for cleaning and to allow for expansion, a floating-head exchanger is used, as shown in Figure 15.4. Here, one of the tube sheets is independent of the shell so that the entire tube bundle can be removed for cleaning of the shell and of the outside of the tubes. This also allows for the differential expansion between the tubes and the shell.

Selection of the type of shell-and-tube exchanger depends upon a number of factors. A satisfactory choice and design will depend upon a compromise of such

stresses between tubes and shell may cause loosening of the tube joints. In addition, shell-side cleaning is very difficult without removing the tube bundle. Floating heads eliminate stresses but add to cost of fabrication.

The nature of the shell-side fluid is also important and will influence the selection of the type of exchanger. Since the shell side of the exchanger is difficult to clean, the least corrosive and cleanest fluid should be placed in the shell. To retard or eliminate corrosion requires the use of expensive alloys, and so the corrosive fluid should be placed in the tubes to save the cost of an expensive alloy shell. The viscosity of the fluids is important in choosing which fluid should be on the shell side. In the shell, induced turbulence and added form friction can be attained by using baffles. A fluid that would normally be flowing in laminar flow in the tubes should be placed in the shell to improve the heat-transfer characteristics. High-pressure fluids should flow through the tubes to avoid expensive high-pressure shells. Only a complete analysis of local design specifications can answer all the questions of selection.

Extended-Surface Heat Exchangers. If heat exchange is occurring between two fluids where one fluid has a very high resistance to heat transfer in comparison to

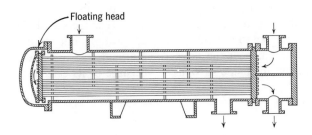

(a) Internal floating-head split-ring type.

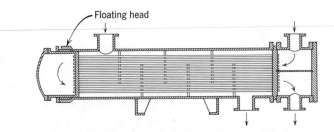

(b) External packed floating-head type.

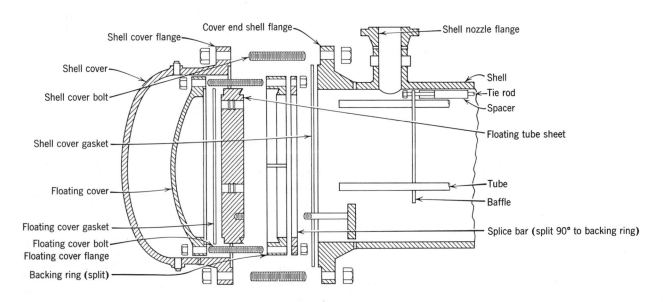

(c) Detail of floating head.

Figure 15.4. Two-tube-pass, one-shell-pass, floating-head heat exchangers. (Courtesy National U.S. Radiator Corp., Heat Transfer Division.)

the other, the higher-resistance fluid "controls" the rate of heat transfer. (See Chapter 14.) Such cases occur, for example, in the heating of air by steam or in the heating of a very viscous oil, flowing in laminar flow, by a molten salt mix. The relative magnitude of the heat-transfer coefficient is about 10 for the oil or air, compared with 2000 for the steam or salt. This poor heat-transfer situation will require much transfer surface for a reasonable flow rate of air or oil.

In order to compensate for the high resistance of the oil or air, the heat-transfer surface exposed to these fluids may be increased by extension of the surface, as in the addition of fins to the outside of the tube, as illustrated in Figure 15.5. The fins are referred to as an *extended surface*; they increase the transfer area substantially in a given amount of space. Some automobile radiators are good illustrations of extended-surface heat exchangers.

Figure 15.5 Extended surface tubing. (Courtesy The Griscom-Russell Co.)

HEAT-TRANSFER COEFFICIENTS

Typical over-all heat-transfer coefficients for industrial heat exchangers are listed in Table 15.1. These are only order-of-magnitude values. Individual film coefficients may be calculated from the empirical correlations of Chapter 13. Individual coefficients may be combined into over-all heat-transfer coefficients by the methods of Chapter 14.

the influence the temperature has upon the thermal conductivity. Wilson found in a study of steam-condenser operations that the sum of the condensate, scale, and tube-wall resistances remained substantially constant for varying cooling-water rates. For the turbulent flow of water, the water-side resistance is an inverse function of the fluid velocity through the tubes, as indicated by Equation 13.77.

Table 15.1. TYPICAL OVER-ALL COEFFICIENTS OF HEAT TRANSFER

Type of Exchanger	Inside or Tube		Outside or Shell		U, Btu/hr sq ft °F	Ref.
	Fluid	Velocity, ft/sec	Fluid	Velocity, ft/sec		
Shell and tube	brine	1–3	water	1–5	50–400	7
Shell and tube	water	2	gas oil	3.0	50–70	7
Shell and tube	water	2	lube oil	0.2	15	7
Shell and tube	water	5	gasoline	condensing	90	7
Shell and tube	crude oil	2	gasoline	condensing	20–30	7
Shell and tube	crude oil	10	gas oil	6.0	80–90	7
Shell and tube	water	4–6	steam	condensing	400–800	7
Double pipe	water	3–8	brine	3–8	150–300	7
Coil in box	gas oil	condensing	water	natural convection	8–20	7
Tube bank	steam	condensing	air	10	9	7
Basket evaporator	brine	boiling	steam	condensing	150–225	7
Vertical tube evaporator	water	boiling	steam	condensing	400–1000	7
Vaporizer	steam	condensing	organic	boiling	300	14
Vaporizer	steam	condensing	acetic acid	boiling	450	14

Evaluation of Heat-Transfer Coefficients. The heat-transfer coefficient most easily measured experimentally is the over-all coefficient. The over-all temperature difference and the total heat transfer can usually be measured directly for a heat exchanger of known area. By the relationship $q = -UA(\Delta T)_{lm}$, the over-all coefficient (U) may be calculated. Heat-exchangers are often designed using over-all coefficients rather than individual coefficients. The determination of individual coefficients is rather difficult because of the uncertainty regarding the measurement of surface temperatures.

A method of calculating film coefficients was proposed by Wilson (37) for condensing vapors. It is based on the over-all resistance ($1/UA$) being equal to the sum of the individual resistances,

$$\Sigma R = \frac{1}{UA} = R_c + R_w + R_d + R_L \quad (15.1)$$

where R_c = condensate resistance
R_w = wall resistance
R_d = scale or dirt resistance
R_L = liquid-side resistance

The vapor-side resistance depends upon the temperature driving force and upon the temperature of the condensate. The dirt resistance and tube-wall resistance also depend upon their respective temperatures and upon

Thus

$$\frac{1}{UA} = R_c + R_d + R_w + \frac{1}{C_2(\bar{v})^{0.8}} = C_1 + \frac{1}{C_2(\bar{v})^{0.8}} \quad (15.2)$$

where C_1, C_2 = empirical constants depending upon geometry and physical properties
$\bar{v}$ = average water velocity, ft/sec

A plot of $1/UA$ as a function of $1/(\bar{v})^{0.8}$ on rectangular coordinates will determine the constants in Equation 15.2. If the average temperature of the water varies widely, C_2 will vary enough that the analysis is not dependable. This relation is indicated in Figure 15.6.

If the straight line is extrapolated to an infinite velocity (zero water-side resistance) the intercept C_1 of Equation 15.2 will be determined. For the moment, consider that the condenser tube is clean so that $R_d = 0$. Therefore the value of C_1 equals the sum of the condensate and tube-wall resistances. The tube-wall resistance, which is a function of thermal conductivity and wall thickness, can be calculated and subtracted from C_1 to give the condensate resistance; its reciprocal is the condensate heat-transfer coefficient.

The slope of this line equals $1/C_2$, and, at a water velocity of 1 ft/sec, the water-side resistance also equals $1/C_2$. The data for a Wilson plot must be obtained over a varying range of water velocities with both condensate and water temperatures, on the average, remaining as nearly constant as possible, so that the Prandtl number remains essentially constant. Also, as high a water velocity as possible should be used to give a short extrapolation.

Now, suppose that the same condenser tube is tested after several months service, during which time the tube has become fouled. Equation 15.2 is still valid but the straight line resulting from the test will now lie above the clean-tube case and will give a larger intercept. The new intercept will include in its value a dirt or scale resistance which would be equal in magnitude to the difference between intercepts for clean and dirty tubes.

HEAT-EXCHANGER CALCULATIONS

The design calculations for heat exchangers are based upon the application of Equation 14.62. The equation as written is perfectly general but may present difficulty of integration in some applications. Consider again the exchange of heat between two fluids within a heat exchanger. At a particular point in the exchanger, the rate of heat transfer may be expressed by a modification of Equation 14.62, or

$$dq = -U \, \Delta T \, dA \qquad (15.3)$$

In most heat exchangers, the temperature of at least one of the fluids will vary, and hence the driving force between the hot and cold fluids will vary. Not only will ΔT vary throughout the exchanger but so will q and U. The over-all coefficient of heat transfer will vary as a result of variations in physical properties and in the flow regime brought about by the transfer of heat. Consequently, the integration of Equation 15.3 is complex. Solving Equation 15.3 for the area gives

$$\int_0^A dA = -\int_0^q \frac{dq}{U \, \Delta T} \qquad (15.4)$$

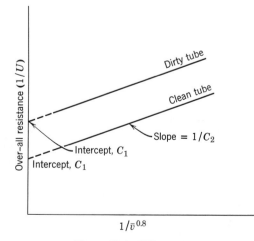

Figure 15.6. Wilson plot.

The temperature variations of the fluids within a heat exchanger are shown in Figure 15.7, where the temperature of each stream is plotted as a function of the length of the exchanger.

In a true parallel-flow heat exchanger, Figure 15.7a, both the hot and cold fluids enter the exchanger at the same end and flow through the exchanger in the same direction. At the entrance of the exchanger a large driving force prevails, giving a relatively large heat-transfer rate. As the fluids progress through the exchanger, the temperature driving force becomes less and less so that the rate drops off asymptotically as the streams approach some limiting temperature. The net result of this type of variation in ΔT is that the exchanger is much more effective for a unit area of heat-transfer surface at the entrance than it is near the exit. Simply extending the length of the exchanger will result in little improvement in the amount of heat transferred. On the other hand, Figure 15.7b illustrates true counter-current operation of a heat exchanger. In this type of operation, the two fluids exchanging heat pass each other in opposite directions. Here, the driving force is much more nearly constant throughout the length of the exchanger. The net result is that a unit surface gives about the same rate of exchange through the entire exchanger. In general, the variations in driving force and resistances necessitate the rigorous integration of Equation 15.4 to size heat-exchange equipment.

Illustration 15.1. (8). It is desired to heat 1000 bbl/day of cottonseed oil from 70°F to 180°F by means of steam condensing at atmospheric pressure in a shell-and-tube heat exchanger. Ten copper pipes (0.622 in. I.D. and 0.840 in. O.D.) 6 ft long will be used in parallel in each pass. Assuming a steam heat-transfer coefficient of 2000 Btu/hr sq ft °F, how many passes in series will be required?

The properties of cottonseed oil are as follows:
Density = $0.8836 - 0.0002879T°F$, gm/cu cm
Specific heat = $0.45 + 0.000625(T°F - 60)$, Btu/lb °F
Thermal conductivity = 0.0808 Btu/hr sq ft (°F/ft)
Viscosity = 11.2 centipoises at 100°F; 2.5 centipoise at 210°F
(Assume log viscosity versus temperature is linear.)

SOLUTION. The heat-transfer equation is

$$\int_0^A dA = -\int_0^q \frac{dq}{U \, \Delta T}$$

but dq may be evaluated by the heat gain of the oil,

$$dq = [w_i c_P \, dT]$$

Thus,

$$A = \int_{T_1}^{T_2} \frac{w_i c_P \, dT}{-U \, \Delta T} = w_i \int_{T_1}^{T_2} \frac{c_P \, dT}{-U \, \Delta T} \qquad (a)$$

The transfer area per pass may be evaluated from the pipe data. Because of the physical characteristics of the cottonseed oil, it will offer great resistance to heat transfer in comparison to the steam or pipe-wall resistances. Therefore,

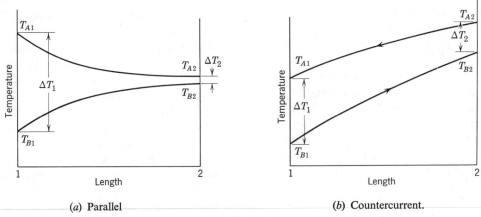

(a) Parallel. (b) Countercurrent.

Figure 15.7. Temperature in parallel and countercurrent heat exchangers.

the inside pipe area is associated with the controlling resistance and will be used in these calculations.

There are 10 pipes per pass, each 0.622 in. I.D. by 6 ft long

$$\frac{\text{Inside area}}{\text{Pass}} = \pi\left(\frac{0.622}{12}\right)(6)(10) = 9.78 \text{ sq ft/pass}$$

Thus the total heat-transfer area becomes $9.78N$ where N is the number of passes.

Therefore,

$$9.78N = w_i \int_{T_1}^{T_2} \frac{c_P \, dT}{-U_i \Delta T} \tag{b}$$

In many simple cases in heat transfer, the right-hand side of Equation b could be integrated analytically, assuming constant values of c_P and U_i. However, in this problem both c_P and U_i vary with temperature to an appreciable extent, and it is necessary to evaluate them at several temperatures between T_1 and T_2. Equation b is then integrated graphically using the point values of $c_P/U_i \Delta T$.

To aid in the solution of this problem, it will be advantageous to plot the physical properties of the cottonseed oil as a function of its temperature. This is done in Figures 15.8, 15.9, and 15.10.

The over-all resistance based on the inside surface area is

$$\frac{1}{U_i A_i} = \frac{1}{h_i A_i} + \frac{\Delta x}{k A_m} + \frac{1}{h_o A_o} \text{*} \tag{c}$$

or

$$\frac{1}{U_i} = \frac{1}{h_i} + \frac{\Delta x A_i}{k A_m} + \frac{A_i}{h_o A_o}$$

For circular cross sections, A is proportional to D; therefore,

$$\frac{1}{U_i} = \frac{1}{h_i} + \frac{\Delta x D_i}{k D_m} + \frac{D_i}{h_o D_o}$$

---

* It is common practice in heat-transfer calculations to base the over-all heat-transfer coefficient upon either the inside or the outside surface area of the tube when cylindrical geometry is considered. Throughout this chapter the subscripts i or o will denote inside or outside surface. Previously, subscripts 1, 2, etc., were used to denote phase boundaries. Thus sub 1 now becomes sub i.

$D_i = 0.622$ in. $\Delta x = 0.109$ in.
$D_o = 0.840$ in. $k_{\text{Cu}} = 220$ Btu/hr sq ft (°F/ft)
$D_m = 0.731$ in.

(for thin-walled tubes the arithmetic mean is satisfactory) so that

$$\frac{1}{U_i} = \frac{1}{h_i} + \frac{(0.109/12)(0.622)}{(220)(0.731)} + \frac{(0.622)}{(2000)(0.840)}$$

or
$$\frac{1}{U_i} = \frac{1}{h_i} + 0.000405 \tag{d}$$

Equation d is a general equation for this problem expressing the over-all resistance in terms of the inside resistance. In this case, both the steam resistance and copper-pipe resistance are constant.

The mass-flow rate of cottonseed oil is computed as follows:

$$w_i = \left(1000 \frac{\text{bbl}}{\text{day}}\right)\left(42 \frac{\text{gal}}{\text{bbl}}\right)\left(\frac{1 \text{ day}}{24 \text{ hr}}\right)$$
$$\times (8.34) [0.8836 - 0.0002879(70)] \text{ lb/gal}$$

$w_i = 12,600$ lb/hr feed to exchanger (constant for steady state and independent of temperature)

It will be convenient to express the flow rate as a mass velocity so as to avoid calculations of linear velocities. Linear velocities are dependent upon the specific gravity of the oil which is temperature sensitive.

Thus,

$$G = \frac{(12,600 \text{ lb/hr})(144 \text{ sq in./sq ft})}{(0.304 \text{ sq in./tube})(10 \text{ tubes})} = 596,000 \text{ lb/hr sq ft}$$

At any point within the exchanger, $-(\Delta T) = (212 - T)°\text{F}$, where T is the bulk average temperature of the oil. Thus, all terms in Equation b have been determined in a general manner so that it is now possible to evaluate them explicitly for specific points within the exchanger. For this illustration, points in the exchanger corresponding to cottonseed-oil bulk average temperatures of 70, 100, 130, 160, 180°F will be selected. Only the 70°F point will be completely calculated.

For 70°F. First, the flow character must be determined through the use of a Reynolds number. Since frequent computations of N_{Re} are necessary, it will be expressed as a function of oil viscosity, or

$$N_{Re} = \frac{DG}{\mu} = \frac{(0.622/12)(596,000)}{(2.42\mu')} = \frac{12,800}{\mu'}$$

where μ' is in centipoises.

The same technique will be used for the Prandtl number.

$$N_{Pr} = \frac{c_P \mu}{k} = \frac{(c_P)(2.42\mu')}{0.0808} = 29.9(c_P)(\mu')$$

The j-factor plot (Appendix C-5) for heat transfer is convenient to use in this case because it eliminates the need for concern regarding the appropriate coefficient correlation to use. Since use of this plot may involve an L/D,

$$\frac{L}{D} = \frac{6 \text{ ft}}{(0.622/12) \text{ ft}} = 116$$

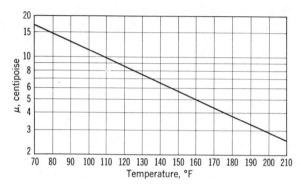

Figure 15.8. Viscosity of cottonseed oil.

Now the sample calculations for 70°F will be completed. From Figures 15.8 and 15.9,

At 70°F, $\mu' = 16.8$ centipoises, $c_P = 0.456$ Btu/lb °F

$$(N_{Re})_{70} = \frac{12,800}{16.8} = 762 \text{ (definitely laminar flow)}$$

$$(N_{Pr})_{70} = 29.9(0.456)(16.8) = 229; \quad (N_{Pr})^{2/3} = 37.2$$

Before the oil coefficient can be evaluated, it is necessary to know the oil-side pipe-surface temperature; it must be assumed for a preliminary calculation and checked by the series-resistance concept after completion of the preliminary calculation.

Assume $T_{is} = 208°$; then $\mu_{is}' = 2.6$ centipoises and $\mu_{is}/\mu = 2.6/16.8 = 0.155$ and $(\mu_{is}/\mu)^{0.14} = 0.77$. From Appendix C-5, at a Reynolds number of 762 and an $L/D = 116$, $j_q = 0.0046$ and so from the definition of j_q (Chapter 13)

$$j_q = 0.0046 = \left(\frac{h_i}{c_P G}\right)(N_{Pr})^{2/3}\left(\frac{\mu_{is}}{\mu}\right)^{0.14}$$

and

$$h_i = j_q c_P G/(N_{Pr})^{2/3}\left(\frac{\mu_{is}}{\mu}\right)^{0.14}$$

or

$$h_i = \frac{(0.0046)(0.456)(596,000)}{(37.2)(0.77)} = 43.7 \text{ Btu/hr sq ft °F}$$

and from Equation d

$$\frac{1}{U_i} = \frac{1}{h_i} + 0.000405 = \frac{1}{43.7} + 0.000405 = 0.0223$$

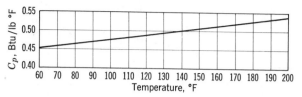

Figure 15.9. Specific heat of cottonseed oil.

Now check the assumed T_{is} by proportioning resistances and driving forces as in Chapter 14.

$$\frac{212 - T_{is}}{0.000405} = \frac{212 - 70}{0.0223}$$

The left side of the equation is the ratio of the temperature drop across the steam and pipe to the sum of their resistances, and the right side is the ratio of the total ΔT to the total resistance. Solving for T_{is} gives

$$T_{is} = 209°F$$

which is a satisfactory check.

The same sequence of calculations must be repeated for each of the other temperatures. The results are tabulated below along with the quantities necessary for the solution of Equation b.

T_i	$-\Delta T = (212 - T)$	h_i	U_i	c_P	$\dfrac{c_P}{U_i(-\Delta T)}$
70	142	43.7	44.8	0.456	0.0000732
100	112	42.5	42.0	0.475	0.000101
130	82	36.4	35.8	0.493	0.000168
160	52	53.5	52.4	0.512	0.000188
180	32	72.3	70.3	0.525	0.000234

It is apparent from the above table that neither the total resistances $1/U$, nor c_P is constant. A graphical integration of $c_P \, dT/U \, \Delta T$ from Equation b will therefore be required since there is no analytical relationship between all the variables.

$c_P/U_i \, \Delta T$ is plotted as a function of T_i and the area under the curve is determined between the limits of $T_i = 70$ and $T_i = 180$. A value of 0.0157 is found. The number of passes may be determined from Equation b,

$$N = \frac{12,600 \times 0.0157}{9.78} = 20.3 \approx 21 \text{ passes}$$

Logarithmic-Mean Temperature Difference. The solution to the previous illustration recognizes all variations

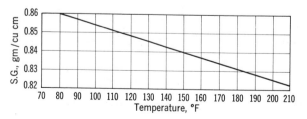

Figure 15.10. Specific gravity of cottonseed oil.

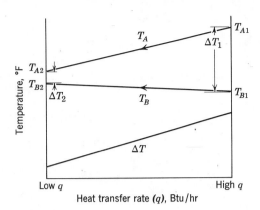

Figure 15.11. Heat-transfer rate as a function of driving forces.

in properties and therefore is as rigorous as accuracy in graphical integration will permit. The procedure is necessarily tedious, but, fortunately, under some conditions simplifying assumptions lead to negligible error. Equation 15.3 may be solved using the following assumptions, which are very often valid.

1. The over-all coefficient (U) is constant. This is not rigorously valid, but, for fluids whose physical properties are not too sensitive to temperature variations, the over-all coefficient based upon average fluid properties leads to little error.

2. The specific heats of the two fluids are independent of temperature.

3. The mass rates of flow of both fluids are constant.

If the second and third assumptions are valid, a simpler integration of the design equation may be used. The development that follows is for true parallel-flow operation but applies equally well to countercurrent operation. Figure 15.11 shows the temperature history as a function of the heat-transfer rate throughout such an exchanger. Because of constant specific heats the temperature lines are linear with heat transferred. The hot fluid flows at a constant rate of (w_A) lb/hr and changes in temperature from T_{A1} to T_{A2}. The cold fluid will flow at a steady rate of w_B lb/hr and be heated from T_{B1} to T_{B2}. The temperature driving force, in the transfer direction, at any point in the exchanger is

$$-\Delta T = (T_A - T_B) \tag{15.5}$$

Differentiating Equation 15.5 gives

$$-d(\Delta T) = dT_A - dT_B \tag{15.6}$$

Now, the heat given up by the hot fluid as it passes through the exchanger is

$$-dq_A = -w_A c_{PA}\, dT_A \tag{15.7}$$

and the heat gained by the cold fluid is

$$dq_B = w_B c_{PB}\, dT_B \tag{15.8}$$

From Equation 15.7

$$dT_A = \frac{dq_A}{w_A c_{PA}} \tag{15.7a}$$

and from Equation 15.8

$$dT_B = \frac{dq_B}{w_B c_{PB}} \tag{15.8a}$$

Combining Equations 15.7a and 15.8a with Equation 15.6 yields

$$-d(\Delta T) = \frac{dq_A}{w_A c_{PA}} - \frac{dq_B}{w_B c_{PB}} = dq_A\!\left[\frac{1}{w_B c_{PB}} + \frac{1}{w_A c_{PA}}\right] \tag{15.9}$$

since $dq_A + dq_B = 0$ at steady state.

By the assumptions previously stated, the terms within the brackets of Equation 15.9 are constant. Therefore Equation 15.9 may be integrated between the limits of ΔT_1 and ΔT_2 and between 0 and q to give

$$-(\Delta T_2 - \Delta T_1) = q_A\!\left[\frac{1}{w_B c_{PB}} + \frac{1}{w_A c_{PA}}\right] \tag{15.10}$$

It would be advantageous in many calculations to evaluate the integral of Equation 15.3 directly when the simplifying assumptions hold. This would be possible if a rigorous average ΔT were known. Derivation of a rigorous mean value follows:

Combining Equation 15.9 with Equation 15.3 for steady state yields

$$dq_A = -U\,\Delta T\, dA = -\frac{d(\Delta T)}{\dfrac{1}{w_B c_{PB}} + \dfrac{1}{w_A c_{PA}}} \tag{15.11}$$

Rearrangement of Equation 15.11 to separate variables for integration, with U being assumed constant or appropriately averaged, gives

$$-\left[\frac{1}{w_B c_{PB}} + \frac{1}{w_A c_{PA}}\right] U\!\int_0^A dA$$
$$= -\int_{\Delta T_1}^{\Delta T_2} \frac{d(\Delta T)}{\Delta T} \tag{15.12}$$

Upon integration, Equation 15.12 becomes

$$\left[\frac{1}{w_B c_{PB}} + \frac{1}{w_A c_{PA}}\right] UA = \ln\frac{\Delta T_2}{\Delta T_1} \tag{15.13}$$

But from Equation 15.10

$$-\left[\frac{1}{w_B c_{PB}} + \frac{1}{w_A c_{PA}}\right] = \frac{\Delta T_2 - \Delta T_1}{q_A} \tag{15.10a}$$

Therefore, substituting Equation 15.10a into Equation 15.13 gives

$$-\left[\frac{\Delta T_2 - \Delta T_1}{q_A}\right] UA = \ln\frac{\Delta T_2}{\Delta T_1} \tag{15.14}$$

Rearranging,

$$q_A = -UA\left[\frac{\Delta T_2 - \Delta T_1}{\ln\dfrac{\Delta T_2}{\Delta T_1}}\right] = -UA(\Delta T)_{lm} \quad (15.15)$$

Equation 15.15 is the simplified design equation for cases where the aforementioned assumptions are valid. The bracketed term of Equation 15.15 defines the logarithmic-mean temperature driving force (ΔT_{lm}). Thus, the variation in driving force is handled very simply by using the log mean of the driving forces at the terminals of the exchanger. Equation 15.15 applies also to countercurrent flow as well as to a situation where one of the fluids is at constant temperature such as a condenser. It is rigorous for any case where ΔT is linear with q.

Colburn (9) has suggested that a more accurate calculation can be made using a logarithmic-mean value of the combination $U \Delta T$. It should be emphasized that this is the product of coefficient at one terminal and ΔT at the other as indicated by the subscript.

Accordingly,

$$q = -A\left[\frac{U_1 \Delta T_2 - U_2 \Delta T_1}{\ln\dfrac{U_1 \Delta T_2}{U_2 \Delta T_1}}\right] \quad (15.16)$$

where U_1, U_2 = overall coefficients at exchanger terminals
$\Delta T_1, \Delta T_2$ = temperature driving forces at terminals

In Equation 15.16 it is assumed that the over-all coefficient varies linearly with q through the exchanger, as well as that specific heats are constant and that steady state exists.

It must be re-emphasized that in the general case where the simplifying assumptions are not valid, heat-exchanger area must be evaluated by graphical integration of Equation 15.4.

Illustration 15.2. Solve Illustration 15.1 using a logarithmic-mean temperature difference and by Equation 15.16. Base calculations on an average oil temperature.

SOLUTION. The total heat load on the exchanger will be the same as Illustration 15.1.

Therefore $q = wc_P(T_2 - T_1) = (12{,}600)(0.48)$
$$\times (180 - 70) = 665{,}000 \text{ Btu/hr}$$

$$(\Delta T)_{lm} = \frac{(212 - 70) - (212 - 180)}{\ln\dfrac{212 - 70}{212 - 180}} = 74.3°\text{F}$$

The average oil temperature is $(180 + 70)/2 = 125°\text{F}$. Therefore, $U_{125°} = 37.0$ Btu/hr sq ft °F from the table in Illustration 15.1,

and $A = \dfrac{q}{U(\Delta T)_{lm}} = \dfrac{665{,}000}{(37)(74.4)} = 244$ sq ft

Thus $N = \dfrac{244}{9.78} = 24.9$ passes

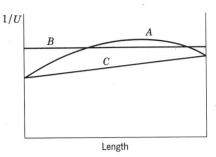

Figure 15.12. Comparison of calculations for Illustrations 15.1 and 15.2.

Here the total heat exchanger is about 20 per cent larger than that of detailed solution.

If Equation 15.16 is used, the following area results:

$$A = \frac{q \ln\left(\dfrac{U_1 \Delta T_2}{U_2 \Delta T_1}\right)}{U_1 \Delta T_2 - U_2 \Delta T_1}$$

or $A = 665{,}000 \dfrac{\ln\left(\dfrac{43.9 \times 32}{70.3 \times 142}\right)}{(43.9)(32) - (70.3)(142)} = 153$ sq ft

Thus $N = 153/9.78 = 15.6$ passes, or about 30 per cent lower than that of Illustration 15.1.

Figure 15.12 shows one reason for the difference between Illustrations 15.1 and 15.2.

Curve A represents the actual over-all resistance as a function of position in the exchanger. The reason for the maximum can be seen by reference to the Reynolds number–j-factor relationships. The oil enters the exchanger at a $N_{Re} = 762$ which is laminar flow and leaves at a $N_{Re} = 3420$, so that the flow character goes from laminar into the transition zone, and the heat-transfer coefficient will go through a minimum. Examination of the tabulated data of Illustration 15.1 indicates also that U will have a minimum, and hence the total resistance becomes a maximum at some point in the apparatus. Curve B is the constant resistance selected as that for $T = 125°$. In this case the resistance is relatively high and constant; therefore, a larger area is required. Curve C is the resistance assumed in the Colburn relationship. This curve shows the resistance to vary linearly with temperature of the oil but to be considerably lower than either of the other two cases, thereby indicating a smaller heat-exchange area.

The general conclusion from this illustration is that the simplified procedures are not appropriate when the flow character of either fluid is such that the fluid passes from laminar flow into moderate or extreme turbulence which causes a variable over-all heat-transfer coefficient. Thus, if one examines the Reynolds number of the fluid at the terminals of the exchanger and if it brackets $N_{Re} = 2100$, according to Appendix C-5 the rigorous graphical solution is required. If both Reynolds numbers exceed 10,000 the resistance may be nearly enough constant so that the logarithmic-mean temperature difference may be used. For both Reynolds numbers between 2100 and 10,000 or for both less than 2100, about a 10 per cent error is likely if either Equation 15.15 or 15.16 is used. The recommended procedure is to examine both terminal Reynolds numbers, the specific heat, and the over-all heat-transfer coefficient. If the Reynolds number exceeds 10,000 and the specific heat is constant an over-all resistance based upon the average fluid temperature may be used in conjunction with the log mean ΔT. If these conditions are not met, a rigorous point-to-point solution of the design equation is required for an accurate result.

Illustration 15.3. A double-pipe heat exchanger is to be designed to cool benzene from 180°F to 100°F. Water enters the annular space at 70°F and flows countercurrent to the benzene at a velocity of 5 ft/sec. The inside pipe is Sch. 40,* $1\frac{1}{4}$-in. steel pipe; the outside pipe is Sch. 40, 2-in. steel pipe. The pipe comes in 20-ft sections. How many sections in series must be used to cool 7500 lb of benzene per hour? Although pipe is available in 20-ft sections, the jacket must be shorter than the inner tube as can be seen by reference to Figure 15.1.

SOLUTION. The solution of this problem is clearly one of evaluation of the various resistances that are encountered along the transfer path. In this particular case the resistances are: (1) the benzene resistance (2) the pipe wall, and (3) the water resistance. Fluid properties will be evaluated at the average bulk temperatures.

The benzene resistance (inside pipe)

The inner diameter of the Sch. 40, $1\frac{1}{4}$-in. pipe $= \dfrac{1.38}{12}$

$$= 0.115 \text{ ft}$$

The cross-sectional flow area of the inner pipe = 0.0104 sq ft

The average bulk temperature of the benzene $= \dfrac{180 + 100}{2}$

$$= 140°$$

Density of benzene at 140° = 52.3 lb/cu ft

Viscosity of benzene at 140° = 0.39 centipoise

Mean velocity of benzene $= (7500 \text{ lb/hr})\left(\dfrac{1 \text{ hr}}{3600 \text{ sec}}\right)$

$$\times \left(\dfrac{1 \text{ cu ft}}{52.3 \text{ lb}}\right)\left(\dfrac{1}{0.0104 \text{ sq ft}}\right) = 3.83 \text{ ft/sec}$$

$(N_{Re})_i = \dfrac{D\bar{v}\rho}{\mu} = \dfrac{(0.115)(3.83)(52.3)}{(0.39 \times 6.72 \times 10^{-4})} = 8.78 \times 10^4 = 87,800$

It is quite apparent from the magnitude of N_{Re} that the benzene is flowing in complete turbulence.

Specific heat of benzene at 140° = 0.45 Btu/lb °F

---

* The schedule number for a pipe is a convenient method for characterizing the strength of the pipe. More will be said of nominal pipe size and schedule number in Chapter 21.

Thermal conductivity of benzene at 140° = 0.085 Btu/hr sq ft °F/ft.

$$N_{Pr} = \frac{c_P\mu}{k} = \frac{(0.45)(0.39 \times 2.42)}{0.085} = 5.0$$

Thus the benzene coefficient can be estimated using Equation 13.78

$$\frac{h_i D}{k} = 0.023(N_{Re})^{0.8}(N_{Pr})^{0.3}$$

$$\frac{h_i D}{k} = 0.023(87,800)^{0.8}(5)^{0.3} = 337$$

$$h_i = \frac{(337)(0.085)}{0.115} = 249 \text{ Btu/hr sq ft °F}$$

The area associated with h_i is $(3.14)(0.115)(1) = 0.361$ sq ft/ft.

Thus the benzene resistance becomes

$$R_i = \frac{1}{h_i A_i} = \frac{1}{(249)(0.361)}$$

$$R_i = 0.0111 \text{ hr ft °F/Btu}$$

The water resistance (in annulus)

An equivalent diameter of the annulus will be required to obtain N_{Re} and h_o

The inner diameter of the outer pipe $= \dfrac{2.067}{12} = 0.172$ ft

The outer diameter of the inner pipe $= \dfrac{1.66}{12} = 0.1385$ ft

The equivalent diameter for this geometry is defined as $D_{eq} = \dfrac{4S}{P}$ (see p. 159)

The annular area, $S = \dfrac{\pi}{4}[(0.172)^2 - (0.1385)^2] = 0.00816$ sq ft

The wetted perimeter, $P = \pi(0.172 + 0.1385) = 0.977$ ft Thus

$$D_{eq} = \frac{4S}{P} = \frac{(4)(0.00816)}{0.977} = 0.0356 \text{ ft}$$

The exit temperature of the water may be obtained by an over-all heat balance.

$$\int_{T_{A1}}^{T_{A2}} w_A c_{PA}\, dT_A = \int_{T_{B1}}^{T_{B2}} w_B c_{PB}\, dT_B$$

$$w_A c_{PA}(T_{A2} - T_{A1}) = w_B c_{PB}(T_{B2} - T_{B1})$$

Water at 5 ft/sec = (5)(3600)(62.3)(0.00816) = 9180 lb/hr Thus

$$(7500)(0.45)(100 - 180) = (9180)(1)(70 - T_{B1})$$

$$(70 - T_{B1}) = \frac{-(7500)(0.45)(80)}{9180} = -29.4$$

$$-T_{B1} = -29.4 - 70 = -99.4° \quad \text{or} \quad T_{B1} = 99.4°$$

Average water temperature $= \dfrac{99.4 + 70}{2} = 84.7 \approx 85°F$

Viscosity of water at 85°F = 0.8 centipoise
Thermal conductivity of water at 85°F = 0.342

$$(N_{\mathrm{Re}})_o = \frac{D_{\mathrm{eq}}\bar{v}\rho}{\mu} = \frac{(0.0356)(5)(62.3)}{(0.8 \times 6.72 \times 10^{-4})} = 20{,}600$$

The water is also flowing in well-developed turbulence. Again, Equation 13.77 is used, but in this case the exponent on N_{Pr} is 0.4 because the water is being heated.

$$N_{\mathrm{Pr}} = \frac{c_P \mu}{k} = \frac{(1)(0.8 \times 2.42)}{(0.342)} = 5.67$$

$$\frac{h_o D_{\mathrm{eq}}}{k} = 0.023(N_{\mathrm{Re}})^{0.8}(N_{\mathrm{Pr}})^{0.4}$$

$$\frac{h_o D_{\mathrm{eq}}}{k} = 0.023(20{,}600)^{0.8}(5.67)^{0.4} = 130.5$$

$$h_o = \frac{(130.5)(0.342)}{(0.0356)} = 1250 \ \text{Btu/hr sq ft °F}$$

The area associated with the water resistance is the outside surface area of the inner pipe or

$$A_o = (3.14)(0.1385)(1) = 0.435 \ \text{sq ft/ft}$$

Thus the water-side resistance becomes

$$R_o = \frac{1}{h_o A_o} = \frac{1}{(1250)(0.435)} = 0.001805 \ \text{hr ft °F/Btu}$$

Resistance of pipe wall

Pipe-wall thickness $= \dfrac{0.140}{12} = 0.0117 \ \text{ft}$

Thermal conductivity of steel = 26 Btu/hr sq ft (°F/ft)
Arithmetic mean diameter of wall

$$= \frac{0.1385 + 0.115}{2} = 0.127 \ \text{ft}$$

The arithmetic average is justified since the wall is thin. The mean wall area is

$$\pi D L = (3.14)(0.127)(1) = 0.398 \ \text{sq ft/ft}$$

Thus the pipe-wall resistance is

$$R_w = \frac{\Delta x}{k A_m} = \frac{0.0117}{(26)(0.398)} = 0.00113 \ \text{hr ft °F/Btu}$$

The total resistance to heat transfer based upon a preliminary estimation of the surface coefficient is

$$\Sigma R = R_i + R_P + R_o$$
$$= 0.0111 + 0.00113 + 0.001805 = 0.01404$$

Log-mean driving force

$$(\Delta T)_{lm} = \frac{\Delta T_1 - \Delta T_2}{\ln \dfrac{\Delta T_1}{\Delta T_2}} = \frac{(180 - 99) - (100 - 70)}{\ln \dfrac{(180 - 99)}{(100 - 70)}} = 51°$$

The total heat requirement is

$$q = (7500)(0.45)(180 - 100) = 270{,}000 \ \text{Btu/hr}$$

The total resistance $\Sigma R = 0.01404$ hr ft °F/Btu.

Therefore, the total length of heat exchanger is found from

$$q/\text{ft} = \frac{(\Delta T)_{lm}}{\dfrac{1}{UA}} = \frac{51}{0.01404} = 3640 \ \text{Btu/hr ft}$$

or the total length is 270,000/3640 or 76.4 ft

Therefore, four lengths will have to be used, which allows about one foot per length for jacket couplings as ineffective for heat transfer.

This illustration was solved using the ΔT_{lm} concept which necessitates the validity of several assumptions. The assumption of constant values of c_P for both benzene and water is rather good in this case since only a slight variation may be expected over the temperatures involved. However, the use of ΔT_{lm} also requires the constancy of the total resistance. To check the validity of this assumption, U may be calculated at the terminals of the exchanger, and from these values the appropriateness of the use of ΔT_{lm} may be determined.

Calculation of U at the terminals

Hot end of exchanger

For benzene:
   Temperature of benzene = 180°F
   Specific heat of benzene at 180° = 0.47 Btu/lb °F
   Viscosity of benzene at 180° = 0.3 centipoise
   Density of benzene at 180° = 50.8 lb/cu ft
   Thermal conductivity at 180° = 0.082 Btu/hr sq ft (°F/ft)

$$(N_{\mathrm{Re}})_i = \frac{D\bar{v}\rho}{\mu} = \frac{(0.115)(3.83)(50.8)}{(0.3 \times 6.72 \times 10^{-4})} = 111{,}000$$

$$(N_{\mathrm{Pr}})_i = \frac{c_P \mu}{k} = \frac{(0.47)(0.3 \times 2.42)}{(0.082)} = 4.16$$

$$\frac{h_i D}{k} = 0.023(N_{\mathrm{Re}})^{0.8}(N_{\mathrm{Pr}})^{0.3}$$
$$= 0.023(111{,}000)^{0.8}(4.16)^{0.3} = 381$$

$$h_i = \frac{(381)(0.082)}{(0.115)} = 272 \ \text{Btu hr sq ft °F}$$

For water:
   Temperature of water = 99°
   Specific heat of water at 99° = 1.0 Btu/lb °F
   Viscosity of water at 99° = 0.7 centipoise
   Density of water at 99° = 62.3 lb/cu ft
   Thermal conductivity at 99° = 0.350 Btu/hr sq ft (°F/ft)

$$(N_{\mathrm{Re}})_o = \frac{D_{\mathrm{eq}}\bar{v}\rho}{\mu} = \frac{(0.0356)(5)(62.3)}{(0.7 \times 6.72 \times 10^{-4})} = 23{,}500$$

$$(N_{\mathrm{Pr}})_o = \frac{c_P \mu}{k} = \frac{(1)(0.7 \times 2.42)}{(0.350)} = 4.84$$

$$\frac{h_o D_{\mathrm{eq}}}{k} = 0.023(N_{\mathrm{Re}})^{0.8}(N_{\mathrm{Pr}})^{0.4}$$
$$= 0.023(23{,}500)^{0.8}(4.84)^{0.4} = 135$$

$$h_o = \frac{(135)(0.350)}{(0.0356)} = 1330 \ \text{Btu/hr sq ft °F}$$

The total resistance at the hot end of the heat exchanger is

$$\Sigma R = \frac{1}{(0.361)(272)} + 0.00113 + \frac{1}{(0.435)(1330)}$$
$$= 0.01306 \ \text{hr sq ft °F/Btu}$$

This value of 0.01306 is about 7 per cent lower than that based upon the average properties.

Cold end of exchanger

For Benzene: Temperature = 100°F

$$c_P = 0.43 \text{ Btu/lb °F}$$
$$\mu = 0.52 \text{ centipoise}$$
$$\rho = 53.8 \text{ lb/cu ft}$$
$$k = 0.09 \text{ Btu/hr sq ft (°F/ft)}$$

$$(N_{\text{Re}})_i = \frac{D\bar{v}\rho}{\mu} = \frac{(0.115)(3.83)(53.8)}{(0.52 \times 6.72 \times 10^{-4})} = 67,700$$

$$(N_{\text{Pr}})_i = \frac{c_P\mu}{k} = \frac{(0.43)(0.52 \times 2.42)}{(0.09)} = 6.02$$

$$\frac{h_i D}{k} = 0.023(67,700)^{0.8}(6.02)^{0.3} = 290$$

$$h_i = \frac{(290)(0.09)}{0.115} = 227 \text{ Btu/hr sq ft °F}$$

For water: Temperature = 70°F

$$\rho = 62.3 \text{ lb/cu ft}$$
$$c_P = 1 \text{ Btu/lb °F}$$
$$\mu = 0.94 \text{ centipoise}$$
$$k = 0.334 \text{ Btu/hr sq ft (°F/ft)}$$

$$N_{\text{Re}} = \frac{D_{\text{eq}}\bar{v}\rho}{\mu} = \frac{(0.0356)(5)(62.3)}{(0.94 \times 6.72 \times 10^{-4})} = 17,500$$

$$N_{\text{Pr}} = \frac{c_P\mu}{k} = \frac{(1.0)(0.94 \times 2.42)}{(0.334)} = 6.81$$

$$\frac{h_o D_{\text{eq}}}{k} = 0.023(17,500)^{0.8}(6.81)^{0.4} = 123.0$$

$$h_o = \frac{123.0 \times 0.334}{0.0356} = 1160 \text{ Btu/hr sq ft (°F/ft)}$$

$$\Sigma R = \frac{1}{(0.361)(227)} + 0.00113 + \frac{1}{(0.435)(1160)}$$

$$= 0.01521 \text{ hr sq ft °F/Btu}$$

Thus the cold-end total resistance is about 9 per cent higher than that based upon the average properties. Thus, it seems justifiable in view of the comparison made between U at the terminals and U based upon average properties that the average temperature is satisfactory and the $(\Delta T)_{lm}$ assumptions are valid. Actually the use of $(\Delta T)_{lm}$ is usually justified whenever the flow character is fully turbulent *and physical properties do not vary widely*, so that checking terminal values is rarely necessary.

SHELL-AND-TUBE HEAT EXCHANGER CALCULATIONS

Under certain restricted conditions, the heat-transfer design equation may be written in the form of Equation 15.15 as $q = -UA(\Delta T)_{lm}$. It is a useful equation despite its many restrictions. In most commercial heat exchangers of the shell-and-tube variety, however, the flow is neither cocurrent nor countercurrent. On the contrary, reference to Figure 15.3 shows that the flow is really a complex pattern of mixed flows. Note however that, if either stream is at a constant temperature as in a condenser or vaporizer, the heat transfer behaves as in a simple exchanger.

If the same restrictions that are applicable for simple-flow exchangers are assumed to hold, the $(\Delta T)_{lm}$ may be used only in conjunction with a suitable correction factor to account for the geometry of the exchanger and for the fact that flow is neither countercurrent nor cocurrent. Thus for complex exchangers,

$$q = -UAY(\Delta T)_{lm} \tag{15.17}$$

where Y is the geometry factor and $(\Delta T)_{lm}$ is the temperature driving force evaluated from terminal ΔT's based upon countercurrent operation.

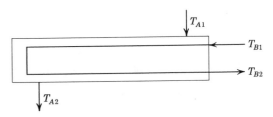

Figure 15.13. Schematic representation of a one shell pass two tube pass exchanger.

Bowman, Mueller, and Nagle (5) have correlated the Y-factor in terms of two dimensionless ratios.

Figure 15.13 schematically indicates a one shell pass two tube pass exchanger with phase A entering the shell at temperature T_{A1} and leaving at T_{A2}. Phase B enters the tubes at temperature T_{B1} and leaves at T_{B2}. Two dimensionless ratios are defined. One is a temperature ratio,

$$Z = \frac{T_{A1} - T_{A2}}{T_{B2} - T_{B1}} \tag{15.18}$$

Equation 15.18 is in reality a ratio of the hourly heat capacities, that is, the heat required to raise the hourly flow 1°F for each of the two phases. The second dimensionless ratio, which is also a temperature ratio expressing the effectiveness of the exchanger, is defined as

$$X = \frac{T_{B2} - T_{B1}}{T_{A1} - T_{B1}} \tag{15.19}$$

Figures 15.14 and 15.15 show the graphical correlation proposed by Bowman for several basic configurations. In the derivations of these diagrams, it was assumed that the shell-side fluid is so well mixed that an average temperature at any cross section is applicable. Also, the assumptions of constant over-all resistance, constant specific heat, no phase change, and equal transfer area in each pass are implied. In the event that these assumptions are not valid, reference should be made to Gardner (11, 12) who treats more complex situations.

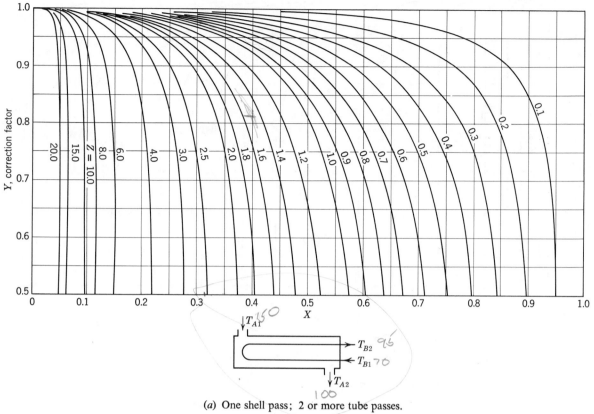

(a) One shell pass; 2 or more tube passes.

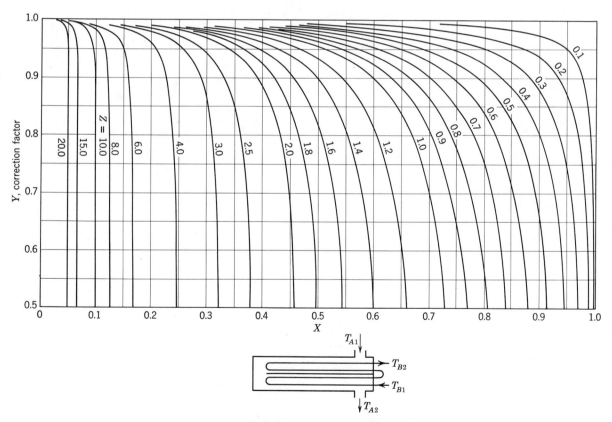

(b) Two shell passes; 4 or more tube passes.

Figure 15.14. $(\Delta T)_{lm}$ correction factor for mixed-flow heat exchangers. (Tubular Exchange Mfg. Assoc.)

$$X = \frac{T_{B2} - T_{B1}}{T_{A1} - T_{B1}} \qquad Z = \frac{T_{A1} - T_{A2}}{T_{B2} - T_{B1}}$$

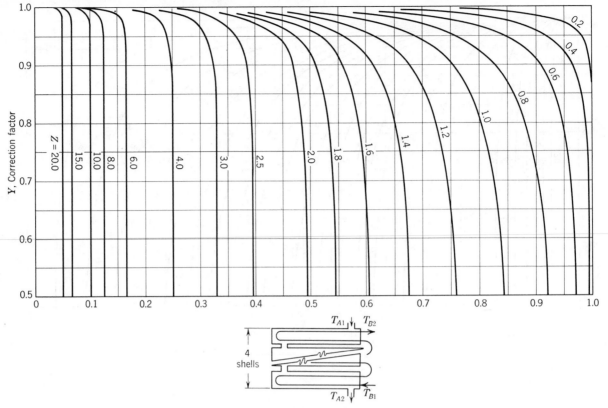

(c) Four shell passes; 8 or more tube passes.

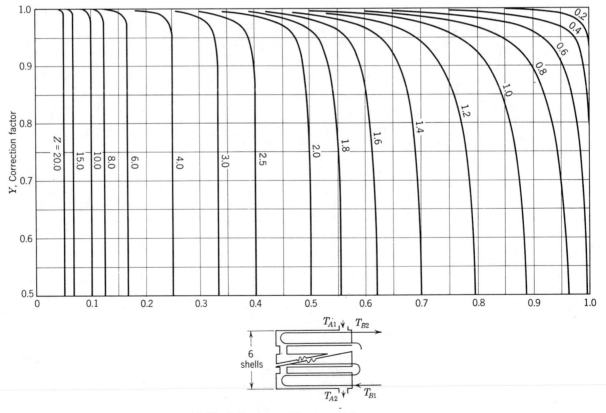

(d) Six shell passes; 12 or more tube passes.

Figure 15.14 (*continued*). $(\Delta T)_{lm}$ correction factor for mixed-flow heat exchangers.   (Tubular Exchange Mfg. Assoc.)

$$X = \frac{T_{B2} - T_{B1}}{T_{A1} - T_{B1}} \qquad Z = \frac{T_{A1} - T_{A2}}{T_{B2} - T_{B1}}$$

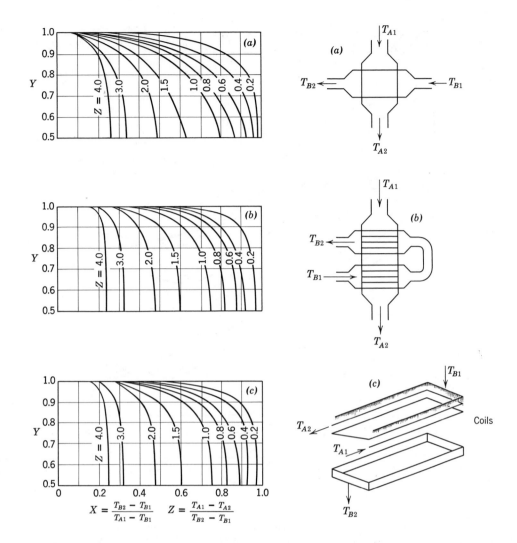

Figure 15.15. $(\Delta T)_{lm}$ correction factor for crossflow heat exchangers. (*a*) Crossflow, both fluids unmixed, one tube pass. (*b*) Crossflow, shell fluid mixed, two tube passes. (*c*) Crossflow drip, helical coils with two turns. (4)

Illustration 15.4. A shell-and-tube heat exchanger having two shell passes and four tube passes is being used for cooling. The shell-side fluid enters at 400°F and leaves at 200°F and the tube-side fluid enters at 100°F and leaves at 200°F. What is the mean temperature difference between the hot fluid and the cold fluid?

SOLUTION. A simple sketch will aid in this solution. The 2–4 exchanger can be represented as follows

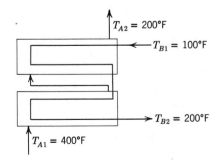

First, the $(\Delta T)_{lm}$ based upon true countercurrent operation is evaluated. This evaluation may be represented by

$$400° \rightarrow 200°$$
$$200° \leftarrow 100°$$
$$\Delta T_1 = 200° \quad \Delta T_2 = 100°$$

Therefore

$$\Delta T_{lm} = \frac{\Delta T_1 - \Delta T_2}{\ln \dfrac{\Delta T_1}{\Delta T_2}} = \frac{200 - 100}{\ln \dfrac{200}{100}} = 144°$$

The Y factor can be obtained from Figure 15.14*b*.

$$Z = \frac{T_{A1} - T_{A2}}{T_{B2} - T_{B1}} = \frac{400 - 200}{200 - 100} = 2$$

$$X = \frac{T_{B2} - T_{B1}}{T_{A1} - T_{B1}} = \frac{200 - 100}{400 - 100} = 0.33$$

and from Figure 15.14*b*, $Y = 0.96$

Thus $(\Delta T)_{\text{corrected}} = Y(\Delta T)_{lm} = 0.96 \times 144 = 138°F$

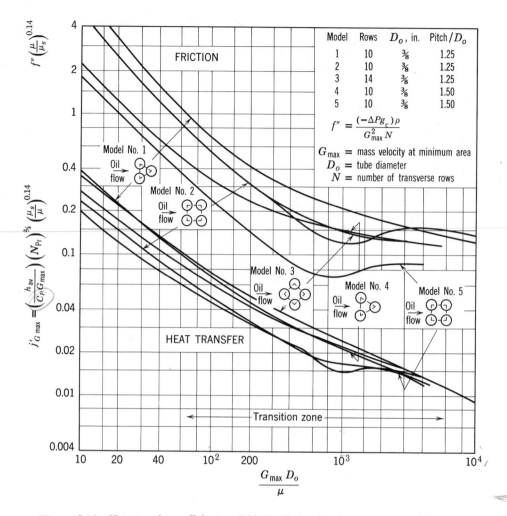

Figure 15.16. Heat-transfer coefficients and friction factors for flow normal to tube banks.
(From Bergelin, O. P., Brown, G. A., Doberstein, S. C., *Trans. ASME* **74,** 953 (1952). With permission.)

Tube-Side Coefficients. Heat-transfer coefficients for the fluid flowing through the tubes are calculated in accordance with correlations presented in Chapter 13. The tube-side coefficient should be evaluated first and analyzed to ascertain whether the tube-side resistance is dominant. This should be done considering the nature of the shell-side fluid. Calculation of tube-side coefficients first may eliminate the complex procedures necessary in evaluation of shell-side coefficients. In any event, the usual heat-transfer-coefficient correlations apply for the tube-side fluid.

Shell-Side Coefficients. The heat-transfer coefficients for the shell-side fluid cannot be accurately calculated by any of the correlations presented thus far. A continually changing path of the shell-side fluid prevails as a result of the baffles influencing its flow direction and because the cross-sectional flow area will vary as the fluid flows among the tubes, the heat-transfer coefficient must be determined entirely from experimental data. In addition, leakage past baffles and short circuits of flow paths

also limit the effectiveness of the exchanger. It is not the purpose to present here the myriad of techniques available to account for leakage, variable areas, and other corrections in the evaluation of the shell-side coefficient. The reader is referred to Tinker (35) for a more detailed analysis.

For most exchangers, a reasonable approximation can be made by simple procedures. The simplest procedure is to use 60 per cent of the coefficient calculated for flow normal to rectangular tube banks, assuming no leakage around baffles and no ineffective transfer area of tubes.

Heat-transfer coefficients have been investigated (3, 4, 15) for fluids flowing normal to banks of tubes in several different types of heat exchangers. The resulting correlation is shown in Figure 15.16 along with the friction-factor correlation to enable one to predict pressure drop through the shell.

Tube-Sheet Layout. Several different tube-bundle arrangements are shown in Figure 15.17.

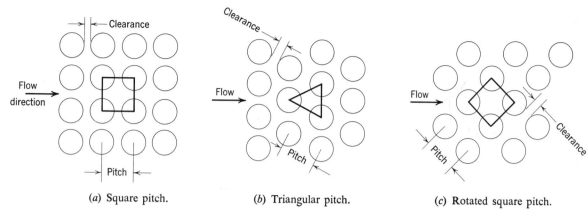

| (a) Square pitch. | (b) Triangular pitch. | (c) Rotated square pitch. |

Figure 15.17. Common tube-sheet layouts.

Each tube arrangement presents certain advantages. For instance, the square pitch makes the tubes more available for external cleaning, and the triangular pitch increases turbulence of the shell-side fluid. Accordingly, the triangular pitch will give a higher coefficient but also a higher pressure drop.

The *tube pitch* is defined as the shortest center-to-center distance between adjacent tubes. The *tube clearance* is the shortest distance between adjacent tubes.

Shell-Side Coefficients for Liquid Metals. The significance of the Prandtl number in heat-transfer coefficients for liquid metals has already been discussed in Chapter 13. Lyon (22) has shown that, for fluids of low Prandtl number, flowing inside tubes, the heat transfer is an unique function of the Peclet number. On the same basis, Rickard, Dwyer and Dropkin (32) postulated and experimentally verified that a similar relationship is valid for flow of fluids of low Prandtl numbers across banks of tubes. Their correlation is

$$\frac{\bar{h}_o D_o}{k} = 4.03 + 0.228 \left[\frac{D_o v_{max} \rho c_P}{k}\right]^{0.67} \quad (15.20)$$

where $\bar{h}_o$ = average heat-transfer coefficient, Btu/hr sq ft °F

D_o = outside diameter of tube, ft

v_{max} = shell-side velocity of fluid based on minimum flow area, ft/hr

This correlation is valid for Reynolds numbers between 20,000 and 200,000.

Fouling. The build-up of dirt or scale on heat-transfer surfaces is a problem which must be considered in heat-exchange equipment. As a result of the formation of scale, another resistance is added to the heat-transfer path causing a reduction in the rate of exchange. If the thickness and thermal conductivity of the scale can be established, the magnitude of the scale resistance can be calculated. Unfortunately, the scale continually builds up so that its resistance is a varying quantity.

The deposit which increases resistance may be simple dirt, mechanically deposited; it may be a true crystalline scale deposited because of the inverted solubility curve of some dissolved constituent present; or it may be a deposit originating from chemical reactions within the fluid or between the fluid and the tube. The rate of deposit is a complex function of heat duty, flow rate, geometry, and the particular characteristics of the fluid.

Since scale build-up varies with time and its increase is accompanied by a drop in the transfer rate, some optimum time exists for operation before cleaning. True scale will be shown in Chapter 19 to be a parabolic function of time. An exponential decrement in coefficient has been used to express build-up in other cases. Scales may approach an asymptote, or they may "mushroom" rapidly once they start (27).

The absence of a definitive theory regarding scale formation usually requires actual performance data to help in determining length of service before cleaning. The limit in permissible fouling may be expressed in resistance units or as a percentage of the clean coefficient below which operation becomes impossible or uneconomical.

Some fouling factors recommended by the Tubular

Table 15.2. TYPICAL FOULING FACTORS

Fouling Factor, $R_d = 1/h_d$
hr °F sq ft/Btu

	Water Velocity,	
	3 ft/sec or less	3 ft/sec or more
Sea water (up to 125°F)	0.005	0.0005
Well water	0.001	0.001
Delaware, Lehigh River waters	0.003	0.002
Brine	0.001	0.001
Fuel oil	0.005	0.005

Exchange Manufacturer's Association (34) are found in Table 15.2. These factors are treated as resistances and added to the "clean" tube resistance give the operating resistance that might be expected.

Illustration 15.5. A steel mill desires to cool 3000 gal/hr of an inorganic liquid from 250°F to 180°F by Lehigh River water flowing through a shell-and-tube exchanger at a rate of 5000 gal/hr. The liquid is somewhat acidic in nature and so an alloy exchanger is probably required.

Within the plant is an unused shell-and-tube, floating-head exchanger having one shell pass and two tube passes. The shell is mild steel, and the tubes are type 304 stainless. Specifications of the surplus exchanger reveal a 20-in. I.D. shell with 158 1-in. 14 BWG tubes 16 ft long laid out on a $1\frac{1}{4}$-in. square pitch. Baffles are placed 6 in. apart.

Will the surplus exchanger be suitable for the cooling requirements? Assume the inorganic liquid has the physical properties of water.

SOLUTION. Since the liquor to be cooled is corrosive, it should be placed in the tubes. The tubes of 304 stainless will probably be acceptable for this purpose. The river water will in all probability be dirty, and the exchanger should have a removable tube bundle. The square pitch of the tube layout is good for cleaning the outside of the tubes. Consequently, from a point of view of materials of construction and ease of maintenance, the surplus exchanger seems adequate. However, this is not enough, since the heat load on the exchanger will demand a certain transfer area. It will be necessary therefore to estimate the cooling surface required and check this against the available area. The heat load can be evaluated from the temperature reduction of the inorganic material.

$$w = (3000 \text{ gal/hr})(8.34 \text{ lb/gal}) = 25000 \text{ lb/hr}$$

$$q = wc_P(T_{A2} - T_{A1})$$

$$= (25000 \text{ lb/hr})(1 \text{ Btu/lb °F})(250 - 180°F)$$

$$= 1{,}750{,}000 \text{ Btu/hr}$$

From this heat duty, the temperature rise of the cooling water can be calculated

$$(T_{B2} - T_{B1}) = \frac{1{,}750{,}000 \text{ Btu/hr}}{(5000 \text{ gal/hr})(8.34 \text{ lb/gal})}(1 \text{ Btu/lb °F}) = 42°F$$

To be safe, the cooling-water entrance temperature should be based on the maximum river-water temperature, say 80°F. Thus the water will enter at 80°F and leave at 122°F. Therefore, the true countercurrent $(\Delta T)_{lm}$ for this exchanger is

$$(\Delta T)_{lm} = \frac{(250 - 122) - (180 - 80)}{\ln \dfrac{(250 - 122)}{(180 - 80)}} = 113.5°$$

For a one shell pass, two tube pass exchanger, Figure 15.14a is used

$$X = \frac{T_{B2} - T_{B1}}{T_{A1} - T_{B1}} = \frac{180 - 250}{80 - 250} = 0.41$$

$$Z = \frac{T_{A1} - T_{A2}}{T_{B2} - T_{B1}} = \frac{80 - 122}{180 - 250} = 0.60,$$

Therefore from Figure 15.14a, at $X = 0.41$ and $Z = 0.60$, $Y = 0.97$

$$(\Delta T)_{corr.} = Y(\Delta T)_{lm} = (0.97)113.5 = 110$$

Tube-Side Coefficient (Inorganic Fluid)

1-in. 14 BWG tubes:

$$D_i = 0.834 \text{ in. (Appendix C-7)}$$

$$S_i = \frac{\pi}{4} D_i^2 = (0.785)\left(\frac{(0.834)^2}{144}\right) = 0.00381 \text{ sq ft/tube}$$

Therefore, the total flow area

$$= \frac{(\text{no. of tubes})(\text{cross-sectional area per tube})}{(\text{no. of passes})}$$

$$S_i = \frac{158 \times 0.00381}{2} = 0.301 \text{ sq ft}$$

Mass velocity $G = \dfrac{25{,}000}{0.301} = 83{,}000 \text{ lb/hr sq ft}$

The viscosity of the liquid at $T_{av} = (250 + 180)/2 = 215°F$ is 0.25 centipoise. The flow character is determined by evaluating N_{Re}

$$N_{Re} = \frac{DG}{\mu} = (0.834/12)\frac{83{,}000}{(0.25 \times 2.42)} = 9550$$

$$\frac{L}{D} = \frac{(16)}{(0.834/12)} = 230$$

Thus j_q (Appendix C-5) $= 0.0038$

To calculate the Prandtl number,

$$c_P = 1.0 \text{ Btu/lb °F}, k = 0.402 \text{ Btu/hr sq ft (°F/ft)}$$

Therefore, $N_{Pr} = \dfrac{c_P\mu}{k} = \dfrac{(1)(0.25 \times 2.42)}{0.402} = 1.5; N_{Pr}^{2/3} = 1.31$

and
$$j_q = \left(\frac{h_i}{c_P G}\right)\left(\frac{c_P\mu}{k}\right)^{2/3}\left(\frac{\mu_{is}}{\mu}\right)^{0.14} = 0.0038$$

as a first trial, assume $(\mu_{is}/\mu)^{0.14} = 1$, then solving the above equation for h_i gives

$$h_i = \frac{(0.0038)(1)(83{,}000)}{(1.31)(1)} = 240 \text{ Btu/hr sq ft °F}$$

Shell-Side Fluid (Cooling Water)

Flow rate $= 5000 \times 8.34 = 41{,}600 \text{ lb/hr}$

Flow area: 1-in. O.D. tubes on a $1\frac{1}{4}$ in. square pitch gives a clearance for fluid flow of 0.25 in. The number of tubes at the center of the exchanger may be evaluated by dividing the shell diameter by the tube pitch, and subtracting one from the resulting number, or

$$\text{Tubes at center} = \frac{20}{1.25} - 1 = 16 - 1 = 15$$

Taking the length of the shell-side flow area as equal to the baffle spacing, the minimum flow area for the shell-side fluid thus becomes

$$S_{shell} = (15)\left(\frac{0.25}{12}\right)\left(\frac{6}{12}\right) = 0.1566 \text{ sq ft}$$

Therefore, $G_{max} = \dfrac{41,600}{0.1566} = 266,000$ lb/hr sq ft

μ at T_{av} of 101°F = 0.68 centipoise

and $N_{Re} = \dfrac{D_o G_{max}}{\mu} = (1/12)\dfrac{(266,000)}{(0.68 \times 2.42)} = 13,460$

From Figure 15.16 $j'_{Gmax} = 0.0085$

Therefore $\left(\dfrac{h_o}{c_P G}\right)\left(\dfrac{c_P \mu}{k}\right)^{2/3}\left(\dfrac{\mu_{os}}{\mu}\right)^{0.14} = 0.0085$

at 101°, $c_P = 1$, $k = 0.350$, assume $(\mu_{os}/\mu)^{0.14} = 1$

$N_{Pr} = \dfrac{c_P \mu}{k} = \dfrac{(1)(0.68 \times 2.42)}{0.350} = 4.7; \ (N_{Pr})^{2/3} = 2.81$

and

$h_o = \dfrac{(0.0085)(1)(266,000)}{(2.81)(1)} = 805$ Btu/hr sq ft °F

Take 60 per cent: $h_o = 0.60 \times 805 = 483$ Btu/hr sq ft °F

The tube-side resistance is the larger of the two fluid resistances, consequently the over-all resistance will be based upon the inside tube area. Therefore, summarizing the resistances on this basis and neglecting the fouling factor for the moment give

			Resistance
Tube-side fluid,	$\dfrac{1}{h_i} = \dfrac{1}{240}$		= 0.0042
Tube,	$\dfrac{\Delta x (A_i)}{k\bar{A}}$	$= \dfrac{0.083(0.218)}{(12)(9)(0.240)}$	= 0.0007
Shell-side fluid,	$\dfrac{A_i}{h_o A_o}$	$= \dfrac{1(0.218)}{483(0.262)}$	= 0.00173
Total resistance	$= \dfrac{1}{U_i}$		= 0.00663
			hr sq ft °F/Btu

and $U_i = \dfrac{1}{0.00663} = 150$ Btu/hr °F sq ft inside area

Using this over-all coefficient, the surface coefficients may be readjusted to account for $(\mu_s/\mu)^{0.14}$ not being unity. The surface temperature may be calculated by the usual manner of proportioning resistances and driving forces based on coefficients calculated in the first trial. The outside-surface temperature is determined from

$$\dfrac{\text{Shell-side resistance}}{\text{Total resistance}} = \dfrac{\text{shell-side driving force}}{\text{total driving force}} = \dfrac{0.00173}{0.00663}$$

$$= \dfrac{\Delta T_0}{(215 - 101)}$$

$\Delta T_o = 31$°F. Therefore, the average outside surface temperature = 132°F, and $\mu_{os} = 0.5$ centipoise.

$$\left(\dfrac{\mu_{os}}{\mu}\right)^{0.14} = \left(\dfrac{0.5}{0.68}\right)^{0.14} = 0.958$$

Adjusting, $h_o = \dfrac{483}{0.958} = 505$ Btu/hr sq ft °F

Tube-side surface temperature is

$$\dfrac{0.0042}{0.00663} = \dfrac{\Delta T_i}{114}$$

$$\Delta T_i = 72.0°F$$

and the surface temperature is $215 - 72 = 143°F$, and $\mu_{is} = 0.45$ centipoise

$$\left(\dfrac{\mu_{is}}{\mu}\right)^{0.14} = \left(\dfrac{0.45}{0.25}\right)^{0.14} = 1.085$$

Adjusting, $h_i = \dfrac{240}{1.085} = 221$ Btu/hr sq ft °F

The new total resistances becomes

Tube side, $\dfrac{1}{h_i} = \dfrac{1}{221}$	= 0.00453
Tube, $\dfrac{\Delta x A_i}{k\bar{A}}$	= 0.00070
Shell side, $\dfrac{A_i}{h_o A_o} = \dfrac{0.834}{505}$	= 0.00165
Total, $\dfrac{1}{U_i}$	= 0.00688 hr sq ft °F/Btu

Thus the corrected over-all resistance (dirt free) is

$$(U_i)_{clean} = \dfrac{1}{0.00688} = 145.0$$ Btu/hr sq ft °F

The required area is,

$$A_i = \dfrac{q}{U_i Y(\Delta T)_{lm}} = \dfrac{1,750,000}{(145)(0.97)(113.5)} = 111$$ sq ft

The total surface actually available in the exchanger is

$$A_i = (158)(16)(3.14)\left(\dfrac{0.834}{12}\right) = 550$$ sq ft

Thus, the required area is considerably less than the available area so that, even with scale build-up as operation continues, the heat exchanger should perform the necessary cooling. To verify the allowable dirt factor, calculate $1/U$ using the actual surface area, or

$$\left(\dfrac{1}{U_i}\right)_{dirty} = \dfrac{A_i Y(\Delta T)_{lm}}{q} = \dfrac{(550)(0.97)(113.5)}{1,750,000}$$

$$= 0.0346 \text{ hr sq ft °F/Btu}$$

The permissible scale resistance is then

$$R_d = \dfrac{1}{U_{id}} - \dfrac{1}{U_{ic}} = 0.0346 - 0.00688$$

$$= 0.0277 \text{ hr sq ft °F/Btu}$$

This allowable scale resistance is significantly greater than the value for Lehigh River water reported in Table 15.2. In the event that the required area and available area were closer together, the dirt resistance could be much more important because frequent cleaning would be necessary to keep the exchanger performing as specified. This solution is not complete in that pressure-drop requirements should also be examined, as well as the economics of the operation.

Condensers. Heat exchangers whose prime purpose is to bring about a change of phase from a vapor to a liquid by means of a coolant are commonly called condensers. Chapter 13 has introduced the condensation coefficient and the mechanism of dropwise and film-type condensation. Most commonly encountered condensation operations in the process industry are of the film type.

The same basic design equation used for heat exchangers is valid, but the nature of the coefficient correlations is somewhat different depending upon whether the condensation process takes place on vertical or horizontal tubes.

For *vertical* surfaces and film-type condensation, McAdams (24, 25) recommends

$$h = 1.13\left[\frac{k_f^3 \rho_f^2 g(\Delta H)}{L\mu_f(T_{sv} - T_s)}\right]^{1/4} = 1.18\left[\frac{k_f^3 \rho_f^2 g\pi D}{\mu_f w}\right]^{1/3} \tag{15.21}$$

where ΔH = latent heat of condensation, Btu/lb
g = gravitational acceleration, ft/hr²
L = length of tube or vertical surface, ft
T_{sv} = saturated-vapor temperature, °F
T_s = surface temperature, °F
D = outside diameter of tube, ft
w = pounds of condensate per hour

Either part of Equation 15.21 is usable, depending upon the information at hand. For film condensation on *horizontal* tubes (23),

$$h = 0.725\left[\frac{k_f^3 \rho_f^2 g(\Delta H)}{ND\mu_f(T_{sv} - T_s)}\right]^{1/4} = 0.95\left[\frac{k_f^3 \rho_f^2 g(\Delta H)}{\mu_f w}\right]^{1/3} \tag{15.22}$$

where N = the number of tubes in a vertical row of tubes.

The condensate properties are evaluated at the average film temperature (T_f), recognizing that the variation in condensate thickness around the tube, or down it in the case of vertical tubes, will cause a considerable variation in the tube-wall temperature, McAdams (23) suggests for T_f,

$$T_f = [T_{sv} - \tfrac{3}{4}(T_{sv} - T_s)] \tag{15.23}$$

Equation 15.22 is written on the assumption that the condensate flows on or over the tube surface in laminar flow. Equation 15.21 includes a 28 per cent increment above the theoretical equation to account for the presence of turbulence. No conclusive correlation for turbulent-flowing films is yet at hand, but Kirkbride (21) and Badger (1, 2), investigating the condensation of diphenyl and "Dowtherm" on vertical nickel tubes, suggest

$$h = 0.0077\left[\frac{4w}{\mu_f \pi D}\right]^{0.4}\left[\frac{k_f^3 \rho_f^2 g}{\mu_f}\right]^{1/3} \tag{15.24}$$

if the flow at the bottom of the tube is turbulent.

Vaporizers. A vaporizer is a heat exchanger specifically designed to supply latent heat of vaporization to a fluid. If the vapor formed is steam, the exchanger is commonly called an evaporator. If the exchanger is used to supply the heat requirements at the bottom of a distillation tower, it is called a reboiler. Vaporizers satisfy the latent-heat requirements to a boiling fluid, and the heat duty for this type of exchanger is easily computed. Many different design arrangements are available depending upon the service.

Heat-transfer coefficients for vaporizers are a complex function of such items as flow rates, fraction of liquid vaporized, transfer area, physical design, etc. Practical-design heat fluxes are of the order of magnitude of 20,000 Btu/hr sq ft. Design procedures for vaporizers and reboilers may be found in Kern (19).

EXTENDED SURFACES FOR HEAT TRANSFER

The desirability of extending the surface of a tube by addition of fins or spines has already been discussed. Design calculations for finned heat exchangers are complex because of uncertainties in knowing the temperature driving force distribution over the resistances between the fluids exchanging heat.

Longitudinal Fins. The simplest type of fin is the longitudinal metal strip extending lengthwise down a tube. To understand the mechanism of this type of exchanger, one may consider the steady flow of heat along a thin fin extending from a tube wall. Figure 15.18 illustrates a portion of such an exchanger.

The mathematical solution to the heat-transfer equation for this configuration has been offered by Gardner (13), based upon the following assumptions:

1. Steady state exists so that the temperature distribution is constant and heat flow through the fin is uniform.
2. No heat source exists within the fin.
3. The fin has a constant thermal conductivity.

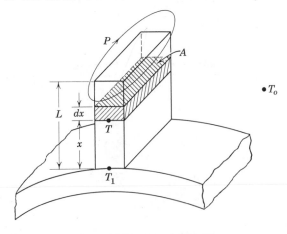

Figure 15.18. Longitudinal fin.

4. The heat-transfer coefficient is uniform over the entire fin surface.

5. Ambient temperature is constant.

6. The temperature at the base of the fin is constant.

7. The fin is thin with respect to its height and so no temperature gradient exists across the width of the fin.

8. The heat transferred from the fin tip is negligible compared with the heat leaving the fin sides.

9. Good contact exists between fin and tube.

Referring to Figure 15.18, let

L = the height of the fin

T_1 = temperature at the base of the fin

T = temperature of the fin at a distance x from the base

T_o = temperature of fluid surrounding fin, $< T_1$

A = transfer area of fin

P = fin perimeter

h_f = heat-transfer coefficient for fin

Consider the element of the fin (dx) as shown in Figure 15.18.

The net heat flow into this elemental slab by conduction from the base of the fin is

$$q_{\text{in}} = -kA\left(\frac{d^2T}{dx^2}\right) dx \qquad (15.25)$$

The heat flow from the surface of (dx) by convection must be

$$q_{\text{out}} = h_f(T - T_o)P\, dx \qquad (15.26)$$

But to maintain a constant temperature in dx requires that $q_{\text{in}} = q_{\text{out}}$. Adding Equation 15.25 and Equation 15.26 gives

$$-kA\left(\frac{d^2T}{dx^2}\right)dx + h_f(T - T_o)P\, dx = 0 \quad (15.27)$$

Substituting the defined factors $\theta = T - T_o$ and $\lambda = \sqrt{h_f P / Ak}$ into Equation 15.27 and rearranging gives

$$\frac{d^2\theta}{dx^2} - \lambda^2\theta = 0 \qquad (15.28)$$

The solution to Equation 15.28 is

$$\theta_x = C_1 \cosh \lambda x + C_2 \sinh \lambda x \qquad (15.29)$$

The constants C_1 and C_2 can be evaluated by substituting the boundary conditions.

At $x = 0$, $\theta = \theta_1$, and $C_1 = \theta_1$

At $x = L$, $\dfrac{d\theta}{dx} = 0$ by assumption 8 of Gardner;

therefore, from Equation 15.29

$$\frac{d\theta}{dx} = \lambda C_1 \sinh \lambda L + \lambda C_2 \cosh \lambda L = 0 \quad (15.30)$$

and $\qquad C_2 = -\theta_1 \dfrac{\sinh \lambda L}{\cosh \lambda L}$

The temperature distribution is given by

$$\theta_x = \theta_1 \cosh \lambda x - \theta_1 \tanh \lambda L \sinh \lambda x \quad (15.31)$$

or

$$\theta_x = \theta_1 \frac{\cosh \lambda(L - x)}{\cosh \lambda L} \qquad (15.31a)$$

Rewriting Equation 15.31a gives

$$\frac{(\Delta T)_x}{\Delta T_1} = \frac{\cosh \lambda(L - x)}{\cosh \lambda L} \qquad (15.31b)$$

where $(\Delta T)_x$ = temperature difference between the fin and the surrounding fluid at a distance x from the base of the fin = $T_x - T_o$

ΔT_1 = temperature difference at the base of the fin = $T_1 - T_o$

A fin efficiency may be defined as

$$\eta_F = \frac{(\Delta T)_m}{\Delta T_1} = \frac{\tanh \lambda L}{\lambda L} \qquad (15.32)$$

where $(\Delta T)_m$ is the temperature difference between the fin and surrounding fluid averaged over the entire length of the fin. Equation 15.32 applies only to the fin and not to the unfinned surface of the tube. The fin efficiency will always be less than 100 per cent since a unit area of fin surface is not as effective as a unit area of tube surface. A temperature gradient within the fin must exist to cause conduction through the fin.

An empirical correlation for evaluating fin-side coefficients for longitudinal fins in double-pipe heat exchangers has been developed by Kern (20) for numerous heating and cooling experiments on different varieties of equipment arrangements. The correlation is shown in Figure 15.19.

Transverse Fins. Figure 15.5 illustrates several types of transverse fins. The fin-efficiency equation for transverse fins is somewhat more difficult to develop than that for the longitudinal fin. The mathematical development of transverse fins has also been accomplished by Gardner (13). Consider the most general type of transverse fin, that of varying cross section. The simplest fin of this type is sketched in Figure 15.20.

In this type of fin, the surface available for heat transfer is a function of L.

The net heat entering the differential section (dx) by conduction from the base is

$$\frac{d}{dx}\left(kA\frac{dT}{dx}\right)$$

Heat will leave this elemental section by convection in amount $dq = h_f \Delta T\, dA$. For thermal equilibrium to exist in dx, the decrease in conduction through the fin at that point must equal the convection from the fin, or

$$\frac{d}{dx}\left(kA\frac{dT}{dx}\right) - h_f \Delta T\, dA = 0 \qquad (15.33)$$

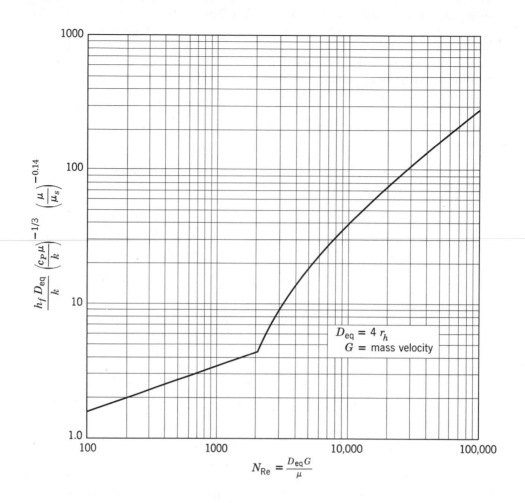

Figure 15.19. Heat transfer in longitudinally finned heat exchangers.
(Kern, D. Q. *Process Heat Transfer*, McGraw-Hill Book Co., New York, 1950, p. 525.)

If the thermal conductivity is assumed constant, Equation 15.33 becomes

$$kA \frac{d^2T}{dx^2} + k\left(\frac{dA}{dx}\right)\left(\frac{dT}{dx}\right) - h_f \Delta T \, dA = 0 \quad (15.34)$$

or

$$\frac{d^2T}{dx^2} + \frac{1}{A}\left(\frac{dA}{dx}\right)\left(\frac{dT}{dx}\right) - \frac{h_f}{kA} \Delta T = 0$$

$$(15.34a)$$

This second-order differential equation has been solved by Gardner using Bessel functions. Efficiencies calculated from Equation 15.34a may be found elsewhere (18). It is generally advisable to base design calculations on experimental values of h_f for the particular fin design being used. This information may usually be obtained from the manufacturer.

THERMAL-ENERGY TRANSFER BY RADIATION

So far, the transfer of thermal energy has been considered when it occurs by the mechanisms of conduction and convection or turbulent diffusion. In these cases, the transfer rate can be determined by integration of the usual rate equation, and equipment sizes can be determined by combining the rate equation with appropriate material and energy balances and the proper design

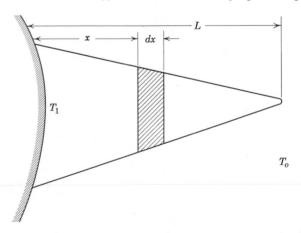

Figure 15.20. Varying-cross-section transverse fin.

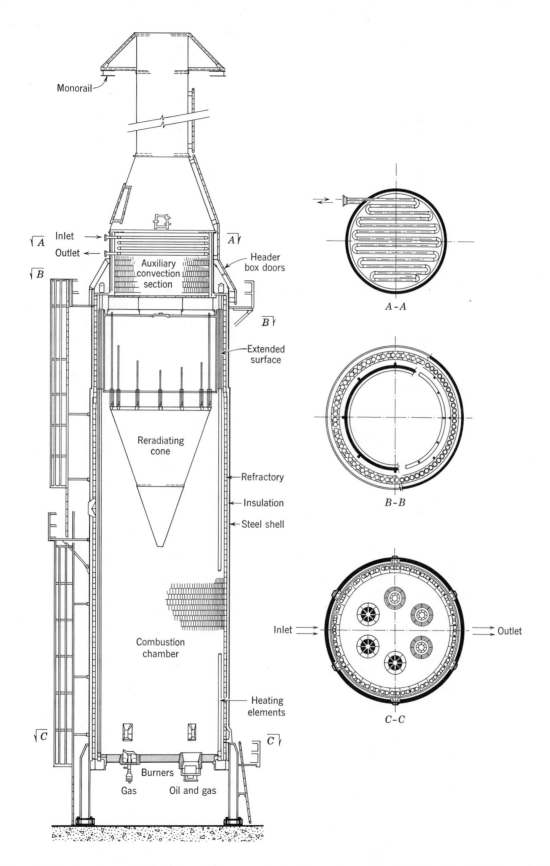

Figure 15.21. Petroleum pipe still with auxiliary convective heating sections. (Courtesy Petro-Chem Development Co.)

equation. Simultaneously and independently, energy is transferred by radiation. Thus, the relation

$$q = q_c + q_k + q_r \qquad (15.35)$$

where q = total rate of energy transfer as heat

q_c = rate of thermal-energy transfer by the mechanism of convection

q_k = rate of thermal-energy transfer by the mechanism of conduction

q_r = rate of thermal-energy transfer by the mechanism of radiation

is applicable in every case. Of course in many cases, one of the mechanisms may transfer a negligible amount of thermal energy compared to one or more of the others. In this section, the phenomenon of energy transmission by radiation will be considered in some detail. The fundamental mechanism involved and the quantitative relations describing this mechanism will be developed, for the transfer of radiant energy does not seem to be adequately described in terms of rate equations of the form considered in the preceding chapters.

Equipment Transferring Energy Mainly by Radiation. Radiation is the major energy-transfer mechanism in a large variety of chemical-process equipment. Electric heaters, direct-fired kettles, steam boilers, rotary kilns, blast furnaces, and petroleum pipe stills are examples of such units. Figure 15.21 is a schematic drawing of a petroleum pipe still. Furnaces like this are built for heat-release rates up to 160,000,000 Btu/hr and are a common unit in most petroleum refineries.

The major heat exchange occurs from the flame to the single row of vertical tubes around the wall of the furnace. The refractory-brick chamber walls and the metal reradiating cone reflect the radiant energy back to the heating tubes. In this furnace, convective heat transfer is obtained between the flue gases confined in an annular ring around the reradiating cone and the heated tubes, which are finned in this section. A final convective-heating section is installed at the base of the stack. These vertical furnaces require an unusually short stack to supply the necessary draft since the flue-gas travel is uniformly vertical, eliminating the flow-direction changes that cause much of the pressure drop in horizontal furnaces. The monorail attached to the top of the stack permits the removal of the heating-element tubes vertically through the header-box doors.

The Nature of Radiation. Although the mechanism of radiant-energy transfer is not completely understood, the associated phenomena are explainable in terms of a dualistic theory. This theory deals separately with the emission and reception of radiation, and with its transmission. Radiation is emitted and received in discrete particles or pulses called photons. The energy transmitted in each photon is a function of the frequency of

emission but is otherwise constant. The emission of energy must lower the energy of the emitting body. In terms of the Bohr model of the atom, which pictures electrons orbiting around the atom nucleus at various energy levels, emission would occur as electrons jump from a high-energy orbit to a lower-energy orbit. Emission also occurs by lowering the vibrational energy of the nucleus with respect to other nuclei in the molecule or by decreasing the rotational or translational motion of the molecule. The frequency of emission is a function of the energy levels of the emitting atoms, and, since any finite body contains atoms in a wide range of energy states, the emitted radiation will have a complete spectrum of frequencies. In transmission, radiation has all the properties of an electromagnetic wave, that is, a vibrating electric field and a similarly vibrating magnetic field. The planes of vibration of these fields are perpendicular to each other, and the directions of vibration are perpendicular to the direction of radiant transmission. The speed of the beam is fixed by the medium through which it travels. In a vacuum, the speed is about 186,000 miles/sec, whereas in denser media it is slightly less. In any case:

$$C = \lambda n$$

where C = the velocity of the electromagnetic beam

λ = the wave length, or distance from energy peak to energy peak

n = the frequency of radiation, or the number of energy peaks passing any point per unit time

The problem of reconciling the wave and particle properties of radiant energy is identical to that which exists for electrons and atoms. It has gradually become understood that these properties are different aspects of the same phenomenon and that the aspects are necessary and complementary. The uncertainty principle of Heisenberg states that it is impossible to determine simultaneously the velocity (or associated energy or momentum) and position of a particle. Thus, experiments designed to determine particle velocity find that radiant energy has the characteristics of waves, whereas experiments designed to determine particle position find that the energy has particle properties. The wavelike appearance of energy probably arises from the identical form of equations describing wave motion and those describing the probability of finding a photon at a given location.

Upon striking a receiver, the photons travel into it until they strike an electron or nucleus which is susceptible to the energy level of the photon. The collision results in an increase in energy of the receiving atom, as an increased amplitude of vibration of the nucleus, as an increasing rotational or translational energy of

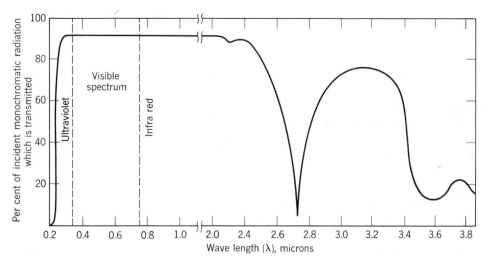

Figure 15.22. Transmission of radiant energy by window glass (31).

the molecule, or as an increase in energy level of one of the orbiting electrons. Since solids have closely spaced atoms, radiation is usually absorbed very close to the surface. For liquids, the penetration would be greater, whereas, for gases much greater penetrations are obtained. The depth of penetration will depend upon the characteristics of the incident photons as well as upon the characteristics of the receiving body.

Thermal Radiation. Thus, all solids, liquids, and most gases receive and give off energy in the form of electromagnetic waves, which are similar whether they are perceived in the form of thermal energy, light, X-rays, radio waves, or radar signals. These various forms differ only in wave length. Light waves, those perceived by the human eye, have wave lengths ranging between 0.35 micron and 0.75 micron, whereas electromagnetic waves received as thermal energy have wave lengths between 0.3 micron and a value greater than 10 microns (10^3 microns = 1 mm). The spectrum of wave lengths radiated from any body will depend upon its temperature as well as upon its surface characteristics, with lower temperatures producing longer wave-length radiation.

Since thermal and light radiation are fundamentally identical, the laws of optics apply to both these radiation forms. Radiation travels in straight-line paths from any emitter. It can be received directly only by bodies that can be "seen" by the emitter. Receivers shade the area away from the emitter, so that objects in this space receive no radiant energy directly from the emitter under consideration. Of course, objects so shaded receive radiation from all the other objects which they can see.

Such objects may also receive reflected radiant beams, so that an object is rarely shaded completely by another body. As a radiant beam travels through space from object to object it may be partly absorbed or deflected by the gas through which it passes. Many common gases

such as water, carbon dioxide, and hydrocarbons absorb and generate appreciable amounts of radiant energy. On the other hand, monatomic and symmetrical diatomic gases such as hydrogen and nitrogen are almost completely transparent to thermal radiation as their nuclei and electrons cannot be energized by photons of the available energy levels. Deflection of the radiant beam occurs when the gas space carries dust or other finely divided solid particles.

The Reception of Thermal Radiation, Kirchhoff's Law. As with optical waves, thermal radiation falling on an object will be partly absorbed, partly reflected, and partly transmitted. The proportions of the incident energy that are absorbed, reflected, and transmitted depend primarily upon the characteristics of the receiver, but also to a lesser extent upon the wave length of the radiation and the temperature of the receiver. For example, most solids absorb or reflect all the radiant energy that strikes them, but glass is transparent to radiation of short wave length, whereas for long-wave-length radiation it is almost opaque. This is shown in Figure 15.22. These considerations can be stated more concisely as

$$\rho + \alpha + \tau = 1 \qquad (15.36)$$

where ρ = reflectivity, the fraction of the total incident radiation that is reflected

α = absorptivity, the fraction of the total incident radiation that is absorbed

τ = transmissivity, the fraction of the total incident radiation that is transmitted through the body

This equation is obviously only a form of energy balance. It traces the possible means of dispersal of the radiant energy arriving at a point. For most solids and liquids, it simplifies to

$$\rho + \alpha = 1 \qquad (15.37)$$

However for glass, the situation is quite different. For energy of some wave lengths striking the glass, as for example light in the near infrared range, τ would approach one, and therefore ρ and α would be very small. The beam of infrared radiation would pass through virtually unchanged. On the other hand, for radiation in the far ultraviolet frequency range, τ would be nearly zero, and almost all the incident radiation would be either absorbed or reflected. Here, Equation 15.37 applies. In this case, most of the incident radiation is absorbed, so that α is almost one and ρ is small. Another example of unexpected behavior is that of snow which, despite the blinding reflection of visible light from its surface, absorbs almost all the thermal radiation striking it. Thus for snow, α approaches one, whereas ρ is quite small.

As with other forms of radiation, thermal radiation penetrates the surface of any receiver to some extent before the energy is completely converted to heat. In metals, the penetration is about 1 micron, whereas, for electric nonconductors, it may be as much as 1000 microns (10). Radiation may be reflected diffusely from a surface, that is, in all directions, or it may be specularly reflected as are light beams from a mirror.

The Ideal Radiation Receiver. A body can be imagined which has $\tau = 0$ and $\rho = 0$; all the incident radiant energy would be absorbed. Such a body is called a "black body," but its surfaces need not be black in color. "Black" bodies do not exist in nature, although many materials approach black behavior. It can be shown as follows that a black body emits radiation at the maximum rate possible of any body of equal size and shape at the same temperature. Consider two bodies identical in every characteristic except that one is black whereas the other has an absorptivity (α) less than one. If each body is put into a furnace held at a constant temperature (T_0), both bodies will eventually reach T_0. This will occur because thermal energy will be transferred to each of them by radiation from the furnace until such time as there is no longer a temperature difference between the furnace and the enclosed bodies. Once they reach T_0, the bodies must maintain this temperature, growing neither hotter nor cooler than the furnace. This is necessary by the second law of thermodynamics which explicitly prohibits the spontaneous generation of a temperature difference between two bodies that can transfer thermal energy only to each other. At this condition of thermal equilibrium then, each body must be absorbing and emitting thermal radiation at equal rates, otherwise the temperature would not stay constant. The rate of absorption and emission for the black body will, however, be different than that for the nonblack body. The situation may be

clarified by writing thermal-energy balances around each body. For the black body,

$$\frac{q_{\text{furnace-to-black body}}}{A} = e_b$$

$$\frac{q_{\text{black body-to-furnace}}}{A} = e_b$$

and

$$e_b = e_b$$

where e_b is the rate at which thermal energy is transferred to and from the black body, Btu/hr sq ft. For the nonblack body,

$$\frac{q_{\text{furnace-to-body}}}{A} = e_b$$

$$\frac{q_{\text{absorbed by body}}}{A} = \alpha e_b$$

$$\frac{q_{\text{reflected by body}}}{A} = (1 - \alpha)e_b$$

and

$$e_b = \alpha e_b + (1 - \alpha)e_b \qquad (15.38)$$

If e is defined as the actual rate at which radiant energy is absorbed and consequently emitted by the nonblack body per unit of surface (Btu/hr sq ft),

$$e = \alpha e_b = \epsilon e_b \qquad (15.39)$$

which defines ϵ, the emissivity, as the ratio of the rate at which energy is radiated from a unit surface and that at which it would be radiated from a black surface of equal area. From Equation 15.39,

$$\alpha = \epsilon \qquad (15.40)$$

which is known as *Kirchhoff's law*. Since α is temperature dependent, ϵ must be also. Thus, Kirchhoff's law holds only at constant temperature, or, in other words, thermal equilibrium between emitter and receiver is required.

Though black-body characteristics cannot be found in naturally occurring surfaces, they are very closely approximated by a cavity in a solid substance which is open to the atmosphere only through a pinhole in the solid surface. Any radiant beam coming to this opening will pass into the cavity. There it will bounce from wall to wall, at each contact being partly absorbed. Since its chances of finding the opening again are almost nil, the beam's energy will be completely absorbed, even though only a modest fraction of this energy is absorbed at each contact. The surface pinhole thus absorbs all the radiant energy striking it.

Under the condition that the pinhole and its surroundings are at thermal equilibrium, the emission from the pinhole must be at a rate equal to that at which radiant energy strikes it from the surroundings. Then, since e_b is both absorbed and radiated, ϵ as well as α must be equal to one. This argument can be followed regardless of the temperature chosen for thermal equilibrium.

The Stefan-Boltzmann Relation. The relation between emitting black-body temperature and the rate of radiant-energy emission was first deduced empirically by Stefan and later proved by Boltzmann (6) through considering the analogy between black-body radiation and perfect-gas behavior.

The photons of energy can be treated in the same way as the molecules of a model gas. The differences are that the photons all move at a uniform rate, and obey much more closely the criteria of no interaction and of elastic collisions. Since radiant beams surely travel at finite velocity, a volume of space through which radiant energy flows will have a certain concentration of this energy. Consider a system consisting of an enclosed isothermal volume of space in which the concentration of radiant energy is U_r. This concentration must be a function of temperature alone, or

$$\frac{E_r}{V} = U_r = \phi(T) \qquad (15.41)$$

where E_r = internal energy due to radiation, Btu
V = volume of the system
ϕ = some unspecified function

The insignificance of such variables as the properties of the boundary walls or the size and shape of the system can be shown through the second law of thermodynamics by inserting, reversibly, a new wall. If such an insertion creates a change in the system, the change must be spontaneous and must be accompanied by an increase in entropy. On reversibly removing the new wall, the system, consisting of the enclosed space, must return to its original state. This change also must be spontaneous, occurring with increasing entropy. A contradiction has appeared, for a reversible cyclical process has produced a net increase in entropy. Therefore, the insertion of new walls cannot change the thermodynamic state of the system.

A pressure effect of radiation must result from the momentum change occurring at collision between photons and the enclosing walls of the system. The derivation is almost identical to that for the pressure exerted by a model gas. By definition pressure is the change in momentum resulting from all the collisions against unit area of wall in unit time. For a single photon, the momentum change per impact is $2\,mu$, where u is the velocity component normal to the wall and m is the mass of the photon. The factor 2 occurs because of the elastic nature of the impact. The number of photons striking unit area of the surface per unit time will be half of all the photons within the distance u from the surface. Within the unit time, the photons will move toward the surface, strike it, and rebound. The factor $\frac{1}{2}$ occurs because half the photons will be moving away from the surface, Thus, the number of impacts

per unit time per unit surface area is $\frac{1}{2}uN/V$, where N is the number of photons in the system. Multiplying this impact rate by the momentum change per impact, and defining the pressure effect of radiation (P_r) as momentum change per unit time per unit surface

$$P_r = \frac{N}{V}\,mu^2 \qquad (15.42)$$

where P_r is the pressure effect of radiation.

Since $\qquad c^2 = u^2 + v^2 + w^2 \qquad (15.43)$

where $\quad c$ = velocity of light
u, v, w = components of the velocity of light in the x, y, z-directions,

and since $\qquad u^2 = v^2 = w^2 \qquad (15.44)$

because of the random paths of the photons within the system,

$$c^2 = 3u^2 \qquad (15.45)$$

Combining Equations 15.42 and 15.45,

$$P_r = \frac{1}{3}\frac{N}{V}\,mc^2$$

From Einstein's relation and Equation 15.41,

$$Nmc^2 = E_r$$

and $\qquad\qquad P_r = \frac{U_r}{3} \qquad (15.46)$

Equations 15.41 and 15.46 can be inserted into the first two laws of thermodynamics to give the relation between absolute temperature and the rate of radiant-energy emission. The first and second laws of thermodynamics may be written,

$$dE = T\,dS - P\,dV \qquad (15.47)$$

where S = entropy
E = internal energy

Since only radiant energy affects the system, all the thermodynamic properties can be written with the subscript r.

Thus $\qquad dE_r = T\,dS_r - P_r\,dV \qquad (15.47a)$

Dividing by dV at constant T,

$$\left(\frac{\partial E_r}{\partial V}\right)_T = T\left(\frac{\partial S_r}{\partial V}\right)_T - P_r \qquad (15.48)$$

but $\qquad (\partial S_r/\partial V)_T = (\partial P_r/\partial T)_V$

(a Maxwell relation based on Helmholtz free energy), and then

$$\left(\frac{\partial E_r}{\partial V}\right)_T = T\left(\frac{\partial P_r}{\partial T}\right)_V - P_r \qquad (15.49)$$

From Equation 15.41 $E_r = VU_r$

and
$$\left(\frac{\partial E_r}{\partial V}\right)_T = U_r \qquad (15.50)$$

From Equation 15.46 $P_r = \dfrac{U_r}{3} = \dfrac{E_r}{3V}$

and
$$\left(\frac{\partial P_r}{\partial T}\right)_V = \frac{1}{3}\frac{dU_r}{dT} \qquad (15.51)$$

Placing Equations 15.50 and 15.51 into Equation 15.49

$$U_r = \frac{1}{3}T\frac{dU_r}{dT} - \frac{U_r}{3}$$

or
$$4U_r = T\frac{dU_r}{dT} \qquad (15.52)$$

Separating variables and integrating, this equation becomes

$$\ln U_r = 4\ln T + \ln b \qquad (15.53)$$

where b is an arbitrary constant of integration. From this equation,

$$U_r = bT^4 \qquad (15.54)$$

This equation is the desired relation between rate of energy emission and temperature, though perhaps not the most useful form of the relation. Before converting this equation to a more useful form, it is well to note that b arose as a constant of integration and therefore is a true constant independent of temperature and radiant-energy concentration. Surface properties cannot affect b of course, as long as the original black-body restriction is maintained.

Since radiant-energy transfer is being considered, Equation 15.54 should be written in terms of q_{rb} rather than U_r. Since q_{rb} is the amount of radiant energy emitted by a black surface per unit time

$$\frac{q_{rb}}{A} = (\text{a constant}) \ cU_r \qquad (15.55)$$

where again c is the velocity of light. The constant here can be obtained by integration of the radiant energy emitted by a small area to the whole visible hemisphere. Such integrations show this constant to be $\frac{1}{4}$. Then

$$\frac{q_{rb}}{A} = \frac{cb}{4}T^4 = \sigma T^4 \qquad (15.56)$$

Equation 15.56 is the Stefan-Boltzmann equation which is basic to all radiant-energy-transfer calculations. The constant σ, the Stefan-Boltzmann constant, is a true constant with a value of 1.73×10^{-9} Btu/sq ft hr °R⁴. The total radiant energy emitted from a unit area of black-body surface per unit time is q_{rb}/A. No account is taken here of the fact that radiant energy is absorbed as well as emitted or that all the emitted energy may not strike a given receiver. The Stefan-Boltzmann equation also gives no information on the wave-length distribution of the emitted radiation.

Wave-Length Distribution. The distribution of wave lengths of radiation emitted by a black body can only be found by considering the mechanism of radiant transfer. Attempts to arrive at a satisfactory relation for this distribution finally led to the quantum theory (29, 30). Using this theory, a satisfactory relation between emitting-body temperature, wave length, and the intensity of radiation is obtained. Calculations based on this theory led to Figure 15.23. Here, the increase in radiant-energy emission for a differential wave-length range, or the monochromatic-radiation intensity, is given as a function of wave length for various constant temperatures. For any given temperature of the emitting black body, the monochromatic-radiation intensity increases with increasing wave length until a maximum is reached and then decreases. The maxima shift toward shorter wave lengths as the temperature increases. Wien first showed this shift to be necessary by developing his displacement law (36), from considerations similar to those of Boltzmann

$$B = \lambda_{\max}T_{\max} \qquad (15.57)$$

Here B is a constant found by experiment to be 5193 microns °R. As indicated on Figure 15.23, the area under any one of the isothermal curves gives the total rate of radiant-energy emission of a black body at that temperature. Thus,

$$\frac{q_{rb}}{A} = \int_0^\infty \left(\frac{d\left(\dfrac{q_{rb}}{A}\right)}{d\lambda}\right) d\lambda \qquad (15.58)$$

Analytical integration at constant temperature of the equation from which Figure 15.23 comes must also give q_{rb}. As would be expected this integration gives the Stefan-Boltzmann equation.

$$q_{rb} = \sigma A T^4 \qquad (15.59)$$

Net Radiant-Energy Transfer. Though the Stefan-Boltzmann equation is the fundamental relation for radiant-energy transfer, it only gives the amount of radiation *emitted* by a black body without regard to that which it *absorbs*. In practice, interest always centers on the *net* energy flow. This must, of course, be the difference between the radiant energy emitted and that absorbed per unit time. Considering only black-body radiation,

$$(q_{rb})_{\text{net}} = (q_{rb})_{\text{emitted}} - (q_{rb})_{\text{absorbed}} \qquad (15.60)$$

But since

$$(q_{rb})_{\text{emitted}} = \sigma A T_1^4$$

and
$$(q_{rb})_{\text{absorbed}} = \sigma A T_2^4$$

the net rate of energy flow is

$$(q_{rb})_{\text{net}} = \sigma A T_1^4 - \sigma A T_2^4 = \sigma A(T_1^4 - T_2^4) \qquad (15.61)$$

where T_1 = temperature of emitting body

T_2 = temperature of surrounding surfaces

A = surface area of body which is emitting and absorbing

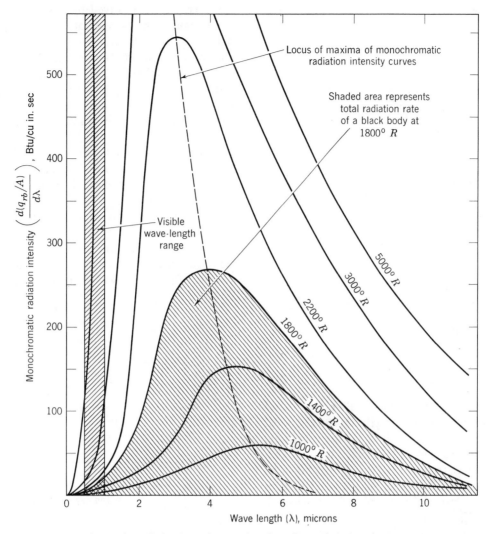

Figure 15.23. Monochromatic-radiation intensity as a function of wave length and temperature for black bodies.

In differential form,

$$\frac{d(q_{rb})_{net}}{dA} = \sigma(T_1^4 - T_2^4) \qquad (15.62)$$

For Equations 15.61 and 15.62 to be valid, very special circumstances must exist. A black body must be emitting energy only to its surroundings and receiving black-body radiation from its surroundings. Thus, the radiating body cannot "see" any of itself; that is, all its faces must be flat or convex. Moreover, the surroundings must be black and isothermal. One case would be that of two parallel infinite planes, both black and at constant temperature. Another possible case would be that of a small black sphere or cube or cylinder enclosed in a large isothermal space. In this case, all radiant energy emitted by the black body would be ultimately absorbed by the walls of the space even though the walls are not black, since the original radiant beam would have negligible chance of being reflected back to the body. Moreover, the walls of the space are so large

that even though they are imperfect radiators the small black body receives over its entire surface radiation characteristic of the wall temperatures.

Actually, even with the case of a small convex-surfaced body radiating to a large enclosure, Equation 15.61 will not apply if the body is not black. The situation for a real body is diagrammed in Figure 15.24.

The small body radiates energy at a rate characteristic of its temperature, area, and surface emissivity (symbolized by beam 1). This energy strikes the walls, where it is partly absorbed and partly reflected by the wall. The reflected energy continues to pass from surface to surface of the enclosure, being partly absorbed at each contact, until it is completely absorbed. Since the originating body is so small in relation to the walls, there is a negligible chance that any significant quantity of the reflected radiation strikes it. At the same time, the walls emit radiation at a rate characteristic of their temperature, area, and surface emissivity (beam 2). Most of this radiant energy merely ricochets from wall

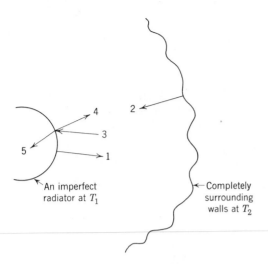

Figure 15.24. Radiant-energy transfer from and to an imperfect radiator.

(1) Radiation from small body: $(q_r)_{\text{emitted}} = \sigma A T_1^4 \epsilon$
(2) Radiation from wall: $(q_r)_{\text{wall emitted}} = \sigma A_W T_2^4 \epsilon_W$
(3) Radiation from wall striking body:
$\quad (q_r)_{\text{wall to body}} = \sigma A T_2^4$
(4) Radiation reflected from body: $(q_r)_{\text{reflected}} = \sigma A T_2^4 \rho$
(5) Radiation absorbed by body: $(q_r)_{\text{absorbed}} = \sigma A T_2^4 \alpha$

to wall, but part of it strikes the small body. Since the wall area is so large compared to the area of the body, the energy striking the body will equal the maximum radiation that could be emitted by a black body of area A and temperature T_2 (beam 3). This can be shown by a thermal equilibrium mechanism such as that used in defining α and ϵ. Part of the energy striking the small body is reflected, part absorbed (beams 4 and 5). Under these conditions, the net rate of radiant-energy transfer must be

$$(q_r)_{\text{net}} = (q_r)_{\text{emitted}} - (q_r)_{\text{absorbed}} = (\text{beam 1}) - (\text{beam 5})$$

or

$$(q_r)_{\text{net}} = \sigma A T_1^4 \epsilon - \sigma A T_2^4 \alpha$$

which can be written as

$$(q_r)_{\text{net}} = \sigma A (T_1^4 \epsilon - T_2^4 \alpha) \qquad (15.63)$$

In the relations given for the individual radiation rates and the net rate of transfer given by Equation 15.63, α, ρ, ϵ, and A are properties of the small body, and A_w and ϵ_w are properties of the surrounding walls.

Constants for Imperfect Radiators. The values of α, ρ, and ϵ used in these equations require further consideration. The radiation emitted by the small body is characteristic of its temperature (T_1). Therefore, ϵ will be that emissivity applicable to this particular surface at T_1. The radiation striking the body will,

however, be at a wave-length spectrum characteristic of an emitting black body at T_2, the temperature of the surroundings. Therefore, ρ and α must apply to the surface of the small body receiving energy characteristic of T_2. By Kirchhoff's law, $\alpha = \epsilon$ at thermal equilibrium, and hence the absorptivity (α) will equal the emissivity (ϵ) of the small body itself at T_2. This is true despite the fact that the temperature of the small body is not T_2.

Kirchhoff's law implies that α, ϵ, and ρ are functions of temperature and therefore of the wave length of radiant energy. This dependence is illustrated for a few materials in Figure 15.25. The behavior shown for polished aluminum is typical of most metals, which have higher absorptivity for short-wave-length radiation. The curves shown for other materials are typical of nonmetal surfaces. For them, the absorptivity is a very irregular function of wave length but generally increases

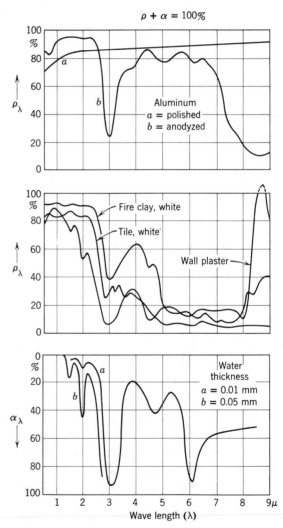

Figure 15.25. Reflectivity and absorptivity as a function of wave lengths for different materials (33). [From W. Sieber, *Z. Tech. Physik*, **22**, 139 (1941) by permission.]

with increasing wave length. For some materials, the variation of absorptivity with wave length is small, so that α and ϵ may be considered to be independent of wave length and temperature. These materials are called *grey bodies*. Many materials are considered to be grey bodies for engineering calculations even when the assumption is not justified. The reason for this practice is that, since ϵ is a surface property, it will change markedly with small changes in surface condition. Therefore, values of ϵ are seldom known to high accuracy.

In engineering applications the radiant energy being transferred is not monochromatic but rather of a range of wave lengths characteristic of the temperature of emission. Then, in order to use Equation 15.63, or any similar equation, emissivities and absorptivities integrated over the entire range of wave lengths must be used. Since the wave-length distribution is given in Figure 15.23, the proper emissivity or absorptivity may be obtained by integrating the monochromatic-emissivity curve, like those of Figure 15.25, while applying the wave-length distribution of Figure 15.23, as is done in Illustration 15.7. More usually, the total emissivity is directly measured experimentally. Such total emissivities are tabulated in Appendix D-15 and shown graphically in Figure 15.26 (33). In normal use, the ϵ versus T data of Appendix D-15 may be interpolated linearly.

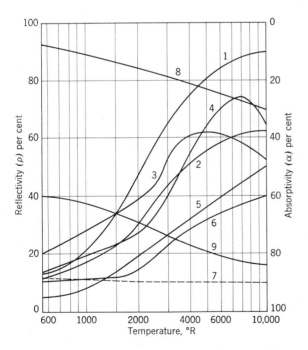

Figure 15.26. Reflectivity and absorptivity of various materials (33). $\rho + \alpha = 100\%$. (1) White fire clay; (2) asbestos; (3) cork; (4) wood; (5) porcelain; (6) concrete; (7) roof shingles; (8) aluminum; (9) graphite. [From E. Eckert, *Introduction to the Transfer of Heat and Mass*, McGraw-Hill Book Co., 1950, by permission.]

Illustration 15.6. An uninsulated steam pipe runs through a dark warehouse room. The pipe is 2-in. Sch. 40 steel pipe; the steam is saturated at 100 psia; and the warehouse is kept at 40°F. How much steam condenses per hour per foot of pipe? The coefficient of heat transfer by natural convection from the outside pipe surface may be taken as 2.0 Btu/hr sq ft °F, and the pipe surface may be treated as a grey body.

SOLUTION. The total heat loss will be due to radiation and convection. These mechanisms operate in parallel and independently to transfer energy from the pipe surface to the warehouse. Thus,

$$q = q_r + q_c \qquad (a)$$

where
 q = total loss of thermal energy, Btu/hr
 q_c = loss of thermal energy by convection from the pipe surface, Btu/hr

or,

$$q = \epsilon A \sigma (T_p^4 - T_w^4) + h_c A (T_p - T_a) \qquad (b)$$

where

T_p, T_w, T_a = temperature of pipe surface, warehouse wall, and warehouse air respectively, °R
 ϵ = surface emissivity of pipe at average of emitting and receiving temperatures from Appendix D-15
 A = area of outside surface of pipe

The pipe surface temperature (T_p) should be calculated by setting up the series-resistance equation relating temperature driving force to resistance along each step of the total path—steam to inside pipe wall, inside pipe wall to outside pipe wall, and outside pipe wall to warehouse—by the parallel mechanisms of convection and radiation. However, here the condensing-film resistance and pipe-wall resistance are so low compared to the outside-surface-to-warehouse resistance that the total ΔT may safely be approximated as equal to the ΔT from pipe surface to warehouse. Then, the pipe surface temperature (T_p) will equal the condensing-steam temperature of 328°F. Moreover, the temperature of the warehouse walls and air will be identical at 40°F. Then,

$$q = 0.173 \times A \times 0.95 \left[\left(\frac{T_p}{100} \right)^4 - \left(\frac{T_w}{100} \right)^4 \right]$$
$$+ 2.0 \times A \times (T_p - T_w) \quad (c)$$

$$= 0.173 \times 0.622 \times 0.95[(7.88)^4 - (5.00)^4]$$
$$+ 2.0 \times 0.622(788 - 500)$$

$$= 329 + 358 = 687 \text{ Btu/hr ft of pipe}$$

where the Stefan-Boltzmann constant has been used in the form $0.173 \times (1/100)^4$. Rate of condensation = $q/\Delta H$ = 687/888.8 = 0.78 lb of steam per hour per foot of pipe. Note that even at these relatively low temperatures radiation has accounted for almost 50 per cent of the total heat transfer.

Illustration 15.7. Determine the total emissivity of anodyzed aluminum at 1800°R.

SOLUTION. This problem requires integration of the wavelength-distribution function of Figure 15.23 with monochromatic emissivities taken from Figure 15.25. From Equation 15.58,

$$q_{rb} = A \int_0^\infty \left[\frac{d\frac{q_{rb}}{A}}{d\lambda} \right] d\lambda$$

$$\epsilon = \frac{q_r}{q_{rb}} = \frac{\int_0^\infty \epsilon_\lambda \left[\frac{d\frac{q_{rb}}{A}}{d\lambda} \right] d\lambda}{\int_0^\infty \left[\frac{d\frac{q_{rb}}{A}}{d\lambda} \right] d\lambda}$$

The integration can be most conveniently done graphically using finite wave-length intervals.

λ	$\Delta\lambda$	$\dfrac{d(q_{rb}/A)}{d\lambda}$	ϵ_λ	$\left[\dfrac{d(q_{rb}/A)}{d\lambda}\right]\Delta\lambda$	$\epsilon_\lambda\left[\dfrac{d(q_{rb}/A)}{d\lambda}\right]\Delta\lambda$
0					
	1	15.7	0.12	15.7	1.88
1					
	1	47.1	0.05	47.1	2.35
2					
	1	182.0	0.30	182.0	54.6
3					
	1	188.5	0.35	188.5	66.0
4					
	1	135.0	0.20	135.0	27.0
5					
	1	104.0	0.20	104.0	20.8
6					
	2	67.5	0.42	135.0	56.5
8					
	2	37.6	0.90	75.2	67.7
10					
	10	15.7	0.95	157.0	149.0
20				$\Sigma = 1039.5$	$\Sigma = 445.83$

$$\epsilon = \frac{445.83}{1039.5} = 0.43$$

Geometric Factors. Thus far, the geometric relationships between emitting and receiving surface have been kept very simple by arranging that the emitting surface sees only the receiving surface. All the radiation emitted struck the receiver. Moreover, it was always necessary that none of the emitted radiation returned to the emitter. In most cases, these restrictions cannot be maintained, and a more general approach is necessary. This approach uses a *geometric factor* which relates the radiant energy striking a surface to the total radiant energy emitted.

The situation illustrated in Figure 15.24 can be re-examined with attention focused on the surrounding walls. In this consideration, the walls will be presumed to be black. Thus, the total rate of radiation emission from the walls will be

$$(q_r)_{\text{emitted from wall}} = \sigma A_w T_2^4$$

However, all of this radiant energy will not strike the small body. Most of it will strike other parts of the wall surfaces and will be absorbed there. Therefore,

$$(q_r)_{\text{net emitted by wall}} = (q_r)_{\text{emitted from wall}}$$
$$- (q_r)_{\text{radiated from wall to itself}}$$

or

$$(q_r)_{\text{net emitted by wall}} = \sigma A_w T_2^4 - b\sigma A_w T_2^4$$
$$= (1 - b)\sigma A_w T_2^4$$

where b is the fraction of the total radiation emitted that is received by the walls. In another form

$$(q_r)_{\text{net emitted by wall}} = F_{w\to1}\sigma A_w T_2^4 \quad (15.64)$$

where $F_{w\to1}$ is called the *geometric* or *view factor* for radiation from w to 1. Since b is a fraction of the originally emitted energy, $F_{w\to1}$ must be a number from 0 to 1 representing the fraction of the total radiation from surface w, the wall, that strikes surface 1, the small body.

The net rate of energy emission from the wall equals the rate of energy emission from the wall to the body. Then,

$$(q_r)_{\text{emitted from wall}} = (q_r)_{\text{emitted from wall to itself}}$$
$$+ (q_r)_{\text{emitted from wall to body}}$$

or

$$\sigma A_w T_2^4 = F_{w\to w}\sigma A_w T_2^4 + F_{w\to1}\sigma A_w T_2^4$$
$$1 = F_{w\to w} + F_{w\to1} \quad (15.65)$$

This equation is a restricted example of the general result that all the energy emitted by a surface must strike some receiver.

$$1 = \Sigma F = F_{1\to1} + F_{1\to2} + F_{1\to3} + F_{1\to4} + \ldots + F_{1\to n} \quad (15.66)$$

In the interchange of radiant energy between two black bodies

$$(q_{rb})_{1\to2} = \sigma A_1 F_{1\to2} T_1^4$$
$$(q_{rb})_{2\to1} = \sigma A_2 F_{2\to1} T_2^4$$
$$(q_{rb})_{\text{net }1\leftrightarrow2} = \sigma A_1 \left(F_{1\to2} T_1^4 - \frac{A_2}{A_1} F_{2\to1} T_2^4 \right) \quad (15.67)$$

At thermal equilibrium $(q_{rb})_{\text{net }1\leftrightarrow2} = 0$, and $T_1^4 = T_2^4$. Then

$$F_{1\to2} T_1^4 = \frac{A_2}{A_1} F_{2\to1} T_2^4$$

$$F_{1\to2} = \frac{A_2}{A_1} F_{2\to1} \quad \text{or} \quad F_{1\to2} A_1 = F_{2\to1} A_2 \quad (15.68)$$

Since the view factors are functions of geometry only, Equation 15.68 is general and not restricted to thermal equilibrium. Equation 15.67 can thus be rewritten as

$$(q_{rb})_{\text{net } 1 \leftrightarrow 2} = \sigma A_1 F_{1 \to 2}(T_1{}^4 - T_2{}^4)$$
$$= \sigma A_2 F_{2 \to 1}(T_1{}^4 - T_2{}^4) \qquad (15.69)$$

The exact relation between the view factor and the system geometry will now be determined. This will be done by determining the view factor for a perfectly general geometry. The general case of a differential surface area radiating to another surface in space is illustrated in Figure 15.27. Here the differential area

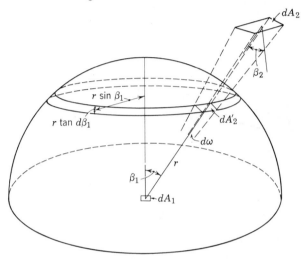

Figure 15.27. Radiation between two differential surfaces in space.

(dA_1) is radiating to surface dA_2. A reference hemisphere has been drawn at a fixed radius (r) through which all radiant energy must pass. The geometric factor will then be the ratio of the energy flux through $dA_2{}'$, the projected area of dA_2 located on the hemisphere surface, to that through the entire hemisphere. If the hemisphere were drawn through the center of dA_2 then

$$dA_2{}' = dA_2 \cos \beta_2$$

where β_2 is the angle between the perpendicular to dA_2 and the radial beam between dA_1 and dA_2. Under this same condition, r is the distance between dA_1 and dA_2.* The amount of radiation emitted from any single point on dA_1 that strikes dA_2 must be proportional to the area of the projected surface on the hemisphere $(dA_2{}')$ and inversely proportional to the square of the distance (r^2) between the point and the projected area. Similarly, the area of dA_1 as seen from $dA_2{}'$ must be the projected

---

* As drawn in Figure 15.27, r is the distance from dA_1 to $dA_2{}'$ but is shorter than the distance from dA_1 to dA_2. This is done for convenience in visualization and does not alter the basic development.

area $dA_1{}'$ on the plane perpendicular to the connecting beam. As before,

$$dA_1{}' = dA_1 \cos \beta_1$$

The fraction of the total energy flux emitted by dA_1 that strikes dA_2 is the view factor. This fraction must be proportional to the projected areas of dA_1 and dA_2 and inversely proportional to the square of the distance between them. The numerator of this fraction must be

$$(q_r)_{dA1 \to dA2} = \frac{i \, dA_1{}' \, dA_2{}'}{r^2} = \frac{i \, dA_1 \cos \beta_1 , \, dA_2 \cos \beta_2}{r^2}$$
$$(15.70)$$

where

$(q_r)_{dA1 \to dA2}$ = radiation from dA_1 which is intercepted by dA_2, Btu/hr

i = a proportionality constant, Btu/hr sq ft

$dA_1{}', dA_2{}'$ = projected areas of dA_1 and dA_2 on the planes normal to the radiant beam, sq ft

r = distance from dA_1 to dA_2

β_1, β_2 = angles between the perpendiculars to the centers of dA_1 and dA_2 and the ray connecting them, radians

Equation 15.70 is known as the *cosine law*. It can be written in terms of the solid angle, $d\omega$, over which a point on dA_1 sees dA_2. Since a solid angle is defined as the area subtended on the surface of a sphere divided by the square of the sphere radius,

$$(q_r)_{dA1 \to dA2} = i \, dA_1{}' \frac{(dA_2{}')}{(r^2)} = i \cos \beta_1 \, dA_1 \, d\omega \quad (15.71)$$

The denominator of the fraction that is the geometric factor must be the total rate of radiation emission from dA_1. This rate can be found by integrating Equation 15.70 or 15.71 over the entire hemisphere.

$$(q_r)_{dA1 \to \text{hemisphere}} = \int i \cos \beta_1 \, dA_1 \, d\omega$$
$$= i \, dA_1 \int_{\beta_1 = 0}^{\beta_1 = \pi/2} \cos \beta_1 \frac{2\pi r \sin \beta_1 \, r \, d\beta_1}{r^2} \quad (15.72)$$

where the integration is done in rings of $r \tan d\beta_1$ in width which sweep outward from the normal axis of the hemisphere to the horizon. Each of the rings is $r \sin \beta_1$ in radius and $2\pi r \sin \beta_1$ in circumference. Since $d\beta_1$ is very small, the $\tan d\beta_1$ has been taken as equal to $d\beta_1$ in determining the width. Integrating,

$$(q_r)_{dA1 \to \text{hemisphere}} = i \, dA_1 2\pi [\tfrac{1}{2} \sin^2 \beta]_0^{\pi/2} = i \, dA_1 \pi \quad (15.73)$$

Equation 15.73 gives the total radiation from a blackbody surface. By the Stefan-Boltzmann equation,

$$(q_r)_{dA1 \to \text{hemisphere}} = \sigma \, dA_1 \, T_1{}^4 \quad (15.74)$$

Combining Equations 15.73 and 15.74, and solving for i

$$i = \frac{\sigma T_1{}^4}{\pi}$$

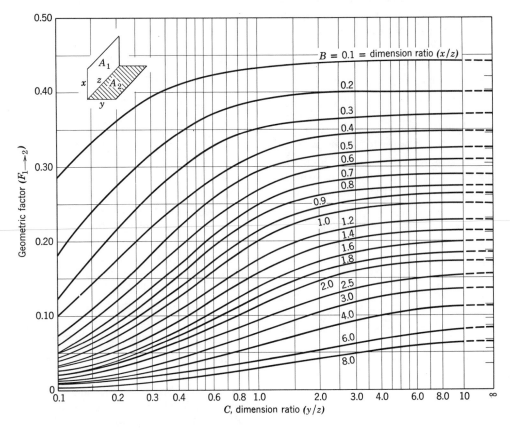

Figure 15.28. Geometric shape factor for radiation between adjacent rectangles in perpendicular planes (16).

(From C. O. Mackey, L. T. Wright, Jr., R. E. Clark, and N. R. Gay, *Cornell University Engineering Experiment Station Bulletin 32*, Ithaca, N.Y., August, 1943, by permission.)

B = ratio (Length of unique side of that rectangle on whose area the heat-transfer equation is based/length of common side) = x/z in sketch

C = ratio (Length of unique side of other rectangle/length of common side) = y/z in sketch

Since both numerator and denominator of the fraction that is the view factor have been obtained, the view factor itself is

$$F_{dA1 \to dA2} = \frac{(q_r)_{dA1 \to dA2}}{(q_r)_{dA1 \to \text{hemisphere}}}$$

$$= \frac{i\, dA_1 \cos \beta_1 \, dA_2 \cos \beta_2 / r^2}{i\, dA_1 \, \pi}$$

$$F_{dA1 \to dA2} = \frac{\cos \beta_1 \cos \beta_2 \, dA_2}{\pi r^2} \qquad (15.75)$$

The view factor applicable for transfer from dA_1 to a finite receiver (A_2) can then be found by integration.

$$F_{dA1 \to A2} = \frac{1}{\pi} \int_{A_2} \frac{\cos \beta_1 \cos \beta_2 \, dA_2}{r^2} \qquad (15.76)$$

In general, engineering applications involve transfer between two finite bodies. A view factor is desired that applies to the entire geometry present. Such an over-all view factor could be simply used to calculate the entire energy-transfer rate.

$$(q_r)_{A1 \to A2} = \sigma A_1 F_{1 \to 2} T_1^4 \qquad (15.77)$$

This energy-transfer rate must be the integral of the transfer rates from all the differential sections of A_1.

$$(q_r)_{A1 \to A2} = \Sigma (q_r)_{dA1 \to A2}$$

but

$$(q_r)_{dA1 \to A2} = \sigma \, dA_1 F_{dA1 \to A2} T_1^4 \qquad (15.78)$$

and therefore

$$(q_r)_{A1 \to A2} = \int \sigma F_{dA1 \to A1} T_1^4 \, dA_1 \qquad (15.79)$$

Combining Equations 15.76, 15.77, 15.78, and 15.79,

$$F_{1 \to 2} = \frac{1}{A_1} \int_{A_1} F_{dA1 \to A2} \, dA_1 \qquad (15.80)$$

or

$$F_{1 \to 2} = \frac{1}{A_1} \int_{A_1} \left[\frac{1}{\pi} \int_{A_2} \frac{\cos \beta_1 \cos \beta_2 \, dA_2}{r^2} \right] dA_1 \qquad (15.81)$$

Equation 15.81 is perfectly general and may be integrated to any desired geometry.

Such integrations have been carried out for certain fixed geometries and presented in graphical form. Some of these solutions are given in Figures 15.28 and 15.29.

Other solutions are available (17). For cases not previously solved, Equation 15.81 must be integrated. Usually the exact integration cannot be done. Then, an approximate answer may be obtained by evaluating the radiant interchange between portions of each body as follows:

1. Divide each surface into roughly equal parts.

2. Using Equations 15.78 and 15.76 directly, evaluate the interchange between representatively selected parts of surface 1 and all visible portions of surface 2.

3. Average the interchange energies so obtained and multiply by the number of portions into which surface 1 was originally divided. The averaging procedure must be consistent with the original division of the total surface. The appropriate weighting procedure should be used for unequal portions of area.

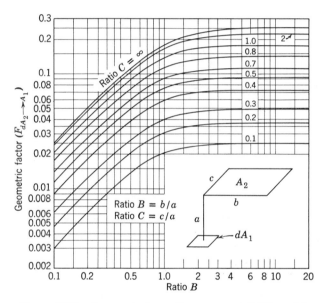

Figure 15.29. Geometric factor for a system of differential surface and a finite rectangular surface parallel to it (16). (From C. O. Mackey, L. T. Wright, Jr., R. E. Clark, and N. R. Gay, *Cornell University Engineering Experiment Station Bulletin 32*, Ithaca, N.Y., August, 1943, by permission.)

Illustration 15.7. Determine the geometry factor to be used for radiation between a wall and adjoining floor. Both surfaces are black; the wall is 10 ft by 20 ft and the floor is 20 ft by 40 ft. The other walls and ceiling may be considered to be transparent to radiation.

SOLUTION. The physical situation is shown in Figure 15.30. This geometric arrangement must be used in defining the terms of Equation 15.81 before integration.

From Figure 15.30, the following substitutions can be made:

$$dA_1 = dy\, dz_1$$

$$dA_2 = dx\, dz_2$$

$$\cos \beta_1 = \frac{x}{\sqrt{x^2 + y^2 + \Delta z^2}}$$

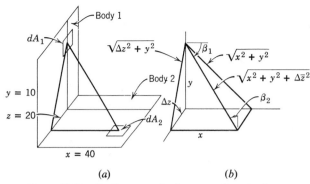

Figure 15.30. Radiant transfer from a wall to the floor.

(a) Physical arrangement.

(b) Geometric relation of terms.

where $\Delta z = z_2 - z_1$

$$\cos \beta_2 = \frac{y}{\sqrt{x^2 + y^2 + \Delta z^2}}$$

$$r = \sqrt{x^2 + y^2 + \Delta z^2}$$

$$F_{1\to 2} = \frac{1}{\pi A_1} \int_{A_1} \int_{A_2} \frac{xy\, dx\, dz_2\, dy\, dz_1}{(x^2 + y^2 + \overline{\Delta z^2})^2}$$

$$F_{1\to 2} = \frac{1}{200\pi} \int_{z_1=0}^{z_1=20} \int_{y=0}^{y=10} \int_{z_2=0}^{z_2=20} \int_{x=0}^{x=40}$$

$$\times \frac{xy\, dx\, dz_2\, dy\, dz_1}{[x^2 + y^2 + (z_2 - z_1)^2]^2}$$

This equation has been integrated with respect to each of the variables successively by Hottel (16) to give

$$F_{1\to 2} = \frac{1}{4\pi}$$

$$\times \ln \frac{(1 + B^2 + C^2)^{B-(1/B)+(C^2/B)}(B^2)^B(C^2)^C}{(1 + B^2)^{B-(1/B)}(1 + C^2)^{(C^2/B)-(1/B)}(B^2 + C^2)^{B+(C^2/B)}}$$

$$+ \tan^{-1}\frac{1}{B} + \frac{C}{B}\tan^{-1}\frac{1}{C} - \sqrt{1 + \frac{C^2}{B^2}}\tan^{-1}\frac{1}{\sqrt{B^2 + C^2}} \quad (a)$$

where $B = x/z$, $C = y/z$, and where z is the common edge and x and y are the other edges of the surface. For this case, $x/z = 2.0$ and $y/z = 0.5$. Putting these values in Equation (a) gives $F_{1\to 2} = 0.079$. This result can also be read from Figure 15.28.

Illustration 15.8. Determine the net heat flux between a furnace door 2 ft by 2 ft and a wall located as shown below:

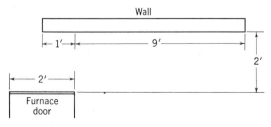

Top view of geometry for Illustration 15.8.

The wall is 8 ft high, and the bottom of the furnace door is 3 ft from the floor. Assume all surfaces are black. The furnace door is at 600°F, and the wall is 100°F.

SOLUTION. The net heat flow can be determined from Equation 15.69.

$$(q_r)_{\text{net } 1\leftrightarrow 2} = \sigma A_1 F_{1\rightarrow 2}(T_1^4 - T_2^4)$$

The determination is readily made once the view factor is determined. This factor can be found by using the general procedure given on p. 259 coupled with the geometric factors read from Figure 15.29. The geometric arrangement of furnace door and wall is probably too unsymmetrical to expect that already-calculated view factors like those of Figure 15.28 would be available. In cases where already-calculated view factors are available they are, of course, convenient. However the procedure to be shown here is powerful enough to allow calculation of many cases for which view factors are not available.

A front view of the system geometry is shown below:

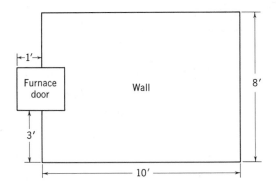

The general procedure calls for dividing the furnace door into a number of representative sections and finding the view factor from several of these sections. Let the door be divided as follows:

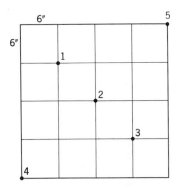

Since, from Figure 15.29, only differential areas are used, the five points shown will be taken as representative. The use of so few points can only be justified because of the symmetry of the system. The geometric factor will now be determined for each of these points.

Point 1. The spatial relation is as shown:

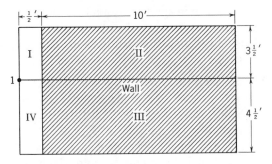

The desired view factor will be

$$F_{1\rightarrow w} = F_{1\rightarrow(\text{I}+\text{II})} + F_{1\rightarrow(\text{III}+\text{IV})} - F_{1\rightarrow\text{I}} - F_{1\rightarrow\text{IV}}$$

For $F_{1\rightarrow(\text{I}+\text{II})}$ $B = \dfrac{b}{a} = \dfrac{10\frac{1}{2}}{2} = 5.25$

$$C = \frac{c}{a} = \frac{3\frac{1}{2}}{2} = 1.75$$

where the factors b, c, B, and C are as defined on Figure 15.29, and $F_{1\rightarrow(\text{I}+\text{II})} = 0.22$.

For $F_{1\rightarrow(\text{III}+\text{IV})}$: $B = 5.25$ $C = 2.25$ $F_{1\rightarrow(\text{III}+\text{IV})} = 0.24$
For $F_{1\rightarrow\text{I}}$: $B = 0.25$ $C = 1.75$ $F_{1\rightarrow\text{I}} = 0.057$
For $F_{1\rightarrow\text{IV}}$: $B = 0.25$ $C = 2.25$ $F_{1\rightarrow\text{IV}} = 0.06$

Then $F_{1\rightarrow w} = 0.22 + 0.24 - 0.057 - 0.06 = 0.343$.

Point 2. Here the spatial arrangement is slightly simpler:

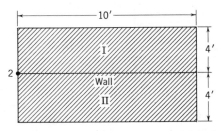

Again using Figure 15.29 to determine partial view factors gives

$$F_{2\rightarrow w} = F_{2\rightarrow\text{I}} + F_{2\rightarrow\text{II}} = 0.23 + 0.23 = 0.46$$

Point 3. Here the spatial arrangement is:

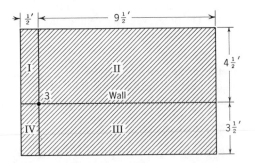

$$F_{3\rightarrow w} = F_{3\rightarrow\text{I}} + F_{3\rightarrow\text{II}} + F_{3\rightarrow\text{III}} + F_{3\rightarrow\text{IV}}$$
$$= 0.06 + 0.24 + 0.22 + 0.057 = 0.575$$

By similar constructions $F_{4\rightarrow w} = 0.22$ and $F_{5\rightarrow w} = 0.615$. Averaging the five geometric factors gives $F_{av\ dA1\rightarrow w} = 0.443$. When the average geometric factor thus obtained is used, Equation 15.80 gives

$$F_{1\rightarrow 2} = \frac{1}{A_1}\int_{A_1} F_{dA1\rightarrow A2}\,dA_1 = \frac{1}{A_1}F_{av\ dA1\rightarrow A2}\cdot A_1$$

so that $\qquad F_{\text{door}\rightarrow\text{wall}} = 0.443$

Then $(q_r)_{\text{net}\ 1\rightarrow 2} = 0.173 \times 4 \times 0.443[(10.60)^4 - (5.60)^4]$

$\qquad\qquad = 3580$ Btu/hr transferred from the furnace door to the wall

Allowance for Real Surfaces. When the bodies are not black but are real liquids and solids, further difficulties arise in evaluating the proper emissivity. If the surface is opaque to thermal radiation, any radiant beam striking it is partly absorbed and partly reflected. The reflected beam may return to the original source, may strike another portion of the receiver, or may strike a third object. In this way, every beam bounces from surface to surface, at each contact losing part of its intensity, until it is completely absorbed. Moreover, the characteristics of the surface influence the manner of travel of the beam, for from some surfaces reflection is specular whereas from others it is diffuse.

The "effective" emissivity under such conditions will be found for the case of diffuse radiation with parallel plates that are essentially infinite in extent. Here, the geometric factor (F) is unity. The effective emissivity would also be unity if the surfaces were black. With nonblack surfaces, the geometric factor will remain as with black surfaces, but an interchange emissivity must be found. The situation for parallel plates is shown schematically in Figure 15.31. There, the amount of heat absorbed by surface 2 of that emitted by surface 1 per unit time per unit of area 1 is

$$\frac{q_{1\rightarrow 2}}{A_1} = (1 - \rho_2\alpha_1 - \rho_2{}^2\rho_1\alpha_1 - \rho_2{}^3\rho_1{}^2\alpha_1$$
$$- \rho_2{}^4\rho_1{}^3\alpha_1 - \ldots)\epsilon_1\sigma T_1{}^4$$

which shortens to

$$\frac{q_{1\rightarrow 2}}{A_1} = \left(1 - \frac{\alpha_1\rho_2}{1 - \rho_1\rho_2}\right)\epsilon_1\sigma T_1{}^4 \qquad (15.82)$$

since ρ_1 and ρ_2 are less than 1.

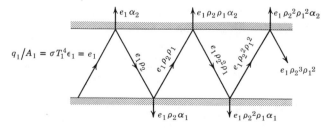

Figure 15.31. Radiant interchange between parallel nonblack surfaces.

Considering the surfaces to be grey and opaque, we can apply Kirchhoff's law and replace the α and ρ terms by ϵ's.

$$q_{1\rightarrow 2} = \left[1 - \frac{\epsilon_1(1 - \epsilon_2)}{\epsilon_1 + \epsilon_2 - \epsilon_1\epsilon_2}\right]\epsilon_1\sigma A_1 T_1{}^4$$

$$= \frac{\epsilon_2\epsilon_1}{\epsilon_1 + \epsilon_2 - \epsilon_1\epsilon_2}\sigma A_1 T_1{}^4 \qquad (15.83)$$

Applying the same reasoning to a beam originally leaving surface 2, we get

$$\frac{q_2}{A_1} = \sigma T_2{}^4 \frac{\epsilon_1\epsilon_2}{\epsilon_1 + \epsilon_2 - \epsilon_1\epsilon_2}$$

The net interchange then is

$$\frac{q_1 - q_2}{A_1} = \sigma \frac{\epsilon_1\epsilon_2 T_1{}^4 - \epsilon_2\epsilon_1 T_2{}^4}{\epsilon_1 + \epsilon_2 - \epsilon_1\epsilon_2} = \frac{q_r}{A_1}$$

or

$$\frac{q_r}{A_1} = \frac{\sigma}{\dfrac{1}{\epsilon_1} + \dfrac{1}{\epsilon_2} - 1}(T_1{}^4 - T_2{}^4)$$

Thus, the interchange emissivity (ϵ_{1-2}) is given by

$$\epsilon_{1-2} = \frac{1}{\dfrac{1}{\epsilon_1} + \dfrac{1}{\epsilon_2} - 1} \qquad (15.84)$$

for the case of parallel plates. For concentric cylinders or concentric spheres with diffusely reflecting surfaces, a similar development leads to

$$\epsilon_{1-2} = \frac{1}{\dfrac{1}{\epsilon_1} + \dfrac{A_1}{A_2}\left(\dfrac{1}{\epsilon_2} - 1\right)} \qquad (15.85)$$

where 1 refers to the smaller body. The interchange emissivity may be used in one of the following two equations:

$$q_r = \sigma A_1\epsilon_{1-2}(T_1{}^4 - T_2{}^4) \qquad (15.86)$$

for cases where the geometric factor is unity, or

$$q_r = \sigma A_2\epsilon_{1-2}F_{2\rightarrow 1}(T_1{}^4 - T_2{}^4) \qquad (15.87)$$

for cases in which the geometric factor is less than unity.

Effect of Zero-Flux Surfaces. In industrial-furnace calculations, the engineer must usually account for the fact that the radiation emitter and ultimate receiver are not the only surfaces present but are connected by refractory surfaces. These surfaces are insulated; consequently, any incident radiant energy must be reradiated. They usually have high absorptivities and act as a shield preventing radiant energy from being lost. Thus, the geometry factors and interchange emissivities must be altered to account for the additional surfaces. Geometry factors for black bodies which include the effect of the refractory surfaces are designated $\bar{F}_{1\rightarrow 2}$. Such

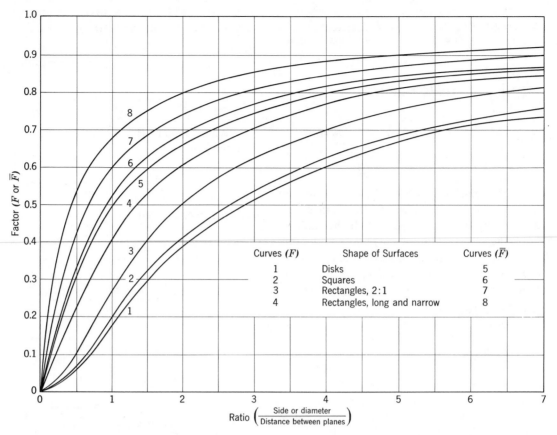

Figure 15.32. Geometric factors F and $\bar{F}$ for radiation between identical parallel opposed black planes (26).

(Curves 1, 2, 3, 4—no refractory surfaces. Curves 5, 6, 7, 8—opposed planes of an enclosure
completed by refractory surfaces.)
(Courtesy of H. C. Hottel, from *Heat Transmission*, 3rd ed., 1954, by W. H. McAdams,
McGraw-Hill Book Co., New York.)

factors for parallel planes enclosed by refractory walls are given in Figure 15.32. It can be seen that the presence of refractory walls increases the rate of heat transfer markedly by preventing radiation leakage.

For a single radiation source and a single receiver both of which are black and are connected by refractory walls, the $\bar{F}$ can be determined from the values of $F_{1\rightarrow2}$ and $F_{2\rightarrow1}$ that apply.

From the meaning of $\bar{F}_{1\rightarrow2}$, by tracing the possible paths of radiant energy leaving the source

$$\bar{F}_{1\rightarrow2} = F_{1\rightarrow2} + F_{1\rightarrow R}F_{R\rightarrow2} + F_{1\rightarrow R}F_{R\rightarrow R}F_{R\rightarrow2}$$
$$+ F_{1\rightarrow R}\bar{F}_{R\rightarrow R}^2 F_{R\rightarrow2} + F_{1\rightarrow R}\bar{F}_{R\rightarrow R}^3 F_{R\rightarrow2} + \cdots$$

$$\bar{F}_{1\rightarrow2} = F_{1\rightarrow2} + F_{1\rightarrow R}F_{R\rightarrow2}(1 + F_{R\rightarrow R} + \bar{F}_{R\rightarrow R}^2$$
$$+ \bar{F}_{R\rightarrow R}^3 + \cdots)$$

$$\bar{F}_{1\rightarrow2} = F_{1\rightarrow2} + \frac{F_{1\rightarrow R}F_{R\rightarrow2}}{(1 - F_{R\rightarrow R})} \qquad (15.88)$$

where the subscript 1 refers to the emitting surface, the subscript 2 to the receiving surface, and the subscript

R to the refractory surfaces. This relation may be put in more convenient form by applying the general restrictions on F factors given in Equations 15.66 and 15.68. As indicated above, the thermal-equilibrium restriction also requires that the net heat flux from source to refractory walls must equal that from refractories to receiver. Thus,

$$q_{1\rightarrow R} = q_{R\rightarrow2}$$

or

$$F_{1\rightarrow R}A_1\sigma(T_1^4 - T_R^4) = F_{R\rightarrow2}A_R\sigma(T_R^4 - T_2^4) \quad (15.89)$$

These considerations and the restriction that both receiver and emitter are flat or convex ($F_{1\rightarrow1} = F_{2\rightarrow2} = 0$) may be used with Equation 15.88 to give

$$\bar{F}_{1\rightarrow2} = \frac{A_2 - F_{1\rightarrow2}^2 A_1}{A_1 + A_2 - 2F_{1\rightarrow2}A_1} \qquad (15.90)$$

This relation requires no knowledge about the refractory surfaces except that they are at a constant temperature and that they receive and re-emit all the radiant energy that does not go directly from A_1 to A_2.

The same relations and restrictions can be used to determine the uniform refractory temperature, giving

$$T_R{}^4 = \frac{(A_1 - F_{1\to 2}A_1)T_1{}^4 + (A_2 - F_{1\to 2}A_1)T_2{}^4}{(A_1 - F_{1\to 2}A_1) + (A_2 - F_{1\to 2}A_1)} \quad (15.91)$$

Geometric Factors for Imperfect Radiators. If the surfaces are imperfect radiators, the geometric factors ($\bar{F}_{1\to 2}$ and $F_{1\to 2}$) must be combined with the proper interchange emissivities. These factors are frequently combined into the single factor $\mathscr{F}_{1\to 2}$ so that the simple equation

$$(q_r)_{1\to 2} = A_1\mathscr{F}_{1\to 2}\sigma(T_1{}^4 - T_2{}^4) \quad (15.92)$$

may be used to determine the net heat flux from source to sink including the effects of reradiating walls and emissivities less than one. If the enclosure has only a single source and single sink both of which are grey bodies, the factor $\mathscr{F}_{1\to 2}$ becomes (26)

$$\mathscr{F}_{1\to 2} = \frac{1}{\dfrac{1}{\bar{F}_{1\to 2}} + \left(\dfrac{1}{\epsilon_1} - 1\right) + \dfrac{A_1}{A_2}\left(\dfrac{1}{\epsilon_2} - 1\right)} \quad (15.93)$$

Illustration 15.9. In an electrically heated dryer the stock is moved on a belt conveyor through the drying chamber. There, it is heated by convection and radiation from glow bars along the top of the drying chambers. The glow bars are 1 in. in diameter and are spaced in a row 4 in. from the roof of the dryer and 4 in. from center to center. The stock surface is 8 in. from the bar surfaces in a bed wide enough to be considered infinite compared to the rest of the system dimensions. If the glow bars are at 2800°F with an emissivity of 0.9 while the stock surface is at 200°F with $\epsilon = 0.5$, determine the rate of heat transfer to the stock surface by radiation alone. The refractory walls of the dryer may be considered black.

SOLUTION. From Figure 15.33 $\bar{F}_{1\to 2}$ may be read. Here the ratio of center-to-center distance of tubes in a row to the outside tube diameter is $B/D = 4/1 = 4$; and a single row of tubes is present. From the proper curve on Figure 15.33, $\bar{F}_{1\to 2} = 0.6$ where 1 refers to the stock surface. Since the surfaces are not all black, the $\mathscr{F}$ factor must be used, or interchange emissivities combined with the view factor. For this purpose, Equation 15.93 will be used, for, though there are a multitude of sinks, these sinks are all at the same temperature, are geometrically similar, and hence can be treated as a single surface.

$$\mathscr{F}_{1\to 2} = \frac{1}{\dfrac{1}{\bar{F}_{1\to 2}} + \left(\dfrac{1}{\epsilon_1} - 1\right) + \dfrac{A_1}{A_2}\left(\dfrac{1}{\epsilon_2} - 1\right)} \quad (15.93)$$

$$\mathscr{F}_{1\to 2} = \frac{1}{\dfrac{1}{0.6} + \left(\dfrac{1}{0.5} - 1\right) + \dfrac{4/12}{\pi \times 1/12}\left(\dfrac{1}{0.9} - 1\right)}$$

$$= \frac{1}{1.667 + 1.0 + 1.29(0.111)} = \frac{1}{2.808} = 0.356$$

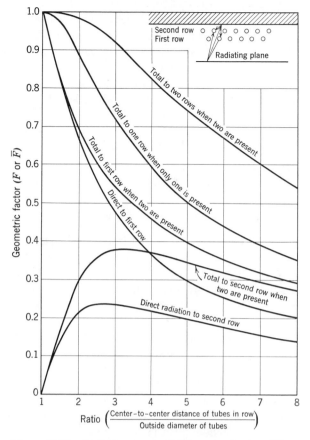

Figure 15.33. Radiation to a bank of staggered tubes (16). (Courtesy of H. C. Hottel, from *Heat Transmission*, 3rd ed., 1954, by W. H. McAdams, McGraw-Hill Book Co., New York.)

Here the unit length of bed has been chosen as 4 in., which corresponds to one complete glow bar. Then

$$(q_r)_{1\to 2} = 1 \times 0.356 \times 0.173(\overline{6.60}^4 - \overline{32.60}^4)$$

$$= 0.0616 \times (1900 - 1{,}132{,}000)$$

$$= -69{,}600 \text{ Btu/hr sq ft}$$

The minus sign indicates that thermal energy is being transferred to the stock surface, and the unit of area is 1 sq ft of stock.

Radiation through Absorbing Media and from Luminous Flames. As mentioned earlier certain unsymmetrical gas molecules absorb significant quantities of radiant thermal energy. The most common of them are water vapor and CO_2, though most organic vapors and oxides of nitrogen and of sulfur absorb radiant energy also. In all cases, the absorptivities of these gases are significant through narrow wave-length bands but are nearly zero through the rest of the wave-length spectrum. The pattern of the absorption bands corresponds to the natural vibration frequencies of the atoms within the molecule and hence is characteristic of the individual molecules. In other wave-length ranges (infrared,

X ray, etc.) the strength and position of the absorption bands serve as analytical devices. Beer's law states that for a given wave length the strength of energy absorption (or emission) is logarithmically proportional to the number of molecules struck by the energy beam. For perfect gases, the strength will be proportional to the partial pressure of absorbing gas and the length of beam travel through the gas. Thus,

$$\ln(1 - \alpha_\lambda) = \gamma pl \qquad (15.94)$$

where p = partial pressure of absorbing gas
l = length of beam travel through the gas
γ = a constant
α_λ = absorptivity for radiation of wave length λ

The constant γ depends upon the wave length of the incident radiation and on the monochromatic absorptivity of the gas at fixed values of p and l. Thus, in considering the total absorptivity of a gas receiving radiant energy, the temperature of emitting source, the atomic and molecular properties of the gas, the partial pressure of receiving gas, and the length of travel of an average beam will all influence the result.

Design charts showing these relations have been obtained experimentally for CO_2, H_2O, and SO_2 and are given in standard texts (26, 28). Also given are factors by which a characteristic dimension of various shapes may be multiplied in order to obtain the length of a mean radiant beam. With water vapor at $pl = 1.0$ ft atm the emissivity varies from 0.1 to 0.3 as the temperature varies from 3500 to 500°F. Over the same temperature range the emissivity varies from 0.3 to 0.6 when pl has been increased to 20 (26).

From the emissivities and absorptivities so obtained the radiant-energy interchange between a black surface and the gas will be

$$q_{rb} = \sigma A_s (\epsilon_G T_G^4 - \alpha_G T_s^4) \qquad (15.95)$$

where ϵ_G = gas emissivity evaluated at the gas temperature
α_G = gas absorptivity evaluated at the surface temperature

For a grey surface, the interchange must be evaluated by tracing typical beams as discussed previously. For surfaces of high emissivities ($\epsilon \geq 0.7$), the resulting surface emissivity factor is approximately $(\epsilon_s + 1)/2$. Thus Equation 15.95 becomes

$$q_r = \sigma A_s \frac{1 + \epsilon_s}{2} (\epsilon_G T_G^4 - \alpha_G T_s^4) \qquad (15.96)$$

The soot particles in most luminous flames also emit and absorb radiant energy. The strength of such emission can be obtained from an empirical design chart based on optical pyrometer measurements of the apparent flame temperature. Such a chart is given by Perry (28). The resulting emissivity can be used to obtain radiant-energy interchange through Equation 15.96.

Radiant-Heat-Transfer Coefficients. In some cases where there are many repetitive calculations of radiant-heat transfer to be made, or where heat transfer by convection and radiation must be considered, it is convenient to reduce the calculations to the form of the heat-transfer coefficient used in convection calculations. Thus for interchange between black surfaces

$$q_{rb} = \sigma A(T_1^4 - T_2^4) = h_{rb} A(T_1 - T_2) \quad (15.97)$$

from which

$$h_{rb} = \sigma \frac{(T_1^4 - T_2^4)}{T_1 - T_2}$$

where h_{rb} = radiation "coefficient" for black-body interchange. For nonblack surfaces the proper emissivities and F factors can be inserted into Equation 15.97. Such coefficients can obviously be used only for the conditions applied in their development, but they do significantly reduce the labor of some calculations of radiant-heat transfer.

REFERENCES

1. Badger, W. L., *Ind. Eng. Chem.*, **22**, 700 (1930).
2. Badger, W. L., *Trans. Am. Inst. Chem. Engrs.*, **33**, 441 (1937).
3. Bergelin, O. P., G. A. Brown, and S. C. Doberstein, *Trans. ASME*, **74**, 960 (1952).
4. Bergelin, O. P., A. P. Colburn, and H. L. Hull, *Eng. Expt. Sta.*, Bull. 2, University of Delaware, 1950.
5. Bowman, R. A., A. C. Mueller, and W. M. Nagle, *Trans. ASME*, **62**, 283 (1940).
6. Boltzmann, L., *Wiedemann's Annalen*, **22**, 291 (1884).
7. Brown, G. G., and Associates, *Unit Operations*, John Wiley and Sons, New York, 1950, p. 425.
8. *Chemical Engineering Problems*, American Institute of Chemical Engineers, 1946.
9. Colburn, A. P., *Ind. Eng. Chem.*, **25**, 873 (1933).
10. Eckert, E., *Forsch. Gebiete Ingenieurw.*, **9**, 251 (1938).
11. Gardner, K. A., *Ind. Eng. Chem.* **33**, 1215 (1941).
12. Gardner, K. A., *Trans. ASME*, **67**, 33 (1945).
13. Gardner, K. A., *Trans. ASME*, **67**, 621 (1945).
14. Gilmour, C. H., Preprint of paper presented at 2nd National Heat Transfer Conference, AIChE-ASME, Chicago, 1958.
15. Grimison, E. D., *Trans. ASME*, **59**, 583 (1937); **60**, 381 (1938).
16. Hottel, H., *Trans. ASME* FSP **53–19b**, 265 (1931).
17. Jakob, M., *Heat Transfer*, Vol. 2, John Wiley and Sons, New York, 1957.
18. Kern, D. Q., *Process Heat Transfer*, McGraw-Hill Book Co., New York, 1950, p. 542.
19. Kern, D. Q., *Process Heat Transfer*, McGraw-Hill Book Co., New York, 1950, Chapter 15.
20. Kern, D. Q., *Process Heat Transfer*, McGraw-Hill Book Co., New York, 1950, p. 525.
21. Kirkbride, C. G., *Ind. Eng. Chem.*, **26**, 425 (1934).
22. Lyon, R. N., Report ORNL-361, Oak Ridge National Laboratory, 1949.

23. McAdams, W. H., *Heat Transmission*, 3rd ed., McGraw-Hill Book Co., New York, 1954, p. 330.

24. McAdams, W. H., *Heat Transmission*, 3rd ed., McGraw-Hill Book Co., New York, 1954, p. 335.

25. McAdams, W. H., *Heat Transmission*, 3rd ed., McGraw-Hill Book Co., New York, 1954, p. 337.

26. McAdams, W. H., *Heat Transmission*, 3rd ed., McGraw-Hill Book Co., New York, 1954, p. 82–124.

27. McCabe, W. L., and C. Robinson, *Ind. Eng. Chem.*, **16**, 478 (1924).

28. Perry, J. H., *Chemical Engineer's Handbook*, 3rd ed., McGraw-Hill Book Co., New York, 1950.

29. Planck, M., *Verhandl. deut. physik Ges.*, **2**, 202 and 237 (1900).

30. Planck, M., *Ann. Physik*, **4**, 553 (1901).

31. Polentz, L. M., *Chem. Eng.*, **65**, No. 7, 137 (1958).

32. Rickard, C. L., O. E. Dwyer, and O. Dropkin, *Trans. ASME*, **80**, 646 (1958).

33. Sieber, W., *Z. Tech. Physik*, **22**, 139 (1941).

34. *Standards of Tubular Exchanger Mfg. Assoc.*, 2nd ed., New York, 1949.

35. Tinker, T., *Proceedings of the General Discussions on Heat Transfer*, Inst. of Mech. Engrs., London and Am. Soc. Mech. Engrs., 1951, pp. 89–116; and *Trans ASME*, **80**, 36 (1958).

36. Wien, W., *Ben. preuss, Akad. Wiss.*, **55**, (1893).

37. Wilson, R. E., *Trans. ASME*, **37**, 47 (1915).

PROBLEMS

15.1. It is proposed to heat 140,000 lb/hr of a nonvolatile oil from 60°F to 400°F by means of a vapor condensing at 500°F. The heat exchanger will consist of a bundle of 1 in. I.D. copper tubes inside a steel shell with the oil flowing through the tubes. Calculate the heat-transfer area required for the exchanger.

Values of the over-all heat-transfer coefficient and the specific heat of the oil are tabulated below. (Normally, the over-all coefficients would have to be calculated using methods of Chapter 13.)

Temp., °F	c_P, Btu/lb/°F	U, Btu/hr sq ft °F
60	0.48	470
100	0.50	540
200	0.56	680
300	0.62	880
400	0.68	1184

15.2. A gear oil is flowing through a 2 in. Sch. 40 steel pipe at a velocity of 3 ft/sec and is heated from 80°F to 200°F by steam condensing at 5 psig. The steam-side coefficient may be assumed to be 1500 Btu/hr sq ft °F. A scale resistance of 0.003 hr sq ft °F/Btu is predicted. How many 20-ft lengths of pipe are necessary for the exchanger?

Oil Properties

$\mu_{100} = 40$ centipoises

$\mu_{200} = 5.4$ centipoises (assume $\ln \mu$ vs T. °F is linear)

Temp., °F	k, Btu/hr sq ft (°F/ft)	ρ, lb/cu ft	c_P, Btu/lb °F
80	0.081	53.3	0.473
100	0.0803	52.0	0.485
150	0.0792	50.8	0.512
180	0.0785	50.1	0.530
200	0.0780	49.7	0.541

15.3. A straw oil flows through a horizontal uninsulated 1 in. Sch. 40 steel pipe which is exposed to quiet room air at 50°F. The oil flows steadily through the pipe entering at an inlet Reynolds number of 3000. The average oil inlet temperature is 200°F. How far from the entrance will the oil reach a temperature of 130°F?

Oil Properties

$\rho = 50$ lb/cu ft

$c_P = 0.50$ Btu/lb/°F

$k = 0.08$ Btu/hr sq ft (°F/ft)

T, °F	70	80	95	110	130	160	200	240
μ, centipoises	22	18	13.8	10.8	8.1	5.7	3.8	2.7

15.4. 100,000 lb/hr of water are to be cooled from 200°F to 100°F by a coolant entering the exchanger at 60°F and leaving at 95°F. Assuming an over-all heat-transfer coefficient of 400 Btu/hr sq ft °F, determine the heat-transfer surface required for:

(a) A double pipe heat exchanger with parallel flow.

(b) A double pipe heat exchanger with countercurrent flow.

15.5. For Problem 15.2, calculate the heating area required by the method indicated:

(a) Find μ at 80°F and 200°F, and determine an arithmetic average. For the temperature corresponding to this average viscosity, calculate U based on properties evaluated at this temperature. Use $(\Delta T)_{lm}$.

(b) Average 80°F and 200°F, and use this temperature to determine physical properties and coefficients. Use $(\Delta T)_{lm}$.

(c) Use arithmetic average of h at 80° and h at 200°F for the exchanger. Use $(\Delta T)_{lm}$.

(d) Compare answers with the rigorous solution of Problem 15.2.

15.6. Process steam at atmospheric pressure is condensed in a shell-and-tube condenser and the condensate cooled to 80°F in a simple double pipe exchanger. The exchanger is a 1⅛ in., 16 BWG, copper tube inside a 3 in. Sch. 40 steel pipe. Cooling water enters the annulus at 60°F and leaves at 190°F. If the steam rate is 2500 lb/hr to the condenser, how much coolant is required (gal/min), and how long must the exchanger be?

15.7. Nitrogen at 30 psig is flowing in the annulus of a double pipe heat exchanger made from a Sch. 40, 2-in. steel pipe and a 1-in., 16 BWG copper tube. The gas flow is 6500 cu ft/hr, measured at 60°F, 760 mm Hg. Water flows in the tube at a velocity of 2 ft/sec, entering at 50°F. The nitrogen is to be cooled from 250°F to 60°F. How long should the exchanger be?

15.8. Butyric acid is being cooled from 180° to 140°F by heat exchange with benzene entering a double pipe heat exchanger at 100°F and leaving at 150°F. A 100 ft long exchanger is fabricated from a 3-in. Sch. 40 outer pipe and a 2-in., Sch. 40 inner pipe. The flow rate of the butyric acid is 20,000 lb/hr. $k_{acid} = 0.09$ Btu/hr sq ft (°F/ft).

(a) What is the dirt factor if the hot stream is placed in the annulus?

(a) What is the dirt factor if the cold stream is in the annulus?

15.9. 3000 lb/hr of a lube oil flow through a smooth ¾-in. I.D. copper tube whose inner surface is maintained at a constant temperature of 250°F by steam condensing on the outside. The tube is 20 ft long, and the oil enters at 100°F. What will be the exit temperature of the oil?

Oil Properties

Specific gravity $= 0.82 - 0.0004T$ where $T = $ °F

Viscosity $= 5.5 \left(\dfrac{60}{T}\right)^{1.25}$ centipoises

$c_P = 0.5$ Btu/lb °F

$k = 0.1$ Btu/hr sq ft (°F/ft)

15.10. It is desired to preheat 30,000 lb of air from 70°F to 200°F at constant pressure of 1 atm by steam condensing at 200°F inside vertical 1-in., Sch. 40 steel pipes. The heater consists of a staggered bank of pipes, each 4 ft long, spaced on 2.5-in. equilateral centers. The heater is 40 pipes wide. What is the number of pipe rows required?

15.11. A horizontal ¾-in., 16 BWG copper condenser tube is 20 ft long and has water entering at a velocity of 5 ft/sec and 70°F. Dry saturated steam at 175°F is fed to a well-insulated jacket surrounding the pipe. Assuming film-type condensation, how many pounds per hour of steam can be condensed?

15.12. Benzene flows through the annulus of a double pipe heat exchanger made up of a 2-in., Sch. 40 steel pipe inside a 3-in., Sch. 40 steel pipe. This exchanger has been operating for several months, and the exit temperature of the benzene has been decreasing with time. Analysis of the benzene shows that some acid impurity is present. Tests on the steam indicate no reason for excessive corrosion inside the inner pipe. From the test data, calculate the magnitude of the scale resistance.

> Flow rate = 4400 lb/hr benzene
> Benzene temperature = inlet 70°F, outlet 130°F
> Steam temperature = 240°F
> Steam-heat-transfer coefficient = 2000 Btu/hr sq ft °F
> Effective heat-transfer length = 15 ft

15.13. A shell-and-tube heat exchanger is to be used to cool 200,000 lb/hr of water from 100°F to 90°F. The exchanger is a one shell pass, two tube pass unit with the water flowing through the tubes. River water flows through the shell and enters at a temperature of 75°F and leaves at 90°F. The heat-transfer coefficient of the shell-side fluid is 1000 Btu/hr sq ft °F. Design specifications state that the tube-side fluid shall have a pressure drop as close to 10 psia as possible and that the tubes be 16 ft long made from 18 BWG copper tubing. For these specifications, what diameter tube and how many tubes are needed?

15.14. 20,000 lb/hr of air is to be cooled from 150°F to 100°F by water flowing in a shell-and-tube heat exchanger. Cooling water is available at 70°F and should leave the exchanger at 95°F. Available to do this cooling assignment is a one shell pass, two tube pass exchanger consisting of fifty 1-in. O.D. 18 BWG copper tubes, each 10 ft long. The tubes are arranged inside a 10-in. shell on a staggered 1¼-in. triangular pitch with baffles spaced every 6 in. Is this exchanger satisfactory?

15.15. A housewife's electric iron is rated at 1000 watts and has a ¼ sq ft heated surface of stainless steel.

(a) What is the rate of radiant-energy transfer from the heated surface when it is at 450°F?

(b) What is the net radiant-energy-transfer rate (in watts) from the iron at 450°F to the room in which it sits if the room is at 60°F?

15.16. If the thermostat on the iron of Problem 15.15 were inoperative, what equilibrium temperature would be attained by the heated surface? Assume all the energy loss is by radiation.

15.17. On a clear night, objects on the surface of the earth exchange thermal energy with the sky as if the sky were a black body at −40°F. What will be the steady-state temperature of a galvanized iron roof on such a night when the air temperature is 80°F? A convection-heat-transfer coefficient of 5 Btu/hr sq ft °F from roof to the air may be used. Assume the roof is perfectly insulated from the room below.

15.18. An asbestos block 4 in. by 4 in. by 12 in. is put in a furnace at 3000°F. What is the rate of temperature rise of the block because of radiation when it is at 600°F? Assume uniform temperature in the block. The specific heat of asbestos is 0.25 cal/gm °C, and its density is 120 lb/cu ft.

15.19. A horizontal steam line of 3-in. Sch. 80 pipe 300 ft long carries steam at 250 psig through an unheated warehouse. The line is insulated with ½ in. of 85 per cent magnesia and painted black with shiny lacquer of negligible thickness. If the steam is dry and saturated as it enters, how much liquid water should be trapped out hourly at the other end of the warehouse on a winter day when the warehouse temperature is 10°F?

15.20. White fire-clay tiles ½ in. by 6 in. by 6 in. are removed from the kiln at 3000°C and cooled in a room at 100°F. If these shapes are well separated from each other during cooling, how long will it take them to cool to 500°C? Use a convection coefficient of 2 Btu/hr sq ft °F to all surfaces and ignore temperature gradients within this tile.

15.21. Repeat Problem 15.20 for the case in which the tiles are placed on edge 2 in. apart on a table to cool.

15.22. It is sometimes desirable from an architectural and safety standpoint to shield steam radiators. With steam at 212°F and room temperature fixed at 70°F, determine the heat transfer to the room for the case of a shielded and unshielded radiator. Use a convection coefficient of 1.5 Btu/hr sq ft °F for both cases at all surfaces. State and justify any assumptions.

15.23. Determine the heat loss by radiation from an oven wall to the wall of the room within which it is located. The oven wall is square, 4 ft on a side, and is 2 ft from the room wall and parallel to it. The oven wall is of steel sheet and is at 400°F, and the wall is at 80°F and is of asbestos board.

15.24. Determine the total absorptivity of white tile for black-body radiation at 2200°F.

15.25. A thermocouple probe used to measure flue-gas temperature from a cement kiln reads 800°F. The probe extends 2 ft into the flue from a fitting in the wall. The probe is made of ½ in. Sch. 40 pipe with the thermocouple tip and pipe end welded together. The ½-in. pipe is insulated with 1-in. thick magnesia pipe insulation out to 1 in. from the tip. Duct wall temperature is measured to be 300°F. The flue gas is flowing at 25 ft/sec at which velocity its convective coefficient to the probe tip is calculated to be 18 Btu/hr sq ft °F. Determine the flue-gas temperature, ignoring heat transfer through the insulation to the probe.

15.26. Liquid N_2 boiling at −320°F is held in a glass Dewar flask. This flask is made of a double layer of silvered glass 4 in. I.D. and ½ in. between the concentric glass walls. The interwall space is completely evacuated. The flask bottom is hemispherical. The top, covered with a 1½ in. thick black rubber stopper with one small hole for vapor removal, is 6 in. above the boiling N_2 surface. The N_2 liquid is 6 in. deep at the center. Determine the rate of evaporation in liters per hour.

chapter 16

Mass Transfer

In mass transfer one or more components of two discrete phases are transferred between the phases. In Part I those mass-transfer operations utilizing *discontinuous* contact of the two phases in stage equipment were analyzed. The concept of the equilibrium stage assumed that the two phases leaving the stage were in equilibrium. This chapter will consider mass transfer in equipment which provides *continuous* contact of the two phases. In continuous-contact equipment it is necessary to consider the *rate* of transfer and the *time* of contact of the phases. Use will be made of the concepts of driving force and resistance in mass transfer.

The design equation for mass transfer which was derived in Chapter 14

$$dA = - \frac{d(Vc_a)}{K(c_a - c_a{}^*)} = - \frac{d(Lc_a)}{K(c_a - c_a{}^*)} \quad (14.65)$$

is perfectly general whether the operation is absorption, distillation, extraction, or adsorption, and it will be the objective of this chapter to discuss in detail its specific application. In application of the transfer equation (14.65), the following information must be available: (1) flow quantities and compositions to permit the establishment of the material balance; (2) equilibrium relationships between the two phases in contact; and (3) knowledge concerning the rate of transfer in the form of a mass-transfer coefficient.

EQUIPMENT FOR MASS TRANSFER

Thorough contact over large areas between the two phases exchanging solute is a prime requirement for equipment used in the mass-transfer operations. For this reason, packed towers or spray columns are frequently used in the process industries.

Packed towers used in continuous countercurrent contacting of two phases are vertical shells filled with a suitable material having a large surface area. Figure 16.1 illustrates the main components of a packed tower.

Phase L enters the top of the column and is distributed over the surface of the packing either by nozzles or distribution plates. In most of the mass-transfer rate operations, phase L will be liquid. Thus, the liquid, upon good initial distribution, will flow down through the packing by tortuous routes, thereby exposing a large surface area for contacting with the rising V-phase. The V-phase will enter the bottom of the tower and rise upward through a similarly tortuous path. The V-phase for absorption and distillation is a gas or vapor, and in extraction it is a liquid.

The column itself is nothing more than a simple cylinder, such as a length of pipe. Care must be exercised in choosing the proper materials of construction depending upon the chemicals being handled. The simplicity in shell and in the entire assembly is one advantage of packed columns.

Tower Packings. Many types of packing materials have been used ranging from simple, readily available solids such as stones or broken bottles to expensive complex geometrical shapes. In general, the packing material should have the following characteristics:

1. It should have a large wetted surface area per unit volume of packed space so as to present a potentially large interfacial area for phase contacting.
2. It should have a large void volume. This will allow reasonable throughputs of phases without serious pressure drop.
3. It should have good wetting characteristics.
4. It should be corrosion resistant.
5. It should have a low bulk density. In large packed towers, the weight of the packing can be quite large resulting in serious support problems.
6. It should be relatively inexpensive.

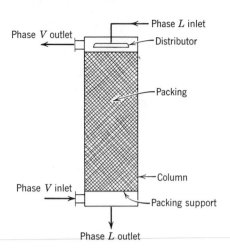

Figure 16.1. Packed tower
components.

Figure 16.2 illustrates several common packing shapes. Raschig rings are widely used in the process industries because of their low cost, although they may not be as efficient as some of the newer packing materials. The wall thickness of the Raschig ring is an important factor because, as the wall thickness is decreased, mechanical strength decreases. A greater wall thickness will result in an increased pressure drop, lower free space, and reduced surface area. Best results are obtained when walls are relatively thin. The diameter and the height of the Raschig ring are equal. Raschig rings may be fabricated from porcelain, clays, carbon, or metals. The "Intalox" saddle gives a greater degree of randomness than the Raschig ring. It is designed to give a more effective nesting of pieces so that good wetting characteristics are obtained. This packing is more expensive in initial cost than the Raschig ring. Berl saddles are costly to produce but do have some advantage over other packings. They can be packed with more randomness than rings, and they give a relatively large amount of surface area per unit volume. The Pall ring has sections of the ring wall stamped and bent inward to give better circulation of the contacting phases.

For many packed-tower operations using rings in large diameter towers, the packing is usually arranged in a stacked configuration. The cyclohelix is good for use in stacked-packing installations. The internal helix affords good contact between gas and liquid. The cross-partition and Lessing rings are simple modifications of the Raschig ring to improve operating characteristics.

Actually, no one packing possesses all the above desirable qualities and so compromises must be made. Although the packing material itself may be designed to give excellent phase contacting, the method of packing into the tower must also be considered because, if the liquid and gas do not contact everywhere within the tower, the packing is not completely effective.

There exists considerable experimental study of the problems of wetting characteristics of packing and liquid distribution within packed towers. Wetted areas of packing have been measured (22) by use of heavy-paper Raschig rings which were contacted with a dye, and the dyed areas of the paper rings were measured. It was found that the external surface of the ring showed somewhat more wetting than the inside surface, and, more significantly, the inside wall area of the column was wetted to a larger extent than the packing. In general the downflowing liquid tends to select preferred

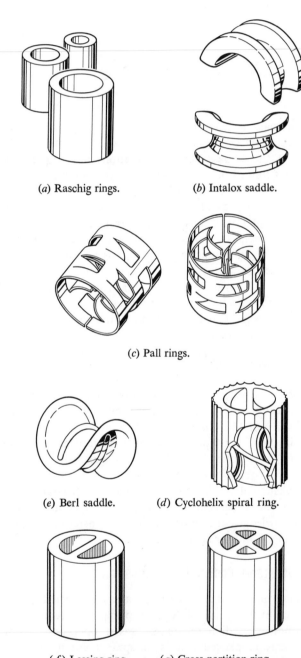

(a) Raschig rings. (b) Intalox saddle.

(c) Pall rings.

(e) Berl saddle. (d) Cyclohelix spiral ring.

(f) Lessing ring. (g) Cross-partition ring.

Figure 16.2. Common packing shapes. (Courtesy U.S. Stoneware.)

paths, or channels, for its flow down the packing. This tendency is referred to as *channeling*. As may be expected, the fluid tends to move towards the region of greatest void space, which is the region near the wall, because the packing material cannot nest so tightly with the plane wall as it can with itself. Consequently, in large-diameter columns, it is customary to employ liquid redistributors at spacings of several diameters (16) to re-direct the flow towards the center of the packing. With good initial liquid distribution and a ratio of tower diameter to packing size greater than 8 to 1, the tendency of the liquid channeling toward the wall is diminished (3) but not eliminated, except with liquid redistributors. Channeling is a principal contribution towards poor performance in packed columns. Leva (18) found that random packing produced less channeling than stacked packing.

Table 16.1 lists some physical characteristics of various types of industrial packing materials.

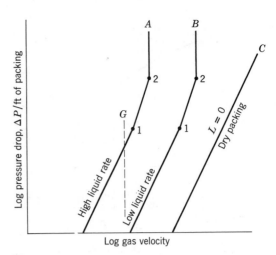

Figure 16.3. Pressure-drop characteristics in packed towers.

Table 16.1. PHYSICAL CHARACTERISTICS OF DRY COMMERCIAL PACKINGS*

Packing	Per Cent Voids (ϵ)	Specific Surface (a_v) sq ft/cu ft	Pieces per cu ft	Dumped Weight lb per cu ft
Ceramic Raschig rings				
$\frac{1}{4}$ in.	73	240	88,000	46
$\frac{1}{2}$	63	111	10,500	54
1	73	58	1350	40
2	74	28	162	38
Carbon Raschig rings				
$\frac{1}{4}$ in.	55	212	85,000	46
$\frac{1}{2}$	74	114	10,600	27
1	74	57	1325	27
2	74	28.5	166	27
Berl saddles				
$\frac{1}{4}$ in.	60	274	113,000	56
$\frac{1}{2}$	63	142	16,200	54
1	69	76	2200	45
2	72	32	250	40
Pall rings				
1 in.	93.4	66.3	1520	33
2	94.0	36.6	210	27.5
Cyclohelix and single spiral				
$3\frac{1}{4}$ in.	58	40	63	60
4	60	32	31	61
6	66	21	9	59

* Courtesy of U.S. Stoneware.

Pressure Drop in Packed Columns. The pressure drop in a packed tower is influenced by both the gas- and liquid-flow rates in a manner shown in Figure 16.3. Flow of the gas through the tower packing is usually turbulent; consequently, the slope of curve C is about 2, as would be predicted by a turbulent-flow pressure-drop equation. For a constant gas velocity, the pressure drop increases with increasing liquid rate as can be observed from line G in Figure 16.3. Each type of packing material has a fixed void volume for liquid passage so that, as the liquid rate increases, the voids fill with liquid, thus reducing the cross-sectional area available for gas flow.

The nature of random packing is such that a myriad of expansion and contraction losses and considerable turbulence is created by the flow of the two fluids around the individual solid packing elements. The pressure drop is a combination of skin friction and form drag with form drag predominant at the higher velocities. By the analogy equations of Chapter 13, there is a direct relationship between heat- and mass-transfer surface coefficients and skin friction, so that it is advantageous to have a high percentage of the total pressure drop attributable to skin friction. It has been estimated (4) that not over 10 per cent of the pressure drop is the result of skin friction.

Returning to Figure 16.3, consider line A, a line of constant liquid rate, and imagine that a transparent column is being used. Up to point 1 on this curve, the pressure-drop characteristic is quite similar to that of the dry packing (C). The slope of this portion of the line is about the same as that of the dry curve, but the pressure drop is greater. The larger pressure drop is due to "blocking" of part of the voids by the liquid and roughening of the surfaces by waves. Visual observation of the packing would indicate orderly trickling of the liquid downward through the packing with no observable liquid build-up. At point 1, a change in slope occurs,

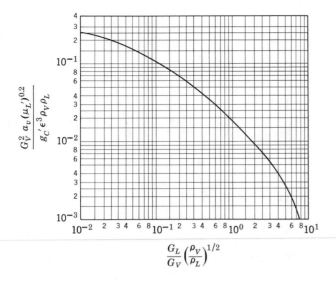

Figure 16.4. Flooding velocities in packed towers (21).

G_L = mass velocity of liquid, lb/hr sq ft
G_V = mass velocity of vapor, lb/hr sq ft
μ_L' = liquid viscosity, centipoises
g_c' = 4.17 × 10⁸ ft-lb/lb$_f$ hr²
ρ_V = gas density, lb/cu ft
ρ_L = liquid density, lb/cu ft
a_v = sq ft/cu ft of packing (Table 16.1)
ϵ = porosity

indicating the pressure drop increases more rapidly with an increase in gas velocity. This point may not be distinct enough to allow observation of any change in the flow pattern or characteristics. Perhaps, it might be possible to observe an increase in the quantity of liquid retained in the packed section. The retained liquid is referred to as hold-up. Point 1 is called the *loading point*, and the increased dependence of the pressure drop on the gas velocity is a consequence of drag between the phases. As point 1 is passed, visual observation shows a greater amount of liquid hold-up. Perhaps observation will show a liquid layer at the top of the packing and a gradual filling of the packing voids with liquid. The liquid now has filled a large portion of the packing, and the gas must bubble through it. This condition is sometimes called "visual flooding." At a gas rate somewhat greater than that corresponding to "visual flooding," a second change in the slope of the pressure-drop line occurs (point 2). This point is more reproducible than the visual observation and is generally taken as the *flooding point*. At this point, the drag force of the gas bubbling through the liquid is quite important. Each liquid rate has its own loading and flooding points.

Operation of packed towers is not practical above the loading point. Colburn (6) presents methods for evaluation of the optimum gas velocity through a packed column. For simplicity and safety packed towers are designed using gas velocities of about 50–75 per cent of the flooding velocity at the expected liquid rate. This

design will normally ensure a stable operating condition somewhere below loading and should provide for thorough wetting of the packing surface.

Numerous investigators have determined flooding velocities for a variety of fluids and packing materials. The results are correlated in Figure 16.4.

Illustration 16.1. Calculate the flooding velocity for a column packed with 1-in. ceramic Raschig rings used for the absorption of ammonia from air by water. Operating conditions are:

Temperature of air–NH₃ mixture = 70°F; total pressure = 1 atm; $G_L/G_V = 1.0$

SOLUTION. Figure 16.4 requires gas and liquid densities. Gas density (assume perfect-gas behavior and negligible effect of NH₃) = (29)(492)/(359)(530) = 0.075 lb/cu ft.

Liquid density = 62.3 lb/cu ft

Thus the abscissa of Figure 16.4 is

$$\frac{G_L}{G_V}\sqrt{\frac{\rho_V}{\rho_L}} = 1\sqrt{\frac{0.075}{62.3}} = 0.0347$$

For 1-in. ceramic Raschig rings, from Table 16.1,

$$a_v = 58 \text{ sq ft/cu ft}$$

$$\epsilon = 0.73$$

From Figure 16.4,

$$\frac{G_V^2\left(\dfrac{a_v}{\epsilon^3}\right)(\mu_L')^{0.2}}{g_c'\rho_V\rho_L} = 0.18$$

$$G_V = \left[\frac{(0.18)\left(4.17 \times 10^8 \dfrac{\text{ft - lb}}{\text{lb}_f\text{hr}^2}\right) \times \left(0.075 \dfrac{\text{lb}}{\text{cu ft}}\right)\left(62.3 \dfrac{\text{lb}}{\text{cu ft}}\right)(0.73)^3}{\left(58 \dfrac{\text{sq ft}}{\text{cu ft}}\right)(1 \text{ centipoise})^{0.2}}\right]^{1/2}$$

$$G_V = 1530 \text{ lb/hr sq ft}$$

The tower would be designed for a gas velocity 50 per cent of the flooding, or

$$G_V = (0.5)(1530) = 765 \text{ lb/hr sq ft}$$

and the corresponding liquid rate is also 765 lb/hr sq ft.

For gas flow below loading (irrigated flow), Leva (19) reports the pressure drops for flow through packings whose diameter is less than one-eighth the size of the shell. Treybal (31) has correlated these as:

$$\frac{\Delta P}{z} = m(10^{-8})(10^{nG_L/\rho_L})\frac{G_V^2}{\rho_V} \tag{16.1}$$

where
z = packed height, ft
m, n = packing constants (see Table 16.2)
G_L = liquid rate, lb/hr sq ft empty tower
G_V = gas rate, lb/hr sq ft empty tower

Table 16.2. CONSTANT FOR TOWER PACKING

Type	m	n
Raschig rings		
$\frac{1}{2}$ in.	139	0.00720
1 in.	32.1	0.00434
2 in.	11.13	0.00295
Berl saddles		
$\frac{1}{2}$ in.	60.4	0.00340
1 in.	16.01	0.00295

DESIGN OF PACKED TOWERS FOR MASS TRANSFER

Mass-transfer operations are carried out in plate towers and in packed columns. The developments that follow will be devoted exclusively to packed columns, but a qualitative comparison between the two types of contacting devices is warranted.

1. For the identical gas-flow rates, less pressure drop occurs through a plate tower than through a packed tower.

2. Packed towers are cheaper and easier to assemble than plate columns for corrosive chemicals.

3. Plate columns can handle greater liquid loads without flooding.

4. Liquid hold-up is less in a packed tower than in a plate tower.

5. For the same duty, plate columns weigh less than packed towers.

6. Large-diameter packed columns are not too satisfactory because of maldistribution of liquid.

7. Plate columns are more easily cleaned.

It has already been established that, in packed-tower contactors used for mass-transfer operations, the two fluids are in continuous contact throughout the tower. This is in direct contrast to stage contactors where the two fluids are brought together, mixed, and then separated; the two streams leaving the stage approach equilibrium; the rate of approach may be rapid. In a continuous-contacting apparatus, the two streams in contact never reach equilibrium, and the point-to-point relationships of coexisting concentrations in the two streams are exceedingly important in establishing driving forces that determine the rate of mass transfer. The relationship between the phase concentrations throughout the tower is the operating line. Just as in stage operations, the operating line is established by a material balance, but unlike the stage operations every point on the operating line has physical meaning. The operating line for stage calculations has significance only with respect to stream compositions to or from a particular stage. The design equation (Equation 14.60)

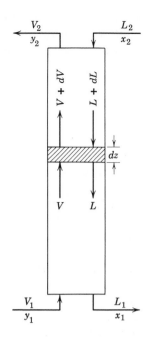

Figure 16.5. Packed tower.

will now be rewritten and applied with particular emphasis on equimolar counterdiffusion although the findings will be translatable to all mass-transfer operations. In application of this equation, the following convention shall be adopted: *transfer of mass to the V-phase will be taken as positive and the integration of the design equation will always be carried out between the limits corresponding to the bottom of the tower and the top.* The V-phase is always the one rising through the column.

The Design Equation. Consider a packed tower used for a mass-transfer operation such as distillation or gas absorption. This tower is shown schematically in Figure 16.5. For this sketch, the following nomenclature is employed:

V = molar flow rate of phase V, lb moles/hr
L = molar flow rate of phase L, lb moles/hr
y = more volatile component in phase V, mole fraction
x = more volatile component in phase L, mole fraction
dz = differential tower height, ft

Subscript 1 = tower bottom
Subscript 2 = tower top

For steady-state operation, a material balance over the differential tower section yields

$$dV = dL \tag{16.2}$$

and a component balance over the same section gives

$$d(Vy) = d(Lx) \tag{16.3}$$

Equation 16.3 is the component material balance equation. It relates the bulk compositions of the two phases in contact at any point throughout the length of the tower. Upon integration between the tower bottom and any point within the tower Equation 16.3 becomes

$$Vy - V_1y_1 = Lx - L_1x_1$$

or

$$Vy + L_1x_1 = Lx + V_1y_1 \qquad (16.4)$$

Equation 16.4 is the operating-line equation and is valid for all values of x between x_1 and x_2 and all values of y between y_1 and y_2. In the general case in which V and L vary with change in concentration or position, this equation describes a *curved* operating line. For certain cases of extreme dilution in which the change in concentration has negligible effect upon V and L, the operating line as defined in Equation 16.4 is substantially linear.

In accordance with the development of Equation 14.60, the rate of change of a component within a phase must be equal to the rate of transfer to the phase. Thus, for the V-phase in equimolar counterdiffusion

$$d(Vy) = k_y'(y_i - y)\,dA = K_y'(y^* - y)\,dA \quad (16.5)$$

where dA is the interfacial transfer area associated with the differential tower length. This transfer area in a packed column is extremely difficult to measure and is more conveniently expressed as

$$dA = aS\,dz \qquad (16.6)$$

where a = interfacial area per unit volume of packing, sq ft/cu ft

  S = empty tower cross-sectional area, sq ft

The term a in Equation 16.6 is not to be confused with a_v in Table 16.1 which is written for dry packing. This new term is a function of a combination of packing characteristics and retained liquid and is therefore dependent upon the flow rates of the two phases. The quantity a is influenced by the flow rates in the same manner as described in flooding phenomena. Since a is usually unknown in packed columns, it is combined with the surface coefficient to give *a composite coefficient* $k_y'a$ with units of amount transferred per unit of (time) (driving force) (volume of packing). Thus Equation 16.5 may be combined with Equation 16.6 to give

$$d(Vy) = k_y'a(y_i - y)S\,dz = K_y'a(y^* - y)S\,dz \quad (16.7)$$

Equation 16.7 may be used to solve for the tower height required by integrating over the total change in concentration between the tower terminals.

$$\int_0^z dz = \int_{y_1}^{y_2} \frac{d(Vy)}{k_y'aS(y_i - y)} = \int_{y_1}^{y_2} \frac{d(Vy)}{K_y'aS(y^* - y)} \quad (16.8)$$

In a similar manner, the tower height can be evaluated using L-phase changes in composition along with the corresponding rate, with all terminology based on the liquid phase.

$$\int_0^z dz = \int_{x_1}^{x_2} \frac{d(Lx)}{k_x'aS(x - x_i)} = \int_{x_1}^{x_2} \frac{d(Lx)}{K_x'aS(x - x^*)} \quad (16.9)$$

To use Equation 16.8 or 16.9 requires knowledge of the following:

1. The value of $k_x'a$, $k_y'a$, $K_x'a$, or $K_y'a$ as a function of the flow rates of V and L and the transport properties of these two phases.

2. The relationship between point compositions in either phase and point driving forces, $(x - x_i)$, $(x - x^*)$, $(y_i - y)$ or $(y^* - y)$. Bulk compositions in either phase may be calculated from Equation 16.4 and driving forces may be determined by methods described in Chapter 14.

The design equation for the case of diffusion through a stationary component may be obtained in the same manner as described for Equations 16.8 and 16.9 by employing the appropriate mass-transfer coefficient to account for the difference in mechanisms as explained in Chapter 9. Thus, for diffusion through a stationary component such as is encountered in absorption,

$$\int_0^z dz = \int_{y_1}^{y_2} \frac{d(Vy)}{k_yaS(y_i - y)} = \int_{y_1}^{y_2} \frac{d(Vy)}{K_yaS(y^* - y)} \quad (16.10)$$

and in terms of the L-phase

$$\int_0^z dz = \int_{x_1}^{x_2} \frac{d(Lx)}{k_xaS(x - x_i)} = \int_{x_1}^{x_2} \frac{d(Lx)}{K_xaS(x - x^*)} \quad (16.11)$$

Thus, Equations 16.8, 16.9, 16.10, and 16.11 represent rigorous design equations for mass-transfer operations involving equimolar counterdiffusion and diffusion through a nondiffusing phase. These equations are the exact counterpart to Equation 14.62, the design equation for heat transfer. The equations will now be examined in more detail in an effort to attach some physical significance to them.

Equimolar Counter Diffusion. Equimolar counter diffusion occurs in distillation since there is an equal rate of interchange of the more and less volatile constituent between phases. If the assumptions of McCabe and Thiele are valid (Chapter 7), Equation 16.8 may be written as

$$\int_0^z dz = V\int_{y_1}^{y_2} \frac{dy}{k_y'aS(y_i - y)} = V\int_{y_1}^{y_2} \frac{dy}{K_y'aS(y^* - y)} \quad (16.12)$$

where the molar flow rate (V) is now considered constant through the enriching section of the tower. It will also be constant in the stripping section but of a different magnitude depending upon the feed quantity and conditions. For the same assumptions, Equation 16.9 becomes

$$\int_0^z dz = L \int_{x_1}^{x_2} \frac{dx}{k_x'aS(x - x_i)} = L \int_{x_1}^{x_2} \frac{dx}{K_x'aS(x - x^*)} \quad (16.13)$$

Some simplifications may be made to put Equation 16.12 or 16.13 into a more usable form. The phase flow rates (L and V) have already been assumed constant. Since the individual and over-all mass-transfer coefficients are functions of these flow rates, the coefficients must also be constant providing physical properties do not vary appreciably. Therefore,

$$\int_0^z dz = \frac{V}{k_y'aS} \int_{y_1}^{y_2} \frac{dy}{(y_i - y)} = \frac{V}{K_y'aS} \int_{y_1}^{y_2} \frac{dy}{(y^* - y)} \quad (16.14)$$

and

$$\int_0^z dz = \frac{L}{k_x'aS} \int_{x_1}^{x_2} \frac{dx}{(x - x_i)} = \frac{L}{K_x'aS} \int_{x_1}^{x_2} \frac{dx}{(x - x^*)} \quad (16.15)$$

Tower heights required for a given separation can be obtained by use of either Equation 16.14 or Equation 16.15, provided that the assumptions stated are valid. Note that in both equations the integral term is equal to the total change of composition for the particular phase divided by the available driving force. This is a measure of the difficulty of separation and upon integration gives a quantity which Chilton and Colburn (5) have defined as the *number of transfer units* (N). The physical significance of this term will be considered later in the chapter. The quantity outside the integral is called the *height of a transfer unit* (H). The tower height is determined by multiplying the number of transfer units by the height of a transfer unit,

$$z = H_G N_G = H_{OG} N_{OG} = H_L N_L = H_{OL} N_{OL} \quad (16.16)$$

where

$$H_G = \frac{V}{k_y'aS} \text{ ft}, \qquad H_L = \frac{L}{k_x'aS} \text{ ft}$$

$$H_{OG} = \frac{V}{K_y'aS} \text{ ft}, \qquad H_{OL} = \frac{L}{K_x'aS} \text{ ft}$$

$$N_G = \int_{y_1}^{y_2} \frac{dy}{y_i - y}, \qquad N_L = \int_{x_1}^{x_2} \frac{dx}{x - x_i}$$

$$N_{OG} = \int_{y_1}^{y_2} \frac{dy}{y^* - y}, \qquad N_{OL} = \int_{x_1}^{x_2} \frac{dx}{x - x^*}$$

Before discussing the significance and uses of Equations 16.14 through 16.16, it is important to develop the similar equations for diffusion through a nondiffusing phase.

Diffusion through a Stationary Component. Important in this classification of mass-transfer operations are gas absorption, liquid-liquid extraction, ion exchange, and adsorption. The starting point for the development of working equations will again be Equation 16.10 or 16.11. However, unlike distillation, the assumption of constant V and L is not valid. For instance in gas absorption, both phases will have variable flow rates as a result of the transfer of solute to or from the phases. This complication can be circumvented by recognizing that the quantity of solute-free gas, which is essentially insoluble, is a constant. Therefore let V' be the molar flow rate of the solute-free gas and $V' = V(1 - y)$; thus

$$d(Vy) = V' d\left(\frac{y}{1 - y}\right) = V' \frac{dy}{(1 - y)^2} = V \frac{dy}{(1 - y)} \quad (16.17)$$

Consequently, in accordance with Equation 16.7, the design equation may be written as

$$d(Vy) = V \frac{dy}{(1 - y)} = k_y a(y_i - y)S \, dz$$
$$= K_y a(y^* - y)S \, dz \quad (16.18)$$

Solving Equation 16.18 for the tower height,

$$\int_0^z dz = \int_{y_1}^{y_2} \frac{V}{k_y aS} \cdot \frac{dy}{(1 - y)(y_i - y)}$$
$$= \int_{y_1}^{y_2} \frac{V}{K_y aS} \cdot \frac{dy}{(1 - y)(y^* - y)} \quad (16.19)$$

In terms of the L-phase, the design equation is

$$\int_0^z dz = \int_{x_1}^{x_2} \frac{L}{k_x aS} \cdot \frac{dx}{(1 - x)(x - x_i)}$$
$$= \int_{x_1}^{x_2} \frac{L}{K_x aS} \cdot \frac{dx}{(1 - x)(x - x^*)} \quad (16.20)$$

Equations 16.19 and 16.20 are rigorous design equations that are applicable to diffusion through stationary nondiffusing phases. These equations are always appropriate although it is often possible to make simplifications in their use. As before, the equations may be expressed in terms of transfer units and the height of a transfer unit, although the latter quantity remains within the integral. Or

$$\int_0^z dz = \int_{y_1}^{y_2} H_G \, dN_G = \int_{y_1}^{y_2} H_{OG} \, dN_{OG} = \int_{x_1}^{x_2} H_L \, dN_L$$
$$= \int_{x_1}^{x_2} H_{OL} \, dN_{OL} \quad (16.21)$$

where H_G, H_{OG} are defined as before. The transfer-unit definition includes the term $(1 - y)$ or $(1 - x)$ in the denominator.

Table 16.3. THE NUMBER OF TRANSFER UNITS (N) AND THE HEIGHT OF A TRANSFER UNIT (H)

Mechanism		Number of Transfer Units	Height of a Transfer Unit		Driving Force
Equimolar counter-diffusion	N_G	$\displaystyle\int_{y_1}^{y_2} \frac{dy}{y_i - y}$	H_G	$\dfrac{V}{k_y'aS}$	$y_i - y$
	N_{OG}	$\displaystyle\int_{y_1}^{y_2} \frac{dy}{y^* - y}$	H_{OG}	$\dfrac{V}{K_y'aS}$	$y^* - y$
	N_L	$\displaystyle\int_{x_1}^{x_2} \frac{dx}{x - x_i}$	H_L	$\dfrac{L}{k_x'aS}$	$x - x_i$
	N_{OL}	$\displaystyle\int_{x_1}^{x_2} \frac{dx}{x - x^*}$	H_{OL}	$\dfrac{L}{K_x'aS}$	$x - x^*$
Diffusion through a stationary component	N_G	$\displaystyle\int_{y_1}^{y_2} \frac{(1 - y)_{lm}\, dy}{(1 - y)(y_i - y)}$	H_G	$\dfrac{V}{k_yaS(1 - y)_{lm}}$	$y_i - y$
	N_{OG}	$\displaystyle\int_{y_1}^{y_2} \frac{(1 - y)_{lm}\, dy}{(1 - y)(y^* - y)}$	H_{OG}	$\dfrac{V}{K_yaS(1 - y)_{lm}}$	$y^* - y$
	N_L	$\displaystyle\int_{x_1}^{x_2} \frac{(1 - x)_{lm}\, dx}{(1 - x)(x - x_i)}$	H_L	$\dfrac{L}{k_xaS(1 - x)_{lm}}$	$x - x_i$
	N_{OL}	$\displaystyle\int_{x_1}^{x_2} \frac{(1 - x)_{lm}\, dx}{(1 - x)(x - x^*)}$	H_{OL}	$\dfrac{L}{K_xaS(1 - x)_{lm}}$	$x - x^*$

It would be convenient for ease of solution if the terms equivalent to H_G or H_{OG} could be removed from the integral sign and treated as a constant. However, as the equation has been developed, this is not possible. As shown in Chapter 13 (Equation 13.83a),

$$\frac{k_c'D}{\mathscr{D}} = 0.023\,(N_{\text{Re}})^{0.83}(N_{\text{Sc}})^{0.44} \qquad (13.83a)$$

Although the above equation is for the mass-transfer coefficient k_c', it will be shown later that a similar correlation for the composite coefficient $k_c'a$ exists. Thus, for the time being it is postulated:

$$\frac{k_y'aD}{\mathscr{D}} = C(N_{\text{Re}})^a(N_{\text{Sc}})^b \qquad (16.22)$$

For a given system of specified flow conditions, geometry, and physical properties, $k_y'a$ is a constant. The relationship between equimolar counterdiffusion and diffusion of a solute through a stationary component is, from Chapter 13, p. 170

$$k_y'a = k_ya \frac{y_{blm}}{y_t} \qquad (16.23)$$

Since $k_y'a$ is constant for a specified system, $k_ya\, y_{blm}/y_t$ must be independent of concentration. This fact will be used to simplify the integral term of Equation 16.19,

the design equation by manipulation of equations.

The numerator and denominator of the second term of Equation 16.19 are divided by y_{blm}/y_t. This ratio will be expressed in terms of component a by expressing y_b as $(1 - y)$, and by taking the logarithmic mean of $(1 - y)$ in the bulk stream and at the interface. This results in

$$\int_0^z dz = \int_{y_1}^{y_2} \frac{V}{k_yaS(1 - y)_{lm}} \cdot \frac{(1 - y)_{lm}\, dy}{(1 - y)(y_i - y)}$$

$$= \int_{y_1}^{y_2} \frac{V}{K_yaS(1 - y)_{lm}} \cdot \frac{(1 - y)_{lm}\, dy}{(1 - y)(y^* - y)} \qquad (16.24)$$

By Equations 16.22 and 16.23, the term $k_ya(1 - y)_{lm}$ in the denominator is equivalent to $k_y'a$ and is independent of concentration. It is dependent on the geometry of the apparatus and on the flow rate of the V-phase.

Correlations for mass transfer in general involve flow rates, frequently expressing a proportionality between k and $V^{0.8}$. As a reasonable, though obviously not rigorous, approximation, k_ya will be assumed proportional to V over the range involved in one application. On the basis of these assumptions the group of terms $V/k_yaS(1 - y)_{lm}$ and $V/K_yaS(1 - y)_{lm}$ are assumed constant for a particular contactor, usually at an average of the terminal values, and may be removed from the integral in Equation 16.24.

$$\int_0^z dz = \left[\frac{V}{k_y aS(1-y)_{lm}}\right]_{av} \int_{y_1}^{y_2} \frac{(1-y)_{lm}\, dy}{(1-y)(y_i - y)} \quad (16.25)$$

and

$$\int_0^z dz = \left[\frac{V}{K_y aS(1-y)_{lm}}\right]_{av} \int_{y_1}^{y_2} \frac{(1-y)_{lm}\, dy}{(1-y)(y^* - y)} \quad (16.26)$$

Colburn (7) suggests for the L-phase:

$$\int_0^z dz = \left[\frac{L}{k_x aS(1-x)_{lm}}\right]_{av} \int_{x_1}^{x_2} \frac{(1-x)_{lm}\, dx}{(1-x)(x - x_i)}$$

$$\int_0^z dz = \left[\frac{L}{K_x aS(1-x)_{lm}}\right]_{av} \int_{x_1}^{x_2} \frac{(1-x)_{lm}\, dx}{(1-x)(x - x^*)}$$

$$(16.27)$$

In either Equation 16.26 or 16.27 the height of a transfer unit is defined by the quantity outside the integral sign, and the number of transfer units is defined by the integral term. Either equation is appropriate for sizing mass-transfer equipment. Although no great simplification of Equation 16.19 or 16.20 seems to have resulted, the transfer-unit height has been removed from within the integral with little loss in rigor.

Evaluation of Transfer Units. The transfer unit has already been introduced as a measure of the difficulty of separation. Table 16.3 lists the most general definitions of the transfer unit, along with the height of the transfer unit for the two general mechanisms of mass transfer. It should be noticed that, if the value of the integral is unity, the meaning of the transfer unit is obvious: It is the amount of contacting necessary to accomplish an enrichment of one phase equal to the driving force in the same phase.

The number of transfer units and the height of a transfer unit may be expressed as functions of driving forces in any set of units. Some authors contend that since the height of a transfer unit has a single dimension of feet, it presents an advantage over the use of the many-dimensioned mass-transfer coefficient. In reality, this is only a superficial advantage because the definition of the height of a transfer unit includes the mass-transfer coefficient, which in turn must be consistent with the units of driving force employed. Correlations for heights of transfer units will be presented later.

A single transfer unit may be defined as giving an enrichment of one of the phases equal to the average driving force producing this enrichment. With this definition, it is of interest to compare the separation obtained using one transfer unit with the separation associated with an equilibrium stage. Figure 16.6 shows a conventional xy-diagram with portions of an equilibrium curve and operating line.

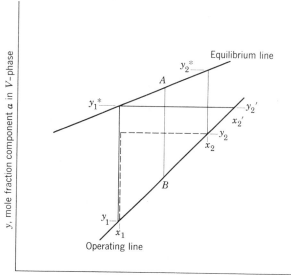

Figure 16.6. Comparison of a transfer unit and an equilibrium stage.

The equilibrium stage is shown by the solid line to give a change in V-phase composition of $y_2' - y_1$ which is equal to the driving force $(y_1^* - y_1)$ based upon the liquid composition *leaving* the stage (x_1). The transfer unit shown by the dotted line gives a change in composition $y_2 - y_1$ equal to the *average* driving force in the transfer unit. The driving force changes as the composition changes. The driving force is $y_1^* - y_1$ at (x_1, y_1) and $y_2^* - y_2$ at (x_2, y_2). In the most general terms, the average driving force for the gas phase is

$$(y^* - y)_m = \frac{\int_{y_1}^{y_2} dy}{\int_{y_1}^{y_2} \frac{dy}{y^* - y}} \quad (16.28)$$

Liquid phase concentrations and driving forces may also be used to show the difference between an equilibrium stage and a transfer unit. Solution of Equation 16.28 depends upon the relationship between y^* and y. If both the equilibrium curve and operating lines are straight or nearly so, solution of Equation 16.28 gives the arithmetic average,

$$(y^* - y)_m = \frac{(y_1^* - y_1) + (y_2^* - y_2)}{2} \quad (16.29)$$

This equation requires a trial-and-error solution to evaluate y_2, but the solution is not very complex. Referring to Figure 16.6, the dotted line indicates one transfer unit with the separation in this unit equal to $(y_2 - y_1)$ which is determined by trial such that $(y_2 - y_1) = AB$, the average driving force of y_2 and y_1. The use of Equation 16.29 will offer a convenient tool

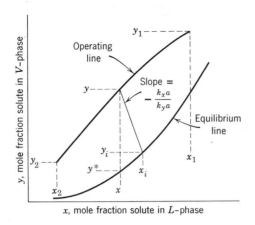

Figure 16.7. xy-diagram for absorption.

toward a graphical determination of the number of transfer units required for a given separation.

Determination of the Number of Transfer Units. Consider the evaluation of the number of transfer units required for a gas-absorption problem where, for example,

$$N_{OG} = \int_{y_1}^{y_2} \frac{(1-y)_{lm}\,dy}{(1-y)(y^*-y)} \qquad (16.30)$$

$$N_G = \int_{y_1}^{y_2} \frac{(1-y)_{lm}\,dy}{(1-y)(y_i-y)} \qquad (16.30a)$$

Several methods are available for the solution of Equation 16.30 or Equation 16.30a.

1. *Graphical Integration.* A rigorous solution of Equation 16.30 or 16.30a is not completely possible in the absence of the necessary functional relationships between the various variables. Nevertheless, a good answer may be obtained by using the techniques of graphical integration. Thus, one should obtain the necessary information required for Equation 16.30 or Equation 16.30a as follows:

(a) Plot the equilibrium data on xy-coordinates and place the operating line on the same graph. Thus, in Figure 16.7, are shown the equilibrium curve and the operating line with the terminal compositions indicated.

(b) Progressing upward through the column, local driving forces $(y_i - y)$ or $(y^* - y)$ and nondiffusing component concentrations $(1 - y)$ may be determined for any value of y. Repeating this operation at several values of y between y_1 and y_2 through the tower permits the plotting, as in Figure 16.8, of the necessary information for evaluation of the number of transfer units as stated by Equation 16.30 or 16.30a. This procedure is rigorous and is limited only by the precision of the graphical techniques.

This graphical procedure was demonstrated for gas absorption but applies equally well to any mass-transfer operation with appropriate modification for the mecha-

nism and the definition of the transfer unit. Either over-all driving forces or driving forces through a single resistance may be employed to evaluate the number of transfer units required for a given enrichment.

2. *Simplification of the Integral.* For many instances of diffusion through a stationary component such as gas absorption or extraction, Wiegand (33) showed that $(1-y)_{lm}$ may be approximated by an arithmetic average of $(1-y)$ and $(1-y^*)$. If the small error introduced by this approximation is acceptable Equation 16.30 integrates to

$$N_{OG} = \int_{y_1}^{y_2} \frac{dy}{(y^*-y)} + \tfrac{1}{2}\ln\frac{1-y_2}{1-y_1} \qquad (16.31)$$

Normally the integral term must be evaluated graphically, but it is much simpler than the integration of Equation 16.30. In addition this integral term will hold for equimolar diffusion. In many instances, the last term of Equation 16.31 is negligible, but the term must be examined for each case. Often for dilute solutions $(1-y)$ and $(1-y)_{lm}$ are approximately equal, so that they may be assumed to cancel in Equation 16.30, again resulting in the simplified integral. In any event, it should be emphasized that the simplifications cannot be made without regard for the accuracy of the assumptions and the error to be encountered.

To summarize the results of the simplification of the integral in Equation 16.30,

$$N_G = \int_{y_1}^{y_2} \frac{dy}{y_i - y} + \tfrac{1}{2}\ln\frac{1-y_2}{1-y_1} \qquad (16.32)$$

$$N_{OG} = \int_{y_1}^{y_2} \frac{dy}{y^* - y} + \tfrac{1}{2}\ln\frac{1-y_2}{1-y_1} \qquad (16.33)$$

$$N_L = \int_{x_1}^{x_2} \frac{dx}{x - x_i} - \tfrac{1}{2}\ln\frac{1-x_2}{1-x_1} \qquad (16.34)$$

$$N_{OL} = \int_{x_1}^{x_2} \frac{dx}{x - x^*} - \tfrac{1}{2}\ln\frac{1-x_2}{1-x_1} \qquad (16.35)$$

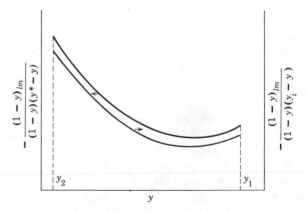

Figure 16.8. Graphical integration for N_G or N_{OG}.

Equations 16.32 through 16.35 are in terms of mole-fraction units, but other units may be used. For instance, in terms of mole ratios,

$$N_G = \int_{Y_1}^{Y_2} \frac{dY}{Y_i - Y} + \frac{1}{2} \ln \frac{1 + Y_2}{1 + Y_1} \qquad (16.36)$$

$$N_{OG} = \int_{Y_1}^{Y_2} \frac{dY}{Y^* - Y} + \frac{1}{2} \ln \frac{1 + Y_2}{1 + Y_1} \qquad (16.37)$$

The choice of equation to use is arbitrary, but, just as in heat transfer where the over-all coefficient is usually based upon the area associated with the largest resistance, it seems logical for purposes of accuracy to use the equation which involves driving forces in terms of the principal resistance. That is, if the gas phase can be said to be controlling, then the driving force across the gas phase is large and can be determined with good precision. In this case Equation 16.32 or Equation 16.36 would be recommended.

3. *Graphical Method of Baker.* From Equation 16.28, one transfer unit may be defined as

$$N_{OG} = 1 = \frac{y_2 - y_1}{(y^* - y)_{\text{mean}}} \qquad (16.38)$$

A simple graphical procedure for evaluating the number of transfer units needed to accomplish a certain change has been proposed by Baker (2) based upon the use of an arithmetic average of the driving forces and the assumption that, over the length of one transfer unit, both operating line and equilibrium line are essentially linear. The procedure is illustrated in Figure 16.9. A construction line *AB* is located midway (vertically) between the operating and equilibrium lines, and the evaluation is carried out as follows: Starting at point *F*, a line is drawn horizontally to the halfway line (*AB*) at point *G*, and is extended a distance equal to *FG* beyond to *H*. From *H*, a vertical line to *M* on the operating line is drawn. *FHM* is one transfer unit, since it gives a change in gas-phase composition of $y_M - y_F$ which is equal to the average driving force *KP*. The procedure is now continued so that $MN = NO$, etc.

4. *Log-Mean Driving Force.* Another simplification that can be made of Equation 16.30 is of importance in the case of dilute concentrations. It is of interest because of the similarity with the simplified heat-transfer design equation using log-mean driving forces. In a dilute mixture, Equation 16.33 reduces to

$$N_{OG} = \int_{y_1}^{y_2} \frac{dy}{y^* - y} \qquad (16.39)$$

Now assume that over the range of concentrations involved, both equilibrium and operating lines are linear. This assumption is analogous to the assumption of constant c_P in heat-exchange design which yielded a linear *T*- versus- *q* curve.

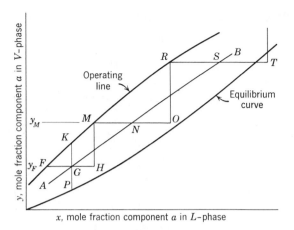

Figure 16.9. Graphical evaluation of N_{OG}.

For the equilibrium curve, let

$$y^* = mx + C \qquad (16.40)$$

Then $$y_2{}^* = mx_2 + C$$
and $$y_1{}^* = mx_1 + C$$
or, eliminating *C*

$$m = \frac{y_2{}^* - y_1{}^*}{x_2 - x_1} \qquad (16.41)$$

The equation for the operating line is

$$L(x - x_1) = V(y - y_1)$$

or

$$x = x_1 + \frac{V}{L}(y - y_1) \qquad (16.42)$$

Therefore,

$$y^* = m\left[x_1 + \frac{V}{L}(y - y_1)\right] + C \qquad (1640a)$$

Substitution for y^* in Equation 16.39 from Equation 16.40a gives

$$N_{OG} = \int_{y_1}^{y_2} \frac{dy}{y^* - y} = \int_{y_1}^{y_2} \frac{dy}{m\left[x_1 + \dfrac{V}{L}(y - y_1)\right] + C - y} \qquad (16.43)$$

After the denominator is rearranged to put Equation 16.43 into the form $dx/(a + bx)$, the equation may be integrated to give

$$N_{OG} = \frac{1}{\dfrac{mV}{L} - 1} \ln \frac{m\left[x_1 + \dfrac{V}{L}(y_2 - y_1)\right] + C - y_2}{mx_1 + c - y_1} \qquad (16.44)$$

Equation 16.44 may be simplified by substituting appropriate forms of Equations 16.40, 16.41, and 16.42.

$$N_{OG} = \frac{1}{\dfrac{y_2{}^* - y_1{}^*}{x_2 - x_1} \cdot \dfrac{x_2 - x_1}{y_2 - y_1} - 1} \ln \frac{y_2{}^* - y_2}{y_1{}^* - y_1} \qquad (16.45)$$

Further rearrangement gives

$$N_{OG} = \frac{y_2 - y_1}{(y_2{}^* - y_2) - (y_1{}^* - y_1)} \ln \frac{y_2{}^* - y_2}{y_1{}^* - y_1} \quad (16.46)$$

But by definition

$$(y^* - y)_{lm} = \frac{(y_2{}^* - y_2) - (y_1{}^* - y_1)}{\ln \dfrac{y_2{}^* - y_2}{y_1{}^* - y_1}}$$

Therefore, Equation 16.46 reduces to

$$N_{OG} = \frac{y_2 - y_1}{(y^* - y)_{lm}} \quad (16.47)$$

Equation 16.47 is restricted to dilute solutions and straight equilibrium and operating lines. Unfortunately, unlike the use of $(\Delta T)_{lm}$, the applicability of $(y - y^*)_{lm}$ is not so general. Many times however, for dilute-solution situations, plotting of the equilibrium curve and operating line in mole-ratio units will make these lines linear and validate the use of Equation 16.47 in mole ratio units.

5. *Rapid Evaluation of Transfer Units When Equilibrium or Operating Lines Are Slightly Curved.* Approximations to the integration of the various equations describing the number of transfer units are given by Colburn (8) for operating conditions where both the operating line and the equilibrium curve are nearly linear. These solutions are useful for mass-transfer operations such as gas absorption, desorption, distillation, stripping operations, and liquid extraction. They give a rapid evaluation of the number of transfer units. Colburn's result is also reported in Perry (26).

Illustration 16.2. A gas stream containing a valuable hydrocarbon (MW = 44) is to be scrubbed with a non-volatile oil (MW = 300) in a tower packed with 1-in. Raschig rings. The entering gas analyzes 20 mole percent hydrocarbon and 95 per cent of this hydrocarbon is to be recovered. The gas stream enters the column at 5000 lb/hr sq ft and hydrocarbon-free oil enters the top of the column at 10,000 lb/hr sq ft. Determine the N_{OG} for this operation.

SOLUTION. It is appropriate to write the design equation even though only the number of transfer units is required. Thus, the depletion rate of the hydrocarbon in the gas stream is equal to the rate of transfer, or

$$d(Vy) = K_y a(y^* - y) \, dzS \quad (16.18)$$

This equation has already been reduced to

$$z = \frac{V}{K_y a(1 - y)_{lm}S} \int_{y_1}^{y_2} \frac{(1 - y)_{lm} \, dy}{(1 - y)(y^* - y)} \quad (16.26)$$

For this illustration, only the integral term is sought, and, to obtain this quantity, it is first necessary to determine the limits of integration. This is done by a material balance over the entire column.

Basis: 1 lb mole of entering gas (solute-free gas has a molecular weight of 29)

Therefore,

Gas in   = (1)(0.8) = 0.8 lb moles = 23.2 lb
Hydrocarbon =   0.2 lb moles = 8.8 lb
Average mole weight of gas   = 32.0 lb

The *total* molar gas rate entering the bottom of the absorber is

$$V_1 = \frac{5000}{32.0} = 156.2 \text{ lb moles/hr sq ft}$$

To calculate V_2, the total molar gas rate out of the column, requires the recognition that all the solute-free gas entering the column must also leave the top of the absorber in the gas stream if no solubility in the oil is assumed. Therefore,

Solute-free gas in = solute-free gas out = 0.8 lb moles
Hydrocarbon out = (0.2)(0.05) = 0.01 lb moles

$$\text{Therefore } y_2 = \frac{0.01}{0.01 + 0.8} = 0.01235$$

and

$$V_2 = \frac{(156.2)(0.8)}{(1 - 0.01235)} = 126.6 \text{ lb moles/hr sq ft}$$

Thus, the flow rates and compositions of the gas stream are determined. For the oil,

$$L_2 = \frac{10,000}{300} = 33.3 \text{ lb moles/hr sq ft}$$

and

$$x_2 = 0$$

Pound moles of hydrocarbon in exit oil = (0.2 − 0.01) = 0.19 moles/mole gas in
Therefore,

Hydrocarbon in exit oil = 0.19 × 156.2 = 29.7 lb moles
Solute-free oil   =   = 33.3 lb moles

  Total moles   =   = 63.0

and

$$x_1 = \frac{29.7}{63.0} = 0.472$$

A summary of the material balance is made by reference to the following sketch.

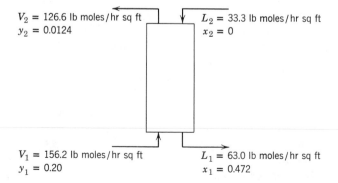

V_2 = 126.6 lb moles/hr sq ft
y_2 = 0.0124

L_2 = 33.3 lb moles/hr sq ft
x_2 = 0

V_1 = 156.2 lb moles/hr sq ft
y_1 = 0.20

L_1 = 63.0 lb moles/hr sq ft
x_1 = 0.472

It is quite evident that the phase-flow rates are not constant throughout the column so that the treatment used in the development of Equation 16.17 is needed. To plot the operating line, write a material balance from the tower bottom to any point within the tower, or

$$V_1 y_1 + Lx = Vy + L_1 x_1 \qquad (a)$$

This equation may be rewritten in terms of the nondiffusing component where

$$V' = V(1 - y) \qquad (b)$$

and

$$L' = L(1 - x) \qquad (c)$$

substituting Equations b and c into Equation a gives

$$V'\left(\frac{y_1}{1 - y_1}\right) + L'\left(\frac{x}{1 - x}\right) = V'\left(\frac{y}{1 - y}\right) + L'\left(\frac{x_1}{1 - x_1}\right) \qquad (d)$$

Equation d is valid throughout the column, and simply by knowing V', L', x_1, and y_1 any point (x, y) intermediate between the terminals may be obtained. Substituting the known values of V', L', x_1, and y_1 into Equation d gives

$$125.2\frac{(0.2)}{(0.8)} + 33.3\frac{(x)}{(1 - x)} = 125.2\frac{(y)}{(1 - y)} + 33.3\frac{(0.472)}{(0.528)} \qquad (e)$$

where $V' = (156.2)(0.8) = 125.2$ and $L' = 33.3$. Thus the equation for the operating line becomes

$$(x/1 - x) = 3.76(y/1 - y) - 0.045 \qquad (f)$$

Data for the operating line are calculated from Equation e and are tabulated below and plotted in Figure 16.10.

x	0.0	0.0307	0.1325	0.271	0.383	0.472
y	0.0124	0.02	0.05	0.10	0.15	0.20

On the same diagram is plotted the equilibrium curve for this system. Having now satisfied the material balance and taken into account the equilibrium relationships, several procedures for evaluating N_{OG} are available. The graphical evaluation of the integral term in Equation 16.26 is always valid since it is substantially free of any assumptions. This solution, of course, is somewhat tedious so that the solution to the problem should be judged from the point of view of the simplifications that can be made and the accuracy of the result required.

First, examination of the equilibrium line and operating line show them both to be curved over the range of concentrations involved. This curvature would introduce serious error if the log-mean driving force were employed. Furthermore, because the concentration of the hydrocarbon in the gas stream is relatively high throughout most of the column, the assumption that $(1 - y)$ is equal to $(1 - y)_{lm}$ is not completely sound. The graphical procedure of Baker might be adequate, but it works best when the curvature of both operating line and equilibrium curve is slight. It would appear that the best solution is the detailed evaluation of the integral of Equation 16.26. Each problem needs to be examined with regard to the assumptions that may be made within the accuracy desired.

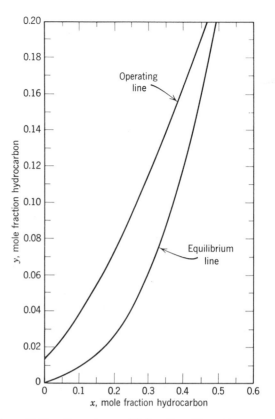

Figure 16.10. Operating-equilibrium line, Illustration 16.2.

The following tabulation indicates the data necessary for the graphical evaluation of N_{OG}. Values of y and y^* are read from Figure 16.10.

y	y^*	$1 - y$	$1 - y^*$	$(1 - y)_{lm}$	$y^* - y$	$\dfrac{(1 - y)_{lm}}{(1 - y)(y^* - y)}$
0.0124	0	0.9876	1.0	0.9938	−0.0124	−81.0
0.03	0.003	.97	0.997	0.9885	−0.027	−37.7
0.05	0.0117	0.95	0.9883	0.9692	−0.0383	−26.6
0.08	0.03	0.92	0.97	0.9450	−0.050	−20.6
0.10	0.0475	0.90	0.9525	0.9265	−0.0475	−21.7
0.14	0.0945	0.86	0.9055	0.8825	−0.0455	−22.6
0.15	0.106	0.85	0.893	0.8715	−0.044	−23.3
0.18	0.1445	0.82	0.8555	0.8352	−0.0355	−28.7
0.20	0.1730	0.80	0.8270	0.8135	−0.0270	−37.6

The last column in the above tabulation is plotted against y in Figure 16.11, and the area under the curve between the limits of y_1 and y_2 gives the N_{OG}. Since y_2 is less than y_1, the integral is positive. In this case, $N_{OG} = 4.91$. Note that the quantities $(1 - y)_{lm}$ and $(1 - y)$ are nearly equal. For dilute concentrations this assumption of equality is usually valid.

Heights of Transfer Units. The relationship between individual and over-all mass-transfer coefficients was shown in Chapter 14 to be

$$\frac{1}{K_y} = \frac{1}{k_y} + \frac{m}{k_x} \qquad (16.48)$$

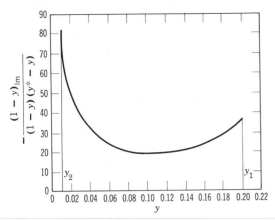

Figure 16.11. Graphical integration for Illustration 16.2.

Earlier it was demonstrated that, for packed-tower operation, the interfacial area per unit volume was combined with the coefficient to give a composite value. It would appear logical to expect that an equation similar to Equation 16.48 can be written using composite coefficients, since the same variables that affect the mass-transfer coefficient, with the exception of diffusivity, also affect the quantity a. Therefore,

$$\frac{1}{K_y a} = \frac{1}{k_y a} + \frac{m}{k_x a} \qquad (16.49)$$

Equation 16.49 may also be written as

$$\frac{V}{K_y a (1 - y)_{lm}} = \frac{V}{k_y a (1 - y)_{lm}}$$
$$+ \frac{mV}{L} \cdot \frac{L(1 - x)_{lm}}{(1 - x)_{lm}} \cdot \frac{1}{(1 - y)_{lm}} \cdot \frac{1}{k_x a} \qquad (16.50)$$

Each of the terms of Equation 16.50 has been defined earlier as a height of a transfer unit (H_{OG}, H_G, or H_L). Thus, Equation 16.50 becomes

$$H_{OG} = H_G + \frac{mV}{L} H_L \qquad (16.51)$$

The ratio $(1 - x)_{lm}/(1 - y)_{lm}$ in the right term approximates unity for the most commonly encountered mass-transfer operations. Equation 16.51 relates H_{OG} to mV/L as a straight line of slope equal to H_L and intercept of H_G. Then if H_{OG} were expressed as a function of mV/L, Equation 16.51 could be used to determine H_G and H_L, the relative magnitudes of resistances encountered in the transfer process. Unfortunately this is not so simple since a constant value of m is not usually encountered outside the range of dilute solutions. The slope of the equilibrium curve for almost all systems is not constant.

Liquid Phase (H_L). Sherwood and Holloway (28) experimentally correlated *absorption* and *desorption* data for cases where the liquid phase was the dominant resistance. For various packing materials,

$$H_L = \beta \left(\frac{G_L}{\mu_L}\right)^n (N_{Sc})^{0.5} \qquad (16.52)$$

where β, n = constants (see Table 16.4)
G_L = mass velocity of the liquid, lb/hr sq ft
μ_L = liquid viscosity, lb/ft hr
N_{Sc} = Schmidt number for the liquid

Table 16.4. VALUES OF CONSTANTS FOR EQUATION 16.52 (28)

Packing	β	n	Range of G_L
Raschig rings			
$\frac{1}{2}$ in.	0.00357	0.35	400–15,000
1 in.	0.01	0.22	400–15,000
$1\frac{1}{2}$ in.	0.0111	0.22	400–15,000
2 in.	0.0125	0.22	400–15,000
Berl saddles			
$\frac{1}{2}$ in.	0.00666	0.28	400–15,000
1 in.	0.00588	0.28	400–15,000
$1\frac{1}{2}$ in.	0.00625	0.28	400–15,000
Spiral rings			
3 in.	0.00909	0.28	400–15,000

Equation 16.52 has been developed solely from experimental observations on absorption and desorption operations. For equimolar transfer, there are not sufficient data to permit a generalized correlation.

Gas Phase (H_G). The correlations for H_G are not as well established as those for H_L because of the difficulty in achieving experimental conditions in which the gas phase comprises nearly all the resistance to transfer. Extensive ammonia–water data of Fellinger (10) have been used to give a useful correlation for H_G. The data are found in Perry (25) in graphical form. The translation of the data to other systems is most easily accomplished by use of the mass-transfer coefficient.

From the extensive *ammonia–water* data, an empirical equation of the following form is suggested:

$$k_G a = b(G_V)^p (G_L)^r \qquad (16.53)$$

where b, p, and r are constants dependent upon the specific flow rates and geometry of the packing used in the operation. Hensel and Treybal (15) question the use of the exponents p and r as true constants, and Shulman and DeGouff (29) suggest that they are variable because k_G and a are not influenced by the same variables. At present, however, Equation 16.53 is the best available tool. Table 16.5 tabulates the values of the constants in the equation.

Table 16.5. AMMONIA–AIR–WATER ABSORPTION DATA
CONSTANTS FOR EQUATION 16.53

Packing Material	b	p	r	Reference
Raschig rings				
$\frac{1}{2}$ in.	0.0065	0.90	0.39	(9)
1 in.	0.036	0.77	0.20	(9)
$1\frac{1}{2}$ in.	0.0142	0.72	0.38	(9)
2 in.	0.048	0.88	0.09	(23)
Berl saddle				
1 in.	0.0085	0.75	0.40	(23)

The correlation for the ammonia–water system may now be applied to other gas-liquid systems by the following method. Assume that a correlation of the type for mass-transfer coefficients as Equation 13.84 may be written for the composite coefficient $k_G a$. Then, for the ammonia system,

$$(k_G a)_{NH_3} = [\text{constant }(N_{Re})^w (N_{Sc})^h]_{NH_3} \quad (16.54)$$

and for any system

$$k_G a = \text{constant }(N_{Re})^w (N_{Sc})^h \quad (16.54a)$$

If the operating conditions for the unspecified system are similar with respect to N_{Re} and tower geometry, constants in the above equations are the same for both systems; then, dividing Equation 16.54a by Equation 16.54 gives

$$\frac{k_G a}{(k_G a)_{NH_3}} = \left[\frac{N_{Sc}}{(N_{Sc})_{NH_3}}\right]^h \quad (16.55)$$

Some doubt exists as to the value of the exponent on the Schmidt group, but a value of $-2/3$ is suggested for the present.

Unfortunately, correlations for mass-transfer coefficients or heights of transfer units for equimolar transfer are not available. Instead experimental data must be available for tower design. It has been suggested (31) that in the absence of specific data for a given system, Equations 16.52 and 16.55 may be used for approximate answers.

Illustration 16.3. A gas mixture containing 6.0 per cent SO_2 and 94.0 per cent dry air from a sulfur burner is to be scrubbed with fresh water in a tower packed with 1-in. Raschig rings to remove the SO_2 so that the exit gas will contain no more than 0.1 mole percent SO_2. The tower must treat 1000 lb/hr of gas and is to be designed using 50 per cent of flooding velocity. The water flow is to be twice the minimum required to achieve this separation in a tower of infinite height. Operating conditions will be isothermal at 30°C and 1 atm pressure. Determine the required tower diameter and height. The liquid-phase Schmidt number may be taken as 570.

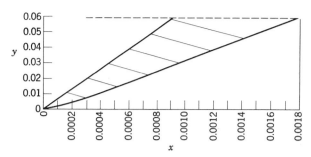

Figure 16.12. Operating-equilibrium line, Illustration 16.3.

EQUILIBRIUM DATA AT 30°C

p_{SO_2}, mm Hg	0.6	1.7	4.7	8.1	11.8	19.7	36.0	52.0	79.0
c. grams SO_2 per 100 grams H_2O	0.02	0.05	0.10	0.15	0.20	0.30	0.50	0.70	1.00
Density of solution lb/cu ft	62.16	62.17	62.19	62.21	62.22	62.25	62.32	62.38	62.47

SOLUTION. First, the equilibrium data are plotted after the tabulated data are converted to mole-fraction units.

y	0.00079	0.00224	0.0062	0.0107	0.01566
x	0.0000565	0.000141	0.000281	0.000422	0.000564

y	0.0259	0.0474	0.0684	0.104
x	0.000844	0.00141	0.00197	0.0028

where

$$y = \frac{p_{SO_2}}{760} \quad \text{and} \quad x = \frac{\dfrac{0.18}{64}c}{1 + \dfrac{0.18}{64}c}$$

Before the material balance can be completed, it will be necessary to determine the minimum liquid flow rate. The concentration of the gas leaving the tower will be $y_2 = 0.001$, and the concentration of the liquid entering will be $x_2 = 0$. The operating line equation may be written, in terms of solute-free streams, as

$$V'\left[\frac{y_2}{1-y_2} - \frac{y_1}{1-y_1}\right] = L'\left[\frac{x_2}{1-x_2} - \frac{x_1}{1-x_1}\right]$$

Now $y_1 = 0.06$, $y_2 = 0.001$, and $x_2 = 0$, and the minimum liquid flow will correspond to a line through (x_2, y_2) and (x_1^*, y_1) since the equilibrium line is concave towards the operating line. From Figure 16.12, $x_1^* = 0.001740$.

$$V_1 = \frac{1000}{(0.06 \times 64) + (0.94 \times 29)}$$

$$= 32.1 \text{ lb moles/hr } \textit{total} \text{ gas flow}$$

$$V_1' = V_1(1 - y_1) = 32.1(0.94)$$

$$= 30.1 \text{ lb moles/hr } SO_2\text{-free gas}$$

Therefore the minimum water rate is

$$30.1\left(\frac{0.001}{0.999} - \frac{0.06}{0.94}\right) = L'\left(0 - \frac{0.00174}{0.99826}\right)$$

$$L' = 1085 \text{ lb moles/hr } SO_2\text{-free water}$$

For a design water rate of $2L'_{min}$

$$L' = 2 \times 1085 = 2170 \text{ lb moles/hr}$$

The design flow rate of $L' = 2170$ lb moles/hr can now be used in the equation for the operating line to determine the exit-liquid composition.

$$30.1\left(\frac{0.001}{0.999} - \frac{0.06}{0.94}\right) = 2170\left(0 - \frac{x_1}{1 - x_1}\right)$$

$$x_1 = 0.000880$$

All terminal compositions and flow rates are now established.

$$x_1 = 0.000880 \qquad y_1 = 0.06$$
$$x_2 = 0.0 \qquad y_2 = 0.001$$
$$L' = 2170 \text{ lb moles/hr} \quad V' = 30.1 \text{ lb moles/hr}$$

The entire operating line may now be evaluated.

$$30.1\left(\frac{y}{1-y} - \frac{0.06}{0.94}\right) = 2170\left(\frac{x}{1-x} - \frac{0.00088}{0.99912}\right)$$

Concentration of the liquid phase is dilute enough so $(1 - x) \cong 1$; therefore

$$x = 0.01388\frac{y}{1-y} - 0.00000655$$

Neglecting the last term, the operating line becomes

y	0.06	0.04	0.03	0.02	0.01	0.001
x	0.00088	0.000591	0.000430	0.000283	0.000141	~0

These data are plotted in Figure 16.12, and the operating line is substantially linear.

It is necessary to calculate the mass-transfer coefficients for both phases; in the absence of any information to the contrary, no assumption as to the controlling phase can be made. To get these values requires a knowledge of mass velocities of both phases since both phase velocities are quantities which normally appear in the correlations. At the moment, only mass rates of flow are known; therefore, it is necessary to determine the tower area first. To do this requires a knowledge of flooding velocities obtained from Figure 16.4.

First, *total* flow rates must be determined.

$$V_1 = 1000 \text{ lb/hr} = 32.1 \text{ lb moles/hr}$$

$$SO_2 \text{ entering} = 32.1 \times 0.06 \times 64 = 123.2 \text{ lb/hr}$$

$$SO_2 \text{ leaving} = \frac{0.001}{0.999} \times 30.1 \times 64 = 1.93 \text{ lb/hr}$$

Therefore, SO_2 absorbed by the water = $123.2 - 1.93 =$ 120.3 lb/hr.

$$\text{Fresh water entering} = 2170 \times 18 = 39,000 \text{ lb/hr}$$

$$\text{Total liquid leaving} = 39,120 \text{ lb/hr.}$$

Assume the density of the gas stream to be essentially that of air at 1 atm and 30°C.

$$\rho_V = \frac{29}{359} \times \frac{273}{303} = 0.0726 \text{ lb/cu ft}$$

$$\rho_L = 62.2 \text{ lb/cu ft}$$

$$\frac{G_L}{G_V}\sqrt{\frac{\rho_V}{\rho_L}} = \frac{39,120}{1000}\sqrt{\frac{0.0726}{62.2}} = 1.335$$

From Figure 16.4 at a value of $\dfrac{G_L}{G_V}\sqrt{\dfrac{\rho_V}{\rho_L}} = 1.335$

$$\frac{G_V{}^2 a_v (\mu_L')^{0.2}}{g_c \epsilon^3 \rho_V \rho_L} = 0.015$$

From Table 16.1, $a_v = 58$ sq ft/cu ft and $\epsilon = 0.73$. Therefore

$$\frac{G_V{}^2 (58)(1.0)^{0.2}}{4.17 \times 10^8 \times (0.73)^3 (0.0726)(62.2)} = 0.015$$

$$G_V = 435 \text{ lb/hr sq ft}$$

Specifications are to use a velocity 50 per cent of flooding; therefore,

$$G_V = 218 \text{ lb/hr/sq ft}$$

Since the gas flow at the bottom of the tower is greatest, base the area on this flow rate, or

$$S = \frac{V}{G_V} = \frac{1000}{218} = 4.59 \text{ sq ft}$$

The tower diameter is therefore

$$D = \sqrt{\frac{4(4.59)}{\pi}} = 2.41 \text{ ft}$$

Consequently,

$$G_V = 218 \text{ lb/hr sq ft}$$

$$G_L = \frac{(2170)(18)}{4.59} = 8500 \text{ lb/hr sq ft}$$

With the above quantities, values of H_G, H_L or $k_y a$, $k_L a$ may be determined.

$$H_L = \beta\left(\frac{G_L}{\mu_L}\right)^n (N_{Sc})^{1/2}$$

$$H_L = 0.01\left(\frac{8500}{2.42}\right)^{0.22}(570)^{1/2} = 1.43 \text{ ft}$$

Since L is essentially constant through the tower, the mass-transfer coefficient $(k_x a)$ can be calculated from the definition of H_L.

$$H_L = \frac{L}{k_x a S} = 1.43$$

$$k_x a = \frac{2170}{(4.59)(1.43)} = 332 \text{ lb moles/hr cu ft } \Delta x$$

The mass-transfer coefficient $k_y a$ may be calculated by Equation 16.53; again, mass velocities of the phases are essentially constant.

$$k_y a = k_g a = b(G_V)^p (G_L)^r. \quad \text{Assume } N_{Sc} = (N_{Sc})_{NH_3}$$

$$k_y a = 0.036(218)^{0.77}(8500)^{0.20} = 13.8 \text{ lb moles/hr cu ft } \Delta y$$

The tower height may be found from Equation 16.24.

$$z = \frac{V}{k_y a S(1-y)_{lm}}\int_{y_1}^{y_2}\frac{(1-y)_{lm}\, dy}{(1-y)(y_i - y)}$$

To get the interfacial compositions requires the knowledge of the ratio of phase resistances. The ratio is obtained by

$$-\frac{k_x a}{k_y a} = -\frac{332}{13.8} = -24.1$$

For several positions along the operating line, the bulk-conditions-to-interfacial-conditions relationships are found by plotting tie lines of slope $-k_x a/k_y a$ and reading interfacial compositions from the equilibrium curve. In this particular problem, $(1 - y)_{lm}$ and $(1 - y)$ are approximately equal and will not be taken into consideration.

y	y_i	$y_i - y$	$\dfrac{1}{y_i - y}$
0.06	0.0475	−0.0125	−80.0
0.05	0.0392	−0.0108	92.5
0.04	0.0308	−0.0092	−109
0.03	0.0224	−0.0076	−132
0.02	0.0140	−0.0060	−167
0.01	0.0065	−0.0035	−288
0.001	0.0005	−0.0005	−2000

The area under the curve of $[1/(y_i - y)]$ versus y is equal to the N_G. From Figure 16.13 by an approximation,

$$N_G = 14.3$$

$$(1 - y)_{lm} \text{ at the bottom} = \frac{0.94 + 0.953}{2} = 0.942$$

$$(1 - y)_{lm} \text{ at the top} = 0.999$$

An average of these two quantities, or 0.97, and also an average V of 31.1 may be used to obtain H_G,
Therefore,

$$H_G = \frac{V_{av}}{k_y a S (1 - y)_{lm}} = \frac{31.1}{(13.8)(4.59)(0.97)} = 0.508 \text{ ft}$$

Finally,

$$z = H_G N_G = 0.508 \times 14.3 = 7.25 \text{ ft}$$

The tower diameter is 2.41 ft and the tower packed height is 7.25 ft.

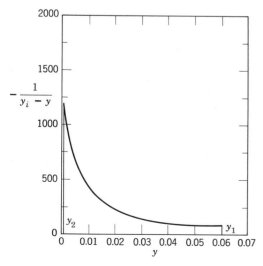

Figure 16.13. Graphical integration for Illustration 16.3.

OTHER MASS-TRANSFER OPERATIONS

The design equation has been applied in considerable detail to the operations of gas absorption and distillation in packed columns. The solution of the design equation requires knowledge of a material balance, an equilibrium relationship of the chemical system, and a specific rate in the form of a mass transfer coefficient. Other unit operations involving the contacting of two liquid phases as in liquid-liquid extraction, or the contacting of solid-liquid or solid-gas phases as in ion exchange or adsorption may also be carried out in packed columns. The same basic design equation that was developed earlier applies to these operations. The fundamental problems in its application are the lack of general information regarding phase equilibrium relationships, mass-transfer coefficients, and the lack of an understanding of the exact nature of the transfer mechanism. Consequently, equipment for these operations is often designed using data applicable only to the particular chemical system and apparatus contemplated.

Liquid Extraction. The mass-transfer operation of liquid extraction has already been introduced in Part I with particular emphasis on extraction carried out in equilibrium stages. A commonly encountered extraction operation is the continuous contacting of two liquid phases in a manner similar to that of gas absorption. However some differences in operating techniques must be employed as a result of the density characteristics of the two liquid phases.

Continuous contacting of two phases may be carried out in packed towers and spray towers. The spray column is the simplest to build. It consists of a vertical shell, free of packing, with provisions for introducing and removing the liquids. For example, the heavy liquid will enter the top, flow through the tower, and exit from the bottom. The light liquid will enter the bottom and rise countercurrent to the heavy liquid. The particular phase which completely fills the tower is called the continuous phase, and the phase which is in the form of droplets is called the dispersed phase. If the light liquid is the dispersed phase, an interface will exist at the top of the tower. If the heavy phase is dispersed the interface will be at the bottom. Either the heavy or the light liquids may be the dispersed phase.

Packing may be placed within the tower shell to increase the physical mixing of the continuous phase and to improve contacting between the dispersed droplets and the continuous phase. Under similar conditions of operation, packed towers give a better rate of transfer than spray towers.

Since liquid-liquid extraction is an operation involving diffusion of a solute through a stationary component, Equations 16.26 and 16.27 apply. The techniques for evaluating the number of transfer units and the height of a transfer unit are the same as before.

Many data pertinent to mass-transfer coefficients and operating characteristics of extraction equipment have been accumulated, but no satisfactory general correlation has yet been evolved. Instead, design of extraction equipment depends upon acquiring performance data from laboratory equipment operating under conditions similar to design specifications. Treybal (32) summarizes some of the performance data available for continuous countercurrent extraction.

Solid-Fluid Operations. The unit operations involving gas-solid or liquid-solid contacting include adsorption, leaching, ion exchange, and drying. Continuous contacting of a solid phase with a liquid or gas phase presents serious problems in equipment design. The continuous countercurrent movement of solids and fluids calls for specially designed equipment to allow for the steady introduction of solids in one end of the apparatus and their removal at the other end. This must be accomplished with little loss of the fluid phase. Although the equipment used for solid-fluid operations is not as simple as a packed tower, the same basic fundamentals apply to these operations as to the operations of absorption and distillation.

Adsorption. Adsorption, as the term is used here, applies to the physical transfer of a solute in a gas or liquid to a solid surface where the solute is held as a result of intermolecular attraction with the solid molecules. The adsorbed solute does not dissolve in the solid but remains on the solid surface or in the pores of the solid. The adsorption process is often reversible so that by changing pressure or temperature the solute may be readily removed from the solid. At equilibrium the adsorbed solute has a partial pressure equal to that of the contacting fluid phase, and by simply changing pressure or temperature of the operation the solute may be removed from the solid.

The selection of adsorbents is quite important. The solids should offer low pressure-drop characteristics, and good strength to withstand the rigors of handling. In addition adsorbents are selective in their ability to adsorb specific solutes. Consequently, the chemical nature of the solid must be considered to assure satisfactory performance. Commercial adsorbents include bentonite, bauxite, alumina, bone charcoal, Fuller's earth, carbon, and silica gel.

Equilibrium data for adsorption are usually presented in the form of adsorption isotherms (See Chapter 3). These data are necessary before the design equation can be applied. In addition, knowledge of adsorption transfer coefficients is necessary. These coefficients are usually available only for specific conditions rather than as general correlations. Design of equipment, therefore, must be made using data which closely parallels design conditions.

Ion Exchange. Another operation of solid-liquid

contacting is that of ion exchange. Ion-exchange operations are normally thought of in connection with water treatment, but over the past few years many other applications of ion exchange have been developed (17, 24, 30).

It is convenient to think of an ion-exchange resin as a homogeneous gel throughout which is distributed a network of hydrocarbon chains. Attached to these chains are ionic groups which are immobile. Their ionic charges are balanced by diffusible ions of opposite charge. For example, a cation-exchange resin has incorporated into it immobile anionic groups bonded to each other through cross linkage. The charges of these fixed anionic groups are balanced by diffusible cations. It is possible to manufacture ion-exchange resins with many groups of fixed ions. Cation-exchange resins may contain such fixed ionic groups as phenolic, carboxylic, sulfonic, and phosphonic ions. Anion exchange resins may have fixed ionic groups of primary, secondary, or tertiary amines, phosphonium, and arsonium. Many types of exchange materials in various physical states have been synthesized. For example, spherical beads, membranes, fibers, and liquids have been prepared. The "zeolites" are naturally occurring clays that are capable of exchanging cations.

Typical of the operation that occurs during ion exchange is that illustrated by water softening. The reaction is very simply indicated by the following equation, in which $[RSO_3]^-$ represents the resin which has as a component a fixed ionic group. In this case, it is the sulfonic group, and the resin is said to be a cation type.

$$2\,[RSO_3]^-Na^+ + Ca^{++} \leftrightarrows (RSO_3)_2{}^- \cdot Ca^+ + 2Na^+$$
$$\text{(solid)} \qquad \text{(solution)} \qquad \text{(solid)} \qquad \text{(solution)}$$

In this process, the hardness ions, such as Ca^{++} and Mg^{++}, are removed from water by passing the water through a bed of resin particles. Since the reaction is reversible, the resin may be returned to the Na^+ form by simply passing a strong salt solution over the resin. In this manner, it is ready for reuse in the next softening cycle.

Water can also be deionized by ion exchange so that nearly all dissolved solids are removed. Deionization requires that the water first be passed through a cation exchanger and then an anion exchanger. The reactions are

$$[RSO_3]^-H^+ + Na^+Cl^- \rightleftharpoons [RSO_3]^-Na^+ + H^+Cl^-$$
$$\text{(solid)} \qquad \text{(solution)} \qquad \text{(solid)} \qquad \text{(solution)}$$

and

$$[RN(CH_3)_3]^+OH^- + H^+Cl^- \rightleftharpoons [RN(CH_3)_3]^+Cl^- + HOH$$
$$\text{(solid)} \qquad \text{(solution)} \qquad \text{(solid)} \qquad \text{(solution)}$$

The cation exchanger is regenerated usually with sulfuric acid, and the anion exchanger with sodium hydroxide.

The expression for phase equilibrium relationships in ion exchange is similar in type to that of adsorption isotherms. Another widely used method for ion-exchange equilibria is based upon the mass-action relationships. For example, consider the reaction

$$[RSO_3]^-H^+ + Na^+Cl^- \rightleftharpoons [RSO_3]^-Na^+ + H^+Cl^-$$
$$\text{(solid)} \qquad \text{(solution)} \qquad \text{(solid)} \qquad \text{(solution)}$$

The equilibrium constant for this reaction is

$$K_{\text{equil.}} = \frac{(\text{conc. } [RSO_3]^-Na^+)(\text{conc. } H^+Cl^-)}{(\text{conc. } [RSO_3]^-H^+)(\text{conc. } Na^+Cl^-)}$$

For most engineering applications, this equilibrium constant may be expressed as

$$K_{\text{equil.}} = \frac{1 - x_a}{x_a} \cdot \frac{y_a}{1 - y_a}$$

where x_a = *equivalent* fraction of solute (a) in liquid phase

y_a = *equivalent* fraction of solute (a) in resin phase

The rate of transfer in ion exchange is complicated by the complex mechanism of the transfer process. A postulated mechanism of ion exchange consists of the following steps:

1. Diffusion of ions from the bulk liquid phase to the surface of the resin.

2. Diffusion of the ions from the surface of the resin into the resin to the site of exchange.

3. Exchange of the ions at the active site.

4. Diffusion of the replaced ions from within the resin to its surface.

5. Diffusion of the replaced ions from the resin surface to the bulk liquid.

Any one of the five steps may be a rate limiting factor, although too few data exist to permit any generalizations. The rate for step 3 is best described by a kinetic reaction rate, whereas the rate for the diffusion steps may be described by the usual rate equation for counter-diffusion as indicated by Equation 16.5. Once again a suitable mass-transfer coefficient is needed. If the exchange of ions (step 3) is very rapid compared to the diffusion steps, then ion-exchange equipment may be designed by employing Equation 16.12. If the rate controlling step is the exchange step, then kinetic procedures are needed.

STAGE EFFICIENCIES

In stage operations, the efficiency of an individual stage depends upon many factors, including the rate of mass transfer. Within a single stage, the contact between the two phases is continuous from the time a phase enters the stage until it leaves. The flow patterns within the stage are usually very complex, and often they cannot be accurately predicted or described. Because of this, it is difficult to predict the contact time and the interfacial area between phases. Therefore, any attempt to predict stage efficiences from the rate of mass transfer involves many approximations. Because knowledge of stage efficiencies is essential in the design of stage-contacting equipment, much research has been directed toward the theoretical and empirical prediction of stage efficiencies. These correlations are usually based upon a particular type of stage equipment, such as the bubble-cap tray. In addition, the correlations are often limited to specific chemical species, such as hydrocarbons.

This section will consider the relationship of the rate of mass transfer to the stage efficiency using a simple model for the phase flow pattern. The complications of more complex flow patterns will be discussed, but no attempt will be made to cover in detail the numerous empirical correlations. Since distillation efficiencies have been most thoroughly investigated, the following discussion will be based upon a gas-liquid model for bubble-cap or sieve trays.

Point Efficiency. The *point* efficiency refers to a single point on the horizontal surface of the contacting tray. At this point the model visualizes a single stream of gas bubbles rising through the liquid, as shown in Figure 16.14. The bubbles form at the opening, such as in the sieve tray as shown or at the slot in a bubble cap (Figures 2.1 and 2.2). The bubbles rise through the liquid and finally break through the liquid surface into the gas above the plate. Mass is transferred between the gas and liquid as the bubble rises. The interfacial area of the bubble is A_B, and its time of contact with the liquid is θ. The change in quantity of the component in the gas bubble will be equal to the transfer across the interface; that is, for the time interval $d\theta$

$$d(V_B y) = K_y' A_B (y^* - y) \, d\theta \qquad (16.56)$$

where V_B = the total quantity of gas in the bubble, lb moles

y = vapor composition, mole fraction a

y^* = composition of the vapor which would be in equilibrium with the liquid *at the point under observation*, mole fraction

K_y' = over-all mass-transfer coefficient, lb moles/hr sq ft (mole fraction)

A_B = bubble interfacial area, sq ft

If the total quantity of gas does not change, as would be the case in equimolar counterdiffusion or in dilute gases, Equation 16.56 becomes

$$V_B \, dy = K_y' A_B (y^* - y) \, d\theta$$

or

$$\frac{dy}{y^* - y} = \frac{K_y' A_B \, d\theta}{V_B} \qquad (16.57)$$

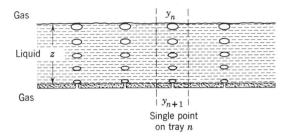

Figure 16.14. Point efficiency in gas–liquid contacting.

This equation may be integrated over the path of travel of the bubble through the liquid.

$$\int_{y_{n+1}}^{y_n} \frac{dy}{y^* - y} = \frac{K_y' A_B}{V_B} \int_0^\theta d\theta$$

$$-\ln \frac{(y^* - y_n)}{(y^* - y_{n+1})} = \frac{K_y' A_B \theta}{V_B} \qquad (16.58)$$

The use of a constant value of y^* implies that the liquid composition is constant in the vertical bubble path. Rearrangement of Equation 16.58 gives

$$\frac{y^* - y_n}{y^* - y_{n+1}} = e^{-(K_y' A_B \theta / V_B)}$$

If both sides of the equation are subtracted from 1,

$$1 - \frac{y^* - y_n}{y^* - y_{n+1}} = 1 - e^{-(K_y' A_B \theta / V_B)}$$

or

$$\frac{y_n - y_{n+1}}{y^* - y_{n+1}} = 1 - e^{-(K_y' A_B \theta / V_B)} \qquad (16.59)$$

The left-hand side of Equation 16.59 is the Murphree *point* efficiency for the V-phase ($E_V{}^0$) as originally expressed by Equation 7.17. Therefore,

$$E_V{}^0 = 1 - e^{-(K_y' A_B \theta / V_B)} \qquad (16.60)$$

The point efficiency ($E_V{}^0$) is not equal to the plate efficiency (E_V) for reasons discussed below. The molar flow rate of the bubble (V_B / θ) may be replaced by the total molar flow rate (V), assuming that the gas flow is uniform across the plate. The bubble interfacial area (A_B) is difficult to evaluate accurately. As in packed columns, it may be replaced by the average interfacial surface area per unit volume.

$$A_B = aSz \qquad (16.61)$$

where a = bubble interfacial surface area per unit volume of liquid sq ft/cu ft
S = cross-sectional area of column, sq ft
z = effective height of liquid, ft

Equation 16.60 may then be written as

$$E_V{}^0 = 1 - e^{-K_y' az/(V/S)} \qquad (16.62)$$

where V/S is the molar mass velocity of gas through the column, lb moles/hr sq ft.

Equation 16.62 shows that the point efficiency is an exponential function of the mass-transfer coefficient, the interfacial area, and the gas mass velocity. The utility of the point efficiency is limited because of the difficulty of evaluating point compositions.

Murphree Stage Efficiency. The Murphree stage efficiency for the gas phase is defined by Equation 7.17.

$$E_V = \frac{y_n - y_{n+1}}{y_n^* - y_{n+1}} \qquad (7.17)$$

In this case y_n^* is the gas composition in equilibrium with the liquid *leaving* the stage. This definition is significantly different from that for the point efficiency because the liquid composition varies as the liquid flows across the typical plate. The variation in liquid composition across the plate depends upon the flow pattern on the plate, which in turn depends upon the construction of the plate. It is also possible for the entering gas composition to vary across the plate if the gas is not thoroughly mixed between plates. Since many different flow patterns occur, no general relationship between point efficiency and plate efficiency can be written.

Typical flow patterns are shown in Figures 2.1 and 2.4. These two cases are diagrammed schematically in Figure 16.15.

In a crossflow stage (Figure 16.15a), the liquid flows perpendicular to the gas and in a reversed direction on each stage. The liquid composition changes as it flows across the stage. The gas may or may not be of uniform composition as it enters the stage. The crossflow pattern is similar to the flow pattern in a baffled heat exchanger, where the flow through the tubes is identical to that of the V-phase and the shell-side flow pattern is similar to that of the L-phase.

The radial flow pattern in the disk-and-doughnut stage (Figure 16.15b) also results in liquid flow perpendicular to the gas flow. This case gives a flow pattern

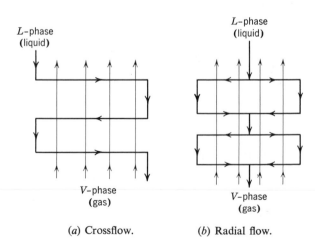

(a) Crossflow.   (b) Radial flow.

Figure 16.15. Typical flow patterns in stage equipment.

similar to that in a heat exchanger with disk-and-doughnut baffles.

These two and other flow patterns are sufficiently complex to make a complete theoretical analysis impossible, although certain simplified cases have been studied. If the gas is thoroughly mixed between stages and if the liquid is sufficiently agitated on the stage so that its composition is uniform, the point and stage efficiencies are equal as indicated by their definitions. Lewis (20) has considered a number of simple cases where mixing is incomplete (20, 27). He also has developed expressions for relating the Murphree stage efficiency to the over-all column efficiency, which is defined as the ratio of the number of equilibrium stages required to the number of actual stages required to make a separation.

Studies of Stage Efficiencies. A summary of empirical and semiempirical correlations of plate efficiencies for distillation and gas absorption is given in reference 27. Recent work by Gerster and associates (11, 12, 13, 14) on gas-liquid transfer has been directed toward predicting stage efficiencies, by considering the resistance to mass transfer offered in each phase. A study of bubble-cap tray efficiencies in distillation columns sponsored by the American Institute of Chemical Engineers is reported in annual reports (1)

Relatively few data are available on stage efficiencies in extractions, adsorption, and some other stage operations.

REFERENCES

1. AIChE Research Committee, *Tray Eff. in Dist. Column*, Annual Reports 1953–1959.
2. Baker, T. C., *Ind. Eng. Chem.*, **27**, 977 (1935).
3. Baker, T. C., T. H. Chilton, and H. C. Vernon, *Trans. Am. Inst. Chem. Engrs.*, **31**, 296 (1935).
4. Chilton, T. H., and A. P. Colburn, *Ind. Eng. Chem.*, **23**, 913 (1931).
5. Chilton, T. H., and A. P. Colburn, *Ind. Eng. Chem.*, **27**, 255 (1935).
6. Colburn, A. P., Collected Papers, ASEE Summer School, 1936.
7. Colburn, A. P., *Trans. Am. Inst. Chem. Engrs.*, **29**, 174 (1933).
8. Colburn, A. P., *Ind. Eng. Chem.*, **33**, 459 (1941).
9. Dwyer, O. E., and B. F. Dodge, *Ind. Eng. Chem.*, **33**, 485 (1941).
10. Fellinger, L., Sc.D. thesis in chemical engineering, M.I.T. (1941).
11. Foss, A. S., and J. A. Gerster, *Chem. Eng. Progr.*, **52**, 18-J (1956).
12. Gerster, J. A., A. P. Colburn, W. E. Bonnet, and T. W. Carmody, *Chem. Eng. Progr.*, **45**, 716 (1949).
13. Gerster, J. A., W. E. Bonnet, and I. Hess, *Chem. Eng. Progr.*, **47**, 523, 621 (1951).
14. Grohse, E. W., R. F. McCastney, H. I. Haver, and J. A. Gerster, and A. P. Colburn, *Chem. Eng. Progr.*, **45**, 725 (1949).
15. Hensel, S. L., and R. E. Treybal, *Chem. Eng. Progr.*, **48**, 362 (1952).
16. Kirschbaum, E., *Z. Ver. deut. Ing.*, **75**, 1212 (1931).
17. Kunin, R., *Ion Exchange Resins*, 2nd ed., John Wiley and Sons, New York, 1958.
18. Leva, M., *Chem. Eng.*, **56**, 115 (1949).
19. Leva, M., *Chem. Eng. Progr. Symposium Ser. No.* **50** (10), 51, (1954).
20. Lewis, W. K., Jr., *Ind. Eng. Chem.*, **28**, 399 (1936).
21. Lobo, W. E., et al., *Trans. Am. Inst. Chem. Engrs.*, **41**, 693 (1945).
22. Mayo, F., T. G. Hunter, and A. W. Nash, *J. Soc. Chem. Ind.*, **54**, 375T, (1935).
23. Molstad, M. C., J. F. Kinney, and R. G. Abbey, *Trans. Am. Inst. Chem. Engrs.*, **39**, 605 (1930).
24. Nachod, F., and J. Schubert, *Ion Exchange Technology*, Academic Press, New York, 1936.
25. Perry, J. H., *Chemical Engineers Handbook*, McGraw-Hill Book Co., New York, 1950, pp. 688–690.
26. Perry, J. H., *Chemical Engineers Handbook*, McGraw-Hill Book Co., New York, 1950, p. 554.
27. Robinson, C. S., and E. R. Gilliland, *Elements of Fractional Distillation*, 4th ed., McGraw-Hill Book Co., New York, 1950.
28. Sherwood, T. K., and F. A. L. Holloway, *Trans. Am. Inst. Chem. Engrs.*, **36**, 39 (1940).
29. Shulman, H. L., and J. J. DeGouff, *Ind. Eng. Chem.*, **44**, 1915 (1952).
30. *Symposium on Ion Exchange and Its Applications*, Soc. Chem. Ind., London, 1955.
31. Treybal, R. E., *Mass Transfer Operations*, McGraw-Hill Book Co., New York, 1955, p. 330; also p. 143.
32. Treybal, R. E., *Liquid Extraction*, McGraw-Hill Book Co., New York, 1951, Chapter 10.
33. Wiegand, J. H., *Trans. Am. Inst. Chem. Engrs.*, **36**, 679 (1940).

PROBLEMS

16.1. A packed tower is to be designed to distill continuously an equimolar mixture of benzene and toluene to obtain products of 95 mole percent purity. The feed and reflux are at their boiling points, and a constant value of $\alpha = 2.58$ can be assumed. If, for the conditions stated $H_G = 0.5$ ft and $H_L = 1.0$ ft, calculate:

(a) The relative importance of the liquid-phase resistance, expressed as per cent of the total resistance, at the top and the bottom of the tower for conditions of total reflux.

(b) Repeat part (a) for each end of the stripping and enriching sections, for a reflux ratio 25 per cent greater than the minimum.

(c) For the reflux ratio of part (b) determine N_L, N_G, and N_{OG}.

(d) For the conditions of part (b) determine the tower height.

16.2. It is desired to design a packed tower to scrub ammonia gas from air by means of ammonia-free water fed to the top of the column. Under anticipated conditions, the equilibrium conditions are given by the expression $Y = 0.8X$. Two gas streams are to be treated: (1) 126 moles/hr of a concentrated gas containing 4.76 mole percent ammonia to be fed to the bottom of the tower, and (2) 133.25 moles/hr of a dilute gas containing 2.44 mole percent ammonia to be introduced at the proper point. The tower is to be tall enough to have an exit-gas concentration of 0.005 mole of ammonia per mole of ammonia free air. Calculate the total packed height required. Use a water flow rate of 200 moles/hr. The maximum velocity of the air stream at any point is to be 30 moles/hr sq ft empty tower. For the packing used $K_Y a = 2.1(V/S)^{0.57}$

16.3. Ammonia is to be removed from a 13 per cent mixture of air and ammonia by scrubbing with water in a tower packed with 1-in. Raschig rings. If 99.9 per cent of the entering ammonia is to be removed, how tall must the column be? Entering-gas rate is 1000 lb/hr sq ft while the water rate is 700 lb/hr sq ft, both at 20°C.

16.4. In the paper industry, large amounts of H_2SO_3 solution are used as a bleach in some processes. Bleach is prepared as it is used by absorbing the flue gas from a sulfur burner in water to produce the bleach liquor.

In a particular plant, 11 per cent SO_2 gas at 1 atm is made in the burner and is scrubbed to recover 97 per cent of the SO_2. Absorbing liquid is pure water, and the bleach is to be 0.75 per cent SO_2 by weight. Operation is assumed isothermal at 30°C.

15,000 SCFH (dry basis) of gas is fed to the absorber, which is 6 ft in diameter, and packed with 3-in. spiral tile rings to a depth of 10 ft.

(a) How much bleach solution is made per hour?

(b) What is $K_g a$ for this operation?

(c) What error (expressed as percentage of the integrated value) is introduced by using a log mean driving force?

16.5. Air at 80°F is used to dry a plastic sheet. The solvent in which the plastic is dissolved is acetone. At the end of the drier, the air leaves containing 0.020 mole fraction acetone. The acetone is to be recovered by absorption with water in a packed tower. The gas composition is to be reduced to 0.0005 mole fraction. The absorption will be isothermal because of the cooling coils within the tower. For the conditions of the absorption, equilibrium relationships may be expressed as $y = 1.8x$, with compositions in mole fractions. The rich gas enters the tower at a flow rate of 1000 lb/hr sq ft, and the water enters the top of the tower at a rate of 1350 lb/hr sq ft. The tower is packed with 1-in. Berl saddles. How tall should the tower be?

16.6. A mixture of benzene and a nonvolatile absorption oil containing 3.8 per cent by weight benzene is preheated under vacuum to 300°F and fed to the top of a packed column where the pressure is 120 mm Hg. Just before entering the column the flashed vapor is removed and separately condensed so the column is fed only with liquid. The molecular weight of the oil is 220. The waste is to contain 0.15 weight percent benzene. The column is fed with live steam at 300°F, and 0.02 mole of steam per mole of benzene-free oil is used. The column is kept at 300°F by a suitable heater. The pressure drop through the column is 60 mm Hg. Assume Raoult's law is valid and that benzene has a vapor pressure of 4220 mm Hg at 300°F. Determine:

(a) The composition of the feed to the column.

(b) The per cent of the benzene vaporized in the preheater.

(c) The composition of the overhead from the tower.

(d) The number of transfer units required.

16.7. The purge gas from an ammonia-synthesis plant contains 4.0 per cent ammonia; the remainder is assumed substantially air. The gas is to be scrubbed at 5 atm pressure with water in a tower packed with 2-in. Raschig rings. The tower temperature will be essentially constant at 70°F. The water rate is to be 600 lb/hr sq ft and the inert gas rate will be 250 lb/hr sq ft. Estimate the tower height needed if 99.9 per cent of the NH_3 entering the tower is absorbed.

16.8. 10,000 lb/hr of a benzene–ethylene dichloride solution containing 50 weight percent benzene is to be continuously rectified in a tower packed with 2-in. Raschig rings to give an overhead and bottom product of at least 95 per cent purity. Feed and reflux will enter the column at their bubble points, and a reflux ratio 1.5 times the minimum is specified. Design the packed tower with regard to tower diameter and tower height.

16.9. Acetic acid is to be extracted from an aqueous solution by continuous contacting with benzene in a packed tower. The aqueous solution is fed to the top of the column and contacted with a rising benzene phase. The inlet aqueous phase enters at a rate of 1000 lb/hr sq ft and analyzes 0.04 lb mole acetic acid/cu ft and leaves the unit analyzing 0.001 lb mole acetic acid/cu ft. Pure benzene enters the tower bottom at a rate of $1.2(G_v)_{min}$. Experiment indicates that a mass-transfer coefficient of $K_c a = 2.5$ lb mole/hr cu ft (lb mole/cu ft) based upon the benzene phase. The equilibrium relationship for this system is given by

$$c_V = 0.025\, c_L$$

where c_V = concentration of acid in benzene, lb mole/cu ft

c_L = concentration of acetic acid in water, lb mole/cu ft

Determine the packed height needed.

16.10. The height of a 6-in. diameter spray tower is to be determined for the extraction of diethylamine (DEA) from water by contacting the aqueous solution with a dispersed toluene phase. The aqueous solution analyzes 0.02 lb mole DEA/cu ft and enters the column at a mass velocity of 200 lb/hr sq ft. Pure toluene enters the bottom of the tower at a mass velocity of 200 lb/hr sq ft. If 90 per cent of the DEA is to be removed from the water phase, how tall a column is needed? Operation is isothermal at 31°C.

For a system of this type and the contemplated equipment, Liebson and Beckmann [*Chem. Eng. Prog.* **49**, 405 (1953)] report

$$H_{OL} = 3.9 + (18)\,\frac{mL}{V}$$

The distribution coefficient for this system is given by

$$\frac{\text{Concentration of DEA in water}}{\text{Concentration of DEA in toluene}} = 1.156$$

chapter 17

Simultaneous Heat and Mass Transfer I: Humidification

In every case involving mass transfer, heat must also be transferred. When a component is transferred from a gas phase to solution in a liquid phase, the latent heat associated with the condensation is evolved. When a component is transferred from solution in one solvent to solution in a second, as in liquid-liquid extraction, the difference between the heats of solution of the solute in the two solvents is evolved. Similar heat effects are present in distillation, adsorption, leaching, drying, etc. In every case, the interfacial temperature will adjust itself so that at steady state the rate of heat transfer will balance the equivalent rate of heat transfer associated with the mass transfer. In operations where mass transfer proceeds by equimolar counterdiffusion, as in distillation, or in operations where the latent-heat effects are small, as in liquid-liquid extraction, gas absorption in dilute solutions, and leaching, heat transfer is of minor importance as a rate-limiting mechanism. In others, particularly where there is a net transfer of mass from the gas phase to a condensed phase, or vice versa, the heat-transfer rate is important. In these cases, it significantly limits the rate at which mass can be transferred. In still other operations, such as boiling, condensation, evaporation, and crystallization, mass and heat transfer occur simultaneously in large amounts, but the rates at which the simultaneous transfer of mass and heat occur can be determined by merely considering the rate of heat transfer from an external source.

Of those operations where both heat transfer and mass transfer affect the rate, humidification and dehumidification are the simplest and are also the most direct applications of the theory. Here, only two components and only two phases are involved. The liquid phase, most often water, is a single component, and the gas phase consists of a noncondensable gas, usually air, in which some vapor from the liquid phase is present. Because of the inherent simplicity, the basic relations for operations involving simultaneous heat and mass transfer will be discussed in this chapter in relation to humidification and dehumidification.

The principles of simultaneous heat and mass transfer as developed here for evaporation of a liquid without an external heat source are also fundamental to reaction kinetics. The problems of mass transfer to a point of reaction and of heat transfer away from it are exactly parallel. The chemical reaction does not, however, allow thermodynamic equilibrium to exist at the interface, whereas equilibrium does exist at the interface in vaporization cases. The rate of the chemical reaction must be included in an evaluation of a reaction kinetics situation.

HUMIDIFICATION: INDUSTRIAL APPLICATIONS AND EQUIPMENT

Humidification processes may be carried out to control the humidity of a space or, more usually, to cool and recover water by contacting it with low-humidity air. The water that has lost heat to the atmosphere can then be reused in heat exchangers throughout a plant. Alternatively, the water could be cooled in surface heat exchangers. The choice is one of economics, with the designer balancing the loss of cooling water inherent in the air-water contact cooler against the cost of supplying and handling the cooling source to the surface cooler and the higher cost of the surface units.

Dehumidification is practiced most commonly as a step in an air-conditioning system. It may also be used as part of a solvent-recovery system, but in these applications the condensable vapor is not water but rather a solvent such as trichlorethylene, benzene, or methanol.

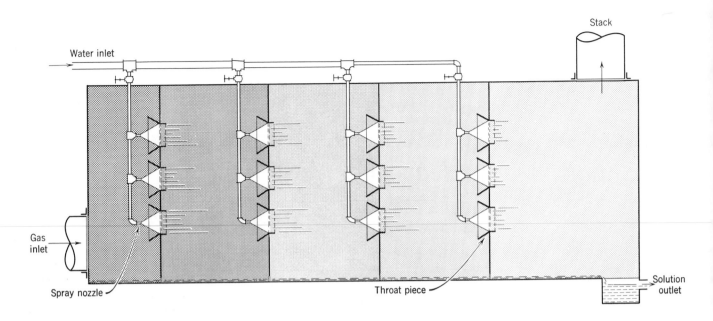

Water inlet

Stack

Gas
inlet

Spray nozzle

Throat piece

Solution
outlet

Figure 17.1. Spray chamber type of fume scrubber. (Courtesy Schutte & Koerting Co.)

Figure 17.2. A spray pond located in a protected area so that louvered fencing is unnecessary.
(Courtesy Schutte & Koerting Co.)

Any one of these applications can, in principle, be carried out in similar pieces of equipment. The direction of mass transfer and that of heat transfer are determined by the relation between the humidity and temperature of the inlet gas phase and the temperature of the contacting liquid. However, the unit size, the convenience of recovering gas or liquid phases, and the materials of construction used all limit the applicability of a single piece of apparatus.

The most obvious form of humidification equipment is the spray chamber. Here, the contacting liquid is sprayed as a mist into the gas stream. Gas velocity is kept low so that the contact time is high and so that there will be only a small amount of liquid physically entrained in the gas stream. These units are usually restricted to small-scale operations and are frequently used in humidity control of a room or plant where either the humidification or dehumidification of the inlet air is required.

The fume scrubber shown in Figure 17.1 is a variation of a spray chamber. Here, throat pieces are used to assure close contact of gas and liquid streams, and the alternate open areas allow separation of the two phases. Nozzle capacities are normally 5 to 10 gal/min of water with throat pieces handling 300 cfm of gas/per gallon per minute of liquid. The scrubbers are often applied to removing dusts from a gas stream or for reacting the liquid with a component of the gas stream as well as for removal of a condensable component from the gas phase.

Where the water-recirculation rate is low or where a very large land area is available, water cooling may be accomplished by spraying the water in fountains over a shallow pool. At best, the cooling effect is small since the water is only sprayed once (5), whereas losses due to entrainment of water in the air passing over the pond, called windage loss, may be large compared to other cooling methods. Figure 17.2 pictures a spray pond in a protected location. Typical spray nozzles used in this application deliver 50 gal/min each at about 7 psig. The nozzles are located about 6 ft above the pond surface and are spaced about every 4 ft on the water headers. In exposed locations, spray ponds are surrounded with a louvered fence about 12 ft high to reduce windage losses. Empirical correlations are available that allow estimation of the size of spray pond necessary for a given duty at known air conditions and typical spray-nozzle spacing and design (5).

More typically, and especially for larger cooling duties, cooling towers are used. These towers are most often of wood construction with multiple wood-slat decks. However, aluminum, steel, brick, concrete, and asbestos-board casings have also been used. The water is sprayed above the top deck and trickles down through the various decks to a bottom collection basin.

Corrosion is prevented by construction entirely of inert materials such as redwood, stainless steel, and porcelain. Air flow may be by natural draft in which case the sides of the tower are of louver construction with air flowing across the decks. Natural-draft towers may be used only where ample open space is available and where a natural wind velocity greater than about 3 mph can be relied upon to carry away the humid air. More often forced- or induced-draft towers are used. Induced-draft towers are preferred since they prevent the recirculation of the humidified air. In these towers, fans are located on the top of the tower. Air is pulled into louvers around the bottom of the tower and up through the decks countercurrent to the water flow. In the largest towers (capacities up to 100,000 gal/min of water) air may be pulled in through two open sides of the tower and up to a central fan through inclined baffles. Empirical correlations are available to facilitate cooling-tower selection and design (1).

Figure 17.3 is a cutaway drawing of a double-flow, induced-draft cooling tower. Here air enters both sides of the tower and flows across the path of the falling water to a central duct. A fan pulls the air up the duct and blows it out the top of the tower. Water is distributed initially by flooding a distributor plate containing porcelain nozzles through which the water flows onto the packing. Packing is of wood-slat construction assembled without nails. The drift eliminators, which are wood baffles through which the air passes before entering the fan, reduce entrainment of water and thus reduce mist content in the air discharged from the fan. Such a tower would be about 48 ft wide by 30 ft high with a fan 14 ft in diameter. Several units may be constructed end to end to give a single long tower. In somewhat smaller cooling towers, directly countercurrent flow may be used. Figure 17.4 pictures this type of tower. Internal construction is similar to that shown in Figure 17.3, though the arrangement is different.

Various other contacting devices are used to dehumidify air or to remove a single component from a noncondensable gas. Often packed towers are used or wash columns in which contacting is essentially stagewise. A contactor of this sort is shown in Figure 17.5. These devices are used to contact the gas stream with several liquids, keeping the liquid streams separate. Water jet ejectors are frequently used to condense or "scrub" a vapor from a gas stream. The ejectors handle large volumes of water and produce intimate cocurrent contacting of the two phases.

PHASE RELATIONS AND DEFINITIONS

As in previous consideration of mass-transfer and heat-transfer processes, in the simultaneous transfer of

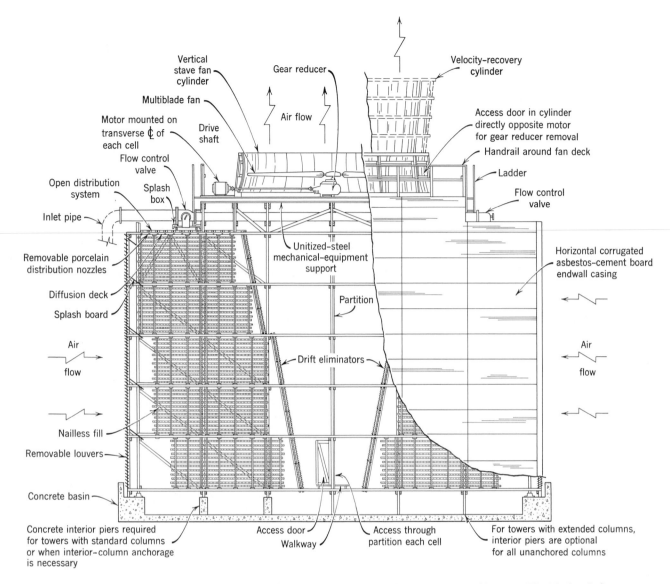

Figure 17.3. Transverse cross-section view of double-flow induced-draft cooling tower. (Courtesy The Marley Co.)

heat and mass the direction of transfer and the extent of this transfer are controlled by the equilibrium condition toward which the transfer tends. Therefore, a statement of phase equilibrium is necessary as a basis for application of the rate equation.

The phase conditions of gas-vapor mixtures are conveniently shown on a concentration-temperature diagram plotted for a single constant pressure. On this diagram, the concentration of saturated gas is plotted against temperature from data which must in general be obtained experimentally.

For the special case of ideal solutions in both phases, the saturation locus can be plotted from the results of Raoult's and Dalton's law calculations.

$$p_a = P y_s = P_a x \qquad (17.1)$$

where p_a = partial pressure of condensable component in the gas phase

P = total gas-phase pressure

P_a = vapor pressure of condensable component

y_s, x = the saturation mole fraction of condensable component in the equilibrium vapor and liquid phases respectively

In this case $x = 1$ since the liquid phase is a single component. Thus

$$y_s = \frac{P_a}{P} \qquad (17.2)$$

Lines of constant per cent saturation can also be plotted on this diagram, where *per cent saturation* is defined as 100 times the weight ratio of condensable component to noncondensable component present divided by the weight ratio of condensable component to noncondensable component present at saturation. Thus, by definition,

$$\text{Per cent saturation} = \frac{100 \, Y \cdot \dfrac{M_a}{M_b}}{Y_s \dfrac{M_a}{M_b}} = 100 \frac{Y}{Y_s} \quad (17.3)$$

where Y = mole ratio of condensable to noncondensable component present

$\quad Y_s$ = mole ratio present at saturation

M_a, M_b = molecular weights of condensable and noncondensable components respectively

Note that the per cent saturation can be defined in terms of mole ratios as well as in terms of mass ratios. On the concentration-temperature diagram, lines of constant per cent saturation will be shaped like the saturation curve but will be at proportionally lower concentrations.

An alternative concentration expression is the *relative saturation*, which is defined as the partial pressure of condensable component present divided by the partial pressure of condensable component present at saturation.

$$\text{Per cent relative saturation} = \frac{100 p_a}{(p_a)_s} = \frac{100 y}{y_s} \quad (17.4)$$

where again the subscript *s* refers to saturation. Relative saturation is the measure normally used in reporting weather data. As shown by Equations 17.3 and 17.4 relative saturation is a ratio of mole fractions, in distinction from per cent saturation which is a ratio of

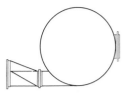

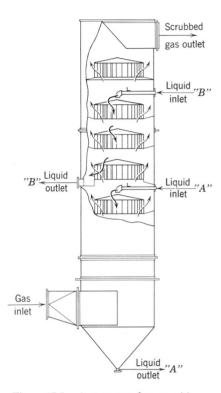

Figure 17.5. A contactor for use with multiple-wash liquid. (Courtesy Claude B. Schneible Co.)

mole ratios. Since $Y = y/(1 - y)$,

$$\text{Per cent saturation} = \frac{\dfrac{100 y}{1 - y}}{\dfrac{y_s}{1 - y_s}} = \frac{100 y}{y_s} \cdot \frac{(1 - y_s)}{(1 - y)}$$

or the per cent saturation is equal to the per cent relative saturation multiplied by $(1 - y_s)/(1 - y)$ (17.5)

Since $y_s > y$, the per cent saturation will always be less than, or in the limit equal to, the per cent relative saturation.

Traditionally mole-ratio units are preferred since they are based on the unchanging unit quantity of noncondensable gas, and hence, they will be used here. However, most practical applications involve the air–water system at 1 atm total pressure and temperatures between 50 and 130°F. For these cases, concentrations

Figure 17.4. Wood counterflow induced-draft cooling tower. (Courtesy The Marley Co.)

are low enough so that $y \approx Y$, and the per cent saturation and per cent relative saturation are practically equal. In systems other than air–water or at high temperatures with the air–water system, the concentrations y and y_s may be significantly higher, so that Y is not equal to y. Therefore, the generality that $y \approx Y$ should be used cautiously.

For most of the systems of industrial importance, the gas phase is dilute, the pressure is low, and Raoult's law is followed. Equation 17.2 can be used to calculate the saturation locus, and partial saturation can be expressed in terms of vapor pressures and partial pressures.

$$\text{Per cent saturation} = \frac{100\,Y}{Y_s} = \frac{100 p_a/(P - p_a)}{P_a/(P - P_a)}$$

$$= 100 \frac{p_a(P - P_a)}{P_a(P - p_a)} \quad (17.6)$$

$$\text{Per cent relative saturation} = \frac{100y}{y_s} = \frac{100 p_a}{P_a} \quad (17.4a)$$

When the air–water system is discussed, it has become common practice to call the weight ratio of water vapor to dry air the *humidity*. The mole ratio of these components is called the *molal humidity*, whereas *relative humidity* and *per cent humidity* are synonymous with *relative* and *per cent saturation*. This nomenclature has frequently been used for systems that do not contain water, so that it is not uncommon to hear of the *humidity* of ethanol in air or to see a *humidity* diagram for the benzene–N_2 system (10).

In humidification calculations, it is frequently necessary to know the volume and specific heat of the gas-vapor phase. For this use, the *humid volume* is defined as the volume of 1 lb of dry gas plus its contained vapor; similarly the molal humid volume is the volume of 1 lb mole of dry gas plus its contained vapor. Thus,

$$V_h = (1 + Y) \times 359 \times \frac{T}{492P} \quad (17.7)$$

applies to those conditions where the perfect-gas law is valid. Here, V_h is the *molal humid volume*. Similarly, the *molal humid heat* is defined as the heat capacity of 1 lb mole of dry gas plus that of its associated vapor. That is,

$$c_h = c_b + Yc_a \quad (17.8)$$

where c_h, c_b, and c_a are the molal humid heat and the molal heat capacities of components b (noncondensable gas) and a (condensable vapor) respectively.

The *enthalpy* of the vapor-gas mixture is not generally given any special name. However it frequently appears in calculations. By analogy to Equation 17.8,

$$H = H_b + YH_a \quad (17.9)$$

where H is the enthalpy of 1 lb mole of dry gas plus that of its contained vapor, and H_b and H_a are the molal enthalpies of components b and a respectively. These enthalpies must, of course, be computed relative to the enthalpy of some arbitrary base condition at which $H = 0$. The base conditions cannot be identical for b and a since different components are being considered. However, they may be at the same temperatures, pressures, and phase conditions, though this is in no way necessary. For the air–water system, it is common practice to take the enthalpy to be zero for liquid water at the triple-point. The base for the enthalpy of air is often taken as the dry gas at 32°F, 1 atm pressure. These bases will be used here. On these bases, the enthalpy of moist air may be related to the appropriate latent and specific heats.

$$H = c_b(T - T_0) + Y[\lambda_0 + c_a(T - T_0)] \quad (17.10)$$

where T_0 = the base temperature, here 32°F for both components
λ_0 = the latent heat of evaporation of water at the base temperature, Btu/lb mole

and the effect of pressure on the enthalpy of the liquid is ignored. In this equation, both c_b and c_a have been taken as constant. If this assumption is not tenable, these specific heats must be the integrated mean values applicable between T and T_0.

Other terms used to describe vapor-gas mixtures are the dew point, wet-bulb temperature, and adiabatic-saturation temperature. The *dew point* is that point at which condensation begins when the pressure or temperature is changed over a mixture of fixed composition. Usually the temperature is lowered at constant pressure, so that the dew-point temperature is obtained. The *wet-bulb temperature* is the steady-state temperature attained by a wet-bulb thermometer exposed to a rapidly moving stream of the vapor-gas mixture. The bulb of the wet-bulb thermometer is coated with the same liquid that forms the vapor in the vapor-gas mixture. The wet-bulb temperature is related to the humidity of the gas phase as will be shown below. This relation allows the use of the wet-bulb temperature coupled with the temperature itself as a measure of humidity.

The *adiabatic-saturation temperature* is the temperature that the vapor-gas mixture would reach if it were saturated through an adiabatic process. Many of the processes discussed in this chapter and in Chapter 18 occur approximately adiabatically, so that the adiabatic-saturation temperature is a particularly useful and important quantity. The relation between this temperature and the gas-phase humidity will also be shown below.

For convenience these definitions are collected in Table 17.1. The saturation locus and partial-saturation curves on the temperature-concentration diagram are shown schematically on Figure 17.10 and quantitatively in Appendix D-14, the humidity chart for the air–water system.

Adiabatic-Saturation Process. The relation between the adiabatic-saturation temperature and the gas-phase humidity can be developed by considering the process by which the gas can be saturated. A general humidification process is shown in Figure 17.6. Here the subscripts 1 and 2 refer to the bottom and top of the column respectively; subscripts L and V refer to liquid and vapor phases respectively; L and V are the molal rates of flow of liquid and vapor respectively; and V' is the molal flow rate of noncondensable gas, lb moles/hr. A material balance around the tower gives

$$(L_2 - L_1) = V'(Y_2 - Y_1) \qquad (17.11)$$

and an enthalpy balance gives

$$q + L_2 H_{L2} - L_1 H_{L1} = V'(H_{V2} - H_{V1}) \qquad (17.12)$$

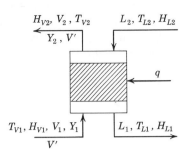

Figure 17.6. General humidification process.

For this general process to be specialized and to be made adiabatic, several restrictions must be imposed. First, no heat can be transferred to the tower, or $q = 0$. Second, the liquid stream will be recycled. In this way, at steady state $T_{L1} = T_{L2}$. As the process continues, the liquid temperature will become constant, and thus no net sensible heat will be brought to the tower or taken

Table 17.1. DEFINITIONS OF HUMIDITY TERMS

Term	Meaning	Units	Symbol
1. Humidity	vapor content of a gas	lb vapor/lb noncondensable gas	$Y' = Y \dfrac{M_a}{M_b}$
2. Molal humidity	vapor content of a gas	moles vapor/mole noncondensable gas	Y
3. Relative saturation or relative humidity	ratio of partial pressure of vapor to partial pressure of vapor at saturation	atm/atm, or mole fraction/mole fraction, often expressed as per cent	$100 \dfrac{y}{y_s}$
4. Per cent saturation or per cent humidity	ratio of concentration of vapor to the concentration of vapor at saturation with concentrations expressed as mole ratios	mole ratio/mole ratio, expressed as per cent	$100 \dfrac{Y}{Y_s}$
5. Molal humid volume	volume of 1 lb mole of dry gas plus its associated vapor	cu ft/lb mole of dry gas	$V_h = (1 + Y) \times 359 \dfrac{T}{492P}$
6. Molal humid heat	heat required to raise the temperature of 1 lb mole of dry gas plus its associated vapor 1°F	Btu/lb mole of dry gas, °F	$c_h = c_b + Yc_a$
7. Adiabatic-saturation temperature	temperature that would be attained if the gas were saturated in an adiabatic process	°F or °R	T_{sa}
8. Wet-bulb temperature	steady-state temperature attained by a wet-bulb thermometer under standardized conditions	°F or °R	T_w
9. Dew-point temperature	temperature at which vapor begins to condense when the gas phase is cooled at constant pressure	°F or °R	T_d

from it by the liquid. The only effect passage through the tower will have on the liquid stream is that some of it will be vaporized into the gas stream. Under these conditions, Equation 17.12 becomes

$$H_L(L_2 - L_1) = V'(H_{V2} - H_{V1}) \qquad (17.13)$$

Combining Equations 17.11 and 17.13 to eliminate $(L_2 - L_1)$,

$$H_L(Y_2 - Y_1) = H_{V2} - H_{V1} \qquad (17.14)$$

The enthalpies can now be expressed in terms of molal latent heat and humid heat as given by Equation 17.10.

$$c_L(T_{L2} - T_0)(Y_2 - Y_1) = [c_{h2}(T_{V2} - T_0) + \lambda_0 Y_2] - [c_{h1}(T_{V1} - T_0) + \lambda_0 Y_1] \qquad (17.15)$$

Now, if the further restriction is made that the tower be so tall that the gas and liquid phases reach equilibrium at the top of the tower, the gas phase will be saturated, and $T_{L2} = T_{L1} = T_{V2} = T_2$. The temperature at the top of the tower (T_2) will then be the adiabatic-saturation temperature, and Y_2 will be the molal humidity of gas saturated at T_2. The gas-phase temperature at the bottom of the tower (T_{V1}) may be simply designated as T_1. Applying these restrictions, Equation 17.15 becomes

$$c_L(T_2 - T_0)(Y_2 - Y_1) = [c_{h2}(T_2 - T_0) + \lambda_0 Y_2] - [c_{h1}(T_1 - T_0) + \lambda_0 Y_1] \qquad (17.16)$$

Either of the humid heats may be replaced in terms of the other since

$$c_{h1} = c_b + Y_1 c_a \quad \text{and} \quad c_{h2} = c_b + Y_2 c_a$$

or

$$c_{h2} = c_{h1} - Y_1 c_a + Y_2 c_a \qquad (17.17)$$

so that

$$c_L(T_2 - T_0)(Y_2 - Y_1) = [c_{h1}(T_2 - T_0) - Y_1 c_a(T_2 - T_0) + Y_2 c_a(T_2 - T_0) + \lambda_0 Y_2] - [c_{h1}(T_1 - T_0) + \lambda_0 Y_1] \qquad (17.18)$$

Rearranging and collecting terms

$$c_{h1}(T_2 - T_1) = (Y_2 - Y_1)[c_L(T_2 - T_0) - \lambda_0 - c_a(T_2 - T_0)] \qquad (17.19)$$

The bracketed terms in this equation are equal to $-\lambda_2$, as can be shown by the following enthalpy balance:

$$H_{L2} - H_{a2} = -\lambda_2 = (H_{L2} - H_{L0}) + (H_{L0} - H_{a0}) + (H_{a0} - H_{a2}) = c_L(T_2 - T_0) + (-\lambda_0) + [-c_a(T_2 - T_0)] \qquad (17.20)$$

Equation 17.20 expresses the result obtained when determining $H_{a2} - H_{L2} = \lambda_2$ by using a path in which the system, here one mole of component a, is first cooled as

a liquid to T_0, is then vaporized at T_0, and is finally heated as a vapor back to T_2. The summation of enthalpy changes along the entire path will equal the difference between final and initial enthalpies because enthalpy is a state function. Making the indicated substitution,

$$c_{h1}(T_2 - T_1) = \lambda_2(Y_1 - Y_2) \qquad (17.21)$$

Since in this development T_2 and Y_2 were conditions at the adiabatic-saturation temperature, Equation 17.21 can be written

$$c_{h1}(T_{sa} - T_1) = \lambda_{sa}(Y_1 - Y_{sa}) \qquad (17.21a)$$

where the subscript sa refers to the adiabatic-saturation condition corresponding to the initial gas-phase conditions designated by the subscript 1.

Equation 17.21a gives the relation between the temperature and humidity of a gas at any entering condition and the corresponding conditions for the same gas at its adiabatic-saturation temperature. Thus, if both the adiabatic-saturation temperature and the actual temperature of a gas are known, the gas humidity is obtainable. The humidity at the adiabatic-saturation condition can be found from the saturation locus, and the gas humidity can then be found through Equation 17.21a. Note also that Equation 17.21a was developed from *over-all material* and *enthalpy balances* between the initial gas condition and the adiabatic-saturation condition. Thus, it is applicable only at these two points but may not describe the path followed by the gas as it becomes saturated.

Wet-Bulb Temperature. One of the oldest, and still the most common, methods of measuring the humidity of a gas stream is to measure its "wet-bulb temperature" as well as its temperature. This is done by passing the gas rapidly past a thermometer bulb which is kept wet with the liquid that forms the vapor in the gas stream. The usual physical arrangement is shown in Figure 17.7. During this process, if the gas is not saturated,

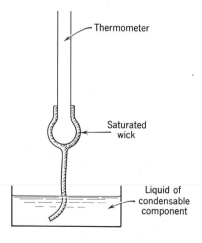

Figure 17.7. Wet-bulb thermometer.

some liquid is evaporated from the saturated wick into the moving gas stream, carrying with it the associated latent heat. The latent-heat removal results in a lowering of the temperature of the thermometer bulb and wick, and therefore sensible heat will be transferred to the wick surface by convection from the gas stream and by radiation from the surroundings. At steady state, the *net* heat flow to the wick will be zero, and the temperature will be constant. The wet-bulb temperature is the temperature attained at steady state by the thermometer exposed to rapidly moving gas. Thus, the rate of heat transfer to the wick is

$$q = (h_c + h_r)A(T_1 - T_w) \qquad (17.22)$$

and the rate of mass transfer from the wick is

$$N_a = k_Y A(Y_1 - Y_w) \qquad (17.23)$$

where the subscripts w and 1 refer to the wick surface at the wet-bulb temperature and to the bulk-gas temperature. The rate of sensible heat transfer to the wick (q) follows the usual convection and radiation mechanisms. The rate of mass transfer of component a toward the wick (N_a) also follows the usual gas-phase mass-transfer mechanism. The coefficients h_c and h_r are those for heat transfer by convection and radiation to the bulb. By using h_r, it is assumed that the radiant-heat transfer may be approximated by

$$q_r = h_r A(T_1 - T_w) \qquad (17.24)$$

At steady state all the heat transferred to the wick is used to vaporize N_a moles of liquid, or

$$q = -N_a \lambda_w \qquad (17.25)$$

which expresses the condition that the rate of sensible-heat transfer to the wick is exactly equal to that of the latent heat carried from the wick by the mass (N_a).

Combining Equations 17.22, 17.23, and 17.25,

$$(h_c + h_r)A(T_1 - T_w) = -k_Y A(Y_1 - Y_w)\lambda_w$$

or, if the areas available for heat and mass transfer are equal,

$$T_1 - T_w = \frac{k_Y \lambda_w}{(h_c + h_r)}(Y_w - Y_1). \qquad (17.26)$$

Equation 17.26 relates the wet-bulb temperature to the gas humidity, just as Equation 17.21a did with the adiabatic-saturation temperature, provided that predictable values of the term $k_Y \lambda_w/(h_c + h_r)$ can be obtained. This ratio of transfer coefficients obviously depends upon the particular flow, boundary, and temperature conditions encountered. In measuring wet-bulb temperature, several precautions are made to ensure that reproducible values of $k_Y \lambda_w/(h_c + h_r)$ are obtained. The radiation coefficient is minimized by shielding the wick from sight of the surrounding surfaces if the surfaces are hotter than the wet-bulb temperature.

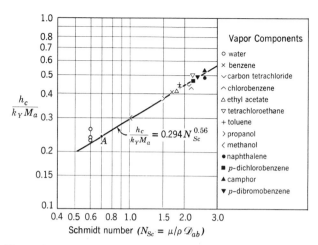

Figure 17.8. The ratio h_c/k_Y for wet-bulb thermometry in air-vapor systems (2). [From Treybal, R. E., *Mass Transfer Operations*, McGraw-Hill Book Company, Inc., New York (1955), by permission.]

Gas movement past the bulb is made rapid, often by swinging the thermometer through the gas as is done with the sling psychrometer or by inserting the wet-bulb thermometer in a constriction in the gas-flow path. Under these conditions Equation 17.26 reduces to

$$T_1 - T_w = \frac{k_Y \lambda_w}{h_c}(Y_w - Y_1) \qquad (17.26a)$$

where the k_Y/h_c ratio is that obtained under turbulent-flow conditions.

For turbulent flow past a wet cylinder such as the wet-bulb thermometer, the accumulated experimental data give (2)

$$\frac{h_c}{k_Y} = 8.50\left(\frac{\mu}{\rho \mathscr{D}_{ab}}\right)^{0.56} \qquad (17.27)$$

when air is the noncondensable gas, and

$$\frac{h_c}{k_Y} = c_h\left(\frac{N_{Sc}}{N_{Pr}}\right)^{0.56} \qquad (17.28)$$

for other gases. The data on which Equation 17.27 is based are summarized on Figure 17.8. Equation 17.28 is based on heat- and mass-transfer experiments with various gases flowing normal to cylinders.

Point A on Figure 17.8 is for pure air for which the $N_{Sc} \cong N_{Pr} \cong 0.70$. For this case, both Equations 17.27 and 17.28 give $h_c/k_Y = c_h = 6.94$. Experimental data for the air–water system give h_c/k_Y values ranging between 7.82 and 6.83. The latter figure is recommended. Thus, for the air–water system, the h_c/k_Y value can be reasonably replaced by c_h within moderate ranges of temperature and humidity, provided that the flow is turbulent. As a result Equation 17.21a, which relates the humidity to the adiabatic-saturation conditions, and

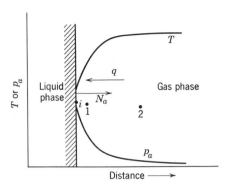

Figure 17.9. Transfer from a saturated surface.

Equation 17.26a, which relates the humidity to the wet-bulb temperature, are identical. Therefore, the adiabatic-saturation temperature equals the wet-bulb temperature for the air–water system. For systems other than air–water, this coincidence does not occur, as can be seen from the psychrometric charts given by Perry (10).

For systems other than air–water, the behavior in a saturating contactor is quite complex. If the column over-all conditions satisfy the adiabatic model, the liquid and the gas entering and leaving must be related by the adiabatic-saturation equation. This equation does not reveal anything of the humidity-enthalpy path of either the liquid phase or the gas phase at various points within the contacting device. Each point within the system must conform to the wet-bulb relation, which requires that the heat-transferred be exactly consumed as latent heat of vaporization of the mass transferred. Consideration of the adiabatic-saturation tower for such a system indicates that the liquid entering and leaving must be at the temperature of adiabatic saturation, as must also the gas leaving. Speculation on the path taken indicates that the gas could follow the straight line joining initial conditions and adiabatic-saturation conditions, with consequent swinging of the temperature of the liquid to follow the required wet-bulb temperature, or that the liquid could remain constant in temperature at the adiabatic-saturation point, with the gas following a path during its humidification that would satisfy the wet-bulb relations at every point within the column. More probably, there is some change in the temperature of the liquid and some difficultly predictable path of gas temperature and composition so that at all points within the system the wet-bulb relation is satisfied and so that at the ends of the system the adiabatic-saturation conditions are satisfied.

The Lewis Relation. The identity of h_c/k_Y with c_h was first found empirically by W. K. Lewis and hence is called the Lewis relation. The basis for its general validity may be determined by examining more closely the heat- and mass-transfer processes occurring at a saturated surface. The gradients of interest are shown in Figure 17.9.

For the transport of heat and of mass occurring between the interface and a point in the main bulk of a turbulently flowing gas stream such as point 2 of Figure 17.9, the general transport equation (Equation 13.61) can be written for the two mechanisms. For heat transfer this becomes

$$\frac{q}{A} = -h_c\,\Delta T = \frac{-4(\alpha + \bar{E}_q)}{\gamma_q D}\,\rho c_p\,\Delta T \quad (17.29)$$

and for mass transfer Equation 13.61 is written

$$\frac{N_a}{A} = k_g\,\Delta p = \frac{-4(\mathscr{D} + \bar{E}_N)}{\gamma_N D}\,\Delta c_a = \frac{-4(\mathscr{D} + \bar{E}_N)}{\gamma_N D}\,\frac{\Delta p_a}{RT} \quad (17.30)$$

where the nomenclature is as given in Chapter 13. Equation 17.29 may be divided by Equation 17.30 to obtain, after simplification, the value of the ratio of transfer coefficients.

$$\frac{h_c}{k_g} = \frac{(\alpha + \bar{E}_q)}{(\mathscr{D} + \bar{E}_N)}\,\frac{\gamma_N}{\gamma_q}\,\rho c_P RT \quad (17.31)$$

For completely turbulent transport, such as that between points 1 and 2 of Figure 17.9, α and $\mathscr{D}$ are negligibly small compared to $\bar{E}_q$ or $\bar{E}_N$. Then

$$\frac{h_c}{k_g} = \frac{\bar{E}_q}{\bar{E}_N}\,\frac{\gamma_N}{\gamma_q}\,\rho c_P RT \quad (17.32)$$

The terms γ_q and γ_N represent the ratio of the difference in transferent property between the interface and the bulk fluid to the difference in this property between the interface and the maximum value in the fluid. Thus their values usually are slightly less than 1.0 and $\gamma_N \approx \gamma_q$.

From this

$$\frac{h_c}{k_g P} = c_P\,\frac{\bar{E}_q}{\bar{E}_N} \quad (17.33)$$

for turbulent transport in an ideal gas system under any conditions of N_{Sc} or N_{Pr}.

For the Lewis relation to hold $\bar{E}_q$ must equal $\bar{E}_N$. This requires equality of N_{Sc} with N_{Pr} even in the regime of complete turbulence.

For equal Prandtl and Schmidt numbers $\gamma_q = \gamma_N$ and also $\bar{E}_q = \bar{E}_N$ as shown in Chapter 13 (see Equation 13.109). Also at low concentration $k_g P \approx k_g p_{bm} = k_Y$. From this

$$\frac{h_c}{k_g P} = c_P \approx \frac{h_c}{k_Y} \quad (17.33a)$$

Thus the Lewis relation holds for transfer within a completely turbulent regime of $N_{Pr} = N_{Sc}$. Near the fluid boundary, as for instance with transport between points i and 1 of Figure 17.9, laminar flow prevails, and

α and $\mathscr{D}$ are large in comparison to $\bar{E}_q$ and $\bar{E}_N$. In this region Equation 17.31 reduces to

$$\frac{h_c}{k_g} = \frac{\alpha}{\mathscr{D}} \frac{\gamma_N}{\gamma_q} \rho c_P R T \qquad (17.34)$$

As before, $\gamma_N = \gamma_q$ when $N_{\mathrm{Sc}} = N_{\mathrm{Pr}}$, but this equality of Prandtl and Schmidt numbers also demands that $\alpha = \mathscr{D}$. Thus again

$$\frac{h_c}{k_Y} = c_P \qquad (17.34a)$$

The Lewis relation then has been found to hold for each part of the transfer path from i to 2, or from the interface to any point in the main gas stream, and thus for the entire transfer path, if $N_{\mathrm{Pr}} = N_{\mathrm{Sc}}$. If $N_{\mathrm{Pr}} \neq N_{\mathrm{Sc}}$ the Lewis relation must be modified to include $\bar{E}_q/\bar{E}_N$ for turbulent transport, and $\alpha/\mathscr{D}$ for laminar transport. This equality of N_{Pr} with N_{Sc} is found to hold true for simple monatomic and diatomic gases, but it is not valid for more complicated fluids. Of course, the most important system to which it applies is air with small concentrations of water vapor.

Where it applies, the Lewis relation implies that the mechanisms of heat and of mass transfer are identically dependent upon the flow conditions. Both the eddy and molecular diffusivities for heat transfer equal the eddy and molecular diffusivities for mass transfer. Hence, the relative importance of eddy to molecular transport of heat is the same as the relative importance of eddy to molecular transport of mass (see Chapter 13). The simplifications of concept and of calculational procedures that result for those systems to which the Lewis relation applies are major ones. The resulting identity of the wet-bulb-temperature equation (Equation 17.26a) and the adiabatic-saturation equation (Equation 17.21a) has already been shown. It will be shown below that this is one of the simplifications permitting the combination of temperature difference and mole-ratio difference driving forces for heat and mass transfer into a single driving force expressed as a specific enthalpy difference. With this simplification, calculations involving equipment in which heat and mass are transferred simultaneously are no more cumbersome than are those for equipment where only heat or mass is transferred.

The Humidity Chart. The previously described concepts and definitions are normally presented graphically on a "humidity chart." Such a chart applies to a single noncondensable and a single condensable component at a fixed total pressure. From the chart, information on the humidity, relative or per cent saturation, humid volume, and humid heat or total enthalpy should be obtainable from conveniently measured quantities such as the dry- and wet-bulb temperatures. The saturation locus and lines of constant per cent

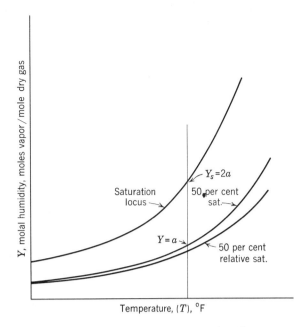

Figure 17.10. Partial and total saturation lines on a humidity chart.

or relative humidity are of primary importance on the humidity diagram. As discussed previously, the saturation locus is directly related to the vapor-pressure curve, and it will appear roughly as shown in Figure 17.10. Of similar curvature will be lines of constant per cent saturation or of constant relative saturation. If lines of constant per cent saturation are plotted, they will fall at ordinates directly proportional to the ordinate of the saturation locus at the same temperature. For example if $Y_s = 0.10$ at T_1, Y at T_1 for the 30 per cent saturation curve will be 0.03. If lines of constant relative saturation are plotted, no such direct proportion exists. This is evident from the definition of relative saturation, Equation 17.4. The relative-saturation curves will always be lower than the equivalent per cent saturation curves. Only one of the measures of partial saturation is normally presented on a single chart.

In order to allow immediate determination of the absolute humidity from wet-bulb-temperature measurements, lines of constant wet-bulb temperature are usually plotted using Equations 17.26 and 17.27. If the N_{Sc} is constant through the range of temperature and humidity covered by a single line of constant wet-bulb temperature, Equation 17.27 shows that the line will be straight. In Chapter 9 it was shown that for gases of fixed composition both $\mathscr{D}$ and ν vary approximately as $T^{3/2}$. The mass diffusivity ($\mathscr{D}$) varies only slightly over a wide concentration range. Thus the variation in N_{Sc} is slight and may be considered negligible when saturation occurs at low molal humidity. For the lines to be straight and parallel, the ratio $k_Y \lambda_w/h_c$ must be constant throughout the entire range of T and Y covered

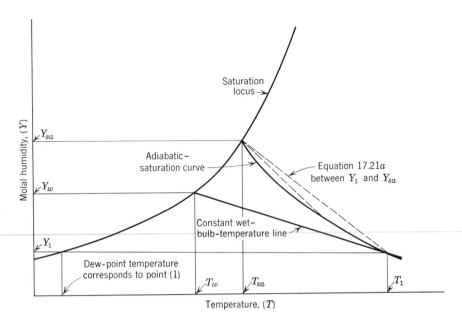

Figure 17.11. Constant wet-bulb-temperature and adiabatic-saturation-temperature curves on a humidity chart.

by the chart. Thus, N_{Sc} and λ_w must either be constant or vary in a compensating way. For the air–water system, the lines are approximately straight and parallel and are often so represented.

Curves connecting all points having the same adiabatic-saturation temperature are also plotted on the humidity chart and are useful in calculations concerning humidification and drying processes. From Equation 17.21a, the slope of the line connecting any point and its adiabatic-saturation temperature will be $-c_{h1}/\lambda_{sa}$. If c_h is constant over the range of temperature and humidity covered by the locus of all points having the same adiabatic-saturation temperature, c_h/λ_{sa} will also be the slope of this locus. Generally, c_h is not constant but increases as the humidity increases. Then, the locus of points of fixed adiabatic-saturation temperature, usually called an "adiabatic-saturation line," will be concave upwards as shown in Figure 17.11. In Figure 17.11, the adiabatic-saturation temperature and the wet-bulb temperature are shown for a single point (T_1, Y_1). For the air–water system or any other system for which the Lewis relation holds, the constant adiabatic-saturation-temperature line and the constant wet-bulb temperature line would be identical, i.e., $T_w = T_{sa}$.

Information on humid volume cannot be presented on coordinates of humidity versus temperature, and so usually a separate volume-versus-temperature plot is made. On this plot are shown the specific volume of noncondensable component and also the specific volume of saturated vapor-gas mixture. From these curves, the humid volume of any vapor-gas mixture can be determined from the per cent saturation.

$$V_h = V_{dry} + (V_s - V_{dry}) \cdot (\text{per cent saturation}) \quad (17.35)$$

where V_h = humid volume of moist gas at T_1
V_{dry} = specific volume of dry gas at T_1
V_s = humid volume of saturated gas at T_1

For convenience, these curves are usually superimposed on the humidity chart using a common temperature scale but an independent volume ordinate.

Enthalpy information is plotted separately in the same way as is the volume information. As with humid volume,

$$H = H_{dry} + (H_s - H_{dry}) \cdot (\text{per cent saturation}) \quad (17.36)$$

where the subscripts "dry" and s refer to dry and saturated conditions respectively and all enthalpy values must be read at a single temperature. As with the volume plots, the enthalpy curves for dry and for saturated gas are superimposed on the humidity chart using the same temperature scale but an independent enthalpy ordinate.

Both Equations 17.35 and 17.36 apply rigorously only to ideal solutions, that is, solutions for which there is no change in volume and no heat effect when the two components are mixed regardless of the mixture concentration. In most cases where Equations 17.35 and 17.36 are used, the gas phase is a very dilute solution of components that are chemically inert to each other, and the assumption of ideal-solution behavior is closely followed. However, gaseous solutions do not always follow ideal-solution laws, and the reader should be especially careful when the vapor concentration becomes large.

The enthalpy of a vapor-gas mixture may also be determined more directly from the humidity chart without use of Equation 17.36. This method uses the fact that the lines of constant adiabatic-saturation temperature are also lines of nearly constant enthalpy.

This fact can be shown by rearranging Equation 17.21a.

$$c_{h1}T_1 + \lambda_{sa}Y_1 = c_{h1}T_{sa} + \lambda_{sa}Y_{sa} \qquad (17.21b)$$

and, adding $-c_{h1}T_0$ to both sides of the equation,

$$c_{h1}(T_1 - T_0) + \lambda_{sa}Y_1 = c_{h1}(T_{sa} - T_0) + \lambda_{sa}Y_{sa} \qquad (17.21c)$$

The right side of Equation 17.21c equals the enthalpy at the adiabatic-saturation point (H_{sa}) if c_{h1} equals the mean value of c_h between T_0 and T_{sa} and if $\lambda_{sa} = \lambda_0$. Similarly, the left side of Equation 17.21c equals the enthalpy at any point on the constant-adiabatic-saturation-temperature curve (H_1) if c_{h1} equals the mean c_h between T_0 and T_1 and if again $\lambda_{sa} = \lambda_0$. Thus,

$$H_1 \approx H_{sa} \qquad (17.37)$$

and the adiabatic-saturation curve would appear as a straight, horizontal line when plotted on enthalpy-temperature coordinates as shown in Figure 17.12. The fact that an adiabatic process is being considered itself indicates that $\Delta H = 0$.

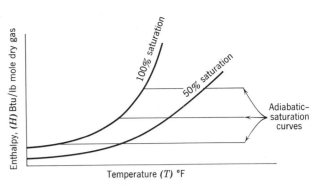

Figure 17.12. Adiabatic-saturation curve on enthalpy-temperature coordinates.

The approximation in this equation arises from the same causes and is of the same order of magnitude as the inaccuracy in assuming the adiabatic-saturation curves are straight and parallel. Using this relation, the enthalpy of any vapor-gas mixture can be determined directly from the saturated-vapor enthalpy curve plotted on the humidity diagram. Errors are insignificant at low vapor concentrations.

A humidity chart for the air–water system at $P = 1$ atm is given in Appendix D-14 and is shown in skeleton form in Figure 17.13. The saturation locus, lines of fixed per cent humidity, and lines of constant adiabatic-saturation temperature are plotted on coordinates of

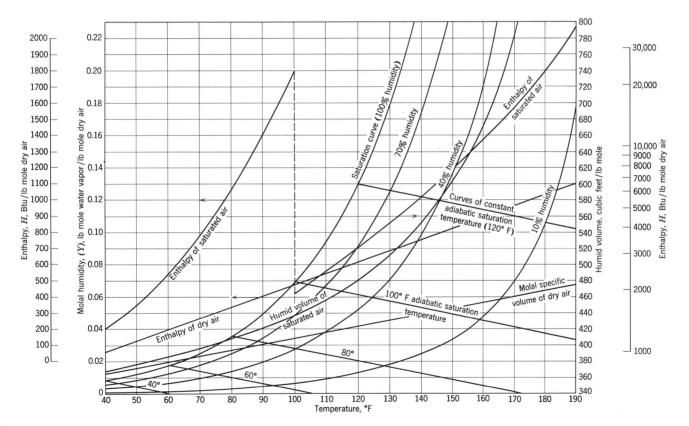

Figure 17.13. Psychrometric chart for air–water system, 1 atm total pressure.

molal humidity versus temperature in this chart. Since the chart is for the air–water system, the lines of constant adiabatic-saturation temperature are also lines of constant wet-bulb temperature. The lines are nearly straight and parallel and are spaced at 5°F temperature intervals along the saturation curve. Overlaying the basic chart are two others: one with coordinates of humid volume versus temperature on which are plotted the volumes of dry and of saturated air, and one with coordinates of enthalpy versus temperature on which are plotted the enthalpies of dry and of saturated air. The temperature scale is identical for all three plots.

As with any presentation of thermodynamic data, the details of the chart depend on the system represented, the units chosen, the bases used, and the ranges of P, T, and Y shown. The chart given in Appendix D-14 applies at a total pressure of 1 atm for the binary air–water system. Significant changes in total pressure or quantities of a third component such as CO_2 will prohibit the use of the chart. Units are given based on 1 lb mole of dry air, and enthalpies are calculated using bases of $H = 0$ for dry air at 32°F and 1 atm total pressure and for liquid water at 32°F and at its triple point pressure. Humidity charts have been published for systems other than air–water (10), with bases different from those chosen here (11), with significant quantities of a third component present (12), and with different variables used as major coordinates.

Various representations of the information in the humidity chart have been developed for special convenience in doing particular calculations which must be performed repeatedly. The broad applicability of the arrangement used here to a variety of chemical-engineering calculations has made this chart a favorite since Grosvenor proposed it (3). In many humidification and drying problems, the convenience of the unit "moles of vapor per mole of dry gas" and the ease with which material balances may be evaluated for changing quantities of vapor in a constant quantity of dry gas overshadow other considerations. The use of the humidity chart is illustrated below.

Illustration 17.1. Air enters the drying chamber of a tray dryer at 210°F after having been heated from an ambient condition of 70°F and 50 per cent relative humidity. If the air leaves the drying chamber at 80 per cent humidity as the result of an adiabatic-saturation process within the dryer, what is the temperature and humidity of this exhaust air?

SOLUTION. The humidity of ambient air must first be determined, since this humidity will not change in the heating process. From the humidity chart (Appendix D-14) $Y_s = 0.0262$ at 70°F. Then from Dalton's law,

$$0.0262 = \frac{P_a}{1 - P_a} ; \qquad P_a = 0.0256 \text{ atm}$$

and from Equation 17.4a, $p_a = 0.5 \times 0.0256 = 0.0128$ atm. Then from Equation 17.6,

$$\text{Per cent saturation} = \frac{0.0128/0.9872}{0.0256/0.9744} \times 100 = 49.3 \text{ per cent}$$

From the humidity chart, this saturation gives $Y = 0.0129$ which will be constant through the heating process. At the entrance to the drying chamber, $Y = 0.0129$ and $T = 210°F$. The adiabatic saturation temperature will be 93°F, and the constant-adiabatic-saturation-temperature curve passes through the 80 per cent saturation curve at 99°F and $Y = 0.0540$. These conditions will be those of the air leaving the drying chamber.

Illustration 17.2. Determine the enthalpy, relative to dry air at 32°F and saturated water at 32°F, of air at 190°F, 10 per cent humidity.

SOLUTION. The desired answer can be read directly from the humidity chart. At the conditions given, the adiabatic-saturation temperature is 134°F. At 134°F, the enthalpy of saturated air is read to be 4700 Btu/lb mole of dry air.

This answer can be checked by finding the enthalpy using the proper latent and specific heats. For any vapor-gas mixture.

$$H = c_b(T - T_0) + Y[c_a(T - T_0) + \lambda_0] \qquad (17.10)$$

Here, $Y = 0.177$ lb mole H_2O per pound mole dry air. Then c_b, the specific heat of air, may be taken as 6.95; c_a, the specific heat of water vapor, as 8.10; and λ_0, as 19,350. This gives

$$H = 6.95(190 - 32) + 0.177[8.10 (190 - 32) + 19,350]$$
$$= 1100 + 0.177(20,630) = 4760 \text{ Btu/lb mole dry air}$$

Note that the difference between these two answers is only a little greater than 1 per cent, even though the conditions were chosen in the high-humidity region where Equation 17.37 is no longer rigorously correct.

Still a third method of determining this enthalpy is through use of the curves for the enthalpy of dry and of saturated air as indicated by Equation 17.36. At 190°F, the enthalpy of saturated air is 37,700 Btu/lb mole dry air, and that of dry air is 1100 Btu/lb mole dry air. The enthalpy of air at 10 per cent saturation is then,

$$H = 1100 + (37,700 - 1100) \times 0.10$$
$$= 4760 \text{ Btu/lb mole dry air}$$

CALCULATIONS FOR HUMIDIFICATION AND DEHUMIDIFICATION OPERATIONS

The design of equipment to carry out humidification and dehumidification operations depends on the concepts and equations developed in Chapters 15 and 16. As discussed there, the direction of the transfer processes and their rates can be determined by writing enthalpy- and material-balance equations and rate equations and by combining these equations into the applicable design equations. In this section these concepts will be applied to the special conditions surrounding the operation of humidification and dehumidification equipment. The development is exactly parallel to that used earlier for heat-transfer and mass-transfer equipment.

In a dehumidification process a warm vapor–gas mixture is contacted with a cool liquid. Vapor condenses from the gas phase; the gas phase cools; and the liquid is warmed. Both sensible and latent heat are transferred toward the liquid phase. These conditions are shown schematically in Figure 17.14. In the normal convention, the gas phase is chosen as the system, and heat- and mass-transfer rates and quantities are positive when transfer is toward the gas phase. This sign convention is equivalent to measuring distance in the direction toward the gas phase. Then, heat- and mass-transfer quantities for this physical situation will be negative.

In a water-cooling process (Figure 17.15), warm water is contacted by a cooler gas-vapor mixture. Mass and heat are transferred toward the gas phase. As a result, the signs of the driving-force terms and transfer rates are positive based upon the convention described above.

Cases may be met where the sensible heat and latent heat transfer in opposite directions. This often occurs at the lower end of water-cooling towers where the water has cooled below the bulk-gas temperature. This overcooling is possible because the latent-heat transfer far overshadows the sensible-heat transfer. The liquid cannot cool below the wet-bulb temperature of the gas, as was shown above. This case is shown in Figure 17.16.

Since in these operations, rates and quantities of both heat and mass transfer are significant, enthalpy-balances and heat-transfer-rate equations must be written parallel to the material-balances and mass-transfer rate equations. Therefore, the analysis of a general transfer process previously done in Chapters 15 and 16 will be repeated here, but transfer of both heat and mass will now be considered simultaneously.

The nomenclature and physical arrangement, are shown in Figure 17.17. The conditions here differ from

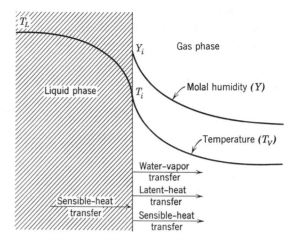

Figure 17.15. Conditions in a cooling tower (humidifier).

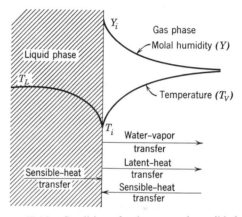

Figure 17.16. Conditions for latent- and sensible-heat transfer in opposing directions.

those of Figure 16.5 in that the liquid phase is presumed to be a single component, identical with the solute in the gas phase. As before, the subscripts 1 and 2 refer to the bottom and top of the column.

L_2 = liquid flow rate into the top of the column, lb mole/hr

V_1 = flow rate of gas phase entering the column, lb mole/hr

V' = flow rate of solvent, or "dry" gas, lb mole/hr

Y_2 = mole ratio of solute to solvent gas at the top of the column

H_{V1} = enthalpy of gas phase entering the column, Btu/lb mole of "dry" gas

H_{L2} = enthalpy of liquid phase entering the top of the column, Btu/lb mole of liquid

q = heat transferred to the column from its surroundings, Btu/hr

T_L, T_V = temperature of liquid and gas phases respectively

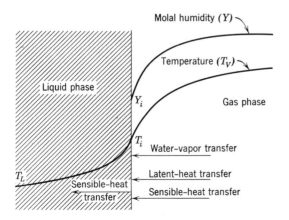

Figure 17.14. Conditions in a dehumidifier.

dz = a differential height of column packing, ft
A = interfacial surface, sq ft
a = interface area, sq ft/cu ft of column volume
S = tower cross section, sq ft

On this basis, for a tower of constant cross section, an over-all material balance gives

$$L_1 - L_2 = V_1 - V_2 \qquad (17.38)$$

A material balance for the condensable component gives

$$V'(Y_2 - Y_1) = L_2 - L_1 \qquad (17.39)$$

and an enthalpy balance gives

$$L_2 H_{L2} + V' H_{V1} + q = L_1 H_{L1} + V' H_{V2} \qquad (17.40)$$

Most commonly, the column will operate nearly adiabatically with $q \to 0$. The approach to adiabatic operation will be closer the larger the column diameter. For this situation, similar balances for the differential height (dz) will be written. The condensable component balance becomes

$$V' \, dY = dL \qquad (17.41)$$

The corresponding enthalpy balance is

$$V' \, dH_V = d(LH_L) \qquad (17.42)$$

If the rate of solute transfer between phases is small compared to the total flow stream, an average value of L may be used, and the change in enthalpy of the liquid phase may be expressed as if it resulted solely from the change in temperature at constant specific heat. Thus

$$d(LH_L) = L_{av} c_L \, dT_L \qquad (17.43)$$

where

$$L_{av} = \frac{L_1 + L_2}{2}$$

For the change in gas-phase enthalpy, the expression in terms of temperature is rigorous if c_h is constant.

$$V' \, dH_V = V' d[c_h(T_V - T_0) + Y\lambda_0]$$
$$= V' c_h \, dT_V + V' \lambda_0 \, dY \qquad (17.44)$$

Rate equations for heat and mass transfer can also be written. Here, however, complications arise because of the fact that heat is transferred from the bulk of the liquid phase to the liquid-gas interface entirely as a result of the temperature potential, but, from the interface to the bulk of the gas phase, heat is transferred by two mechanisms. On the gas-phase side of the interface, heat is transferred as the result of a temperature potential, and the latent heat associated with the mass transfer is transferred as the result of a concentration driving force. The quantities of heat transferred by these two mechanisms are separated as the two terms on the right side of Equation 17.44.

Applying these ideas, equations for the heat-transfer processes indicated in Equations 17.43 and 17.44 can be written separately. For the liquid-phase transfer,

$$\frac{L_{av}}{S} c_L \, dT_L = h_L a(T_L - T_i) \, dz \qquad (17.45)$$

where T_i = the interfacial temperature, °F

For the gas-phase sensible-heat transfer,

$$\frac{V'}{S} c_h \, dT_V = h_c a(T_i - T_V) \, dz \qquad (17.46)$$

and, for the gas-phase latent-heat transfer,

$$\frac{V'}{S} \lambda_0 \, dY = \lambda_0 k_Y a(Y_i - Y) \, dz \qquad (17.47)$$

where Y_i = the gas-phase mole ratio of solute to solvent at the interface

Development of the Design Equation. The enthalpy-balance equations and the rate equations given above can now be combined to give a design equation in the form of Equation 14.61. These equations will relate the change in gas-phase temperature and molal humidity to the rates of heat and of mass transfer to, or from, the gas phase. Thus combining Equation 17.44 with Equations 17.46 and 17.47,

$$\frac{V'}{S} dH_V = h_c a(T_i - T_V) \, dz + \lambda_0 k_Y a(Y_i - Y) \, dz$$

$$(17.48)$$

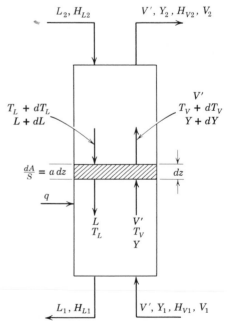

Figure 17.17. Nomenclature for the general humidification process.

for the gas phase. Separating $k_Y a$ from the right side of the equation and designating $h_c/k_Y c_h$ as r, the psychrometric ratio, give

$$\frac{V'}{S} dH_V = k_Y a[(c_h r T_i + \lambda_0 Y_i) - (c_h r T_V + \lambda_0 Y)] dz$$

(17.49)

In putting r into this equation for $h_c a/k_Y a c_h$, the assumption is made that a, the area per unit volume of tower, is the same for heat transfer as it is for mass transfer. This will be true only at such high liquid rates that the tower packing is completely wet. If r is equal to 1, as it is for the air–water system under normal conditions, the terms within parentheses in Equation 17.49 are enthalpies as defined by Equation 17.10.

$$\frac{V'}{S} dH_V = k_Y a(H_i - H_V) dz$$

(17.50)

or

$$\int_{H_{V1}}^{H_{V2}} \frac{V' dH_V}{S k_Y a(H_i - H_V)} = \int_0^z dz = z$$

(17.51)

Equation 17.51 is a design equation similar to Equations 14.62 and 14.63 except that the pertinent driving force is expressed as an enthalpy difference. Enthalpy is an extensive thermodynamic property. As such, it cannot be a driving force for any transfer operation. Thus, the mathematical treatment that leads to Equation 17.51 should be examined. First, in the energy balances, the total flow per unit time is fixed. The enthalpy terms are then "specific enthalpy," based on a fixed mass of material. Second, in the basic design equation, Equation 17.49, the driving force is a function of T and Y, the quantities that would be expected to control rates of heat and of mass transfer. Only in the fortuitous case that $r = 1$ can H be substituted for these T and Y functions. In all other cases, Equation 17.51 would be written as

$$\int \frac{V' d(c_h T_V + \lambda_0 Y)}{S k_Y a[(c_h r T_i + \lambda_0 Y_i) - (c_h r T_V + \lambda_0 Y)]} = \int dz$$

(17.51a)

Integration of the Design Equation. The integration indicated by Equation 17.51 is usually performed by using values of V' and $k_Y a$ averaged over the column height. This introduces small error in light of the low concentration of water vapor in the gas stream. Beyond this, knowledge of the relation between the enthalpy in the main gas phase and that at the gas-liquid interface is necessary. Such a relation can be obtained by now considering the transfer process on the liquid side of the interface. Combining the enthalpy balance (Equation

17.42) with the liquid-transfer rate (Equation 17.45) gives

$$\frac{V'}{S} dH_V = h_L a(T_L - T_i) dz$$

(17.52)

and combining this equation with Equation 17.50,

$$-\frac{h_L a}{k_Y a} = \frac{H_V - H_i}{T_L - T_i}$$

(17.53)

Equation 17.53 applies at any point in an air–water contacting device. From it, the temperature and the enthalpy of the interface can be determined at any point for which the liquid temperature (T_L), the gas enthalpy (H_V), and the ratio of the liquid-phase heat-transfer coefficient to the gas-phase mass-transfer coefficient based on mole-ratio driving forces are known.

The interface conditions can be obtained through Equation 17.53 using a graphical method. A plot is drawn with coordinates of liquid-phase temperature versus enthalpy of the gas phase. On it, the locus of interface H_i and T_i values can be plotted by realizing that at the interface the vapor phase will be saturated at the interface temperature if the assumption that equilibrium exists at a phase boundary is tenable. From the saturation curve on the air–water humidity chart, the saturation molal humidity can be obtained for any desired temperature. The saturation, or interface, enthalpy can then be calculated or read from the humidity chart.

On the same plot, an "operating line" of H_V versus T_L can be plotted by combining Equations 17.42 and 17.43 and integrating. This curve represents the path of bulk-phase conditions as the fluids pass through the unit. Thus,

$$\int_{H_{V1}}^{H_{V2}} V' dH_V = \int_{T_{L1}}^{T_{L2}} L_{av} c_L dT_L$$

(17.54)

where the limits again refer to the bottom and top of the column. Integrating,

$$V'(H_{V2} - H_{V1}) = L_{av} c_L(T_{L2} - T_{L1})$$

(17.55)

and rearranging,

$$\frac{H_{V2} - H_{V1}}{T_{L2} - T_{L1}} = \frac{L_{av} c_L}{V'}$$

(17.56)

This equation gives the slope of the H_V versus T_L "operating" line as $L_{av} c_L/V'$. For the air–water system and for most other dilute gas-solution systems, this ratio is constant over a moderate humidity range, and the straight H_V-versus-T_L line can be determined from knowledge of the liquid- and gas-phase flow rates and the conditions of both streams at one end of the column, or alternately the line can be located from the conditions at both ends of the column.

Figure 17.18 shows such a diagram for a humidification operation. The equilibrium curve represented

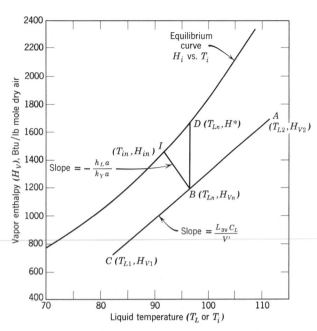

Figure 17.18. Graphical representation of adiabatic gas-liquid contacting operation.

on it was obtained from data on the air–water psychrometric chart, Appendix D-14. On this curve are located all possible conditions of T_i, H_i throughout the column. Line ABC is the operating line containing all values of H_V corresponding to liquid temperature (T_L) throughout the column. This line could have been obtained from knowledge of the two end conditions, (T_{L1}, H_{V1}) and (T_{L2}, H_{V2}), or from either one of these two plus the slope ($L_{av}c_L/V'$). On this line, point B represents an arbitrary point in the column at which the liquid temperature and gas enthalpy have the values of T_{Ln} and H_{Vn}. The interface conditions at this point can be found by using Equation 17.53. Thus, a tie line starting at point B and having a slope equal to $-h_La/k_Ya$ will intercept the equilibrium curve at the interface conditions corresponding to point B. Point I represents the interphase conditions designated by (T_{in}, H_{in}). In this way, interface conditions can be found corresponding to any point between A and C on the operating line. Of course, values of k_Ya and h_La applicable to the column conditions are necessary. Usually, these values must be obtained experimentally.

From this construction, Equation 17.51 can now be integrated graphically. The necessary driving force in terms of enthalpy is determined for representative points throughout the column. For point B on Figure 17.18, this would be ($H_{in} - H_{Vn}$). Values of $V'/Sk_Ya(H_i - H_V)$ are then plotted against H_V and the area under this curve determined between the limits of H_{V1} and H_{V2} in order to find the column height (z). Solutions of this sort in terms of temperature and concentration driving forces have been carried out in Chapters 14, 15, and 16.

Illustration 17.3. A water-cooling tower of the wood-slat, forced-draft, countercurrent-flow design cools water from 110°F to 80°F when the ambient air is at 90°F with a wet-bulb temperature of 60°F. Previous experience with towers of this design leads the engineer to predict that $h_La/k_Ya = 145$ and that $k_Ya = 0.2V'$. A liquid rate of 1500 lb/hr sq ft of tower cross section will be used.

(a) Determine the minimum gas rate. At this rate, there will be a zero driving force somewhere in the column, and an infinitely high tower will be necessary. Usually, at minimum gas rate, the exit gas will be in equilibrium with the inlet water.

(b) Determine the necessary tower height if a gas flow rate is used which is twice the minimum found in part (a).

SOLUTION. This problem can be solved using the graphical method outlined above. For this, an enthalpy-temperature plot must be made, and the locus of interface conditions must be plotted on this diagram. This plot is made from Appendix D-14 by replotting the curve of saturated-air enthalpy as a function of temperature.

The inlet and outlet known conditions are shown on Figure 17.19. The enthalpy of the moist air can be determined from the humidity diagram by taking advantage of the fact that adiabatic-saturation curves are lines of nearly constant enthalpy. Thus, the inlet-air enthalpy is the same as that of saturated air at 60°F. From Appendix D-14, $H_V = 543$ Btu/lb mole dry air. The conditions at the bottom of the tower (H_{V1}, T_{L1}) can now be plotted and are shown at point C on Figure 17.20.

The H_V, T_L operating line will be straight and have a slope of $L_{av}c_L/V'$. For the minimum gas rate called for in part (a), this line will be tangent to the equilibrium curve, at some point between inlet- and outlet-column conditions or will intersect the equilibrium curve at one of these end conditions. Here, tangency with the equilibrium curve occurs at T_{L2}, the inlet water temperature. Thus, at the top of the tower, the gas stream is saturated at the liquid temperature. This results in line CB' on Figure 17.20.

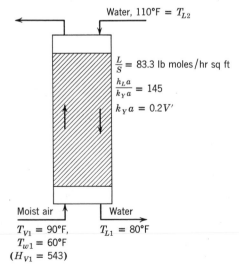

Figure 17.19. Column conditions for Illustration 17.3.

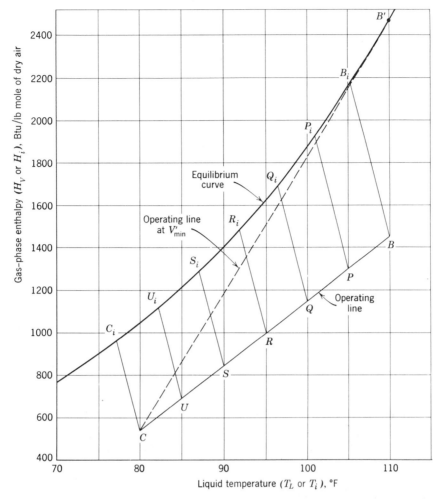

Figure 17.20. Graphical representation of driving forces, Illustration 17.3.

From the slope of this line

$$\frac{L_{av}c_L}{V'_{min}} = \frac{1850}{30} = 61.9$$

and

$$\frac{V'_{min}}{S} = \frac{83.3 \times 18}{61.9} = 24.2 \text{ lb moles/hr sq ft}$$

which is the answer of part (a).

For a gas-flow rate $V' = 2V'_{min} = 48.4$, the slope of the operating line is

$$\frac{L_{av}c_L}{V'} = \frac{61.9}{2} = 30.9$$

and the plotted line CB results. The position of point B gives $H_{V2} = 1450$, but tells nothing about T_{V2}. From H_{V2} the value of $(T_w)_{V2}$, the exit-air wet-bulb temperature, equals 92°F.

Since the operating-line position is fixed, tie lines running from any point on the operating line to the equivalent interface condition on the equilibrium curve can be plotted. From Equation 17.53, these tie lines will have the slope $-h_L a/k_Y a$, which here equals -145. Such tie lines are

shown as CC_i, BB_i, and PP_i, QQ_i, RR_i, SS_i, and UU_i in Figure 17.20.

Corresponding values of H_V and H_i can now be taken from the diagram, and the design equation integrated graphically as shown on Figure 17.21.

T_L	H_V	H_i	$(H_i - H_V)$	$\dfrac{V'}{Sk_Y a(H_i - H_V)}$
80	543	960	417	0.0120
85	690	1110	420	0.0119
90	840	1290	450	0.0111
95	990	1480	490	0.0102
100	1150	1680	530	0.00945
105	1300	1920	620	0.00807
110	1450	2185	735	0.00680

From which $z =$ area under curve 1–2 = 9.27 ft.

In Illustration 17.3 the specification of a hotter inlet water temperature would have complicated the solution of part (a), the determination of minimum vapor rate. Suppose the inlet water temperature had been 120°F,

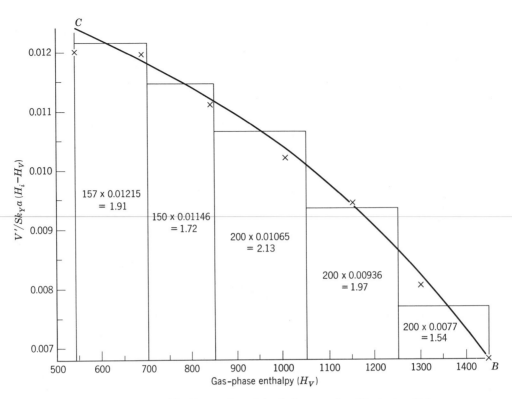

Figure 17.21. Graphical integration of the design equation, Illustration 17.3.

and consider the situation as the vapor rate is decreased from 1500 lb/ hr sq ft with the tower bottom conditions fixed as given in Illustration 17.3 (point C on Figure 17.22). Initially, the operating line would be almost identical to line CB. Of course, it would extend past B to 120°F. As point D, the top end of this new operating line, is moved toward the equilibrium curve, an increasing column height would be necessary to balance the decreasing transfer driving force. Eventually, the line CD would become tangent to the equilibrium curve at some point such as E. Point D would not yet be coincident with F, but an infinitely tall column would be required because of the zero driving force at the point of tangency. This slope fixed the minimum gas rate, for with any steeper operating line the specified end conditions could not be obtained, even with an infinitely high column. This condition of minimum vapor rate is shown schematically in Figure 17.22.

In Illustration 17.3, integration of Equation 17.51 to get the required tower height was done graphically. In contrast to other rate operations, the analytical integration of the design equation for humidification is not very difficult. The reason is that all information necessary for the solution is given in equation form except the equilibrium curve. Since this curve is always the same for the air–water system, it can be represented by one empirical equation to any desired degree of accuracy for all humidification problems. For instance, the relation

$$H_i = 2295 - 63.85T_i + 0.598T_i^2 \qquad (17.57)$$

represents the equilibrium curve between $60 < T_i < 120$ within about ±3 per cent. Using Equation 17.57 along with the tie-line equation,

$$-\frac{h_L a}{k_Y a} = \frac{H_V - H_i}{T_L - T_i} \qquad (17.53)$$

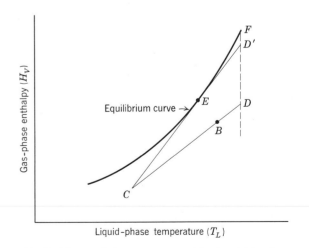

Figure 17.22. Minimum gas rate with zero driving force between the column ends.

and the equation for the operating line,

$$\frac{H_V - H_{V1}}{T_L - T_{L1}} = \frac{L_{av}c_L}{V'} \qquad (17.56)$$

the design equation,

$$\Delta z = \int_{H_{V1}}^{H_{V2}} \frac{V' \, dH_V}{Sk_Ya(H_i - H_V)} \qquad (17.51)$$

can be written in terms of a single variable and formally integrated since four equations are available in the variables H_i, H_V, T_i, and T_L. The most convenient solution is obtained by writing the design equation in terms of T_i and then integrating. In contrast, in gas absorption and heat transfer, rarely does the same equilibrium condition apply to more than one problem. Thus, the graphical solution is usually easier than the derivation of an analytical expression for the equilibrium curve.

Illustration 17.4. Solve Illustration 17.3b analytically using the results obtained in Illustration 17.3a.

SOLUTION. The conditions imposed by Illustration 17.3 result in the following numerical substitutions into Equations 17.53, 17.56, and 17.51:

$$-145 = \frac{H_V - H_i}{T_L - T_i} \qquad (a)$$

$$H_V - 543 = 30.9(T_L - 80) \qquad (b)$$

$$z = \int_{H_{V1}}^{H_{V2}} \frac{dH_V}{0.2(H_i - H_V)} \qquad (c)$$

Combining Equations a and b and solving for H_i,

$$H_i - H_V = 145(T_L - T_i) \qquad (a)$$

$$H_V - 543 = 30.9T_L - 2472$$

$$T_L = \frac{H_V + 1929}{30.9} \qquad (b)$$

$$H_i - H_V = 145\left(\frac{H_V + 1929}{30.9} - T_i\right) \quad (a) \text{ and } (b)$$

$$H_i = 9052 + 5.693H_V - 145T_i \qquad (d)$$

Now this equation can be combined with the equation of the equilibrium curve (Equation 17.57) to obtain H_V as a function of T_i.

$$2295 - 63.85T_i + 0.598T_i^2 = 9052 + 5.693H_V - 145T_i$$

Collecting terms,

$$H_V = -1187 + 14.25T_i + 0.1050 \, T_i^2 \qquad (e)$$

From Equations e and 17.57

$$H_i - H_V = 3482 - 78.10T_i + 0.493T_i^2$$

and differentiating Equation e gives

$$dH_V = (14.25 + 0.210T_i) \, dT_i \qquad (f)$$

Substituting the last two equations into the design equation results in

$$z = \frac{1}{0.2} \int_{H_{V1}=543}^{H_{V2}=1450} \frac{(14.25 + 0.210T_i) \, dT_i}{3482 - 78.10T_i + 0.493T_i^2} \qquad (g)$$

The indicated integration can be performed after the limits are expressed in terms of T_i. Solving Equation e, $T_{i1} = 77.10°F$, and $T_{i2} = 104.4°F$.

The integration of Equation g can be done in two parts by breaking the integral into two separate integrals of the forms

$$\int \frac{dx}{a + bx + cx^2} = \frac{2}{\sqrt{4ac - b^2}} \tan^{-1} \frac{2cx + b}{\sqrt{4ac - b^2}}$$

and

$$\int \frac{x \, dx}{a + bx + cx^2} = \frac{1}{2c} \ln(a + bx + cx^2) - \frac{b}{2c} \int \frac{dx}{a + bx + cx^2}$$

Inserting the proper numbers and solving gives $z = 9.285$ ft.

The answers obtained for the column height in Illustrations 17.3 and 17.4 differ by about 2 per cent. This percentage is the order of accuracy obtainable with the crude representation of the equilibrium curve used and the method of graphical construction. Uncertainties in the values of coefficients, nonuniformity of flow distribution, inaccuracy of humidity determinations, etc., do not show up here, though they may contribute errors many times greater than this.

Over-all Coefficients. As shown earlier (see Equations 14.49 and 14.50, p. 211) the individual mass-transfer coefficients may be converted to over-all coefficients if the equilibrium curve is straight. In this case, an over-all driving force must also be used. In Figure 17.18 the vertical distance BD represents such an over-all driving force. Point $D(T_L, H^*)$ represents saturated vapor at the temperature of the bulk liquid and is identical in concept to the "equilibrium equivalent" used in mass-transfer work. The appropriate rate equation is

$$\frac{V'}{S} dH_V = K_Ya(H^* - H_V) \, dz \qquad (17.58)$$

for constant V' and K_Ya

$$\int_{H_{V1}}^{H_{V2}} \frac{dH_V}{H^* - H_V} = \frac{SK_Ya}{V'} z \qquad (17.59)$$

As with the equilibrium equivalent for mass transfer, point $D(T_{Ln}, H^*)$ is not, in general, the interface condition at a point in the column where the bulk-phase conditions give point $B(T_{Ln}, H_{Vn})$; nor does it represent the liquid-phase conditions equivalent to some known gas-phase condition. It represents an *imaginary* interface condition where the interface temperature is that of the bulk-liquid phase at the same point in the column T_{Ln}. Of course, the enthalpy is that of saturated air at this temperature.

If the liquid-phase resistance to heat transfer is very small compared to the gas-phase resistance to mass transfer, the actual interface temperature will approach the bulk-liquid temperature. The slope $-h_L a/k_Y a$ approaches $-\infty$, and point I on Figure 17.18 approaches the equilibrium equivalent of B located at point D. For this special situation, $K_Y a = k_Y a$, and the use of the over-all mass-transfer coefficient is rigorously correct.

Where a significant liquid-phase resistance to heat and mass transfer exists, the interface temperature will differ significantly from the bulk-liquid-phase temperature. In other words, points I and D will be separated as shown on Figure 17.18. For this case, the use of an over-all mass-transfer coefficient with an appropriate enthalpy-difference driving force $(H^* - H_V)$ is correct only if the interface locus between I and D is straight.

Usually, the use of $K_Y a$ brings in negligible error, since $h_L a$ is usually large compared to $k_Y a$ and the curvature of the interface locus is not great. The over-all coefficient $(K_Y a)$ has the practical advantage of being easily measured.

The left-hand side of Equation 17.59 is the number of transfer units defined on p. 273 and is exactly analogous to similar parts of Equations 16.14 and 16.15. Rearranging Equation 17.59 gives

$$z = \frac{V'}{SK_Y a} \int \frac{dH_V}{H^* - H_V} = H_{OG} N_{OG} \quad (17.60)$$

which is identical to Equation 16.14 except that an enthalpy, rather than a mole-fraction, driving force is used.

Equation 17.51 is the equivalent equation in terms of gas-phase resistance and driving force. Thus,

$$\int \frac{V'\, dH_V}{Sk_Y a(H_i - H_V)} = H_G N_G = z \quad (17.51b)$$

Determination of Bulk-Gas-Phase Temperature. It was shown earlier (Illustration 17.3 and Figure 17.22) that for simultaneous-heat-and-mass-transfer operations, as well as with mass-transfer and with heat-transfer operations, tangency or intersection of operating and equilibrium lines represents a point of zero driving force and hence can be attained only with an infinitely large contactor. This zero driving force condition then forms one limit on the range of feasible operating conditions.

Another, and often more limiting, restriction on the range of practical operating conditions occurs when fog forms in the vapor phase. Fog will form when the bulk-gas phase becomes supersaturated. The carrying of water droplets by the gas phase disrupts the basic material and energy balances with respect to V' and L' and hence makes the mathematical procedures outlined here useless. Physically, the fog represents a serious inconvenience. If fog forms, separation of the droplets from the main gas stream is costly and inconvenient. Water losses are high in a water-cooling operation, and, in a dehumidification operation, the major process aim is thwarted. With systems other than the air–water system, the presence of fog may be a serious health hazard or economic loss. Prevention of fog formation is the main reason that air fed to a contact sulfuric acid plant must be dried.

On the H_V–T diagram used here, the bulk-gas-phase temperature has not been shown. If a curve of T_V versus H_V could be plotted on a diagram such as Figure 17.18, the intersection of this curve with the interface locus would represent a fogging condition. That is, the bulk gas phase would reach saturation. If no intersection occurs, the column would be operable. Any point on the T_V–H_V curve would indicate a gas-phase condition occurring in the column corresponding to a liquid-phase temperature read from the T_L–H_V curve at the same value of H_V. Without plotting such a T_V–H_V curve a design may be inadvertently developed in which the column may fail as a result of fogging conditions even though it would otherwise be operable.

To determine the gas-phase T_V–H_V plot Mickley

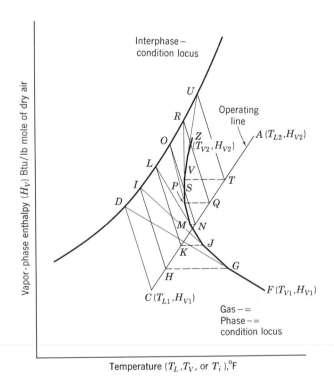

Figure 17.23. Graphical construction of gas-phase-condition locus.

(8) developed a convenient graphical method. Dividing Equation 17.46 by Equation 17.50 gives

$$\frac{V' c_h \, dT_V}{V' \, dH_V} = \frac{h_c a (T_i - T_V) \, dz}{k_Y a (H_i - H_V) \, dz} \qquad (17.61)$$

or

$$\frac{dT_V}{dH_V} = \frac{h_c a}{k_Y a c_h} \left(\frac{T_i - T_V}{H_i - H_V} \right) \qquad (17.61a)$$

but, by the Lewis relation, $h_c a / k_Y a c_h = 1$, and therefore

$$\frac{dT_V}{dH_V} = \frac{T_i - T_V}{H_i - H_V} \approx \frac{\Delta T_V}{\Delta H_V} \qquad (17.62)$$

where the Δ's refer to a small but finite difference. If the gas-phase conditions at either end of the column are known, a stepwise method can be used to draw the locus of gas-phase conditions throughout the tower. The procedure is illustrated in Figure 17.23 where the situation is the same as was shown in Figure 17.18. Warm water enters the tower at T_{L2} and leaves cooled to T_{L1}. This is done using a countercurrent air flow which enters the bottom of the column at H_{V1}, T_{V1}. The air temperature may be greater or less than T_{L1}, but here $T_{V1} > T_{L1}$. It must be greater than T_i at H_{V1}. By Equation 17.62, the slope of the T_V–H_V curve will equal the slope of the straight line from point F to the interface condition corresponding to points C and F. The interface condition can be determined from Equation 17.53, using the ratio of coefficients to get the tie line from a point on the T_L–H_V curve to the corresponding interface point. The stepwise procedure is as follows, with the letters referring to points on Figure 17.23:

1. From Equation 17.53 get point D. Draw DF. As long as the interface conditions are constant FGD will represent the path of the gas-phase conditions as indicated by Equation 17.62. Let this suffice for an arbitrary short distance FG.

2. The operating-line conditions corresponding to G will be at H. Thus T_H will be the liquid temperature at the point in the column where T_G is the gas-phase temperature. From H, determine I by Equation 17.53. Draw line IG, and arbitrarily presume this to be the gas-condition curve to point J.

3. Repeat. The construction determines the points in alphabetical order. Points C, H, K, N, Q, T, and A fall on the operating line, whereas points F, G, J, M, P, S, V, and Z are on the gas-phase-condition locus. Point Z concludes the gas-phase condition locus at the gas outlet enthalpy.

Obviously, this construction accumulates errors due to the approximation inherent in using finite differences rather than differentials in Equation 17.62. Each line segment of the curve FZ represents a line tangent to the true gas-phase condition locus.

As the line segments FG, GJ, JM, etc., are shortened,

becoming closer to true differentials, the errors become less, and the correct gas-condition locus is approached. In any case, the curvature of the gas-condition locus found by using finite differences is less than the true curvature of this line when the calculations are started from one end of the column and is also less than the true curvature when the calculations are started from the opposite column end. Thus, one way to closely approach the true gas-condition locus is to carry out the stepwise solution from both ends of the column and then draw a final locus with a shape curving between that of the two stepwise solutions.

Equations 17.61a and 17.62 state that the slope of a tie line extending from a point on a phase-condition locus (here the vapor phase) to the equivalent interface condition is equal to a constant times the slope of the phase-condition line itself. This statement is markedly different from that given by previous equations involving the slope of a tie line, as for example Equations 17.53 and 14.48. In each of the previous cases, the slope was related to the ratio of appropriate coefficients. For instance, in Equation 17.53, the enthalpy driving force for the gas phase divided by the temperature driving force in the liquid phase equaled the inverse ratio of gas-phase mass-transfer coefficient to liquid-phase heat-transfer coefficient. Equation 17.61 shows no such ratio. If it did the gas-phase operating line would have to be straight, since both $h_c a$ and $k_Y a$ are usually nearly constant throughout the column. The basic difference lies in the quantities related by these different equations. Equation 17.53 relates the *total* heat-transfer rate through the vapor phase by the mechanisms of convective heat transfer *and* mass transfer to the *total* heat-transfer rate through the liquid phase. Equation 14.48 for a mass-transfer situation equates the total mass-transfer rate through the liquid phase to that through the gas phase. In contrast, Equation 17.61 or 17.62 relates the rate of heat transfer through the vapor phase by the mechanism of convection to the total rate of heat transfer through the vapor phase by convection and by latent-heat transfer associated with the mass transfer. The vapor-phase coefficients are involved but are themselves related through the psychrometric ratio $(h_c a / k_Y a c_h = r)$. Thus the gas-phase condition locus differs fundamentally from the operating lines used in heat transfer, mass transfer, and simultaneous-heat-and-mass-transfer calculations.

Illustration 17.5. A packed tower is to be used to dehumidify air as a part of an air-conditioning system. Air enters the tower at 100°F, 90 per cent humidity, and is contacted with water entering at 60°F and leaving at 90°F. Flow is countercurrent with a liquid rate of 1000 lb/hr sq ft and a gas rate of 27,000 std cu ft/hr sq ft of column cross-section. The ratio of coefficients $(h_L a / k_Y a)$ is expected to be 200. Determine the outlet air conditions.

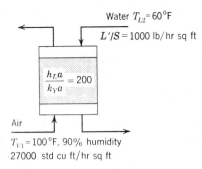

Figure 17.24. Conditions given for Illustration 17.5.

SOLUTION. The given conditions are shown on Figure 17.24. From Appendix D-14, the humidity of inlet air is $Y_1 = 0.0629$ mole/mole dry air and $H_{V1} = 1738$ Btu/lb mole dry air. Also

$$\frac{V'}{S} = \frac{27,000}{359} \times \frac{1}{1.0629} = 70.6 \text{ moles/hr sq ft}$$

and

$$\frac{Lc_L}{V'} = \frac{1000 \times 18}{70.6 \times 18} = 14.1$$

These conditions are plotted on Figure 17.25. The location of point (T_{L1}, H_{V1}) and the slope (Lc_L/V') determine the

location of the operating line, and with the inlet water temperature given they also fix $H_{V2} = 1325$. Thus the air-outlet wet-bulb temperature is 88°F. The air-outlet temperature is determined with a Mickley solution starting from (T_{V1}, H_{V1}). Note that, in this case, the gas-phase-condition locus actually crosses the equilibrium curve. Fogging would occur in this column, and an exact answer is impossible. The outlet air would probably be saturated at 88°F, giving a humidity of $Y_2 = 0.0467$ lb mole/lb mole dry air.

The method outlined above breaks down for the case of an adiabatic saturator. In this case the operating line reduces to a point located on the equilibrium curve. The gas phase T_V–H_V line becomes a straight line at constant enthalpy moving from the inlet air condition toward the equilibrium curve. Under these conditions Equation 17.46 can be integrated directly since T_i is constant and equal to T_L. The integration gives

$$\ln \frac{T_L - T_{V2}}{T_L - T_{V1}} = \frac{h_c az}{c_h V'/S} \qquad (17.46a)$$

Determination of Coefficients in Operating Equipment. The stepwise construction of Mickley may be done in reverse to determine the rate constants, ($k_Y a$, $h_c a$, and

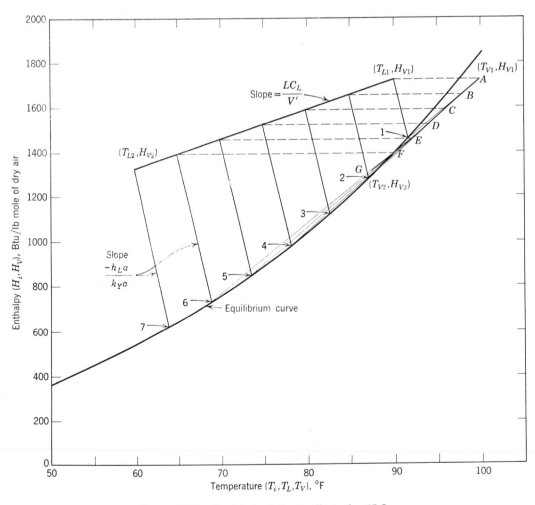

Figure 17.25. Graphical solution to Illustration 17.5.

$h_L a$) from a single set of test data. From the inlet and outlet bulk gas- and liquid-phase temperatures and gas-phase humidities, the end points of the operating line and the gas-phase condition line are fixed. The operating line can be immediately drawn. The gas-phase-condition curve can be obtained by assuming a value of $-h_L a/k_Y a$ and plotting the curve stepwise. If this curve does not meet the experimental end condition, a new value of $-h_L a/k_Y a$ must be chosen. Once a proper $-h_L a/k_Y a$ ratio has been found the "driving force" $(H_i - H_V)$ is read from the construction, and the design equation, in the integrated form

$$\int \frac{dH_V}{H_i - H_V} = \frac{S k_Y a z}{V'} \qquad (17.51)$$

is solved directly to give $k_Y a$. This value of $k_Y a$ then gives $h_L a$ through the previously determined ratio $-h_L a/k_Y a$. Finally, the Lewis relation can be used to get $h_c a$. As before, these solutions presume that the areas for mass and heat transfer are equal.

Illustration 17.6. A dehumidifying tower packed with wood slats operates with water entering at 82°F and leaving at 100°F. The air comes in at a temperature of 125°F dry bulb, 111°F wet bulb. It leaves with a 96°F dry-bulb temperature and a 95°F wet-bulb temperature. The liquid flow rate is 900 lb/hr sq ft, and the tower is 8 ft high.

(a) Determine the values of $h_L a$, $h_c a$, and $k_Y a$ obtained in the column.

(b) If the gas inlet temperature rose to 140°F with a 111°F wet-bulb temperature, what would be the air outlet conditions? Water inlet temperature and air and water flow rates would not change.

SOLUTION. (a) The column conditions are shown in Figure 17.26. From the gas wet-bulb temperatures, the enthalpies of inlet and outlet gas streams can be found. From Appendix D-14, the enthalpy of inlet air is $H_{V1} = 2500$ Btu/lb mole of dry air, and the enthalpy of outlet dry air is $H_{V2} = 1600$ Btu/lb mole of dry air.

Points A and B on Figure 17.27 can now be plotted, and the T_I–H_V operating line can be drawn between them.

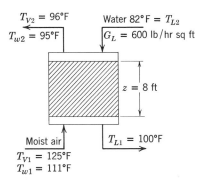

Figure 17.26. Column conditions for Illustration 17.6

The slope of this line (AB) is 900/18 = 50 Btu/lb mole. Since

$$\text{Slope} = 50 = \frac{L c_L}{V'}; \quad \frac{V'}{S} = \frac{\dfrac{900}{18} \times 18}{50} = 18 \text{ lb moles/hr sq ft}$$

or

$$G_{V'}' = 18 \times 29 = 521 \text{ lb of dry air/hr sq ft of column}$$

Now points C and D, representing inlet and outlet air temperature and enthalpy, can be plotted, and the bulk-gas-condition locus connecting points C and D can be found. This requires a trial-and-error construction in which various slopes of the tie lines are tried. A convenient starting point might be to try $h_L a/k_Y a = \infty$ giving vertical tie lines such as AF'. Using these vertical tie lines with the Mickley stepwise construction, the curve CD' is obtained. Since D' is not coincident with D, the ratio $h_L a/k_Y a$ is not ∞. Trying $h_L a/k_Y a = 1.0$, the curve CD'' is obtained from tie lines parallel to AF''. The correct solution is obtained with tie lines parallel to AF. These tie lines give curve CD with the construction shown. The slope of AF is -275. Thus $h_L a/k_Y a = +275$.

Interphase conditions corresponding to the bulk conditions throughout the tower are now known. That is, where the liquid temperature plots at M the interphase temperature plots at G. This permits the solution of Equation 17.51.

T_L	H_V	H_i	$\dfrac{1}{H_V - H_i}$
100	2500	1940	0.00178
97	2350	1770	0.00173
94	2200	1630	0.00176
91	2050	1500	0.00182
88	1900	1370	0.00189
85	1750	1250	0.00200
82	1600	1150	0.00223

The integral of Equation 17.51 [$\int dH_V/(H_V - H_i)$] is the area under the curve of $1/(H_V - H_i)$ versus H_V, as shown on Figure 17.28. This is found, and

$$N_G = \int_{1600}^{2500} \frac{dH_V}{H_V - H_i} = 1.821 = \frac{S k_Y a \Delta z}{V'}$$

from which

$$k_Y a = \frac{1.821 \times 18}{8} = 4.10 \text{ lb moles/hr (mole ratio) cu ft of column volume}$$

From this, $h_L a$ and $h_c a$ are

$$h_L a = 275 \times 4.10$$
$$= 1170 \text{ Btu/hr (cu ft of column volume) °F}$$

$$h_c a = c_h \times k_Y a$$
$$= 7.60 \times 4.10$$
$$= 31.2 \text{ Btu/hr (cu ft of column volume) °F}$$

where $h_c a$ has been found using the Lewis relation.

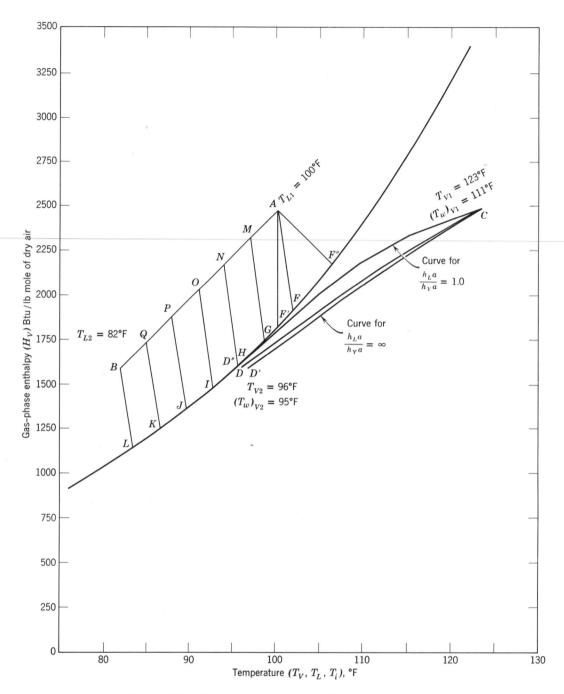

Figure 17.27. Mickley method solution of Illustration 17.6(a).

(b). Part (b) requires a second, almost independent solution. The known conditions are shown on Figure 17.29. The data available do not fix the operating line since liquid temperature and gas enthalpy are not both available at either end of the column. However, the number of transfer units is fixed as well as the slope of the operating line and the slope of the tie lines. Thus, the operating line can be located by a trial-and-error procedure. Laying down parallel lines of slope $Lc_L/V' = 50$ between $T_{L2} = 82°F$ and $H_{V1} = 2500$ Btu/lb mole, the N_G is found as was done in part (a). The correct operating line will be found when $N_G = 1.821$. This line will then determine T_{L1} and H_{V2}. A Mickley stepwise solution will then fix T_{V2}.

Figure 17.30 shows the temperature-enthalpy plot. The equilibrium line is plotted on it, as are limiting lines at $H_{V1} = 2500$ and $T_{L2} = 82°F$. The operating line shown is a first trial. This line runs from T_{L2} to H_{V1} at a slope of $Lc_L/V' = 50$. Tie lines of slope $-h_La/k_Ya = -275$ are drawn in. Resulting determination of N_G requires graphical integration.

T_L	H_V	H_i	$H_V - H_i$	$\dfrac{1}{H_V - H_i}$
82	1550	1130	420	0.00238
86	1750	1290	460	0.00217
89	1900	1420	480	0.00208
92	2030	1560	490	0.00204
95	2200	1700	500	0.00200
98	2350	1850	500	0.00200
101	2500	2000	500	0.00200

The actual integration is carried out on Figure 17.31. The resulting area is $N_G = 1.980$, which is too large. Subsequent locations of the operating line will be parallel to that shown but further from the equilibrium curve.

The final operating line is also shown on Figure 17.30. From it $H_{V2} = 1600$, $T_{L1} = 100°F$. The Mickley stepwise construction results in an air outlet temperature of 98°F. From H_{V2}, the outlet wet-bulb temperature is found to be 95°F.

Thus, the Mickley construction forms a tool by which rate coefficients may be easily determined for full-size, operating cooling towers and dehumidifiers. Unfortunately, almost no data of this sort have been reported in the technical literature.

If coefficients obtained by measurements on operating-plant equipment are unavailable, the designer must use coefficients obtained on experimental or pilot-plant equipment or coefficients obtained on the basis of analogies between heat, mass, and momentum transfer. Despite the simplicity of humidification systems where all the mass-transfer gradient occurs in the gas phase, data are few. The lack of data is mostly the result of some rather formidable experimental difficulties. In long columns with large contacting areas, the outlet vapor will be essentially saturated. Therefore, the driving forces are small, and small errors in concentration are magnified several fold. The remedy is to use short columns. Here though, the end effects get large and can only be corrected by taking several sets of data with varying column height and extrapolating the result

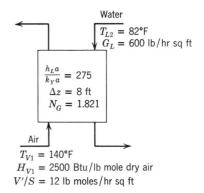

Water
$T_{L2} = 82°F$
$G_L = 600$ lb/hr sq ft

$\dfrac{h_L a}{k_Y a} = 275$
$\Delta z = 8$ ft
$N_G = 1.821$

Air
$T_{V1} = 140°F$
$H_{V1} = 2500$ Btu/lb mole dry air
$V'/S = 12$ lb moles/hr sq ft

Figure 17.29. Conditions of Illustration 17.6(b).

back to zero. The method is rife with areas for error.

If the end-effect correction is not made, commercial towers based on the experimental results will be oversized if they are shorter than the experimental column but undersized if they are longer. This is probably one of the reasons for the generally held belief that increasing the height of a packed tower decreases its efficiency. Another cause of this effect is the fact that bed porosity at the walls is always higher than that at the center of the tower. Therefore, the liquid tends to flow down the column wall rather than to be distributed evenly throughout the packing. This channelling effect is greater in tall columns, and elaborate redistribution systems have sometimes been used.

The mass-transfer coefficients for humidification are not fundamentally different from those obtained on absorption equipment. Only gas-phase mass-transfer coefficients are pertinent. As with other gas-liquid mass-transfer coefficients, they will depend upon the gas flow rate, viscosity, and diffusivity and on the geometry of the packing. They will also be dependent on liquid flow rate, primarily because at low liquid rates the packing is not completely wet. Thus, gas-absorption mass-transfer rates, such as those calculated from Equation 16.59, should be usable for humidification or dehumidification operations if the geometries are identical.

Part of the mass-transfer data available from experimental work is summarized in Table 17.2 (16). Some understandable inconsistencies among data taken from a variety of sources are present. Also, since the relations shown are empirical results of experimental data, they should *not* be extrapolated beyond the range of the data without full consciousness of the uncertainty that results.

A similar situation exists in relation to appropriate heat-transfer coefficients. For the air–water system at high liquid rates, $h_c a$ may be calculated through the Lewis relation. At low liquid rates, incomplete wetting of the packings makes the area for heat transfer (a_h)

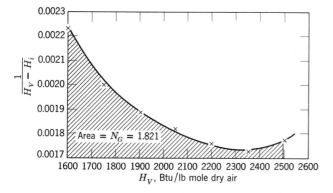

Figure 17.28. Determination of transfer units in dehumidification, Illustration 17.6(a).

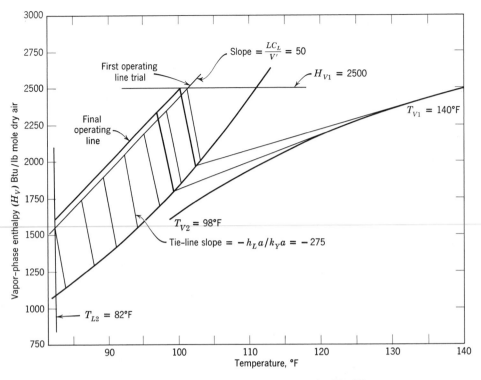

Figure 17.30. Graphical solution to Illustration 17.6(*b*).

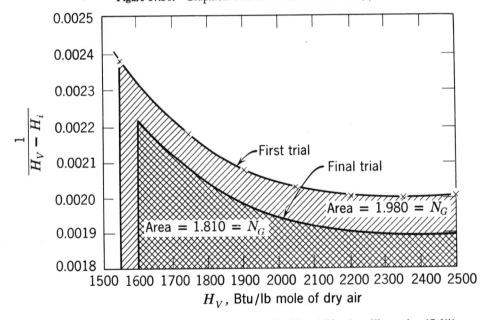

Figure 17.31. Determination of operating-line location in dehumidification, Illustration 17.6(*b*).

considerably larger than that for mass transfer (a_M). The result is that measured values of $h_c a_h/k_Y a_M$ may be two or three times as large as c_h depending upon the liquid flow rate (4). The liquid-phase heat-transfer coefficient may not be so readily calculated. Here, experimental data on the packing to be employed must be used, or analogies between heat and mass transfer must be depended upon.

The methods discussed here apply to the air–water

system or to any system where $h_c a/k_Y a c_h = 1$. If this restriction cannot be met but $h_c a/k_Y a c_h = r$, the development must be based on Equation 17.51a. Instead of enthalpy, the graph would have ($c_h r T + \lambda_0 Y$) plotted against T and used as the "driving force" in the rate-equation solution. Simplified procedures for this general case are available (6). Such procedures retain the graphical method used here but adjust the interface-locus curve to account for the fact that r is not unity.

Table 17.2. EXPERIMENTAL MASS-TRANSFER RATES FOR CONTACT OF A PURE LIQUID WITH A GAS

Item	Equipment	Process and Notes	Range of Flow Rates	Equation	Reference
1	8-in.-diam. tower, 1-in. Raschig rings.	Liquid cooling with air and water, methanol, benzene, ethyl butyrate $k_g a$ corrected for end effects and independent of G_L for $G_L = 1600$–5000.	$G_L = 1600$ $G_{V'} = 150$–500	$k_g a = 0.486 D^{0.15} G_{V'}^{0.72}$	15
2	10-in.-diam. tower; 12.5 in. depth of 15-, 25-, and 35-mm Raschig rings.	Humidification cooling, liquid cooling, and dehumidification with air–water.	$G_L = 200$–4160 $G_{V'} = 137$–586	$k_Y a = 0.0155 G_{V'} G_L^{0.2}$	19
3	21.5-in.-square tower, $1\frac{1}{2}$-in. Berl saddles.	Humidification cooling with air–water. Corrected for end effects.	$G_L = 120$–6800 $G_{V'} = 100$–700	$k_Y a = 0.0431 G_{V'}^{0.39} G_L^{0.48}$	4
4	4-in.-diam. tower, $\frac{1}{2}$-in. spheres.	Humidification cooling with air–water. Corrected for end effects.	$D_p G_{V'}/\mu = 304$–927 $G_L = 85$–1895	$H_G = \dfrac{13.4}{G_L^{0.5}} \left(\dfrac{D_p G_{V'}}{\mu}\right)^{0.1}$	17
5	6-ft-square tower, 11-ft-3-in. packed height. Wood slats, $\frac{3}{8}$ by 2 in., spaced parallel, 15 in. between tiers.	Liquid cooling with air–water.	$G_L = 350$–3000 $G_{V'} = 664$–1680	$K_Y a = 0.0068 G_L^{0.4} G_{V'}^{0.5}$	7
6	$41\frac{5}{8}$- by $23\frac{7}{8}$-in. tower, packed height $41\frac{3}{8}$ in, Wood slats, $\frac{1}{4}$ by 2 by 23.5 in., bottom edge serrated, on $\frac{5}{8}$ in. horiz. centers, $3\frac{5}{8}$ to $2\frac{5}{8}$ in. vert. centers. 18 spray nozzles.	Liquid cooling with air–water.	$G_L = 880$–1500 $G_{V'} = 700$–1500	$K_Y a = 0.00001\, G_L G_{V'}$ $- 0.00396 G_{V'}$ $- 0.00458 G_L + 10.7$	14
7	$6\frac{3}{8}$-in. square tower by 6 ft. Carbon slats, $6 \times 1 \times \frac{1}{8}$ in., bottom edge serrated, on $\frac{3}{4}$-in. horiz. centers, $1\frac{1}{4}$ in. vert. centers. Alternate tiers at right angles, $a = 23.6$ sq ft/cu ft.	Liquid cooling with air–water.	$G_{V'} = 1000$–3000 $G_L = 930$–2100	$K_G a = \left(\dfrac{0.0222}{1000}\, G_L + 0.0526\right) G_{V'}^{0.8}$	9
			$G_L = 2100$–2810	$K_G a = 0.0992 G_{V'}^{0.8}$	
8	Spray tower, 31.5 in. diam. by 52 in. high. 6 solid cone spray nozzles.	Liquid cooling and dehumidification with air–water.	$G_L = 300$–800 $G_{V'} = 200$–750	$N_{OG} = \dfrac{0.0526 G_L}{G_{V'}^{0.58}}$	13
9	Perforated plate (sieve tray). 83 $\frac{1}{8}$-in. holes on $\frac{3}{8}$-in. triangular centers.	Humidification cooling with air–water.	$G_{V'} = 670$–1920	Liquid depth, in. N_G $\quad$ 0.5 $\qquad$ ca. 1.5 $\quad$ 1 $\qquad$ 2 $\quad$ 2 $\qquad$ 2.5	18

From Treybal, R. E., *Mass Transfer Operations*, McGraw-Hill Book Co., New York (1955) by permission.

REFERENCES

1. Anonymous, The Fluor Corp., Ltd., *Bulletin T337* (1939). See also Perry, J. H., *Chemical Engineers' Handbook*, 3rd ed., McGraw-Hill Book Co., New York, 1950, p. 791.
2. Bedingfield, C. H., and T. B. Drew, *Ind. Eng. Chem.*, **42**, 1164 (1950) as reported by Treybal, R. E., *Mass Transfer Operations*, McGraw-Hill Book Co., New York, 1950, p. 169.
3. Grosvenor, W. M., *Trans. Am. Inst. Chem. Engrs.*, **1**, 184 (1908).
4. Hensel, S. L., and R. E. Treybal, *Chem. Eng. Progr.*, **48**, 362 (1952).
5. Kelley, R. C., paper presented before the California Natural Gasoline Association, December 3, 1942.
6. Lewis, J. G., and R. R. White, *Ind. Eng. Chem.*, **45**, 486 (1953).
7. Lichtenstein, J., *Trans. ASME*, **65**, 779 (1943).
8. Mickley, H. S., *Chem. Eng. Progr.*, **45**, 739 (1949).
9. Norman, W. S., *Trans. Inst. Chem. Engrs.*, London, **29**, 226 (1951).
10. Perry, J. H., *Chemical Engineers' Handbook*, 3rd ed., McGraw-Hill Book Co., New York, 1950, pp. 813–816.
11. Perry, J. H., *Chemical Engineers' Handbook*, 3rd ed., McGraw-Hill Book Co., New York, 1950, pp. 759, 766.
12. Perry, J. H., *Chemical Engineers' Handbook*, 3rd ed., McGraw-Hill Book Co., New York, 1950, p. 767.
13. Pigford, R. L., and C. Pyle, *Ind. Eng. Chem.*, **43**, 1649 (1951).
14. Simpson, W. M., and T. K. Sherwood, *Refrig. Eng.*, **52**, 535 (1946).
15. Surosky, A. E., and B. F. Dodge, *Ind. Eng. Chem.*, **42**, 1112 (1950).
16. Treybal, R. E., *Mass Transfer Operations*, McGraw-Hill Book Co., New York, 1955, p. 190.
17. Weisman, J., and E. F. Bonilla, *Ind. Eng. Chem.*, **42**, 1099 (1950).
18. West, F. B., W. D. Gilbert, T. Shimizu, *Ind. Eng. Chem.*, **44**, 2470 (1952).
19. Yoshida, F., and T. Tanaka, *Ind. Eng. Chem.*, **43**, 1467 (1951).

PROBLEMS

17.1. An air-conditioning system is to be built to change completely the air in a laboratory every 10 min with no recirculation. The laboratory building has two floors, each 50 by 200 by 10 ft and is to be supplied with air at 75°F, 40 per cent saturation. The air conditioner takes outside air, refrigerates it and separates out the condensed water, and then reheats it in an exchanger against condensing 50 psig steam.

(a) When the outside air is at 100°F, 90 per cent relative humidity, determine the volume of air brought in to the air conditioner and the temperature to which it must be cooled.

(b) For the same conditions determine the tons of refrigeration required (1 ton of refrigeration = 12,000 Btu/hr removed) and the pounds per hour of saturated steam used.

17.2. A tray drier takes in ambient air at 90°F with a 75°F wet-bulb temperature, heats it to 220°F along with some recycled air in a steam-heated exchanger, and passes it over the wet stock. The stock dries from 50 per cent moisture by weight to 10 per cent moisture and may be assumed to enter and leave the drier at the wet-bulb temperature of the drying air. After leaving the drying chamber, part of the air is recycled back to the steam-heater inlet; the rest discharged. If 20 lb of dry air are used per pound of dry stock handled and the exit air is at 135°F, determine the fraction of moist air that is recycled and the per cent humidity of the outlet air.

17.3. A mixture of air and acetone vapor at 1000 mm Hg total pressure and 100°F is 60 per cent saturated. Calculate (a) the molal humidity, (b) the humidity, (c) the partial pressure of acetone, (d) the relative humidity, (e) the volume percent acetone, (f) the humid heat, (g) the humid volume, and (h) the dew point. Assume the mixture is an ideal solution obeying the perfect-gas law.

17.4. For the mixture of Problem 17.3 calculate the wet-bulb temperature.

17.5. Prepare a psychrometric chart for the air–ethylene dichloride system at 1000 mm Hg total pressure. Cover temperatures ranging between 0 and 200°F; mass ratios between 0 and 1 lb vapor per pound dry air. Use air and liquid $C_2H_4Cl_2$ at 0°F as reference states. The following curves should be plotted against temperature:

(a) Absolute humidity at 100, 75, 50, and 10 per cent humidity.

(b) Absolute humidity at constant adiabatic-saturation temperatures of 70, 100, and 130°F.

(c) Absolute humidity at a constant wet-bulb temperature of 100°F.

(d) Dry and saturated humid volumes.

(e) Dry and saturated humid heat.

(f) Enthalpy of dry and saturated mixtures expressed as Btu/lb dry gas.

17.6. Prepare a psychrometric chart for the air–acetone system at 1 atm total pressure. Show lines of constant relative humidity, adiabatic-saturation temperature, and wet-bulb temperature. Use reasonable ranges of temperature and humidity.

17.7. A pan of water sits on the ground exposed to the sun.

(a) In the early morning the air is at 70°F with a 65°F wet-bulb temperature. The water is at 50°F.

(b) At 10 A.M. the air is at 80°F with a 70°F wet-bulb temperature. The water is at 70°F.

(c) At noon the air is at 90°F with a 75°F wet-bulb temperature. The water is at 120°F.

For each case draw a diagram showing temperature and humidity on both sides of the air-water interface (like Figure 17.14.). Show the direction of water-vapor transfer and of sensible- and latent-heat transfer.

17.8. (a) In Table 17.2, item 8, data are given on the performance of a spray humidification tower. Assuming that mass- and heat-transfer rates are uniform throughout the length of the tower, convert the equation given to $K_Y a = f(G_L, G_{V'})$. (Probably this assumption of uniform rate is a poor one. Indications are that mass transfer is most rapid in the drop-formation area.)

(b) A spray tower is to be used in a water-cooling operation. Air at 140°F with 80°F wet-bulb temperature will be contacted countercurrently with a water spray initially at 140°F to cool the water to 100°F. G_V will be 400 lb/hr sq ft and G_L will be twice this. If the $K_Y a$ relation obtained above is valid, what column height will be required?

17.9. If the operation of Problem 17.8b were to be carried out in the exact column used by Pigford and Pyle (item 8, Table 17.2), what would be the exit-water temperature?

17.10. A dehumidifier consisting of a tower packed with 1-in. Raschig rings contacts air entering at 120°F, 80 per cent saturated, countercurrently with water entering at 70°F. The air is to leave at a wet-bulb temperature of 80°F. The 6-ft-diameter tower handles 1000 std cu ft/min of air and is operated at 1.5 times minimum water flow. In a similar tower, h_L/k_Y has been found to be 60. Determine, for a height of packing of 1.5 ft:

(a) Outlet water temperature.

(b) Tower height.

(c) Water flow rate.

(d) Outlet air temperature and humidity.

(e) Whether fogging will occur in the tower. If fogging occurs how should the conditions be changed to eliminate it?

17.11. On a particularly hot day, a forced-draft cooling tower cooled water from 113°F to 90°F by countercurrent contact with air at 120°F and 70°F wet-bulb temperature. Outlet air was found to be 101°F with a 96°F wet-bulb temperature. Water flow rate was 1000 lb/hr sq ft of tower cross-section.

Under more normal operating conditions, inlet air would be at 100°F with a 50 per cent relative humidity. To what temperature could 113°F water flowing at 1000 lb/hr sq ft be cooled under these conditions?

17.12. An engineer is called upon to design a coke-packed dehumidifier which will handle 2000 cu ft of 70 per cent saturated air per minute. The air stream will enter the tower at a temperature of 130°F and should be cooled until the humidity of the exit-air stream is 0.046 lb water per pound dry air. The operation is to be conducted at atmospheric pressure. Cooling water is available at 73°F and can rise to 90°F. It is agreed to use a gas velocity of 1200 lb of dry air per hour per square foot of tower cross section and a water feed rate of 249 lb/min. Under these flow conditions, $h_c a = 300$ and $h_L a = 1500$.

(*a*) Determine the height and diameter of the tower required to do the dehumidification.

(*b*) After the tower of part (*a*) was constructed, it is discovered that the engineer used the incorrect entering air humidity. Instead of the 70 per cent saturation as specified, he read the humidity chart at a 60 per cent saturation.

1. How will this error affect the tower design?

2. Can the constructed tower be made to operate to do the specified dehumidification? How?

17.13. A guided-missile research center located on a desert operates a countercurrent cooling tower to permit reuse of their cooling and process water. An engineer at this center takes the following data on the cooling tower in order to learn more about its operating potential:

Ambient air conditions:
  Temperature = 120°F
  Wet-bulb temperature = 70°F
Air conditions at discharge from the cooling tower:
  Temperature = 101°F
  Wet-bulb temperature = 96°F
Water conditions:
  Inlet temperature = 113°F
  Outlet temperature = 90°F
  Flow rate = 1000 lb/hr sq ft
Tower height = 20 ft

What are the values of $h_L a$, $k_Y a$, and $h_c a$ for this tower?

17.14. A cooling tower with wood-slat packing cools 1000 gal/min of water from 105°F to 90°F using a countercurrent forced draft of air entering at 110°F and 20 per cent saturation. Measurements indicate that the air leaves at 96°F with a 94°F wet-bulb temperature.

The plant manager wishes to cool the water to as cold a temperature as possible. One possible means would be to increase the air flow rate, and toward this end it is found that fan speed can be increased without overloading the fan motors so that air flow is 1.5 times that previously used. Tower flooding will not occur at this higher gas rate.

What will be the outlet water temperature attained with the higher air rate, and what will be the air outlet condition?

chapter 18

Simultaneous Heat and Mass Transfer II: Drying

As used in this chapter, the term drying applies to the transfer of liquid from a wet solid into an unsaturated gas phase. The removal of moisture from gases, also frequently referred to as drying, has been discussed under the names of dehumidification and adsorption.

The factors discussed in relation to humidification and dehumidification and the mass- and heat-transfer rate equations written in Chapter 17 apply also to drying operations. The drying process, then, is identical with a humidification process except for the influence exerted by the solid itself on the drying process. This influence is considerable. The study of drying and the calculation of required dryer size must take into account a host of problems in the areas of fluid mechanics, surface chemistry, and solid structure as well as the transfer-rate problems dealt with in the discussion of humidification. In many cases, these physicochemical phenomena are so complicated and so incompletely understood that the quantitative design of the dryer is impossible. For example, in the drying of timber some of the liquid is held within the wood fibers. This moisture can migrate into the drying air only by diffusion through the fiber walls. Since this moisture diffusion through the wood is slow, the wood may dry nearly completely at the surface before the liquid escapes. The uneven drying can cause cracking and warping of the wood. The whole problem has been attacked empirically so that conditions can be set to dry timber successfully, but the basic mechanism of liquid movement remains doubtful. Again, in spray drying of detergents and of many other materials, the required drying time depends to a large extent upon the size of droplets sprayed into the hot chamber. Any atomizing device produces a range of droplet sizes which, of course, dry to different extents under the same conditions. In the detergent drying, exposure of the droplet to the drying medium produces a tough, partly dry, shell around the droplet which inhibits moisture escape. Heat flows readily through the shell, however, and evaporates the liquid inside the droplet. The vapor produced swells the droplet, often rupturing it and sometimes even blowing new "balloons" on the sides of the original droplet. Thus, the dried particles are mixtures of hollow spheres, sphere fragments, and clusters of hollow spheres. The size distribution cannot be readily predicted from the size distribution of the spray itself. The rate of fall of the particles through the hot air and therefore the drying time are very difficult to predict, and the initial droplet-size distribution is itself only occasionally calculable. Again, in drum drying of such products as milk and baby cereal, the slurry coats a hot drum and adheres to it during drying. The thickness of the drying layer is a function of the surface tension and cohesive characteristics of the slurry, but exactly what this relationship is, is still unknown. Of course, the thickness controls the extent of drying produced and hence the final moisture content. These factors take on added importance since, in general, the dried solid is the valuable product. Its shape, color, stability, stickiness, and hence its salability all depend on the drying process to which it has been subjected.

GENERAL DRYING BEHAVIOR

In drying a wet solid with a gas of fixed temperature and humidity, one general pattern of behavior always appears. Immediately after contact between the sample and the drying medium, the solid temperature adjusts until it reaches a steady state. The solid temperature and the rate of drying may increase or decrease to reach the steady-state condition. At steady state, a temperature probe would find the temperature of the wet-solid surface to be the wet-bulb temperature of the drying medium. Temperatures within the drying solid would also tend to equal the wet-bulb temperature of the gas, but here agreement would be imperfect because of lag in movement of mass and heat. Once these stock temperatures reach the wet-bulb temperature of the gas, they are found to be quite stable, and the drying rate also remains constant. This is the so-called *constant-rate drying period*. The period ends when the solid reaches the *critical moisture content*. Beyond this point, the surface temperature rises, and the drying rate falls off rapidly. The *falling-rate period* may take a far longer time than the constant-rate period even though the moisture removal may be much less. The drying rate approaches zero at some *equilibrium moisture content* which is the lowest moisture content obtainable with this solid under the drying conditions used. Figures 18.1 and 18.2 show typical drying curves, one on a moisture-content-versus-time basis and the other on a rate-of-drying-versus-moisture-content basis. The moisture-content-versus-time plot (Figure 18.1) is the form in which drying test data might be obtained. Figure 18.2, the rate-of-drying-versus-moisture-content plot, is much more descriptive of the drying process. However, it is obtained by differentiating data in the form of Figure 18.1 and thus is subject to considerable scattering of data and resulting uncertainty.

These typical drying curves are related to the mechanism by which drying occurs. The drying period represented by segment *AB* of the curves of Figure 18.1 and Figure 18.2 is the unsteady-state period during which the solid temperature reaches its steady-state value. Although the shape shown is typical, almost any shape

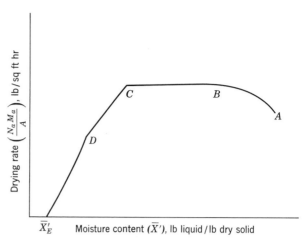

Figure 18.2. Typical drying rate curve for constant drying conditions, drying rate as a function of moisture content.

is possible, and *AB* may occur at decreasing rate as well as the increasing rate shown. During the constant-rate period (segment *BC* of the drying curves of Figures 18.1 and 18.2), the entire exposed surface is saturated with water. Drying proceeds as from a pool of liquid with the solid not directly influencing the drying rate. It is possible that the roughness of the solid surface over which the liquid film extends may increase mass- and heat-transfer coefficients, but this effect has not been firmly established. The surface temperature reaches the wet-bulb temperature as would be expected. The constant-rate drying regime continues with the mass that is transferred from the surface continuously replaced by movement of liquid from the interior of the stock. The mechanism of liquid movement and consequently the rate of this movement vary markedly with the structure of the solid itself. With solids having relatively large open void spaces, the movement is likely to be controlled by surface tension and gravity forces within the solid. With solids of fibrous or amorphous structures, liquid movement is by diffusion through the solid. Since the diffusion rates are much slower than the flow by gravity and capillarity, solids in which diffusion controls the liquid movement are likely to have short constant-rate periods, or even to dry without a measurable constant-rate period. At point *C*, the moisture content of the solid is barely adequate to supply the entire surface.

During the drying period between points *C* and *D* of Figure 18.2, called the "first falling-rate period," the surface becomes more and more depleted in liquid because the rate of liquid movement to the surface is slower than the rate of mass transfer from the surface, until at point *D* there is no significant area of liquid-saturated surface. The part of the surface that is saturated dries by convective transfer of heat from and of mass to the drying gas stream. Vapor from lower levels in the sample diffuses to the part of the surface

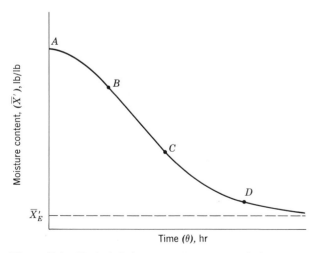

Figure 18.1. Typical drying curve for constant drying conditions, moisture content as a function of time.

that is not saturated and then continues its diffusion into the gas stream. This mechanism is very slow compared to the convective transfer from the saturated surface.

At moisture contents lower than that at point D of Figure 18.2, all evaporation occurs from the interior of the solid. As the moisture content continues to fall, the path for diffusion of heat and mass grows longer, and eventually the concentration potential decreases until at $\bar{X}_E{}'$, the equilibrium moisture content, there is no further drying. The equilibrium moisture content is reached when the vapor pressure over the solid is equal to the partial pressure of vapor in the *incoming* drying gas. This period is called the "second falling-rate period."

Classes of Materials according to Drying Behavior. Materials may be divided into two major classes on the basis of their drying behavior. Granular or crystalline solids which hold moisture in the interstices between particles or in shallow, open surface pores constitute the first of these classes. In these materials, moisture movement is relatively unhindered and occurs as a result of the interplay of gravitational and surface tension or capillary forces. The constant-rate period continues to relatively low moisture contents. Though the falling-rate period divides into the two regions mentioned above, it usually approximates a straight line on a rate-versus-moisture-content basis. The solid itself, which is usually an inorganic, is relatively unaffected by the presence of the liquid and therefore is unaffected by the drying process. As a result, drying conditions can be chosen on a basis of convenience and economic advantage with little concern over the effect of the conditions on the properties of the dried products. In the case of hydrates, the drying conditions will affect the product by changing the hydrate obtained, but otherwise the materials are not affected by the drying conditions over wide temperature and humidity ranges. Examples of this class of materials would be crushed rock, titanium dioxide, chrome yellow, catalysts, zinc sulfate monohydrate, and sodium phosphates. For these substances equilibrium moisture contents are usually very close to zero.

Most organic solids are either amorphous, fibrous, or gel-like and constitute the second of the major classes. These materials hold moisture as an integral part of the solid structure or trapped within fibers or fine interior pores. In these substances, moisture movement is slow and probably occurs by the diffusion of the liquid through the solid structure. As a result, the drying curves of the substances show only very short constant-rate periods, ending at high values of critical moisture content. For the same reasons, the first falling-rate period is much reduced, and most of the drying process is liquid diffusion controlled; that is, the drying rate is controlled by the rate of diffusion of liquid through the solid. The bulk of the drying occurs in the second falling-rate period. The equilibrium moisture contents are generally high, indicating that a significant quantity of water is held so intimately in the solid structure, or in such fine pores, that its vapor pressure is significantly reduced. Since the water present is such an intimate part of the solid structure, such solids are affected by moisture removal. The surface layers tend to dry more rapidly than the interior. If the drying rate is high, it may cause such differences in moisture content through the sample that warping or cracking occurs. In other cases, it may cause formation of a relatively impervious partly dry shell which further inhibits interior drying and may accentuate the unevenness of moisture content through the sample and consequent tendency for solid deterioration. Because of these reactions, the conditions under which drying occurs are critical. The conditions must be chosen with primary consideration for the effect of the conditions on product quality, and process economy or operating convenience must be subordinated. Examples of such materials include eggs, detergents, glues, soluble coffee extract, cereals, starch, animal blood, and soy-bean extract.

Moisture Movement—Diffusion Mechanism. In relatively homogeneous solids such as fibrous organics, gel-like substances, or porous cakes, moisture probably moves toward the surface mainly by molecular diffusion. The rate of moisture movement is then expressed by Fick's law (Equation 11.11) altered to apply to this particular case.

$$\frac{d\bar{X}'}{d\theta} = \mathscr{D}_L{}^* \frac{\partial^2 \bar{X}'}{\partial x^2} \tag{18.1}$$

where $\mathscr{D}_L{}^* =$ liquid-phase diffusion coefficient applicable for movement through the solid phase, sq ft/hr

The integration of this equation requires that boundary conditions be chosen and that the characteristics of $\mathscr{D}_L{}^*$ be specified. For the simplest case, $\mathscr{D}_L{}^*$ would be considered constant, and drying would occur from one face of a slab, the sides and bottom of which are insulated. Applying these restrictions and assuming the initial moisture content to be evenly distributed through the slab, Sherwood (23) and Newman (17) obtained

$$\frac{\bar{X}' - \bar{X}_E{}'}{\bar{X}_C{}' - \bar{X}_E{}'} = \frac{8}{\pi^2}\left\{e^{-\mathscr{D}_L{}^*\theta\left(\frac{\pi}{l}\right)^2} + \frac{1}{9}e^{-9\mathscr{D}_L{}^*\theta\left(\frac{\pi}{l}\right)^2}\right.$$
$$\left. + \frac{1}{25}e^{-25\mathscr{D}_L{}^*\theta\left(\frac{\pi}{l}\right)^2} + \dots\right\} \tag{18.2}$$

where l = distance from face to center of a slab drying from both faces, or total thickness of a slab drying from one face, ft

$\bar{X}_E'$ = equilibrium moisture content, lb of liquid/lb of dry solid

$\bar{X}'$ = moisture content at time θ, lb of liquid/lb of dry solid

$\bar{X}_C'$ = moisture content at the start of the period during which drying rate is diffusion controlled, lb of liquid/lb of dry solid.

Since the liquid movement by diffusion is relatively slow, the drying-rate curve may show almost no constant-rate period. In any case, $\bar{X}_C'$ will be the moisture content at the end of the constant-rate period and is identical to the critical moisture content. Equation 18.2 then gives the moisture-content–time curve during the falling-rate period.

Since Equation 18.1 differs from Equation 11.11 only in the definition of $\mathscr{D}_L^*$ and the consistent units used for concentration, the methods of solution of Equation 11.11 outlined in Chapter 11 are also usable with Equation 18.1. Specifically, the Gurney-Lurie charts are useful graphical solutions to Equation 18.1, and Figure 11.12 may be used to solve Equation 18.2.

Even if diffusion does control the moisture movement through the solid, Equation 18.2 does not adequately fit the measured drying-rate curve. Many solids change their pore characteristics during drying, so that $\mathscr{D}_L^*$ is seldom constant. Moreover, the moisture distribution at the critical moisture content is seldom uniform. For some substances such as wood and clay, the distribution has been found to be nearly parabolic, and solutions of Equation 18.1 for this boundary condition have been given (23).

Moisture Movement—Capillarity Mechanism. For beds of particulate solids or for substances with a large open-pore structure, the molecular diffusion mechanism is obviously incorrect. For these materials, movement of liquid within the solid results from a net force arising from differences in hydrostatic head and in surface-tension effects (1). Surface tension causes the pressure under a curved liquid surface to be different from that of

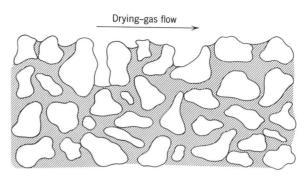

Drying-gas flow →

Figure 18.4. Moisture distribution in a particulate-solid bed during the first falling-rate period.

a flat surface. For a sphere of radius r, it can be shown (9) that

$$-\Delta P = \frac{2\gamma}{r} \qquad (18.3)$$

where $-\Delta P$ = decrease in pressure caused by surface-tension effect, lb_f/sq in

γ = surface tension for contact between liquid and gas phases, lb_f/in

r = radius of curvature of sphere, in.

The radius here is positive for a bubble surrounded by liquid, and negative for a liquid droplet in a gas. If a small tube is inserted in a liquid, as shown in Figure 18.3, the rise of liquid in the tube can be determined from a force balance at point A. The liquid surface in the tube has a radius of curvature equal to the radius of the tube only if the liquid wets the tube so thoroughly that the contact angle at the tube wall is zero. If this occurs, the force balance gives

$$-\Delta P = \Delta z\, g_c(\rho_L - \rho_V) = \frac{2\gamma}{r}$$

$$\Delta z = \frac{2\gamma}{r g_c(\rho_L - \rho_V)} \qquad (18.4)$$

and r may be taken as the inside-tube radius. In any drying solid, pore sizes are not uniform, and wetting may not be complete, but the mechanism of movement is as described.

At lower moisture contents (those between C and D of Figure 18.2), the liquid interface begins to retreat from the surface. The retreat is not uniform, as the radii of curvature of the liquid menisci at the surface are not uniform. Liquid in larger pores is pulled down into the sample to supply the menisci in smaller pores by flow through the surface-tension driving force. As drying proceeds, liquid in the larger pores continues to retreat until it either reaches a narrow "waist" in the pore and thus takes on a curvature matching that in the filled

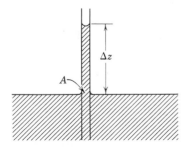

Figure 18.3. Capillary effect.

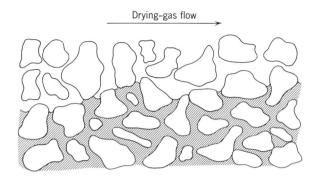

Figure 18.5. Moisture distribution in a particulate-solid bed during the second falling-rate period.

pore or until it retreats far enough so that the unbalance in surface-tension force is matched by an unbalance in gravitational head. As the moisture is depleted, more and more surface pores lose their moisture in this way so that, between C and D, the proportion of the total surface that is saturated becomes less and less. Figure 18.4 shows the solid phase during this period. Drying proceeds from the saturated surface at the same rate as that observed in the constant-rate period. The over-all drying rate is reduced since heat and mass must both diffuse through the top layers of the solid. By the time point D, the "second critical point," is reached, moisture has retreated from all the surface pores. Further drying involves a longer and longer diffusion path for both heat and mass. The physical situation for a granular-solid bed is shown in Figure 18.5. During the later drying stages, the solid surface approaches the temperature of the drying medium, but the surface from which evaporation actually occurs remains at the wet-bulb temperature. The second critical point is hard to find experimentally, and frequently the drying-rate curve may be smooth from C to E. The curves vary greatly in shape and slope depending upon the structure of the solid and the ease of movement of moisture within it. Toward the end of the drying process, the moisture present exists

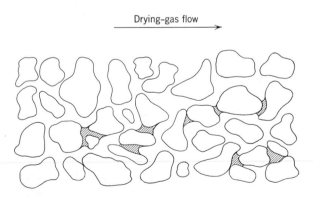

Figure 18.6. Moisture distribution in a solid bed toward the end of the drying process.

in small pockets in pore corners scattered throughout the solid as shown in Figure 18.6. The actual drying surface is scattered and discontinuous, and the mechanism controlling the drying rate is that of diffusion of heat and mass through the porous solid.

CALCULATION OF DRYING-TIME

In calculations involving drying, the drying-rate curve must be considered in its major sections, for the controlling factors differ along different parts of the curve. The drying rate is defined as

$$R = \frac{-W_s \, d\bar{X}'}{A \, d\theta} = \frac{N_a M_a}{A} \qquad (18.5)$$

where R = drying rate, lb of liquid evaporated per hr and sq ft of solid surface
W_s = weight of dry solid, lb
$\bar{X}'$ = bulk moisture content of the solid, lb of liquid/lb of dry solid

which can be rearranged and integrated to obtain the drying time.

$$\int_0^\theta d\theta = \frac{-W_s}{A} \int_{\bar{X}_1'}^{\bar{X}_2'} \frac{d\bar{X}'}{R} \qquad (18.6)$$

where $\bar{X}_1'$ = bulk moisture content at time 0
$\bar{X}_2'$ = bulk moisture content at time θ

The Constant Rate Period. For the constant-rate period R will be constant at R_c, and Equation 18.6 may be readily integrated to

$$\theta_C = \frac{-W_s}{AR_C}(\bar{X}_C' - \bar{X}_1') \qquad (18.7)$$

where $\bar{X}_C'$ = moisture content at the end of the constant rate period, lb of water/lb of dry solid
$\bar{X}_1'$ = moisture content at start of drying process
θ_C = time of constant-rate drying, hr

R_C will depend upon the heat- and mass-transfer coefficients from the drying medium to the solid surface, for

$$R_C = k_Y(Y_i - Y_V)M_a = \frac{h_V}{\lambda}(T_V - T_i) \quad (18.8)$$

where the transfer coefficients apply from the evaporating interface to the bulk-gas phase.

Since the total amount of heat will be transferred by the mechanisms of conduction through the solid, convection from the gas, and radiation from the surroundings, all to the evaporating surface, the total rate of heat transfer (q_T) is

$$q_T = h_V A(T_V - T_i) = h_c A(T_V - T_i)$$
$$+ h_r A(T_W - T_i) + U_k A(T_V - T_i) \quad (18.9)$$

where h_c = coefficient for heat transfer by convection from gas to solid surface

h_r = coefficient for radiant-heat transfer between material surface and walls of drying chamber

U_k = over-all coefficient for heat transfer to drying surface by convection and conduction *through the bed* to the evaporating surface

T_W = temperature of the walls of the drying space

T_V, T_i = temperature of the drying gas and the liquid-gas interface, respectively

If the drying-chamber walls are at the bulk-gas temperature,

$$h_V = h_c + h_r + U_k \qquad (18.10)$$

In most cases, the heat transferred by radiation and by conduction through the bed is minor. Then the total heat-transfer coefficient is essentially a convection coefficient and may be correlated by a relation in the usual *j*-factor form. Thus,

$$\left(\frac{h_V}{c_P G_V}\right)(N_{\text{Pr}})^{2/3} = b\left(\frac{DG_V}{\mu}\right)^n \qquad (18.11)$$

Few data are available to allow the constants in Equation 18.11 to be fixed with certainty. Almost all reported experimental results use air as a drying medium and hence cannot be used to substantiate the exponent on the Prandtl number. The 2/3 exponent proved satisfactory for drying in superheated steam (27) and for the evaporation of butyl alcohol, water, and benzene into their superheated vapors (2). The data result in the equation

$$\left(\frac{h_V}{c_P G_V}\right)(N_{\text{Pr}})^{2/3} = 26.6\left(\frac{DG_V}{\mu}\right)^{-0.7} \qquad (18.11a)$$

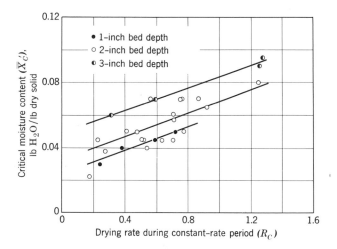

Figure 18.7. Influences of constant drying rate and bed thickness on critical moisture content when drying sand in superheated steam (27). (Courtesy of American Chemical Society.)

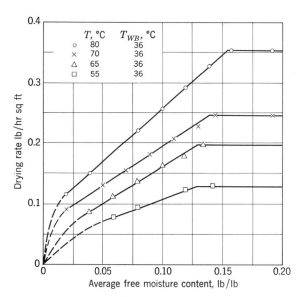

Figure 18.8. Drying of asbestos pulp in air (15). (Courtesy of Am. Inst. Chem. Engrs.)

For air drying, the extensive experimental results have been correlated (21) by the relation

$$h_V = 0.0128 G_V^{0.8} \qquad (18.12)$$

Equation 18.12 is recommended for determining the coefficient during the constant-rate period when air is the drying medium. In this case, the surface temperature (T_i) may be taken as the wet-bulb temperature of the air.

Critical Moisture Content. The moisture content existing at the end of the constant-rate period is called the critical moisture content. At this point, the movement of liquid to the solid surface becomes insufficient to replace the liquid being evaporated. Therefore, the critical moisture content depends upon the ease of moisture movement through the solid, and hence upon the pore structure of the solid in relation to the drying rate. The complicated and uneven structure of porous solids makes predicting the critical moisture content very difficult. At present, the engineer must depend upon experimental measurements made under simulated production conditions in order to determine $\bar{X}_C{}'$. The critical moisture content may be expected to depend upon the pore structure of the solid, sample thickness, and drying rate. Thus, the test conditions should use production widths of the actual solid to be dried. The dependence of critical moisture content on drying rate is weak and may often be ignored, though this dependence has been illustrated repeatedly (27). Drying conditions should be fixed so that the drying rate during the constant-rate period will match that expected in the plant. Figure 18.7 illustrates the dependence of the critical moisture content upon bed depth and drying rate during the constant rate period for the drying of sand with superheated steam (27). Figure 18.8 illustrates the

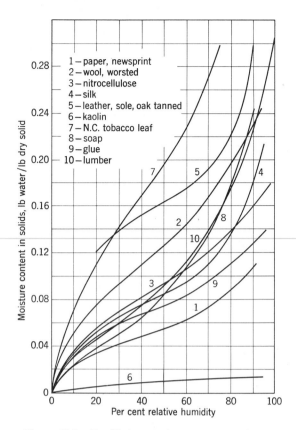

Figure 18.9. Equilibrium moisture content of some solids at 25°C (11).

dependence of critical moisture content on constant drying rate when asbestos pulp is dried into moist air (15).

Falling-Rate Period. The shape of the drying curve during the falling-rate period is as difficult to predict as is the critical moisture content. The shape will depend primarily on the structure of the solid being dried and also upon the drying rate during the constant-rate period and on the critical moisture content. The gas flow rate, which so strongly influences the constant drying rate, becomes less and less important as the drying rate falls. This decreasing importance of the gas flow rate results from the fact that the drying rate is controlled more and more completely by diffusion of heat and mass through the porous solid. In many cases, the drying-rate–free-moisture-content curve during the falling-rate period approximates a straight line from the critical moisture content to the origin (8). Then,

$$\frac{R}{\overline{X}'} = \frac{R_C}{\overline{X}_C'} \qquad (18.13)$$

where the subscript C refers to the critical moisture content. Substituting in Equation 18.6,

$$\int_{\theta_C}^{\theta} d\theta = \frac{-W_s}{A} \int_{\overline{X}_C'}^{\overline{X}'} \frac{d\overline{X}'}{\frac{R_C}{\overline{X}_C'} \overline{X}'}$$

and integrating,

$$(\theta - \theta_C) = \frac{-W_s}{A} \frac{\overline{X}_C'}{R_C} \ln \frac{\overline{X}'}{\overline{X}_C'} \qquad (18.14)$$

The parallel to logarithmic mean (ΔT) can be seen; accordingly, a logarithmic-mean drying rate for a straight-line falling-rate period can be used in Equation 18.6. This logarithmic-mean drying rate appears to be a good approximation when the material being dried is a rigid particulate solid such as sand. It is not suitable for use in drying fibrous material such as leather or gel-like material such as soap.

Equilibrium Moisture Content. For many materials, such as the leather and soap mentioned above, the solid will contain a significant moisture content when the drying rate has dropped to zero. The moisture remains no matter how long drying continues if the drying-gas conditions do not change. It is in equilibrium with the vapor contained in the drying medium and is called the equilibrium moisture content.

The amount of moisture held by a solid in equilibrium with humid gas depends upon the structure of the solid, the temperature of the gas, and the moisture content of the gas. For many materials, it also depends upon whether the solid originally held more or less liquid than the equilibrium value.

A few typical equilibrium-moisture-content–humidity curves (11) are shown in Figure 18.9. The data are for the equilibrium water content of solids in contact with moist air. The graph shows that, for example, a sample of wool containing 0.2 lb of H_2O per pound of dry wool placed in air at 25°C and 60 per cent relative humidity will gradually dry till a moisture content of 0.145 lb of H_2O per pound of dry wool is reached. Further exposure will produce no further decrease in moisture

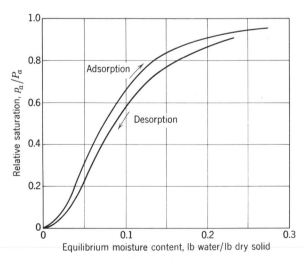

Figure 18.10. Equilibrium moisture content of sulfite pulp showing hysteresis (21). [From Treybal, R. E., *Mass Transfer Operations*, McGraw-Hill Book Co., New York (1955), by permission.]

content. Upon increasing the air humidity to 90 per cent saturation at 25°C, the wool will absorb moisture till it reaches a water content of about 0.23 lb H$_2$O per pound dry wool. However, in contact with air at zero relative humidity, the equilibrium moisture content is zero. The same pattern holds for all the materials shown. The zero equilibrium moisture content in contact with completely dry air indicates that none of the water content is held chemically.

Figure 18.10 shows the equilibrium moisture content curve for sulfite pulp (21). This material is an example of those in which the equilibrium-moisture-content curve shows a hysteresis effect. In drying operations the desorption curve is always the one of interest. In Figure 18.11, the effect of temperature on the equilibrium moisture content of raw cotton is shown.

If the curves of Figure 18.9 were extended to 100 per cent relative humidity, they would show the *maximum* moisture content that could exist *in equilibrium* with moist air. At this point, the water would exert its full vapor pressure. Any additional water would only continue to exert the same vapor pressure and would be in equilibrium with saturated air. The added moisture acts like free water and is called "unbound moisture." The moisture contained in the solid in equilibrium with partly saturated air is called "bound moisture." It exerts a vapor pressure lower than that of pure water for several possible reasons. It may be held in fine capillary pores with highly curved surfaces, or it may contain a large concentration of dissolved solids, or it may be in physical combination in a natural organic structure. The extent of vapor-pressure lowering due to these mechanisms can vary widely with variations in moisture

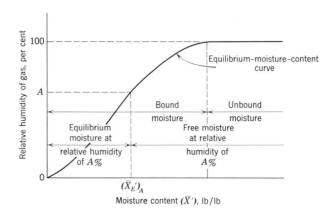

Figure 18.12. Types of moisture involved in the drying of solids.

content as shown. Moisture held in excess of the equilibrium moisture content is called "free moisture." Figure 18.12 shows the relation among the various terms. On this figure at any gas humidity, such as point A, the corresponding equilibrium moisture content ($\overline{X}_E'$) can be read.

To this point, the discussion of equilibrium moisture content has mainly concerned itself with the moisture held in the structure of insoluble solids. If the solids are soluble in water, all the various phase relations are possible that are found in the normal two-component condensed-phases system. These phase relations are discussed at length in texts devoted to the phase rule (4).

Figure 18.13 shows schematically the equilibrium-moisture-content relation for CuSO$_4$. Here, several different hydrates form as well as a CuSO$_4$–H$_2$O solution. The product obtained in drying this material would depend markedly on the humidity of the air used for drying. For example, with air containing between 10 and 20 mm Hg partial pressure of water as the drying medium, CuSO$_4$·5H$_2$O would be obtained at 25°C. If the partial pressure of water dropped to 7.8 mm Hg, an undetermined mixture of CuSO$_4$·5H$_2$O and CuSO$_4$·3H$_2$O would result. If the water partial pressure were only 2 mm Hg, CuSO$_4$·H$_2$O would be the product. Here too, a zero equilibrium moisture content is found in contact with completely dry air.

APPLICATION TO DRYING-EQUIPMENT DESIGN

The application of basic principles to the design of drying equipment is not straightforward. In addition to the difficulty in predicting the drying-rate curve, questions arise concerning variation of drying conditions through the dryer, difference of area for heat transfer and for mass transfer, gas flow pattern, and effect of operating variables and choice of equipment on the condition of the dried product. Also, the usual economic factors of processing cost must be considered in relation

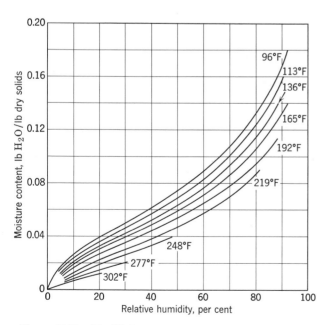

Figure 18.11. Equilibrium moisture content of raw cotton, desorption data (25).

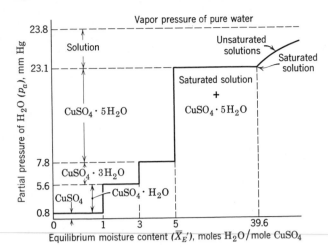

Figure 18.13. Equilibrium moisture content for $CuSO_4$ at 25°C.

to the most desirable product condition from a sales point of view. For these reasons, the choice of dryer is usually based on preliminary tests, drying the material under simulated production conditions. Most dryer manufacturers maintain service laboratories in which pilot-plant-sized dryers are available. The customer may dry samples of his material on these dryers at various operating conditions to attempt to find the optimum combination of equipment type and operating conditions.

Here, some major types of dryers will be described and their operation discussed emphasizing the mechanisms controlling the process. The dryers will be compared so that their relative advantages may be seen.

Tray Dryers. The simplest of dryers is the tray dryer, which is shown schematically in Figure 18.14. The tray dryer is essentially a cabinet into which the material to be dried is placed on trays. It is basically a batch unit used for small-capacity operation. The dryer may have space for ten, twenty, or more trays. They may

be solid-bottomed pans with the air circulating across the top and bottom of the pan, or they may have a screen base with the air circulation controlled so that it passes through the tray and the solids contained on it. (Figure 18.15). The material to be dried may even be hung on racks or hooks if it exists in the form of a sheet. The drying conditions are simply controlled and readily changed so that this dryer is particularly well suited for laboratory operations, or for drying materials that require altering the drying conditions as the drying progresses. If the external controllable conditions are constant, the drying conditions will be constant for any given tray of moist solids. Thus, the drying history at any given point will be just as discussed above. However, the trays nearest to the air inlet will be subjected to conditions markedly different from those located near the end of the air-flow path. As a result, material on some pans dries more rapidly, and that on others dries less rapidly than the average. This is of little concern with material that is not heat sensitive, but, if overdrying can cause scorching, the pans must be removed at varying times, or the air temperature must be reduced at the approach of the end of the drying process. In some dryers, the manufacturers have at least partially overcome this difficulty by arranging for reversal of the air-flow path.

As mentioned above, for some loosely packed material or material that can be formed into small discrete shapes before drying, the tray dryer may be equipped with screen-bottomed trays. The air flow is then directed so that it passes through the material placed on the tray. This arrangement results in much shorter drying times, but usually a smaller quantity of material can be placed in the dryer than would be the case with a solid-tray dryer. Figure 18.15 is a schematic diagram of this kind of dryer.

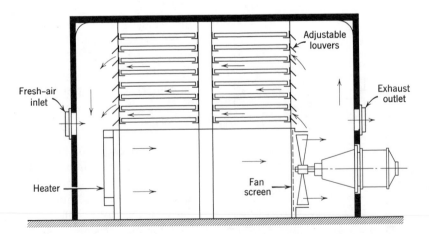

Figure 18.14. A typical 20-tray dryer. This unit has 30 by 40 in. trays on 4 in. centers. Heating is normally by heat exchange with steam, although gas-fired or electrically heated dryers are often used. (Courtesy Proctor & Schwartz, Inc.)

In tray dryers, steam, gas, or electrically heated air is usually used as the drying medium. Energy cost is a major part of the total processing cost. To conserve energy and also to control the air humidity at that level which results in the best product, part of the drying air may be recycled. In this case of air recirculation, the humidity-temperature path will be as shown in Figure 18.16. The air mixture containing both fresh and reheated air enters the dryer at point A and, after passing over the trays, leaves at point B. For A–B to be part of an adiabatic-saturation curve, the solids must be at the wet-bulb temperature of the air. Otherwise, curve A–B will have a slope different from that of an adiabatic-saturation curve but not markedly different from that which is shown. The moist air is then mixed with fresh air at condition E to give air at point C. Air of condition C is reheated to condition A where again it enters the dryer. Had none of the air been recycled the air path would have been E–D–B' with E–D being the air-heating step and D–B' the solids-drying step. The net result of air recycling, aside from control of inlet-air humidity, is that discharge-air humidity has been increased to point B rather than to B' as would have been true for drying without recycle. Air discharged from the dryer has received more energy as latent heat but no more energy as sensible heat. The result is a direct saving of fuel energy, though a proportionately longer drying time is required.

Some materials cannot be satisfactorily dried in air at normal atmospheric pressure because they deteriorate at temperatures necessary to obtain reasonable drying rate, or because they react with the oxygen in air. Tray dryers can be built which exclude atmospheric air and use other media such as superheated organic vapor or evacuated air for drying. The freeze dryer shown in Figure 18.17 dries material under vacuum so that in many cases the materials being dried lose liquid by sublimation from the solid state at very low temperatures.

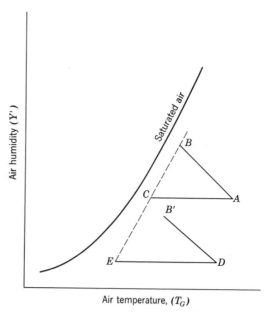

Figure 18.16. Air reheat for a dryer.

This is fundamentally an expensive drying method practicable only for products of high unit price. Thus, its major use has been with pharmaceuticals.

Illustration 18.1. Raw cotton of 0.7 gm/cu cm density when dry is to be dried in a batch tray dryer from a moisture content of 1 lb H_2O/lb dry solids to 0.1 lb H_2O/lb dry solids. Trays are 2 ft square and $\frac{1}{2}$ in. thick and are arranged so that drying occurs from the top surface only with the bottom surface insulated. Air at 160°F with a 120°F wet-bulb temperature circulates across to the pan surface at a mass flow of 500 lb/hr sq ft. Previous experience under similar drying conditions indicates that the critical moisture content will be 0.4 lb H_2O/lb dry solids, and that the drying rate during the falling-rate period will be proportional to the free moisture content. Determine the drying time required.

SOLUTION. For this batch dryer, the drying-rate-versus-moisture-content curve will be of the typical shape. The problem gives no information on initial stock temperature, so that the initial period during which the stock temperature approaches the wet-bulb temperature of the air cannot be specified. Thus the constant-rate period will be presumed to start at the initial moisture content. This is equivalent to assuming the initial stock temperature to be 120°F.

The drying-rate curve can be completely fixed once the equilibrium moisture content is determined and the rate during the constant-rate period is calculated. The equilibrium moisture content depends on the relative humidity of the air, as shown in Figure 18.11. From the humidity chart, $Y = 0.114$ mole H_2O/mole dry air, and the humidity is 24 per cent for the conditions given. Thus, for saturated air at 160°F, $Y_s = 0.114/0.24 = 0.475$. Since $Y = y/(1 - y)$, $y = 0.1023$ and $y_s = 0.322$ for the actual and the saturated air respectively. Then,

$$\text{Relative humidity} = \frac{0.1023}{0.322} \times 100 = 31.8 \text{ per cent}$$

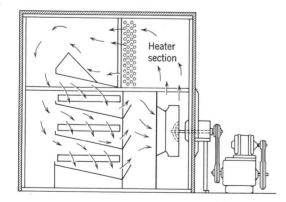

Figure 18.15. Through-circulation tray dryer. In the standard size, the dryer is about 9 ft high, 9 ft long, and 6 ft wide. It has room for six trays with a face area of 44 sq ft. (Courtesy Proctor & Schwartz, Inc.)

Figure 18.17. Large freeze dryer which is used to dry material which is highly heat sensitive such as some pharmaceuticals. The entire drying operation is accomplished under vacuum. Devices are available to seal bottles automatically within the freeze dryer before breaking the vacuum. (Courtesy F. J. Stokes Corp.)

and, from Figure 18.11, the equilibrium moisture content, $(\bar{X}_E{}')$ is 0.04 lb H_2O/lb dry cotton.

For the constant drying rate the heat-transfer coefficient may be calculated from Equation 18.12.

$$h_V = 0.0128\, G_V{}^{0.8} = 0.0128(500)^{0.8} = 1.86 \text{ Btu/hr sq ft }°F$$

and

$$R_C = \frac{1.86}{\lambda}(T_V - T_i) = \frac{1.86}{1025.8}(160 - 120) =$$

$$0.0724 \text{ lb/hr sq ft}$$

Using these values of $\bar{X}_E{}'$ and R_C, the drying-rate curve is as shown in Figure 18.18. From the pan dimensions and the density of dry raw cotton, the total weight of dry solids is

$$W_s = 2 \times 2 \times \frac{1}{2} \times \frac{1}{12} \times 62.4 \times 0.7 = 7.27 \text{ lb of dry solids}$$

The drying time during the constant rate period is then

$$\theta_C - \theta_0 = \frac{(1 - 0.4) \times 7.27}{0.0724 \times 4} = 15.1 \text{ hr}$$

The drying time during the falling-rate period can be obtained by modifying Equation 18.14.

$$(\theta_f - \theta_C) = \frac{-W_s(\bar{X}_C{}' - \bar{X}_E{}')}{A R_C} \ln \frac{\bar{X}_f{}' - \bar{X}_E{}'}{\bar{X}_C{}' - \bar{X}_E} \quad (18.14a)$$

$$= \frac{-7.27}{4} \times \frac{(0.4 - 0.04)}{0.0724} \ln \frac{(0.1 - 0.04)}{(0.4 - 0.04)} = 16.2 \text{ hr}$$

Total drying time, $\theta_f - \theta_0 = 15.1 + 16.2 = 31.3$ hr

Conveyor and Tunnel Dryers. Tray dryers may be made continuous by moving the moist solids through the drying chamber continuously. This may be done by mounting the trays on dollies, by conveying the feed through the dryer on a belt as in the dryer of Figure 18.19, or, for material in sheet form, by moving the moist sheet through the dryer over rollers. Air flow may be directed perpendicular to the flow of material, countercurrent to it, or cocurrent with it. Usually, the flow path is not simple but successively takes several of these directions. In any case, the stock is subjected to a drying medium of varying conditions along the drying path. Consequently, the characteristic drying curve is greatly altered. For example, the "constant-rate" period no longer shows a constant drying rate. The rate decreases as the drying-air temperature decreases even though the surface temperature of the stock stays constant. The time required for the constant-rate period may be calculated through use of the heat-transfer-rate equation, but an appropriate mean-temperature driving force must be used. The falling-rate period may be calculated by successively treating incremental decreases in moisture content during which the drying-medium temperature and humidity are considered to be constant.

The calculation of drying rate and time in continuous dryers requires enthalpy and material balance calculations as well as use of the heat- and mass-transfer rate equations. Complete and rigorous calculations involve successive approximation solutions of the proper equations written for each small segment of the dryer. To reduce the labor involved, a number of approximations are usually made, though they are not essential to dryer calculations. The first of the approximations is the assumption that all the heat transferred is transferred from air to stock by convection mechanisms. This assumption requires that the apparatus operate nearly adiabatically and that heat transfer by conduction and radiation be negligible. Second, the area for mass transfer is assumed equal to that for heat transfer.

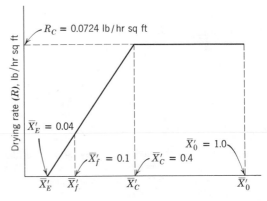

Figure 18.18. Drying rate diagram for Illustration 18.1.

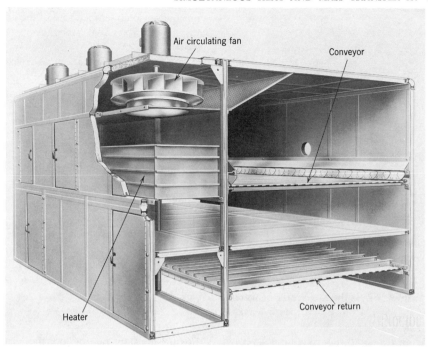

Figure 18.19. Single conveyor dryer. The fans pull the air through the heater section located beneath them. The air then passes through the wet stock carried on the conveyor and back into the bottom of the heater. The conveyor belt is made of perforated-metal plate or woven wire, and the heater is usually of steam-heated finned tubes. Conveyor widths vary between 15 in. and 9 ft and lengths vary up to 160 ft. Preforming of the feed to increase drying rate and obtain a more desirable product is often practiced. It may be accomplished by extrusion or granulating steps or merely by scoring the cake removed from a continuous filter so that it breaks into reasonably uniform pieces. Thickness of material on the belt will depend on the material being processed, but it can vary between $\frac{1}{2}$ and 6 in. (Courtesy of Proctor & Schwartz, Inc.)

This approximation is much less satisfactory than was the first one, because some heat is always transferred to the sides and bottom of any pan or conveyor system used. Only in through dryers, where the drying air flows through the solid being dried, is this assumption relatively good. Third, no evaporation occurs during the initial warm-up or cool-down period. This assumption brings in a varying error depending on the condition of the feed relative to its steady-state temperature maintained during the constant-rate period. By using it, the condition of stock and drying medium at the start of the constant-rate period may be more readily calculated. Fourth, moisture removal occurs entirely at the adiabatic-saturation temperature. The heating of the sample that occurs during the falling-rate period is thus assumed to occur after the last of the moisture has been removed. This final condition is closely followed in the drying of a granular sandlike material. For porous or fibrous materials, it may be seriously in error. The net result of the assumptions is that the drying process occurs entirely at a constant temperature. This temperature is the adiabatic-saturation temperature of the drying air which follows a constant-adiabatic-saturation-temperature curve during moisture removal.

Illustration 18.2. The raw cotton dried in a tray dryer in Illustration 18.1 is to be dried in a tunnel dryer through which it passes on a conveyor belt 2 ft wide in a layer $\frac{1}{2}$ in. deep. Sixty pounds of air per pound of dry solids flow through the dryer countercurrent to the stock at a mass rate of 500 lb/hr sq ft and enter at 200°F with a 120°F wet-bulb temperature. The stock enters at 80°F, and as before it contains 1 lb of moisture per pound of dry solids. It leaves at 150°F with a moisture content of 0.1 lb of water per pound of dry solids. The specific heat of dry raw cotton may be taken as 0.35 Btu/lb °F. Determine the drying time required.

SOLUTION. Although the raw cotton specified here is not one of the granular materials for which the assumptions stated above apply most accurately, the assumptions will be used in this solution. Using the assumptions and applying the data developed in Illustration 18.1, the problem can be attacked directly.

Writing an over-all material balance based on 1 lb of dry stock gives the humidity of the outlet air.

$$\bar{X}_o' - \bar{X}_f' = (Y_0 - Y_f)\frac{M_a}{M_b} \times \frac{VM_b}{W_s/\theta} \qquad (a)$$

$$1.0 - 0.1 = (Y_0 - 0.098)\frac{18}{29} \times 60; \quad Y_0 = 0.122$$

where the subscripts "0" and "f" refer to the conditions at the ends of the dryer where stock enters and exits, respectively.

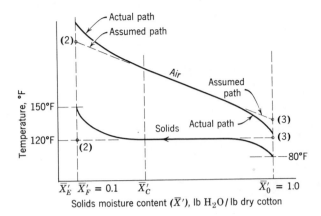

Figure 18.20. Temperature–moisture-content history through the continuous dryer of Illustration 18.2.

These subscripts apply whether air or stock properties are being considered; for example, Y_0 is the mole ratio of water to dry air at the end of the dryer from which the air leaves, but in which the stock enters.

The path of stock and air temperature through the dryer is shown in Figure 18.20. Heat balances can be made from the stock-outlet end of the dryer (air-inlet end).

For the dried-stock heating step,

$$\Delta H_{\text{solids} + \text{moisture}} = -\Delta H_{\text{air}} \times \frac{W_b}{W_s} \times \frac{1}{M_b} \qquad (b)$$

where the subscript b refers to dry air.

Assuming a 120°F wet-bulb temperature,

$$1 \times 0.35 \times 30 + 0.1 \times 1 \times 30 = \frac{60}{29} \times (-\Delta H_{\text{air}}) \quad (b')$$

$$\Delta H_{\text{air}} = -6.52 \text{ Btu/lb mole dry air}$$

Since at the inlet condition (120°F wet-bulb temperature, 200°F dry-bulb temperature) the air enthalpy is 3240 Btu/lb mole, $H_2 = 3234$ Btu/lb mole. For air at a wet-bulb temperature of 119°F, $H = 3154$ Btu/lb mole. Thus, the 120°F wet-bulb temperature is still substantially correct for the air at point 2; and the 30°F ΔT inserted in Equation b' above is substantially correct, since the stock reaches the wet-bulb temperature of the drying air.

The wet-bulb temperature of the air will not change between points 2 and 3 of Figure 18.20, so that $H_3 \approx 3234$ Btu/lb mole dry air.

Now, writing a heat balance over the initial solids warm-up part of the cycle, assuming no evaporation,

$$1 \times 0.35 \times 40 + 1 \times 1 \times 40 = \frac{60 \times (-\Delta H_{\text{air}})}{29}$$

$$-\Delta H_{\text{air}} = \frac{1.35 \times 40 \times 29}{60} = 26.1 \text{ Btu/lb mole dry air}$$

$$H_{\text{air leaving the drier}} = 3207 \text{ Btu/lb mole}$$

The wet-bulb temperature is still substantially 120°F, and, since the humidity is 0.122 lb mole H_2O per pound mole dry air, the temperature is 144°F. Since neither the humidity

nor the wet-bulb temperature of the drying air changes during the initial and final warm-up periods, $T_2 = 200°F$ and $T_3 = 144°F$.

For the entire operation, the heat-transfer coefficient is fixed and can be calculated from Equation 18.12.

$$h_V = 0.0128(500)^{0.8} = 1.86 \text{ Btu/hr sq ft °F}$$

On the basis of 1 sq ft of conveyor area,

$$\text{Solids held} = 1 \times \frac{1}{2} \times \frac{1}{12} \times 62.4 \times 0.7 = 1.82 \text{ lb dry cotton}$$

For the final period during which the dried solids are heated,

$$q = 1.82(1 \times 0.35 \times 30 + 0.1 \times 1.0 \times 30) = 1.82 \times 13.50$$
$$= 23.6 \text{ Btu/sq ft of conveyor}$$

$$\frac{23.6}{\theta_f - \theta_2} = h_V A \, \Delta T_{lm} = 1.86 \times 1.0$$

$$\times \frac{(200 - 120) - (200 - 150)}{\ln \frac{80}{50}}$$

$$\theta_f - \theta_2 = \frac{23.6}{1.86 \times 63.9} = 0.199 \text{ hr}$$

For the constant-rate period, the heat-transfer-rate equation can be used directly.

$$dq = h_V A \, \Delta T = \frac{-W_s \lambda' \, d\bar{X}'}{d\theta} \qquad (c)$$

$$\int_{\theta_3}^{\theta_C} h_V A \, \Delta T \, d\theta = -\int_{\bar{X}_3'}^{\bar{X}_C'} W_s \lambda' \, d\bar{X}'$$

$$h_V A \, \Delta T_{lm} (\theta_C - \theta_3) = W_s \lambda' (\bar{X}_3' - \bar{X}_C') \qquad (c')$$

As before, the critical moisture content ($\bar{X}_C'$) is 0.4 lb H_2O/lb dry stock. To get ΔT_{lm}, the air temperature at the point where the solids reach $\bar{X}_C'$ must be known. At this point, the wet-bulb temperature is still 120°F, but the humidity is

$$(1.0 - 0.4) = -(Y_C - Y_0)\frac{18}{29} 60$$

$$Y_C = 0.106 \text{ moles of water/mole of dry air}$$

From this value of Y_C, T_C is fixed as 180°F. Now, substituting in Equation c',

$$(\theta_C - \theta_3) = \frac{1.82 \times 0.6 \times 1025.8}{1.86 \times 1.0 \times \frac{60 - 24}{\ln(60/24)}} = 15.3 \text{ hr}$$

The falling-rate period can be attacked in the same way. Here, the evaporating liquid is assumed to stay at 120°F, but the effective heat-transfer coefficient ($h_{V \text{eff}}$), decreases as the drying progresses. Again,

$$dq = h_{V \text{eff}} A \, \Delta T = \frac{-W_s \lambda' \, d\bar{X}'}{d\theta}$$

$$\int h_{V \text{eff}} A \, \Delta T \, d\theta = -\int W_s \lambda' \, d\bar{X}' \qquad (d)$$

If it is assumed that the falling-rate period exhibits a linear-rate-versus-moisture-content curve, $h_{V \text{eff}}$ will also be a linear function of $\bar{X}'$. Also, $h_{V \text{eff}} = h_V = 1.86$ at $\bar{X}' = \bar{X}_C' = 0.40$, and $h_{V \text{eff}} = 0$ at $\bar{X}' = \bar{X}_E'$.

The inlet air, which enters at the end of the drying path, has a relative humidity of 3 per cent. Then, from Figure 18.11, $\bar{X}_E' = 0.008$, and

$$\frac{\Delta h_{V\,\text{eff}}}{\Delta \bar{X}'} = \text{constant} = \frac{1.86}{0.4 - 0.008} = 4.75 = \frac{1.86 - h_{V\,\text{eff}}}{0.40 - \bar{X}'}$$

or

$$h_V = 4.75\bar{X}' - 0.04$$

Rearranging Equation d and solving,

$$A\,\Delta T_{lm}\int_{\theta_C}^{\theta_2} d\theta = -W_s\lambda'\int_{\bar{X}_C'}^{\bar{X}_f'}\frac{d\bar{X}'}{4.75\bar{X}' - 0.04}$$

$$\theta_2 - \theta_C = \frac{-W_s\lambda'}{A\,\Delta T_{lm}}\frac{1}{4.75}\ln\frac{4.75\bar{X}_f' - 0.04}{4.75\bar{X}_C' - 0.04}$$

$$= \frac{-1.82 \times 1025}{1 \times \dfrac{80 - 60}{\ln(80/60)} \times 4.75}$$

$$\times \ln\frac{0.475 - 0.04}{1.90 - 0.04} = 8.22\ \text{hr}$$

Finally, the time required for the initial warm-up period can be determined as was that for the final warming period.

$$q = 1.82(1 \times 0.35 \times 40 + 1 \times 1.0 \times 40)$$
$$= 98.4\ \text{Btu/sq ft of conveyor area}$$

$$\frac{98.4}{\theta_3 - \theta_0} = h_V A\,\Delta T_{lm} = 1.86 \times 1.0$$

$$\times \frac{(144 - 80) - (144 - 120)}{\ln\dfrac{64}{24}}$$

from which,

$$\theta_3 - \theta_0 = \frac{98.4 \times \ln\dfrac{64}{24}}{1.86 \times 40} = 1.3\ \text{hr}$$

Total drying time = 1.3 + 15.30 + 8.22 + 0.20 = 25.02 hr. This is an extremely long processing period for continuous operation. If the conveyor is 200 ft long the rate of movement is 200/25.02 = 7.98 ft/hr, and the production rate from the 2 ft wide conveyor belt would be

$$1.82 \times 2 \times 7.98 = 29.0\ \text{lb/hr of dry raw cotton}$$

or

$$29.0 \times 1.1 = 31.9\ \text{lb/hr of dried product } (\bar{X}' = 0.1)$$

This value requires an air flow rate of 29.0 × 60 = 1740 lb/hr (dry basis) and permits a free dryer cross-section around the conveyor belt of 1740/500 = 3.48 sq ft.

The results of Illustration 18.2 indicate that, for continuous production at high throughput, drying by passing air over the surface of a bed of wet solids is likely to be impractical. Circulation of drying gas *through* the bed as is done in the dryer of Figure 18.19 is one possible solution. As indicated there, the feed must be made into such a form that the gases can pass through the bed, and the particles must be large enough to be retained on the conveyor belt. It is also obvious

that the temperature of inlet air should be as high as possible in order to increase ΔT. Since the stock remains at the wet-bulb temperature during constant-rate drying, heat-sensitive materials may be dried with air entering at temperatures somewhat above the safe stock temperature.

Rotary Dryers. Free-flowing, particulate material may be difficult to hold on a woven-wire or perforated-metal conveyor belt. Such material can be dried in a rotary dryer where the solids are tumbled in a continuous spray through the center of a rotating drum and the air is blown through the spray. Internal flights lift the solid and control its cascade through the air stream. The dryer is tilted so that the solids gradually work their way from feed end to discharge end. Flue gases as well as superheated steam or even electrically heated air may be used directly as the drying medium. In some dryers, steam-heated tubes run through the drying cylinder to maintain the air temperature and to act as drying surfaces. Figure 18.21 shows a rotary dryer. These dryers are built in sizes up to 9 ft in diameter in standard models with lengths up to 80 ft.

Since a much higher solid surface is exposed in rotary dryers than is exposed in tray or tunnel dryers, the drying rate will be much higher. The high rate is an advantage only if the air can be kept unsaturated. Thus, the engineer must use a high air rate or must heat the air as it passes through the dryer. It is for this reason that steam-heated tubes may be inserted into the rotary dryer, as has been done in the dryer shown in Figures 18.22 and 18.23. The tubes usually are inadequate to keep the air temperature constant, though they make the air-temperature–humidity curve considerably steeper than an adiabatic-saturation curve. In some cases the internal heating may even be strong enough to increase the air temperature as it passes through the dryer. Thus, the air path may take any of the directions indicated by Figure 18.24.

In designing rotary dryers, the retention time of solids passing through the dryer must be estimated. The retention time depends on the density and angle of repose of the solid, the arrangement of flights in the dryer, the slope of the dryer, and the mass of material present in the dryer.

The movement of solid through the dryer is influenced by three distinct mechanisms. First, as the dryer turns, each particle is lifted up by the flights and dropped again. At each drop, the particle advances a distance Ds where D = dryer diameter and s = dryer slope, ft/ft. Thus with a dryer of length l rotating at N rpm the time of passage of a particle would be proportional to l/sDN. This is called "flight action." In addition, the particles striking the bottom of the drum bounce, and those striking other particles roll over them, while solids that are not lifted by the flights still move forward

Figure 18.21. A 4-ft diameter by 14 ft long rotary dryer before installation. This unit is for ultimate use in drying bread crumbs. The solids discharge end is in the foreground showing the finned-tube exchanger used to heat the air. (Courtesy of Patterson Foundry and Machine Company.)

Figure 18.22. An internal-steam-tube rotary dryer. The foreground end is the feed-inlet end. The steam enters the tubes through a central inlet at the far end of the dryer. It is distributed to a header and then into the tubes. Condensate is discharged at the near end of the dryer into the collection ring shown in the foreground. From this ring, it flows back to the far end of the dryer through an internal tube for ultimate removal. (Courtesy of Struthers-Wells, Inc.)

in rolling over others on the bottom of the dryer. These effects are called "kiln action" and significantly alter the holding time of a particle in the dryer. Finally, the drying gas blowing through the dryer either advances or hinders the travel of solid depending on whether gas flow is cocurrent or countercurrent. The carrying of particles by the gas stream can only be estimated empirically since the settling velocity is hindered by the large number of particles present. On these considerations and on data obtained in pilot-sized dryers, Friedman and Marshall (6) offer the following relation:

$$\theta = \frac{0.35l}{sN^{0.9}D} \pm 0.6 \frac{BlG_V}{G_F} \qquad (18.15)$$

where θ = average time of passage, min
   l = dryer length, ft
   s = dryer slope, ft/ft
   N = rate of rotation, rpm
   D = dryer diameter, ft
   B = a constant dependent on particle size of material handled, $B = 5.2 (D_p)^{-0.5}$
   D_p = weight average particle size, microns
   G_V = gas mass velocity, lb/sq ft hr
   G_F = feed rate to dryer, lb dry material/hr sq ft of cross section

and where the plus sign is for counter-current flow, and the minus sign for cocurrent flow.

Permissible air velocities are limited by the tendency of the material being dried to "dust." This tendency obviously varies greatly with the properties of the material handled. For instance, Friedman and Marshall report

Figure 18.23. Internal view of rotary dryer of Figure 18.22, from the feed end. The internal arrangement of steam tubes is shown. Each tube has a transverse spiral fin that roughly doubles the external surface area. (Courtesy of Struthers-Wells, Inc.)

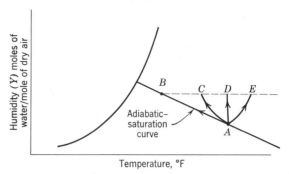

Figure 18.24. Possible air-condition paths in an internally heated rotary dryer.
   Path *AB*—No internal heating. Point *B* may be off of the adiabatic-saturation curve through *A*, because of heating or cooling of solids.
   Path *AC*—Internal heating inadequate to maintain air-temperature constant.
   Path *AD*—Enough internal heating to maintain isothermal operation.
   Path *AE*—Excess internal heating so that air temperature increases as it passes through the dryer.

that sawdust begins to dust excessively at low rates of about 250 lb/hr sq ft, but Ottawa sand (a white sand of very round particle shape) showed no significant dusting at 1000 lb/hr sq ft air flow.

The loading of moist solids in rotary dryers markedly affects the operation. Too low a quantity of solids will reduce the production rate. Too great a quantity of solids will result in uneven and incomplete flight action with some of the solid merely rolling on the bottom of the dryer, and it may result in a moist product. As a result, drying will be uneven, and the power required to turn the dryer will increase. There is more danger of overloading a dryer than underloading it, with experience indicating a hold-up of 3 to 10 per cent of the dryer volume as giving satisfactory operation.

Heat-transfer coefficients in such dryers are obviously indeterminable since the interfacial area is unknown. However, Friedman and Marshall (6) determined Ua from tests in a 1 ft by 6 ft rotary dryer and compared their results with those previously published. The equation used to correlate their results was

$$Ua = \frac{0.63G_V^{0.16}}{D} [G_F + BG_V\rho_b]^{0.5} \qquad (18.16)$$

where Ua = product of over-all coefficient and surface area, Btu/hr °F cu ft of dryer volume
   D = dryer diameter, ft
   G_F = feed rate to dryer, lb/hr sq ft of dryer cross section
   ρ_b = bulk density of dry solids, lb/cu ft
   G_V = air flow rate, lb/hr sq ft
   $B = 5.2D_p^{-0.5}$, where D_p is the weight average particle diameter, microns

This equation is far from exact, with deviations reported in the order of ± 100 per cent, but it will give an indication of required dryer size. The materials dried were granular in characteristic, and hence, most of the drying corresponded to constant-rate conditions. The solid temperature remained constant near the air wet-bulb temperature until it suddenly rose to approach the air temperature. One unexplained observation was that particularly for cocurrent operation, the constant solids temperature was not always the wet-bulb temperature but was sometimes as much as 5°F above it. This could be a result of radiation transfer.

Illustration 18.3. A rotary dryer 5 ft in diameter by 60 ft long is to be used to dry titanium dioxide from a moisture content of 0.3 lb water per pound dry solids to 0.02 lb water per pound dry solids. The TiO_2 has a mean particle size of 50 microns, and a density of 240 lb/cu ft. The dryer operates at 4 rpm with a slope of 0.5 in. in 10 in. of length. Determine inlet-air rate and required temperature, and rate of production of dried TiO_2 if drying air is obtained by heating room air originally at 90°F with a 60°F wet-bulb temperature.

SOLUTION. This problem is not stated so completely that only one answer is possible. The answer obtained will depend upon gas-flow rate and dryer-loading values used. They can be chosen arbitrarily provided reasonable limits are not exceeded. The air-flow rate must be low enough to prevent excessive dusting. The value of 1000 lb/hr sq ft given above probably applies to much larger particle sizes than are involved here, so that another method of estimating the limit of possible air flow must be used. Dusting must occur when the particles are picked up by the air stream. Actually, the air stream does not flow directly opposite to the particles as they fall, but a safe limit may be obtained by keeping the air flow below the velocity at which it could entrain the solid particles if it were flowing directly opposite them.

The relation giving the maximum settling velocity of particles in a fluid when these particles do not interfere with each other is derived in Chapter 22. The equation

$$v_t = \frac{(\rho_s - \rho)gD^2}{18\mu} \qquad (22.15)$$

is derived by combining a force balance on a particle with the relation for a drag force, Equation 13.50. The force balance specifies zero acceleration when the buoyant and drag forces balance any external force on a particle. Simplification is obtained by assuming the particle is spherical and the relative particle-to-fluid velocity is low enough to give laminar flow. The student is referred to the development beginning with Equation 22.1 for the complete derivation.

In Equation 22.15, v_t is the maximum rate at which a spherical particle would fall in still gas if the fall were slow enough that the Reynold's number (N_{Re}) would be less than about 0.5. Conversely, of course, v_t is the vertical gas velocity necessary to suspend a spherical particle in the gas in laminar flow.

Substituting in Equation 22.15,

$$v_t = \frac{1}{18}(240 - 0.05)32.2$$

$$\times \left(\frac{50 \times 10^{-4}}{2.54 \times 12}\right)^2 \times \frac{1}{0.02 \times 0.000672}$$

$$= \frac{240 \times 32.2}{18} \times \frac{2.69 \times 10^{-8}}{0.135 \times 10^{-4}} = 0.860 \text{ ft/sec}$$

at which velocity $N_{Re} = \dfrac{1.64 \times 10^{-4} \times 0.860 \times 0.05}{0.02 \times 0.000672}$

$$= 0.524$$

Thus, if $G_V = (0.860 \times 3600) \times (14.7 \times 29)/(10.73 \times 1000) = 123$ lb/hr sq ft, there would be a chance of entraining significant amounts of solid into the air stream. Here an air temperature of 1000°R is assumed. Of course, some entrainment may occur at lower gas-flow rates because of the complex flow pattern and the interference of particle against particle. Therefore, a gas-flow rate (G_V) of 50 lb/hr sq ft should be appropriate.

The time of passage may now be calculated from Equation 18.15 and from the dryer loading. Substituting in Equation 18.15,

$$\theta = \frac{0.35 \times 60}{0.05 \times (4)^{0.9} \times 5} + 0.6 \frac{5.2 \times 60 \times 50}{(50)^{0.5} \times G_F}$$

$$\theta = 24.2 + \frac{1320}{G_F} \qquad (a)$$

The loading of the dryer may be chosen within the 3 to 10 per cent range noted above. Choosing 5 per cent of the dryer volume,

$$G_F \times \frac{1}{\rho} \times \frac{\theta}{60} = 0.05l$$

$$\theta = \frac{0.05 \times 60 \times 60 \times 240}{G_F} = \frac{43200}{G_F} \qquad (b)$$

Combining Equations a and b,

$$\frac{43200}{G_F} = 24.2 + \frac{1320}{G_F}$$

and $\qquad G_F = 1770$ lb/hr sq ft

The inlet-air temperature must be high enough to give an adequate ΔT so that the required drying occurs in the available dryer volume and still remains unsaturated when it leaves the dryer. From the previous conditions, 1 lb of air must dry 35.4 lb of solids, or it must absorb $35.4(0.3 - 0.02) = 9.9$ lb of water. The air might possibly absorb this much water if it were heated as it passed through the dryer. With a simple rotary dryer with no internal heating, as this dryer must be presumed to be, such a moisture pick-up is absurd. Since the gas-flow rate cannot be controlled by the designer, the dryer loading must be altered. To obtain an outlet humidity in the air stream of about 1 lb of water per pound of dry air, the required feed rate would be $1770/9.9 = 179$ lb/hr of dry solids. Using this feed rate, the required inlet-air temperature can be found by a trial-and-error procedure. At $Y' = 1.0$, the saturation temperature is 189°F. Since the adiabatic-saturation curve, along which the drying air will move, is approximately represented by Equation 17.21a,

$$C_{h1}(T_a - T_1) = \lambda_a(Y_1 - Y_a) \qquad (17.21a)$$

the path of the drying air is known. Also, Y_1 is fixed by the problem statement as 0.0173 mole H_2O/mole dry air. Then, by Equation 17.21a,

$$(T_1 - T_a) = \left(1 \times \frac{29}{18} + 0.0173 - 0.0173\right)\frac{985 \times 18}{7.13}$$

$$= 4000°F$$

This temperature is beyond practicality. Using the moisture pickup as 0.1 lb of H_2O per pound of dry solids, the outlet humidity is $0.1 \times 29/18 + 0.0173 = 0.1783$ mole H_2O/mole dry air. The saturation temperature for this humidity is 130°F.

$$T_1 - T_a = \frac{0.1 \times \dfrac{29}{18} \times 1019 \times 18}{7.13} = 414°F$$

$$T_1 = 544°F$$

This inlet temperature can be attained only if the exit air is saturated; thus, an infinitely long dryer would be necessary. The heat load on each pound of inlet air is

$$q = 0.1 \times 1019 = 101.9 \text{ Btu}$$

Figure 18.25. A vacuum pan dryer. This unit is 96 in. I.D. with a nominal capacity of 875 gal. Heating is effected through steam coils imbedded in the cast-iron walls of the vessel. The kettle can be operated at pressures ranging between a full vacuum and 25 psig, whereas the steam coils are designed for 110 psig. An agitation speed of 9.6 rpm is maintained by a 15-hp drive. The unit was designed for use in dye-stuff intermediate manufacture to drive off the solvent from a precipitated organic solid. (Courtesy of Bethlehem Foundry & Machine Co.)

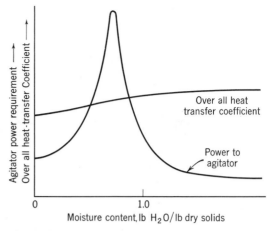

Figure 18.26. Performance of a pan dryer drying a material that passes through a paste or gel stage.

which gives a total heat transfer rate of

$101.9 \times 50 = 5095$ Btu/hr sq ft of dryer cross section

Since the material being dried is granular and is being dried as individual particles, it is safe to assume that all drying occurs in the constant-rate period. The volumetric over-all heat-transfer coefficient can be calculated from Equation 18.16. Thus,

$$Ua = \frac{0.63(50)^{0.16}}{5}\left[17.90 + \frac{5.2 \times 50 \times 240}{(50)^{1/2}}\right]^{0.5}$$

$$= 0.236(17.90 + 8820)^{0.5} = 22.8 \text{ Btu/hr °F cu ft}$$

The required ΔT_{lm} then is

$$\Delta T_{lm} = \frac{5095}{22.8 \times 60} = 3.72°F$$

Since ΔT_1 is about 414°F, ΔT_2 will be very close to zero. Thus, for a solid-feed rate of 17.9 lb/hr of dry solid, the inlet-air temperature would have to be 544°F, and 50 lb/hr sq ft of air would have to be used.

In this illustration, a dryer was used which was not suitable for the operation. An internally heated dryer would have allowed a much greater feed rate. The low permissible gas-flow rate resulted from the very fine particles handled. Here, pelletizing the particles in a preliminary step might be economically attractive. Finally, better throughput and utilization of equipment would have been obtained had the dryer been larger in diameter but shorter.

Pan Dryers. Pan dryers, in which the material being dried is moved by an agitator or scraper over the hot walls of the containing vessel, provide a means for batch drying of small volumes of pastes or slurries. Agitator speeds are usually below 10 rpm with the intention of merely preventing the solids from forming a hard cake on the heating surface and of keeping the charge moving and broken up. These units can be designed to run under vacuum where necessary Figure 18.25 shows such a dryer. These dryers are built in sizes up to 10 ft in diameter with capacities up to 1000 gal and heating areas up to 125 sq ft.

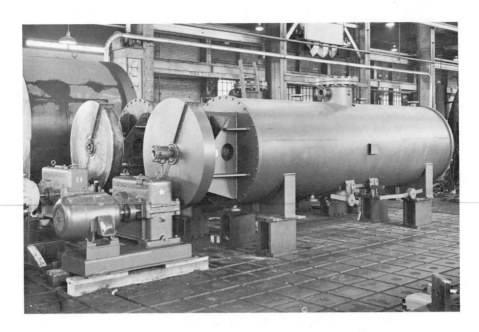

Figure 18.27. Vacuum rotary dryer 4 by 15 ft. (Courtesy of Blaw-Knox Co., Buflovak Equipment Division.)

Figure 18.28. Interior view of dryer shown in Figure 18.27. The helical agitator with individually adjustable straight-faced scraper blades is shown. This agitator moves the charge during drying and assists in discharging the dried solids. (Courtesy of Blaw-Knox Co., Buflovak Equipment Division.)

Very few quantitative performance data are available for use in the design of these dryers. Over-all heat-transfer coefficients varying between 5 and 175 Btu/hr sq ft °F have been reported (13), but for design purposes, where no pilot-plant information is available, coefficients between 10 and 15 Btu/hr sq ft °F should be used.

The variation in agitator power requirement versus solids moisture content (13) is rather interesting. A typical plot is shown schematically in Figure 18.26. At high moisture contents, the slurry is relatively free moving, but, as drying proceeds, it becomes more viscous and pasty. It finally becomes a heavy, adhesive mass which is very difficult to move. If drying continues, the mass breaks into lumps which move much more readily, and the lumps continually break up until a free-flowing granular product is obtained. The enormous changes in physical properties of the material being dried seem to affect the heat-transfer coefficient only to a minor extent. The moisture content at which the power peak occurs and the height of the power peak must depend upon the slurry being dried and the dryer design, but no studies have been reported along this line.

An alternate form of batch slurry dryer is the horizontal vacuum dryer such as that shown in Figures 18.27 and 18.28. The dryer consists of a jacketed shell in which the charge is placed, and in which agitator scrapers keep it moving while the liquid is evaporated. The agitator shaft and arms, as well as the shell, may be heated to increase the drying rate.

Drum Dryers. Slurries and pastes may be dried continuously on drum dryers, provided that the dried product is not hard and gritty enough to score the drum surface. Thus, drum dryers are used most often to dry

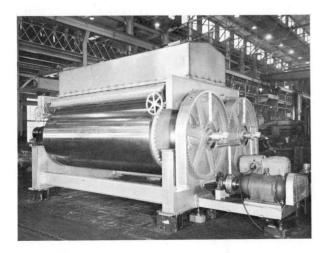

Figure 18.29. An atmospheric double-drum dryer. Each roll is 5 ft in diameter by 12 ft long and has a chromium-plated surface. Discharge screw conveyor has been removed to show the drum. Drum separation and rate of rotation can be adjusted to offset changes in feed properties. (Courtesy of Blaw-Knox Co., Buflovak Equipment Division.)

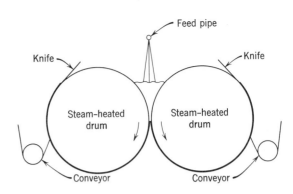

Figure 18.30. Schematic drawing of double-drum dryer showing feed arrangement, drum-rotation direction, and product-discharge system.

organics. These dryers are rotating horizontal cylinders heated with steam on the inside. The slurry being dried is spread over the outside surface of the dryer, clinging to it and drying as the hot drum rotates. When the slurry has made about three-quarters of a complete rotation on the drum surface, it is scraped off with a doctor knife. Rotation rates of 1 to 10 rpm are normal. The rotation rate, drum-surface temperature, and slurry-layer thickness are adjusted to give the desired moisture content in the dried cake scraped from the roll. The product comes off in flaked form.

Various drum arrangements and feed methods are available. In *double-drum dryers*, the slurry is fed into the nip between two rotating drums. The thickness of slurry layer held on the drum will depend upon the characteristics of the slurry and of the drum surface as well as the spacing between drums. Figure 18.29 shows such a dryer, and Figure 18.30 shows the construction arrangement of the dryer. Here, the feed is sprayed into the nip from a perforated pipe. The feed may also be admitted through a pipe which swings like a pendulum from end to end of the dryer.

Twin-drum dryers operate on the same principle as double-drum dryers except that they rotate in the opposite direction, that is, away from each other at the top. Drum spacing no longer influences the thickness of the drying layer but can be adjusted to break up bubbles and smooth out the coating of slurry on the drums where the drums are dip fed. The simplest feed device is merely to have the rolls dip into the feed slurry. Other feed devices are shown in Figures 18.31 and 18.32. The top feeding of Figure 18.31 permits a thick coat to be formed on the roll surfaces, which is advantageous when the material is granular and easily dried. The splash feed of Figure 18.32 forces the slurry against the dryer rolls, thus helping to prevent it from falling off as drying proceeds.

Single-drum dryers and vacuum single- and double-drum dryers are also available and operate on the same

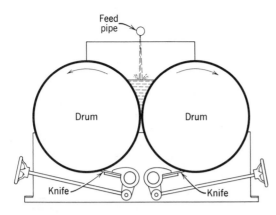

Figure 18.31. Twin-drum dryer with top feed. The feed is held in place by end boards and by the closeness of the two rolls. (Courtesy of Blaw-Knox Co., Buflovak Equipment Division.)

principle as those described above. These drum dryers are shown schematically in Figures 18.33 and 18.34.

The operation of drum dryers and the design of a drum dryer for a particular service are still more art than science. Usually, the design must be based on pilot-plant tests under the proposed operating conditions. Even then, scale-up methods are uncertain. Experiments with drum dryers have yielded a considerable amount of information on drying particular materials under a few conditions. The results cannot be directly extrapolated to plant-sized units, nor can they be applied to the drying of other materials.

Problems arising in operation of drum dryers are concerned with obtaining an even, unbroken sheet of drying material on the roll, with adhesion to the roll, and with obtaining the proper conditions to give a product of desired properties. The need for a high production rate can be met only with a thick, dense, and continuous layer of drying solid. For this purpose, the feed slurry should be as concentrated as possible. Of course, as the concentration increases so does the viscosity, so that an even distribution along the roll length is hard to maintain, and the slurry may not wet and adhere to the roll surface. Adjustment of roll spacing and revolutions per minute may be required by minor changes in feed conditions. Production rates in the order of 5 lb of dry solids/hr sq ft of drum surface are normal when steam at 50 psig is fed to the drums.

Spray Dryers. Spray dryers will handle slurries and solutions at relatively high production rates, whereas drum dryers are most conveniently used at low or moderate throughputs. The product is obtained in the form of small beads which are reasonably uniform in size and relatively dust free and which may be quite rapidly redissolved. For these reasons, spray drying is used in the production of dried milk, coffee, soaps, detergents, and many other products for home use. In a spray-drying operation, the slurry is pumped to a nozzle or rotary-disk atomizer which sprays the feed in fine droplets. The droplets are subjected to a stream of hot air flowing either countercurrently or cocurrently in relation to the falling droplets, or in a complex mixture of the two paths. After being dried, the solid particles are separated from the air by gravity. The exhaust air carries the fines from the drying chamber, and so the air is passed through cyclone separators and perhaps bag filters and wet scrubbers before issuing to the atmosphere.

Figure 18.35 shows a spray-dryer installation used to dry clay slip, and Figure 18.36 is a schematic drawing of this system showing its major constituents.

Any spray-drying unit has as its major parts a feed delivery and atomization system, a hot-gas production

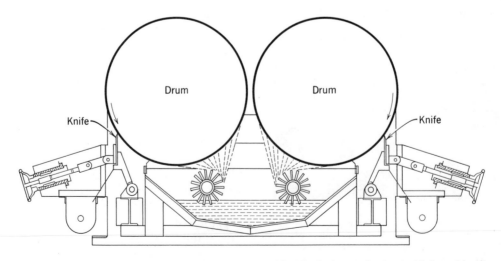

Figure 18.32. Twin-drum dryer with splash feed. Splashing the feed unto the drum aids in making it adhere to the drums and produces a dense coating. (Courtesy of Blaw-Knox Co., Buflovak Equipment Division.)

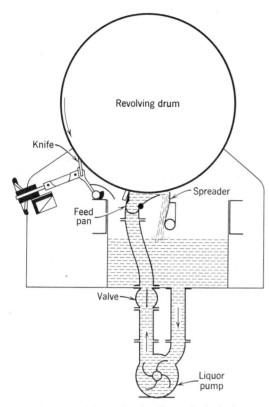

Figure 18.33. Schematic drawing of single-drum dryer with pan feed. The dryers may also be dip fed, in which case the drum rolls through the pan holding the feed slurry, or splash fed as shown in Figure 18.32. (Courtesy of Blaw-Knox Co., Buflovak Equipment Division.)

Figure 18.35. Spray dryer for drying clay slip. (Courtesy of Proctor and Schwartz, Inc.)

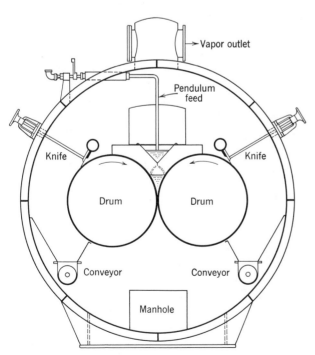

Figure 18.34. Schematic diagram of a vacuum double-drum dryer. (Courtesy of Blaw-Knox Co., Buflovak Equipment Division.)

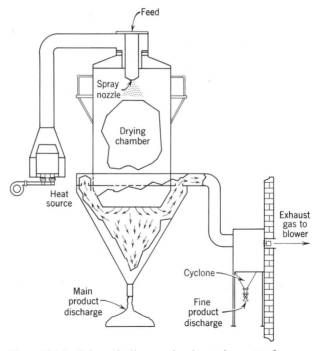

Figure 18.36. Schematic diagram showing major parts of a spray-drying system. (Courtesy of Proctor & Schwartz, Inc.)

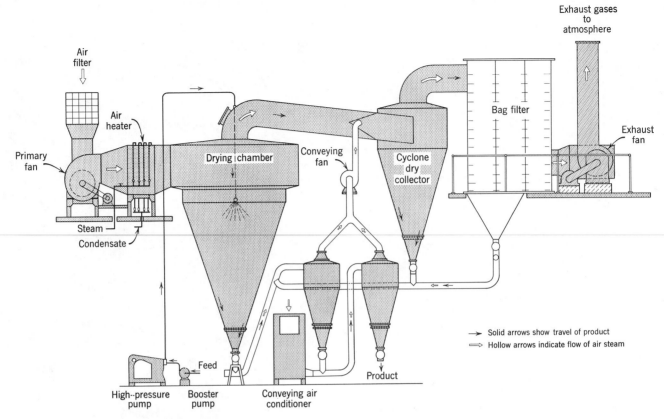

Figure 18.37. A steam-heated spray dryer with a dry collection system. This dryer was designed for drying coffee solubles, but a similar arrangement can be used for drying pharmaceutical products, organic dyes, pigments, resins, and other high-value products that cannot be processed with wet collection equipment and that require lower drying temperatures. (Courtesy Swenson Evaporator Co., a Division of The Whiting Corp.)

and delivery system, a drying chamber, a solids-gas separation system, and a product-discharge system. The design of each of these systems depends upon the material being dried and is influenced by the design of the rest of the unit. Thus, the form of the finished spray dryer may vary enormously from product to product and even from installation to installation for the same product.

Moreover, the properties of the product depend to a great extent on the conditions under which it has been dried. The fineness and uniformity of the spray, the behavior of the sprayed droplets during drying, and the temperature, humidity, mass-flow rate, and flow pattern of the drying air all influence the properties of the dried product. In general, the designer is called upon to make a dryer that will produce a product of fixed bulk density, particle-size distribution, moisture content, and color at a fixed production rate. At present, although spray dryers have been studied intensively, the system design can be achieved only on the basis of trial runs under exact production conditions. Even pilot-plant runs using the same feed sprayed through the same nozzle into air of identical conditions are unreliable because air-flow patterns cannot be reproduced.

The wide variation in component design from one spray-dryer system to another, and the different selection and arrangement of these components, can be noticed in comparing Figures 18.36 and 18.37. The spray dryer of Figure 18.37 is treating a heat-sensitive material, and one which must be fit for human consumption. Thus the temperature must be kept lower, and standards of cleanliness must be much more stringent than in the drying of clay slip. Also both the value of the product and the nuisance that would be caused by letting it escape into the atmosphere lead to a much more complicated and effective product collection system for the coffee dryer than is required by the clay-slip dryer.

Perhaps the most important part of the spray-dryer system is the feed atomizer. Three classes of atomizers are normally used in spray dryers. *Two-fluid nozzles*, such as that shown in Figure 18.38, are used in low-production-rate drying, particularly if a fine particle size is desired. The mechanism of atomization in these nozzles is by air shattering the liquid streams. At low air pressure, the gas blows a bubble of liquid which collapses into droplets. At higher pressures, the liquid issues from the nozzle as ligaments which are then torn into droplets by the gas stream. The average droplet size decreases as the pressure of both streams to the nozzle increases (14).

Single-fluid pressure nozzles operate at higher throughput and give larger and more uniform droplet size than does the two-fluid nozzle shown in Figure 18.38. Thus, they are used in production-sized spray dryers, whereas pneumatic, or two-fluid, nozzles are seldom used except in laboratory or pilot units. A pressure nozzle built for spray-dryer service is shown in Figure 18.39. The nozzle produces a high-velocity tangential motion in the liquid being sprayed. The resulting centrifugal force makes the fluid swirl around the circumference of the nozzle hole forming an air core along the axis of the hole. The fluid then swirls out into a hollow cone-shaped sheet which finally breaks into droplets. Figure 18.40 shows this form of atomization from a swirl-type pressure nozzle operating at 30 psig (14) spraying water. To get the same degree of atomization with a viscous liquid, a pressure of several thousand pounds per square inch might be necessary.

Both pneumatic and pressure nozzles require the fluid being sprayed to flow through narrow passages. Thus, any lumps, crystals, or other solids suspended in the fluid will clog the spray nozzle. Moreover, even the finest grit will wear the nozzle tip, causing enlargement and uneven atomization. For these reasons, the material delivered to the nozzles must be completely homogeneous. If the feed is formed by mixing several ingredients, agitation must be thorough enough to eliminate any of these small particles. Even so, fine screens are usually placed in the feed line ahead of the nozzle and may be actually built into the nozzle. Homogenizers are often incorporated in the high-pressure

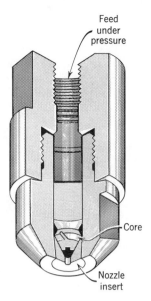

Figure 18.39. Single-fluid pressure nozzle. The nozzle is of the grooved-core design. Nozzles of this design will spray 60 to 1600 gal/hr of water. Throughput for a single nozzle varies by a factor of 2.5 to 3 as the pressure changes from 1000 to 7000 psig. The nozzle body is normally stainless steel, whereas the replaceable core and nozzle insert are of hardened steel or tungsten carbide. (Courtesy of Spraying Systems Co.)

positive-displacement pumps used to feed pressure nozzles. Homogenizers force the feed through short capillary tubes at high velocity, thus breaking up the feed stream and violently remixing it.

Centrifugal disk atomizers may be used to spray fluids that cannot be made homogeneous enough to pass through a nozzle. Moreover, they produce an extremely uniform droplet size and do not require a high-pressure feed or impart an axial velocity to the sprayed droplets. Such an atomizer is shown in Figure 18.41. In the atomizer, the feed liquid is delivered to a wheel rotating at 6000 to 20,000 rpm. The fluid is accelerated to a high centrifugal speed on the disk and forced off the disk. At very low feed rates and slow rates of revolution, the fluid issues from the disk as droplets; at higher flow rates and higher revolutions per minute, as ligaments that then break into droplets; and at high flow rates and revolutions per minute, as a continuous film which breaks into droplets. The three regimes of droplet formation from a spinning disk are shown in Figure 18.42 (10).

Most spinning-disk atomizers used for spray drying operate with high feed rate and high speed so that they are in the film-formation region. In application, only one disk atomizer is used in a spray dryer, whereas

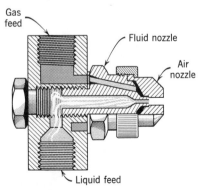

Figure 18.38. Two-fluid nozzle. Here the fluid to be sprayed is delivered to the nozzle at low pressure, while air, steam, or gas is supplied at higher pressure (up to 70 psig) to supply energy for atomization. In the nozzle internal mixing of liquid and gas occurs. In other nozzles, the gas and liquid are discharged separately at the nozzle face, producing external mixing. (Courtesy of Spraying Systems Co.)

Figure 18.40. Break-up of liquid issuing from a lucite model of a swirl pressure nozzle operating at 30 psi. The hollow liquid cone in which waves break into droplets is clearly shown as is the hollow core within the nozzle (14). (Courtesy of American Institute of Chemical Engineers.)

several nozzles may be used simultaneously. In this way, eroded or plugged nozzles may be replaced with only a minor reduction in throughput for the entire dryer. This ease of maintenance compensates for the fact that the nozzles are more likely to require replacement than is the disk atomizer.

After leaving the atomizer, the liquid droplets fall through the hot gas in the drying chamber. If the feed were pure water, the droplet would evaporate at the wet-bulb temperature of the drying air until it completely disappeared. In drying solutions or emulsions, the drying particle reaches a temperature somewhat higher than the wet-bulb temperature as drying proceeds. Initially, the liquid evaporates from the droplet surface. The relatively dry surface may form a tough shell through which liquid from the interior of the droplet must diffuse in order to escape. Since this diffusion is a much slower process than is the transfer of heat through the droplet shell to the interior, the liquid tends to

evaporate in place. As a result, the droplet swells making the shell thinner and diffusion through it faster. If the shell is relatively inelastic and impermeable, the internal evaporation is usually great enough to cause rupture of the shell either producing fragments or making a new hollow bulb beside the original one. Thus, the typical spray-dried product consists of ruptured hollow spheres and whole spheres. Figure 18.43 shows a photomicrograph of a spray-dried organic acid. Here, the rupturing of droplets was relatively nonviolent, producing secondary hollow spheres attached to the original ones. In Figure 18.44, a similar photomicrograph is shown of a spray-dried molding powder. Here, rupture of the droplets resulted in the production of fragments.

It is obvious that the time required for drying cannot be calculated by the simple mechanism which applies for drying beds of porous material. However, the time will depend upon the temperature, humidity, and flow

(a) Centrifugal atomizing assembly showing atomizing disk in place. (Courtesy, Bowen Engineering, Inc.)

(b) Disassembled view of centrifugal atomizing wheel. Feed enters in the annular ring around the central shaft. It is accelerated to disk speed as it flows down the central cone and out toward the ejection ports. Note the erosion-resistant facing at each ejection port. (Courtesy, Nichols Engineering and Research Corp.)

Figure 18.41

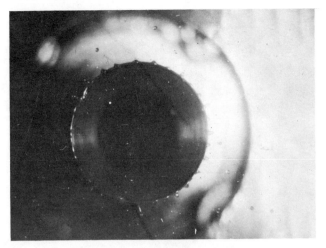

(a) Direct drop formation from a spinning disk.

(b) Ligament formation from a spinning disk.

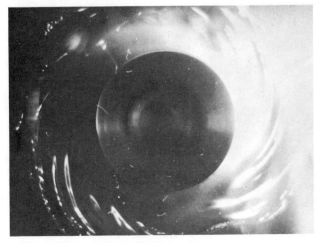

(c) Film formation and break-up from a spinning disk.

Figure 18.42. Photographs of drop formation from a spinning disk (10).

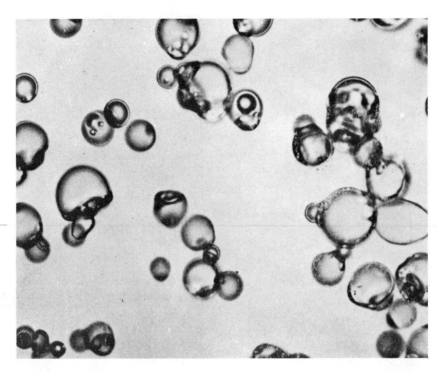

Figure 18.43. Photomicrograph of an organic acid spray dried from a true solution. (Courtesy of Swenson Evaporator Co., a Division of The Whiting Corp.)

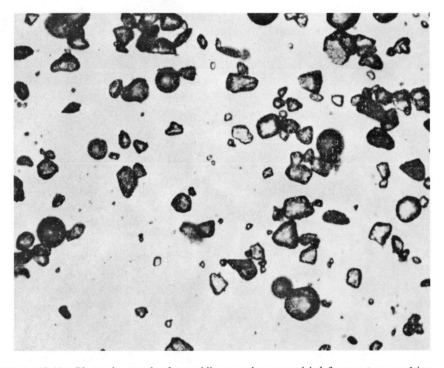

Figure 18.44. Photomicrograph of a molding powder spray dried from a true emulsion. (Courtesy of Swenson Evaporator Co., a Division of The Whiting Corp.)

conditions of the drying gas, the size of droplets produced by the atomizer, and the properties of the material being dried. Moreover, the properties of the finished product will depend upon the same factors. The importance of these factors has long been recognized, and a considerable amount of theoretical and experimental work has been done to try to relate them to each other and to the design of the dryer. Most of this work has pointed toward relating the drop-size distribution from the atomizer to its design, the properties of the feed, and the throughput. Since the mechanism of drop formation differs in the three major classes of atomizers, the correlations obtained have had to differ for the different atomizers.*

Two-fluid atomizers form a spray by impinging cocurrently a high-velocity gas jet against the liquid issuing from a nozzle as discussed above. For the break-up of alcohol–glycerine mixes in an air jet, Nukiyama and Tanasawa (18) obtained

$$D_{vs} = \frac{1410\sqrt{\sigma}}{v_a\sqrt{\rho_L}} + 191\left(\frac{\mu}{\sqrt{\rho_L\sigma}}\right)^{0.45}\left(\frac{1000Q_L}{Q_A}\right)^{1.5} \quad (18.17)$$

where D_{vs} = Sauter mean diameter, microns
  v_a = relative velocity between liquid and air, ft/sec
  σ = surface tension, dynes/cm
  μ = liquid viscosity, centiposies
  ρ_L = liquid density, lb/cu ft
  Q_L, Q_A = volumetric flow rates of liquid and air

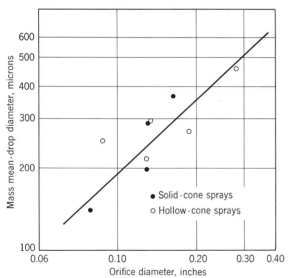

Figure 18.45. Effect of orifice diameter on mean drop size from pressure nozzles. $P = 50$ psig, fluid H_2O (19). (Courtesy American Chemical Society.)

---

* The description of a statistically large sample of small particles so that the particle-size distribution may be known is discussed along with other properties of particulate-solid systems in Appendix B. In the following paragraphs the reader is expected to be familiar with the material in this appendix.

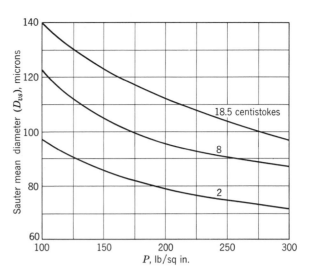

Figure 18.46. Influence of pressure and viscosity on the Sauter mean diameter of droplets from a swirl pressure nozzle (12). (Courtesy Inst. Fuel.)

They also proposed to correlate drop size through the empirical distribution equation

$$\frac{dn}{dD_p} = bD_p{}^m e^{-cD_p{}^\delta} \quad (18.18)$$

where dn/dD_p is the number of particles with dimensions between D_p and $D_p + dD_p$. They reported the exponents $m = 2$ and $\delta = 1$. Later work substantiated $m = 2$ but found δ to vary markedly with nozzle design.

In pressure nozzles, the spray is usually formed by forcing a centrifugally swirling liquid through an orifice. Normally the swirl produces an air core through the center of the nozzle to the end of the swirl chamber. In such a nozzle the mean drop size, usually expressed as a Sauter mean, is a function of nozzle diameter, pressure applied, liquid viscosity, and, to a lesser extent, surface tension. Figures 18.45 and 18.46 give some indication of the mean drop sizes expected from pressure nozzles. Empirical correlations have been developed by assuming that the Sauter mean diameter is independently a function of the tangential and the axial velocity components at the nozzle tip (24). The measured distributions were found to give a linear distribution on probability paper. The drop-size uniformity was correlated in terms of the standard deviation as measured from these lines. A modified standard deviation was found to be a function of tangential velocity and orifice diameter. Other empirical correlations for drop-size distribution have been offered, based on more easily measured properties of the nozzle, but they are usually applicable only to the nozzle design studied. In most of the studies, water was the primary fluid atomized. The effect of fluid properties, particularly viscosity, on the drop size produced has not been adequately determined.

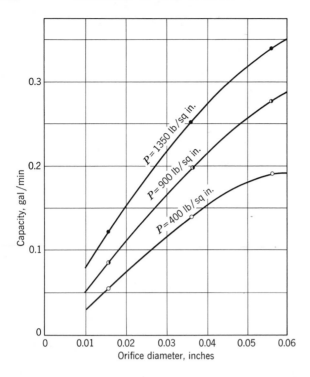

Figure 18.47. Effect of nozzle diameter on the capacity of grooved-core pressure nozzles (16). Groove area 7.7×10^{-4} sq in. Viscosity 50 centipoises. (Courtesy Am. Inst. Chem. Engrs.)

The capacity of the nozzles depends upon nozzle diameter, pressure, internal-nozzle design, and fluid viscosity. The importance of internal design is in the variation in tangential velocity component imparted to the swirling liquid and on pressure drop through the nozzle for a given size. The influence of viscosity is minor, but, in general, capacity increases slightly with increasing viscosity at constant feed pressure as long as atomization occurs. This is because the air-core diameter decreases as the viscosity increases. Nozzle capacities are frequently given in terms of a "flow number" which is pressure insensitive. Defining the flow number by the equation

$$K' = \frac{Q}{\sqrt{P}} \qquad (18.19)$$

where Q = capacity, vol/unit time
  K' = flow number, which is related to the efficiency of conversion of pressure to kinetic energy in the nozzle

it is found that K' is relatively independent of pressure but varies somewhat with liquid viscosity and markedly with nozzle design. Thus, K' can be used to obtain an estimate of the capacity of a nozzle at P_2 if its capacity at P_1 is known. Figure 18.47 shows the relation between orifice diameter, pressure, and capacity of grooved-core nozzles with fixed grooved area spraying a fluid of 50-centipoise viscosity.

In spinning-disk atomization, the fluid leaves the disk in streams which break into droplets from friction as they pass through the relatively still gas. The feed rate can be controlled independent of the disk. Although the hydrodynamics of this break-up is complicated, the correlation of mean drop sizes from spinning disks and the generalizing of drop-size distribution have been more successful than they were for two-fluid atomizers or for pressure nozzles. Friedman, Gluckert, and Marshall (5) correlated Sauter mean diameter through the dimensionless equation

$$\frac{D_{vs}}{r} = 0.4 \left(\frac{G}{\rho_L Nr}\right)^{0.6} \left(\frac{\mu}{G}\right)^{0.2} \left(\frac{\sigma \rho_L L_w}{2}\right)^{0.1} \qquad (18.20)$$

where r = disk radius
  G = mass velocity, lb/(ft of L_w) (unit time)
  σ = surface tension
  L_w = wetted periphery, ft
  N = disk speed, rpm

The spray distribution was found to form a straight line on log-normal probability paper. Thus, the entire spray pattern could be determined through the Sauter mean diameter and one other point. The point chosen was the maximum drop size. Friedman found the maximum particle size $D_{pm} = 3D_{vs}$ through a dimensionless correlation, but other workers found more uniform spray distributions (14).

The rates of heat and mass transfer to a falling drop and hence the drying rate have been studied using boundary-layer theory. The results show that these rates should be a maximum at the side of the drop facing the oncoming air, should decrease to a minimum near the boundary-layer separation point behind the drop equator, and should increase again to a higher value at the rear center of the drop. From these considerations, combined with heat- and mass-transfer equations, a set of simultaneous partial-differential equations results from which the rates of heat, mass, and momentum transfer can be determined. The general solution of these equations involves so many assumptions as to be of no practical value. In the limit for a finely dispersed system, the N_{Re} approaches zero, and, for this case, the solution gives (14)

$$\frac{h_c D_p}{k} = \frac{k_g \bar{M} D_p P}{\mathscr{D} \rho} = 2 \qquad (18.21)$$

where h_c = convective heat-transfer coefficient, Btu/hr sq ft °F
  k = gas-phase thermal conductivity, Btu/hr ft °F
  k_g = gas-phase mass-transfer coefficient, lb moles /hr sq ft atm
  $\bar{M}$ = mean molecular weight of gas phase
  $\mathscr{D}$ = gas-phase diffusivity sq ft/hr

which is of practical and theoretical importance as it fixes the conditions during evaporation to dryness of a liquid drop. Experimental studies (20) give

$$\frac{k_g \bar{M} D_p P}{\mathscr{D} \rho} = 2 + K_1 (N_{Sc})^m (N_{Re})^n \qquad (18.22)$$

where K_1, m, and n are constants. This equation should give by analogy (7, 14).

$$\frac{h_c D_p}{k} = 2 + K_2 (N_{Pr})^p (N_{Re})^q \qquad (18.23)$$

where $\quad K_1 = K_2 = 0.6$
$\qquad p = m = \frac{1}{3}$
$\qquad q = n = \frac{1}{2}$

At high N_{Re}, the effect of the constant 2 becomes negligible so that Equations 18.22 and 18.23 convert into the usual j-factor relation.

The time of exposure of a drop to the spray-drying atmosphere depends upon the drop size, particle shape and density, and gas-flow path and velocity. The usual equations for free fall of a particle through a fluid (see Chapter 22) will apply after the initial jet action from the nozzle has been spent. Of course, the numerical values employed must change throughout the particle-size range that is present, and will also change as the particle changes shape and density during fall.

The time required to dry a droplet can be estimated from the usual drying-rate equations if they can be solved. The constant-rate period will end when the drop surface becomes solid. This may occur while the interior of the drop is still a relatively dilute solution. If swelling and rupture do not then occur, the rate during the falling-rate period will be controlled by diffusion through the shell.

The production engineer must know how variations in feed or operating conditions affect his spray-dried product. The effects differ markedly from product to product, and only a few generalizations can be made concerning them. Duffie and Marshall (3) studied the effect of feed temperature, air temperature, and feed concentration on the bulk density of product for typical spray-dried materials and compared their results with other published results. Only in the case of air temperature was the behavior consistent for all the materials studied, which ranged between inorganic salts and corn syrup. In all cases, increased inlet-air temperature decreased the bulk density. In the case of feed concentration and temperature, effects on atomization and on drying behavior both influenced the product bulk density but in conflicting ways. Increasing feed concentration put more solid in each droplet of fixed size but also increased the average particle size. The net effect on most materials studied was a slight decrease in product bulk density with increased feed concentration.

Increasing the feed temperature decreased its viscosity, which decreased the pressure drop through a nozzle and resulted in a slightly smaller sprayed droplet. On the other hand, the initial warm-up period may be faster, the particle swelling greater, and final moisture content lower. On most materials sprayed, the final bulk density decreased slightly with increased feed temperature, though gelatin appeared to act in a reverse manner.

Qualitatively, the normal result of other operating changes may be inferred from experience. Increasing the pressure to a spray nozzle should increase the throughput and the droplet size. The increase should result in lower air temperatures and a product of higher moisture content, larger particle size, and higher bulk density. Increasing the throughput to a disk atomizer should have similar effects. If, at the same time, the air temperature is increased, the net result may be a product of larger particle size but relatively unchanged moisture content or bulk density.

The dryers discussed above are only a few of the many commercial types. They have been discussed here because they illustrate a wide variety of dryers and because they are among the most common. Under special conditions, dryers will be met which differ fundamentally from them, as, for instance, infrared dryers where most of the heat is transferred by radiation. But, most dryers differ only in minor detail from the ones discussed here. For example, cylinder dryers such as those used in drying paper are essentially drum dryers to which the material is fed in sheet form.

REFERENCES

1. Ceaglske, N. H., and O. A. Hougen, *Trans. Am. Inst. Chem. Engrs.*, **33**, 263 (1937); *Ind. Eng. Chem.*, **29**, 805 (1937).
2. Chu, J. C., A. M. Lowe, and D. Conklin, *Ind. Eng. Chem.*, **45**, 1586 (1953).
3. Duffie, J. A., and W. R. Marshall, *Chem. Eng. Progr.*, **49**, 480 (1953).
4. Findlay, A., A. N. Campbell, and N. O. Smith, *The Phase Rule and Its Applications*, 9th ed. Dover Publications, New York, 1951.
5. Friedman, S. J., R. A. Gluckert, and W. R. Marshall, *Chem. Eng. Progr.*, **48**, 181 (1952).
6. Friedman, S. J., and W. R. Marshall, *Chem. Eng. Progr.*, **45**, 482 573 (1949).
7. Frössling, N., *Gerlands Beitr Geophys.*, **52**, 170 (1938).
8. Gilliland, E. R., *Ind. Eng. Chem.*, **30**, 506 (1938).
9. Glasstone, S., *Textbook of Physical Chemistry*, 2nd ed., D. Van Nostrand Co., Princeton, 1946, 487 pp.
10. Hinze, J. O., and H. Mulborn, *J. Appl. Mechanics*, **17**, 145 (1950).
11. *International Critical Tables*, Vol. 2, pp. 322–325.
12. Joyce, J. R., *J. Inst. Fuel*, **22**, 150 (1949).
13. Laughlin, H. G., *Trans. Am. Inst. Chem. Engrs.*, **36**, 345 (1940).
14. Marshall, W. R., Jr., "Atomization and Spray Drying," *Chem. Eng. Progr. Monograph Series No.* **50** (2) (1954).

15. McCready, D. W., and W. L. McCabe, *Trans. Am. Inst. Chem. Engrs.*, **29**, 131 (1933).
16. McIrvine, J. D., M. S. Thesis, University of Wisconsin (1953).
17. Newman, A. B., *Trans. Am. Inst. Chem. Engrs.*, **27**, 203 (1931).
18. Nukiyama, S., and Y. Tanasawa, *Trans. Soc. Mech. Engrs.* (*Japan*) **4**, No. 14, 86 and No. 15, 138 (1938); **5**, No. 18, 63, and 68 (1939); **6**, No. 22 II-7 and No. 23 II-8 (1940).
19. Pigford, R. L., and C. Pyle, *Ind. Eng. Chem.*, **43**, 1649 (1951).
20. Ranz, W. E., and W. R. Marshall, Jr., *Chem. Eng. Progr.*, **48**, 141, 173 (1952).
21. Seborg, C. O., *Ind. Eng. Chem.*, **29**, 169 (1937).
22. Shepherd, C. B., C. Hadlock, and R. C. Brewer, *Ind. Eng. Chem.*, **30**, 388 (1938).
23. Sherwood, T. K., *Ind. Eng. Chem.*, **21**, 12, 976 (1929); *Trans. Am. Inst. Chem. Engrs.*, **32**, 150 (1936).
24. Tate, R. W., and W. R. Marshall, Jr., *Chem. Eng. Progr.*, **49**, 169, 226 (1953).
25. Toner, R. K., C. F. Bowen, and J. C. Whitwell, *Textile Research J.*, **17**, 7 (1947).
26. Wenzel, L. A., Ph.D. Thesis, University of Michigan (1949).
27. Wenzel, L. A., and R. R. White, *Ind. Eng. Chem.*, **43**, 1829 (1951).

PROBLEMS

18.1. The following data were obtained when drying sand in superheated steam. (26):

Drying Time, hr	Sample Weight, lb	Drying Time, hr	Sample Weight, lb
0	43.72	5.00	37.93
0.25	43.32	5.25	37.70
0.50	42.95	5.50	37.48
0.75	42.54	5.75	37.28
1.00	42.21	6.00	37.12
1.25	41.85	6.25	36.90
1.50	41.52	6.50	36.73
1.75	41.20	6.75	36.58
2.00	40.89	7.00	36.42
2.25	40.57	7.50	36.22
2.50	40.30	8.00	36.05
2.75	40.03	8.50	35.83
3.00	39.81	9.00	35.69
3.25	39.59	9.50	35.61
3.50	39.36	10.00	35.50
3.75	39.08	10.50	35.39
4.00	38.84	11.00	35.35
4.25	38.60	11.50	35.33
4.50	38.40	12.00	35.31
4.75	38.16		

The sample was 2 in. thick, it weighed 35.28 lb when dry. It was dried in steam at 50 psia, superheated 53.5°F, and flowing at 1000 lb/hr sq ft. Drying was from the top face only, and this face measured 8 in. by $29\frac{3}{4}$ in. Determine:

(a) The critical moisture content.

(b) The drying rate during the constant-rate period.

(c) The heat-transfer coefficient observed during the constant-rate period.

Compare the answer for part (c) with that obtained by using Equation 18.12.

18.2. A granular solid with dry bulk density of 100 lb/cu ft is being dried in a batch drier in air at 150°F with a humidity of 0.005 lb

H₂O/lb dry air. The solids, containing 0.5 lb H_2O/lb dry solids are in 1-in. pans insulated so that heat and mass transfer occur to the top surfaces only. The solids are to be dried to a final moisture content of 0.02 lb H_2O/lb dry solids and have a critical moisture content of 0.10 lb H_2O/lb dry solids. Air passes over the pans at a mass velocity (G_Y) of 1200 lb/hr sq ft. Heat transfer by conduction and radiation may be neglected. For this granular material, $\bar{X}_E' = 0$.

(a) Calculate the drying time required.

(b) What would be the drying time if the air flow rate was raised to 1800 lb/hr sq ft?

(c) What would be the drying time if the air temperature was raised to 200°F?

18.3. Brick clay of bulk density $\rho_B = 110$ lb/cu ft is being dried in a conveyor dryer with a 3 ft wide belt. The brick-clay layer is 1 in. thick, and air flows countercurrent to the belt. The space for air flow above the belt is 1 ft high and is equal to the belt width. Air enters the dryer at 80°F with a 50°F wet-bulb temperature. It is mixed with an equal amount (on a weight-of-dry-air basis) of recycle air, and the combined air stream is heated to 250°F in a steam-tube heater. After passing over the sample, half the air is discharged from the dryer; the other half is reheated. Exit air is at 80 per cent relative humidity, product is at $\bar{X}_f' = 0.2$ lb H_2O/lb dry solids, and feed enters at $\bar{X}_0' = 1.5$ lb H_2O/lb dry solids. For this material the critical moisture content, ($\bar{X}_c'$), is 0.15 lb H_2O/lb dry solids. Assuming that heat and mass are only transferred to the top surface of the clay and that the clay enters the dryer at the adiabatic-saturation temperature of the exit air, determine:

(a) The mass flow rate of air, lb/hr sq ft, passing over the belt if this belt moves at a velocity of 1 ft/min.

(b) The length of belt required.

18.4. A conveyor dryer is to be used to dry wood chips and sawdust from an initial moisture content of 0.3 lb/lb of dry solids. The critical moisture content for this material is found to be 0.1 lb of H_2O/lb of dry solids. The drying air enters and flows countercurrent to the solids flow at a rate of 10,000 cu ft/min at 240°F with a humidity of 0.005 lb H_2O/lb dry air. The conveyor belt is 3 ft wide and is loaded to a wet feed weight of 26 lb/sq ft of belt. It moves at a velocity of 1 ft/min with air continuously passing through it and through the material held on it. Under these conditions heat-transfer coefficients of 100 Btu/hr °F (sq ft of bed area) can be expected in the constant-rate period. The feed enters the dryer at 95°F. The dryer is 100 ft long. What is the final moisture content and temperature of the wood chips? What is the outlet air condition?

18.5. The following data on the drum drying of sodium acetate on a trough-fed double-drum dryer has been reported (Harcourt, a paper presented at the Niagara Falls Meeting, ASME, September 17–23, 1936):

Moisture Content, weight percent		Steam Pressure, psia	Drum Speed, rpm	Feed Temperature, °F	Capacity, lb product/ hr sq ft
Feed	Product				
39.5	0.44	70	3	205	1.57
40.5	10.03	67	8	200	5.16
63.5	9.53	67	8	170	3.26

It is proposed to drum-dry sodium acetate from an initial 60 per cent solids by weight to 95 per cent solids in a double-drum dryer 10 ft long by 3 ft in diameter. Steam at 70 psia will be used as heating medium, and the feed will be preheated to 200°F. Recommend a drum speed to give the desired product concentration, and estimate the production rate in pounds per hour of dried product. Discuss the areas of uncertainty in your solution, list any assumptions you have made, and suggest any additional physical or chemical data that would decrease the uncertainty in your answer.

18.6. In a pilot process, mashed potatoes are dried in pans $\frac{1}{2}$ in. deep which are insulated on the bottom. Drying air is at $180°F$ with a 10 per cent relative humidity. Drying from an initial moisture content of 0.6 lb H_2O/lb of dry solids to a final moisture content of 0.15 lb/lb of dry solids requires 6 hr. All the drying is in the falling-rate period and is diffusion controlled. The equilibrium moisture content is 0.1 lb/lb of dry solids when in contact with air at this temperature and humidity. Bulk density of the product is 0.75 gm/cu cm.

In the plant process, the potatoes are to be dried from the same initial moisture content as in the pilot process to a final moisture content of 0.25 lb/lb. The process will take place in a pan dryer with pans 2 in. deep but with perforated metal bottoms so that drying occurs from both faces. If the drying-air conditions are controlled to duplicate those in the pilot dryer, what drying time will be required?

18.7. A steam-tube rotary dryer is to be used to dry crushed dolomite ($\rho_B = 180$ lb/cu ft) of $-14 + 20$ mesh particle size range ($\bar{D}_p = 0.039$ in). The dryer is 4 ft in diameter and can be purchased in 5-ft increments of length. The 40 steam tubes are 3 in. in diameter located in two concentric rings around the inside of the shell. They are of the low-finned variety with a total surface area of 40 sq ft/ft of dryer length. Previous experience indicates that the over-all coefficient of heat transfer from the condensing $325°F$ steam to the air in the dryer will be about 10 Btu/hr sq ft °F. Air enters the dryer at $300°F$ with an initial humidity of 0.003 lb H_2O/lb dry air, and at a flow rate of 1500 lb/hr sq ft of dryer cross section. Solids, flowing countercurrent to the air, enter at 0.4 lb H_2O/lb dry solids and leave at 0.02 lb H_2O/lb dry solids. Product is obtained at a rate of 1500 lb/hr. Throughout the dryer the solids exist at the wet-bulb temperature of the drying air.

If the dryer is sloped at 1 in./ft of length and rotates at 5 rpm, determine the length of dryer required.

18.8. A fibrous material is dried in a countercurrent, adiabatic, tunnel dryer in sheets 4 ft by 8 ft by 1 in. in size. The sheets are carried through the drier on racks holding 20 sheets separated so that each sheet dries from both sides. 50 lb of air at $260°F$ with a $105°F$ wet-bulb temperature enters the dryer per pound of dry stock and flows around the stock at a rate of 800 lb/hr sq ft of free area. The stock enters at $80°F$ with a moisture content of 0.80 lb H_2O/lb of dry stock and leaves with a moisture content of 0.04 lb H_2O/lb of dry stock. The stock has a bulk dry density of 40 lb/cu ft and a heat capacity of 0.3 cal/gm °C. Pilot tests indicate that a critical moisture content of 0.25 lb H_2O/lb of dry stock can be expected, but that there is negligible equilibrium moisture content. Determine the drying time required and the condition of the exhaust air.

18.9. Sand is continuously dried by conveying it at a rate of 1 ton of dry solids per hour through a drying chamber countercurrent to the drying air. The sand enters at $80°F$ with a moisture content of 0.6 lb H_2O/lb of solids and leaves at $180°F$ with a moisture content of 0.03 lb H_2O/lb dry solids. Air initially at $80°F$ and 70 per cent humidity is heated to $200°F$ in contact with steam-heated tubes and fed to the dryer at a rate of 12,000 scfm. After passing through the drying chamber, the air is exhausted to the atmosphere. Assume that each of four parallel conveyor belts carries a 2-in. thick, 3-ft wide layer of sand and that drying occurs from the top surface only. Determine the holding time required for the 4-ft diameter unit. The bulk density of dry sand is 100 lb/cu ft.

18.10. A single-shell, direct, countercurrent rotary dryer is needed to dry 10 tons/hr of manganese ore at $80°F$ with an average particle diameter of 0.04 in. from an initial moisture content of 0.1 lb H_2O/lb dry solids to a final moisture content of 0.005 lb H_2O/lb dry solids. At these conditions, the critical moisture content will be less than 0.005 lb/lb. The drying medium will be air initially at $80°F$ with a $70°F$ wet-bulb temperature heated to $260°F$ before entrance to the dryer. Specify dryer length, diameter, pitch, rpm, and air-flow rate.

18.11. Soluble coffee is spray dried from a 25 per cent solids solution having a density of 0.95 gm/cu cm and a viscosity of 5 centipoises. It is sprayed from a swirl-type pressure nozzle at 250 psia. For the purposes of this problem, consider the drops to be uniform in size. Drying occurs countercurrently. Assume that the sprayed particles neither swell nor shrink and that the drying history follows a constant-rate-linear-falling-rate pattern. Critical moisture content is 0.5 lb H_2O/lb dry solids, and equilibrium moisture content is 0.01 lb H_2O/lb dry solids. Air enters the dryer at $450°F$ and a humidity of 0.01 lb H_2O/lb dry air, and leaves at $200°F$.

(a) Determine the height of dryer required to dry to an average moisture content of 0.03 lb H_2O/lb dry solid, and the required air flow rate and tower cross section for a production rate of 1000 lb/hr of dried product. Use Equation 22.15, p. 451 for settling velocity.

(b) Why does the answer obtained above differ so radically from the height of commercial spray dryers?

18.12. In the spray described in Problem 18.11 the drop-size distribution is found to follow the normal probability function, that is

$$f(D_p)\sigma = \frac{1}{\sqrt{2\pi}} e^{\frac{-(D_p - \bar{D}_p)^2}{2\sigma^2}}$$

where $f(D_p)$ is the population density of drops of diameter D_p, $\bar{D}_p$ is mean drop diameter, σ is standard deviation, $\sigma = \sqrt{\Sigma(D_p - \bar{D}_p)^2/n}$, and n is the total number of drops. Plot this drop-size distribution when $2\sigma = \bar{D}_{vs}$, the Sauter mean diameter. Outline the steps needed to solve Problem 18.11 using this drop-size distribution.

chapter 19

Simultaneous Heat and Mass Transfer III:
Evaporation and Crystallization

In an evaporation operation, a solution is concentrated by boiling off the solvent. Usually, the desired product is the concentrated solution, but occasionally the evaporated solvent is the primary product as, for example, in the evaporation of sea water to make potable water. The concentrating may be continued until the solution becomes saturated and further until the solute precipitates as a crystalline solid. In this case, the operation is often called *crystallization.*

In either case, several rate processes occur. First, heat is transferred from the heating medium to the solution. The transfer usually occurs through a solid surface but may be a direct transfer from combustion gases to the evaporating solution. Second, mass and heat are simultaneously transferred from the liquid to the vapor phase. For this transfer process, all the rate, balance, and equilibrium equations which were discussed for a humidification or a distillation process apply. Finally, for a crystallization operation, there is also the simultaneous transfer of heat and mass between solutions and solid phases. Here again, all the considerations mentioned above apply. This process sequence differs from that of the operations discussed in Chapters 17 and 18 mainly in that it is *not* nearly adiabatic.

The *evaporation* process is really equivalent to a single-stage separation. However, the components normally separate so sharply that the composition equilibrium is frequently forgotten. In applications where the vapor phase is the valuable product, this equilibrium may be important. For example, if the vapor is to be condensed and recycled to a boiler for reevaporation, the stream must usually be specially treated to adjust pH and remove trace impurities. Figure 19.1 is a temperature-composition diagram for the sodium hydroxide–water system at 1 atm total pressure. Equilibrium concentrations of vapor and liquid phases occur at the intersection of saturated liquid and vapor boundary lines with a line of constant temperature. Note that a discernible concentration of sodium hydroxide in the vapor phase is not obtained until the liquid phase reaches 95 per cent NaOH and boils at some 700°F.

Despite the multitude of rate processes occurring, the engineer concerned with these operations can usually consider the entire process in terms of the heat transfer from heater to solution. It is this step which controls the over-all rate and which is most easily analyzed. The contributions of the other transfer processes are minor in their effect on the over-all operation. Moreover, so little is presently known, outside of the few companies that build evaporators, of most of the refinements in design that there is no reason to treat here more than the heat and material balances and the simple formulation of heat-transfer rate through the heating surface.

The initial heat-transfer process usually requires the transfer of heat from a heating fluid through a dividing wall to the evaporating solution. Thus, the series-resistance concepts developed in Chapter 14 apply. Equation 14.26 may be modified to

$$q = \frac{T_1 - T_2}{\dfrac{1}{h_1 A_1} + \dfrac{\Delta x}{k A_m} + R_d + \dfrac{1}{h_2 A_2}} \qquad (19.1)$$

where R_d is resistance of wall fouling, 1 refers to the heating fluid, and 2 refers to the evaporating solutions.

The heating fluid is usually condensing steam so that h_1 is the steam-condensing coefficient, though some small fraction of the heat may be transferred to the solution from superheated steam before the steam reaches the condensing temperature. The wall fouling depends upon the solution being evaporated, the time in service since cleaning, and the temperature difference across the surface. The coefficient h_2 is that for boiling solution though some heat may be transferred to the solution before it reaches the boiling point. Obviously all three of the terms can be estimated only very roughly from theoretical equations or empirical correlations. The over-all coefficient that results is then extremely uncertain. As a result, the heat-transfer rate is usually expressed by the simple equation

$$q = U_0 A_0 (-\Delta T) \qquad (14.31a)$$

where U_0, the over-all heat-transfer coefficient based on outside surface area, is determined experimentally in pilot equipment or from previous experience. Experience has shown U_0 to depend upon the properties of the solution, the heating medium, and the surface geometry and type. The solution properties are fixed only by specifying the components and their concentrations, the local pressure, and the details of fluid motion. The usual phase-rule considerations would allow the substitution of local temperature for either the pressure or one of the component concentrations. Similar specification is necessary to determine the heating-medium conditions. The surface conditions would include cleanliness and smoothness of surface and composition and thickness of metal. Some experimentally determined over-all coefficients are available (11), but

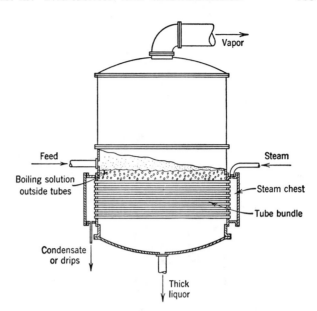

Figure 19.2. Cross-sectional diagram of horizontal-tube evaporator. (Courtesy Swenson Evaporator Company, a Division of The Whiting Corporation.)

they are seldom adequate as a basis for commercial evaporator design.

EVAPORATION

As discussed above, evaporation is the operation of concentrating a solution by boiling away the solvent. The concentration is normally stopped before the solute begins to precipitate from solution. Basically, then, an evaporator must consist of a heat exchanger capable of boiling the solution and a device to separate the vapor phase from the boiling liquid. In its most simple form it might be a pan of liquid sitting on a hot plate. The surface of the hot plate is a simple heat exchanger, and vapor disengaging is obtained by the large area for vapor flow and its consequent slow rate of flow. In industrial operation the equipment is usually arranged for continuous operation; the heat-exchange surface is vastly increased; boiling is much more violent, and vapor evolution is rapid. Such problems as foaming, scaling, heat sensitivity, corrosion, and space limitations are met. These problems have resulted in variations and refinements in evaporator design to meet different combinations of solution properties and economic conditions.

Evaporator Construction. Evaporator-body construction falls into a few general categories. They are illustrated in Figures 19.2 through 19.9.

The *horizontal-tube evaporator* of Figure 19.2 is one of the classic construction types and has been widely used for many years. The solution to be evaporated boils outside of horizontal tubes within which steam condenses.

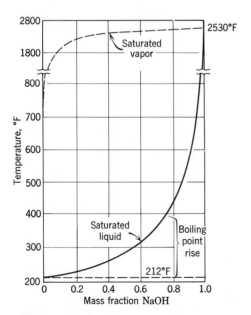

Figure 19.1. Phase equilibrium for H_2O–NaOH system at 1 atm total pressure.

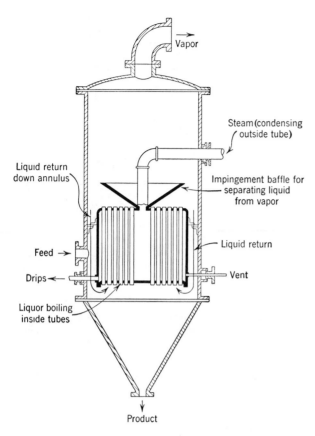

Figure 19.3. Cross-sectional diagram of basket evaporator. (Courtesy Swenson Evaporator Company, a Division of The Whiting Corporation.)

The horizontal tubes interfere with the natural circulation of the boiling liquid and thus minimize liquid agitation. As a result, the over-all heat-transfer coefficient is lower than in other forms of evaporators, especially if the solution is viscous. No provision is made for breaking foam that occurs because of the boiling action. Over-all coefficients of 200 to 400 Btu/hr sq ft °F result, depending upon the over-all temperature difference, boiling temperature, and solution properties. Moreover, fouling from the evaporating solution builds up on the outside of the tubes where it cannot be as easily removed as it could be from the inside of the tubes. Traditionally, the horizontal tubes have been inserted into the steam chests with packed glands rather than by rolled or welded joints. For these reasons, horizontal tubes are presently used mainly in small installations where the solution to be treated is dilute and neither foams nor deposits solids on the evaporator tubes, or where materials of construction preclude welding or rolling of tubes.

The *vertical-tube evaporators* of the *basket* and *standard vertical* varieties, which are shown in Figures 19.3 and 19.4 respectively, are distinct improvements over the horizontal-tube evaporator. In both of them the

solution boils inside vertical tubes with the heating medium, usually condensing steam, held in a chest through which the tubes pass. In the basket evaporator, the steam chest forms a basket hung in the center of the evaporator. Boiling or heating of the liquid in the tubes causes flow upward through the tubes, and unevaporated liquid flows downward through the annulus around the basket. In the standard vertical-tube evaporator, the steam chest is doughnut shaped. Liquid flows upward through the tubes and downward through the central hole. In large installations, there may be several return ducts rather than the single central one shown here. In both types, the tubes are inserted into tube sheets by rolling or welding, which considerably reduces the cost compared to the packed glands traditionally used in the horizontal-tube evaporator.

These evaporators overcome most of the operational disadvantages of the horizontal-tube evaporator. Natural circulation is promoted, having been measured at between 1 and 3 ft/sec in the tubes (4). As a result, coefficients are somewhat higher than in horizontal-tube evaporators and range between 200 and 500 Btu/hr sq ft °F, depending again on solution properties, over-all ΔT, and boiling temperature. Any solid deposit will build up inside the tubes where it is readily removed by mechanical cleaning. Foam breaking is ineffective, though entrainment separators and impingement baffles, which are usually supplied, will reduce the foam build-up. Viscous liquids can be handled, but circulation is

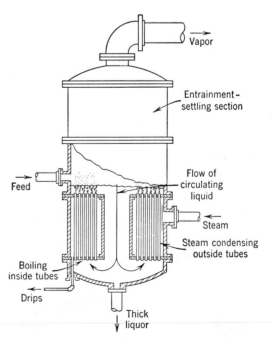

Figure 19.4. Cross-sectional diagram of standard vertical-tube evaporator with natural circulation. (Courtesy Swenson Evaporator Company, a Division of The Whiting Corporation.)

slow, and the coefficients obtained are poor. Thus, vertical-tube evaporators are completely satisfactory for most evaporation demands and are impractical only where the liquid being evaporated is very viscous, foams markedly, or may be subjected to evaporator temperatures for very short periods only.

Forced-circulation evaporators are shown in Figures 19.5 and 19.6. In these evaporators, the evaporating liquid is pumped through a heat exchanger where the heating medium surrounds the tubes carrying the solution. Pressure drop and hydrostatic head in combination are frequently great enough to prevent the solution from boiling in the exchanger tubes, so that the vapor generated is flashed off as the liquid enters the disengaging space. Since the velocity of the flashing mixture is high, impingement baffles are important to minimize entrainment. A properly designed

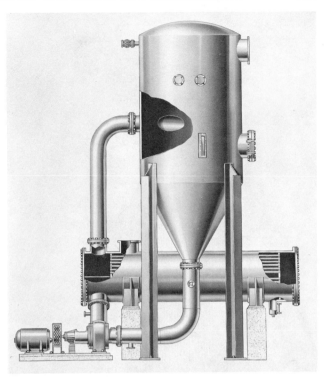

Figure 19.6. Cross-sectional diagram of forced-circulation evaporator with an external horizontal heater. (Courtesy Chicago Bridge and Iron Company.)

baffle will promote coalescence of small bubbles as well as merely changing the direction of flow. Modern forced-circulation evaporators are usually equipped with external heaters, as shown in Figure 19.6, rather than with heat-transfer surface built into the evaporator body. This makes the cleaning of tubes and the replacement of any corroded or eroded tube much simpler than it is with the internal heating element. It also permits the construction of a more compact unit, so that it can be installed in spaces with low headroom. In evaporating some solutions, it is important to prevent boiling in the tubes in order to reduce solids deposition. In the evaporator with external heating, boiling can be easily prevented by lowering the heater relative to the disengaging space. It cannot be prevented so easily when the heat-exchange surface is inside the evaporator body.

In forced-circulation evaporators, the heat-transfer coefficient will depend upon circulation rate as well as over-all ΔT, boiling temperature, and system properties. At low circulation rates, boiling occurs throughout much of the tube length. Boiling increases the turbulence and may make the boiling-side coefficient twice as large as it would be without boiling. The fraction of the liquor vaporized while passing through the tube is very small; hence, total circulation through the tubes is many times larger than the feed rate. At circulation rates over about 4 ft/sec, boiling in the tube is almos

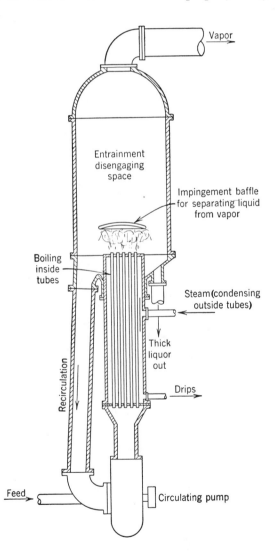

Figure 19.5. Cross-sectional diagram of vertical-tube evaporator with forced circulation. (Courtesy Swenson Evaporator Company, a Division of The Whiting Corporation.)

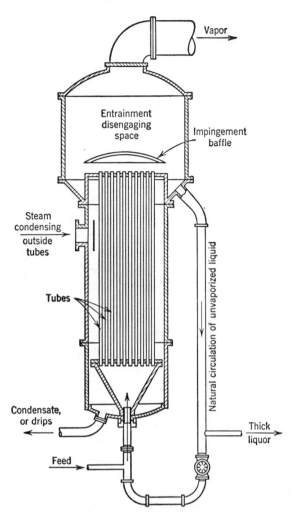

Figure 19.7. Cross-sectional diagram of long-tube vertical evaporator. (Courtesy Swenson Evaporator Company, a Division of The Whiting Corporation.)

be kept from boiling in the tubes to minimize solids deposition, the choice of a forced-circulation evaporator may be almost mandatory.

The *long-tube vertical evaporator*, shown in Figure 19.7, obtains reasonably high liquor flow through the tubes by natural convection. The tubes are usually 12 to 20 ft long. The vapor-liquid mixture issues from the top of the tubes and impinges on a baffle. The liquor velocity is high enough so that the baffle acts as an effective foam breaker. Little published information is available on coefficients obtained in these evaporators, but over-all coefficients ranging between 200 and 800 Btu/hr sq ft °F can be expected. Thus, among natural-circulation evaporators, the long-tube vertical evaporator

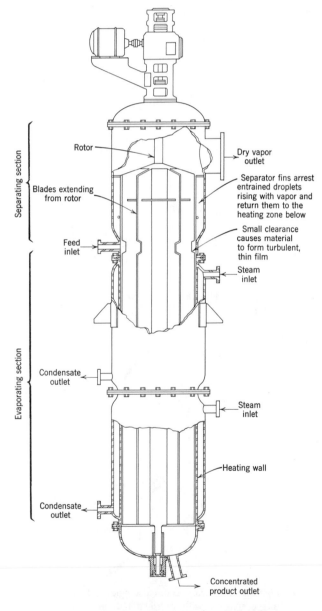

Figure 19.8. Cross-sectional view of turbulent-film evaporator. (Courtesy Rodney-Hunt Machine Co.)

completely suppressed, and the liquor-side coefficient can be reasonably predicted from Equation 13.79 for forced convection inside tubes. Over-all coefficients varying between 200 and 1200 Btu/hr sq ft °F have been reported, with normal experience giving values of 500 to 1000 Btu/hr sq ft °F. Such higher coefficients permit reduction in the size of unit required for a given application, but the saving in first cost is balanced by the power cost of pumping the solution in the recirculation loop. Thus, the decision to use a forced-circulation evaporator depends upon a favorable economic balance which includes the evaporator first cost, the power costs of operating the recirculating pump, the cost of maintaining the pump, and the comparative cost of cleaning the evaporator tubes. High circulation velocities may be attractive when moderate- or high-pressure steam is to be used, with the steam used first in a turbine to drive the pumps and then used as the heating medium. In the case of viscous fluids or of solutions that must

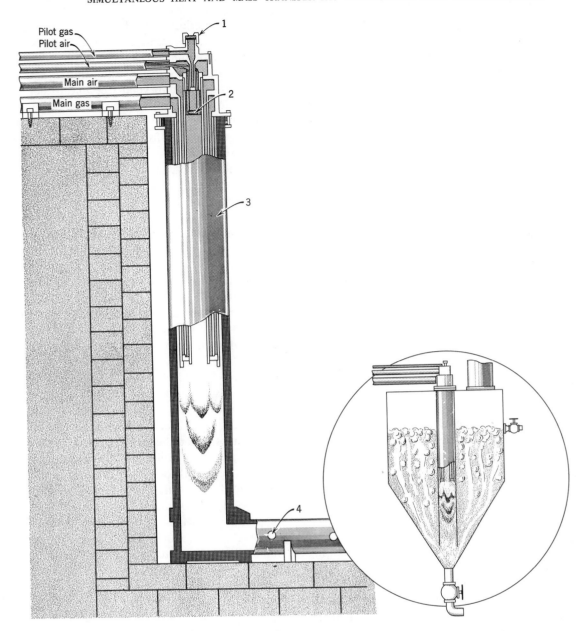

Figure 19.9. Cross-sectional diagram of submerged-combustion evaporator. (Courtesy Submerged Combustion Company.)

1 Sight port.
2 Safety type igniter, temperature of which controls operation of gas valves.
3 Combustion casing controls discharge of exhaust gases as desired and protects burner parts from corrosion during operation.
4 Exhaust ports to distribute gases as desired.

competes most favorably with the forced-circulation evaporator for application in large modern installations. The coefficients are lower than they may be for the forced-circulation evaporator, but no circulating pump is required. The fraction evaporated per pass is usually larger than for forced-circulation operation. If desired, the exchanger surface can be located external to the evaporator body to permit easy cleaning. Although viscous fluids cannot be handled, foaming materials

can be, and, for certain heat-sensitive materials, this evaporator has the advantage of being readily operated without recirculation, i.e., as a once-through unit. Evaporation per pass through the tubes is normally far higher than in other types of natural-circulation evaporators and can be increased still more by further lengthening the exchanger tubes or by coiling them.

Modern evaporator developments have centered on the problem of successfully handling viscous and corrosive materials. The *turbulent-film evaporator* shown in Figure 19.8 will handle viscous liquids and can even be adapted to the evaporation of a solution or slurry to dryness. The unit consists of a vertical tube heated over the bottom two-thirds with a steam jacket, and containing a central rotor. Mounted on the rotor are blades which extend almost to the heated walls. The top third of the cylinder is of larger diameter and is unheated. In this region, the vertical rotor blades carry horizontal baffles which collect entrained droplets and return them to the evaporator walls. Feed enters at the top of the heated section. In the evaporator, the feed is hurled against the heated walls by the action of the rotor blades. The blades also keep the heating surface free of solid deposits. The concentrated solution gradually works its way to the bottom of the evaporator, being continuously forced against the heated walls and thoroughly stirred by the rotor. These units are built in sizes up to 42 in. in diameter by 33 ft long with heat-transfer areas up to 198 sq ft. Over-all coefficients of 40 to 400 Btu/hr sq ft °F are claimed when evaporating materials of viscosities up to 20,000 centipoises. An additional advantage lies in the low hold-up and consequent short time of residence of the fluid in the evaporator.

Another unit which has been successfully used with both viscous materials and corrosive fluids is the *submerged combustion burner* illustrated in Figure 19.9. With this burner, no metal heat-transfer surface is required, since the products of combustion bubble up through the process fluid. The small amount of submerged equipment and its simple design reduce replacement costs to a minimum. Moreover, the parts can be made of ceramic or other resistant material, even though these materials are virtually thermal insulators, since they are not used as heat-transfer surface. Transfer of energy from the combustion gases to the process fluid is complete enough so that substantial savings in fuel are claimed compared with the use of steam generated on the site. Units are built for energy release rates up to 22×10^6 Btu/hr in applications such as heating dilute H_2SO_4, oxidizing asphalt, and evaporating arsenic acid, muriatic acid, and clay slurry.

The selection of the appropriate evaporator is ultimately a matter of economics. However, the properties of the material being evaporated may sharply limit the choice. For instance, a highly viscous solution might not move readily past the heating surface by the action of natural convection alone. In this case use of the forced-circulation or turbulent-film evaporator might be necessary. In the evaporation of detergent solutions, foaming occurs in certain concentration ranges, whereas gelling occurs in others. Thus, the evaporator body should contain as nearly a uniform concentration as possible. This is achieved by using forced-circulation evaporators where the circulation rate is large compared

to the evaporation rate. In the concentration of fruit juices, the food value and taste depreciate rapidly upon exposure to heat. Therefore, no recirculation can be tolerated since some of the solution would remain in the evaporator beyond the mean holding time. Here, a long-tube vertical evaporator is used without recirculation, frequently with downward flow to superimpose gravitational force and vapor drag to speed the flow of the liquid film down the tube. In evaporating spent pickle liquor, the corrosive nature of the solution prohibits the use of a solid heating surface. In this case, submerged combustion is used. Because of the better effectiveness of heating surface and the resulting reduction in evaporator size, most evaporators installed now are of the long-tube vertical or forced-circulation types.

Prediction of Solution Boiling Temperature. The vaporization of a pure liquid presents no particular problems from a physicochemical viewpoint. The temperature of the boiling liquid is fixed by the pressure under which it exists, and it may be readily calculated once the relation between vapor pressure and temperature is known. If the liquid exists in depth, the pressure that must be used in calculating the boiling point at any depth must include the hydrostatic head of liquid above that point. It is very important in long-tube vertical evaporators, where liquid depths of 20 ft are common. In such a case, the liquid enters the heated tube subcooled by flow through the downcomer and addition of the feed. As it flows up the tube, it is heated to its boiling point at some particular depth. Superheating is probable as the solution flows upwards and the pressure drops. When boiling starts, the superheat is rapidly lost, and thereafter the liquid temperature drops with the decreasing pressure. The acceleration of the mixture imposes additional pressure beyond the static ones; this pressure may be the major one if the flow approaches the velocity of sound in the two-phase mixture. A resulting temperature-height curve is shown in Figure 19.10.

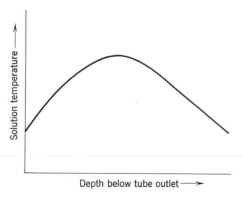

Figure 19.10. Temperature traverse in long, vertical evaporator tubes.

When a *solution* is evaporated, not only are the effects of liquid depth and acceleration present, but the effect of concentration on the boiling point must also be considered. For ideal solutions, the concentration effect may be estimated through Raoult's and Dalton's laws, for

$$P = p_a + p_b = P_a x_a + P_b x_b \qquad (19.2)$$

where a and b refer to solute and solvent respectively. If the solute is nonvolatile, as is normally the case for evaporation operations, then

$$P = P_b x_b \qquad (19.3)$$

With P^0 defined as the total pressure of the vapor over a single-component liquid phase, then, for the pure solvent $P^0 = P_b$ since $x_b = 1$, and the vapor pressure lowering due to the presence of solute becomes

$$\frac{P^0 - P}{P^0} = \frac{1 - x_b}{1} = x_a \qquad (19.4)$$

Thus, for solutions obeying Raoult's law, the fractional decrease in vapor pressure is equal to the concentration of nonvolatile solute. For these solutions then, the vapor pressure-temperature curves for fixed concentrations will be parallel. If the curves are assumed to be straight parallel lines in the region of the boiling point, the vapor pressure lowering $(P^0 - P)$ will be proportional to the boiling-point rise, or

$$T_B - T_B{}^\circ = k x_a \qquad (19.5)$$

where k is a constant for a given solvent in ideal solutions and $(T_B - T_B{}^\circ)$ is the difference between the boiling point of the solution and that of the pure solvent at the same total pressure. This simple relation holds only under stringent restrictions. The range of boiling point must be narrow, and the solution must obey Raoult's law. This restriction implies that the solution be dilute and that the solute be such that ionization, complex formation, etc., do not occur. Obviously, when solutions of 20 to 50 per cent solute are to be obtained by evaporation, only qualitative agreement can be expected. However, one useful generalization results. For most solutions, the boiling-point rise, $(T_B - T_B{}^\circ)$ is a function of the solution constituents and the concentration. In other words, the boiling-point rise (abbreviated BPR) is relatively insensitive to pressure. Hence, for a solution of fixed concentration, the BPR will not change significantly over wide pressure ranges. This generalization was first noted by Dühring in 1878. It permits the boiling-point–concentration network for a system to be simply represented and based on a minimum of experimental data. The Dühring chart is a plot of T_B against $T_B{}^\circ$ at the same total pressure for solutions of constant composition. On such a plot, the zero-concentration line is obviously a 45-degree line passing through the origin. Lines for other concentrations will be found to be roughly straight and parallel to this line but displaced above it. Such a plot for the NaOH–H$_2$O system is shown in Figure 19.11. This system is highly nonideal, as indicated by its large heat of solution. Nevertheless, the Dühring lines are reasonably straight and parallel up to 60 mass percent NaOH. Thus, for most systems, the Dühring lines can be plotted if the boiling point is known as a function of concentration at one pressure. For even a nonideal solution like the NaOH–H$_2$O system, it would be adequate to know the boiling-point–concentration data at two widely separated pressures, because the Dühring lines are straight except for the most extremely nonideal systems.

Note that the equilibrium vapor rising from any solution exhibiting a boiling-point rise will exist at a temperature and pressure such that it is superheated. The vapor rises at the solution boiling temperature, but the vapor is free from solute, not concentrated as is the solution, and therefore it condenses only after the BPR has been removed. It is superheated by the extent of the BPR.

Calculations on a Single Evaporator Stage. From the considerations presented above, the calculations necessary to fix the heating-area requirement of a single evaporator from a knowledge of its duty are straightforward. The heat-transfer-rate equation need only be solved simultaneously with the over-all heat balance and the total and solute mass balances. As mentioned above, the over-all heat-transfer coefficient is usually obtained from previous experience.

Illustration 19.1. Determine the heating area required for the production of 10,000 lb/hr of 50 per cent NaOH solution from a 10 per cent NaOH feed entering at 100°F. Evaporation is to be carried out in a single-body standard evaporator for which an over-all coefficient of 500 Btu/hr sq ft °F is expected. Steam is available saturated at 50 psig, and the evaporator can be operated at 10-psi vacuum relative to a barometric pressure of 14.7 psia.

SOLUTION. Figure 19.12 is a diagram of this evaporator with the given information included.

The solution of this problem is based on several assumptions which are typical of evaporator calculations:

(*a*) The solution in the evaporator body is violently mixed and therefore homogeneous in composition. This assumption is realistic except in the case of long-tube vertical evaporators and special types such as the turbulent-film evaporator. Since 50 per cent NaOH solution is being removed from the evaporator, the solution within the evaporator must be uniformly 50 per cent NaOH.

(*b*) The boiling point of the solution is that of 50 per cent NaOH at 4.7 psia. This statement ignores the effects of hydraulic head. Here, this simplification may be allowable

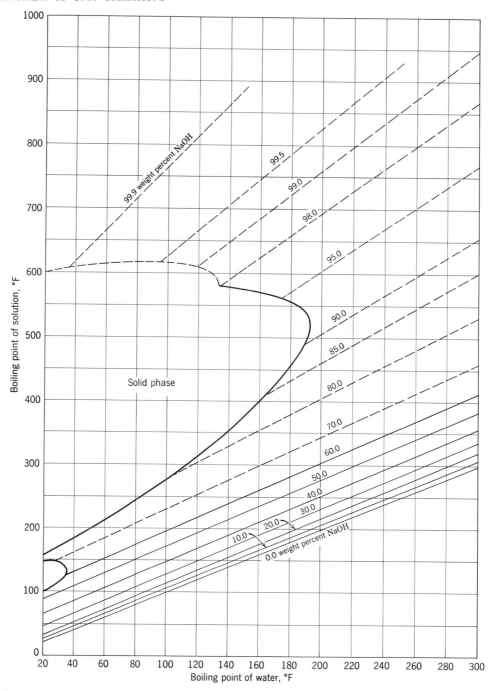

Figure 19.11. Dühring lines for the NaOH–H₂O system.

because the liquid depth will not be great, because acceleration is small and because liquid enters the tubes near its boiling point in the body. In the case of long-tube vertical evaporators, hydraulic and acceleration heads should be considered, at least by using an average depth.

(*c*) The vapor leaving at V_1 is pure water-vapor at the temperature of the boiling solution and the pressure given.

(*d*) The condensate, or drips, leaving at D is a saturated liquid at the steam pressure. Subcooling will occur, but its contribution to the heat flux is minute.

(*e*) There is no heat loss from the evaporator body to the surroundings. This approximation is closer to fact the larger

the unit, since the shell-surface-to-heat-transfer-surface ratio decreases markedly as the unit size increases.

On these assumptions, the area can be obtained through the solution of the rate equation, coupled with heat and material balances.

Over-all material balance: $F + V_0 = V_1 + L_1 + D$

$$(Note:\ D = V_0) \quad \text{(i)}$$

NaOH balance: $Fx_F{}' = L_1x_1{}'$ (ii)

Over-all heat balance: $V_0\lambda_{V0} + Fh_F = V_1H_1 + L_1h_1$ (iii)

Heat-transfer-rate equation: $V_0\lambda_{V0} = UA(-\Delta T)$ (iv) where F, V_0, V_1, L_1, D = quantities of streams entering or leaving the evaporator as noted on Figure 19.12. All these quantities are given on the same basis; here, one hour of operation

H = enthalpy per unit mass of a vapor stream designated by subscript, Btu/lb

h = enthalpy per unit mass of a liquid stream designated by subscript, Btu/lb

λ = latent heat of evaporation per unit mass of stream designated by subscript, Btu/lb

x' = mass fraction of solute in stream designated by subscript, lb solute/lb of total stream

The enthalpy quantities must be determined relative to a constant-base condition. If steam-table data are to be used, all enthalpies must be fixed relative to $H = 0$ for water at its vapor pressure at 32°F. Any convenient base may be used for the enthalpy of the solute such as crystalline solid at 32°F or infinitely dilute solution at 60°F. Here, there are four equations in the four unknowns, F, V_1, V_0, and A, so that the solution of the problem is direct once the necessary data have been supplied. The enthalpy of vapor (H_1) and latent heat (λ_{V_0}) can be obtained from steam tables (6). Liquid-phase enthalpies (h_F and h_1) are obtained from an enthalpy-concentration plot such as that of Figure 19.13. This plot is for 1 atm pressure, but, since the effect of pressure on liquid enthalpy is small, the pressure restriction need not be of concern. The boiling point of the solution is most conveniently obtained from a Dühring chart for the NaOH–H$_2$O system (Figure 19.11). At 4.7 psia, water boils at 160°F (6). From Figure 19.11, the boiling point of a 50 per cent NaOH solution at a pressure such that water boils at 160°F is 233°F, corresponding to a BPR of 73°F.

Solving the material balances, Equation ii gives

$$F(0.1) = 10,000(0.5); \quad F = 50,000 \text{ lb/hr}$$

From Equation i, $V_1 = 40,000$ lb/hr.
Equation iii may be now solved for V_0.

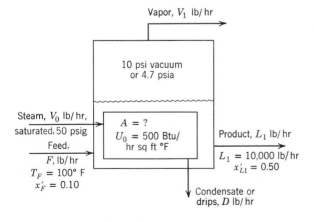

Figure 19.12. Conditions for Illustration 19.1.

$$V_0 \times 920.0 + 50,000 \times 60 = 40,000 \times 1160 + 10,000 \times 245$$

$$V_0 = \frac{45,850,000}{920} = 49,800 \text{ lb/hr}$$

The enthalpy of steam leaving the evaporator (H_1) has here been read from the steam tables at a pressure of 4.7 psia and a temperature of 233°F. The superheat results from the BPR. Finally, from Equation iv

$$45,850,000 = 500A \times (297 - 233) = 64 \times 500A$$

$$A = 1400 \text{ sq ft}$$

In this example, as in any single-effect evaporation, each pound of vapor is produced at the cost of condensing approximately 1 lb of steam. Here, more than 1 lb of steam was required primarily because the feed entered much colder than the solution boiling point. If the feed entered superheated, it is possible that 1 lb of steam would produce more than 1 lb of vapor. The fact that the latent heat of evaporation of water decreases as the pressure increases tends to make the ratio of vapor produced per pound of steam condensed less than unity.

Multiple-Effect Evaporation. In any evaporation operation, the major process cost is in the steam consumed. Therefore, methods of reducing steam consumption (or of increasing *economy*, defined as mass of vapor produced per unit mass of steam consumed) are very attractive. The most common of the available methods is to use the vapor generated in the first evaporator as the heating fluid for a second evaporator. Ideally, this method should produce almost 2 lb of vapor for every pound of steam consumed. The method is feasible if the second evaporator is operated at a lower pressure than the first, so that a positive value of $-\Delta T$ is obtained across the steam-chest surface of the second evaporator. Obviously, several evaporators can be connected in series in this way with the intent being to obtain a number of pounds of vapor for each pound of steam consumed equal to the number of evaporator bodies. The increase in latent heat with decreasing pressure and additional radiation losses result, however, in the economy being increasingly lower than the number of evaporators used as this number is increased. This method of evaporator operation in series is called *multiple-effect evaporation*, and each stage is called an *effect*.

Multiple-effect evaporators may be connected in a variety of ways. In a *forward-feed* system, the flow of process fluid and of steam are parallel. Forward feed has the advantage that no pumps are needed to move the solution from effect to effect. It has the disadvantage that all the heating of cool feed is done in the first effect, so that less vapor is generated here for each pound of steam, resulting in lower economy. It has the further

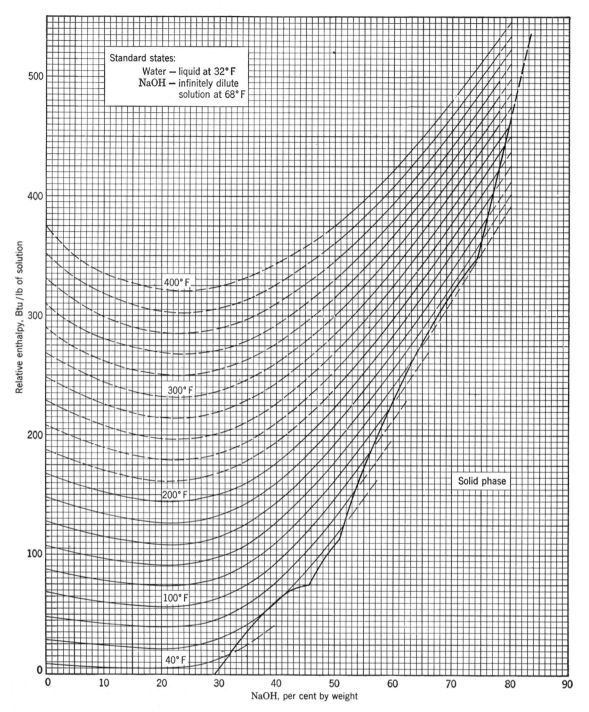

Figure 19.13. Enthalpy-concentration diagram for aqueous solutions of NaOH under a total pressure of one atmosphere. The reference state for water is taken as liquid water at 32°F under its own vapor pressure. This reference state is identical with the one used in most steam tables (6). For sodium hydroxide, the reference state is that of an infinitely dilute solution at 68°F. [From McCabe, W. L., *Trans. Am. Inst. Chem. Engrs.*, **31**, 129 (1935), by permission.]

result that the most concentrated solution is subjected to the coolest temperatures. Cool temperatures may be helpful in preventing decomposition of organics, but the high viscosity that may be found sharply reduces the coefficient in this last effect. In a *backward-feed* system, the process solution flows counter to the steam flow. Pumps are required between effects. The feed solution is heated as it enters each effect, which usually results in better economy than that obtained with forward feed. The viscosity spread is reduced since the concentrated solution evaporates at the highest temperature, but organic materials may tend to char and decompose.

For best over-all performance, evaporators may be operated with flow sequences that combine these two (i.e., *mixed feed*), or they may be fed in *parallel*, with fresh feed evaporating to final concentrate in each effect. In multibody operation, the number of stages may be different for steam and liquor. Three common feed arrangements for multiple-effect evaporation are illustrated by schematic flow sheets in Figure 19.14.

In addition to the economy increase in multiple-effect evaporation, a capacity variation would be expected. Note, however, that the temperature difference from initial steam to the final condenser which was available for a single-effect evaporator will be unchanged by inserting any additional effects between the steam supply and the condenser. For the simplest case, where each effect has area and coefficient equal to that of every other effect and where there are no boiling-point rises,

$$q_t = q_1 + q_2 + q_3 + \ldots$$

where q_t is the total heat-transfer rate in all effects and q_1, q_2, q_3 are the heat-transfer rates in each of the individual effects.

$$q_t = U_1 A_1 (-\Delta T_1) + U_2 A_2 (-\Delta T_2) + U_3 A_3 (-\Delta T_3) + \ldots$$

Since the areas and transfer coefficients are equal,

$$q_t = U_1 A_1 (-\Delta T_1 - \Delta T_2 - \Delta T_3 - \ldots) \quad (19.6)$$
$$= U_1 A_1 (-\Delta T_{\text{total}})$$

This rate of heat transfer is the same as that obtained with a single effect operating between the same ultimate temperature levels. Thus, multiple-effect evaporation using n effects increases the steam economy but decreases the heat flux per effect by a factor of about $1/n$ relative to single-effect operation under the same terminal conditions. Therefore, no increase in capacity is obtained, and, in fact, the additional complexity of equipment usually results in increased heat losses to the surroundings and a reduction in capacity. The increased steam economy must, then, be balanced against the increased equipment cost. The result is that evaporation using more than five or six effects is rarely economical.

When the solution being evaporated has a significant boiling-point rise, the capacity obtained is very much reduced, for the boiling-point rise reduces the $-\Delta T$ in each effect. Consider again the situation of Illustration 19.1. Here the total $-\Delta T$ was $(297 - 160)$ or $137°F$. Of this, the BPR used $73°F$, leaving an effective $-\Delta T$ of $64°F$. If two effects are used for this evaporation, about equal evaporation would occur in each effect. Then the product from the first effect would be $5000/30,000$ or 16.7 per cent NaOH. The pressure in the first effect will depend upon the relative areas and coefficients present. Assuming that they are such that water would boil in this effect at $250°F$, the boiling point of the 16.7 per cent NaOH solution would

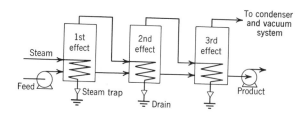

(a) Forward feed.

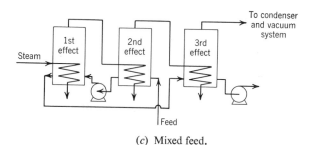

(b) Backward feed.

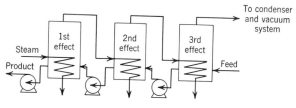

(c) Mixed feed.

Figure 19.14. Schematic flow sheets of feed arrangements for multiple-effect evaporation.

then be about $267°F$ giving a $17°F$ BPR. The resulting temperatures throughout these two evaporator systems are then as listed in Table 19.1.

The total effective $-\Delta T$ is seen to decrease by the amount of the BPR in the first effect. The vapor from the first of the two effects leaves at $267°F$ but at a pressure such that it condenses at $250°F$. This superheat results from the first effect BPR and is lost before the major heat transfer occurs in the second effect. The solution temperature in the final effect is identical

Table 19.1. TEMPERATURE IN SINGLE-EFFECT AND DOUBLE-EFFECT EVAPORATORS OPERATING UNDER IDENTICAL OVER-ALL CONDITIONS

	Single Effect	Double Effect
Steam temperature, 1st effect	$297°F \longleftrightarrow$	$297°F$
Solution temperature, 1st effect	$233°F \longleftarrow$	$267°F$
Effective $-\Delta T$, 1st effect	$64°F$	$30°F$
Steam temperature, 2nd effect		$250°F$
Solution temperature, 2nd effect		$\longrightarrow 233°F$
Effective $-\Delta T$, 2nd effect		$17°F$
Total effective $-\Delta T$	$64°F$	$47°F$

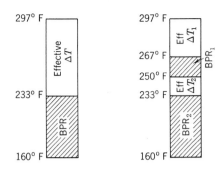

Figure 19.15. Distribution of total temperature difference in (a) single- and (b) double-effect evaporation.

in the two cases. As a result, the capacity is reduced, and the economy is also reduced. Another method of presenting the same information is by a bar graph, as is done in Figure 19.15. Again, the effective $-\Delta T$ applicable to the single-effect evaporator is apportioned to the two effects in the double-effect evaporator. Moreover, in the double-effect evaporator, the BPR in the first effect subtracts from the available $-\Delta T$. This is a general result. The BPR in each effect subtracts directly from the available $-\Delta T$; that is,

$$-\Delta T_{\text{total effective}} = -\Delta T_{\text{total}} - \Sigma \text{BPR} \quad (19.7)$$

Each effect of a multiple-effect evaporator, except possibly the last one, is in essence a vaporizer and a surface condenser, both operating on the same fluid. If the first effect of a multiple-effect evaporator is considered, then the contents of the first effect is the fluid; the steam chest of the first effect vaporizes a portion of the fluid, and the steam chest of the second effect condenses the vapor produced in the first effect. A general analysis of this sort of single element may be examined as follows.

Consider a fluid contained in the unit shown in Figure 19.16, in which the unit is equipped with a vaporizer unit and a condenser unit, arranged so that the vapor that is condensed returns to the pool of boiling liquid. The surface areas and over-all heat-transfer coefficients of the system may be stated as A_v and U_v for the vaporizer and A_c and U_c for the condenser. The heating medium to the vaporizer enters at T_s, and the coolant to the condenser is maintained as an average at T_c. The temperature of the operating fluid may be taken as T_f. If the operating fluid is a single pure component, T_f will be the same in the condenser as it is in the boiler. If this fluid is a solution with significant BPR, the fluid temperature in the boiler will exceed the fluid temperature in the condenser by the value of the BPR. Here for simplicity, the operating fluid will be taken as a pure liquid.

Then, the rate of transfer of heat from the vaporizer

can be written as

$$q_v = -U_v A_v (T_f - T_v)$$

and the rate at which heat is transferred to the condenser is

$$q_c = -U_c A_c (T_c - T_f).$$

For the simplified system in which there are no heat losses through the walls of the vessel, $q_v = q_c$, and therefore

$$T_f = \frac{T_c U_c A_c + T_v U_v A_v}{U_c A_c + U_v A_v} \quad (19.8)$$

In the derivation of Equation 19.8, no reference was made to the nature of the fluid used in the evaporator element, nor is any specification written for the pressure. It should be evident that the pressure in the effect is a function of T_f. According to Equation 19.8 the fluid temperature (T_f) is dependent on the transfer rates and terminal temperatures only. Therefore the unit could be used with any fluid and the temperature T_f would be the same as long as the heat-transfer coefficients did not change. The pressure in the vapor chest would, however, depend on the fluid used, since it is related to T_f through the vapor pressure curve of the fluid.

As stated above, this action controls the pressure in the body of each effect except the last one in a multiple-effect evaporator. The last effect is controlled to a predetermined pressure and related boiling point. The conditions in the other effects adjust themselves so that the resulting temperature drops with existing transfer coefficients accomplish heat-transfer rates called for by the material and energy balances.

With these concepts, the calculation of size of multiple-effect evaporators is a simple extension of the methods used for single-effect evaporators. Material and heat balances and rate equations are written for each effect. They are then solved simultaneously to get the required

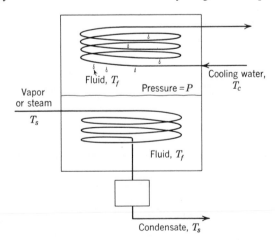

Figure 19.16. A generalized unit of a multiple-effect evaporator.

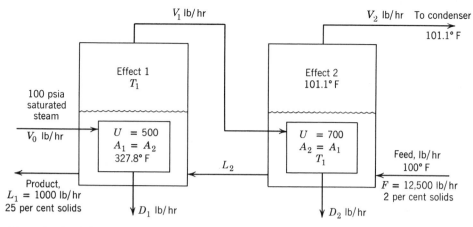

Figure 19.17. Schematic flow sheet for backward-feed double-effect evaporator with conditions given for Illustration 19.2.

information. Thus, for an n-effect evaporation, we have $3n$ independent equations and can solve for $3n$ unknowns. Unknowns might include the feed and steam quantities, the composition and temperature in the first effect, and the heat-transfer area in each effect. The equations can be written and a solution obtained directly if the evaporating liquor has no BPR and if the latent heat of steam can be considered constant. If a significant BPR is found or if more than two effects are used, a successive-approximation solution is involved.

Illustration 19.2. An aqueous solution containing 2 per cent dissolved organic solids is to be concentrated to 25 per cent solids by double-effect, backward-feed evaporation in forced-circulation evaporators of 2000 sq ft heating area each. If the coefficient in the first effect is 500 Btu/hr sq ft °F and that in the second is 700 Btu/hr sq ft °F, what production rate will be obtained? The solution exhibits no boiling-point rise; feed enters at 100°F; steam is available at 100 psia; and the condenser operates at 2 in. Hg absolute.

SOLUTION. The steam tables (6) give the steam-condensing temperature as 327.8°F and the condenser temperature as 101.1°F. Taking 1000 lb/hr of product as a basis, an over-all material balance gives

$$1000 \times 0.25 = F \times 0.02; \qquad F = 12,500 \text{ lb/hr}$$

and

$$V_1 + V_2 = 12,500 - 1000 = 11,500 \text{ lb/hr}$$

With these results included, the flow diagram is shown schematically in Figure 19.17. The procedure will be to determine T_1 so that $A_1 = A_2$. Then, the basis chosen [1000 lb/hr of product (L_1)] will be multiplied by the ratio $2000/A_1$ to get the actual production rate. Since no exact information is available on the thermal properties of the solution, specific heat of the solution will be taken as that of water, and all enthalpies will be calculated from the steam-table basis of liquid water at 32°F and the triple-point pressure. As a result, all thermodynamic properties can be read directly from the steam tables.

The following equations can be written:

Material balance around effect 1:

$$L_2 = L_1 + V_1 \tag{i}$$

Solids balance around effect 1:

$$L_2 x_2' = 1000 \times 0.25 \tag{ii}$$

Material balance around effect 2:

$$F = L_2 + V_2 \tag{iii}$$

Heat balance, effect 1:

$$V_0 \lambda_{V0} + L_2 h_2 = L_1 h_1 + V_1 H_1 \tag{iv}$$

Heat balance, effect 2:

$$V_1(H_1 - h_{D2}) + F h_F = V_2 H_2 + L_2 h_2 \tag{v}$$

Heat-transfer-rate equation, effect 1:

$$V_0 \lambda_{V0} = q_1 = U_1 A_1 (327.8 - T_1) \tag{vi}$$

Heat-transfer-rate equation, effect 2:

$$V_1[H_1 - (T_1 - 32)c_{\text{liq}}] = V_1 \lambda_{V1}$$
$$V_1 \lambda_{V1} = q_2 = U_2 A_2 (T_1 - 101.1) \tag{vii}$$

Here, there are seven independent equations involving the variables F, L_2, L_1, V_0, V_1, V_2, x_2', λ_{V0}, h_1, h_2, h_F, H_1, H_2, T_1, A_1, A_2, U_1, and U_2. Of the eighteen, the variables F, L_2, λ_{V0}, h_2, h_F, H_2, U_1, and U_2 are already fixed by the problem statement. Three other equations are needed. These are $A_1 = A_2$, which has already been written, $H_1 = \phi(T_1)$, and $h_1 = \phi'(T_1)$. The last two of the equations express the fact that the steam tables give H_1 and h_1 immediately, once T_1 is fixed. However, since these steam-table relations are not explicit equations, the final solution will involve a trial-and-error method.

A good estimate of the final condition is obtained by realizing that q_1 and q_2 will be almost equal. Assuming $q_1 = q_2$ gives

$$U_1 A_1 (-\Delta T_1) = U_2 A_2 (-\Delta T_2)$$

or

$$(-\Delta T_1)/(-\Delta T_2) = \frac{U_2 A_2}{U_1 A_1} = \frac{U_2}{U_1} = \frac{700}{500} = 1.4$$

$$(-\Delta T_1) + (-\Delta T_2) = \Sigma(-\Delta T) = 327.8 - 101.1 = 226.7$$

Here, there is no BPR, so that the total $-\Delta T$ is also the $-\Delta T$ which is effective for heat transfer. Then, $(-\Delta T_2) + 1.4(-\Delta T_2) = 226.7$; $-\Delta T_2 = 95°F$; $-\Delta T_1 = 131.7°F$, which gives $T_1 = 196.1°F$, $H_1 = 1144$, and $\lambda_{V1} = 980$ Btu/lb from the steam tables (6).

From Equations iv and v, coupled with the over-all material balance,

$$V_1\lambda_{V1} + Fh_F = (11,500 - V_1) H_2 + L_1h_1 + V_1H_1 - V_0\lambda_{V0} \quad \text{(viii)}$$

Since it has been assumed that $q_1 = q_2$,

$$V_0\lambda_{V0} = V_1\lambda_{V1}$$

and Equation viii becomes

$$Fh_F = (11,500 - V_1) H_2 + L_1h_1 + V_1H_1 - 2V_1\lambda_{V1}$$

This equation can be solved for V_1 to give

$$V_1(2\lambda_{V1} + H_2 - H_1) = 11,500H_2 + L_1h_1 - Fh_F \quad \text{(ix)}$$

$$V_1(1960 + 1105.7 - 1144) = 11,500 \times 1105.7$$
$$+ 1000(196.1) - 12,500 \times 68$$

so that $V_1 = 6250$ lb, $V_2 = 5250$, $L_2 = 7250$. Also, $x_2' = 0.0345$, and, from Equation iv,

$$V_0\lambda_{V0} = q_1 = 6,834,000$$

The validity of the first assumption may now be checked by noting (Equation vii) that $q_2 = V_1\lambda_{V1}$. From this, $V_1 = 7000$ which is 1.12 times the V_1 obtained from Equations iv and v through use of the assumed temperature distribution.

As a next assumption, q_1 might be chosen equal to $1.1q_2$. Then, by the method shown above, $-\Delta T_2 = 89.2°F$, and $-\Delta T_1 = 137.5°F$. From this, $T_1 = 190.3°F$, $H_1 = 1142$, and $\lambda_{V1} = 983.3$. Solving Equation ix with this new value of T_1 gives $V_1 = 5920$. From Equations i and iii, $V_2 = 5580$ and $L_2 = 6920$. Use of Equation vi to check these results gives $V_0\lambda_{V0} = q_1 = 6,439,000$; then, from Equation vii, $V_1 = 5950$. This check is satisfactory. The area then may be found from both Equations vi and vii and averaged. The result gives $A_{av} = A_1 = A_2 = 93.8$ sq ft for producing 1000 lb/hr of product.

The production rate then is $2000/93.8 \times 1000$ or 21,400 lb/hr of product having 25 per cent solids.

This example illustrates typical evaporator calculations. The method is essentially a trial-and-error solution for the intermediate temperature. The initial trial is guided by the physical restriction that each pound of condensing steam evaporates about 1 lb of vapor. Adjustment of the first trial comes from information gained in that trial. For example, in the first trial where $q_1 = q_2$, the V_1 as calculated from the heat and material balances was $1/1.12$ times the V_1 found through q_2 from Equation vii. Therefore, q_2 was decreased by a factor of 1.1 for the second trial. The factor 1.1 was used rather than 1.12 because a change in q_2 will result in an even greater change in the relative values of required transfer areas.

The trial-and-error solution was necessary here because tabular thermodynamic data had to be used for steam. If the system had a significant BPR, a graphical equation, the Dühring line chart or its equivalent, would also have been involved. If the thermodynamic data had been expressed analytically in terms of specific and latent heats, the example could have been solved analytically.

The existence of a boiling-point rise would not have complicated the solution greatly. At the intermediate compositions, the BPR is small and almost independent of temperature. The composition of the intermediate stream could be estimated from the equal evaporation principle, and, from this, the BPR obtained. During the successive approximations, changes in intermediate temperature or composition usually do not require changes in BPR.

The addition of a third or fourth effect results in additional heat and material balances and an additional heat-transfer-rate equation for each effect. The calculations are more tedious but involve the same trial-and-error solution of simultaneous equations. The steps of such a solution are as follows:

1. From the given product concentration and condenser pressure, determine the temperature, boiling-point rise, and enthalpies in the last effect. They may be directly determined if forward feed is used.

2. From an over-all material balance, determine the total amount of evaporation, and apportion it among the effects by assuming equal evaporation in each effect. If the feed is very cold or is very much superheated, the portions may be modified appropriately. The assumed vapor quantities will give an estimate of the concentration in each effect and consequently of the BPR. Only the crudest guess as to the pressure in each effect need be made, for BPR is virtually independent of pressure.

3. Find the $-\Delta T$ available for heat transfer by subtracting the sum of all BPR's from the total $-\Delta T$. The available $-\Delta T$ can be apportioned among the various effects by assuming that $q_1 = q_2 = q_3$ so that $(-\Delta T_1)/U_1A_1 = (-\Delta T_2)/U_2A_2 = (-\Delta T_3)/U_3A_3 = \ldots$.

4. Calculate the amount of evaporation in each effect through energy and material balances. If the amounts differ significantly from the values assumed in step 2, steps 2 and 4 must be repeated with the amounts of evaporation just calculated. Usually, this recalculation will represent a very slight revision in the BPR and enthalpy values previously used.

5. By means of the rate equations for each effect, calculate the surface required for each effect.

6. If the surfaces calculated do not fit the required area distributions, (usually $A_1 = A_2 = A_3 = \ldots$), revise the temperature-difference distributions of step 3.

Unless BPR's are very large, the revision will not affect the values assumed in step 2.

7. Continue adjusting the temperature differences and recalculating surface areas until the areas are distributed satisfactorily.

The mechanism involved in solving evaporator problems should not camouflage the fact that the evaporator operates under a balance of natural physical forces. All the designer does is to try to determine how the interplay of these forces affects the evaporator conditions. The operator of an evaporator can control the steam pressure, the condenser pressure, and the feed flow rate. The evaporator bodies will come to steady-state temperatures and pressures that are determined as much by the physical properties of the evaporating solution and by the physical arrangement of equipment as they are by any control exercised by the operator.

The procedure for determining evaporator operating conditions consists of writing heat and material balances around each effect and solving them in conjunction with the rate equation for heat transfer in each effect. In the procedure outlined above, steps 1, 2, and 3 merely fix a reasonable starting point for the solution. Actually any assumed condition would do as a starting point for calculation. The advantage of the method outlined is only that it shortens the trials necessary by allowing the calculator to start at a point reasonably close to the correct answer.

Illustration 19.3. A triple-effect evaporator system is to be used to concentrate 5 per cent NaOH to 50 per cent NaOH. Forward feed is to be used with the feed entering at 60°F. Over-all coefficients of 800, 500, and 300 Btu/hr sq ft °F are expected for the three effects in the given order. Steam is available at 125 psia, and ejectors capable of maintaining a pressure of 1 psia will be used. The heating areas of each effect are to be equal and large enough to produce 10 tons/hr of concentrate. What heating area is needed for each effect? What will be the steam consumption and the economy?

SOLUTION. A schematic diagram of the evaporator system is shown in Figure 19.18.

By reference to Figure 19.11, the solution and the vapor in the third effect are found to be at 171.7°F as is the product (L_3). Thus the BPR in the third effect is 70°F.

An over-all material balance gives

$$0.05F = 0.5 \times 20,000$$

$$F = 200,000 \text{ lb/hr}$$

and $\qquad V_1 + V_2 + V_3 = 180,000 \text{ lb/hr} \qquad$ (i)

Assuming equal evaporation, $V_1 = V_2 = V_3 = 60,000$ lb/hr. From this, the solution in the first effect is 10,000/140,000 = 7.1 per cent NaOH, and the solution in the second effect is 10,000/80,000 = 12.5 per cent NaOH. If 280°F and 220°F are used as rough guesses for the boiling point of water in the

first and second effect, the solutions boil at 285°F and 230°F respectively, which gives BPR's of 5°F and 10°F for these effects.

The total effective $-\Delta T$ is then

$$(-\Delta T)_{\text{eff}} = (344.3 - 101.7) - (70 + 10 + 5) = 157.6°F$$

The available $-\Delta T$ is apportioned among the various effects, assuming that $q_1 = q_2 = q_3$. Although this assumption is in conflict with the previous $V_1 = V_2 = V_3$ assumption, so that a revision in one or the other of them will be necessary eventually, it is a reasonable starting point. Calculating $-\Delta T$'s gives

$$U_1 A_1 (-\Delta T_1) = U_2 A_2 (-\Delta T_2) = U_3 A_3 (-\Delta T_3)$$

$$\frac{800}{500} = \frac{-\Delta T_2}{-\Delta T_1}; \quad \frac{500}{300} = \frac{-\Delta T_3}{-\Delta T_2};$$

$$(-\Delta T_1) + (-\Delta T_2) + (-\Delta T_3) = 157.6$$

from which $-\Delta T_1 = 30.0°F$, $-\Delta T_2 = 47.9°F$, $-\Delta T_3 = 79.7°F$.

Thus, $\qquad T_{3s} = 171.7 + 79.7 = 251.4; \quad T_2 = 261.4$

$$T_{2s} = 309.3°F; \quad T_1 = 314.3°F; \quad T_{1s} = 344.3°F$$

where the nomenclature is as shown on Figure 19.18.

The guesses as to boiling point of water in the first two effects are seen to be greatly in error, but a recheck of Figure 19.11 shows that the BPR's calculated need not be changed. Temperature in the three evaporator bodies will then be as follows for the first trial solution of the heat and material balances:

Effect 1	Effect 2	Effect 3	Condenser
$T_{1s} = 344.3°F$	$T_{2s} = 309.3°F$	$T_{3s} = 251.4°F$	$T_c = 101.7°F$
$T_1 = 314.3°F$	$T_2 = 261.4°F$	$T_3 = 171.7°F$	

Heat and material balances can now be solved around each effect to find a V_1, V_2, and V_3 consistent with these assumed temperatures. Enthalpy data for water or steam can be taken from the steam tables (6) and that for solutions from Figure 19.13. These balances are

$$V_0 \lambda_{V0} + F h_F = V_1 H_1 + (F - V_1) h_1 \qquad \text{(ii)}$$

$$V_1 H_1 + (F - V_1) h_1 = V_2 H_2 + V_1 h_{D2} + (F - V_1 - V_2) h_2 \qquad \text{(iii)}$$

$$(F - V_1 - V_2) h_2 + V_2 H_2 = V_2 h_{D3} + (180,000 - V_1 - V_2) H_3$$
$$+ L_3 h_3 \qquad \text{(iv)}$$

In these three equations, the only unknowns are V_1, V_2, and V_0. Inserting numerical values from the steam tables and from Figure 19.13 gives

$$875.4 V_0 + 200,000 \times 26 = 1185.1 V_1$$
$$+ (200,000 - V_1) 260 \qquad \text{(ii)}$$

$$875.4 V_0 - 925.1 V_1 = 200,000 \times 234$$

$$1185.1 V_1 + (200,000 - V_1) 260 = 1169.9 V_2 + 279.3 V_2$$
$$+ (200,000 - V_1 - V_2) 208 \qquad \text{(iii)}$$

$$843.8 V_1 - 961.9 V_2 = -200,000 \times 52$$

$$(200,000 - V_1 - V_2) 208 + 1169.9 V_2 = 220.0 V_2$$
$$+ (180,000 - V_1 - V_2) 1137.4$$
$$+ 20,000 \times 200 \qquad \text{(iv)}$$

$$1879.3 V_2 + 929.4 V_1 = 167,200,000$$

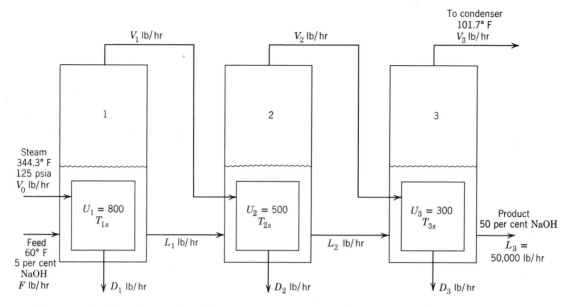

Figure 19.18. Flow sheet of triple-effect evaporator with forward feed with conditions for Illustration 19.3.

Solving simultaneously gives $V_1 = 57,000$ lb, $V_2 = 60,800$ lb, $V_3 = 62,200$ lb, and $V_0 = 110,800$ lb. The check of these results with the $V_1 = V_2 = V_3$ assumption, although not good, is adequate for a first approximation. The large value obtained for V_0 indicates however, that the equal-area requirement will not be met and a temperature adjustment will be needed. Accepting V_1, V_2, and V_3 as calculated, and solving the rate equations for the areas results in

$$A_1 = \frac{q_1}{U_1(-\Delta T_1)} = \frac{V_0 \lambda_{V_0}}{U_1(-\Delta T_1)}$$

$$= \frac{110,800 \times 875.4}{800 \times 30} = 4040 \text{ sq ft}$$

$$A_2 = \frac{V_1(H_1 - h_1)}{U_2(-\Delta T_2)} = 2160 \text{ sq ft}$$

$$A_3 = \frac{V_2(H_2 - h_2)}{U_3(-\Delta T_3)} = 2410 \text{ sq ft}$$

These unequal areas indicate that the original apportionment of the total available $-\Delta T$ among the various effects was improper. In adjusting the $-\Delta T$'s it is common practice to choose them so that

$$-\Delta T_1' = -\Delta T_1 \frac{A_1}{A_m} \tag{v}$$

$$-\Delta T_2' = -\Delta T_2 \frac{A_2}{A_m} \tag{vi}$$

$$-\Delta T_3' = -\Delta T_3 \frac{A_3}{A_m} \tag{vii}$$

where $-\Delta T_1'$, $-\Delta T_2'$, and $-\Delta T_3'$ are the newly assumed effective $-\Delta T$'s for each of the three effects, and $-\Delta T_1$, $-\Delta T_2$, $-\Delta T_3$ are those that were originally assumed. A_m is the mean area of one effect obtained by proper averaging of A_1, A_2, and A_3. Equations v, vi, and vii are corrections on the original assumptions that $q_1 = q_2 = q_3$ which led to the values of A_1, A_2, and A_3 obtained above. The result is that

each of the areas will approach A_m. The equal-evaporation assumption has been discarded, since it has served its purpose.

The simplest method of calculating A_m is merely to average A_1, A_2, and A_3, as was done in Illustration 19.2. This method, however, will give $-\Delta T$'s that will not add up to the total effective $-\Delta T$. One common practice is to calculate A_m by this averaging technique and then arbitrarily adjust the $-\Delta T$'s so that $(-\Delta T_1') + (-\Delta T_2') + (-\Delta T_3') = -\Delta T_{\text{eff}}$.

An alternate, and more satisfactory, method is to apply the restriction of a constant sum.

$$(-\Delta T_1') + (-\Delta T_2') + (-\Delta T_3') = \Sigma(-\Delta T_{\text{eff}})$$
$$= (-\Delta T_1) + (-\Delta T_2) + (-\Delta T_3) \tag{viii}$$

in the calculation of A_m. Combining Equations viii, v, vi, and vii for this purpose,

$$(-\Delta T_1)\frac{A_1}{A_m} + (-\Delta T_2)\frac{A_1}{A_m} + (-\Delta T_3)\frac{A_3}{A_m}$$
$$= (-\Delta T_1) + (-\Delta T_2) + (-\Delta T_3) \tag{ix}$$

from which

$$A_m = \frac{(-\Delta T_1)A_1 + (-\Delta T_2)A_2 + (-\Delta T_3)A_3}{(-\Delta T_1) + (-\Delta T_2) + (-\Delta T_3)} \tag{x}$$

The A_m value obtained from Equation x will give $\Sigma(-\Delta T') = \Sigma(-\Delta T_{\text{eff}})$ without adjustment unless BPR values change. Determining A_m in this way.

$$A_m = \frac{30.0 \times 4040 + 47.9 \times 2160 + 79.7 \times 2410}{157.6}$$
$$= 2645 \text{ sq ft}$$

and $-\Delta T_1' = 30.0 \times \dfrac{4040}{2645} = 45.8°F$

$$-\Delta T_2' = 47.9 \times \frac{2160}{2645} = 39.2°F$$

$$-\Delta T_3' = 79.7 \times \frac{2410}{2645} = 72.6°F$$

$$\Sigma(-\Delta T') = 157.6°F$$

If A_m had been calculated by averaging A_1, A_2, and A_3, a value of 2870 sq ft would have been obtained, from which $-\Delta T_1' = 42°F$, $-\Delta T_2' = 35.9°F$, $-\Delta T_3' = 66.7°F$, and $\Sigma(-\Delta T') = 144.6°F$. Considerable adjustment would have been required to make $\Sigma(-\Delta T') = 157.6$.

Using the $-\Delta T'$ values calculated by Equation x gives the significant temperatures as

Effect 1	Effect 2	Effect 3	Condenser
$T_{1s} = 344.3°F$	$T_{2s} = 293.5°F$	$T_{3s} = 244.3°F$	$T_c = 101.7°F$
$T_1 = 298.5$	$T_2 = 254.3$	$T_3 = 171.7$	

Indication of these temperatures to fractions of a degree should not mislead one as to the precision with which actual operating conditions are predicted.

Equations ii, iii, and iv can again be written using enthalpy values obtained at the new temperatures.

$$875.4V_0$$
$$+ 200,000 \times 26 = 1180.3V_1$$
$$+ (200,000 - V_1)245 \qquad \text{(ii)}$$

$$1180.3V_1$$
$$+ (200,000 - V_1)245 = 1167V_2 + 263V_1$$
$$+ (200,000 - V_1 - V_2)198 \qquad \text{(iii)}$$

$$198(200,000 - V_1 - V_2)$$
$$+ 1167V_2 = 212V_2 + (180,000 - V_1 - V_2)1137.4$$
$$+ 20,000 \times 200 \qquad \text{(iv)}$$

Solving them gives $V_1 = 57,300$, $V_2 = 61,100$, $V_3 = 61,600$, and $V_0 = 111,000$.

Since these values do not coincide with the initial assumption of equal evaporation, the resulting concentrations should be checked to see how the BPR's have been affected. Such a

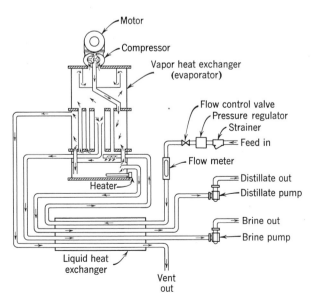

Figure 19.19. Schematic diagram of vapor-recompression evaporator. (Courtesy of Badger Mfg. Company.)

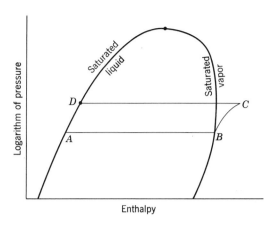

Figure 19.20. Schematic pressure-enthalpy diagram for water showing the operation cycle of a vapor-recompression evaporator.

check gives $x_1' = 0.07$, $x_2' = 0.123$, and a recheck through Figure 19.11 shows no change in the boiling-point rises.

Calculating the required areas as before gives $A_1 = 2640$ sq ft, $A_2 = 2640$ sq ft, and $A_3 = 2720$ sq ft. The values are equal within the accuracy range practically required. Moreover they nearly equal the value of A_m obtained from the first temperature approximation. Final answers to this problem would be 2700 sq ft surface per evaporator body; 111,000 lb/hr of steam required, and a steam economy of 1.62.

Vapor Recompression. Another method of increasing steam economy is by recompressing the vapors from the evaporator and feeding them back into the steam chest of the same evaporator effect. Mechanical compression or steam-jet injection may be used. In the case of the steam-jet compression, a high-pressure steam jet is used to entrain and compress a major part of the vapors from the evaporator. Some of the vapor is separately condensed to compensate for the motive steam added. A schematic flow diagram of a vapor-recompression evaporator in which mechanical compression of the vapor is used is shown in Figure 19.19. Here, the compressor is located on top of the evaporator body to prevent heat losses in any external piping. The principle of operation is shown in Figure 19.20 where the cycle is shown on a schematic pressure-enthalpy diagram for water. Water at A is evaporated in the evaporator to become saturated vapor at B. This vapor is compressed along path B–C, the work done by the compressor being $H_C - H_B$ if the compression is adiabatic. The compressed vapor at C enters the steam chest where it condenses by giving enthalpy to the evaporating liquid. Condensate at condition D leaves the unit through an exchanger in which the feed is heated. The heat released by the condensing vapor $(H_C - H_D)$ must be enough to evaporate an equal quantity of liquid, $(H_B - H_A)$ and to supply all the heat losses. The available $-\Delta T$ in the exchanger is that between the boiling solution at low pressure and the condensing vapor at higher pressure. With this method, evaporation equivalent to fifteen or more evaporator stages can be obtained. In this country, these systems are used only in special applications because of several mechanical and economical disadvantages: (1) The compressor is expensive and is subject to higher maintenance costs than is the remainder of the evaporator system. (2) In reasonable operating ranges, the $-\Delta T$ obtained is small, about $10°F$. As a result the evaporator equipment must be large for any reasonable production rate. Higher $-\Delta T$'s could be obtained with a higher compression ratio, but the efficiency would decrease proportionally. (3) A recompression evaporator cannot be started up without an auxiliary steam or heat supply. (4) If the solution has a large BPR the cost of recompression increases rapidly because the vapor must be compressed to such a pressure that its saturation temperature is above the temperature at which the solution boils. In a situation where fuel costs are high but electrical power is cheap, recompression evaporation is economical and is used. In the United States, fuel costs are low relative to power costs and will probably continue to be so for many years; thus, recompression evaporation is not generally attractive. It was regularly used to supply potable water at advanced military bases and on ships at sea. In Europe, fuel is relatively scarce and expensive, whereas power from hydroelectric stations is quite inexpensive in some areas. As a result, there are several large recompression-evaporation units in operation in European chemical plants.

Integration of Evaporators into the Total Plant Economy. The need for economy in the generation of steam dictates careful integration of the evaporator unit into the over-all plant economy. Such general economic considerations will dictate what steam pressure is available in a plant where steam is raised at high pressure and is used in an extraction turbine for generating electrical energy. Economics may also fix conditions of feed preheating and of product withdrawal temperature and may point to advantages in a particular feeding arrangement or to the withdrawal of some steam at one or more points between effects for ultimate use elsewhere in the plant. This latter procedure allows the evaporator to act as an efficient method of generating low-pressure steam when higher pressure steam is available. The advantages of steam withdrawal from an evaporator rather than mere throttling of the high-pressure steam are in the "free" evaporation which has been obtained by the evaporator. The disadvantage is that the steam generated may contain minute amounts of salt or other impurities that promote the corrosion of equipment in which it is used.

Scaling. In addition to the usual problems of process control and maintenance, the operator of evaporators is sometimes faced with serious loss of capacity resulting from the deposit of solids from the evaporating solution onto the evaporator heat-transfer surfaces. This problem is particularly acute when evaporating materials exhibit "inverse solubility," that is, where the solubility decreases as the temperature increases. When such a solution passes through the tubes of an evaporator, the material nearest the tube walls is heated to a higher temperature than is the material in the center of the tube. If the solution has a bulk temperature close to its saturation temperature, the fluid close to the wall may become hot enough so that the saturation concentration becomes lower than the existing concentration. The precipitated solid will cling to the tube wall, increase in temperature still further, and promote further precipitation. Solutions of sodium sulfate (see Figures 19.24 and 19.25) and of calcium sulfate are among those that exhibit inverse solubility and behave in this way. The scale that forms adheres tightly to the tube surface and must eventually be removed. Methods for removal include opening the evaporator and drilling out the tubes with a special cleaning tool, and boiling out the evaporator with a dilute acid solution. Such scaling can sometimes be greatly reduced by maintaining a suspension of the solid phase in the liquor, thereby providing extensive surface for crystallization within the slurry itself.

Scaling solutions can be best handled where the velocity through the tubes is high. As well as physically obstructing the deposit of scale in the tube, the resulting turbulence reduces the temperature variation across the tube diameter. Because they maintain a high circulation rate, forced-circulation evaporators are most suitable for this service. However, even at high circulation velocities, scale does deposit and must be removed. In one evaporator design, scale is removed by periodically switching the flow of solution and of steam through the two sides of the evaporative exchanger. Thus, surfaces on which scale deposits from the boiling solution are later washed clean by the steam condensate. The exchanger is made with square tubes on a checkerboard pattern so that liquor and steam passages are essentially identical. Many solutions, containing organic matter, cause deposits which are not true scales. If the operating conditions are held constant, the amount of true scale build-up should be proportional to the total amount of heat transferred through the surface since the start of the operation. Also, the over-all heat-transfer coefficient must decrease as the scale builds up since an additional heat-transfer resistance is being formed in series with those already present. On this basis (7),

$$U = \frac{1}{R_0 + R_s} = \frac{1}{\dfrac{1}{U_0} + aQ} \qquad (19.9)$$

where U = over-all coefficient of heat transfer at time θ after the start of evaporation.

U_0 = over-all coefficient of heat transfer at the start of evaporation

R_0 = resistance to heat transfer existing at the start of evaporation

R_s = resistance to heat transfer resulting from scale formation

Q = total amount of heat transferred from time 0 to time θ

a = a constant

The rate of heat transfer at any time (θ) is given by the basic rate equation as

$$q = \frac{dQ}{d\theta} = UA(-\Delta T) \qquad (14.31)$$

The relation between the over-all coefficient (U) and the time (θ) may be obtained by eliminating Q between Equations 19.8 and 14.31. Solving Equation 19.9 for Q and differentiating gives

$$aQ = \frac{1}{U} - \frac{1}{U_0}$$

$$a\,dQ = \frac{-dU}{U^2} \qquad (19.9a)$$

This relation may be substituted into Equation 14.31 to give the desired result in differential form

$$\frac{-dU}{U^3} = A(-\Delta T)a\,d\theta \qquad (19.10)$$

Integrating between the limits of $\theta = 0$ and $\theta = \theta$ gives

$$-\int_{U_0}^{U} \frac{dU}{U^3} = \int_0^\theta A(-\Delta T)a\,d\theta$$

$$\frac{1}{U^2} - \frac{1}{U_0^2} = A(-\Delta T)\,a\theta/2 \qquad (19.11)$$

Equation 19.11 shows that a graph of $1/U^2$ plotted against θ for the evaporation of a true scaling solution should give a straight line. If this is true, the entire path of U as a function of time can be found from values of the over-all coefficient at two points in time after the start of operation. This information, coupled with knowledge of the shutdown time required to clean the tubes, permits the engineer to determine the optimum time of operation between shutdowns.

CRYSTALLIZATION

In many cases, the salable product from a plant must be in the form of solid crystals. Throughout the history of the modern chemical industry, crystals have been produced by crystallization methods ranging between those as simple as setting pans of hot concentrated solution out to cool and those as complex as continuous, carefully controlled, many-step processes tailored to give a product of uniform particle size, shape, moisture content, and purity. Customer demands as to product quality have gradually forced the discontinuance of the simpler crude crystallizers, because a modern crystal product must usually meet very rigid specifications governing the properties listed above plus such others as color, odor, and caking characteristics.

Crystallization Equipment. Crystallizers can be conveniently classified in terms of the method used to obtain deposition of particles. The groups are:

1. Crystallizers that obtain precipitation by cooling a concentrated, hot solution.

2. Crystallizers that obtain precipitation by evaporating a solution.

3. Crystallizers that obtain precipitation by adiabatic evaporation and cooling.

In the first group are found the pan coolers mentioned above, agitated batch crystallizers, and the Swenson-Walker continuous crystallizer. In the second group are those evaporators in which crystallization as well as evaporation takes place, called salting evaporators, and the Oslo crystallizer. In the third group are the vacuum crystallizers.

Of the crystallizers that operate by cooling, the simplest are various forms of batch crystallizers. *Pan crystallizers* need little description being merely pans in which a hot solution is allowed to cool and crystallize. Actually they are seldom met in modern practice, except in small-scale operations, since they are wasteful of floor space and of labor and usually give a low-quality product. *Agitated batch crystallizers* are a simple improvement. They consist of an agitated tank, usually cone bottomed, containing cooling coils. In small-scale or batch processing, such crystallizers are quite convenient because of their low first cost, simplicity of operation, and flexibility. They are too wasteful of manpower and give too uneven a product to be attractive for large-scale continuous processing.

The *Swenson-Walker crystallizer* is a cooling crystallizer designed to operate continuously. It consists of an open round-bottomed trough 24 in. wide by 10 ft long, jacketed for cooling water, and containing a long, ribbon mixer which turns at about 7 RPM. As many as four of these units may be connected together with the agitators driven from a single shaft. If longer lengths are required, a second set of crystallizers may be set up at a slightly lower level and fed by overflow from the end of the first crystallizer set. Figure 19.21 shows a crystallizer consisting of two of these connected units. The hot, concentrated solution is fed continuously at one end of the crystallizer and flows slowly toward

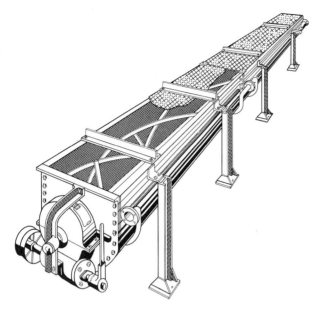

Figure 19.21. Swenson-Walker crystallizer. (Courtesy Swenson Evaporator Company, a Division of The Whiting Corporation.)

the other end while it is being cooled. The function of the agitator is to scrape the crystals from the cold walls of the unit and to pick up the crystals and cascade them down through the solution so that precipitation occurs mainly by build-up on previously formed crystals rather than by the formation of new crystals.

Of the evaporating crystallizers, the *salting evaporator* is the most common. In older forms, this crystallizer consisted of an evaporator below which were settling chambers into which the salt settled. The chambers were connected in tandem so that salt could be removed from one of them while the other was connected to the system. Modern practice no longer has these settling chambers directly connecting to the body. They are most commonly replaced by a classifying settler, through which a stream of the liquor from the evaporator body is pumped continuously. The dimension of the settler and the circulation rate are so related that only the coarser crystals settle out. The fine ones remain in suspension and are returned to the evaporator body for further growth. The thick slurry from the bottom of the settler is pumped continuously to filters or centrifuges or other appropriate processing equipment. Mother liquor may be returned to the evaporator or discarded in whole or in part to eliminate impurities from the system.

The *Oslo crystallizer*, one form of which is the "*Krystal*" crystallizer shown in Figure 19.22, is a modern form of evaporation crystallizer. This unit is particularly well adapted to the production of large-sized, uniform crystals usually somewhat rounded. It consists essentially of a forced-circulation evaporator with an external heater containing a combination salt filter and

particle-size classifier on the bottom of the evaporator body. If desired, the external heater may be used as a cooler, in which case crystallization occurs by solution cooling.

The unique feature of the Oslo crystallizer is that a slightly supersaturated solution is passed upward through a bed of crystals, depositing on them the excess solute above saturation and simultaneously classifying them so that only the larger ones settle against the stream. Fine crystals and saturated solution leave the top of the bed and are recycled. The recycle stream is treated to attain supersaturation in one of three ways:

(1) it may be subjected to evaporation; (2) it may be cooled; or (3) it may be enriched in solute by addition of a strong feed solution. The first will be described.

The two sections of the body and the external heater are clearly shown in Figure 19.22. The lower section of the body contains the bed of crystals through which the liquor flows upward to exert its classifying action. Solute corresponding to the supersaturation is largely deposited on the existing crystals. The saturated solution and fine crystals are then heated in the external heater, which is maintained under enough hydrostatic head to suppress boiling on the heating surface. The heated liquid passes upward to the top section of the body. At the higher elevation, some of the solution flashes to vapor, leaving a supersaturated solution which

Figure 19.22. "Krystal" Crystallizer for production of ammonium sulfate. (Courtesy Struthers-Wells Corp.)

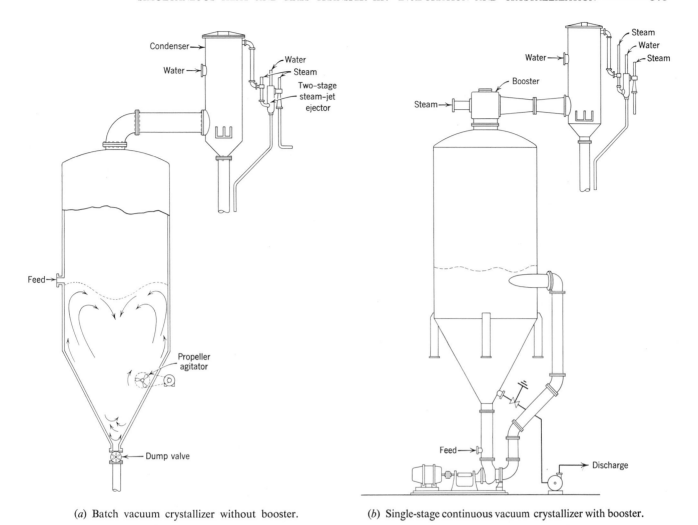

(a) Batch vacuum crystallizer without booster.

(b) Single-stage continuous vacuum crystallizer with booster.

Figure 19.23. Two types of vacuum crystallizers. The batch unit at the left employs mechanical agitation and a two stage jet ejector, whereas the unit on the right recycles the liquor-crystal mix for agitation and has a booster and a two-stage ejector. (Courtesy Swenson Evaporator Company, a Division of The Whiting Corporation)

is led downward through a central duct to the bottom of the crystal bed. The large crystals are discharged through an opening in the bottom of the lower section of the body.

In *vacuum crystallizers*, evaporation is obtained by flashing a hot solution into a low-pressure space. Energy for vaporization is obtained at the expense of sensible heat in the feed. As a result, the temperature of the vapor-liquor mixture after flashing is much lower than that of the liquor before flashing. In Figure 19.23, two designs of a vacuum crystallizer are shown. These units may be operated continuously as one or more stages or batchwise. In batch operation, the hot feed is pumped into the vessel and agitation started. Then, the ejectors are started, and the system pressure and temperature are gradually lowered. The run is finished when the vacuum pump develops the minimum obtainable pressure. During the run, liquid has been vaporzied at all pressures from atmospheric to the final minimum

pressure. Much of this vapor has to be compressed only slightly so that the ejectors can operate at high capacity. Only toward the end of the run is the compression ratio high, and, during this period, a booster ejector may be necessary. The over-all result is that the vapors are removed much more economically than would be possible if the unit operated continuously at the lowest pressure.

For high-capacity units, usually above 10,000 gal/hr of feed, batch operation is impractical and the crystallizer must be operated continuously. To avoid the poor energy economy mentioned above, the continuous units may be operated in stages. As the number of stages increases, the steam consumption approaches that of batch operation, but ultimately the capital costs become excessive. For continuous operation, the feed is put into the first effect which operates at only a slightly reduced pressure. The point of feed injection is selected so that minimum superheating of any significant quantity

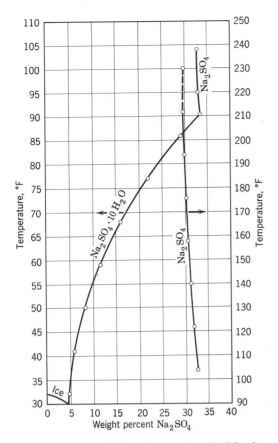

Figure 19.24. Solubility diagram for Na_2SO_4 in H_2O at 1 atm total pressure. (Data from *International Critical Tables*, Vol. 4, page 236.)

for transfer. In crystallization, mass is transferred from the solution to the crystal surface. The concentration necessary for crystals to form and the chemical species which separates can be determined from a temperature-composition phase diagram, which is often called a solubility diagram. Data on the solubility of solid compounds in solvents as a function of temperature can be found in most handbooks, and a very complete collection is given by Seidell (14). In Figure 19.24 data for the Na_2SO_4-H_2O system are plotted to show solubility as a function of temperature for a total pressure of one atmosphere. Over the temperature range shown, only three solid species precipitate. At temperatures between 30°F and 90.5°F and at concentrations above 4.5 per cent, the saturated solution is in equilibrium with the decahydrate. Above 90.5°F, the solid phase is anhydrous Na_2SO_4. Note that the anhydrous salt exhibits an inverse solubility between 90.5°F and about 220°F. At 220°F, a minimum solubility of 29.5 per cent is found. Above this temperature, increasing the temperature increases the solubility. At concentrations below 4.5 per cent Na_2SO_4, the solid phase in equilibrium with saturated solution will be ice. This system exhibits an eutectic at about 4.5 per cent Na_2SO_4, at which concentration the minimum freezing point temperature of 29.84°F occurs. Although the data for this diagram are presented at a total pressure of 1 atm, the diagram can be used over a reasonable pressure range since only liquid and solid phases are involved. For crystallizer calculations in which the final temperature is known and in which there has been no evaporation, the saturated-solution concentration, the weight of crystal per given weight of feed of known concentration, and the species precipitated can be determined from a solubility diagram. These problems require only a material balance for their solution.

For crystallization problems involving energy balances, such as those in which evaporation occurs or in which the final temperature of an adiabatic crystallizer is unknown, enthalpy data as well as solubility data are needed. In such cases, the enthalpy-composition diagram is convenient. Figure 19.25 shows the enthalpy-composition diagram for the Na_2SO_4-H_2O system. This is a relatively simple diagram, and the principles involved in its construction and interpretation should be familiar to the reader from his study of Chapter 3. Within any field, the relative quantities are calculated from the familiar lever-arm ratio. The $CaCl_2$-H_2O diagram shown in Figure 19.26 is much more complicated, and some explanation of its features are in order. This system has four hydrates and an eutectic at point *b*. Points *c*, *d*, *e*, and *f* represent transition points from one hydrate in equilibrium with solution to another. As with the Na_2SO_4-H_2O diagram, the major part of the data used here apply at 1 atm pressure. In addition to the

of solution is obtained. Usually, the superheating is limited to 5°F to prevent caking up of the unit and formation of large numbers of new, small crystals. The product withdrawn from the crystallizer contains both crystals and mother liquor and is pumped to the second effect. The second effect operates at a pressure somewhat below that of the first effect. In each effect, some evaporation takes place; the solution cools and deposits solids, and the pressure is reduced compared to the preceding stage. In the last stages, a booster ejector may be necessary.

In each of these units, attempts have been made to solve the problems of preventing caking on the walls, of separating vapor from liquid and liquid from crystals, of obtaining low operating cost and first cost, of conserving floor space, of promoting the desired crystal growth of pure crystals, and of keeping maintenance costs low. The choice, from among these and other crystallizers, of a unit to do a particular job depends upon the economics of each individual situation and upon the product limitations imposed by the sales situation.

Solubility Relations. As with other two-phase systems in which transfer occurs, an equilibrium statement is necessary in order to set the limit of the driving force

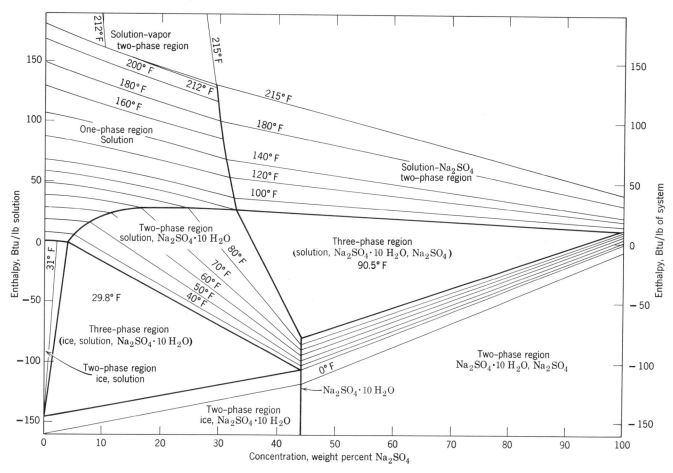

Figure 19.25. Enthalpy-concentration diagram for the Na_2SO_4–H_2O system at 1 atm total pressure. Bases: $H = 0$ for water at $32°F$ and the triple point pressure, $H = 0$ for Na_2SO_4 solid at $32°F$ and at 1 atm pressure.

1-atm data, the vapor-liquid saturation lines for 0.5 atm and 0.2 atm are given. All other lines on the chart apply at 1 atm.

It may be instructive to follow on Figure 19.26 the course of a process in which a mixture of $CaCl_2 \cdot 6H_2O$ and $CaCl_2 \cdot 4H_2O$ having an over-all mixture composition of 60 per cent $CaCl_2$, 40 per cent H_2O is heated under a 1 atm total pressure. At low temperatures the system consists of a mixture of $CaCl_2 \cdot 6H_2O$ and $CaCl_2 \cdot 4H_2O$. Use of the inverse lever-arm principle shows the mix to contain about 90 weight percent $CaCl_2 \cdot 4H_2O$. As the mixture is warmed, no phase change occurs until the temperature reaches $86°F$. At this point, solution of composition and enthalpy represented by point c begins to form. This formation occurs mainly by the dissolving of the $CaCl_2 \cdot 6H_2O$ phase to form saturated solution and some $CaCl_2 \cdot 4H_2O$. As heating continues, the temperature remains at $86°F$, but the $CaCl_2 \cdot 6H_2O$ continues to dissolve until only $CaCl_2 \cdot 4H_2O$ and solution remain. During this period, the solution composition and enthalpy are continuously represented by point c. From this point, further heating increases the system temperature until it reaches $113°F$. At the same time, some of the

$CaCl_2 \cdot 4H_2O$ dissolves to form saturated solution, and the solution concentration and enthalpy move along line cd. Further heating at $113°F$ causes the $CaCl_2 \cdot 4H_2O$ to dissolve and recrystallize forming $CaCl_2 \cdot 2H_2O$ and solution of condition d. This continues until no further $CaCl_2 \cdot 4H_2O$ remains. Still further heating raises the system temperature, dissolves $CaCl_2 \cdot 2H_2O$, and forms more concentrated solution until, at about $190°F$, the last of the $CaCl_2 \cdot 2H_2O$ dissolves leaving only the solution phase. From this point, the temperature rises with no phase change until, at $290°F$, boiling begins. Beyond this point, additional heating raises the temperature, increases the proportion of vapor phase present, and increases the concentration of the boiling liquid phase. The vapor phase may be assumed to be pure water, and its enthalpy may be obtained from the steam tables at the proper temperature and 1 atm total pressure.

Illustration 19.4. One pound each of Na_2SO_4 and H_2O at $50°F$ are mixed and allowed to reach equilibrium at atmospheric pressure. If the system is perfectly insulated so that equilibrium is reached with no gain or loss of enthalpy, what will be the temperature and phase condition of the product?

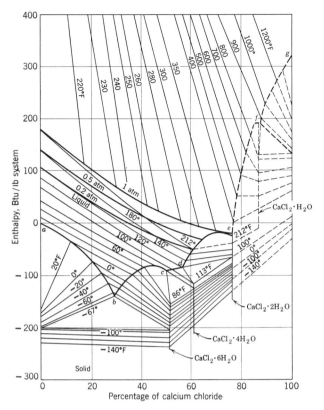

Figure 19.26. Enthalpy-concentration diagram for the $CaCl_2$–H_2O system. (Hougen, Watson, and Ragatz, *Chemical Process Principles*, Part I, 2nd ed., John Wiley and Sons, New York, 1954.)

SOLUTION. The adiabatic mixing process would be represented on the enthalpy-composition diagram, Figure 19.25, by a straight line joining the two components of the mixture. At 50°F, $H_{H_2O} = 18$ Btu/lb, $H_{Na_2SO_4} = 4$ Btu/lb Then, the enthalpy of a 1:1 mix will be 11 Btu/lb, and the concentration will be 50 weight percent Na_2SO_4. These values of H and x put the mix point on the H–x diagram in the 90°F three-phase region. Therefore, the temperature is 90°F, and the mixture consists of Na_2SO_4, $Na_2SO_4 \cdot 10H_2O$, and saturated solution at 32.7 per cent Na_2SO_4.

The quantities of the three phases present can be found using the inverse-lever-arm principle. A tie line is drawn from any corner of the three-phase region to the opposite side through the mix point (J). By the inverse-lever-arm principle, the relative quantity of the phase used in drawing the tie line and of the mixture of the other two phases can be found. The inverse-lever-arm principle will also, then, give the relative proportion of the other two phases present. The method is illustrated schematically in Figure 19.27. On this diagram, the relative distances are as follows:

$$\frac{\overline{AJ}}{\overline{JJ'}} = \frac{1.79}{3.33} \qquad \frac{\overline{BJ'}}{\overline{J'C}} = \frac{4.06}{1.88}$$

From these relative distances,

$$\frac{A}{J'} = \frac{3.33}{1.79} = 1.860; \qquad \frac{B}{C} = \frac{1.88}{4.06} = 0.464$$

also $A + B + C = 2.0$ and $A + J' = 2.0$

Then $\qquad A = 1.860(2.0 - A); \qquad A = 1.300$

$$B + C = 0.700$$

$$B = 0.464(0.700 - B); \qquad B = 0.228$$

and by difference

$$C = 2.00 - 0.228 - 1.300; \quad C = 0.472$$

From this, the final answer is

$T = 90°F \qquad$ Solution (32.7 %) = 1.300 lb

$\qquad\qquad\qquad Na_2SO_4\,10H_2O = 0.228$ lb

$\qquad\qquad\qquad Na_2SO_4 = 0.472$ lb

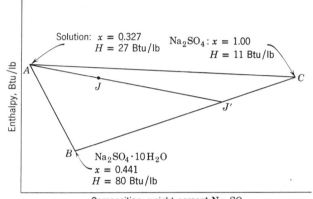

Figure 19.27. Application of the inverse-lever-arm principle to a three-phase region, Illustration 19.4.

Crystallizer Material and Energy Balances. As was done for evaporation operations, material and energy balances and a heat-transfer-rate equation can be written for the crystallization process. The material balance will give the process yield, that is, the mass of crystals formed from a given mass of solution, if the extent of evaporation or of cooling can be determined. The general crystallization process shown schematically in Figure 19.28 gives the nomenclature involved.

A solute material balance gives

Solute in feed = solute in product crystals + solute in product liquor

$$Fx_F{}' = C\frac{M_a}{M_h} + \left[F(1 - x_F{}') - V\right.$$

$$\left. - C\frac{M_h - M_a}{M_h}\right]X' \quad (19.12)$$

which rearranges to

$$C = \frac{[F(1 - x_F{}') - V]X' - Fx_F{}'}{\left(\dfrac{M_h - M_a}{M_h}\right)X' - \dfrac{M_a}{M_h}} \quad (19.12a)$$

where C = mass of crystals in the product magma per unit time

$\qquad M_a$ = molecular weight of anhydrous solute

M_h = molecular weight of hydrate crystal

x_F' = mass fraction of anhydrous solute in feed

X' = solubility of the material at product temperature expressed as a weight ratio of anhydrous salt to solvent

F = total mass of feed per unit time

V = evaporation in pounds mass of solvent per unit time

This relatively simple material balance is applicable to all single-stage crystallization units or to multistage units from which only one liquid-solid product is withdrawn. In the calculation of the solute in the final liquor, account must be taken of the solvent lost by evaporation and of the solvent lost as water of crystallization.

The determination of V or of the final magma temperature depends upon a heat balance and a rate equation. For adiabatic crystallization as in a vacuum crystallizer, the rate equation is unnecessary, for the extent of evaporation or cooling may be fixed by an enthalpy balance. In any case, the enthalpy balance can be written with the aid of an enthalpy-composition diagram and the steam tables. As in evaporation calculations, the rate equation is written in terms of an overall coefficient, and this coefficient must usually be determined on the basis of experience.

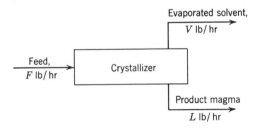

Feed, F lb/hr

Crystallizer

Evaporated solvent, V lb/hr

Product magma L lb/hr

Figure 19.28. Schematic diagram of generalized crystallization process.

Illustration 19.5. A single-stage, continuous, Krystal crystallizer is to be used to obtain $CaCl_2 \cdot 4H_2O$ product from a feed containing 40 weight percent $CaCl_2$ in water at 180°F. The vacuum system on the crystallizer will give an equilibrium magma at 90°F.

(*a*) What range of heat input per pound of feed solution can be used to obtain a product containing only $CaCl_2 \cdot 4H_2O$ crystals?

(*b*) What would be the maximum yield of crystal in the product (pounds of $CaCl_2$ as crystal per pound of $CaCl_2$ in feed)?

SOLUTION. This problem can be solved directly using the graphical heat- and material-balance methods developed in Part I. From Figure 19.26 the enthalpy of the feed is read to be −8 Btu/lb of solution.

The products of the crystallizer will be vapor at 90°F and magma consisting of $CaCl_2 \cdot 4H_2O$ crystals in saturated $CaCl_2$ solution also at 90°F. The enthalpy of 90°F vapor cannot be

read from Figure 19.26 but is nonetheless fixed and can be determined from the steam tables. Note also that the system pressure, though it is much less than 1 atm, cannot be determined from Figure 19.26. This fact does not prevent the determination of the vapor enthalpy since, at these pressures, enthalpy is practically independent of pressure. Thus, from the steam tables, $H_V = 1100.9$ Btu/lb.

The product magma must have a composition and enthalpy lying on the 90°F tie line between saturated solution and $CaCl_2 \cdot 4H_2O$ crystals. Therefore its over-all concentration is limited to 52 to 61 per cent and its enthalpy to a range between −90 and −120 Btu/lb.

Using this information, a graphical or an analytical solution of the energy and material balances can be carried out. The graphical solution gives a less precise answer but illustrates the principles better than does the analytical solution. Such a graphical solution is shown in Figure 19.29. The enthalpy and concentration of the feed put it at point (x_F', H_F). The vapor conditions are plotted at point (y_V', H_V), and the magma conditions are located along the line from (x_S', H_S), the saturated solution at 90°F, to (x_C', h_C), the crystal conditions. The heated feed must then lie in the range of (x_{F+h}', h_{F+h}) indicated. From this information the feed heating is fixed.

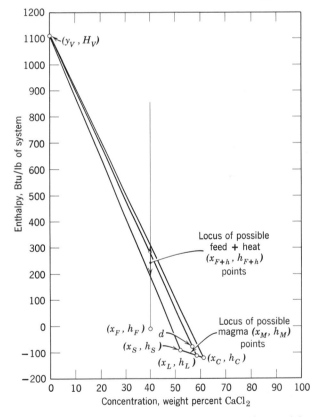

Figure 19.29. Graphical solution of heat- and material-balance equations, Illustration 19.5.

(*a*) Minimum feed heating = 185 − (−8) = 193 Btu/lb of feed

Maximum feed heating = 305 − (−8) = 313 Btu/lb of feed

(*b*) Using the maximum heating, the products would consist only of crystals and vapor. The yield would be 1 lb of $CaCl_2$ as tetrahydrate crystal per pound of $CaCl_2$ in the feed. This yield is impractical because, first, the product could not be removed from the crystallizer and, second, the tie line connecting crystal and vapor passes through the saturation curve to the right of point d. This means that initially $CaCl_2 \cdot 2H_2O$ will precipitate and that, although at equilibrium it will be converted to $CaCl_2 \cdot 4H_2O$, the conversion will be slow and may not be complete by the time the product leaves the crystallizer. If the maximum yield is taken as corresponding to just enough heating so that the vapor-magma tie line passes through point d, the construction line (y_V', H_V)–d–(x_L', h_L) results. Products from the crystallizer now consist of vapor at (y_V', H_V) and magma at (x_L', h_L). Then,

Feed heating required $= 270 - (-8) = 278$ Btu/lb of feed

Mass of vapor generated $= \dfrac{58.5 - 40}{58.5} = 0.316$ lb/lb of feed

Mass of crystals obtained $= (1 - 0.316)\dfrac{58.5 - 52}{61 - 52}$

$$= 0.494 \text{ lb/lb of feed}$$

Yield of $CaCl_2 = \dfrac{0.494 \times \dfrac{111}{183}}{0.40} = 0.75$ lb $CaCl_2$ in crystal per lb of $CaCl_2$ fed

This yield of $CaCl_2$, however, gives a magma of $(0.494/0.684) \times 100 = 72$ per cent crystals, 28 per cent solution, which is more concentrated than most withdrawal systems will tolerate. A magma containing about 55 per cent by weight of crystals is about the most concentrated that is recommended by most manufacturers.

The material and energy balances may not always be made in as direct a way as is indicated above. There is some evidence that the assumption of equilibrium in the interface which is inherent in the use of the energy balance is not accurate in the case of vacuum crystallizers. With these units, rates of mass transfer from the liquid into the vapor are so great that the "absolute vaporization rate" is approached (13). This phenomenon is somewhat analogous to the sonic velocity phenomenon in that increasing the driving force no longer proportionally increases the rate of transfer. As a result of this phenomenon, and of the accumulation of crystals at the interface, the pressure in the vapor space may be significantly less than would be predicted on the basis of the liquid phase temperature. For example, the products from a vacuum crystallizer operating at 40°F may have to be pumped out from a pressure that would correspond to a 35°F liquid temperature This behavior cannot now be predicted without experimental data.

Crystallization Mechanism. The mechanism by which crystallization occurs is of more than passing interest because it influences the conditions within the crystallizer and the properties of the product obtained. As would be expected, the deposition of a solid crystal can occur only as the result of a concentration driving force from the bulk solution to the interface of the solid surface. Since the interface concentration must be the equilibrium concentration for solid and solution coexistence, that is, since the interface concentration is that of saturated solution, the concentration of the bulk fluid must be greater than saturation. The extent of the supersaturation depends upon the number and shape of crystals upon which precipitation occurs, the temperature level, the solution concentration, and the violence of the agitation present.

The crystallization process is usually divided into *nucleation* and *crystal-growth* mechanisms. The actual mechanism of nucleation is still uncertain. However, it probably involves the assembly of solute molecules into a specific concentrated array as a result of their normal molecular motion. The assembly may exist only momentarily, because, in addition to the attractive and repulsive forces between the molecules, there are the disruptive forces of solvent molecule movements and convective eddy currents acting on the entire system. The greater the concentration of solution, the more probable is the existence of such assemblies at any instant, and the more probable that they become stable. Ultimately, some of the molecule groups must grow and stabilize until a solid-liquid interface results. Upon this interface, additional precipitation occurs. Once formed, the crystals have a solubility slightly less than the first associated molecules. Miers (8)(9), on the basis of experimental results, concluded that nucleation could occur only at concentrations with some minimum degree of supersaturation. In the complete absence of solid particles, a "supersaturation" curve roughly parallel to the saturation curve was found below which spontaneous nucleation did not occur. This curve is shown schematically in Figure 19.30. At concentrations above curve CD, nucleation will occur more rapidly as the concentration increases. According to Mier's original theory, crystals will grow between curve CD and curve AB, but no new crystal nuclei will form.

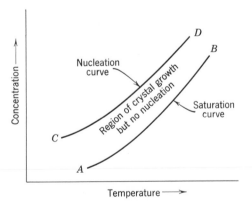

Figure 19.30. Qualitative presentation of Mier's supersolubility curve.

More recent experiments (12) have shown that, with sufficient time, nucleation can occur well below the curve CD, so that at present the supersaturation curve is considered to be more of an area. The principal contribution of Mier's theory is in emphasizing that considerable supersaturation is necessary for nucleation to occur. Experimental evidence that nucleation is more rapid in large volumes of solution than in small ones confirms the hypothesis that this nucleation is the result of random collision of molecules. In all but the most meticulously controlled experimental atmospheres, insoluble solids in suspension and dust particles from the air form ready-made nuclei so that crystallization occurs readily at only slight supersaturation.

The process of crystal growth has been studied extensively, with recent work suggesting that crystal growth occurs at erratic rates (2), that the solution concentration at the crystal surface is not uniform (3), and that crystal growth may take place in a spiral manner at dislocations in the crystal structure (5). Despite the uncertain mechanism, the gross result can be approached through the usual rate-equation forms. Rate equations for crystal growth have been written using the two-film theory. Separate equations were written for the diffusion of solute from the edge of the laminar layer to the crystal face (10) and for the turbulent transfer of solute from the body of the solution to the laminar interface (1). These equations were then combined to give an equation for transfer across the entire path from bulk fluid to solid surface in terms of a liquid phase transfer coefficient. As shown in Chapter 13, equations of this sort can be expressed in a single relation applicable from solid boundary to bulk fluid. Application of Equation 13.80 gives

$$N_A M_A = \frac{-4(\bar{E}_N + \mathscr{D})A(c_s - c_a)}{\gamma_N x} \quad (19.13)$$

where N_A = lb moles of solute deposited per unit time

M_A = molecular weight of solute

$\bar{E}_N$ = integrated mean eddy diffusivity applicable over the entire transfer path, sq ft/hr

$\mathscr{D}$ = diffusivity, sq ft/hr

c_s, c_a = concentrations of solute at the solid surface and in the main bulk of the fluid phase, lb/cu ft

A = area of crystal surfaces, sq ft

γ_N = ratio of concentration gradients, see Equation 13.57

x = length of entire transfer path, ft

The group $4(\bar{E}_N + \mathscr{D})/\gamma_N x$ is the liquid-phase mass-transfer coefficient (k_L) as defined in Chapter 13. Moreover, here there is no transfer through the solid phase, so that $k_L = K_L$. Thus

$$-N_A M_A = K_L A(c_s - c_a) \quad (19.14)$$

Equation 19.14 can be written for a single crystal if A is taken as the area of a single crystal. In this case

$$-N_A M_A = -\frac{dm}{d\theta} = K_L A(c_s - c_a) \quad (19.15)$$

where m = mass of a crystal, lb

θ = time, hr

since steady-state exists for any differential section of crystal surface.

It has been found experimentally that the shape of a crystal does not change during growth, so that

$$\phi \times A \times D \times \rho = m \quad (19.16)$$

where ϕ = a shape factor, dimensionless.

D = a characteristic crystal dimension

For example, the factor ϕ for a cube is $\frac{1}{6}$ since the area is $6D^2$, the volume is D^3, and the mass is $D^3\rho = \frac{1}{6}AD\rho$. Then Equation 19.15 can be written as

$$-\frac{dD}{d\theta} = \frac{K_L}{\phi\rho}(c_s - c_a) \quad (19.17)$$

This equation states that the rate of linear crystal growth is independent of crystal size. Of course, the rate of volume growth does not show such independence, for

$$V = \phi'D^3; \quad dV = 3\phi'D^2\,dD$$

where ϕ' is a shape factor but different from ϕ. Writing Equation 19.17 in terms of increase in volume gives

$$\frac{dV}{d\theta} = \frac{3\phi'D^2 K_L}{\phi\rho}(c_s - c_a) \quad (19.18)$$

Thus $dD/d\theta \propto \Delta c$, whereas $dV/d\theta \propto D^2\,\Delta c$. From these relations, the size distribution of the crystal mass after growth can be estimated from the size distribution of the crystal mass before growth, as will be shown below.

Crystallization can be carried out at very small supersaturation so that nucleus formation is slow. In such cases, it is common practice to "seed" the crystallizer with small particles of crystalline solute and to expect the major precipitation mechanism to be by crystal growth. This practice gives a product containing large crystals, whereas, if carried out at high supersaturation, there will result a mass of fine crystals. The calculation of crystal-size distribution for a seeded crystallizer by the method indicated above gives at best a very crude estimate of the actual distribution. Nucleation cannot be entirely prevented, and often also a classification process takes place in the crystallizer resulting in longer holding times for small crystals. Since agitation is frequently violent, large crystals may break and erode. A reverse effect results from the fact that small crystals are more soluble than large ones as the result of a surface-energy effect. Thus large crystals may grow at the expense of small ones.

Despite the inexactness of the results, such an estimate does give a useful first approximation of the particle size of the product to be expected from a given seed. The seed crystals used are unavoidably in a range of particle sizes. Even so, the relation between seed and product particle sizes may be written as

$$D_p = D_s + \Delta D \qquad (19.19)$$

where D is again a characteristic particle dimension and the subscripts s and p refer to seed and product respectively. In Equation 19.19, ΔD is constant throughout the range of sizes present as shown by Equation 19.17. From this equation, the seed and product masses may be related (15) for

$$m_p = \phi'\rho D_p{}^3 = \phi'\rho(D_s + \Delta D)^3 \qquad (19.20)$$

and

$$m_s = \phi'\rho D_s{}^3 \qquad (19.21)$$

which combine to give

$$m_p = m_s\left(1 + \frac{\Delta D}{D_s}\right)^3 \qquad (19.22)$$

Here, all the crystals in the seed have been assumed to be of the same shape, and the shape has been assumed to be unchanged by the growth process. This assumption is reasonably close to the actual conditions in most cases. Equation 19.22 was written for the entire crystal mass. It may also be written for differential parts of the crystal masses, each consisting of crystals of identical dimensions. The differential equation which results can then be integrated over the entire range of particle sizes.

$$\int_0^{m_p} dm_p = \int_0^{m_s}\left(1 + \frac{\Delta D}{D_s}\right)^3 dm_s = m_p \quad (19.23)$$

The indicated integration can be carried out stepwise for each small but finite particle-size range in the seed crystals, using successively assumed ΔD's until the product-to-seed ratio reaches a desired value, or conversely it can be integrated with known ΔD to get the mass of product. In either case, the product particle-size range will be determined. Since the fractional growth in mass of small particles is larger then for large ones, the shape of the screen analysis curve will change.

Illustration 19.5. A solute which forms cubic crystals is to be precipitated from solution at a rate of 10,000 lb of solid (dry basis) per hour using 1000 lb/hr of seed crystals. If no nucleation occurs and the seed crystals have the following size distribution, determine the product size distribution:

Tyler Sieve Mesh	Weight Fraction Retained
−48 + 65	0.10
−65 + 100	0.30
−100 + 150	0.50
−150 + 200	0.05
−200 + 270	0.05

SOLUTION. Here, the seed crystal-size distribution and the product-to-seed weight ratio are given, and so the product crystal-size distribution can be determined by trial-and-error selection of ΔD values followed by a summation-of-finite-parts approximate solution of Equation 19.23. The characteristic particle dimension may be taken as equal to the average of the smallest screen opening through which a particle passed and that on which it was retained. Thus the seed crystal dimensions are

Weight Fraction (Δm_s)	Average Dimension (D_s), in.
0.10	0.0099
0.30	0.0070
0.50	0.0050
0.05	0.0035
0.05	0.0025

As a first attempt, a $\Delta D = 0.005$ in. will be used. Then,

Δm_s	$\left(1 + \dfrac{\Delta D}{D_s}\right)^3$	Δm_p
0.10	3.37	0.337
0.30	4.94	1.480
0.50	8.00	4.00
0.05	14.45	0.723
0.05	27.0	1.35
1.00		7.89

Since m_p/m_s is specified as 10.0, the ΔD chosen was too small. Succeeding trials lead to a final $\Delta D = 0.0058$ for which

Δm_s	$\left(1 + \dfrac{\Delta D}{D_s}\right)^3$	Δm_p	D_p
0.10	3.95	0.39	0.0157
0.30	6.00	1.80	0.0128
0.50	10.10	5.05	0.0108
0.05	18.70	0.94	0.0093
0.05	36.60	1.83	0.0083
1.00		10.01	

The particle sizes of product crystals can be obtained by adding ΔD to the dimension D of each fraction of the seed crystals. The weight fraction of product crystals (Δm_p), represented by each increment of seeds, is computed for all increments.

Mass Fraction Product $(m_p/\Sigma\Delta m_p)$	$D_p = D_s + \Delta D$	Product Cumulative Fraction Smaller than D_p
0.039	0.0157	0.961
0.180	0.0128	0.781
0.504	0.0108	0.277
0.094	0.0093	0.183
0.183	0.0083	0.000
1.000		

The product and feed size distributions are plotted in Figure 19.31 as cumulative weight percent of the total sample smaller than the size indicated. Note the flattening of the product distribution curve at the small-particle size range. It is caused by the uniform increase in dimension experienced by the vastly larger number of small particles compared to the same size increase experienced by the smaller number of large particles. In other words, the small crystals have a very large surface to volume ratio compared to that of the large crystals. Therefore the small crystals receive a much higher quantity of mass per unit volume of original seed than do the large crystals.

The product size-range increments are not expressed in the screen-size units that were used to describe the seed particles. A conversion to screen sizes can be made through use of Figure 19.31 as follows:

Tyler Sieve Mesh	Screen Opening, in.	Weight Fraction of Product Smaller than Opening
100	0.0058	0.0
65	0.0082	0.10
48	0.0116	0.80
35	0.0164	0.99
28	0.0232	1.00

Tyler Sieve Mesh	Screen Analysis Weight Fraction
−28 + 35	0.01
−35 + 48	0.19
−48 + 65	0.70
−65 + 100	0.10

The flattening of the size distribution mentioned above is again evident.

The results obtained in Illustration 19.5 are typical of those obtained with the so-called ΔD law. Two major assumptions have been made: first, that there is no nucleation and, second, that every particle grows through the same increase in linear dimension. Neither of these assumptions fits the facts of typical crystallization behavior, but the second is much closer than is the first, particularly so when ΔD is small compared to D_s. For relatively large values of ΔD, the crystals may grow unsymmetrically. When large crystals are grown, the larger crystals experience a greater ΔD than do the small ones.

In actual practice nucleation cannot be prevented. In typical recycling crystallizers, a given particle in the crystallizing mass passes cyclically through regions of high supersaturation and regions of low supersaturation. In high-supersaturation regions, such as that immediately after addition of hot feed in a vacuum crystallizer or after flashing into the vapor separation space of an

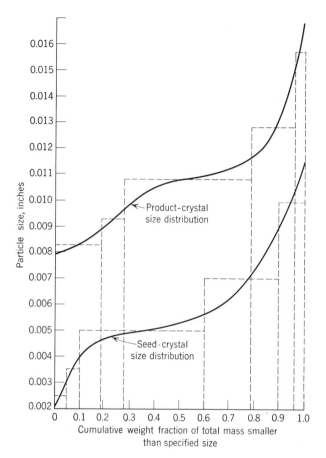

Figure 19.31. Crystallizer-seed and product particle-size distribution, Illustration 19.5.

Oslo crystallizer, nucleation rates will be high. In regions of low supersaturation, there will be virtually no nucleation. If high concentrations of crystals are maintained in the recycling mass, supersaturation can often be kept low enough throughout the crystallizer to minimize nucleation. The presence of relatively few crystals will be unfavorable for crystal growth and thus raise the level of supersaturation throughout the operating cycle. The crystallization rate may be higher, but a large percentage of small crystals will result.

Crystallization from Mixed Solutes. Crystallization from solutions of several solutes generally can be made to result in deposition of crystals of one of the solutes rather than crystals which are a mixture of the two. Thus, separation of one of the components from such a solution is possible by crystallization. The prediction of the products to be obtained from such a system cannot be made based on the two single-solute–water phase diagrams, because the presence of the second solute affects greatly the phase behavior of the first. For example, where the two solutes contain a common ion, such as Na_2SO_4 and Na_2CO_3, the solubility of each solute is greatly reduced.

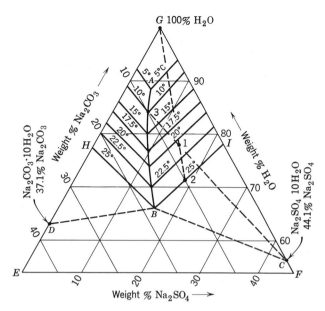

Figure 19.32. Solubility of Na_2CO_3–Na_2SO_4–H_2O system at 5 to 25°C. (Hougen, Watson, and Ragatz, *Chemical Process Principles*, Part I, 2nd ed., John Wiley and Sons, New York, 1954.)

Figure 19.32 is the solubility diagram for the system Na_2CO_3–Na_2SO_4–H_2O at temperatures ranging between 5 and 25°C. Within this temperature range, only the decahydrates of the two salts can crystallize out of solution. At higher temperatures, crystals of Na_2SO_4, Na_2SO_4–Na_2CO_3, Na_2CO_3·$7H_2O$, and Na_2CO_3·H_2O can form depending on the temperature and composition range. In Figure 19.32, point *D* represents Na_2CO_3·$10H_2O$, whereas point *C* represents Na_2SO_4·$10H_2O$. At 25°C, the area bounded by *GHBI* represents unsaturated solutions of the two salts. The rest of the diagram represents heterogeneous mixtures of crystals and saturated solutions. In the area *BIC*, the crystals are pure Na_2SO_4·$10H_2O$; in the area *HBD*, they are pure Na_2CO_3·$10H_2O$; and below line *DBC*, they are mixtures of Na_2CO_3·$10H_2O$ and Na_2SO_4·$10H_2O$. Similar regions exist at other temperatures as indicated on the diagram.

From this diagram, it is quite evident how one or the other of these salts may be fractionally crystallized from any given solution. For example, a solution of composition represented by point 1 at 25°C is in the unsaturated region. If it is evaporated at 25°C, it will move as indicated by line 1–2. At point 2, crystals of pure Na_2SO_4·$10H_2O$ will begin to precipitate and will continue to crystallize as evaporation continues. During this step, the solution concentration will move along line 2–B until point *B* is reached. At this point, Na_2CO_3·$10H_2O$ crystals will begin to form. As another approach, the same original solution could have been cooled to lower temperatures. When the temperature reached about 21°C, precipitation of Na_2SO_4·$10H_2O$ would begin. This precipitation would continue as cooling continued, with solution concentrations moving along line 1–3. When the solution reached point 3, Na_2CO_3·$10H_2O$ would begin to precipitate, but this would not occur until the solution reached about 10°C.

The product from any crystallization separation will consist of the crystals of precipitated solute and the entrapped mother liquor. This trapped solution will carry with it all the impurity to be found in the product. Occlusion of impurities is particularly troublesome with coarse precipitates. However, fine crystals hinder washing and physical removal of solution.

AUXILIARY EQUIPMENT

Evaporation and crystallization processes require a variety of auxiliaries for their successful operation. Slurry pumps, vacuum pumps, mixers, and surface condensers are described elsewhere and hence will not be discussed here. Steam traps, barometric condensers, and entrainment separators are not otherwise discussed, and, since they are vital to evaporation and crystallization operations, they will be dealt with here.

Steam Traps. Steam traps must be placed in the discharge line from any steam-condensing unit to prevent the steam chest from filling with condensate and to prevent live steam from escaping from the discharge line. Some traps are arranged to allow noncondensables to be removed from the process unit. This is particularly important in condensers, for small amounts of noncondensables blanket the surfaces, greatly reducing heat transfer.

Three types are common: mechanical traps, orifice traps, and thermostatic traps.

Mechanical traps may be of the *bucket* or the *float* type. In either case, the mechanical trap operates on a buoyancy principle. In the bucket trap, condensate is admitted under a bucket, flows around it, and discharges from a valve above it. If steam enters the trap, the bucket fills with steam and rises to float. This action closes the discharge valve and prevents further flow. Noncondensables are prevented from floating the bucket by a small vent hole placed in the top of the bucket. Of course a small amount of steam is lost through this vent.

The sequence of operations of an inverted bucket trap is shown in Figure 19.33.

In the *float trap*, a mechanical member is lifted by the collecting condensate until it opens a discharge valve thus allowing flow through the trap. As the condensate layer flows out of the trap, its level drops, thus lowering the float and closing the discharge valve. Figure 19.34 shows a compound-lever float trap.

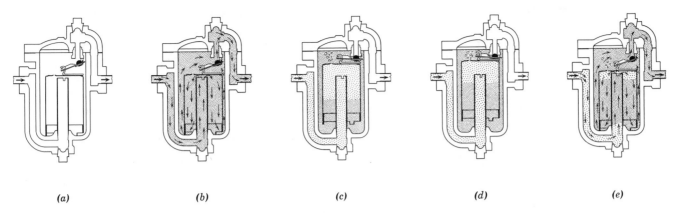

| (a) | (b) | (c) | (d) | (e) |

Figure 19.33. Sequence of operation of an inverted-bucket trap. (Courtesy of Armstrong Machine Works.)

(a) Trap newly installed—steam off, bucket down, valve open.

(b) Steam turned on—condensate reaches bucket and flows through it without lifting the bucket.

(c) Steam reaches trap—condensate is displaced from under the bucket. This condensate displacement floats the bucket and closes the valve. Noncondensables escape through the pinhole in the bucket and collect at the top of the trap.

(d) Condensate flows to trap—the steam under the bucket condenses, and, as more condensate reaches the trap, the bucket begins to fill with water and lose buoyancy.

(e) Condensate flows through trap—as more condensate flows to the trap, the bucket sinks, opening the valve and allowing condensate to pass through. When steam again reaches the trap, the bucket will again rise and close the valve as in (c).

Ball-float traps are operable over wider ranges of pressure and load than are inverted bucket traps and will handle larger quantities of noncondensables than an unmodified inverted-bucket trap. However, the inverted-bucket trap can be supplied with a thermostatic air vent to handle large quantities of noncondensables; it is smaller, and it is less likely to be damaged by air hammer. The ball-float trap has an additional connection so that it may also be used to control liquid level in a tank, or to control the interface level between two liquids. In this case the light liquid would be admitted through

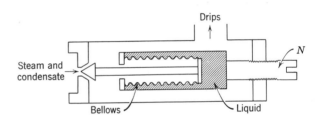

Figure 19.35. Liquid expansion thermostatic trap.

the top of the trap and the heavy liquid through the side opening.

Thermostatic traps operate on a thermal-expansion principle. There are two major types of these traps: the balanced-pressure type and the metal-and-liquid expansion type. In the expansion type, condensate flows around an operating cylinder and out a valve port. The operating cylinder contains liquid held between the cylinder and a metal bellows. Attached to the bellows is a plunger which can close the valve port when the bellows contracts. If steam enters the trap, the liquid around the bellows expands as its temperature rises, thus contracting the bellows and closing the discharge valve. Figure 19.35 shows the principle of operation of the liquid expansion trap. This trap releases subcooled condensate and hence conserves some of the sensible heat in the condensate. However, this means that a significant quantity of condensate may build up in the system which may reduce the heat-transfer area available. The extent of subcooling may be adjusted by turning the knob N so that the condensate leaves at any temperature below 212°F.

The balanced-pressure thermostatic trap is shown

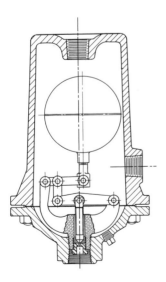

Figure 19.34. Cross-section of a compound-lever, ball-float trap. (Courtesy of Armstrong Machine Works.)

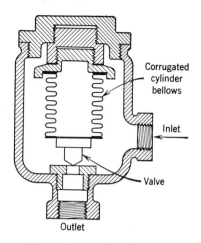

Figure 19.36. Balanced-pressure thermostatic trap.

diagrammatically in Figure 19.36. In this trap, the steam passes around a bellows containing a small quantity of volatile liquid. As this liquid is heated, its vapor pressure increases until it exceeds the pressure of the steam around the bellows, at which point the bellows expands and closes the discharge valve. When the bellows and contained liquid are cooled again by condensate, the vapor pressure of the contained liquid decreases and the valve opens again. The contained liquid is selected so that its vapor pressure is greater than the steam pressure, regardless of how the steam pressure varies.

Thermostatic traps are small and light, and they may be installed in any position. Since they are wide open when cold gas surrounds the bellows, they cannot air bind. They are unaffected by vibration or motion and are not easily damaged by water hammer.

Orifice traps operate on the difference in throttling action of steam and of water passing through a succession of orifices. With steam flow, sonic choking limits the rate of discharge and builds up the intermediate pressure between two orifices in series. With water flow, no such choking occurs, and the intermediate pressure between two orifices in series is the median between the upsteam and downstream pressures. This difference between intermediate pressure when steam flows and when water flows can be used to operate a valve closing and opening the trap. Figure 19.37 is a cross-sectional view of the Yarway Impulse Trap, a patented form of orifice trap. The annular space around the control disk (L) forms the first of the two orifices in series, and the control orifice (O) forms the second one. When steam flows to the trap, the pressure in the chamber (K) builds up forcing the valve stem (F) down upon the seat (G). As long as steam is present, the valve remains closed, but a small steam leak blows through the control orifice (O) and out through the discharge line. When

condensate builds up in the valve approach line, the pressure in the intermediate chamber (K) drops, and the pressure below the control disk (L) lifts the valve stem opening the valve. The impulse trap is small and light. It will operate under a wide range of pressures and handle noncondensables as well as steam. It is, however, affected by corrosion and by clogging of the orifices with dirt or scale.

Other types of steam traps are available which operate on energy balance principles, but they are not as common as the types discussed above.

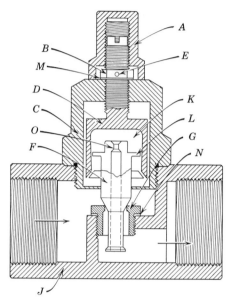

Figure 19.37. Cross-sectional diagram of Yarway impulse trap. (Courtesy Yarnall-Waring Co.) *A*—cap nut, *B*—lock nut, *C*—bonnet, *D*—control cylinder, *E*—lock pin, *F*—valve, *G*—valve seat, *J*—body, *K*—control chamber, *L*—control disc, *M*—washer, *N*—seat gasket, *O*—control-orifice.

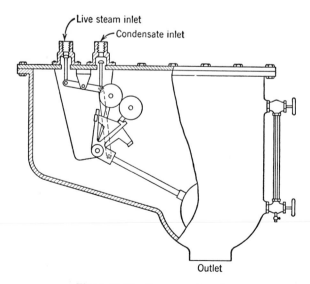

Figure 19.38. Pressure-return trap.

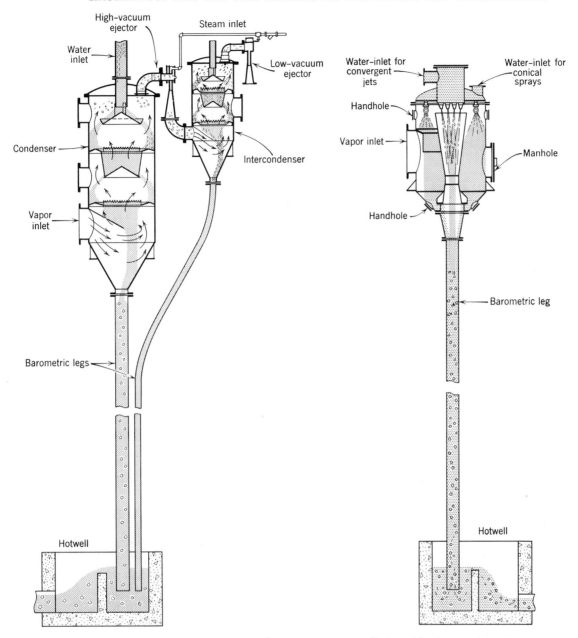

(a) A two-stage countercurrent barometric condenser. (b) A multijet barometric condenser.

Figure 19.39. Barometric condensers. (Courtesy of Acme Coppersmithing and Mfg. Company.)

Condensate Discharging Methods. In multiple-effect evaporators and in vacuum crystallizers, condensate must be discharged from a vacuum space to the atmosphere. The discharging is often done with surface condensers followed by a *pressure return trap* in conjunction with check valves. Figure 19.38 is a schematic diagram of this kind of trap. Condensate enters the trap but is prevented from leaving since the trap is at subatmospheric pressure. As the liquid level rises in the trap, a ball float actuates two valves, closing the condensate-feed valve and opening a live-steam valve. Steam enters the trap, forcing the discharge check valve to open and thus emptying the trap. When the level

drops to a preset point, the float returns the condensate valve to the open position and closes the live-steam valve.

Another, and even more common, solution to the problem is through the *barometric condenser*. Here, a surface, or more commonly a jet, condenser is located on top of a long tail pipe. Condensate stands in the tail pipe high enough so that the hydrostatic head balances the pressure difference between the vacuum space and the atmosphere. Figure 19.39 shows two types of barometric condensers. In the two-stage barometric condenser, vapors are contacted countercurrently with a falling-water shower. Noncondensables are handled in a jet ejector after each stage of the

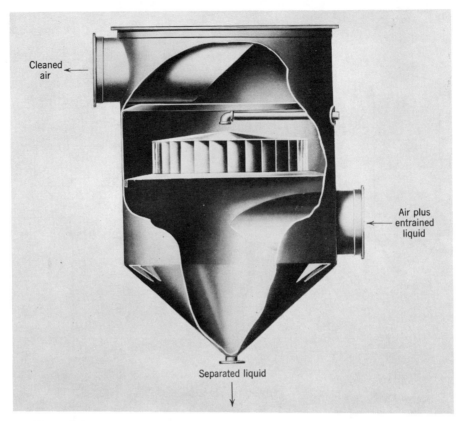

Figure 19.40. An entrainment separator using both centrifugal and impingement mechanisms to effect separation. (Courtesy of Claude B. Schneible Co.)

condenser. In the jet barometric condenser, vapors are entrained into a high-speed water jet. This jet acts as both steam condenser and entrainer of noncondensables. The jet condenser eliminates the steam use of a steam-jet ejector with a barometric condenser, but operates best at pressure greater than 2 in. Hg. The water required by a barometric condenser may be estimated from a simple heat balance. Water discharged to the hot well is usually at 100 to 120°F.

These condensers are always much smaller than surface condensers are for the same duty. They require about a 40-ft headroom and may use more water than a surface condenser would. Because of operating convenience and reduced first cost, they have become almost standard in those applications where the vapors are neither valuable, corrosive, not hazardous to health. Low-level jet condensers are available in which the water is injected at sufficiently high velocity to recompress the noncondensables and discharge both water and gases to atmospheric pressure without a barometric leg.

Entrainment Separators. The entrainment of liquid droplets in the vapor phase can become a serious problem in evaporators and crystallizers as well as in mass-transfer equipment. In standard vertical-tube and horizontal-tube evaporators, the large vapor space in the evaporator body is usually sufficient to let all but the very smallest droplets fall out of the vapor phase.

In forced-circulation and long-tube vertical evaporators and in Oslo crystallizers, the vapor space is seldom sufficient to separate effectively these droplets from the high-velocity vapor-liquid streams. In these cases, auxiliary entrainment separators are often necessary. These separators usually operate on one or both of two principles. Either they operate as cyclone separators on a centrifugal-force mechanism (see Chapter 22), or they coalesce the droplets by impingement with baffles or with metal mesh. Figure 19.40 shows an entrainment separator that operates on a combination of these two mechanisms. The feed enters tangentially so that some droplets are thrown to the walls by centrifugal force. It then passes through a set of baffles that cause further separation by impingement.

REFERENCES

1. Berthoud, A., *J. Chem. Phys.*, **10**, 624 (1912).
2. Bunn, C. W., *Discussions Faraday Soc.*, No. 5, 132 (1949).
3. Butler, R. M., Thesis, University of London (1950).
4. Caldwell, H. B., and W. D. Kohlins, *Trans. Am. Inst. Chem. Engrs.*, **42**, 495 (1946).
5. Frank, F. C., *Discussions Faraday Soc.*, No. 5, 48 (1949).
6. Keenan, J. H., and F. G. Keyes, *Thermodynamic Properties of Steam*, John Wiley and Sons, New York, 1936.
7. McCabe, W. L., and C. S. Robinson, *Ind. Eng. Chem.*, **16**, 478 (1924).

8. Miers, H. A., *J. Inst. Metals.*, **37**, 1, 331 (1927).

9. Miers, H. A., and F. Isaacs, *J. Chem. Soc.*, **89**, 413 (1952).

10. Noyes, A. A., and W. R. Whitney, *J. Am. Chem. Soc.*, **19**, 930 (1897).

11. Perry, J. H., *Chemical Engineers Handbook*, 3rd ed., McGraw-Hill Book Co., New York, 1950.

12. Preckshot, G. W., and G. G. Brown, *Ind. Eng. Chem.*, **44**, 1314 (1952).

13. Schrage, R. W., *A Theoretical Study of Interphase Mass Transfer*, Columbia University Press, New York, 1953.

14. Seidell, A., *Solubility of Inorganic and Metal-Organic Compounds*, D. Van Nostrand Co., Princeton, 1940.

15. McCabe, W. L., *Ind. Eng. Chem.*, **21**, 30, 112 (1929).

PROBLEMS

19.1. The operator of a standard, two-effect evaporator with forward feed, barometric condenser, two-stage jet ejector, and manual controls has been operating at stable conditions for some time. If, with no change in operating controls, he notices each of the following effects, what might be probable causes of each of these effects, and how might these be eliminated?

(*a*) Product rate increases, but concentration falls off proportionately.

(*b*) Steam consumption increases, but process-stream flow rates remain unchanged.

(*c*) Production rate and steam consumption decrease. Steam pressure, vacuum, and feed and product concentrations remain unchanged.

(*d*) Steam consumption increases, and pressure in the first effect increases.

(*e*) Production rate decreases while temperatures increase in both effects.

19.2. Characteristics of liquids which offer problems in evaporation are:

1. Salting.
2. Foaming.
3. Scaling.
4. Corrosive as requiring replacement of tubes.
5. Corrosive as involving expensive materials of construction.
6. Heat sensitivity.
7. High viscosity.
8. High boiling-point rise.

List which of these troublesome characteristics are taken care of in evaporation of the following types:

(*a*) Horizontal tube.

(*b*) Vertical short tube.

(*c*) Long-tube vertical.

(*d*) Forced circulation.

(*e*) Turbulent film.

19.3. Recommend a type of evaporator for each of the following applications, and briefly give the reason for your choice:

(*a*) Concentration of grapefruit juice. The solution is non-foaming. Stainless steel is to be the material of construction.

(*b*) Concentration of glycerol solution from soap making. Water vapor goes to the condenser. Glycerol product is somewhat viscous at the operating condition.

(*c*) Corrosive "black liquor" used in sulfate pulp mills is to be concentrated through a small range. Nonsalting. No foam problems.

19.4. Glycerol–water solutions may be assumed to obey Raoult's law with the vapor pressure of glycerol being negligible. Using water as a reference liquid construct Dühring lines for solutions having 0, 20, 40, 60, 80, 100, and 120 gm of glycerol per 100 gm of water. Plot these lines for reference water temperatures from 80 to 240°F. 1 in. of the plot should not represent more than 20°F.

19.5. The data given below concern NaNO₃ solutions and were taken from the *International Critical Tables*. From it construct a Dühring line plot for the NaNO₃–H₂O system.

VAPOR PRESSURES OF AQUEOUS SOLUTIONS OF NaNO₃

Grams NaNO₃	Vapor Pressures, mm Hg					
100 gm H₂O	0°C	25°C	50°C	75°C	100°C	125°C
0	4.58	23.76	92.54	289.32	760.0	1740.5
5	4.50	23.34	90.9	284.1	746.3	1709
10	4.42	22.93	89.2	278.9	732.5	1677
20	4.28	22.14	86.1	268.8	705.6	1615
30	4.15	21.39	83.1	259.1	679.6	1554
40	4.04	20.69	79.6	249.9	654.8	1496
50	3.93	20.04	77.5	241.1	631.3	1442
60	3.83	19.42	74.9	232.9	609.0	1390
70	3.73	18.83	72.56	225.1	588.1	1341
80	3.64*	18.29	70.25	217.8	568.2	1294
90		17.77	68.1	210.8	549.4	1250
100		17.29*	66.1	204.3	531.6	1209
110			64.2	199.4	514.8	1170
120				192.1	498.9	1135
130				186.5	483.7	1097
140				181.2*	469.4	1064
150					455.7	1032
160					442.6	1002
170					430.1	973
180					418.3*	945
190						919
200						893
210						869
220						846
230						824

* Supersaturated solution.

19.6. Shown in the figure is a boiler connected cyclically to a condenser. Steam supply is at 100 psia, and cooling water is at 80°F average temperature. The boiler has a surface area of 100 sq ft, and the condenser area is 50 sq ft. Over-all coefficients of 600 Btu/hr

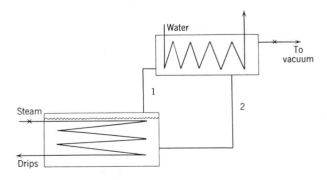

sq ft °F for the boiler and 800 Btu/hr sq ft °F for the condenser are realized.

(*a*) With the vacuum line sealed off and water in the system what are the conditions at points 1 and 2 (*P*, *T*, and flow rate)?

(*b*) What would these conditions be for a 30 per cent NaOH solution?

(*c*) What would they be for a 50 per cent NaOH solution if water *T* = 50°F?

19.7. An evaporator is being installed to concentrate caustic soda solution from 48 per cent NaOH to 80 per cent NaOH by weight with feed at 100°F. A forced-circulation evaporator will be used with Dowtherm "A" (an eutectic of diphenyl and diphenyl oxide) condensing in the external steam chest. The pressure in the evaporator will be 10.4 psia with the Dowtherm "A" condensing at 20 psia (B.P. = 525°F). The circulation rate will be sufficient to give an inside film coefficient of 800 Btu/hr sq ft °F in the external heater. A barometric jet condenser will be used with water supply at 90°F, 50 psig. Heater tubes will be nickel, 1 in. O.D. × 16 BWG, 12 ft long, with 273 of them in parallel. What capacity in Btu/hr will be required for the Dowtherm boiler?

Data for Dowtherm

Temp., °F	μ, lb/ft hr	$\left(\dfrac{\mu^2}{k^3 \rho^2 g}\right)^{1/3}$	λ, Btu/lb
450	0.970	0.001110	129
475	0.930	0.001082	126
500	0.895	0.001062	123
525	0.870	0.001060	120

19.8. A batch evaporator 6 ft in diameter with a steam coil 6 in. above the bottom surface is to be used to concentrate NaOH solution from 2 weight percent to 15 weight percent by condensing 100-psig steam in the coil. The coil surface is 30 sq ft in area, and an over-all heat-transfer coefficient of 200 Btu/hr sq ft °F is realized from coil to solution throughout the boiling period. With the pressure over the solution constantly 1 psia, how long will the evaporation take if the solution is initially 6 ft deep?

19.9. A 5 per cent NaNO₃ solution is to be concentrated to 25 per cent NaNO₃ at the rate of 15,000 lb of concentrate per hour in a standard vertical-tube evaporator. It is planned to run the evaporator at 300 mm Hg absolute pressure and to use saturated steam at 50 psig as the heating medium. The following information concerning this operation is available:

U_0, the over-all heat-transfer coefficient based on outside surface area = 400 Btu/hr sq ft °F

Feed temperature = 100°F

Specific heat of feed = 0.90 cal/gm °C

Condensate from steam chest leaves at steam saturation temperature

Effect of hydrostatic head of solution in the evaporator on ΔT is negligible

The latent heat of vaporization of the solution may be taken from steam tables at the operating pressure

Heat losses amount to 5 per cent of the total heat transferred. How much heat-transfer surface is required? If 1 in. O.D. × 20 BWG tubes are used what would be a reasonable length of tube and number of tubes in the steam chest?

19.10. A double-effect evaporator is to be used to concentrate 10,000 lb/hr of a 10 per cent sugar solution to 30 per cent. The feed enters the second effect at 70°F. Saturated steam at 230°F is fed to the first effect, and the vapors from this effect are used to supply heat to the second effect. The temperature in the final condenser will be 110°F. The over-all coefficients are estimated to

be 400 Btu/hr sq ft °F for the first effect and 300 for the second effect. The heating areas for the two effects are the same. The specific heats of the solutions may be taken as constant and equal to 0.95.

Calculate:

(1) The temperature in each effect.

(2) The heating surface per effect.

(3) Steam consumption in pounds per hour.

(4) Pounds of water evaporated per pound of steam.

Solve this problem, using $\lambda = 1000$ for steam at all pressures and assuming the solutions have no boiling-point rises, by writing appropriate enthalpy balances, material balances, and rate equations and by solving them simultaneously by direct algebraic methods.

19.11. A basket-type double-effect evaporator connected for backward feed is to be used to concentrate a NaOH solution. Each evaporator body has a 2000 sq ft heating area. The caustic solution enters at 80°F and 5 weight percent NaOH. It is to be concentrated to 50 weight percent. Previous operation indicates that over-all coefficients of 400 Btu/hr sq ft °F and 550 Btu/hr sq ft °F may be obtained in the first and second effects respectively. Saturated steam at 150 psia is available, and a vacuum of 2 in. Hg absolute may be obtained with existing ejectors. What is the maximum production rate obtainable?

19.12. A double-effect evaporator which has been idle for some time is to be replaced in service evaporating glycerol solution. The evaporator is piped for backward feed. In previous operation an absolute pressure of 115 mm Hg was maintained on the second effect and superheated steam at 5 psig and 250°F was supplied to the first effect; these conditions will be maintained in the proposed operation. The evaporator bodies are of the standard vertical-tube type with heating areas of 750 sq ft each. Previous operation indicates over-all coefficients of 300 Btu/hr sq ft °F for each of the effects. If a 10 per cent glycerol feed entering at 60°F is to be evaporated to a 40 per cent glycerol product, what throughput can be maintained without altering the operating conditions?

19.13. It has been decided to withdraw 30,000 lb/hr of low-pressure steam from the vapor line leaving the first effect of the evaporators of Problem 19.11. What will be the pressure and temperature of this steam, and what production rate will now be obtained from the evaporators?

19.14. A triple-effect evaporator system is to be used to concentrate a glycerol–water solution from 10 to 40 per cent glycerol by weight at a rate of 1 ton/hr of concentrate. Identical long-tube natural-circulation evaporators are to be used, and over-all heat-transfer coefficients of 600, 500, and 350 Btu/hr sq ft °F are expected for the first, second, and third effects respectively when using forward feed. Feed enters at 80°F. Saturated steam at 300°F is used to heat the first effect. The condenser temperature is kept at 100°F. Specify the evaporator sizes required in terms of steam-chest surface area.

19.15. A triple-effect evaporator is to be used to evaporate 20,000 lb/hr of 10 per cent NaNO₃ solution to obtain a 50 per cent NaNO₃ product. The system is arranged for forward feed, which enters at 70°F. The steam supplied to the first effect is saturated at 50 psig, and a barometric condenser–ejector system maintains the pressure in the last effect at 4 in. Hg absolute. The heat-exchange areas of all these effects are to be equal. For this system determine:

(1) Steam economy, pounds of steam per pound of vapor formed.

(2) Temperatures in each effect.

(3) Heating surface required in each effect.

The following additional information may be helpful:

(*a*) Over-all coefficients for the three effects are 550, 500, and 400 Btu/hr sq ft °F respectively.

(*b*) Specific heats of NaNO₃ solutions can be considered to be independent of temperature and to be given by the equation:

$c_P = 4.175 - 3.742x + 2.166x^2$

x = wt. fraction $NaNO_3$ (1–39 per cent conc.)

c_P = specific heat, joules/gm °C at 20°C and 1 atm pressure

19.16. A double-effect, forward-feed evaporator system is to be used to crystallize NaCl from a 5 weight percent solution originally at 80°F. Each evaporator body has a 1000 sq ft heating area, and over-all heat-transfer coefficients of 600 Btu/hr sq ft °F and 400 Btu/hr sq ft °F are obtained in the first and second effects respectively. Steam saturated at 320°F is available, and a total pressure of 1 psia can be maintained on the second effect. It is desired to obtain 40 lb of NaCl crystals per 1000 lb of feed. Calculate the feed rate under these conditions.

Data on NaCl *Solutions*

1. Boiling-point rise for solutions at 1 atm

Conc., gms NaCl/100 gm water 0 5 10 20 30 40

B P R, °F   0 1 3 7 11 16.5

2. Solubility of NaCl in H_2O, gm NaCl/100 gm H_2O

Temp., °F.	32	50	90	120	160	212
Solubility	35.7	35.8	36.3	37.0	37.8	39.8

3. Specific heat of NaCl solutions in Btu/lb °F $[c_P \neq f(T)]$

NaCl gms/100 gm water	10	20	30	40	solid crystal
Specific heat	0.91	0.86	0.80	0.72	0.37

4. ΔH of crystallization of NaCl $= -1250$ cal/gm mole.

Assume this to be independent of temp. and concentration.

19.17. A vapor recompression still is to be used to generate fresh water for shipboard use. The feed to this unit is sea water at 60°F containing 3.5 per cent solids which may be assumed to be NaCl. This is concentrated to 10 per cent NaCl in the evaporator. The evaporator loses 100 Btu to the surroundings per pound of vapor generated, and the vapor is compressed to a pressure two times the evaporator pressure before being returned to the steam chest.

(*a*) If the unit runs continuously with no added steam required and no excess steam available and with a compressor operating adiabatically at 60 per cent efficiency, determine the evaporator pressure.

(*b*) If the steam-chest area is 1000 sq ft and an over-all coefficient of 400 Btu/hr sq ft °F is realized, what is the potable water production rate?

19.18. What phases are present and what is the relative quantity of each when 45 weight percent $CaCl_2$ solution is cooled from 180°F to

(*a*) 60°F

(*b*) 0°F.

(*c*) An enthalpy of -200 Btu/lb relative to pure H_2O and $CaCl_2$ at 0°F.

(*d*) -140°F.

19.19. 10 lb of ice at 0°F are added to 10 lb of $CaCl_2 \cdot 4H_2O$ also at 0°F. What would be the temperature and phase conditions (species and amount) that would result if this mix ultimately came to equilibrium adiabatically?

19.20. A 20 weight percent solution of Na_2SO_4 at 175°F is pumped continuously to a vacuum crystallizer from which the product magma is pumped at 60°F. What is the composition of this magma, and what percentage of Na_2SO_4 in the feed is recovered as $Na_2SO_4 \cdot 10H_2O$ crystals after this magma is centrifuged?

19.21. A solution of 32.5 per cent $MgSO_4$ originally at 150°F is to be crystallized in a vacuum adiabatic crystallizer to give a product containing 4000 lb/hr of $MgSO_4 \cdot 7H_2O$ crystals from 10,000 lb/hr of feed. The solution boiling-point rise is estimated at 10°F

Determine the product temperature, pressure, and weight ratio of mother liquor to crystalline product.

For data see Perry, J. H., *Chemical Engineer's Handbook*, 3rd ed., McGraw-Hill Book Co., New York, 1950, pp. 1052–1053.

19.22. Tri-sodium phosphate is to be recovered as $Na_3PO_4 \cdot 12H_2O$ from a 35 weight percent solution originally at 190°F by cooling and seeding in a Swenson-Walker crystallizer. From 20,000 lb/hr of feed 7000 lb/hr of product crystals in addition to the seed crystals are to be obtained. Seed crystals fed at a rate of 500 lb/hr have the following size range:

Weight Fraction	Size Range, in.
10%	$-0.0200 + 0.0100$
20%	$-0.0100 + 0.0050$
40%	$-0.005 + 0.0025$
30%	$-0.0025 + 0.0010$

Latent heat of crystallization of tri-sodium phosphate is 27,500 Btu/lb mole. Specific heat for the tri-sodium phosphate solution may be taken as 0.8 Btu/lb °F, and solubility data may be found in Perry (11).

(*a*) Estimate the product particle-size distribution.

(*b*) To what temperature must the solution be cooled, and what will be the cooling duty in Btu/hr?

19.23. A solution of 28 weight percent Na_2SO_4 is to be crystallized to yield Glauber's salt ($Na_2SO_4 \cdot 10H_2O$) in a four-section Swenson-Walker crystallizer. It is fed to the unit at 100°F and is cooled with 50°F water flowing countercurrently. The crystallizer has a cooling surface of 3 sq ft/ft of length and operates with an over-all coefficient of heat transfer of 20 Btu/hr sq ft °F across the cooling surface. The water leaves at 75°F.

(*a*) What feed rate to the crystallizer should be used to obtain 75 percent of the Na_2SO_4 in the feed as $Na_2SO_4 \cdot 10H_2O$ crystals?

(*b*) The crystallizer is seeded with $Na_2SO_4 \cdot 10H_2O$ crystals analyzing as follows:

Tyler Sieve Size	Mass Fraction
$-48 + 65$	0.04
$-65 + 100$	0.18
$-100 + 150$	0.46
$-150 + 200$	0.32

What would be the screen analysis expected for the product if 10 lb of seed crystals are used per 100 lb of feed solution?

19.24. Crystals of $CaCl_2 \cdot 6H_2O$ are to be obtained from a solution of 35 weight percent $CaCl_2$, 10 weight percent inert soluble impurity, and 55 weight percent water in an Oslo crystallizer. The solution is fed to the crystallizer at 100°F and receives 250 Btu/lb of feed from the external heater. Products are withdrawn from the crystallizer at 40°F.

(*a*) What are the products from the crystallizer?

(*b*) The magma is centrifuged to a moisture content of 0.1 lb of liquid per pound of $CaCl_2 \cdot 6H_2O$ crystals and then dried in a conveyor drier. What is the purity of the final dried crystalline product?

19.25. A solution containing 7 weight percent Na_2SO_4, 20 weight percent Na_2CO_3, and 73 weight percent H_2O is to be crystallized to obtain pure $Na_2CO_3 \cdot 10H_2O$. This operation can be carried out in a Swenson-Walker crystallizer for which cooling can be done by brine at 0°C, or it can be carried out in a batch vacuum crystallizer on which the ejector can pull enough vacuum to give a solution boiling at 70°F. What is the maximum yield of pure $Na_2CO_3 \cdot 10H_2O$ obtainable by each of these crystallizers, and which unit would you recommend?

19.26. A 30 weight percent solution of $MgSO_4$ is fed at 100°F to a Swenson-Walker crystallizer where it is cooled to 60°F. As it enters the crystallizer it is seeded with $MgSO_4 \cdot 7H_2O$ crystals of the following size range:

Tyler Screen Sire	Weight Fraction
−65 + 100	20
−100 + 150	55
−150 + 200	25

The products from the crystallizer are wet-screened with the −28 + 35 mesh material retained but with the undersize and oversize material recycled back into the feed solution before it enters the crystallizers. Determine the optimum rate of addition of seed to the crystallizer in terms of pounds of seed per pound of feed solution.

chapter 20

Momentum Transfer I:
The Energy Balance and Its Applications

In any system where a fluid-solid boundary exists, momentum is transferred from the fluid to the boundaries. It is this transfer of momentum that results in the loss in system pressure which is of interest to the practical designer. The details of this transfer were examined in Chapter 13. Here this phenomenon will be applied to practical design situations.

The previous chapters in Part III have emphasized the interrelations of the rate of transfer, driving forces, and resistances and their influence on equipment design. Many process operations also require a knowledge of the energy and material relationships associated with the flow of materials that are involved in the process. For instance, it might be necessary to know the velocity of a fluid stream leaving a reactor through a specified pipe, when the pressure in the reactor is fixed, or the power requirements to pump a given quantity of fluid through a complex piping system might be needed. Two important tools necessary to design process equipment, in addition to the rate equations, are the material balance and the energy balance.

THE MATERIAL BALANCE

In its simplest form a material balance is nothing more than an accounting of mass. The law of conservation of mass states that the total mass of all substances taking part in a process remains constant. Although there are exceptions to this law for nuclear reactions and processes where the reacting materials are moving at or near the velocity of light, for general engineering purposes the law is inviolate.

Semantically this law may be written as

Mass input = mass output + mass accumulation (20.1)

The accumulation term may be positive or negative depending upon the physical situation. For steady-state operation, the accumulation becomes zero. In a like manner, a conservation of mass of a particular component partaking in a reaction must be observed and is referred to as a component balance.

Consider an apparatus of the type shown in Figure 20.1. This apparatus could be anything from a complete piping system to a specific type of process equipment. The apparatus is bounded by rigid walls but has openings through which mass can enter and leave. In general, mass can be accumulated within the apparatus, so that Equation 20.1 is valid. Under the special restriction of steady state, the accumulation is zero so that Equation 20.1 becomes

$$w_1 = w_2 = \frac{\bar{v}_1 S_1}{V_1} = \frac{\bar{v}_2 S_2}{V_2} = G_1 S_1 = G_2 S_2 \quad (20.2)$$

where $\bar{v}$ = the average velocity, ft/sec
S = the cross-sectional flow area, sq ft
V = the specific volume of material, cu ft/lb
G = the mass velocity, lb/hr sq ft
w = the mass rate of flow, lb/hr

Equation 20.2 simply states that the sum of all the masses entering a system must equal the sum of the masses leaving the system at steady state. It is known as the *continuity equation*.

THE ENERGY BALANCE

To obtain a relationship between the various types of energy involved in the flow system depicted in Figure 20.1 requires several assumptions. Consider for simplicity that the material flowing through the system is a

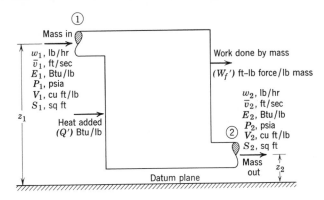

Figure 20.1. Process flow quantities.

fluid. The equation that will be developed is not limited to fluids, however, but is pertinent to any material involved in a flow process.

The definition of an open-flow system in steady-state conditions requires that:

1. The fluid flowing into the system is uniform in properties and velocity and that these items are non-variant with time.

2. The fluid leaving the system is uniform in properties and velocity and that these items are nonvariant with time. The exit conditions will not usually be identical to entering conditions.

3. The physical properties of the fluid at any point within the system are constant with respect to time.

4. Mass rates of flow into and from the system are constant.

5. The rates of addition of heat and production of work are constant.

In accordance with the principle of conservation of energy, the total energy entering the system must equal the total energy leaving the system. If electrostatic and magnetic energies are neglected, the energies under consideration for this system are those carried by the fluid and that transferred between the fluid and surroundings.

The energy carried by the fluid includes:

1. An internal energy (E) which is an intrinsic property of the fluid. The system under study is comprised of molecules which are oriented in a particular geometry in the case of solids or are in random motion in the case of fluids. The orientation and movement of molecules may be likened to *internal* potential and kinetic energy.

2. An external potential energy (zg/g_c) due to the position of the fluid with respect to an arbitrary datum plane.

3. An external kinetic energy $(\bar{v}^2/2g_c\alpha)$ due to the motion of the fluid. The term α must be included in the kinetic-energy term to account for the effect of velocity distribution in the flow channel on the average kinetic

energy. If little velocity gradient exists, such as in fully developed turbulent flow, α approaches unity. For laminar flow, the value of α is not unity and must be included in the kinetic-energy term. Figure 20.2 shows the relationship between α and the Reynolds number for flow of fluids in pipes.

4. An energy of pressure (PV) carried by the fluid as a result of being introduced into the system. In reality, the PV product is a work term at the expense of the energy of the surroundings. This energy is the force exerted by the fluid immediately behind the entrance point times the distance through which it acts. The distance through which the force acts is equal to the specific volume of the material divided by the cross-sectional area at the entrance point. Thus the work done is force times distance, or

$$(PS)\left(\frac{V}{S}\right) = PV \tag{20.3}$$

The energy transferred between the system and surroundings includes:

1. The heat (Q') absorbed by the flowing material from the surroundings. Addition of heat to the fluid may or may not demand that its temperature must rise. It is possible for the fluid to remain in isothermal flow while heat is added, for the added energy can find outlet in other forms. The quantity of heat from the surroundings must include the total net heat passing *through* the boundary of the system. This excludes the heat generated by friction since frictional heat must come from a dissipation of other forms of energy. By convention, the sign of Q' is positive if heat is transferred *from* the surroundings *to* the system.

2. The work (W_f') done on the surroundings by the flowing fluid as it moves through the apparatus. This term is frequently referred to as *shaft work*. This work quantity, like Q', must pass *through* the boundary of the system. For the fluid to do work on the surroundings would require a turbine or some other work-producing apparatus connected from the system to the surroundings. By convention, the sign of W_f' is positive if work is done *by* the fluid and transferred *to* the surroundings.

The balance of all the energies involved in the flow system of Figure 20.1 may be written as

$$wE_1 + \frac{w\bar{v}_1{}^2}{2g_c\alpha} + wz_1\frac{g}{g_c} + wP_1V_1 + wQ'$$

$$= wE_2 + \frac{w\bar{v}_2{}^2}{2g_c\alpha} + wz_2\frac{g}{g_c} + wP_2V_2 + wW_f' \tag{20.4}$$

The summation of the terms on the left side of Equation 20.4 represents the energy transferred *to* the system, and the terms on the right signify the transfer of

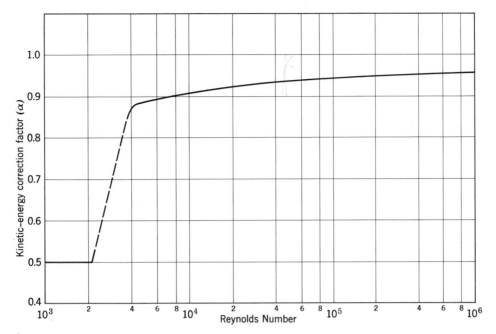

Figure 20.2. Kinetic-energy correction factor as a function of Reynolds number (3).

energy *from* the system to the surroundings. Equation 20.4 is referred to as the total-energy balance. Since it is based only on the concept of conservation of energy, its validity for steady state is rigorous.

The total-energy balance may also be written in the differential form. This is frequently convenient when manipulating mathematically the terms of Equation 20.4, or when applying Equation 20.4 to compressible fluids. Thus, for unit mass, Equation 20.4 becomes

$$dE + d(PV) + d\left(\frac{g}{g_c}z\right) + \frac{\bar{v}\,d\bar{v}}{\alpha g_c} = \delta Q' - \delta W_f' \quad (20.5)$$

The $\delta Q'$ and the $\delta W_f'$ terms in Equation 20.5 are not exact differentials. Their evaluation depends upon the manner in which the heat and work have been added to the system.

The internal energy is an intrinsic thermodynamic property of the system and is obtainable only by difference or by establishing an arbitrary basis from which E can be evaluated. Similarly, the PV term is also an intrinsic property of the system. Furthermore, since E and PV appear as sums on both sides of Equation 20.4, it is sometimes convenient to combine this sum into a single quantity which will also be an intensive property of the system. Thus, the *enthalpy* (H) may be defined

$$H = E + PV \quad (20.6)$$

Then, Equation 20.4 may be written, for unit mass, as

$$H_1 + \frac{\bar{v}_1^2}{2\alpha g_c} + z_1\frac{g}{g_c} + Q' = H_2 + \frac{\bar{v}_2^2}{2g_c\alpha} + z_2\frac{g}{g_c} + W_f' \quad (20.7)$$

or

$$(H_2 - H_1) + \left(\frac{\bar{v}_2^2 - \bar{v}_1^2}{2g_c\alpha}\right) + (z_2 - z_1)\frac{g}{g_c} = Q' - W_f' \quad (20.8)$$

Writing Equation 20.8 as a difference equation gives

$$\Delta H + \Delta\left(\frac{\bar{v}^2}{2g_c\alpha}\right) + \Delta\left(z\frac{g}{g_c}\right) = Q' - W_f' \quad (20.9)$$

where Δ is the increase in going from state 1 to state 2.

Enthalpy, like internal energy, can be evaluated only as a difference. In general, the enthalpy change of a system is a function of the specific heat and the temperature change of the system along a constant-pressure path or where the specific volume is very small as is the case for solids and liquids. Therefore, the change in enthalpy with temperature is

$$\Delta H = \int_{T_1}^{T_2} c_P\, dT \quad (20.10)$$

where ΔH = enthalpy change, Btu/lb
c_P = heat capacity at constant pressure, Btu/lb°F
T_2, T_1 = final and initial temperatures, °F

In using Equations 20.4 through 20.9, each of the terms including heat and work must be dimensionally consistent. The reader is referred to Appendix A for material concerning dimensions.

Illustration 20.1. Air flowing at a constant mass rate of 600 lb/hr is heated from 70°F to 190°F. The heating is accomplished in a vertical heat exchanger of constant cross-sectional area. The exchanger is 12 ft long. The pressure of the air stream entering at the bottom of the exchanger is 18 psia, and a 2-psia pressure drop occurs through the apparatus because of fluid friction. The average linear velocity of the inlet air is 15 ft/sec. Assuming the air to behave as a perfect gas and to have a value of $c_P = 0.24$ Btu/lb °F, calculate the net heat input to the air in Btu per hour.

SOLUTION. The net heat input to the air can be obtained directly from Equation 20.4 although it will be more convenient to use Equation 20.9 in this case. There is no exchange of work between the air and the surroundings and so W_f' is zero. The energy balance is written with each term expressed as Btu per pound. Reference point 1 will be taken at the bottom of the exchanger and point 2 will be taken at the top.

$$(H_2 - H_1) + \frac{\bar{v}_2{}^2 - \bar{v}_1{}^2}{2g_c\alpha} + (z_2 - z_1)\frac{g}{g_c} = Q'$$

From the statement of the problem,

$$T_1 = 70°F = 530°R$$

$$T_2 = 190°F = 650°R$$

$$(z_2 - z_1) = 12 \text{ ft (datum plane at bottom inlet of exchanger)}$$

$$\bar{v}_1 = 15 \text{ ft/sec}$$

$$P_1 = 18 \text{ psia}$$

$$P_2 = 18 - 2 = 16 \text{ psia}$$

In order to solve the energy balance for the net heat input, the quantities $(H_2 - H_1)$ and $\bar{v}_2$ must be evaluated. The outlet velocity can be calculated directly from the perfect-gas law since the velocity is directly proportional to the specific volume for a constant cross-sectional area,

$$\bar{v}_2 = \bar{v}_1 \left(\frac{T_2}{T_1}\right)\left(\frac{P_1}{P_2}\right)$$

$$\bar{v}_2 = (15)\left(\frac{650}{530}\right)\left(\frac{18}{16}\right) = 20.7 \text{ ft/sec}$$

The quantity $(H_2 - H_1)$ can be calculated from the mean specific heat and temperature change of the air. c_P is independent of pressure for a perfect gas.

$$(H_2 - H_1) = c_P(T_2 - T_1) = 0.24(650 - 530) = 28.8 \text{ Btu/lb}$$

Therefore, substituting into the energy equation,

$$Q' = 28.8 \frac{\text{Btu}}{\text{lb}} + \frac{(20.7 \text{ ft/sec})^2 - (15 \text{ ft/sec})^2}{2\alpha\left(32.2 \frac{\text{ft-lb}}{\text{lb}_f \text{ sec}^2}\right)\left(778 \frac{\text{ft-lb}_f}{\text{Btu}}\right)}$$

$$+ \frac{(12 - 0) \text{ ft } (32.2 \text{ ft/sec}^2)}{\left(\frac{778 \text{ ft-lb}_f}{\text{Btu}}\right)\left(32.2 \frac{\text{ft-lb}}{\text{lb}_f \text{ sec}^2}\right)}$$

$$Q' = 28.8 + \frac{0.0041}{\alpha} + 0.0154$$

α will be between 1.0 and 0.5. Since the entire kinetic-

energy term is small compared to the other terms, it can be called zero. Thus

$$Q' = 28.82 \text{ Btu/lb}$$

and for 600 lb/hr

$$Q' = 600(28.82) = 17,290 \text{ Btu/hr}$$

The Energy Balance and Fluid Friction. Consider a flow system in which a fluid that is subject to no shearing stress during motion is flowing under isothermal conditions. This type of fluid is called a perfect fluid. For fluid flow of this type, only mechanical-energy forms are significant, and for this seriously restricted case the energy balance becomes

$$P_1 V_1 + z_1 \frac{g}{g_c} + \frac{\bar{v}_1{}^2}{2g_c\alpha} = P_2 V_2 + z_2 \frac{g}{g_c} + \frac{\bar{v}_2{}^2}{2g_c\alpha} \quad (20.11)$$

Equation 20.11 is commonly referred to as the Bernoulli equation. The significance of this equation is that in the absence of nonmechanical energy the sum of the pressure-energy, potential-energy, and kinetic-energy terms remains constant for a perfect fluid. For example, if the velocity of the fluid varies, it can do so only at the expense of either the potential energy or the pressure energy or a combination of both. Unfortunately, Equation 20.11 is of theoretical interest only, because there is no perfect fluid, and the sum of the three previously mentioned energies appearing at point 2 is always less than the sum of the same energy terms at point 1 because of fluid friction.

Equation 20.4 has been developed considering only the energies that enter and leave the system as depicted in Figure 20.1. Focusing attention on the fluid within the boundaries of the system gives rise to other problems. As a result of flow, fluid friction occurs wherever there is a stress on the fluid. This friction will effectively convert mechanical energy into heat so that all the work done by the fluid will not be transferred to the surroundings. The lost energy will appear in the fluid as heat. Thus,

$$Q = Q' + \Sigma F \quad (20.12)$$

and

$$W_f' = W - \Sigma F \quad (20.13)$$

where
$Q = $ heat absorbed by the fluid
$Q' = $ heat transferred from the surroundings
$W = $ total work done by the fluid
$W_f' = $ work transferred to the surroundings (shaft work)
$\Sigma F = $ total fluid friction

Substituting Equation 20.12 into Equation 20.4 and expressing the result as a difference equation gives

$$\Delta E + \Delta\left(\frac{\bar{v}^2}{2g_c\alpha}\right) + \Delta z \frac{g}{g_c} + \Delta(PV) = Q - \Sigma F - W_f'$$

$$(20.14)$$

By the first law of thermodynamics,

$$\Delta E = Q - W = Q - \int_{V_1}^{V_2} P\,dV \qquad (20.15)$$

Substitution of Equation 20.15 into Equation 20.14 yields

$$\Delta\left(\frac{\bar{v}^2}{2g_c\alpha}\right) + \Delta z\frac{g}{g_c} + \Delta(PV) - \int_{V_1}^{V_2} P\,dV + \Sigma F = -W_f' \qquad (20.16)$$

But

$$\Delta(PV) = \int_{V_1}^{V_2} P\,dV + \int_{P_1}^{P_2} V\,dP \qquad (20.17)$$

Therefore, substituting Equation 20.17 into Equation 20.16 gives

$$\Delta\left(\frac{\bar{v}^2}{2g_c\alpha}\right) + \Delta z\frac{g}{g_c} + \int_{P_1}^{P_2} V\,dP + \Sigma F = -W_f' \qquad (20.16a)$$

Either Equation 20.4, 20.7, 20.14, 20.16, or 20.16a may be used to solve a flow problem. The choice depends solely upon the desired result and upon the information at hand.

Although any one of the energy terms might be calculated from the energy balances, often it is desired to determine the power requirements to move a fluid at a specified rate through a piping system. In this particular case, work will be done *on* the fluid, and the sign of W_f' in Equation 20.16a is negative. Also, as has been established earlier, the units of work will be foot-pounds force per pound mass. Power is the time rate of doing work and so a conversion of W_f' must be made. This is most easily done as follows:

$$w = \bar{v}\rho S \qquad (20.18)$$

where w = mass flow rate, lb/sec
  $\bar{v}$ = average fluid velocity, ft/sec
  ρ = fluid density, lb/cu ft
  S = cross-sectional area of flow, sq ft

Therefore, the theoretical power required becomes

$$\text{Power} = (W_f')(w) \qquad (20.19)$$

where power is expressed as foot-pounds force per second. Horsepower may be calculated by applying the proper conversion factor to Equation 20.19.

Illustration 20.2. A pump takes water at 60°F from a large reservoir and delivers it to the bottom of an open elevated tank 25 ft above the reservoir surface through a 3-in. I.D. pipe. The inlet to the pump is located 10 ft below the water surface, and the water level in the tank is constant at 160 ft above the reservoir surface. The pump delivers 150 gal/min. If the total loss of energy due to friction in the piping system is 35 ft-lb$_f$/lb, calculate the horsepower required to do the pumping. The pump-motor set has an over-all efficiency of 55 per cent.

SOLUTION. The best technique for solving problems of this type is to draw a simple sketch with appropriate data and label the reference points about which the energy balance is to be made. To demonstrate that the reference points may be

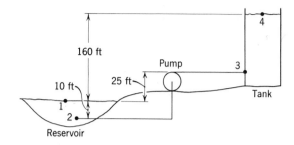

placed anywhere, several different sets will be used. The only requirement is that *all* the energy between the points must be accounted for.

(a) Reference points 1 and 4—the water surface of the reservoir and the water surface in the tank, respectively.

From the statement of the problem,

$$Q' = 0, \quad E_4 = E_1, \quad z_1 = 0, \quad z_4 = 160 \text{ ft}$$

$$P_1 = P_4 = 1 \text{ atm}$$

$$V_1 = V_4 = \frac{1}{62.3} = 0.0161 \text{ cu ft/lb (incompressible, iso-}$$

thermal fluid)

$$\Sigma F = 35 \text{ ft-lb}_f/\text{lb}$$

$$\bar{v}_1 = \bar{v}_4 = 0, \text{ since liquid levels remain approximately}$$
constant

For this problem and the information sought, the most convenient and appropriate energy equation to use is Equation 20.14 or 20.16a.

Using Equation 20.14,

$$\Delta E + \Delta\left(\frac{\bar{v}^2}{2g_c\alpha}\right) + \Delta z\frac{g}{g_c} + \Delta(PV) = Q' - W_f' - \Sigma F$$

$$0 + 0 + 160 + 0 = 0 - W_f' - 35$$

Therefore,

$$-W_f' = 195 \text{ ft-lb}_f/\text{lb}$$

The minus sign connotes work being done *on* the fluid by the pump.

The cross-sectional area of the pipe (S) is $(\pi D^2)/4 = (3.14/4)(3/12)^2 = 0.049$ sq ft. The bulk average fluid velocity in the pipe may be evaluated from the volumetric flow rate, as follows:

$$\left(150\frac{\text{gal}}{\text{min}}\right)\left(\frac{\text{min}}{60 \text{ sec}}\right)\left(\frac{8.34 \text{ lb H}_2\text{O}}{\text{gal}}\right)\left(\frac{\text{cu ft}}{62.3 \text{ lb H}_2\text{O}}\right)$$

$$\times \left(\frac{1}{0.049 \text{ sq ft}}\right) = 6.8 \text{ ft/sec}$$

Thus, the actual power required is found by application of Equations 20.18 and 20.19 and the use of the pump-motor efficiency.

$$\text{Power} = \left(195\frac{\text{ft-lb}_f}{\text{lb}}\right)\left(6.8\frac{\text{ft}}{\text{sec}} \times 0.049 \text{ sq ft}\right)\left(\frac{62.3 \text{ lb}}{\text{cu ft}}\right)$$

$$\times \left(\frac{1 \text{ hp}}{550 \text{ ft-lb/sec}}\right)\left(\frac{1}{0.55}\right) = 13.14 \text{ hp}$$

(b) Reference points 2 and 3—the inlet and exit of the 3-in. pipe respectively.

Application of the energy balance between these two points necessitates the evaluation of the pressures at these points. This evaluation in itself requires an energy balance applied between a point at the reservoir surface and the entrance to the 3-in. pipe. Therefore,

$$P_1 V_1 = P_2 V_2 + z_2 \frac{g}{g_c} + \frac{\bar{v}_2^2}{2 g_c \alpha}$$

where P_2 is the pressure at the pipe inlet, and $\bar{v}_2$ is the fluid velocity at this level. Admittedly, the above equation does not include the frictional loss associated with the fluid entering the pipe. Although the total frictional loss is known, the portions of the ΣF term that comprise the entrance and the exit frictional losses cannot be ascertained at this time. Instead, the ΣF term will be included as a total entity upon application of the energy balance between points 2 and 3.

Rearranging the above equation gives for P_2

$$P_2 = P_1 - \frac{z_2}{V_2} \frac{g}{g_c} - \frac{\bar{v}_2^2}{2 g_c \alpha V_2}$$

and

$$P_2 = (14.7)(144) - \frac{(-10)}{(1/62.3)} - \frac{(6.8)^2}{(2)(32.2)(\alpha)(1/62.3)}$$

$$P_2 = 2120 + 623 - \frac{45}{\alpha}$$

The kinetic-energy correction factor (α) may be evaluated by calculating the Reynolds number and by use of Figure 20.2.

$$N_{\mathrm{Re}} = \frac{D \bar{v} \rho}{\mu} = \frac{(3/12)(6.8)(62.3)}{(1 \times 0.000672)} = 158,000$$

At this Reynolds number $\alpha = 0.94$, and

$$P_2 = 2695 \text{ lb/sq ft absolute}$$

The pressure at 3 is determined in the same manner by taking an energy balance between the liquid surface of the storage tank and the exit of the 3-in. pipe. P_3 is found to be equal to 10,521 lb/sq ft absolute.

Now the total energy balance may be applied between 2 and 3

$$E_2 + P_2 V_2 + z_2 \frac{g}{g_c} + \frac{\bar{v}_2^2}{2 g_c \alpha} + Q' = E_3 + P_3 V_3 + z_3 \frac{g}{g_c}$$
$$+ \frac{v_3^2}{2 g_c \alpha} + W_f' + \Sigma F$$

For these reference points,

$Q' = 0$, $z_2 = 0$, $z_3 = 35$ ft, $E_2 = E_3$

$P_2 = 2695$, $P_3 = 10,521$

$v_2 = v_3 = 6.8$ ft/sec (because the pipe diameter is the same at both reference points)

$$\Sigma F = 35 \text{ ft-lb}_f/\text{lb}$$

or

$$E_2 + 2695 \left(\frac{1}{62.3}\right) + 0 + \frac{(6.8)^2}{(2)(32.2)(\alpha)} + 0 = E_3$$
$$+ 10,521 \left(\frac{1}{62.3}\right) + \frac{(6.8)^2}{(2)(32.2)(\alpha)} + 35 + W_f' + 35$$

and

$$W_f' = -195.4 \text{ ft-lb}_f/\text{lb}$$

This is essentially the same theoretical work as needed before. Note that in this case P_2 and P_3 do not equal atmospheric pressure nor are they equal. Also, the two velocities, although equal, are 6.8 ft/sec instead of zero.

(c) Reference points 2 and 4 (the inlet to the 3-in. pipe and the liquid surface of the storage tank, respectively)

Again, $Q' = 0$, $z_2 = 0$, $z_4 = 170$ ft

$P_2 = 2743$ lb$_f$/sq ft

$P_4 = 14.7 \times 144 = 2120$ lb$_f$/sq ft absolute

$v_2 = 6.8$ ft/sec

$v_4 = 0$

$\Sigma F = 35$

Since the velocities at the two reference points are not equal, the kinetic-energy correction factor will again have to be used.

As before, the energy balance becomes

$$2743 \left(\frac{1}{62.3}\right) + \frac{(6.8)^2}{(2)(32.2)(0.94)} = 2120 \left(\frac{1}{62.3}\right)$$
$$+ 170 + W_f' + 35$$

or

$$W_f' = -194.6 \text{ ft-lb}_f/\text{lb}$$

Again, the same theoretical work requirement results. Other reference points could be used, and exactly the same result would be found, providing all energies at each reference point are considered. It may be concluded that, although any set of reference points may be used, wise selection of reference points can simplify calculations.

PIPE, TUBING, AND FITTINGS

Before considering the ΣF term in Equation 20.14 in more detail, it is important to give some thought to the means of containing the fluid in a given flow system.

Chemical-process applications frequently involve the fluid state. The fluid is usually transferred from one part of the process to another through pipe or tubing of circular cross section. Most chemical-engineering flow problems dictate the use of closed ducts rather than open channels. Pipe or tubing may be manufactured from any available material of construction depending upon the corrosive properties of the fluid being handled and the flow pressure. Such materials as glass, concrete, asbestos, steel, plastics, wood, and many others are frequently used in construction, but iron, steel, copper, and brass are the most common pipe materials encountered in the process industries. As mentioned above, the choice depends upon the application.

Pipe sections may be joined by several techniques. Threaded connections are most common, and emphasis will be placed on this type. Since *pipe* may be manufactured with different diameters and various wall thicknesses, some standardization is necessary. A method of identifying pipe sizes has been established by the American Standards Association. By convention, pipe size and fittings are characterized in terms of a nominal diameter and wall thickness. For steel pipe, nominal diameters may vary between $\frac{1}{8}$ in. and 30 in. The nominal diameter is neither an inside nor an outside diameter but an approximation of the inside diameter. But, regardless of wall thickness, pipes of the same nominal diameter have the same *outside* diameter. This permits interchange of fittings.

The wall thickness of pipe is indicated by a *schedule number*, which is a function of internal pressure and allowable stress. Approximately,

$$\text{Schedule number} \approx 1000 \frac{P}{s} \qquad (20.20)$$

where P = internal working pressure, lb_f/sq in.
s = allowable stress, lb_f/sq in.

Ten schedule numbers are in use: 10, 20, 30, 40, 60, 80, 100, 120, 140, and 160. The pipe-wall thickness increases with the schedule number. For steel pipe, schedule 40 is the number corresponding to "standard" pipe. In Appendix C-6 are tabulated the dimensions and other useful data for schedule 40 and 80 threaded steel pipe. For other schedule numbers and other types of pipe, Perry (6) is recommended.

Some demarcation between pipe and tubing can be set forth. Unlike pipe, tubing is sold on the basis of *actual* outside diameter. The wall thickness may be expressed as Birmingham wire gage (BWG). Appendix C-7 tabulates various dimensions of several sizes of copper tubing.

Fittings. The term fitting refers to a piece of pipe that can:

1. Join two pieces of pipe; for example, couplings, unions.

2. Change pipeline direction; for example, elbows, tees.

3. Change pipeline diameter; for example, reducers.

4. Terminate a pipeline; for example, plugs, valves.

5. Join two streams to form a third; for example, tees, wyes, crosses.

6. Control flow; for example, valves.

Figure 20.3 illustrates typical pipe fittings. Fittings for steel pipe are usually made of cast iron or malleable iron and can be obtained in several wall thicknesses. For pipe sizes greater than 2 in., screw fittings are less frequently encountered. Large-size pipes can be joined by the same types of fittings, but it is usual to install flanged fittings or to employ welded connections. Figures 20.4 and 20.5 show flanged joints and flanged pipe fittings respectively.

Valves. A valve is also a fitting, but it has more important uses than simply to connect pipe. Valves are used either to control flow rate or to shut off the flow of fluid. The basic design of the valve dictates its use as either a stopping device or as a flow-rate control device. Two common valves are the *gate valve* and the *globe valve*.

The *gate valve* (Figure 20.6a) is a simple design possessing a disk which slides at right angles to the flow. It is principally a valve used for stopping flow because small lateral adjustments of the disk give extreme changes in the flow area.

The *globe valve* (Fig. 20.6b) because of its design is more suitable for flow control. In this valve, the fluid passes through an opening whose area is controlled by a disk placed somewhat parallel to the flow direction. Good control of flow can be achieved with a globe valve; however the pressure loss is higher than with a gate valve.

Other common valves include plug cocks used for on-off service, check valves to control the flow direction, safety valves to control pressure and diaphram or bellows valves which eliminate the need for packing glands.

EVALUATION OF FLUID FRICTION

It must be recalled that fluid friction may be grouped into two broad classifications, i.e., form friction and skin friction. In the general flow system, both types of friction are encountered, and

$$\Sigma F = F_s + F_f \qquad (20.21)$$

where F_s = skin friction, ft lb_f/lb
F_f = form friction, ft lb_f/lb

A physical flow system in which only one type of fluid friction is encountered is impossible. Perhaps the closest approach to this is in a long smooth horizontal pipe without change in flow direction. In this case, skin friction predominates. However, as soon as one considers commercial steel pipe in a flow system including many pipe fittings and changes of direction, boundary-layer separation is likely to occur, and form drag will enter the ΣF term to a relatively large extent.

In the usual piping system in the process industries, the fluid friction occurring within the system may be calculated by the methods discussed in Chapter 13. The development of the friction-factor equation, however, was limited to smooth tubes. In the actual case, fluid friction is a function of pipe roughness as well as pipe size, fluid properties, and fluid velocity. A commercial pipe is inherently rough as a result of the manufacturing procedure. This roughness will cause boundary-layer separation and eddy formation within the roughness pockets, resulting in energy dissipation and additional frictional losses. Experiments (5) have confirmed this fact by showing artificially roughened pipes to have a greater frictional pressure drop than the corresponding smooth pipe (3).

In a like manner, the presence of fittings and changes

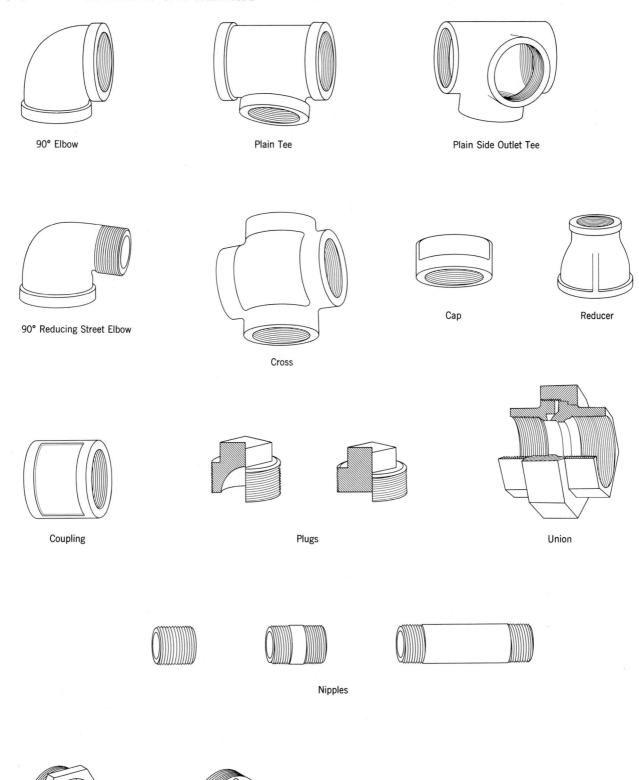

Figure 20.3. Screwed pipe fittings.

(a) Screwed flange.

(b) Slip-on flange.

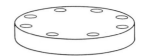

(c) Blind flange.

Figure 20.4. Flanged joints.

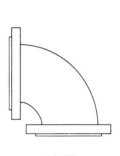

(a) Elbow.

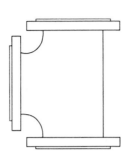

(b) Tee.

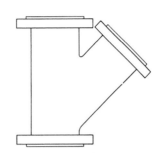

(c) Lateral.

(d) Taper reducer.

Figure 20.5. Flanged fittings.

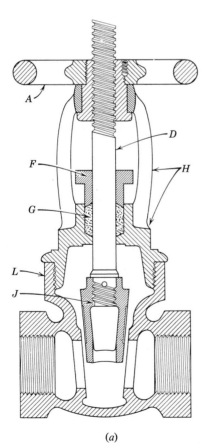

(a)

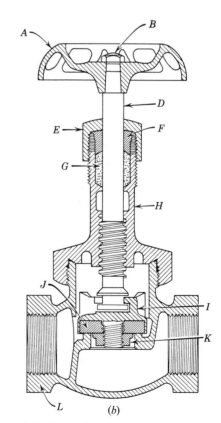

(b)

Figure 20.6. Sectional view of (a) gate and (b) globe valves.

A—Wheel	E—Packing nut	H—Bonnet	K—Disk nut
B—Wheel nut	F—Gland	I—Disc holder	L—Body
D—Spindle	G—Packing	J—Disc	

in flow direction will cause boundary-layer separation which results in form friction. Indeed, the friction encountered in a fitting may be a significant portion of the frictional pressure drop for short flow systems. The friction losses through a fitting and through expansions and contractions are most easily obtained in terms of *equivalent length* of straight pipe. This equivalent length is added to the length of straight pipe to give the total equivalent system length. The advantage of this approach is that both pipe and fittings are expressed in terms of a total equivalent length of pipe of the same relative roughness.

Appendix C-1 shows the variation in roughness factors for various types of pipe with nominal pipe size. In Appendix C-2a are tabulated the equivalent lengths of fittings and of other frequently encountered physical shapes. Equivalent lengths of contractions, expansions, or fittings may also be correlated by a resistance coefficient (K). This coefficient is defined as the number of velocity heads lost because of the fitting. One velocity head equals $\bar{v}^2/2g_c$. Appendix C-2b and c gives K for various expansions and contractions, whereas Appendix C-2d correlates K with equivalent length in terms of pipe diameters.

The friction factor as developed in Chapter 13 should more appropriately be written for application to piping systems as

$$f = \left[\frac{\Sigma F}{\dfrac{\bar{v}^2}{2g_c}}\right]\left[\frac{D}{\Sigma L}\right] = \phi\left[(N_{Re})\left(\frac{\epsilon}{D}\right)\right] \quad (20.22)$$

where f = the friction factor
  D = pipe diameter, ft
  $\bar{v}$ = average fluid velocity, ft/sec
  ΣF = *total* fluid friction loss, ft lb$_f$/lb
  ΣL = *total* equivalent length of fittings, expansions
    and contractions, and straight pipe, ft
  ϵ/D = relative roughness factor

Although the friction factor of Equation 20.22 is a function of pipe roughness in addition to Reynolds number, it is still a ratio of the total momentum transferred to the momentum transferred by the turbulent mechanism. Pipe roughness, in this case, contributes to momentum transfer through form drag.

Equation 20.22 is presented graphically in Appendix C-3. In the event that noncircular cross sections are used, the simple diameter term in Equation 20.22 cannot be used. Instead, the equivalent diameter as previously defined can be substituted, where

$$D_{eq} = 4\left[\frac{\text{cross-sectional area}}{\text{wetted perimeter}}\right] \quad (13.31)$$

This equivalent diameter is applicable, subject to conditions stated in Chapter 13.

Note that to use Equation 20.22 or its graphical representation requires knowledge of the average velocity to determine the frictional pressure drop. There are times when it is necessary to solve for the quantity of fluid flowing to give a specified pressure drop. Since in this case velocity appears in both ordinate and abscissa, the solution of Equation 20.22 would require successive approximations until the stipulated conditions were met. For instance, an assumption of a velocity would have to be made so that a Reynolds number could be calculated. With this Reynolds number and the appropriate roughness factor, a friction factor could be obtained from Appendix C-3. Equation 20.22 could then be solved for ΣF using the f and $\bar{v}$ based upon assumption. If the ΣF calculated meets the requirements, the solution is completed. If not, the process must be repeated.

However, a direct approach is possible. If the Reynolds number is multipled by the $\sqrt{f}$, the Kármán number results. Thus

$$N_{Ká} = \frac{D\bar{v}\rho}{\mu}\left[\frac{1}{\bar{v}}\sqrt{\frac{\Sigma F\left(\dfrac{D}{\Sigma L}\right)}{\dfrac{1}{2g_c}}}\right] = \frac{D\rho}{\mu}\sqrt{\frac{\Sigma F\left(\dfrac{D}{\Sigma L}\right)}{\dfrac{1}{2g_c}}}$$

$$(20.23)$$

No velocity term appears in the Kármán number. If this number is plotted as a function of $1/\sqrt{f}$, as in Appendix C-4, the velocity can be obtained directly for a known pressure drop.

Illustration 20.3. Water is pumped from a reservoir to a storage tank atop a building by means of a centrifugal pump. There is a 200-ft difference in elevation between the two water surfaces. The inlet pipe at the reservoir is 8.0 ft below the surface, and local conditions are such that level is substantially constant. The storage tank is vented to the atmosphere and the liquid level is maintained constant. The inlet pipe to the storage tank is 6.0 ft below the surface. The piping system consists of 200 ft of 6-in. Sch. 40 steel pipe containing two 90° elbows and one open gate valve from the reservoir to the pump; 6-in. pipe follows the pump for a length of 75 ft after which the pipe size is reduced to 4-in. Sch. 40 steel pipe for 250 ft to the tank. The 4-in. pipe contains one gate valve and three 90° elbows. It is desired to maintain a flow of water into the tank of 625 gal/min. Water temperature is 68°F. If the pump-motor set has an over-all efficiency of 60 per cent, what would the pumping costs be in dollars per day if electricity costs $0.02/kwhr?

SOLUTION. For the flow system as sketched, point 1, the reservoir surface, and point 2, the liquid level in the storage tank, will be used as reference points. The datum plane will be at the reservoir surface.

Writing the energy balance in the form of Equation 20.16a, gives

$$\Delta\left(\frac{\bar{v}^2}{2g_c\alpha}\right) + \Delta z\frac{g}{g_c} + \int_{P_1}^{P_2} V\,dP + \Sigma F = -W_f'$$

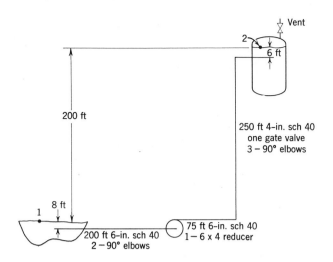

The flow is isothermal and water is incompressible so that $V_1 = V_2$. For the reference points chosen, $P_1 = P_2$ and $\int_{P_1}^{P_2} V\,dP = 0$. Since both liquid levels remain constant, $\bar{v}_1 = \bar{v}_2 = 0$. Thus, the energy balance becomes

$$-W_f' = \Delta z \frac{g}{g_c} + \Sigma F$$

To solve for W_f', only the potential energy and frictional losses need be evaluated. In this case, ΣF will be equal to the friction occurring in both sizes of pipe including all fittings, expansions, and contractions. For both pipe sizes, Equation 20.22 must be employed.

The 6-in. pipe. For this size pipe, the average velocity is

$$\bar{v}_{6\,\text{in.}} = \left(625\,\frac{\text{gal}}{\text{min}}\right)\left(\frac{\text{min}}{60\,\text{sec}}\right)\left(0.1334\,\frac{\text{cu ft}}{\text{gal}}\right)\left(\frac{144\,\text{sq in./sq ft}}{28.89\,\text{sq in.}}\right)$$

$$= 6.92\,\text{ft/sec}$$

The 28.89 sq in. is the cross-sectional area of a 6-in. Sch. 40 pipe obtained from Appendix C-6. The N_{Re} for flow through this pipe can be calculated:

$$D = 6.065\,\text{in.} = 0.506\,\text{ft} \qquad \text{(Appendix C-6)}$$
$$\mu = 1.0\,\text{centipoise} \qquad \text{(Appendix D-9)}$$
$$\rho = 62.3\,\text{lb/cu ft} \qquad \text{(Appendix D-12)}$$

$$N_{\text{Re}} = \frac{D\bar{v}\rho}{\mu} = (6.065/12)\frac{(6.92)(62.3)}{(1\times0.000672)}$$

$$= 325,000$$

To use Equation 20.22 requires the total equivalent length of straight pipe. From Appendix C-2,
Entrance from reservoir to pipe; $K = 0.78$; $L = 27$ ft
90° elbows, $L/D = 30$, $L = 15.2$ ft each   30.4 ft
Gate valve (fully opened) $L/D = 13$ 6.5 ft

Total   63.9 ft

Thus the *total* equivalent length of 6-in. pipe is

$$\Sigma L_{6\,\text{in.}} = 63.9 + 200 + 75 = 339\,\text{ft}$$

From Appendix C-1, the relative roughness factor (ϵ/D) for 6-in. commercial steel pipe is 0.0003.

Rearranging Equation 20.22 to solve for ΣF,

$$\Sigma F = \frac{f\bar{v}^2(\Sigma L)}{2g_c D}$$

The friction factor (f) may be determined from Appendix C-3 using N_{Re} and ϵ/D.

$$f = 0.017$$

So

$$\Sigma F \text{ for the 6-in. pipe} = \Sigma F_{6\,\text{in.}} = \frac{(0.017)(6.92)^2(339)}{(2)(32.2)(0.506)}$$

$$= 8.5\,\text{ft-lb}_f/\text{lb}$$

The 4-in. pipe. Assuming steady state, by material balance the linear velocity of the water in the 4-in. pipe will be inversely proportional to the cross-sectional areas of the 4-in. and 6-in. pipes. Thus,

$$\bar{v}_{4\,\text{in.}} = \bar{v}_{6\,\text{in.}}\left(\frac{S_{6\,\text{in.}}}{S_{4\,\text{in.}}}\right) = 6.92\left(\frac{28.89}{12.73}\right) = 15.7\,\text{ft/sec}$$

So,

$$N_{\text{Re}} = \frac{D\bar{v}\rho}{\mu} = (4.026/12)\frac{(15.7)(62.3)}{(1\times0.000672)} = 489,000$$

The total equivalent length of 4-in. pipe is (from Appendix C-2)

Sudden reduction (6-in. to 4-in. pipe) ($K = 0.24$)   5.6 ft
Straight pipe   250.0
90° elbows, $L/D = 30$, $L = 10$ ft each   30.0
Gate valve (fully opened) $L/D = 13$ 4.4
Sudden enlargement (pipe into tank) ($K = 1.0$)   20.0

Total   310.0 ft

Relative roughness factor for 4-in. pipe, $\epsilon/D = 0.00044$. Therefore, $f = 0.0175$
and

$$\Sigma F_{4\,\text{in.}} = \frac{f\bar{v}^2(\Sigma L)}{2g_c D} = \frac{(0.0175)(15.7)^2(310)}{2(32.2)(4.026/12)} = 62\,\text{ft-lb}_f/\text{lb}$$

Thus the total friction loss in both pipes, including fittings, expansions, and contractions, is $\Sigma F = 8.5 + 62 = 70.5$ ft-lb$_f$/lb.

Returning to the energy balance as written for the problem,

$$-W_f' = (z_2 - z_1)\frac{g}{g_c} + \Sigma F = 200\left(\frac{32.2\,\text{ft/sec}^2}{32.2\,\dfrac{\text{ft-lb}}{\text{lb}_f\,\text{sec}^2}}\right)$$
$$+ 70.5 = 270.3\,\text{ft-lb}_f/\text{lb}$$

This work term may be converted to power as previously described.

$$-W_f' = \left(270.5\,\frac{\text{ft-lb}_f}{\text{lb}}\right)\left(\frac{(\text{hp})(\text{min})}{33,000\,\text{ft-lb}_f}\right)\left(\frac{625\,\text{gal}}{\text{min}}\right)$$
$$\times\left(\frac{8.34\,\text{lb}}{\text{gal}}\right)\left(\frac{0.746\,\text{kwhr}}{\text{hphr}}\right) = 31.7\,\text{kwhr/hr}$$

With 60 per cent efficiency, the actual work required $= \dfrac{31.7}{0.6}$
$= 53.9$ kwhr/hr

Pumping costs per day $= (53.9)(0.02)(24) = \$25.4$ per day

Illustration 20.4. A pump discharges ethanol at 100°F into a 2-in. Sch. 40 steel pipe at a pressure of 40 psig. This pipe is 135 ft long (total equivalent length) and supplies ethanol at 20 psig to a reactor. What is the volumetric flow rate of ethanol to the reactor? There is negligible change in elevation between the pump and reactor.

SOLUTION. A simple sketch is, as usual, warranted.

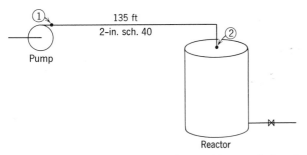

According to the statement of the problem and for the reference points chosen

$$z_1 = z_2 = 0$$
$$P_1 = (40 + 14.7)144 = 7860 \text{ lb}_f/\text{sq ft}$$
$$P_2 = (20 + 14.7)144 = 4990 \text{ lb}_f/\text{sq ft}$$
$$\rho_1 = \rho_2 = 62.3 \times 0.82 = 51.0 \text{ lb/cu ft}$$
$$\bar{v}_1 = \bar{v}_2 = ?$$
$$W_f' = 0$$
$$Q' = 0$$

The energy balance therefore becomes

$$P_1 V_1 = P_2 V_2 + \Sigma F$$

Since the specific volumes are equal for an incompressible fluid,

$$\Sigma F = \frac{(P_1 - P_2)}{\rho} = \frac{7860 - 4990}{51} = 56.4 \text{ ft-lb}_f/\text{lb}$$

For 2-in. Sch. 40 pipe, $\epsilon/D = 0.00086$

$$D = 2.067 \text{ in.} = 0.1724 \text{ ft}$$
$$S = 3.356 \text{ sq in.} = 0.0233 \text{ sq ft}$$

The desired flow rate may be calculated from the average velocity which is most easily obtained from Appendix C-4. From Equation 20.23,

$$\frac{D\rho}{\mu}\sqrt{\frac{2g_c D(\Sigma F)}{\Sigma L}} = \frac{(0.1724)(51)}{(0.9 \times 0.000672)}\sqrt{\frac{(2)(32.2)(0.1724)(56.4)}{135}}$$

$$= 31,200$$

Then, from Appendix C-4,

$$\frac{\bar{v}}{\sqrt{2g_c D(\Sigma F)/\Sigma L}} = 7.1$$

Solving for $\bar{v}$ gives

$$\bar{v} = 15.1 \text{ ft/sec}$$

and the volumetric flow rate is

$$(15.1)(0.0233)(60) = 21.1 \text{ cu ft/min}$$

COMPRESSIBLE FLUIDS

For operations involving the flow of compressible fluids, the solution of the energy balance is more com-plicated because of the variation of specific volume with pressure changes. The integrated equations as previously presented do not apply. Instead, the differential form of the energy balance must be used and integrated over the terminals of the system.

Thus, Equation 20.16a may be written as

$$\int_{v_1}^{v_2} \frac{\bar{v}\, d\bar{v}}{g_c\alpha} + \int_{z_1}^{z_2} \frac{g}{g_c}\, dz + \int_{P_1}^{P_2} V\, dP + (\Sigma F) = -W_f' \quad (20.24)$$

The friction term in this case will be equal to

$$(\Sigma F) = \int_{L_1}^{L_2} \frac{f\bar{v}^2}{2g_c D}\, d(\Sigma L) \quad (20.25)$$

Therefore, substituting Equation 20.25 into Equation 20.24 gives

$$\int_{v_1}^{v_2} \frac{\bar{v}\, d\bar{v}}{g_c\alpha} + \int_{z_1}^{z_2} \frac{g}{g_c}\, dz + \int_{P_1}^{P_2} V\, dP + \int_{L_1}^{L_2} \frac{f\bar{v}^2}{2g_c D}\, dL = -W_f' \quad (20.26)$$

Actually, Equation 20.26 is appropriate for incompressible fluids as well as for gases but is most frequently encountered in gas flow. The solution to Equation 20.26, and in particular $\int V\, dP$, requires a knowledge of the *PVT* relationship for the gas under study. Often this relationship may be obtained by use of the perfect-gas law. In any case, the *PVT* properties may be evaluated using the compressibility factor in conjunction with the gas law.

Since no analytical expression exists for $f\bar{v}^2$ as a function of L, the friction term is more difficult to evaluate. A series of stepwise calculations involving short increments of length using average values of $\bar{v}$ and V over the length may be used with little error.

If steady state is assumed, the mass velocity will be a constant for a pipe of uniform cross section. Or

$$G = \bar{v}\rho = \frac{\bar{v}}{V} \quad (20.27)$$

where G = mass velocity, lb/sec sq ft
   $\bar{v}$ = average linear velocity, ft/sec
   ρ = fluid density, lb/cu ft
   V = specific volume, cu ft/lb

Solving Equation 20.27 for v gives

$$\bar{v} = GV \quad (20.28)$$

and

$$d\bar{v} = G\, dV \quad (20.29)$$

Replacing $\bar{v}$ in Equation 20.26 by the above expressions yields an expression in terms of the specific volume.

$$\int_1^2 \frac{G^2 V\, dV}{g_c\alpha} + \int_1^2 \frac{g}{g_c}\, dz + \int_1^2 V\, dP$$

$$+ \int_1^2 \frac{fG^2 V^2}{2g_c D}\, dL = -W_f' \quad (20.30)$$

Analysis of a particular gas-flow problem will best indicate the appropriate method needed in solving Equation 20.26 or Equation 20.30. For instance, it might be that for nearly isothermal conditions the gas viscosity will vary little, and, if flow is fully turbulent, f will be substantially constant, so that the friction term is easily calculated.

Illustration 20.5. It is desired to have 150×10^6 cu ft of methane, measured at 60°F and 760 mm Hg, per 24-hr day delivered to a synthesis plant located 2 miles from the compressor station. Delivery conditions are to be 60°F and 10 psig at the plant. If the gas leaves the compressor at a temperature of 75°F, what must its pressure be at this point? Assume negligible change in elevation between line terminals. The pipe is steel and is 24 in. I.D.

SOLUTION. If the reference points for the energy balance are chosen so as to have point 1 at the outlet of the compressor and point 2 in the pipe at the synthesis plant, then $-W_f' = 0$. Also by statement of the problem $dz = 0$. For this case, Equation 20.30 will be used.

$$\int_1^2 \frac{GV\,dV}{g_c\alpha} + \int_1^2 V\,dP + \int_1^2 \frac{\bar{f}G^2V^2}{2g_cD}\,dL = 0 \qquad (a)$$

To determine the character of flow, the Reynolds number at the metered conditions is calculated.

$$w = \left(\frac{150,000,000}{24 \times 3600}\right)\left(\frac{1}{359}\right)\left(\frac{492}{520}\right)(16)$$
$$= 73.0 \text{ lb/sec (constant)}$$

Viscosity of methane at 60°F = 0.011cp (Appendix D-9)

Cross-sectional area of 24-in. pipe $= \frac{\pi}{4}(2)^2 = 3.14$ sq ft

$$G = \frac{73.0}{3.14} = 23.3 \text{ lb/sec sq ft}$$

$$(N_{Re})_2 = \frac{DG}{\mu} = \frac{(2)(23.3)}{(0.011 \times 6.72 \times 10^{-4})} = 6.30 \times 10^6$$

$$(N_{Re})_1 = (6.3 \times 10^6)\left(\frac{0.011}{0.012}\right) = 5.78 \times 10^6 \quad \text{(viscosity of}$$

CH$_4$ at 75°F = 0.012 centipoise)
Relative roughness factor (ϵ/D) is 0.00007. Therefore,

$$f_2 = 0.0128 \text{ and } f_1 = 0.0130.$$

These friction factors are so close that an average value of $\bar{f} = 0.0129$ may be used with negligible error. To eliminate V^2 in the friction term, Equation a is divided by V^2.

$$\int_1^2 \frac{G^2\,dV}{g_cV\alpha} + \int_1^2 \frac{dP}{V} + \int_1^2 \frac{\bar{f}G^2\,dL}{2g_cD} = 0 \qquad (b)$$

By the magnitude of N_{Re}, $\alpha = 0.98$ from Figure 20.2. Assume that the perfect-gas law may be used to describe the PVT relationships for methane.

Thus $$PV = \frac{1}{M}RT$$

where M = molecular weight

$$V = \frac{RT}{PM}$$

Substitution of V from above into the second term of Equation b gives

$$\int_1^2 \frac{G^2\,dV}{\alpha g_cV} + \int_1^2 \frac{MP\,dP}{RT} + \int_1^2 \frac{\bar{f}G^2\,dL}{2g_cD} = 0 \qquad (c)$$

Integrating Equation c gives

$$\frac{G^2}{\alpha g_c}\ln\frac{V_2}{V_1} + \frac{M}{2RT_{av}}(P_2{}^2 - P_1{}^2) + \frac{\bar{f}G^2(\Sigma L)}{2g_cD} = 0$$

The use of T_{av} is not in great error in this case. For an ideal gas.

$$\frac{G^2}{\alpha g_c}\ln\frac{P_1}{P_2} + \frac{M}{2RT_{av}}(P_2{}^2 - P_1{}^2) + \frac{\bar{f}G^2(\Sigma L)}{2g_cD} = 0 \qquad (d)$$

Assume as an initial trial the first term of Equation d to be negligible, then

$$P_2{}^2 - P_1{}^2 = -\frac{\bar{f}G^2(\Sigma L)}{g_cD}\left(\frac{RT_{av}}{M}\right)$$

or

$$P_1{}^2 = \frac{\bar{f}G^2(\Sigma L)}{g_cD}\left(\frac{RT_{av}}{M}\right) + P_2{}^2$$

Substituting the known values from the statement of the problem and physical data,

$$P_1{}^2 = \frac{(0.0129)(23.3)^2(2 \times 5280)}{32.2 \times 2.0}\left(\frac{1544 \times 528}{16}\right)$$
$$+ [(10 + 14.7)144]^2$$

$$P_1{}^2 = 71.0 \times 10^6$$

$$P_1 = 8440 \text{ lb/sq ft} = 58.5 \text{ psi}$$

Check the validity of neglecting the velocity term in Equation d

$$\frac{G^2}{\alpha g_c}\ln\frac{P_1}{P_2} = \frac{(23.3)^2}{(0.98)(32.2)}\ln\frac{58.5}{24.7} = 14.5 \text{ lb/sq ft}$$

Therefore, the assumption made above that the velocity term was negligible contributed an error of 0.16 per cent in the final answer.

FLUID METERS

The metering of fluids is an important application of the energy balance. Basically, most flow meters are designed to cause a pressure drop which can be measured and related to the rate of flow. This pressure drop can be brought about by changes in kinetic energy, by skin friction, or by form friction. Some types of meters emphasize one or a combination of these items. The pitot tube for example is a metering device which is based upon a pressure difference resulting almost solely from kinetic energy changes. On the other hand, the pressure change across an orifice plate is due mostly to kinetic-energy change and form friction. In any

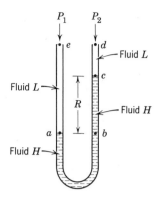

Figure 20.7. Simple manometer.

event, a general equation resulting from an energy balance may be derived to relate flow and pressure drop.

Manometers. Since most fluid meters tend to cause a pressure difference across the metering section, some simple, easy-to-use, pressure-measuring device must be employed to indicate this difference. One of the simplest pressure-measuring instruments is the U-tube manometer.

If a duct is filled with an incompressible fluid and no flow occurs, the energy equation, Equation 20.16a, may be written as

$$\Delta z \frac{g}{g_c} + V(P_2 - P_1) = 0 \qquad (20.31)$$

or

$$\Delta P = -\frac{\Delta z}{V} \frac{g}{g_c} = -\rho \Delta z \frac{g}{g_c} = \rho(z_1 - z_2) \frac{g}{g_c} \ (20.32)$$

Thus the pressure difference can be expressed in terms of a height of a vertical column of fluid of density ρ.

Equation 20.32 may be applied to the U-tube manometer. This instrument is shown schematically in Figure 20.7. Fluid H cannot be identical to, nor miscible with, the flowing fluid. It is sometimes desirable to use a fluid that is heavier than the flowing fluid as is shown and sometimes one that is lighter, in which case the U-tube is inverted. In Figure 20.7, it is desired to measure the difference between P_1 and P_2. The pressure difference will be related to the difference in height between points a and c which is called the *manometer reading*, R.

The pressure at point b can be determined by Equation 20.32; that is,

$$P_b = P_2 + (z_d - z_c)\rho_L + (z_c - z_b)\rho_H \quad (20.33)$$

Since point b is stationary, the pressure as indicated by Equation 20.33 must be balanced by a pressure acting on it in an opposite direction. Or

$$P_b = P_1 + (z_e - z_a)\rho_L \qquad (20.34)$$

Equating P_b as determined from Equations 20.33 and 20.34 gives

$$P_2 + (z_d - z_c)\rho_L + (z_c - z_b)\rho_H = P_1 + (z_e - z_a)\rho_L$$
$$(20.35)$$

But $z_e = z_d$ and $z_a = z_b$; therefore,

$$P_2 - P_1 = [(z_e - z_a) - (z_d - z_c)]\rho_L - \rho_H(z_c - z_b)$$
$$P_2 - P_1 = (z_c - z_a)\rho_L - (z_c - z_a)\rho_H$$

or

$$-\Delta P = P_1 - P_2 = (z_c - z_a)(\rho_H - \rho_L) \quad (20.36)$$

In Equation 20.36, the term $(z_c - z_a)$ would correspond to the manometer reading as indicated by an appropriate scale. This balancing of forces to determine the pressure difference read by a manometer is general. It may be used regardless of the manometer orientation, presence of intermediate fluids, or any other variation in manometer design. Note that the manometer reading is multiplied by the *difference* in density of the two fluids to get the pressure drop.

As indicated above, variations of the manometer shown in Figure 20.7 may be made. It may be inverted so as to use a metering fluid of lighter density than the flowing fluid. For small readings the tube may be inclined at some angle to the horizontal so as to magnify the linear displacement of fluid. Another technique used to magnify small readings is to have one leg of the manometer of significantly larger cross section than the other and to incline the smaller diameter tube. This procedure will allow the pressure to be read in only the smaller diameter tube since the meniscus level in the larger tube will remain substantially constant. A two-fluid manometer as illustrated in Figure 20.8 is often used to indicate low pressure differences.

In the two-fluid manometer, the liquid levels in the top reservoirs remain essentially constant, but the reading can be multiplied a great deal if fluid B and fluid C have nearly the same density. Whatever the type of manometer used, the principles behind Equation 20.32 can be applied to give an equation that can be used

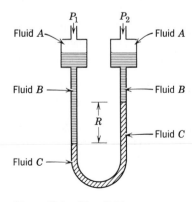

Figure 20.8. Two-fluid manometer.

to meter the pressure drop across a test section of a duct. Manometers are frequently used with flow meters of the orifice-, venturi-, or pitot-tube variety.

The General Meter Equation. Consider a flow system containing a fluid meter about which an energy balance is to be made. This meter must be of the type in which the reading depends upon a pressure difference. Figure 20.9 illustrates a meter of this type. Points 1 and 2 represent locations about which an energy balance is to be made. The two points are separated by a negligible distance compared to the total length of the system.

Assume that for this case Δz, W_f', and Q' are zero. Thus Equation 20.16a becomes, for a fluid of substantially constant density,

$$V(P_2 - P_1) + \frac{\bar{v}_2^2 - \bar{v}_1^2}{2g_c\alpha} + \Sigma F = 0 \qquad (20.37)$$

Equation 20.37 may be rearranged to give

$$\bar{v}_2^2 - \bar{v}_1^2 = -2g_c\alpha[V(P_2 - P_1) + \Sigma F] \qquad (20.38)$$

From the continuity equation,

$$\bar{v}_1\rho_1 S_1 = \bar{v}_2\rho_2 S_2 \qquad (20.39)$$

Since $\rho_1 = \rho_2$, then

$$\bar{v}_2 = \frac{\bar{v}_1 S_1}{S_2} \qquad (20.39a)$$

Therefore, substituting Equation 20.39a into Equation 20.38 results in

$$\bar{v}_1^2\left(\frac{S_1^2}{S_2^2} - 1\right) = -2g_c[V(P_2 - P_1) + \Sigma F]\alpha$$

$$\bar{v}_1 = \sqrt{\frac{2g_c\left[\left(\frac{-\Delta P}{\rho}\right) - \Sigma F\right]\alpha}{\left(\frac{S_1^2}{S_2^2} - 1\right)}} \qquad (20.40)$$

The bracketed term within the square root represents a pressure decrease attributable to changes in kinetic energy between points 1 and 2. The friction term will include losses due to form friction and skin friction as has been shown in Equation 20.21. The manometer connecting points 1 and 2 will give a reading that is indicative of the pressure difference between these two points. This pressure difference is produced by two

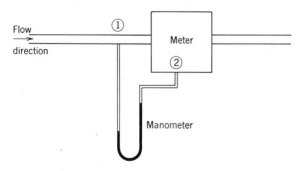

Figure 20.9. Pressure-difference type of fluid meter.

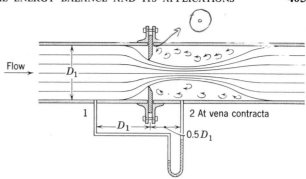

Figure 20.10. Sharp-edged orifice meter.

effects: (1) the change in kinetic energy caused by a change in velocity because of a change in flow area, and (2) permanent pressure loss due to skin and form friction. The manometer indicates only the total pressure difference between these points, and does not differentiate between the two effects. Equation 20.40 is referred to as the general meter equation. It could be written in terms of $\bar{v}_2$ or in terms of the mass rate of flow by the appropriate use of the continuity equation. The general equation will now be applied to specific types of flow meters.

The Orifice Meter. An orifice meter is an extremely simple apparatus; it normally consists of a flat plate with a centrally drilled hole. The drilled plate is inserted perpendicularly to the flow direction and the fluid passes through the hole. A schematic representation of an orifice meter along with the fluid stream lines is shown in Figure 20.10.

By the nature of the streamlines, it is apparent that boundary-layer separation occurs on the downstream side of the orifice plate. Consequently, the pressure loss from form friction is likely to be considerable, and, in fact, the orifice plate is a meter which maximizes form drag. The flow lines actually reach a minimum cross-sectional area one half to two duct diameters downstream from the plate. This point is known as the *vena contracta*. The location of the vena contracta relative to the orifice plate is a function of the fluid velocity as well as the relative diameters of the orifice and the duct. The position of the downstream pressure tap should approximate the vena contracta to ensure a maximum reading on the pressure-difference indicator.

The application of Equation 20.40 to the orifice plate permits the calculation of the fluid velocity and hence of the mass-flow rate through the duct. Some modification of this equation is necessary to make it more useful. Consider again the bracketed term of Equation 20.40. The term $(-\Delta P/\rho)$ represents the *total* pressure difference between points 1 and 2, and the ΣF term includes all fluid friction between the same two points. The difference between these terms represents the kinetic-energy change between points 1 and 2 as indicated by Equation 20.38. This difference can therefore be expressed as a fraction of the total pressure difference $(-\Delta P)$, or

$$- \frac{\Delta P}{\rho} - \Sigma F = C_1^2 \left(-\frac{\Delta P}{\rho} \right) \qquad (20.41)$$

where C_1^2 is a proportionality factor always less than one. Equations 20.40 and 20.41 may be combined to give

$$\bar{v}_1 = C_1 \alpha \sqrt{\frac{2g_c \left(\dfrac{-\Delta P}{\rho} \right)}{\left(\dfrac{S_1^2}{S_2^2} - 1 \right)}} \qquad (20.42)$$

Although the pressure tap at point two can be approximately located at the vena contracta, it is quite difficult to know with any reliability the cross-sectional area of the vena contracta. However, by geometry,

$$S_2 = C_2 S_0 \qquad (20.43)$$

where C_2 = a constant of geometry
S_0 = cross-sectional area of the orifice hole, sq ft

Therefore combining Equation 20.43 with Equation 20.42 gives

$$\bar{v}_1 = C_1 \alpha \sqrt{\frac{2g_c \left(\dfrac{-\Delta P}{\rho} \right)}{\dfrac{S_1^2}{C_2^2 S_0^2} - 1}} \qquad (20.44)$$

Now, define a coefficient C_0 such that

$$C_0 \sqrt{\frac{2g_c \left(\dfrac{-\Delta P}{\rho} \right)}{\dfrac{S_1^2}{S_0^2} - 1}} = C_1 \alpha \sqrt{\frac{2g_c \left(\dfrac{-\Delta P}{\rho} \right)}{\dfrac{S_1^2}{C_2^2 S_0^2} - 1}} \qquad (20.45)$$

This new coefficient C_0 is referred to as the *orifice coefficient*, and the final orifice equation may be written as

$$\bar{v}_1 = C_0 \sqrt{\frac{2g_c \left(\dfrac{-\Delta P}{\rho} \right)}{\dfrac{S_1^2}{S_0^2} - 1}} \qquad (20.46)$$

Bear in mind that the orifice coefficient is a complex function of pressure drop and geometry.

It is sometimes convenient to use mass flow instead of velocity. Equation 20.46 is directly converted into the mass form by applying the continuity equation. Since $\bar{v}_1 = w/S_1\rho$

$$w = \rho C_0 S_1 \sqrt{\frac{2g_c \left(\dfrac{-\Delta P}{\rho} \right)}{\dfrac{S_1^2}{S_0^2} - 1}} \qquad (20.47)$$

Either Equation 20.46 or Equation 20.47 may be used to calculate the flow rate using an orifice meter if the pipe dimensions, orifice dimensions, the pressure drop across the orifice, and the orifice coefficient are known. The orifice meter measures *average* velocity.

The orifice coefficient is most appropriately correlated as a function of the Reynolds number *through the orifice* and the ratio of the diameters of the orifice to the pipe. This practice seems justified in that the orifice equation includes the diameter ratio and the total pressure drop across the orifice, part of which is due to fluid friction. This correlation is shown in Figure 20.11 for sharp-edged orifices. When the Reynolds number through the orifice exceeds 30,000, the orifice coefficient is approximately constant at a value of 0.61, independent of the diameter ratio. Figure 20.11 is correct only if the pressure taps are located as shown in Figure 20.10; for flange taps it is usable. For other locations or for round-edged orifices it should not be used. If the orifice is located in the piping system in such a way as to alter the flow lines through the orifice, the orifice coefficient will be different from that predicted from Figure 20.11. It is recommended that the orifice should be located at least fifty pipe diameters downstream and ten pipe diameters upstream from any disturbances (2). If these undisturbed lengths of piping system are not available, straightening vanes may be used upstream from the orifice.

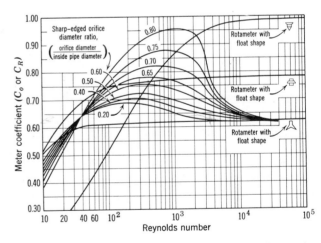

Figure 20.11. Meter coefficients for sharp-edged orifices and for rotameter (1, 9). From Brown, G. G., and Associates, Unit Operations, John Wiley and Sons, New York, 1950, with permission.)

For orifice: $\qquad N_{Re} = \dfrac{D_o \bar{v}_o \rho}{\mu}$

For rotameter: $\qquad N_{Re} = \dfrac{D_{eq} \bar{v}_2 \rho}{\mu}$

D_o = orifice diameter

D_{eq} = equivalent diameter of annular opening between tube and float.

The orifice meter, although simple, has one serious disadvantage in that a large percentage of the pressure drop across the orifice is unrecoverable. The fluid velocity is increased at the orifice opening without much loss of energy, but, as the fluid issues from the opening and starts to slow down, much of the excess kinetic energy is lost. The permanent pressure loss is a function of the ratio of the orifice and pipe diameter as shown in Figure 20.12.

When compressible fluids are metered by an orifice, the pressure drop across the orifice may be a significant fraction of the total pressure of the system. In this case, the specific volume of the fluid and also the fluid velocity will change appreciably between the two pressure taps. For this situation, the orifice equations as developed earlier are inappropriate. Instead, the energy balance must be written in the differential form and integrated between the two pressure taps, accounting for variation in specific volume and velocity. This will entail the use of some thermodynamic equation of state relating the PVT characteristics of the system.

Erroneous results will also be obtained when the orifice equations are used for situations in which pulsating flow is encountered if the frequency of pulsations is considerably greater than the natural frequency of the manometer. The errors are caused by the manometer or other pressure-measuring device indicating an arithmetic average of the differential pressure. The net result is that the indicated differential pressure will be greater than that of the actual flow since the flow varies with the square root of the differential pressure. If the period of pulsations approaches that of the manometer, oscillations will preclude accurate reading.

Illustration 20.6. A lube oil is flowing through a 5-in. Sch. 40 steel pipe at a rate of 300 gal/min. Inserted in this pipe is a 3.5-in. standard sharp-edged orifice to which is attached a mercury manometer. At the flow temperature, the oil has a specific gravity of 0.87 and a viscosity of 15 centipoises. If the manometer tube is inclined at an angle of 30° to the horizontal, what would be the manometer reading measured along the sloping leg?

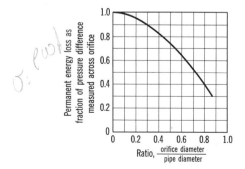

Figure 20.12. Permanent energy loss in sharp-edged orifice (2).

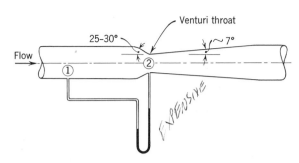

Figure 20.13. Venturi meter.

SOLUTION. Equation 20.46 is appropriate for use in this solution, and its use will require the determination of C_o. To determine the orifice coefficient entails the evaluation of the Reynolds number through the orifice and the use of Figure 20.11.

Orifice area: $\pi D_o^2/4 = 0.785\left(\dfrac{3.5}{12}\right)^2 = 0.0667$ sq ft

Pipe area $= 0.139$ sq ft

Linear velocity through the orifice:

$$\left(300\ \frac{\text{gal}}{\text{min}}\right)\left(0.1338\ \frac{\text{cu ft}}{\text{gal}}\right)\left(\frac{1}{60}\right)\left(\frac{1}{0.0667}\right) = 10.0\ \text{ft/sec}$$

Orifice Reynolds number:

$$(N_{\text{Re}})_o = \frac{D_o \bar{v}_o \rho}{\mu} = \frac{(3.5/12)(10.0)(62.3 \times 0.87)}{(15 \times 6.72 \times 10^{-4})} = 15,650$$

$$\frac{D_o}{D_1} = \frac{3.5}{5.047} = 0.695$$

From Figure 20.11, $C_o = 0.635$. The linear velocity of the oil in the pipe can be evaluated from the continuity equation.

$$\bar{v}_1 = \frac{\bar{v}_o S_o}{S_1} = 10.0(0.695)^2 = 4.83\ \text{ft/sec}$$

Solving Equation 20.46 for $-\Delta P/\rho$ gives

$$\frac{-\Delta P}{\rho} = \frac{\bar{v}_1^2\left(\dfrac{S_1^2}{S_o^2} - 1\right)}{C_o^2 2g_c} = \frac{(4.83)^2\left[\left(\dfrac{0.139}{0.0667}\right)^2 - 1\right]}{(0.635)^2\, 2 \times 32.2}$$
$$= 2.96\ \text{ft of oil}$$

or $-\Delta P = (2.96)(62.3 \times 0.87) = 161\ \text{lb}_f/\text{sq ft}$

From Equation 20.36,

$$R = \frac{-\Delta P}{\rho_H - \rho_L} = \frac{161}{62.3(13.6 - 0.87)} = 0.203\ \text{ft}$$

This manometer reading would result if the manometer were vertical. Since it is inclined 30°, the reading measured along the sloping leg is $2 \times 0.203 \times 12 = 4.86$ in.

The Venturi Meter

A properly designed venturi meter (Figure 20.13) is one in which form friction is minimized. The streamlined shape of the meter virtually eliminates boundary-layer separation so that form drag is negligible. The

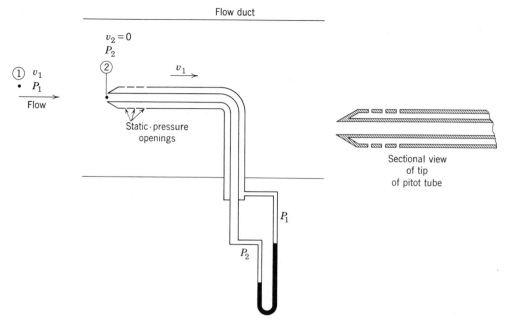

Figure 20.14. Pitot tube.

converging cone is about 25 to 30°, and the diverging cone should not exceed 7°. Venturi meters are difficult and expensive to manufacture and in large sizes are very bulky. The permanent pressure loss in a Venturi is about 10 per cent of the total pressure drop across the meter, substantially less than in an orifice meter. Equations 20.46 and 20.47 are directly applicable to a venturi meter if the coefficient C_0 is replaced by C_v, which has a value of about 0.98 for commonly encountered conditions. A Venturi meter also measures an average velocity.

The Pitot Tube. A pitot tube is a device which measures a point velocity. It normally consists of two concentric tubes arranged parallel to flow. The outer tube is perforated with small holes which lead into the annular space and are perpendicular to the flow direction. The annular space is otherwise sealed except for a manometer lead. The inner tube has a small opening which is pointed into the flow. This tube is connected to the other side of the manometer. There is no fluid motion within the pitot tube. The annular space serves to transmit the static pressure. The flowing fluid is brought to rest at the entrance to the inner tube, and the tube transmits an impact pressure equivalent to the kinetic energy of the flowing fluid. Figure 20.14 illustrates a pitot tube.

Neglecting potential energy changes between points 1 and 2, the energy balance may be written between these two points of Figure 20.14. At point 2, the fluid velocity is zero so the energy balance becomes

$$(v_1)^2 = 2g_c\left[\left(\frac{-\Delta P}{\rho}\right) - \Sigma F\right] \qquad (20.48)$$

Equation 20.48 may be combined with Equation 20.41 and rearranged to give

$$v_1 = C_P\sqrt{2g_c\left(\frac{-\Delta P}{\rho}\right)} \qquad (20.49)$$

The coefficient C_P of Equation 20.49 is usually unity for a well-designed pitot tube. This means that fluid friction between 1 and 2 is very small and that the pressure drop measured by a pitot tube is attributable only to a kinetic-energy change.

The pitot tube measures only a point velocity, and to get accurate values requires a well-designed instrument and perfect alignment of the instrument with the flow. An error of less than 1 per cent is possible with a pitot tube for a wide range of Reynolds numbers. This error may be contrasted with the orifice and venturi meter which in normal practice give results that have an error of about 2 per cent.

To get the average velocity of a flowing fluid with a pitot tube requires a point-to-point traverse across a duct diameter. Consider a circular duct of radius r_1 across which a pitot tube traverse is to be made. The average velocity through this duct may be defined as

$$\bar{v} = \frac{\text{volumetric flow rate}}{\text{cross-sectional area}} = \frac{Q}{\pi r_1{}^2} \qquad (20.50)$$

where Q = volumetric flow rate of the fluid, cu ft/sec. The quantity of fluid flowing through any ring of radius r is $2\pi r v\, dr$, where v is the velocity at radius r. The total flow is then

$$Q = \int_0^{r_1} 2\pi r v\, dr \qquad (20.51)$$

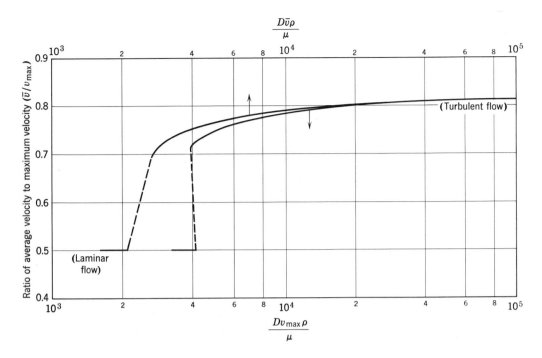

Figure 20.15. Relationship between average velocity and maximum velocity for flow through circular ducts (4).

and

$$\bar{v} = \frac{Q}{\pi r_1^2} = \int_0^{r_1} \frac{2vr\, dr}{r_1^2} \qquad (20.52)$$

If point velocities are measured with the pitot tube at various radii from the center of the duct to the duct wall, the integration of Equation 20.52 will give the average velocity of the fluid flowing through the duct.

If the velocity distribution within the duct is normal, i.e., the velocity distribution is in agreement with the momentum-transfer equations, Figure 20.15 may be used. This graphical representation shows the relationship between average velocity and the maximum velocity for fluids flowing in circular ducts. The maximum velocity would be measured at the center of the duct. To assure normal velocity distribution within a circular duct requires a straight run of at least fifty duct diameters without fittings or other obstructions.

Area Meters. Fluid meters of the previous group indicate a flow by a change in pressure. An area meter, on the other hand, is one in which the pressure drop is constant and the reading is dependent upon a variable flow area. The fluid stream passes through a constriction which accommodates itself to the flow so that a constant pressure difference is maintained. The most important meter in this classification is the rotameter. The rotameter (Figure 20.16) consists of a plummet or float which is free to move inside a vertical tapered glass tube. The fluid enters the bottom of the tube, and as it flows upward it exerts a force on the bottom of the float. When the upward force on the plummet is equal to the gravitational force acting downward on the plummet, the plummet becomes stationary at some point in the tube. The area available for flow is the annulus between the tube walls and the plummet. The constant-pressure drop arises from two factors: the change in kinetic energy and fluid friction. Form friction is of major significance in this instance. Floats may be designed so that form friction may or may not be of significance. The "Stabl-Vis"* float, for example, is a float whose behavior is generally independent of viscosity. The behavior of a spherical float, however, is quite dependent upon viscosity.

Figure 20.17 schematically shows a rotameter with the float at its steady-state position. At this condition, the forces acting on the float must be balanced so that no net force acts to move the float. The forces present are a gravity force, (F_G) acting downward on the float; a buoyant force (F_B) acting so as to raise the float; and a drag force (F_D) resulting from form and skin friction for flow around the float. Thus, at steady state,

$$F_D = F_G - F_B = V_f \rho_f \frac{g}{g_c} - V_f \rho \frac{g}{g_c} = V_f (\rho_f - \rho) \frac{g}{g_c}$$

$$(20.53)$$

where V_f = volume of the float
ρ_f = density of the float
ρ = density of the fluid

---

* Registered trade mark, Fischer-Porter Co., Hatboro, Pa.

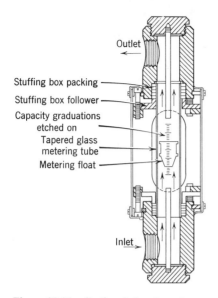

Figure 20.16. Sectional drawing of a rotameter.

The factors that contribute to the drag force can be analyzed by writing an energy balance between point 1 just before the float, and point 2 at the largest float cross section. Between these points, Equation 20.16a reduces to

$$\frac{P_2 - P_1}{\rho} + \frac{\bar{v}_2^2 - \bar{v}_1^2}{2g_c\alpha} + \Sigma F = 0 \qquad (20.54)$$

Using the continuity equation, Equation 20.54 can be rearranged to give

$$\bar{v}_1^2\left(\frac{S_1^2}{S_2^2} - 1\right) = -2g_c\alpha\left[\frac{P_2 - P_1}{\rho} + \Sigma F\right] \qquad (20.54a)$$

As with the orifice, the bracketed term in Equation 20.54a can be represented by Equation 20.41, so that

$$\bar{v}_1 = C_1 \sqrt{\frac{2g_c\left(\dfrac{-\Delta P_{12}}{\rho}\right)}{\left(\dfrac{S_1^2}{S_2^2} - 1\right)}} \qquad (20.55)$$

The pressure difference acting on the maximum float cross section is not identical to $(-\Delta P)_{12}$ because some fraction of this is recovered as the stream returns to the full tube diameter, and the velocity decreases proportionally. The recovery fraction is likely to be small, and, like the permanent pressure loss for the orifice, it is not dependent upon the velocity. Then

$$F_D = (-\Delta P_f)S_f = (-\Delta P_{12})C_f^2 S_f \qquad (20.56)$$

where $(-\Delta P_f)$ is the pressure drop acting on the top of the float and C_f^2 is the fraction of the maximum pressure drop $(-\Delta P_{12})$ that is not recovered to act on the top of the float. Therefore Equation 20.55 when

combined with Equations 20.53 and 20.56 gives

$$\bar{v}_1 = C_R \sqrt{\frac{2g_c V_f(\rho_f - \rho)\dfrac{g}{g_c}}{S_f\rho\left(\dfrac{S_1^2}{S_2^2} - 1\right)}} \qquad (20.57)$$

where C_R is the rotameter coefficient and is equal to C_1/C_f. Since the tube of a practical rotameter tapers very gradually, $S_f = S_1 - S_2$, and Equation 20.57 simplifies to

$$\bar{v}_1 = C_R \frac{S_2}{S_f} \sqrt{\frac{2g V_f(\rho_f - \rho)}{\rho(S_1 + S_2)}} \qquad (20.58)$$

The rotameter coefficient (C_R), like the orifice coefficient, is a function of the Reynolds number through the minimum cross section. It is also a function of the shape of the rotameter float. Coefficients are given in Figure 20.11 for rotameters with floats of three different shapes. These coefficients apply to Equation 20.59 which is a simplified equation obtained by assuming that S_2 is very small in comparison to S_f and by converting Equation 20.58 into the mass-flow form through

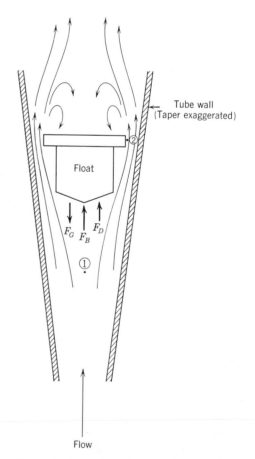

Figure 20.17. Schematic representation of rotameter.

the use of the continuity equation.

$$w = C_R S_2 \sqrt{\frac{2gV_f(\rho_f - \rho)\rho}{S_f}} \qquad (20.59)$$

Proper design of the rotameter float will make the coefficient C_R constant over wide ranges of Reynolds numbers. This constancy of C_R is particularly useful if the fluid being metered is subject to wide viscosity variations while its density is nearly constant.

For a fluid of fixed density in a single rotameter, the terms within the square root symbol of Equation 20.58 are nearly constant and independent of the flow rate. Combining these terms with C_R and S_f to give a new constant (C_R') yields

$$\bar{v}_1 \approx C_R' S_2 \qquad (20.58a)$$

which shows that the flow velocity is a constant times the minimum cross section for flow. Equation 20.58a emphasizes the great advantage of a rotameter relative to an orifice, venturi, or pitot tube. The rotameter calibration curve can be varied in shape at the desire of the designer by varying the tube geometry and float density. The square-root relation no longer applies. Usually the rotameter is designed to give a nearly linear calibration, so that instrument sensitivity is constant throughout the entire operating range.

The major disadvantage of the rotameter is that it becomes prohibitively expensive in the larger sizes, so that rotameters are most often used in installations that have pipe sizes less than 2 in. Another disadvantage is that inherently the flow rate must be read at the instrument rather than at a central control panel. This disadvantage can be overcome by any of several remote-reading methods, but these methods also increase the cost.

Positive Displacement Meters. The meters described above depend upon a pressure drop to measure the flow. There is a group of meters which indicate flow by means of mechanical devices. Metal parts are in contact with a flowing fluid, and the parts integrate in various manners so that a *total* discharge is indicated. This

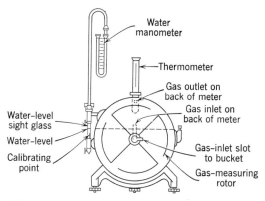

Figure 20.18. Sectional view of wet-test gas meter.

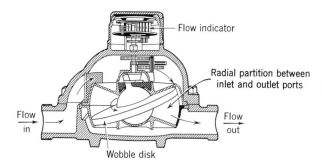

Figure 20.19. Sectional view of a disk meter. (Courtesy Neptune Meter Co.)

discharge can be timed over some convenient interval to determine the flow rate. Prominent among this group are the wet and dry gas meter and the rotary disk meter.

The wet gas meter, Figure 20.18, consists of a segmented drum which is caused to rotate by the flow of gas. Gas enters under the liquid level, filling a segment, which causes the segment to rotate, expelling an equal volume of gas from another segment. The *dry gas meter* uses bellows with the number of oscillations of the bellows being recorded on a series of dials.

The rotary-disk meter that is commonly used to measure the water flow in households is a meter having a cylindrical measuring chamber with a flat diaphragm extending radially across the chamber. This diaphragm is pivoted in such a manner that it wobbles as a result of fluid action upon it. As seen in Figure 20.19, the liquid enters the measuring chamber at one side and forces the disk ahead of it to move in an oscillating manner. The disk then forces the fluid out the discharge end. A counter indicates the number of wobbles which can in turn be converted to volumetric flow.

Mass Meters. A group of meters that has been developed recently depends for its operation upon the mass of the fluid that flows through the meter. There is a distinct advantage in being able to meter directly the mass of material that might be flowing through a system. The direct measurement of mass avoids errors that are likely to be encountered in metering a fluid either volumetrically or by a velocity measurement with the subsequent conversion to volumetric flow rate. Such variables as temperature, pressure, and viscosity can often lead to considerable error in volumetric-flow measurements.

For a mass meter to be independent of all other variables, it must be based upon principles governed only by the mass of the flowing material. One device that does this has as its operating principle that of gyroscopic precession. In this type of meter, the fluid is caused to flow in a circular path. In so doing, the fluid establishes a torque that makes the instrument

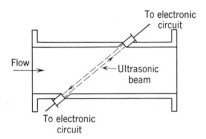

Figure 20.20. Ultrasonic flow meter.

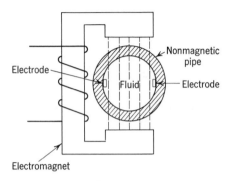

Figure 20.21. Magnetic flow meter.

precess (move about its axis) in proportion to the mass velocity of the flow. The torque that is developed is directly proportional to the mass rate of flow.

Another type of gyro meter depends upon Coriolis acceleration. In this meter, the fluid accelerates radially outward between vanes of an impeller. There is a tendency for the impeller to lag behind a casing that rotates with it, and this lag furnishes a measurable torque that is indicative of the mass-flow rate.

Ultrasonic Flow Meter. The ultrasonic flow meter passes a pulsed pressure from one side of a pipe to the other, and the retardation of the pulse by the flowing material is a measure of the flow rate. Figure 20.20 illustrates the ultrasonic flow meter. This meter is unique in that nothing need be placed within the flow duct for metering purposes. Two sonic beams are used: one transmitted diagonally upstream and the other diagonally downstream. Both beams are transmitted in short bursts. The pulses of energy from the two beams are integrated electronically and can be converted to flow rate. This meter can be used to meter practically all fluids.

Magnetic Flow Meter. As is shown in Figure 20.21, the magnetic flow meter consists of a nonmagnetic pipe through which the fluid flows; this pipe is enclosed by an electromagnet. The fluid flows through the magnetic field. In so doing, the liquid generates a voltage which is directly proportional to the flow through the meter. Its use is restricted to fluids having some electrical conductivity, such as liquid metals.

AGITATION OF LIQUIDS

Agitation is one of the oldest and commonest operations in chemical engineering, yet understanding of the subject is limited. The greatest limitation is the lack of a quantitative definition of agitation. Lacking a definition, no quantitative measurement is possible.

In a mechanical sense, an agitator may be required to do any of several operations. It may be required to disperse a miscible solute equally throughout a mass, such as the solution of a soluble solid in a liquid. In this instance, the agitator must impart motion to the liquid sufficient to maintain a concentration driving force between the solid interface and the bulk of the solution. Thus, the solution of crystals resolves into a mass-transfer phenomenon. Another agitator may be required to blend a miscible liquid into the contents of a tank. The blending of a solution into another miscible solution resolves into a simple matter of circulating the fluid until the blend is effected. A typical agitator for simple mixing is shown in Figure 20.22.

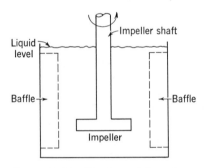

Figure 20.22. Simple paddle mixer, may be used for blending or dissolving. Such mixers may or may not be equipped with baffles.

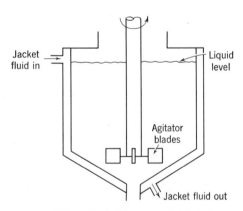

Figure 20.23. Jacketed reactor with four-blade turbine agitator. Hot or cold fluid circulates through the jacket and the agitator mixes the contents and produces forced convection for heat transfer along the walls.

An agitator may be required to produce and maintain a slurry of a solid that would otherwise settle without this motion, as in the storage of certain thin inks and paints. On the other hand, an agitator may be called upon to aid in the compacting or compression of a solid cake, as the rake does in a thickener. Since the two requirements cited are direct opposites, certainly the mechanical features of the agitators will be different.

In batch chemical reactors, especially in the field of organic technology, the proper operation of the reactor depends upon rapid dispersion of reactants and products through the mass and rapid transfer of heat of reaction to or from the surroundings in order to maintain a predetermined temperature in the batch. In this case, the motion imparted to the fluid will simultaneously circulate the batch and move fluid over a heat-transfer surface at sufficient rate to maintain a temperature driving force. An agitated jacketed kettle is shown in Figure 20.23.

In the preparation of emulsions or the dispersion or wetting of pigments, extremely high local mechanical shear may be required to create new liquid-liquid or liquid-solid surface. A high-shear agitator for emulsification and dispersion is shown in Figure 20.24.

The foregoing are just a few of the assignments given to agitation devices, and it should be evident that the mechanical features built into these devices are as varied as the assignments given to them. An excellent description of the varied equipment is available (8).

Fluid Mechanics of Agitation. The agitation of fluids in vessels is an operation based on the principles of fluid mechanics and consequently should be resolvable using principles described in this chapter. An agitator in a baffled tank as shown in Figure 20.25a will maintain a uniform fluid level irrespective of the speed of rotation of the agitator. However, if the tank is not baffled, a *vortex* will form at a high enough speed of rotation as shown in Figure 20.25b. The *vortex* occurs when the fluid level increases along the radius at higher agitator speeds. The liquid level pattern shown in Figure 20.25b is that of a typical vortex. Vortex formation is a phenomenon that can occur at an un-

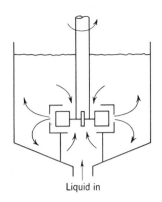

Figure 20.24. Four-blade turbine with stationary deflector ring. High turbine speed and close clearance of the deflector ring produce high shear rates for emulsification or pigment dispersion. Material enters along the shaft and leaves through holes in the deflector ring.

bounded, or free, surface. It is related to wave motion on bodies of water, which can only occur at an unbounded surface.

The analysis of agitation can be approached in terms of the energy equation. For unit mass of fluid

$$\Delta\left(\frac{\bar{v}^2}{2g_c\alpha}\right) + \Delta z\frac{g}{g_c} + \Delta(PV) - \int_{V_1}^{V_2} P\,dV + \Sigma F = -W_f'$$

$$(20.16)$$

In the usual sense, the energy equation relates changes between points in a flowing system. In the over-all sense, the agitator does not constitute a flowing system; therefore, all position-dependent terms go to zero, and the equation becomes

$$-\Sigma F = W_f' \qquad (20.60)$$

and therefore only the friction or work need be known.

The analysis of the *baffled tank* is the simpler of the two cases and will be considered first. The motion of

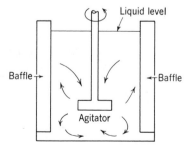

(a) Agitator in a baffled tank.

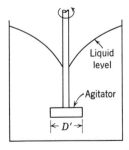

(b) Agitator in an unbaffled tank.

Figure 20.25.

the fluid over the agitator blade and the consequent motion of the fluid past the stationary baffles must have associated with them skin friction and form drag, depending upon the speed and design. Therefore, for a geometrically similar* system of agitator, tank, and baffles, the drag coefficient–Reynolds number relationship may be written

$$C_D = \Phi N_{\text{Re}} \qquad (20.60a)$$

where C_D = the drag coefficient

N_{Re} = the Reynolds number

These terms may be expanded into groups of variables for further consideration.

$$\frac{2Fg_c}{Sv^2\rho} = \Phi \frac{L'v\rho}{\mu} \qquad (20.61)$$

where F = the force applied to the solid surface by the fluid motion

S = the projected area of the solid normal to the fluid motion

L' = a characteristic dimension

v = a characteristic velocity of the fluid

ρ = the density of the fluid

μ = the absolute viscosity of the fluid

The variables in Equation 20.61 with the exception of fluid properties are not in a form that is easily measurable for agitation; therefore, before Equation 20.61 can be applied, some modification is in order. In a geometrically similar system, all length dimensions bear a constant ratio; therefore, any measurable dimension may be used to replace L'. The agitator diameter D' will be used in any instance in which a characteristic length is required.

The term F/S, the force applied to the fluid per unit area, can be used more conveniently if the term is converted to power input. Recall that power is the product of force and velocity of application; then,

$$\frac{vF}{S} = \frac{\mathbf{P}}{S} \qquad (20.62)$$

where $\mathbf{P}$ = power, ft lb$_f$/sec. The area term may be written in terms of the characteristic length; that is, the area is proportional to D'^2.

$$\frac{F}{S} \propto \frac{\mathbf{P}}{vD'^2} \qquad (20.63)$$

The velocity (v) is a linear velocity. In this system, the tangential tip velocity of the agitator is proportional to linear velocity so that

$$v \propto \pi ND' \propto ND' \qquad (20.64)$$

---

* See Appendix A.

where N = speed, rps. Equations 20.63 and 20.64 may be substituted into Equation 20.61.

$$\frac{\mathbf{P}g_c}{N^3 D'^5 \rho} = \Phi \frac{D'^2 N\rho}{\mu} \qquad (20.65)$$

or

$$N_{\text{Po}} = \Phi N_{\text{Re}} \qquad (20.66)$$

where $N_{\text{Po}} = \mathbf{P}g_c/N^3 D'^5\rho$, the drag coefficient for agitation systems referred to as the "power number."

$N_{\text{Re}} = D'^2 N\rho/\mu$, the Reynolds number expressed in variables convenient for agitation

This equation is written expressly for geometrically similar agitator tank systems in which the tanks are *baffled*.

Several typical plots of N_{Po} as a function of N_{Re} are shown in Figure 20.26. Note that, for baffled tanks, the two limits of the curve have characteristics universal for all C_D–N_{Re} plots, namely, at low Reynolds numbers,

$$N_{\text{Po}} = \frac{\text{constant}}{N_{\text{Re}}} \qquad (20.67)$$

and, at high Reynolds numbers,

$$N_{\text{Po}} = \text{constant} \qquad (20.68)$$

The region between that defined by Equations 20.67 and 20.68 represents transition of the boundary layer and development of significance of the accelerative effects associated with form drag.

In *unbaffled tanks*, the formation of vortex introduces an additional mechanism, namely, the fluid forces associated with gravity. In the vortex a part of the tank contents is supported against the gravitational acceleration of the earth, and therefore fluid forces in the system must supply the force to maintain the head of fluid that constitutes the vortex. The nature of the forces can be analyzed if an element of fluid is examined in the vortex, as shown in Figure 20.27. The fluid element is taken at z_0, the lowest point of the vortex, and therefore, at any radial position, a "head" of fluid will exist above the element. At steady state, the pressure on any element in the fluid must be such that the forces on every face of unit area of the element are the same. As elements are examined at z_0, from $r = 0$ to $r = r_1$ the pressure must be increasing because an increasing head of fluid is present. The force associated with the head is represented by F_b, the *body force*, and the radial force is represented by F_c, and surely

$$F_b = F_c \qquad (20.69)$$

The force F_c then is obviously a *centrifugal force* resulting from the motion of the fluid through the volume

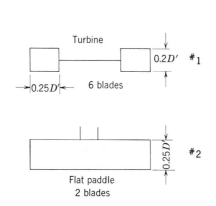

Type of impeller	$\dfrac{D_t}{D'}$	$\dfrac{z_1}{D'}$	$\dfrac{z_i}{D'}$	Baffles No.	Baffles w/D	No.
See #1	3	2.7–3.9	0.75–1.3	4	0.17	1
See #1	3	2.7–3.9	0.75–1.3	4	0.10	2
See #1	3	2.7–3.9	0.75–1.3	4	0.04	4
Same as #1, curved 2 blades	3	2.7–3.9	0.75–1.3	4	0.10	3
Marine propeller 3 blades, pitch = D'	3	2.7–3.9	0.75–1.3	4	0.10	5
See #2	3	2.7–3.9	0.75–1.3	4	0.10	6
Same as 5 but pitch = $2D'$	3	2.7–3.9	0.75–1.3	4	0.10	7

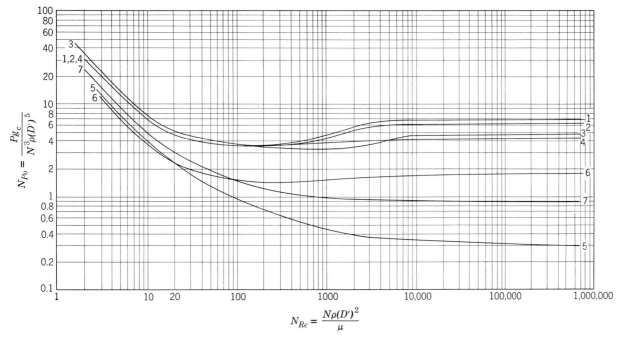

D' = diameter of impeller
D_t = diameter of tank
w = width of baffle
z_i = elevation of impeller
z_1 = height of liquid

Pitch = the axial distance a free propeller would move through a non-yielding liquid during one revolution

Figure 20.26. Plots of N_{Po} as a function of N_{Re} for various agitator-tank systems. Baffled systems (8).

element. Therefore, each of the forces can be related to system variables. The body force on the element $\Delta x\,\Delta y\,\Delta z$ is

$$F_b = \Delta x\,\Delta y\,(z - z_0)\,\rho\,\frac{g}{g_c} \qquad (20.70)$$

and the centrifugal force on the element is

$$F_c = \frac{\Delta x\,\Delta y\,\Delta z\,\rho r\omega^2}{g_c} \qquad (20.71)$$

where $\Delta x, \Delta y, \Delta z$ = the element dimensions, ft
$z - z_0$ = the height of fluid above the element, ft
r = the radial position, ft
ω = the angular velocity, sec^{-1}

These forces can be equated.

$$\Delta x\,\Delta y\,(z - z_0)\rho\,\frac{g}{g_c} = \frac{\Delta x\,\Delta y\,\Delta z\,\rho r\omega^2}{g_c} \qquad (20.72)$$

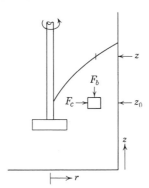

Figure 20.27. Forces associated with an element of fluid in a vortex.

ρ, Δx, Δy, and g_c cancel from the equation so that

$$(z - z_0)g = \Delta z\, r\omega^2 \qquad (20.73)$$

The right-hand term of the equation is related to the kinetic energy of the fluid, and the left-hand side is related to forces associated with gravity. This equation may be written

$$\frac{\Delta z\, r\omega^2}{(z - z_0)g} = 1 \qquad (20.74)$$

The exact numerical value of each term is not known, but, in a geometrically similar system, all length dimensions can be taken proportional, so that

$$\Delta z \propto (z - z_0) \sim r \sim D' \qquad (20.75)$$

and in a dynamically similar system

$$\omega \propto N \qquad (20.76)$$

Therefore, Equation 20.74 may be written as

$$\frac{D'N^2}{g} = \text{constant} \qquad (20.77)$$

for a dynamically similar system. The group $D'N^2/g$ is called the *Froude number*, (N_{Fr}) and represents the ratio

$$\left(\frac{\text{Forces associated with kinetic energy}}{\text{Forces associated with gravity}}\right)$$

and the group is dimensionless.

The additional gravitational effects in an unbaffled agitator system can be included in a correlation equation in terms of the Froude number so that

$$N_{Po} = \Phi\,(N_{Re})\,(N_{Fr}) \qquad (20.78)$$

Equation 20.78 may be expanded into system variables as

$$\left(\frac{P g_c}{N^3 D'^5 \rho}\right) = \Phi\left(\frac{D'^2 N\rho}{\mu}\right)\left(\frac{D'N^2}{g}\right) \qquad (20.79)$$

Typical drag diagrams for unbaffled tanks are presented in Figure 20.28, which is a plot of $N_{Po}/(N_{Fr})^{(a-\log N_{Re})/b}$ as a function of N_{Re} for Reynolds numbers above 300

and N_{Po} versus N_{Re} for Reynolds numbers below 300. The constants a and b are characteristic of each agitator tank system and are given in Figure 20.28.

The design of systems for agitation is usually carried out by the scale-up procedure described in Appendix A. Tests are performed in model or pilot-sized equipment until a satisfactory set of conditions is established, after which a geometrically similar unit is designed according to the same principles as those stated in this chapter. The data presented in graphical form in this chapter can be used directly for geometrically similar systems.

Illustration 20.7. Calculate the power required for agitation for a 3-blade marine impeller of 2-ft pitch and 2-ft diameter operating at 100 rpm in an unbaffled tank containing water at 70°F. The tank diameter is 6 ft, depth of liquid is 6 ft, and the impeller is located 2 ft from the tank bottom.

Data: $\mu = 0.93$ centipoise
$\rho = 62.4$ lb/cu ft

SOLUTION. The data given above fit the conditions for curve 4 in Figure 20.28. The Reynolds number may be calculated

$$N_{Re} = \frac{D'^2 N\rho}{\mu} = \frac{(2)^2(100/60)(62.4)}{(0.93 \times 6.72 \times 10^{-4})} = 6.66 \times 10^5$$

For unbaffled tanks with N_{Re} in excess of 300, the Froude number affects the agitation. Figure 20.28 is entered at $N_{Re} = 6.66 \times 10^5$ and

$$\frac{(N_{Po})}{(N_{Fr})^{(a - \log N_{Re})/b}} = 0.23$$

From the data table on Figure 20.28, $a = 2.1$, $b = 18$

$$\frac{a - \log N_{Re}}{b} = \frac{2.1 - 5.82}{18} = -0.206$$

Then

$$\frac{N_{Po}}{(N_{Fr})^{-0.206}} = 0.23$$

$$N_{Fr} = \frac{N^2 D'}{g} = \frac{(100/60)^2(2)}{(32.2)} = 0.173$$

$$N_{Po} = \frac{0.23}{(0.173)^{0.206}} = \frac{0.23}{0.696} = 0.33$$

$$N_{Po} = \frac{P g_c}{N^3 D'^5 \rho} = 0.33$$

$$P = \frac{N_{Po} N^3 D'^5 \rho}{g_c}$$

$$P = \frac{(0.33)(100/60)^3(2)^5(62.4)}{(32.2)}$$

$$P = \frac{(0.33)(4.63)(32)(62.4)}{32.2}$$

$$P = 94.8 \text{ ft-lb}_f/\text{sec}$$

$$P = \frac{94.8}{550} = 0.173 \text{ hp}$$

Type of Impeller	$\dfrac{D_t}{D'}$	$\dfrac{z_1}{D'}$	$\dfrac{z_i}{D'}$	a	b	No.
Marine propellers, 3 blades, pitch $= 2D'$	3.3	2.7–3.9	0.75–1.3	1.7	18	1
Same as #1 but pitch $= 1.05D'$	2.7	2.7–3.9	0.75–1.3	2.3	18	2
Same as #1 but pitch $= 1.04D'$	4.5	2.7–3.9	0.75–1.3	0	18	3
Same as #1 but pitch $= D'$	3	2.7–3.9	0.75–1.3	2.1	18	4

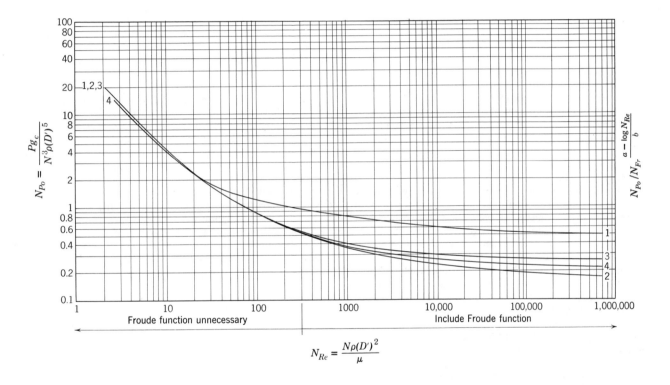

$$N_{Re} = \frac{N\rho(D')^2}{\mu}$$

D' = impeller diameter
D_t = tank diameter
z_i = elevation of impeller
z_1 = elevation of liquid

Figure 20.28. Plot of N_{Po} or $N_{Po}/N_{Fr}^{(a-\log N_{Re})/b}$ as a function of N_{Re} for unbaffled tanks (8).

REFERENCES

1. Fischer and Porter Co., Hatboro, Pa., Catalogs, section 98A, 1947.
2. *Fluid Meters: Their Theory and Application*, 4th ed., ASME, New York, 1937.
3. Kays, W. M., *Trans. ASME*, **72**, 1067 (1950).
4. McAdams, W. H., *Heat Transmission*, McGraw-Hill Book Co., New York, 1933.
5. Nikuradse J., *VDI-Forshungsheft*, 361, (1933).
6. Perry, J. H., *Chemical Engineering Handbook*, 3rd ed., McGraw-Hill Book Co., New York, 1950, section 5.
7. Perry, J. H., *Chemical Engineer's Handbook*, 3rd ed., McGraw-Hill Book Co., New York, 1950, section 17.
8. Rushton, J. H., E. Costich, and H. J. Everett, *Chem. Eng. Progr.*, **46**, 395, 467 (1950).
9. Tuve, G. L., and R. E. Sprenkel, *Instruments*, **6**, 201 (1933).

PROBLEMS

20.1. A 500-watt heating element is used to heat 100 cu ft/min of air flowing into a 1-in., 16 BWG copper-tube heating section at 70°F and 1 atm gage. When the air leaves the heater its pressure is 0.1 atm gage. What is the temperature of the air leaving the heater?

20.2. Air is flowing through a 12-in. I.D. iron pipe which discharges into a 3-in. I.D. iron pipe. At a particular point in the larger pipe, the pressure is 15 psig, the temperature is 80°F, and the air velocity is 5.0 ft/sec, At a point in the smaller pipe, the pressure is 5 psig, the temperature is 80°F. The point in question in the 3-in. pipe is 100 ft above the point in the 12-in. pipe. Assuming perfect-gas behavior, calculate the energy lost due to friction between the two points.

20.3. A closed tank kept partially filled with oil (specific gravity = 0.9) has a pressure in the gas space above the oil of 10 psig. If the oil is discharging through a hose at 50 gal/min, estimate the pressure at the nozzle which is located 10 ft lower than the oil surface and has a 1-in. I.D. discharge. Assume a total frictional loss in the line amounts to 1.0 ft-lb_f/lb.

20.4. A research team is designing a flow system for a nuclear reactor to study corrosion problems. The equipment as constructed consists of 230 ft (total equivalent length) of 1-in. I.D. stainless-steel pipe. Molten bismuth is pumped from a melt tank maintained at 350°C and 5 microns absolute pressure, through a test section included in the 230 ft, and back to the melt tank. If the bismuth velocity is 1.0 ft/sec, how much theoretical power must be supplied to a pump placed in the line? Assume a constant temperature of 350°C is maintained by suitable insulation.

Liquid bismuth properties:
Viscosity = 1.28 centipoise
Density = 613 lb/cu ft.

20.5. A hydro-electric plant is supplied with water by a 4-mile long duct leading from a dam to the turbines. The duct is made of concrete and is 4 ft in diameter. The duct inlet at the dam is 40 ft below the water surface and 200 ft above the turbine entry. The turbine discharges to the atmosphere. What flow of water at 60°F can be expected to the turbine? How much power would the turbine develop for this flow? Pressure at the turbine inlet is 2 psig.

20.6. A summer home is to be supplied with water by a pipe line leading from a spring 600 ft above the home. The line will be comprised of 650 ft of straight, Sch. 40, 1-in. brass pipe, three 90° elbows, and one gate valve. The pipe line discharges into a vented tank. What water flow at 50°F could be expected?

20.7. A large elevated tank is used to supply water at 50°F to a spray chamber. To ensure proper atomization of the fluid, a pressure of at least 40 psig at the nozzle inlet must be maintained to deliver the required 150 gal/min. The line from the tank is 2-in. Sch. 40 steel pipe. In addition to its vertical run, the line has a 10-ft horizontal section and contains four 90° elbows and one gate valve. What is the minimum height above the spray nozzle at which the liquid level in the tank must be maintained?

20.8. Rework Problem 20.7 on the basis that glycerol at 70°F is stored and fed to a reactor through a similar piping system and delivered to the reactor at the same conditions as in Problem 20.7.

20.9. A copper smelter is located in a small city whose water pressure is insufficient to meet the company demands. It is decided that a large elevated tank will be erected to supply water for company needs. Requirements are: water at 65°F and a delivery pressure of 40 psia at 100 gpm. The line leading from the tank to the point of use is standard 2-in. steel pipe. In addition to its vertical run, the line has a 10-ft horizontal run and contains two 90° elbows and a gate valve. The tank is closed but vented to the atmosphere. At what height above the point of use must the bottom of the tank be?

20.10. A paint factory keeps its solvents stored in vented tanks on the second floor of its mixing building. One particular tank contains linseed oil which is fed through a pipe to a mix tank on the first floor through 75 ft of pipe (equivalent length of straight pipe). The bottom of the oil tank is 15 ft above the point of discharge into the open mix tank. Calculate the minimum pipe diameter which will ensure a flow of 10 gal/min to the mixer. Express pipe diameter to nearest commercial size of Sch. 40. pipe. Viscosity of linseed oil = 15 centipoises. Specific gravity of oil = 0.92.

20.11. Water at 55°F is to be pumped from a pond to the top of a tower 60 ft above the level of the water in the storage basin. It is desired to deliver 12 cu ft of water per minute at 15 psig. The transmission line consists of 400 ft of standard 3-in. steel pipe with eight 90° ells and four gate valves. What horsepower is required for pumping? What will be the electrical input to the motor if the motor-pump set has a 40 per cent efficiency?

20.12. A light lube oil (specific gravity = 0.87, μ = 3.0 centipoises is pumped by a rotary gear pump to a header at a rate of 1000 gal/hr. At the header, the flow system is branched into two lines. One is a 4-in. Sch. 40 steel pipe and the other an 8-in. Sch. 40 steel pipe. The smaller pipe is 250 ft long, whereas the larger pipe extends for 100 ft. Neglecting changes in elevation and pressure drop through fittings and the header, what is the volumetric flow rate of oil in each pipe?

20.13. A supply of gasoline at 69°F having a viscosity of 0.667 centipoises and a specific gravity of 0.76 is pumped through a 6-in. standard horizontal pipe at a rate of 500 gal/min. At the end of 200 ft, this pipe branches into three lines consisting of 3-, 2-, and 1-in. standard pipes respectively. If the pipes have respective lengths of 700, 325, and 125 ft and discharge at atmospheric pressure, what is the volume per cent of total flow through each branch?

20.14. Tank A is filled with a 10 per cent NaOH, 10 per cent NaCl solution (specific gravity = 1.10, viscosity = 3 centipoises). The solution is subsequently drained into a reaction tank B located as shown in the following flow diagram. If the outlet gate valve on tank A and the inlet gate valve on tank B are simultaneously fully opened, how long will it take to drain tank A down to a 1-ft level?

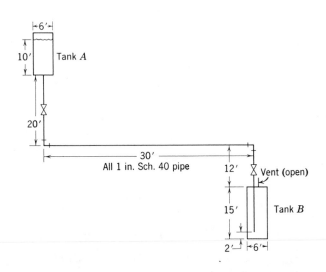

20.15. Two summer cottages receive their water supply from a dammed mountain stream. The piping system is as follows:

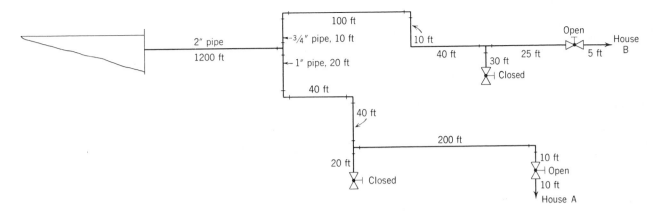

Elbows and tees as shown. All valves are globe valves.

The height difference from water surface to the faucet in house A is 50 ft, and that to the faucet in house B is 30 ft. What rate of flow will be delivered to house A if both faucets are wide open?

20.16. What is the mass velocity of air that can be handled in a $\frac{1}{2}$-in. Sch. 40 horizontal steel pipe, 300 ft long, if the air flows isothermally at 100°F through the pipe? The pressure drops from 50 to 5 psig through the pipe.

20.17. In cases where open liquid manometer columns would be unusually high or when the liquid under pressure cannot be exposed to the atmosphere, the inverted U-tube illustrated is sometimes used. Develop a general expression for the pressure difference between points 1 and 2 in terms of R, a, b, and the fluid densities.

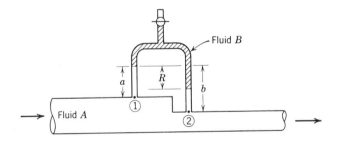

20.18. Show that the pressure-drop reading for the two-fluid manometer pictured in Figure 20.8 is given by

$$\Delta P = R\left[\rho_C - \rho_B + \frac{S_T}{S_B}(\rho_A - \rho_B)\right]$$

where
R = reading
S_T = cross-sectional area of the tubes
S_B = cross-sectional area of the bulbs
ρ_A, ρ_B, ρ_C = density of fluids A, B, and C respectively

20.19. A simple open-end U-tube, using mercury, is used to measure the pressure in a 2-in. pipe carrying CO_2 gas at 70°F. The mercury level at zero reading is 2 ft below the pressure tap.

(a) If the reading on the manometer is 1-in. Hg, what is the pressure in the line?

(b) If water were flowing in the line under the same pressure as the CO_2, what would the manometer reading be?

20.20. Petroleum oil of specific gravity 0.9 and viscosity 13 centipoises flows isothermally through a horizontal Sch. 40, 3-in. pipe. A pitot tube is inserted at the center of the pipe, and its leads are filled with the same oil and attached to a U-tube con-taining water. The reading on the manometer is 3.0 ft. Calculate the volumetric flow of oil in cubic feet per minute.

20.21. A duct traverse is made with a pitot tube of a 20.0-in. I.D. galvanized-iron duct through which air is flowing at 100°F. A water manometer is used in connection with the pitot tube. The following readings were obtained:

Position, (r), in.	ΔP in. water
0.0	3.67
3.0	3.27
5.0	2.67
7.0	1.90
8.0	1.40
9.0	0.80
9.75	0.198

Estimate the average flow rate of the air in cubic feet per minute.

20.22. A sharp-edged orifice in a thin plate has been calibrated with dry air at 70°F and substantially standard atmospheric pressure and a plot prepared from which the volume of air per minute referred to 32°F and normal barometer may be read directly. If this meter were used to measure the flow of dry CO_2 gas at 70°F and normal pressure, would it give high or low results? What correction factor would be necessary to make the plot usable?

20.23. It is desired to meter, by the installation of a sharp-edged orifice, a stream of approximately 500 lb/hr of air flowing at 70°F and 1.0 psig through a standard 4-in. iron pipe line. It is agreed that the orifice will have flange taps and that, for ease in reading the flow, a minimum pressure difference of 2.0 in. H_2O must exist between the two taps. What diameter do you recommend for the orifice which is to be installed? What is your estimate of the static pressure of the air flowing at a point 3 ft downstream from the orifice?

20.24. Brine (specific gravity 1.20) is flowing through a standard 3-in. pipe at a maximum rate of 185 gal/min. In order to measure the rate of flow, a sharp-edged orifice, connected to a simple U manometer is to be installed. The maximum reading of the manometer is to be 400 mm Hg. What size orifice should be installed? Express result to the nearest 1/8 in. Repeat, assuming a venturi meter is used instead of an orifice.

20.25. A standard 0.500-in. orifice is installed in a 2-in. standard steel pipe. Dry air at upstream conditions of 70°F and 15 psig flows through the orifice at such a rate that a U-tube manometer connected across the taps indicates a reading of 35 cm of red oil.

The red oil has a specific gravity of 0.831 referred to water at 60°F.

(a) Calculate the weight rate of air flow in the pipe.

(b) Estimate permanent head loss across the orifice.

(c) What per cent of the power requirement could be saved by using a venturi meter in place of the orifice (assuming a pressure loss from venturi meter equal to 20 per cent of head across venturi).

20.26. An oil of 0.87 specific gravity and 6 centipoises viscosity flows through a pipe line. An orifice with opening diameter of one-half of the inside pipe diameter is used to measure the flow. It is proposed to replace this orifice with a venturi with throat diameter equal to the orifice diameter. If the coefficient of the orifice is 0.61 and that of the venturi 0.98 and the flow rate is unchanged, calculate:

(a) The ratio of the venturi reading to the orifice reading.

(b) The ratio of the net pressure loss due to the venturi installations to that found with the orifice.

20.27. A standard 0.500 in.-flange-tap orifice is installed in a straight horizontal length of standard 2-in. steel pipe. The pipe carries a steady flow of dry air at 85°F. A vertical U-tube manometer indicates a differential head of 70 mm Hg when connected across the orifice. When the upstream leg of the manometer is disconnected from the orifice and left open to the atmosphere, the manometer reading indicates a static head of 10 mm Hg, gage, at the downstream tap. The barometric pressure is 745 mm Hg. Calculate the volumetric flow of dry air in standard cubic feet per minute.

20.28. A venturi meter with a 12-in. throat is inserted in a 24-in. I.D. line carrying chlorine at 70°F. The barometer is 29.5 in. Hg, the upstream pressure 2 in. Hg above atmospheric pressure, and the head measured over the venturi (upstream to throat) is 0.52 in. Hg. Calculate the rate of flow in pounds per hour.

20.29. It is proposed to measure a given flow of water for municipal purposes by the use of either an orifice having a $D_0/D_P = 0.3$ or a venturi. If it is desired to have the same reading for this particular flow on both instruments, how would the power loss compare in the two cases?

20.30. Using the agitator-tank system described in curve 1 of Figure 20.28, calculate the speed for an agitator that imparts 0.01 hp/cu ft of tank contents in which the fluid under test is water at 70°F, and the tank holds 50 cu ft. Calculate the power required for a 10 cu ft tank at the same Reynolds number.

$$\frac{z_1}{D'} = 3.0$$

20.31. Using the agitator-tank system described in curve 6 of Figure 20.26, calculate the speed required for a 20 cu ft tank in which 0.005 hp/cu ft is imparted to the tank. The fluid under test is SAE 10 oil. Calculate the power required for a 2 cu ft tank at the same Reynolds number, and both operating at 100°F.

$$\frac{z_1}{D'} = 3.5$$

20.32. Components for a liquid detergent ($\mu = 10$ centipoises) are blended in the pilot plant in a 10-gal, baffled, flat-bottomed tank 10 in. in diameter. A double-turbine agitator with blades 6 in. in diameter is used. A 1/2 hp motor turns the agitator at 500 rpm for 30 min to attain complete dispersion.

In the plant a geometrically similar unit is planned to blend 200-gal batches of this solution. Determine consistent values of agitator and tank diameter, revolutions per minute, power requirement, and batch time for the plant unit. Base the design on (a) constant N_{Re}, (b) constant agitator peripheral speed, (c) constant rpm.

chapter 21

Momentum Transfer II:
Pumps and Compressors

The energy balance may be used for the evaluation of the energy required for moving a fluid through a complex piping system. It is the purpose of this chapter to describe and discuss some of the equipment that is used for providing energy to transport fluids. In selecting a pump or blower for a specific assignment, the engineer is interested in several specific characteristics of the pump: the capacity, the energy or head supplied to the fluid, the power required to run the pump or blower, and the efficiency of the unit.

To obtain the best characteristics as stated above, the engineer requires certain information pertinent to the particular problem. He needs to know:

1. The nature of the fluid being transported. Is it corrosive? Is it hot or cold, and what is its vapor pressure? Is it viscous or nonviscous? Does it contain suspended solids?

2. The required capacity as well as the range of capacity that the pump or blower might be called upon to deliver.

3. The suction conditions. Is there a suction lift or a suction head?

4. The discharge conditions. What pressure is required? What is the fluid friction that must be overcome?

5. The type of service. Is it continuous or intermittent?

6. The nature of the power available to drive the pump.

7. The pump location. What space is available? What space is required?

CLASSIFICATION OF PUMPS

Pumps may be classed into two main groups, *positive-displacement* pumps and *centrifugal* pumps. Positive-displacement pumps may be either a reciprocating type or a rotary type. The prime feature of a positive-displacement pump is that a definite quantity of liquid will be delivered for each stroke or revolution of the prime mover. Only pump size, design, and suction conditions will influence the quantity of liquid that can be delivered. On the other hand, a centrifugal pump can deliver a variable volume of fluid with varying head for a constant speed.

POSITIVE-DISPLACEMENT PUMPS

Reciprocating Pumps. This type of pump adds energy to a fluid system by means of a piston acting against a confined fluid. The principles of fluid dynamics are of little importance since fluid flow will be determined by pump geometry. The piston may be driven by either a steam engine or an electrical motor. For each stroke of the piston, a fixed quantity of fluid will be discharged from the pump. The quantity of fluid will depend only upon the volume of the cylinder, and the number of times the piston moves through the cylinder. The actual delivery may be less than the swept-out volume of the cylinder because of leakage past the piston and failure to fill the cylinder. Thus, a volumetric efficiency may be defined as the ratio of actual discharge to the discharge based upon piston displacement. The efficiency for well-maintained pumps should be at least 95 per cent.

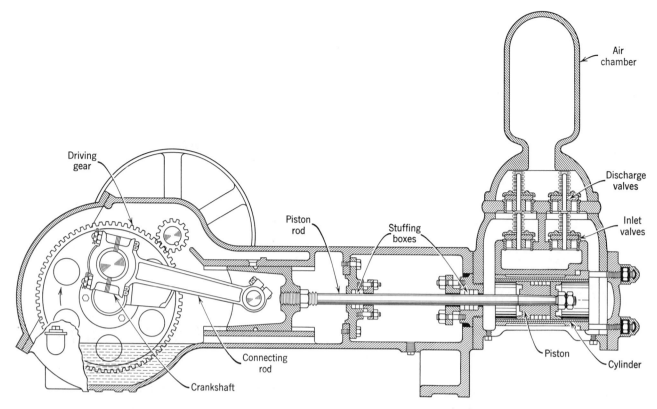

Figure 21.1. Cutaway view of a simplex piston pump. (Courtesy Novo Pump and Engine Co.)

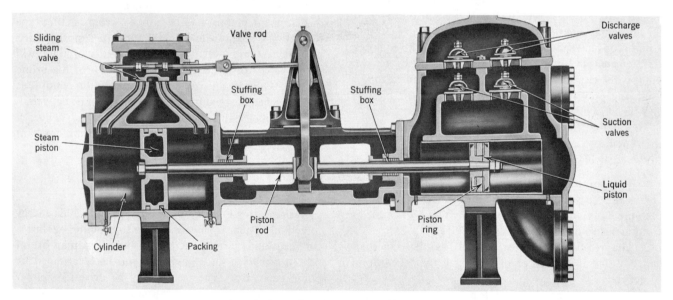

Figure 21.2. Steam-driven double-acting piston pump. (Courtesy Warren Pumps, Inc.)

Another efficiency definition, and one which is perhaps more significant, is the work done on the fluid, divided by the work done on the pump. If an electric motor is used to drive the pump, a pump-motor efficiency is usually employed. It may be defined as the ratio of the work done on the fluid to the electrical energy supplied to the motor.

In a reciprocating pump, while the piston is being withdrawn in the cylinder (liquid intake), the discharge of fluid from the pump ceases. Consequently, liquid delivery is in pulses. The pulses can be lessened by using a double-acting pump or increasing the number of cylinders. A double-acting pump takes advantage of the cylinder volume on both sides of the piston so that delivery will be almost the same for the forward and backward strokes.

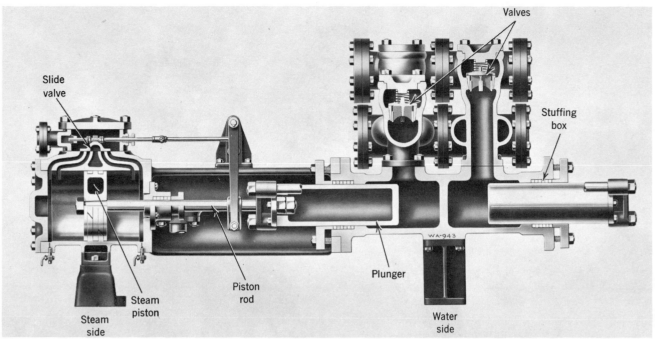

Figure 21.3. Steam-drive duplex plunger pump. (Courtesy Worthington Corp.)

Figure 21.1 illustrates a simplex power-driven piston pump. In this type of pump, the piston is connected to a suitable crank shaft which is driven by an electric motor. Steam-driven pumps are also used in the process industry. Figure 21.2 shows a steam-driven double-acting pump, where a common piston rod connects the steam piston and the liquid piston.

In these pumps, liquid enters the cylinder through a check valve which is opened by an external pressure acting on the fluid. The flow of the fluid through the inlet valves follows the piston movement backward through the cylinder on its inlet stroke. When the piston moves forward, the inlet valve closes, and a second valve is forced open to discharge the liquid. The piston must have a close fit with the cylinder walls in order to minimize liquid slip past the piston. If the piston carries with it its own packing, it is called a piston pump. In contrast to this, a plunger pump uses a close-fitting rod moving through the cylinder past fixed packing. The plunger is simply an extension of the shaft. Figure 21.3 shows a plunger pump.

Reciprocating pumps are particularly well suited for pumping viscous fluids because the high rate of shear acting on the cylinder walls serves as an additional "packing." This pump is also good for attaining high pressures, and, because of its positive-displacement characteristic, it is sometimes used for metering fluids. Liquids containing abrasive solids should not be pumped with a reciprocating pump because of damage to the machined surfaces.

Operating Features of Reciprocating Pumps. Discharge characteristics of reciprocating pumps are indicated in Figure 21.4. Fluid issues from the discharge valves until they close near the end of the stroke, when the piston stops and reverses. During part of the pumping cycle, the flow is zero; however, flow from the discharge line may be nearly constant depending upon pump design. Double-acting pumps will always have a flow from the discharge line. Duplex pumps have the discharge of one cylinder displaced half a stroke from the other. Thus, the total flow from the pump is the addition of the two, to give the solid line of Figure 21.4c. Nearly pulse-free flow can be attained by designing for duplex, triplex, or multiplex operation.

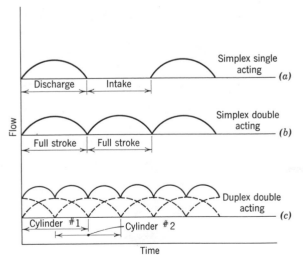

Figure 21.4. Discharge curves for reciprocating pumps. (a) Simplex single acting. (b) Simplex double acting. (c) Duplex double acting.

Figure 21.5. The principle of rotary-gear pumps. (Courtesy Roper Hydraulics, Inc.)

Figure 21.6. Exploded view of rotary-gear pump. (Courtesy Sier-Bath Gear and Pump Co.)

Figure 21.7. Helical-gear pump. (Courtesy Sier-Bath Gear and Pump Co.)

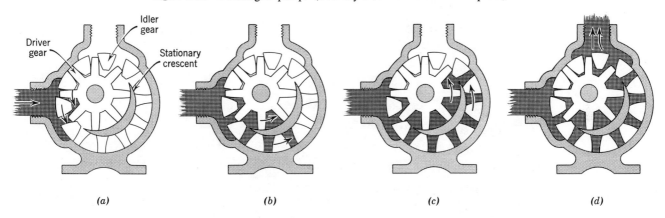

(a)   (b)   (c)   (d)

Figure 21.8. Internal gear pump and its operation. (Courtesy Viking Pump Co.)

The flow capacity of a reciprocating pump varies directly with speed. Current design has speeds between 20 and about 200 strokes per minute. Careful machining and maintenance can give this class of pumps a good

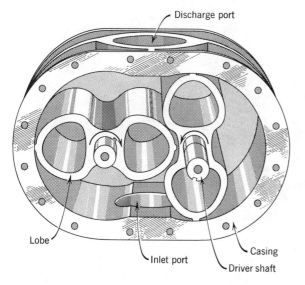

Figure 21.9. Sectional view of lobe pump.

efficiency. Some disadvantages in reciprocating pumps are in their size, high maintenance and initial costs. This type of pump is available in many designs so that a wide selection can be made.

Rotary Pumps. Another group of positive-displacement pumps is the rotary type. This class of pumps can be characterized by the method of taking in and discharging the fluid. Unlike a reciprocating pump, which depends upon check valves for control of entry and discharge, a rotary pump traps a quantity of liquid and moves it along towards the discharge point. This principle is indicated in Figure 21.5. The unmeshed gears at the pump inlet provide a space for liquid to fill. As the gears rotate, the liquid is trapped between the teeth and the pump casing and is eventually released at the discharge line. Rotary pumps can handle almost any abrasive-free liquid and are especially well suited for high-viscosity fluids. Some lubricating action by the fluid decreases wear.

Gear Pumps. Gear pumps are the simplest rotary type. Figure 21.5 is an external-gear pump, and Figure 21.6 shows an exploded view of the same type of pump. Notice its simple construction. Helical gears may be

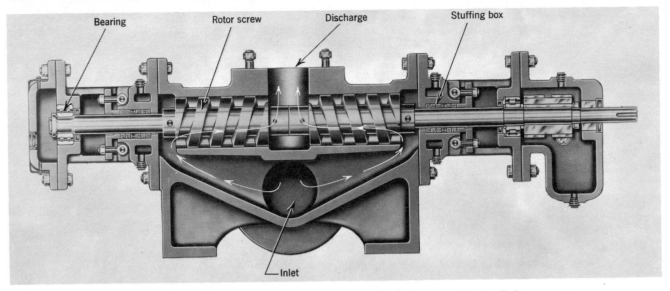

Figure 21.10. Single-screw pump. (Courtesy Sier-Bath Gear and Pump Co.)

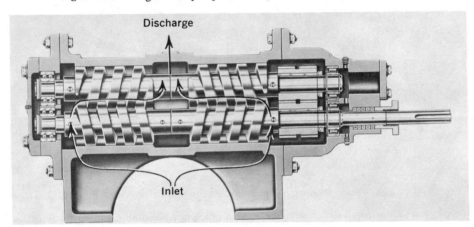

Figure 21.11. Double-screw pump (sectioned behind inlet channel). (Courtesy Sier-Bath Gear and Pump Co.)

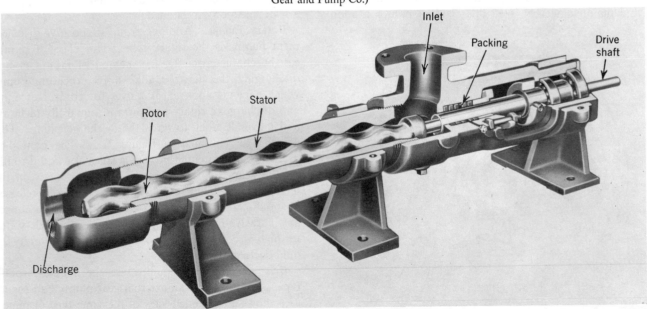

Figure 21.12. "Moyno" pump. (Courtesy Robbins & Myers Inc.)

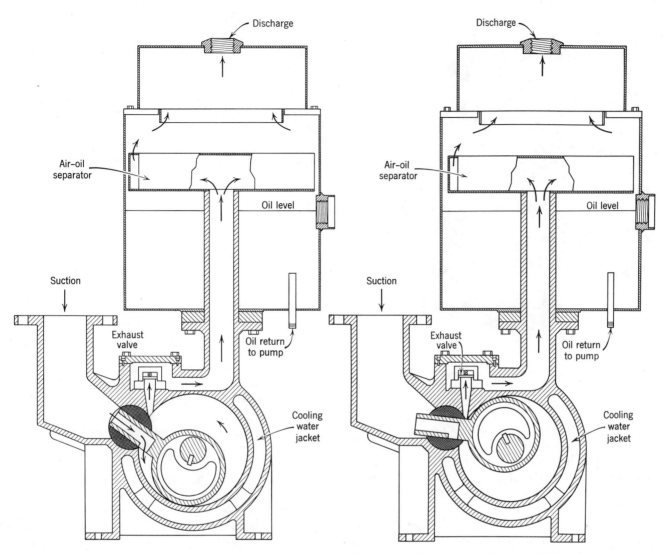

(*a*) Rotating piston creates space for gas in pump chamber.

(*b*) Rotating piston, approaching end of discharge, expels gas from pump chamber.

Figure 21.13. Cutaway view and operation cycle of a rotary-piston pump. (Courtesy Kinney Mfg. Div. New York Air Brake Co.)

used as well as spur gears. A helical-gear pump is shown in Figure 21.7.

The pumps illustrated in Figures 21.6 and 21.7 are referred to as external-gear pumps. Another type of gear pump is the internal-gear pump shown in Figure 21.8. Liquid is drawn into the pump casing and is trapped between the teeth of the rotor and idler (Figure 21.8*a*). The crescent shape on the pump head divides the liquid and serves as a seal between inlet and discharge ports (Figure 21.8*a*). Figure 21.8*c* shows the pump nearly full of liquid, and Figure 21.8*d* shows the pump completely flooded and discharging liquid. The rotor and idler teeth mesh to form a seal midway between inlet and discharge ports.

Lobe Pumps. This pump (Figure 21.9) is similar to the gear pump except that the gears are replaced by two

rotors having two or more lobes. Both rotors are externally driven.

Screw Pumps. Another variation of the gear pump is replacement of the gears with suitable screws turning in a fixed casing. A single-screw pump and a double-screw pump are illustrated in Figures 21.10 and 21.11. Liquid enters the suction chamber of the pump, is divided, and flows to the ends of the pump body. Here it enters the openings between the rotor screw threads, and is moved by the rotors to the discharge port at the center of the pump body. Screw pumps give pulse-free flow and are quite good for handling viscous materials.

An interesting variation of the screw pump is the "Moyno"* pump shown in Figure 21.12. This pump

---

* Registered trade mark, Robbins and Myers, Inc.

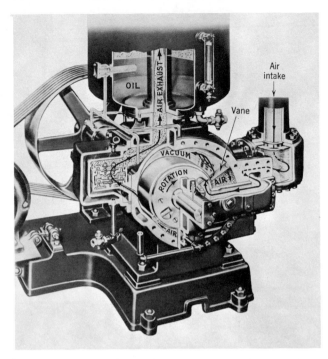

Figure 21.14. Sliding-vane pump. (Courtesy Beech-Russ Co.)

consists of a rotor that revolves within a stator, executing a compound movement: the rotor is revolved about its axis while the axis itself travels in a circular path. The rotor is a true helical screw, and the stator has a double internal helical thread. In each complete revolution of the rotor, the eccentric movement enables the rotor to contact the entire surface of the stator. The voids between the rotor and stator will have entrapped material which is continuously moved toward the discharge. This pumping action gives continuous flow with low, smooth, and uniform velocities. The pump is capable of handling highly viscous materials. Such materials as chocolate, greases, plaster, cake icings, potato salad, and putty are easily pumped.

Rotary-Piston Pump. Instead of gears, this type of rotary pump consists of a circular rotor mounted eccentrically at the center of the pump casing. Figure 21.13 is a cutaway view of a rotary-piston pump. The operating cycle as shown indicates the piston moving in the direction of the arrow. This movement creates space for the fluid in the pump chamber while it also discharges fluid through the exhaust valve. This pump is excellently suited for pumping gases.

Vane Pump. Vane pumps have several sliding vanes fitted into a rotating shaft. Centrifugal force throws the vanes outward, and, as the shaft rotates, the space behind the vane enlarges and draws in fluid. This fluid is trapped between vanes and is eventually forced out the discharge port. The mechanism of this pump is pictured in Figure 21.14.

Operating Characteristics of Rotary Pumps. Rotary pumps are capable of delivering a nearly constant capacity against all pressures within the limits of the pump design. Discharge flow from a rotary pump varies directly as the speed. The discharge is nearly pulse free, particularly for the gear-type pumps. Typical capacity characteristics for an external-gear pump are illustrated in Figure 21.15.

Rotary pumps find a wide range of application. They are capable of pumping fluids of all viscosities with only the restriction that the fluids be free of abrasive materials, lest the closely machined parts be damaged.

CENTRIFUGAL PUMPS

The centrifugal pump is widely used in the process industries because of its simplicity of design, low initial cost, low maintenance, and flexibility of application. Centrifugal pumps have been built to pump as little as a few gallons per minute against a small head and also as much as 605,000 gal/min against a head of 310 ft. In its simplest form, the centrifugal pump consists of an impeller rotating within a casing. Fluid enters the pump near the center of the rotating impeller and is thrown outward by centrifugal action. The kinetic energy of the fluid increases from the center of the impeller to the tips of the impeller vanes. This velocity head is converted to pressure head as the fluid leaves the pump.

The impeller is the heart of the centrifugal pump. It consists of a number of curved vanes or blades shaped in such a way as to give smooth fluid flow between the blades. Common impellers are pictured in Figure 21.16. In the straight vane, single-suction closed impeller (Figure 21.16a), the surfaces of the vanes are generated by straight lines parallel to the axis of rotation. The double-suction impeller (Figure 21.16b) is, in effect, two single-suction impellers arranged back to back in a single casing. For handling liquids containing stringy materials and soft solids, the impellers of Figure 21.16a and 21.16b are likely to become clogged because of restricted flow passages. A nonclogging impeller

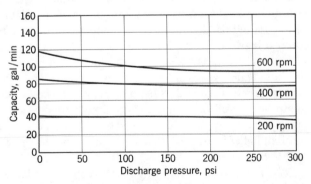

Figure 21.15. Capacity characteristics of rotary-gear pumps.

(a) Straight vane single-suction closed impeller.

(b) Double-suction impeller.

(c) Nonclogging impeller.

(e) Semiopen impeller.

Front

Back

(d) Open impeller.

(f) Mixed-flow impeller.

Figure 21.16. Centrifugal-pump impellers. (Courtesy Worthington Corp.)

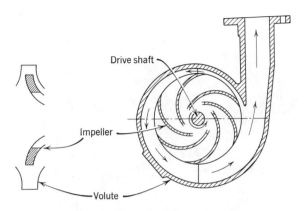

Figure 21.17. Volute centrifugal-pump casing.

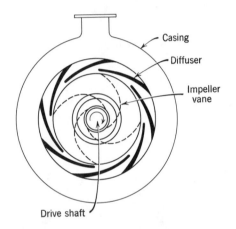

Figure 21.18. Diffuser centrifugal-pump casing.

(Figure 21.16c) is designed to have large flow passages to lessen the possibility of clogging. Open impellers (Figure 21.16d) have vanes attached to a central hub and are well adapted for pumping abrasive solids. The semiopen impeller (Figure 21.16e) has a single shroud, and the closed impeller has shrouds on both sides of the vanes. The semiclosed impeller shown has pump-out vanes located on the back of the shroud whose purpose is to reduce the pressure at the back hub of the impeller. The mixed-flow impeller (Figure 21.16f) is a design in which there are both a radial component and an axial component of flow.

Centrifugal-pump casings may be of several designs, but the main function is to convert the velocity energy imparted to the fluid by the impeller into useful pressure energy. In addition, the casing serves to contain the fluid and to provide an inlet and outlet for the pump. Casings may be of either the volute type or the diffuser type. Figure 21.17 shows a volute casing. Here the impeller discharges into a continuously expanding flow area. This increase in flow area causes the fluid velocity to decrease gradually, thereby reducing eddy formation. By this means, most of the velocity energy is converted to pressure energy with low losses.

The diffuser pump casing (Figure 21.18) has stationary guides which offer the liquid a widening path from impeller to casing. The diffusers serve the same purpose as the volute, and both types of pumps have about the same efficiency. The major applications of the diffuser centrifugal-pump casing are for multistage high-pressure pumps and for use with mixed-flow impellers.

A centrifugal pump with one impeller is referred to as

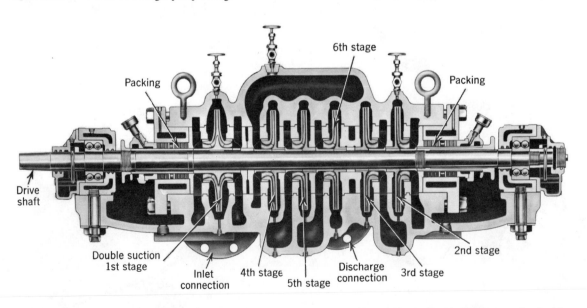

Figure 21.19. Six-stage centrifugal pump. (Courtesy Penna. Pump & Compressor Co.) The fluid enters the double-suction impeller (1st stage) from where it crosses over to the 2nd stage and is discharged to the 3rd stage. There is another cross over to the 4th stage with subsequent passage into the 5th and 6th stages and discharge. The opposed impellers tend to balance forces.

a single-stage pump. If the total head-capacity combination to be developed is greater than that which can be developed from a single impeller, multistage operation may be used. Multistage pumps may be thought of as being several single-stage pumps on one shaft, with flow in series. In effect, discharge from a single-stage pump is fed into the suction side of a second stage where the discharge pressure of the first stage is preserved. The fluid after entering the second stage will have added to it the pressure energy developed in the stage, and so on.

A cutaway view of a multistage centrifugal pump is shown in Figure 21.19. This particular six-stage pump is available in 2 to 4-in. sizes for a capacity range between 50 and 850 gal/min and for heads up to 1400 ft. The size of the outlet pipe fitting is used frequently to characterize the size of a centrifugal pump.

Operating Characteristics of Centrifugal Pumps. A centrifugal pump usually operates at constant speed, and the capacity of the pump depends only on the total head, the design, and the suction conditions. The best way to describe the operating characteristics of a centrifugal pump is through the use of a characteristic curve (Figure 21.20). This figure shows the interrelation of discharge head (H), capacity (Q), efficiency (η), and power input (P), for a given pump at a particular speed. The H-Q curve shows the relation between total head and capacity. The pressure increase created by a centrifugal pump is universally expressed in terms of feet of the fluid flowing. The discharge head, when reported as feet of fluid flowing, is independent of the density of the fluid. In Figure 21.20, the head rises continuously as the capacity is decreased; this type of curve is referred to as a rising characteristic curve. A stable head-capacity characteristic curve is one in

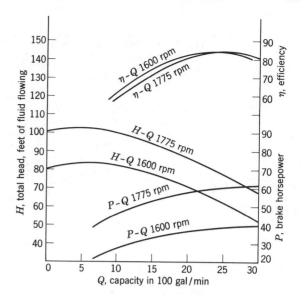

Figure 21.21. Effects of speed change on pump characteristics. (Courtesy Worthington Corp.)

which only one capacity can be obtained at any one head. Pump selection should be made such that stable operating characteristics are available. The P-Q curve of Figure 21.20 shows the relation between power input and pump capacity. The η-Q curve relates pump efficiency to capacity. For a pump having the characteristics of Figure 21.20, maximum efficiency would occur at a capacity of 2500 gal/min and a total head of 80 ft.

When the pump is capable of being operated at variable speeds, characteristic curves such as Figure 21.21 are obtained.

For dimensionally similar pumps changing the impeller diameter changes the area of discharge, thus directly changing the capacity. Changing the diameter also influences the total head directly as the square of the diameters of the impellers, and the power, directly as the cube of the diameters. The effect of impeller diameter on the characteristics of a centrifugal pump is illustrated in Figure 21.22.

Cavitation. When a centrifugal pump is operating at a high capacity, low pressures may develop at the impeller eye or vane tips. When this pressure falls below the liquid vapor pressure, vaporization may occur at these points. The bubbles of vapor formed move to a region of high pressure and collapse. This formation and collapse of vapor bubbles is called cavitation. The bubble collapse is likely to be so quick that the liquid will hit the vane with extreme force and is likely to gouge out small pieces of the impeller. In addition to this pitting of the impeller, noise and vibration will be created. Cavitation may be reduced or eliminated by reducing the pumping rate. If cavitation is not reduced or eliminated, serious mechanical damage to the pump is likely to result.

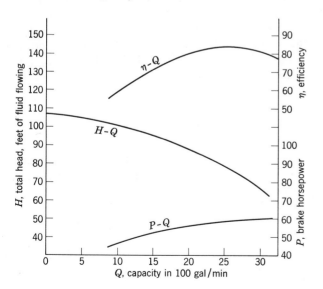

Figure 21.20. Characteristic curves for a centrifugal pump. (Courtesy Worthington Corp.)

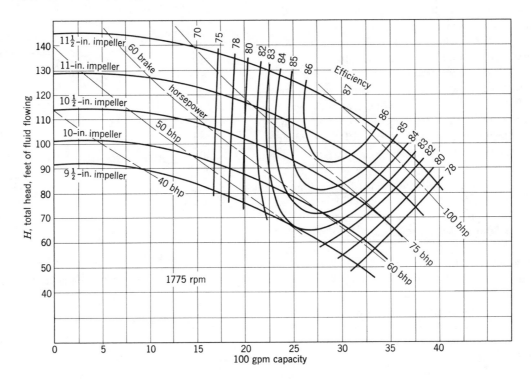

Figure 21.22. Effects of change in impeller diameter on pump characteristics.
(Courtesy Worthington Corp.)

Net Positive Suction Head. Cavitation may be minimized by paying attention to the design of the pump installation on the suction side. Referring to Figure 21.23, let P_1 be the pressure acting on the surface of liquid inside the tank, and z_1 the height of the liquid surface above the pump center line at the suction inlet. If ΣF is the total friction loss in the piping system, application of the energy balance for a unit mass gives

$$\frac{P_1}{\rho} + z_1 \frac{g}{g_c} + \frac{\bar{v}_1^2}{2g_c} = \frac{P_2}{\rho} + z_2 \frac{g}{g_c} + \frac{\bar{v}_2^2}{2g_c} + \Sigma F \quad (21.1)$$

If the datum plane is taken at z_2 and $\bar{v}_1$ is negligible compared with $\bar{v}_2$, the total head at the suction inlet (point 2) is

$$\frac{P_2}{\rho} + \frac{\bar{v}_2^2}{2g_c} = \frac{P_1}{\rho} + z_1 \frac{g}{g_c} - \Sigma F \quad (21.2)$$

From Equation 21.2, the pressure at the suction inlet is

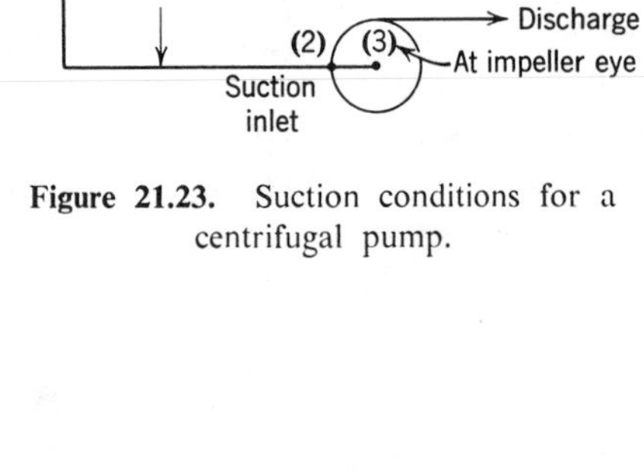

Figure 21.23. Suction conditions for a centrifugal pump.

$$P_2 = P_1 + \rho z_1 \frac{g}{g_c} - \rho \Sigma F - \frac{\rho \bar{v}_2^2}{2g_c} \quad (21.3)$$

At the eye of the impeller, it is reasonable to assume that the pressure will be less than at the suction inlet. This pressure difference can be assumed to be related to the velocity at the eye by the expression $\Delta P = \phi \bar{v}_3^2/2g_c$, where ϕ is a pressure-drop coefficient characteristic of pump geometry and $\bar{v}_3$ is the fluid velocity at the eye.

The *net positive suction head* (NPSH) is defined as the difference between the static head at the suction inlet and the head corresponding to the vapor pressure of the liquid at the pump inlet. Thus combining this definition with Equation 21.2,

$$\text{NPSH} = \left(\frac{P_2}{\rho} + \frac{\bar{v}_2^2}{2g_c}\right) - \frac{P_v}{\rho} = \left(\frac{P_1}{\rho} + z_1 \frac{g}{g_c} - \Sigma F\right) - \frac{P_v}{\rho} \quad (21.4)$$

Cavitation is probable if the total head at the suction minus the acceleration head from there to the eye of the impeller is equal to or less than the vapor pressure head. Thus,

$$\frac{P_1}{\rho} + z_1 \frac{g}{g_c} - \Sigma F - \frac{\bar{v}_2^2}{2g_c} - \phi \frac{\bar{v}_3^2}{2g_c} \lesseqgtr \frac{P_v}{\rho} \quad (21.5)$$

Equation 21.5 may be rearranged to give

$$\left(\frac{P_1}{\rho} + z_1 \frac{g}{g_c} - \Sigma F\right) - \frac{P_v}{\rho} = \frac{\bar{v}_2^2}{2g_c} + \phi \frac{\bar{v}_3^2}{2g_c} \quad (21.6)$$

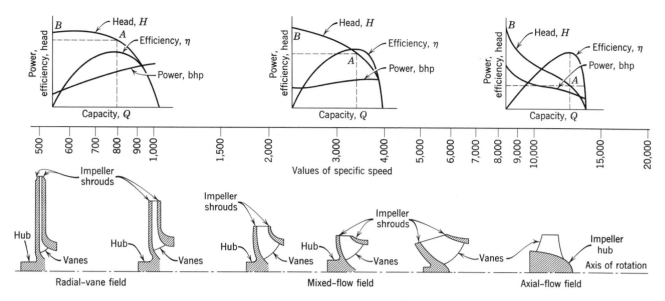

Figure 21.24. Characteristic curves and specific speeds for various impellers. (Courtesy Worthington Corp.)

and from the definition of the net positive suction head

$$\text{NPSH} = \frac{\bar{v}_2^2}{2g_c} + \phi \frac{\bar{v}_3^2}{2g_c} \qquad (21.7)$$

Therefore, to avoid cavitation, the NPSH must be greater than the maximum possible value of $(v_2^2/2g_c) + (\phi\bar{v}_3^2/2g_c)$. The values of $\bar{v}_2$ and $\bar{v}_3$ will depend upon the flow rate so that some control over cavitation may be achieved by variation in flow rates as previously mentioned. It is recommended that the value of the required NPSH for the particular pump being used be obtained from the pump manufacturer.

Specific Speed. Specific speed is the speed in revolutions per minute at which a theoretical pump geometrically similar to the actual pump would run at its best efficiency if proportioned to deliver 1 gal/min against a total head of 1 ft. The specific speed serves as a convenient index of the actual pump type, using the capacity and head obtained at the maximum efficiency point.

Specific speed may be determined from Equation 21.8 for a single-stage pump or one stage of a multistage pump. This equation results from model theory and dimensional analysis.

$$N_s = \frac{n\sqrt{Q}}{H^{0.75}} \qquad (21.8)$$

where N_s = specific speed
   n = actual pump speed, rpm
   H = total head per stage, ft
   Q = pump capacity in gal/min at speed n and total head z

Note that specific speed as defined in Equation 21.8 is a dimensional quantity.

Figure 21.24 shows that the normal range in specific speeds encountered in single-suction pumps for various impeller designs is between 500 and 15,000. On the same figure is shown the type of pump characteristic that is associated with pumps of various specific speeds.

SPECIAL PUMPS

With the advent of nuclear-energy processes, pumping problems that were of minor concern in the chemical-process industry became significant. The most noteworthy of the problems is that of having a pump that has zero leakage. All pumps that have been discussed thus far have a seal problem. Many designs of seals have been developed, but, when handling radioactive materials, no leakage can be tolerated. Many of the flow conditions in the nuclear engineering field are such that common pumps could be used if proper sealing devices were available for them. Many special pumps for nuclear and other applications are currently available.

Canned-Motor Pump. One design of a zero-leak pump is the canned-motor pump shown in Figure 21.25. This pump is suitable for pumping radioactive water which is allowed to fill the motor cavity but which is isolated from the rotor and stator by cans in the magnetic gap. There are no external shaft seals in this pump. The liquid being pumped completely fills all the space inside the sealing cans. This pump can also be designed for pumping liquid metals at temperatures as high as 1600°F.

Magnetic Pumps. An *electromagnetic* pump has no moving parts so that no seals of any type are required. Liquid metals having high electrical conductivity are pumped with this device. The electromagnetic pump

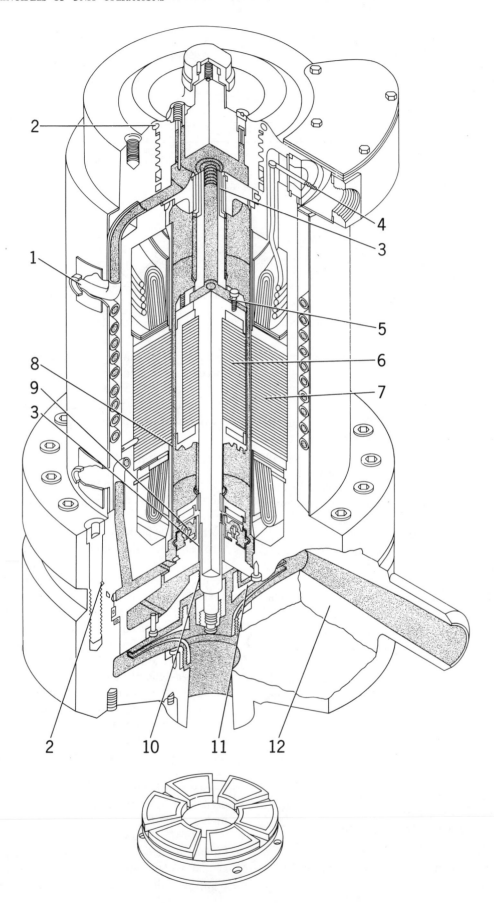

Figure 21.25. Canned motor pump. (Courtesy Atomic Equip. Dept., Westinghouse Elec. Co.)

1. Cooling jacket inlet. Ordinary clean cooling water flowing over the wrap-around coils of the integral heat exchanger removes heat from motor.

2. Seal and welds and gaskets.

3. Radial sleeve bearings. The bearings are made of Graphitar, a carbon graphite compacted material, and can be made self-aligning. Shaft journals are specially hardened metal. The bearings are designed to be lubricated by the pumped fluid only.

4. Motor terminals.

5. Lubrication and cooling. Fluid handled by pump completely fills all spaces inside sealing "can" and provides lubrication for the bearings. Auxiliary impeller provides circulation of fluid around rotor and through wrap-around coils, which dissipates motor heat to secondary cooling water flowing around the cooling coil.

6. Motor rotor. The squirrel-cage rotor is hermetically enclosed by a close-fitting, welded stainless-steel or Inconel sealing "can." The rotor is long and slender thus minimizing fluid friction loss.

7. Motor stator. The stator winding is completely isolated from the pumped fluid by a stainless-steel or Inconel sealing "can." Backup sleeves are used under the end turns to support the "can" against system pressure. Heat from electrical losses is conducted outward and removed directly by the cooling water.

8. Hermetic sealing "can." A sealing "can" within the magnetic gap seals the motor stator from the pumped fluid. The rotor is similarly encased.

9. Thrust bearing.

10. Labyrinth seal. The labyrinth seal together with the thermal barrier minimizes fluid flow and heat transfer between pump casing and motor cavity and thus insulates the motor from high system temperatures.

11. Main impeller. The main impeller is keyed and locked to the motor shaft.

12. Pump casing. The pump casing is arranged so that it can be welded into the piping system. The pump motor and impeller can be easily removed from the casing without disturbing the piping system.

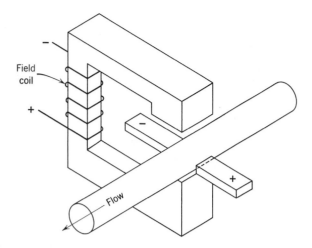

Figure 21.26. Electromagnetic-pump principle.

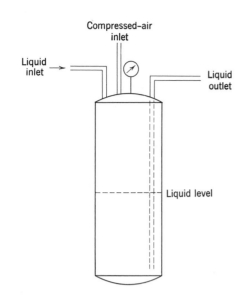

Figure 21.28. Acid egg.

works on the same principle as the induction motor. The liquid metal flows through the field of an electromagnet. An external current passes through the metal in a direction perpendicular to the magnetic field. The force exerted as a result of these interacting fields causes flow to occur. This principle is illustrated in Figure 21.26.

A *magnetic-drive* pump for toxic liquids is shown in Figure 21.27. In this pump, the drive and liquid end are kept separated by a nonmagnetic diaphram, and the liquid itself serves as the pump lubricant.

Acid Egg. A simple inexpensive "pump" involving no moving parts is the acid egg or *monte jus* (Figure 21.28). The principle is simply the displacement of one fluid by another. In the illustration, liquid is fed into a tank by gravity and forced out of the tank by compressed air. The operation can be made continuous by use of several eggs so that one is always discharging while the others are being filled. The acid egg can be made of corrosion resistant materials.

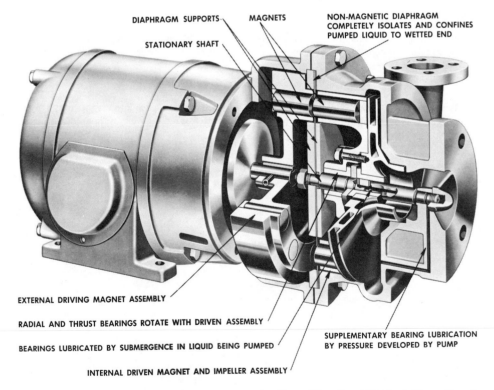

Figure 21.27. Magnetic-drive pump. (Courtesy Peerless Pump Div., Food Mach. & Chem. Corp.)

Table 21.1. PUMP MANUFACTURER'S INFORMATION*

Liquid to be pumped . . . .
. . . .
Capacity required . . . . .gal/min. Lubricating qualities? . . . . .Yes . . . . .No . . . .
Suction:___flooded, or . . . . .in. Hg lift. Discharge . . . . .psig
Pumping Temperature . . . . °F. Viscosity at pumping temp. . . . .
Vapor pressure of the liquid at the pumping temperature . . . . .in. Hg
MAXIMUM viscosity pump may handle for short periods . . . .
MINIMUM viscosity pump may handle for short periods . . . .
Is liquid corrosive?___Yes___No   Is liquid abrasive___Yes___No
What is abrasive substance in liquid? . . . .
. . . .
From your experience, what material best handles this liquid? . . . .
___standard cast iron___cast steel___ . . . .
. . . .
What other particular requirements do you have? (Such as mechanical seals, lantern rings, etc.). . . . .
. . . .
. . . .

* Courtesy Sier Bath Gear & Pump Co.

PUMP SELECTION

Pumps must be selected according to head requirements, capacity, fluid being pumped, and other specifications. Typical of the information that need be supplied to a pump manufacturer to assure satisfactory selection is that found in Table 21.1.

Several steps are necessary in choosing a pump: (1) Sketch the pump and piping system so that total heads may be calculated. (2) Determine the capacity as dictated by local conditions, allowing for some variation. (3) Examine liquid conditions as to specific gravity, viscosity, vapor pressure, etc. (4) Choose class and type of pump based upon the first three items.

The mechanical design of centrifugal pumps is rather complex, and selection of an appropriate pump for a particular assignment demands recognition of the importance of each factor. Probably the most important feature is proper design of the wearing rings which seal between the high-pressure periphery of the impeller and the low-pressure hub area. If these can be made to provide continued close clearances, good efficiency can be attained. If corrosion and erosion are expected to deteriorate the seal, it is necessary to accept fairly large internal circulation of the fluid. The original efficiency will be lowered somewhat, but longer effective life of these parts with less drop-off in performance of the pump may be assured.

In single suction pumps, the back side of the impeller is subjected to approximately the discharge pressure, but the center of the face is above the suction pressure only by the pressure drop through the wearing seal. Because of the variation in fluid dynamics around the casing or volute, neither of these pressures is symmetrical. As a result unbalanced forces play upon the impeller.

In multistage pumps this net force can be reduced by facing some impellers in one direction, and others in the opposite direction, as in Figure 21.19.

Material of construction problems are aggravated in centrifugal pumps for chemical service because the fluid at high velocities removes protective films causing corrosion to occur at an accelerated rate.

COMPRESSORS

Just as in the case of pumps for liquids, compressors for gases may be classed as positive-displacement or centrifugal compressors. Positive-displacement compressors include reciprocating and rotary machines. Gases are propelled by means of blowers and compressors. The distinction between the two is not always clear cut.

Positive-Displacement Machines. *Reciprocating Compressors.* The reciprocating compressor can furnish gas at pressures of a few pounds or at extremely high pressures, such as 35,000 psi. The characteristic features of reciprocating compressors are the same as those of reciprocating pumps—a piston, a cylinder with suitable intake and exhaust valves, and a crankshaft with drive. Single-stage or multistage operation is common, with double-acting cylinder usage being general.

Gas being compressed enters and leaves the cylinder through valves which are set to be actuated when the pressure difference between cylinder contents and outside conditions is that desired. If multistage compression is used, it is general practice to cool the gas between stages. Several pictures of reciprocating compressors are seen in Figures 21.29, 21.30, 21.31, and 21.32.

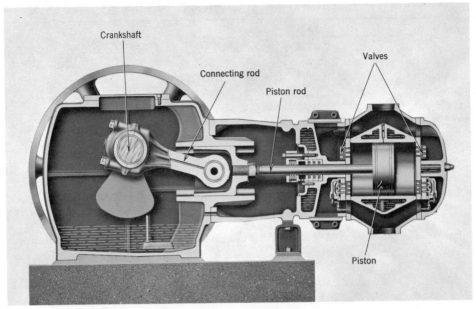

Figure 21.29. Cross section of 7-in.-stroke reciprocating compressor. (Courtesy Ingersoll-Rand.)

Discharge characteristics of reciprocating compressors are similar to those for reciprocating pumps. Compressor operation is fundamentally thought of as being isentropic, and efficiencies are reported relative to this isentropic basis. Thermodynamic losses and fluid friction are grouped together as a compression inefficiency. Mechanical friction losses are termed mechanical inefficiency. An over-all compressor efficiency will be the product of the compression and mechanical efficiencies. The over-all efficiency of most reciprocating compressors is 65 to 80 per cent.

Rotary Compressors. This group of compressors is characterized by a continuous, almost smooth, discharge of gas. Rotary compressors are of the lobe, sliding-vane, and rotating-piston types. The sliding-vane compressor is particularly well suited for evacuation assignments; indeed, this compressor has a wide range of possible pressure, vacuum, and volume conditions.

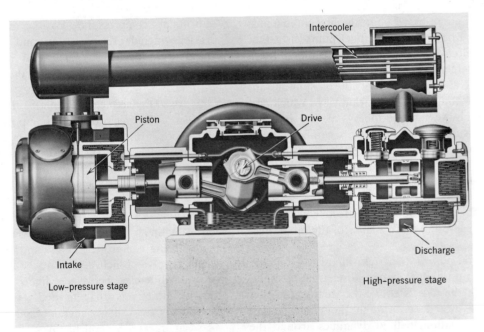

Figure 21.30. Two-stage reciprocating compressor. The basic design for this two-stage unit is for 80 to 125 psi, and it is used in a variety of installations. The compressor is equipped with a shell-and-tube intercooler. (Courtesy Ingersoll-Rand.)

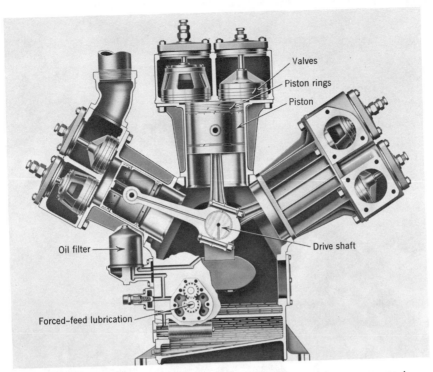

Figure 21.31. Two-stage air-cooled compressor. This two-stage compressor is capable of producing pressures of 80 to 125 psi. It has a piston displacement of 686 cu ft/min and requires 125 hp to drive it. (Courtesy Ingersoll-Rand.)

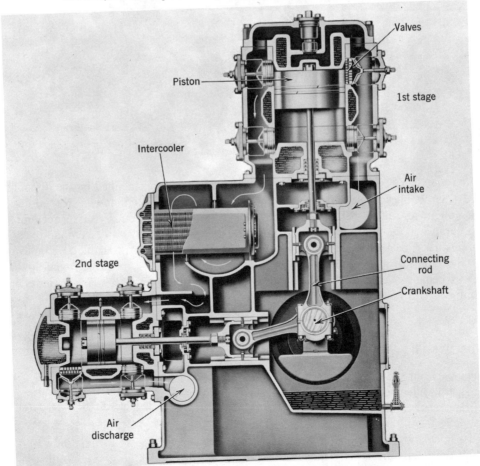

Figure 21.32. Heavy-duty crosshead two-stage compressor. This compressor is available in motor horsepower rating of 125 to 350 hp. It can handle between 800 and 2397 cu ft/min depending upon the cylinder bore and speed and will produce pressures of 100 psi. (Courtesy Ingersoll-Rand.)

Figure 21.33. Inlet end of axial-flow fan. (Courtesy Buffalo Forge Co.)

The lobe compressor pushes gas from the suction inlet to the discharge inlet by action of the lobes. Little compression takes place within the unit but compression occurs when the pump contents are forced into the system against a back pressure from the system. Extremely close clearances of the lobes are necessary; therefore, the gas being moved must be dust- and dirt-free. This compressor is of the same design as that in Figure 21.9.

FANS AND BLOWERS

The difference between a fan and a compressor is not well defined. If any difference can be generalized, it is that a fan operates at pressures low enough so that the compressibility effect on the gas may be neglected. In other words, inlet and outlet volumes for fans are essentially equal, and fans are simply movers of gas.

Fans are classified by the direction of air flow—radial flow and axial flow. Radial-flow fans depend upon centrifugal action for propelling the gas; however, axial-flow fans have comparatively simple flow parallel to the fan shaft. Table 21.2 indicates a simple grouping of fans. The propeller axial-flow fan is a rather loose classification since many varieties are available.

Table 21.2. FAN CLASSIFICATION

Radial or Centrifugal Flow	Axial Flow
Straight blades	Propeller
Forward-curved blades	
Backward-curved blades	
Double-curved blades	

A picture of an axial-flow fan is shown in Figure 21.33, and Figure 21.34 illustrates a centrifugal fan.

Figure 21.34. Centrifugal fan. (Courtesy Buffalo Forge Co.)

Fan Characteristics. The performance of fans can be estimated in terms of generally applicable statements. At a fixed operating condition, corresponding to a specific rating, the volume of gas moved varies directly as the fan speed; the static pressure varies as the square of the fan speed; and the power varies as the cube of the

speed. In practice, exact fan performance can be evaluated only through testing. Fan characteristics will depend to a large extent upon design. For centrifugal fans, the type of blading will influence the performance. For axial-flow fans, the propeller design is a prime variable. Characteristic curves for various fans are shown in Figure 21.35.

The radial-bladed fan has a medium efficiency, and experience with this design has shown it to give a stable pressure response. This type of blading is ideally suited for conveying gases containing suspended solids because centrifugal force tends to keep the blades clean.

The forward-curved fan is a low-speed, large-volume fan. Efficiency is medium. Clean gases are best handled in this fan.

Backward-inclined bladed fans are a recent design and are characterized by a high efficiency and power curve. This fan is most used in clean-gas applications.

Centrifugal Compressors. The main function of a centrifugal compressor is to increase the pressure of the gas flowing through it. This is done by the conversion of velocity energy to pressure energy by accelerating the gas as it flows radially outward from the inlet, in a manner similar to the action of a centrifugal pump. Centrifugal compressors are available in a wide range of capacities—between 200 discharge cu ft/min and 150,000 suction cu ft/min with outlet pressures to over 800 psig.

The centrifugal compressor consists of an impeller and casing similar to the centrifugal pump and is in principle one of the simplest types of machines known for compressing gases. Several types of compressor impellers are shown in Figure 21.36. It can be seen that they are quite similar to pump impellers.

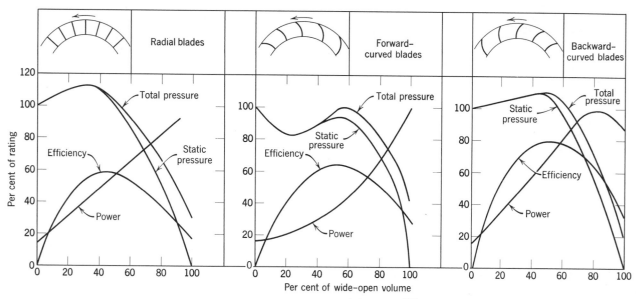

Figure 21.35. Fan-characteristic curves (10).

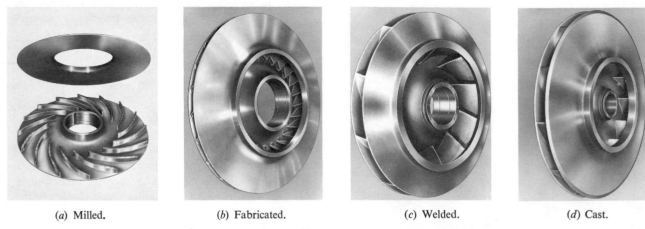

(a) Milled.   (b) Fabricated.   (c) Welded.   (d) Cast.

Figure 21.36. Impellers for centrifugal compressors. (Courtesy Clark Bros., Div. of Dresser Operations, Inc.)

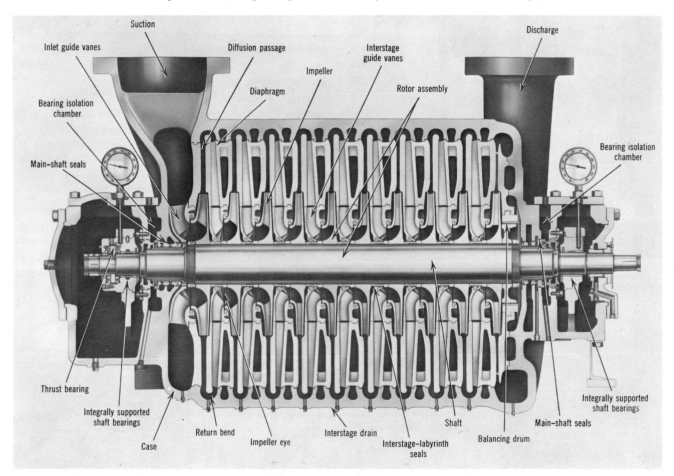

Figure 21.37. Multistage centrifugal compressor. (Courtesy Clark Bros., Div. of Dresser Operations, Inc.)

Gas enters the centrifugal compressor (Figure 21.37) near the eye of the impeller and is ejected at a high velocity and pressure from the tip of the impeller into a diffuser where the remaining conversion of velocity into pressure is accomplished. Centrifugal compressors are usually multistaged to permit the attainment of higher exit pressures.

In multistage operation, the gas leaves the diffuser and enters a diaphragm containing vanes which direct the gas into the eye of the next impeller. The transfer of energy to the gas as it is compressed will cause it to become warm so that cooling channels may be provided between stages.

An interesting centrifugal compressor is shown in Figure 21.38. This compressor has an elliptical casing which is partly filled with liquid, through which the rotor

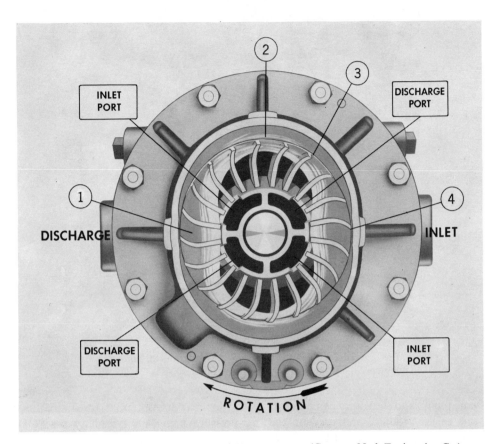

Figure 21.38. Principle of the Nash "Hytor" compressor. (Courtesy Nash Engineering Co.)

blades revolve. The rotor speed is such that liquid is thrown out from the center by centrifugal force creating a liquid ring on the casing wall. The cycle of operation may be seen by reference to Figure 21.38. At (1), the blade "bucket" is full of liquid; as the rotor turns, the liquid follows the casing, moving away from the rotor so gas can enter through the inlet. At (2), the space for gas is at a maximum since the liquid will be at the casing wall. The elliptical wall at (3) is closer to the axis, thereby forcing the liquid back towards the rotor and diminishing the gas space, thereby compressing the gas and discharging it at (4). The cycle is repeated through the second half of the revolution. The liquid is usually supplied continuously to provide cooling; the quantity flowing must be properly controlled for best operation.

Axial-Flow Compressors. Only recently has this type of compressor found general acceptance for industrial applications. For large inlet volumes, the use of an axial-flow compressor is recommended. These machines are capable of handling 860,000 cu ft/min and are about half the diameter of a comparable centrifugal compressor for about the same initial cost. The axial compressor is about 10 per cent higher in efficiency than the comparable centrifugal. Figure 21.39 shows a cutaway view of an axial compressor.

Axial-flow compressors are designed on the basis that half of the pressure rise occurs at the rotor blade and half at the stator blade. The rows of stationary blades serve to increase both static pressure and kinetic energy while they guide the air into the rotor blades. In this regard, they serve as diffusers. Good design practice for an axial-flow compressor calls for air velocities of the order of magnitude of 400 ft/sec. In most compressors of this type, the gas velocity from stage to stage is essentially constant. To achieve constancy, since the pressure increases with each successive stage, requires a smaller annular area. Initial cost of an axial-flow compressor is about the same as that of a centrifugal compressor designed for the same duty. However, the axial compressor is somewhat more efficient than the comparable centrifugal compressor so that a smaller turbine or motor can be used to drive it, resulting in lower initial driver cost.

VACUUM PRODUCERS

The chemical engineer is quite likely to be concerned with processes which operate at subatmospheric pressures. Pressures of $\frac{1}{2}$ in. Hg are easily obtained by use of rotary or reciprocating pumps of the types already discussed. Particularly useful are the rotary-piston

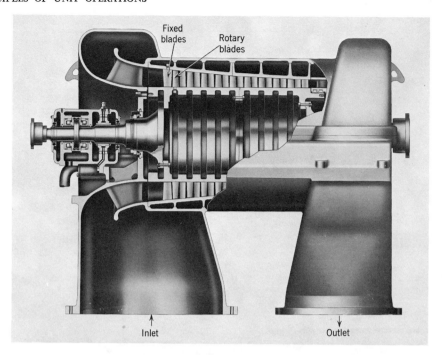

Figure 21.39. Axial-flow blower. Each row of blades on the rotor represents one stage. This is a nine-stage blower. (Courtesy Allis-Chalmers.)

pump, the sliding-vane pump, and the rotary compressor. These machines can easily produce pressures of the magnitude mentioned, and the rotary-piston pump (see Figure 21.13) is capable of producing pressures of 0.001 mm Hg.

Ejectors. A frequently encountered device for producing vacuum of the range above 1 mm Hg is the jet ejector. It is a cheap, simply designed, easily operated piece of equipment. Applications of ejectors are found in nearly every industry.

The ejector is essentially a compressor, but it is unique in that the working fluid mixes with the fluid being compressed. In Figure 21.40 is a sectional view of a single-stage ejector. The working fluid, high-pressure steam or air, is fed through a nozzle into a vapor chamber where it entrains the surrounding vapor or gases. The combined gases or vapors issue from the nozzle at a high velocity and expand through a convergent-divergent nozzle. The diffuser serves to convert velocity energy into pressure energy. The result of this is to deliver a volume of entrained gas at a pressure higher than the pressure in the vapor chamber—effectively, the diffuser is a compressor.

The compression ratio in a single-stage ejector can exceed 10:1, but the capacity per unit of driving fluid becomes uneconomical. For larger compression ratios, it is more economical to use jets in series. As shown in Figure 21.41, the discharge from the low-pressure jet feeds the high-pressure-jet suction chamber. This permits the designer to use each jet at a compression

ratio near its optimum to attain the required over-all pressure increase. As many as six stages have been

Figure 21.39a. Installation view of 860,000 CFM axial compressor showing fixed and rotary blades. (Courtesy Allis-Chalmers.)

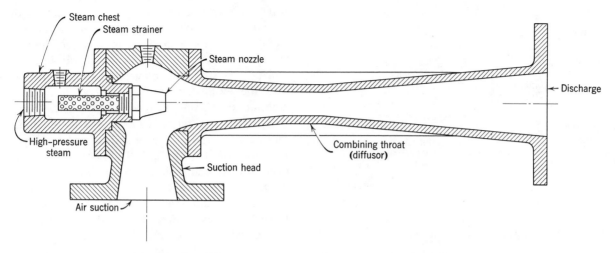

Figure 21.40. Single-stage jet ejector. (Courtesy Croll-Reynolds Co., Inc.)

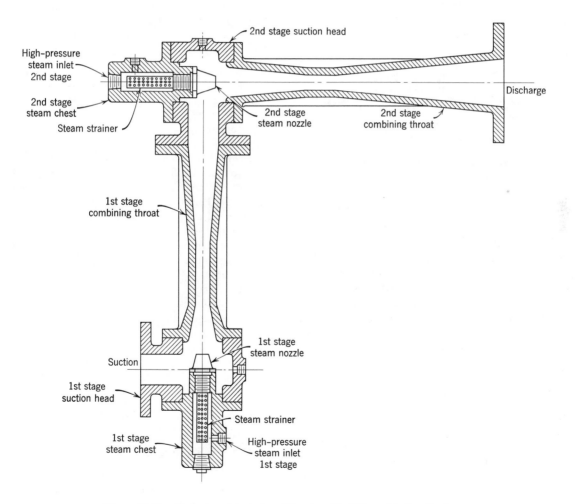

Figure 21.41. Two-stage jet ejector. (Courtesy Croll-Reynolds Co., Inc.)

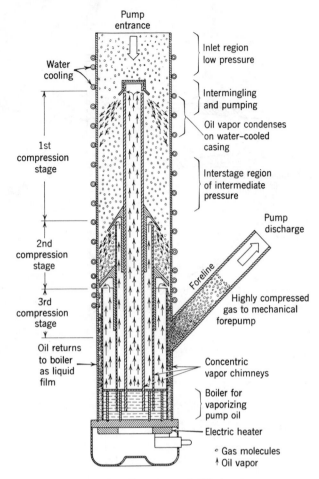

Figure 21.42. Three-stage oil-diffusion pump.

of the pump and the vapor rises upward through the chimneys, as indicated. Gas molecules from the space being evacuated enter the pump by random thermal motion and collide and mingle with molecules of the vaporized pumping fluid. The vaporized-liquid molecules are directed downward through the chimney slots and possess sufficient momentum to keep from leaking into the evacuated space. The gas molecules upon collision are given a downward velocity component away from the pump entrance. The pumping fluid molecules move to the cool surface and condense to be returned to the boiler. The noncondensables are concentrated and pulled out by the forepump. The diffusion pump functions in one direction only—from the high vacuum space to the forepump. The use of three stages, as shown, gives high pumping capacities over a wide pressure range.

Vacuum Pump Selection. The choice of which vacuum pump to use for a specified operation depends primarily upon the degree of vacuum desired. The operating-pressure range of commercially available vacuum pumps is listed in Table 21.4.

Table 21.4. OPERATING RANGES OF VACUUM PUMPS*

Pump Type	Operating Range, mm
Reciprocating	
Single-stage	760–10
Double-stage	760–1
Rotary Piston	
Single-stage Oil-Sealed	$760–10^{-2}$
Two-Stage Oil-Sealed	$760–10^{-3}$
Rotary Lobe	$20–10^{-3}$
Vapor Jet Pump	
Steam Ejector	
One-stage	760–100
Two-stage	760–10
Three-stage	760–1
Four-stage	$760–3 \times 10^{-1}$
Five-stage	$760–5 \times 10^{-2}$
Mercury diffusion with cold trap	
One-stage	$10^{-1}–10^{-6}$
Two-stage	$1–10^{-6}$
Three-stage	$10–10^{-6}$
Oil diffusion	
One-stage	$10^{-1}–5 \times 10^{-6}$
Two-stage	$5 \times 10^{-1}–5 \times 10^{-6}$

* From *Encyclopedia of Chemical Technology*.

used in series. The working fluid accumulates in the gases being compressed, increasing the amount of gas which must be compressed in succeeding stages.

Jet ejectors can use any fluid as the driving fluid, but steam is by far the most economical and is generally used. It has the further advantage of being condensable at reasonable pressures by cooling water and hence can be removed from the gas compressed in the high-pressure stages. This is usually done by an intercooler between stages, which may be either a barometric jet condenser, or, more commonly, a shell and tube condenser. Multi-stage steam jet ejectors can maintain absolute pressures below 1 mm Hg.

Diffusion Pumps. For applications where a very low pressure (high vacuum) is desired, the use of a diffusion pump is dictated. A diffusion pump is capable of producing pressures of 10^{-7} mm Hg or lower. To achieve vacuums of this magnitude requires a diffusion pump backed by a single-stage mechanical forepump. The working fluid in a diffusion pump is a low vapor-pressure liquid, usually mercury or special oils. A typical three-stage diffusion pump is shown in Figure 21.42. The working fluid is vaporized at the bottom

REFERENCES

1. Carter, R., and I. J. Karassik, *Basic Factors in Centrifugal Pump Application*, RP. 477, Worthington Corp., Harrison, New Jersey.
2. Church, A. H., *Centrifugal Pumps and Blowers*, John Wiley and Sons, New York, 1944.

3. Hicks, T. G., *Pump Selection and Application*, McGraw-Hill Book Co., New York, 1957.

4. Hydraulic Institute, New York, Standards.

5. Karassik, I. J., and R. Carter, *Centrifugal Pump Design and Selection*, RP. 736, Worthington Corp., Harrison, New Jersey.

6. Robbins and Myers, Inc., Springfield, Ohio, Catalogs.

7. Geo. D. Roper Corporation, Rockford, Illinois, *How to Solve Pumping Problems*.

8. Stepanoff, A. J., *Centrifugal and Axial Flow Pumps*, 2nd ed., John Wiley and Sons, New York, 1957.

9. Sier-Bath Gear and Pump Company, North Bergen, New Jersey, Catalogs.

10. Compressible Fluids, Editorial Reprints, *Chem. Eng.*, 1956.

PROBLEMS

21.1. Water is to be pumped from a river to a large storage tank for plant service. The piping system shown below consists of 180 ft of 3-in. Sch. 40 pipe on the suction side of the pump and 700 ft of the same pipe in the discharge line. When the level in the storage tank falls below the control point, water is pumped into the tank until the control level is re-established. A centrifugal pump having the characteristics indicated below is used for this intermittent steady-state pumping assignment.

Determine the flow rate through the piping system in gallons per minute and the power required.

Centrifugal Pump Characteristics

Capacity, gal/min	Total Head, ft of water	Efficiency, Per cent
0	280	0
20	260	45
40	220	60
60	160	60
80	110	56
100	63	50
120	28	43
140	10	37
160	5	30

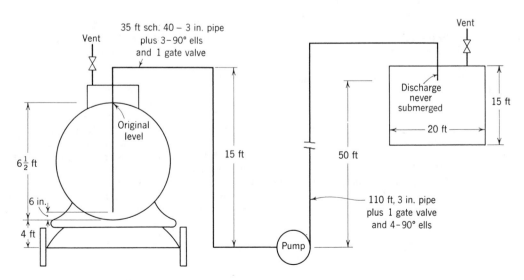

Water pumping system (Problem 21.1)

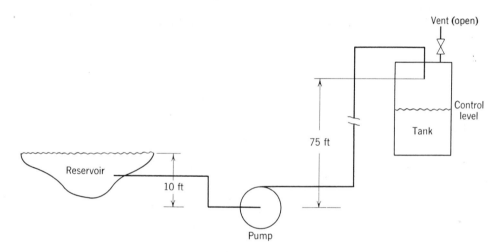

Tank car discharge of benzene (Problem 21.3)

21.2. A hydraulic cylinder, consisting of a 4-in. diameter piston that travels at a rate of 9 ft/min, is actuated by a rotary-gear pump which supplies hydraulic fluid to the cylinder. The length of the stroke is 20 in., and the load on the cylinder is 4 tons. The piping system consists of 30 ft of $\frac{1}{2}$-in. Sch. 80 pipe (total equivalent length), and there is a negligible change in elevation from the fluid reservoir to the cylinder. The hydraulic fluid used has a specific gravity of 0.9 and a kinematic viscosity of 100 centistokes.

A manufacturer's catalog* lists the following gear pumps which are judged satisfactory to meet hydraulic-fluid pumping assignments:

Pump Size	Pipe Size, in.	Minimum Capacity, gal/min				
		150 psi	300 psi	400 psi	800 psi	1000 psi
1	1/2	1.57	1.5	—	—	—
3	1/2	3.35	3.15	—	—	—
5a	1/2	5.47	5.12	—	—	—
5	1/2	—	—	6.73	6.15	5.85
10	3/4	—	—	10.07	9.2	8.75
10a	3/4	11.05	10.4	—	—	—

* Courtesy Roper Hydraulics Inc.

From this information, select the pump which will accomplish the required specifications.

21.3. A tank car is to be emptied of 10,000 gal of benzene at 80°F in 3 hr. (See above figure.) The plant piping system is as indicated below. Available to do the pumping is a centrifugal pump having the following characteristics:

Capacity, gal/min	Total Head, ft	Efficiency, per cent
0	110	0
20	106	29.2
40	90	40.0
60	63	45.0
80	41	47.0
100	22	48.3
120	12	46.5
140	7	40.0

(a) Is this pump satisfactory for the job?
(b) How long will it take to "empty" the tank car?
(c) How much work is necessary?
Note: This problem may be solved by (1) considering the effect of time dependent heads or (2) the use of an average head over the pumping cycle.

21.4. Recommend a pump, and defend this recommendation, for
(a) Pumping cutting oil to a machine tool.
(b) Delivering water from the Hudson River above Albany to New York City.
(c) Delivering H_2 gas at 15000 psi to an ammonia converter.
(d) Pumping concentrated tomato soup to the canning machines (100 GPM).

(e) Metering caustic solution into a reaction stream in response to a pH controller.
(f) Feeding liquid oxygen to the first stage engine of an Atlas rocket.
(g) Pumping detergent slurry to a spray dryer nozzle (1500 psig, 5 GPM).
(h) Recirculating water for a laboratory constant-temperature bath.
(i) Transferring sugar solution from stage to stage of an evaporator system. Feed temp—120°F; flow—30 gpm.
(j) Recirculating molten salt between a storage tank and a heat exchanger. $T = 1000$°F, $\mu = 10$ centipoise; flow = 50 GPM.
(k) Recirculating molten NaK alloy from a nuclear reactor to a steam generator in a nuclear power plant.

21.5. A plastic intermediate is to be pumped from a storage tank into a batch reactor. Pumping time is to be held to a minimum. Preliminary plans call for placing the storage tank on the second floor, the pump being on the ground floor. System specifications are:

Fluid Properties:
$C_p = 0.65$ Btu/lb °F
$\mu = 500$ cp
$\rho = 45$ lb/ft³
$k = 0.36$ Btu/hr ft² (°F/ft)
M. Wt. = 3000
Vapor pressure at 80°F = 100 mm Hg
System geometry:
See sketch

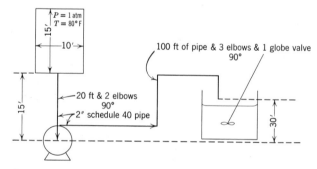

Pump:
Any one of a line of centrifugal pumps of from 10 to 1000 GPM capacity can be purchased. All of these pumps specify a minimum NPSH of 10 ft of water, and develop ample head. What pump capacity should be specified? (Use a friction factor of 0.033).

21.6. Figure 21.22 gives the characteristics of a series of centrifugal pumps. Calculate the specific speed for the $9\frac{1}{2}$, $10\frac{1}{2}$, and $11\frac{1}{2}$ in. impeller sizes.

21.7. The pump of $11\frac{1}{2}$ in. impeller diameter for which characteristics are shown in Figure 21.22 is used in a piping system taking cooling water from a river and delivering it to a stand-pipe. Flow is through 8 in. Sch. 40 steel pipe and the required lift is 70 ft. Equivalent pipe length, including entrance, exit, and fitting losses is 300 ft plus that contributed by a gate valve that controls the flow. The pump is driven by a 75 HP motor. How far can the valve be opened before the pump overloads the motor?

chapter 22

Momentum Transfer III:
Phase Separations Based upon
Fluid Mechanics

This chapter will deal with operations which are useful in the separation of multi-phase mixtures into two or more individual fractions. The separation methods discussed may be classified as mechanical separations, as opposed to those separations requiring vaporization or condensation. For example salt crystals can be separated from their mother liquor by *filtration* or *centrifugation*. Several different sizes of crushed ore can be separated by *screening*, *elutriation*, *jigging*, or *classification*. Sludges can be separated from a liquid by *sedimentation*. The mechanical methods of separation may be grouped into two general classes: (1) those whose mechanism is controlled by fluid mechanics and (2) those whose mechanism is not described by fluid mechanics.

APPLICATIONS OF THE MECHANICS OF PARTICLE MOVEMENT THROUGH A FLUID

General Principles. In Chapter 13 the concept of form drag was introduced. For steady flow of a fluid past a solid, boundary layers are established, and a force is exerted on the solid by the fluid. This force is a combination of boundary-layer drag and form drag, and it can be expressed in terms of a drag coefficient. By Equation 13.50, the drag coefficient is

$$C_D = \frac{2F_D g_c}{v_{fs}^2 \rho S} \tag{13.50}$$

where F is the force acting on the solid, v_{fs} is the free-stream velocity relative to the particle, and S is the projected area of the solid normal to the flow. This equation is important wherever momentum transfer at a fluid-solid

boundary must be examined. Thus it may be applied to the design of particle-separation equipment as well as to that of piping systems. Over the range of flow conditions likely to be encountered, the drag coefficient may be determined from Figure 22.1. Note that the sphericity term is introduced to account for nonspherical particles. (See Appendix B for sphericity determinations.)

Consider a particle moving through a fluid in one dimension only, under the influence of an external force. This external force may be from gravity or from a centrifugal-force field. The basic theory of the flow of solids through fluids is based upon the concept of freely moving bodies,

$$Fg_c = m\frac{dv}{d\theta} \tag{22.1}$$

where F is the resultant force acting on any body, $dv/d\theta$ is the acceleration of the body, and m is the mass of the body.

In Figure 22.2, the forces acting on the falling body are the external force (F_E), a buoyant force (F_B), and the drag force (F_D) due to fluid friction in the direction of the velocity of fluid relative to the particle. Then, by Equation 22.1,

$$(F_E - F_D - F_B)g_c = m\frac{dv}{d\theta} \tag{22.2}$$

The external force (F_E) may be expressed by Newton's law as

$$F_E g_c = ma_E \tag{22.3}$$

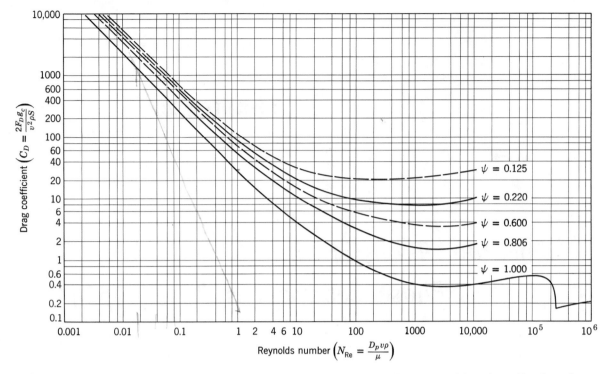

Figure 22.1. Drag coefficient as a function of Reynolds number (43). (From G. G. Brown and Associates, *Unit Operations*, John Wiley and Sons, New York, 1950, with permission.)

where a_E is the acceleration of the particle resulting from the external force. The drag force is obtained from Equation 13.50

$$F_D g_c = \frac{C_D v_{fs}{}^2 \rho S}{2} \qquad (13.50)$$

where ρ is the fluid density and v_{fs} is the velocity of the particle relative to the fluid.

Archimedes principle yields the buoyant force. The mass of the fluid displaced by the solid is $(m/\rho_s)\rho$, where ρ_s and ρ are the solid and fluid densities, respectively. Therefore,

$$F_B g_c = \left(\frac{m}{\rho_s}\right) \rho a_E \qquad (22.4)$$

Substituting Equations 22.3, 13.50, and 22.4 into Equation 22.2 gives

$$\frac{dv}{d\theta} = a_E - \frac{\rho a_E}{\rho_s} - \frac{C_D v_{fs}{}^2 \rho S}{2m} \qquad (22.5)$$

Equation 22.5 is a general equation for the total force acting on a body in any force field. Its solution requires a knowledge of the nature of the external force and the drag coefficient. If the external force is gravity, a_E is equal to the acceleration of gravity g, and Equation 22.5 becomes

$$\frac{dv}{d\theta} = g\left(1 - \frac{\rho}{\rho_s}\right) - \frac{C_D v^2 \rho S}{2m} \qquad (22.6)$$

If the external force is from a centrifugal field, $a_E = r\omega^2$, where r is the radius of path and ω the angular velocity in radians/sec. Equation 22.5 becomes for this case

$$\frac{dv}{d\theta} = r\omega^2\left(1 - \frac{\rho}{\rho_s}\right) - \frac{C_D \rho v^2 S}{2m} \qquad (22.7)$$

Equations 22.6 and 22.7 are both important in solving mechanical-separation problems.

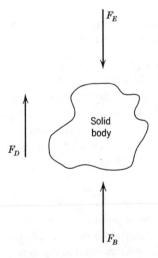

Figure 22.2. Forces on a body.

Terminal Velocity. Consider the particle of Figure 22.2 to be falling in a gravitational field in such a manner that other particles which might be present do not hinder its fall. As the particle falls, its velocity increases and will continue to increase until the accelerating and resisting forces are equal. When this point is reached, the particle velocity remains constant during the remainder of its fall unless the balance of forces is upset. The ultimate constant velocity is called the *terminal velocity.*

For spherical particles the projected area normal to flow is $\pi D_p^2/4$, and the mass is $(\pi D_p^3/6)\rho_s$. Using Equation 22.6 for the gravitational-field case, and substituting for S and m gives

$$\frac{dv}{d\theta} = g\left(1 - \frac{\rho}{\rho_s}\right) - \frac{3C_D v^2 \rho}{4 D_p \rho_s} \qquad (22.8)$$

At the terminal velocity, $dv/d\theta = 0$; therefore,

$$\frac{3C_D v_t^2 \rho}{4 D_p \rho_s} = g\left(1 - \frac{\rho}{\rho_s}\right) \qquad (22.9)$$

Solving Equation 22.9 for the terminal velocity (v_t) yields

$$v_t = \sqrt{\frac{4(\rho_s - \rho)g D_p}{3 C_D \rho}} \qquad (22.10)$$

Equation 22.10 may be used to calculate the terminal velocity in laminar, transition, or turbulent flow if C_D is evaluated from Figure 22.1.

An expression for the terminal velocity independent of C_D may be developed for particles in *laminar flow.* The resisting force due to fluid friction acting on a sphere when the relative motion produces laminar flow has been shown by Stokes (39) to be

$$F_D g_c = 3\pi D_p \mu v \qquad (22.11)$$

This resistance term may be substituted in Equation 22.2 along with the other force terms to give

$$m\frac{dv}{d\theta} = m\left(1 - \frac{\rho}{\rho_s}\right)g - 3\pi D_p \mu v \qquad (22.12)$$

Since $m = (\pi D_p^3/6)\rho_s$ for spheres,

$$\left(\frac{\pi D_p^3}{6}\right)\rho_s\frac{dv}{d\theta} = \frac{\pi D_p^3}{6}(\rho_s - \rho)g - 3\pi D_p \mu v \qquad (22.13)$$

or

$$\frac{dv}{d\theta} = \frac{(\rho_s - \rho)g}{\rho_s} - \frac{18\mu v}{D_p^2 \rho_s} \qquad (22.14)$$

At the terminal velocity, $dv/d\theta = 0$ and

$$v_t = \frac{(\rho_s - \rho)g D_p^2}{18\mu} \qquad (22.15)$$

Equation 22.15 is a statement of Stokes' law which is applicable to the fall of spherical particles in *laminar flow.* It is used, for example, for calculating viscosity using a falling-ball viscosimeter. A ball of known diameter flows through a fluid of unknown viscosity in a tube. The time of fall between two index points is measured, and by Equation 22.15 the viscosity can be determined.

If Equation 22.15 is substituted into Equation 22.10, an expression for the drag coefficient for laminar flow results. First, Equation 22.10 is solved for C_D.

$$C_D = \frac{4(\rho_s - \rho)g D_p}{3 v_t^2 \rho} \qquad (22.10a)$$

If one v_t in Equation 22.10a is replaced by Equation 22.15,

$$C_D = \frac{4(\rho_s - \rho)g D_p}{3 v_t \rho} \cdot \frac{18\mu}{(\rho_s - \rho)g D^2} = \frac{24\mu}{D_p v_t \rho} = \frac{24}{N_{Re}} \qquad (22.16)$$

where N_{Re} is the Reynolds number for the particle. Equation 22.16 is for laminar flow. A particle may be considered to be in laminar flow up to a particle Reynolds number of 0.1, where transition to turbulent flow begins. Above a Reynolds number of 1.0, Equation 22.10 must be used.

Equation 22.10 is referred to as Newton's law and is used to evaluate terminal velocities of falling spherical particles. It may also be used for nonspherical particles if some characteristic dimension is used for D_p, and C_D is evaluated at the proper sphericity (see Appendix B). Note that, in the use of Equation 22.10, C_D is also a function of velocity, resulting in one equation with two unknowns. The second "equation" is Figure 22.1. A technique for simultaneous solution follows.

Equation 22.10 may be solved for C_D and the results expressed in the logarithmic form.

$$\log C_D = \log\left[\frac{4g D_p(\rho_s - \rho)}{3\rho}\right] - 2\log v_t \qquad (22.17)$$

Expressing the Reynolds number at the terminal velocity in logarithmic form yields

$$\log N_{Re} = \log\frac{D_p \rho}{\mu} + \log v_t \qquad (22.18)$$

Eliminating v_t between Equations 22.17 and 22.18 gives

$$\log C_D = -2\log N_{Re} + \log\left[\frac{4g D_p^3 \rho(\rho_s - \rho)}{3\mu^2}\right] \qquad (22.19)$$

Equation 22.19 is the equation for a straight line of slope (-2) passing through the point $N_{Re} = 1$ and $C_D = 4g D_p^3 \rho(\rho_s - \rho)/3\mu^2$. In this equation, v_t does not appear, but it may be determined by plotting Equation 22.19 on Figure 22.1. The intersection of this line with the proper sphericity curve will give the terminal Reynolds number from which v_t can be calculated.

In a manner similar to the development of Equation 22.19, an expression may be derived in which the size of the particle does not appear. The expression is

$$\log C_D = \log N_{Re} + \log \left[\frac{4g(\rho_s - \rho)\mu}{3\rho^2 v_t^3} \right] \quad (22.20)$$

The size of a particle having a specified terminal velocity can be determined, by plotting Equation 22.20 on Figure 22.1. Its intersection with the proper sphericity curve gives the terminal Reynolds number from which D_p can be calculated.

Illustration 22.1. Calculate the terminal velocity for spherical droplets of coffee extract, 400 microns in diameter, falling through air. The specific gravity of the coffee extract is 1.03, and the air is at a temperature of 300°F.

SOLUTION. In this problem, the terminal velocity can be calculated using Equation 22.10; but, since the velocity is unknown, C_D cannot be directly evaluated. This method would require a trial-and-error solution, but the problem is easily solved using Equation 22.19. This equation will be plotted on Figure 22.1 using the specified data. It will pass through the point $N_{Re} = 1.0$ and $C_D = [4g D_p^3 \rho(\rho_s - \rho)/3\mu^2]$ with a slope of -2.

$$D_p = \frac{400 \times 10^{-4}}{30.48} = 1.31 \times 10^{-3} \text{ ft}$$

$$\mu_{air} = 0.026 \times 6.72 \times 10^{-4} = 1.747 \times 10^{-5} \text{ lb/ft sec}$$

$$\rho_s = 1.03 \times 62.3 = 64.1 \text{ lb/cu ft}$$

$$\rho_{air} = \frac{29}{359} \times \frac{492}{760} = 0.0524 \text{ lb/cu ft}$$

$$C_D = \frac{4g D_p^3 \rho(\rho_s - \rho)}{3\mu^2}$$

$$C_D = \frac{(4)(32.2)(0.00131)^3(0.0524)(64.1 - 0.0524)}{(3)(1.747 \times 10^{-5})^2}$$

$$C_D = 611$$

On Figure 22.1a at ($C_D = 611$, $N_{Re} = 1.0$), draw a line of slope of -2. At its intersection with $\psi = 1.0$, $N_{Re} = 14$.

Therefore, $D_p v_t \rho/\mu = 14$,

and

$$v_t = \frac{14\mu}{D_p \rho} = \frac{14(1.747 \times 10^{-5})}{(1.31 \times 10^{-3})(5.24 \times 10^{-2})} = 3.57 \text{ ft/sec}$$

Hindered Fall of Spherical Particles. In the usual operational case, many particles are present, and the surrounding particles interfere with the motion of other individual particles. The influence of the neighboring particles affects the velocity gradients surrounding each particle. For this hindered flow, the settling velocity is considerably less than that which would be calculated from Equation 22.15. The mechanism of fall differs in that the particle is settling through a suspension of particles in a fluid rather than through the fluid itself.

The density of the fluid phase becomes effectively the bulk density of the slurry, which is the quotient of total mass of liquid plus solid divided by the total volume. The viscosity of the slurry is considerably higher than that of the fluid because of the interference of boundary layers around interacting solid particles, and because of the increase of form drag caused by the solid particles. The viscosity of such a slurry is frequently a function of the rate of shear, of the previous history as it affects clustering of particles, and of the shape and roughness of the particles insofar as these factors contribute to a thicker boundary layer.

As would be surmised from the above, generalized predictions of the viscosity of slurries are impossible. Experimental measurements are necessary for accurate values, and extrapolation of any variable should be made with caution. For one particular slurry of nonclustering spheres, the measurements can be correlated into a ratio of the effective bulk viscosity (μ_B) to the viscosity of the liquid expressed as a function of the volume fraction of liquid ($\mathscr{X}$) in the slurry for a reasonable range of compositions (3):

$$\frac{\mu_B}{\mu} = \frac{10^{1.82(1-\mathscr{X})}}{\mathscr{X}} \quad (22.21)$$

For greatest convenience in use, an equation using only the physical properties of the solid and liquid phases would be most desirable. A correction factor (R) incorporating both the viscosity and density effects can be derived for a given slurry, allowing use of a more convenient equation based upon Equation 22.15:

$$v_H = \frac{(\rho_s - \rho)g D_p^2}{18\mu} R \quad (22.22)$$

where v_H is the terminal velocity in hindered settling.

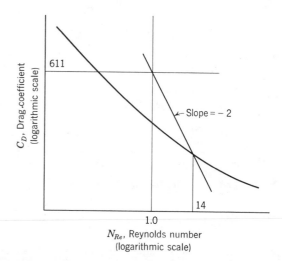

Figure 22.1a. Evaluation of N_{Re} at v_t, Ill. 22.1.

Experimental measurements of effective viscosity as a function of composition may be reported in the form of Figure 22.3, which represents the slurry viscosity fitted to Equation 22.21. For slurries which are non-Newtonian, such a curve must be used only in situations paralleling the measurements. Caution must be exercised that all rheological factors are recognized and controlled in such comparisons.

Two-Dimensional Motion. The previous discussion was limited to simple vertical flow of a particle in one-dimensional flow. However, the particle may have a horizontal velocity component as well. Such would be the case of a liquid being dispersed from a spinning-disk atomizer where the liquid is thrown from the disk in a horizontal path. In this case, a significant initial horizontal-velocity component exists. As the fall continues, the horizontal-velocity component decreases. For a development and solution of the equations of motion for particles having a second velocity component the reader is referred to Lapple and Shepherd (25).

Applications of Particle-Flow Theory. The mechanics of particle motion discussed thus far can be directly applied to industrial operations of particle separation from fluids. The mineral-dressing industries widely use some of the techniques, but the chemical industries also use many of them. For example, the elimination of dust and dirt from flue or stack gases and the removal of solids from liquids prior to discharge as waste are cases common to the chemical-process industries. The emphasis upon clean streams and clean atmosphere makes the chemical engineer more and more concerned with the "cleaning" of fluids. He may also be involved with solid separations as a consequence of one or more steps in a chemical process.

Classification. The separation of solid particles into several fractions based upon their terminal velocities is called *classification.* Suppose for example, that two

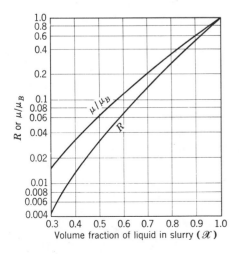

Figure 22.3. Settling factor for hindered settling (38).

particles having different settling velocities are placed in a rising current of water. If the water velocity is adjusted to a value between the terminal velocities of the two particles, a separation will result. The slower-settling particle will move upward with the water, while the faster-settling particle will simultaneously settle out to the bottom. Another way to separate the particles would be to feed the suspension through a tank of large cross-sectional area. When the fluid stream enters the tank, the horizontal velocity component decreases, and the particles start to settle. The faster-settling particles will tend to accumulate near the inlet, but the slower-settling particles will be carried farther and will concentrate nearer the exit.

Suppose two different materials a and b are present in a solid particulate mixture with a more dense than b. If the size range of the two materials is large, no complete separation is likely because the terminal velocities of the largest particles of b may be greater than that of the smallest particles of a. The range of sizes that will result from a separation can be calculated from the ratio of sizes of the particles of a and b which have the same terminal velocities. The equality of terminal velocities occurs when

$$v_t = \sqrt{\frac{4(\rho_a - \rho)gD_a}{3\rho C_{Da}}} = \sqrt{\frac{4(\rho_b - \rho)gD_b}{3\rho C_{Db}}} \quad (22.23)$$

or

$$\frac{D_a}{D_b} = \frac{(\rho_b - \rho)}{(\rho_a - \rho)} \cdot \frac{C_{Da}}{C_{Db}} \quad (22.24)$$

From the shape of the drag coefficient–Reynolds number diagram, it is evident that the value of the drag coefficient is essentially constant at high values of the Reynolds number. Therefore, for particles of the same sphericity settling at high Reynolds numbers, Equation 22.24 becomes

$$\frac{D_a}{D_b} = \frac{\rho_b - \rho}{\rho_a - \rho} \quad (22.25)$$

Under laminar-flow conditions, $C_D = 24/N_{Re}$, and

$$C_{Da} = \frac{24\mu}{D_a v_t \rho} \quad (22.26)$$

$$C_{Db} = \frac{24\mu}{D_b v_t \rho} \quad (22.27)$$

Substituting Equations 22.26 and 22.27 into Equation 22.24 gives, for different-sized particles of different density settling at the same terminal velocity

$$\frac{D_a}{D_b} = \frac{(\rho_b - \rho)}{(\rho_a - \rho)} \cdot \frac{D_b}{D_a} \quad (22.28)$$

or

$$\left(\frac{D_a}{D_b}\right)^2 = \frac{\rho_b - \rho}{\rho_a - \rho} \quad (22.28a)$$

Separation is possible only if a separation ratio, defined as the ratio of the size of the smallest particle of a to that of the largest particle of b, is greater than

$$\frac{D_a}{D_b} > \left[\frac{\rho_b - \rho}{\rho_a - \rho}\right]^n \qquad (22.29)$$

where $n = \frac{1}{2}$ for laminar flow

$n = 1$ for turbulent flow

$\frac{1}{2} < n < 1$ for transition flow

Examination of Equation 22.29 indicates that, by choosing a fluid with a density very nearly equal to that of one of the solid materials, the size ratio of the separation can be made larger or smaller. For example, if ρ approaches ρ_b, the settling ratio approaches zero and particles of any size range can be separated. This is the basis for "heavy-medium separations" which can be much more effective than hydraulic classification. The liquid phase may be made heavier by dissolving a soluble material in it which increases the density. More frequently, the effective density of the liquid is increased by dispersing in it a heavy solid which is so finely ground that its settling velocity is negligible. The "liquid" phase in heavy-medium separations is actually a relatively stable suspension, in which the differences in terminal velocities of particles of different densities are magnified.

Illustration 22.2. A mixture of silica and galena is to be separated by hydraulic classification. The mixture has a size range between 0.008 cm and 0.07 cm. The density of galena is 7.5 gm/cu cm, and the density of the silica is 2.65 gm/cu cm. Assume the sphericity, $\psi = 0.806$. (a) What water velocity is necessary for a pure galena product? Assume unhindered settling of the particles and a water temperature of 65°F. (b) What is the maximum size range of the galena product?

SOLUTION. (a) For equal-sized silica and galena particles, the heavier galena will settle faster. Therefore, the settling velocity of the largest silica particle will determine the water velocity. A water velocity equal to this settling velocity should give a pure galena product. By Equation 22.19, using the metric system of units

$$\log C_D = -2 \log N_{\text{Re}} + \log \left[\frac{4g D_p{}^3 \rho(\rho_{sil} - \rho)}{3\mu^2}\right]$$

The viscosity of water at 65°F = 0.01 poise

$$\log C_D = -2 \log N_{\text{Re}} + \log \left[\frac{(4)(981)(0.07)^3(1.0)(2.65 - 1)}{3(0.01)^2}\right]$$

$$\log C_D = -2 \log N_{\text{Re}} + \log 7400$$

This is a straight line on Figure 22.1 passing through ($N_{\text{Re}} = 1$, $C_D = 7400$) with a slope of -2. This line intersects the $\psi = 0.806$ curve at $N_{\text{Re}} = 28$.

A Reynolds number of 28 corresponds to a settling velocity of

$$v_t = \frac{N_{\text{Re}}\mu}{D_p \rho} = \frac{(28)(0.01)}{(0.07)(1.0)} = 4.0 \text{ cm/sec}$$

This velocity must also be the water velocity to ensure a clean galena product, since it will carry all silica overhead.

(b) Calculation of the size of a galena particle that settles at a velocity of 4.0 cm/sec fixes the smallest galena particle in the galena product. By Equation 22.20,

$$\log C_D = \log N_{\text{Re}} + \log \left[\frac{4g(\rho_{gal} - \rho)\mu}{3\rho^2 v_t{}^3}\right]$$

$$\log C_D = \log N_{\text{Re}} + \log \left[\frac{(4)(981)(7.5 - 1.0)0.01}{(3)(1)^2(4.0)^3}\right]$$

$$\log C_D = \log N_{\text{Re}} + \log 1.33$$

This is a straight line on Figure 22.1 passing through ($N_{\text{Re}} = 1.0$, $C_D = 1.33$) with a slope of $+1$. This line intersects the $\psi = 0.806$ curve at $N_{\text{Re}} = 9.0$. This Reynolds number corresponds to a diameter of

$$D_p = \frac{N_{\text{Re}}\mu}{v\rho} = \frac{(9)(0.01)}{(4.0)(1.0)} = 0.0225 \text{ cm}$$

Thus, the galena product size ranges between 0.0225 cm and 0.07 cm. Galena particles smaller than 0.0225 cm are carried overhead along with all silica.

Classification Equipment. The simplest type of classifier is one which consists of a large tank with provisions for a suitable inlet and outlet. Figure 22.4 illustrates such a *gravity settling tank.*

A slurry feed enters the tank through a pipe. Immediately upon entry, the linear velocity of the feed decreases as a result of the enlargement of cross-sectional area. The influence of gravity causes the particles to settle, with the faster-settling particles falling to the bottom of the tank near the entrance. The slower-settling particles will be carried farther into the tank before settling to the bottom. The placing of vertical baffles within the tank allows for the collection of several fractions. The very fine particles will be carried out of the tank with the liquid overflow. The position in the tank at which a certain size particle may be expected can be calculated on the assumption that the particles quickly reach terminal velocities. The relationships developed earlier apply to the deposition location of any given particle size.

The gravity-settling tank is referred to as a surface-velocity classifier. The resulting separation is not a sharp one, since rather considerable overlapping of size occurs.

The *Spitzkasten* (Figure 22.5) is another type of gravity-settling chamber. It consists of a series of conical vessels of increasing diameter in the direction of flow. The slurry feed enters the top of the first vessel where the larger, faster-settling, particles are separated. The overflow, including unseparated solids, feeds into

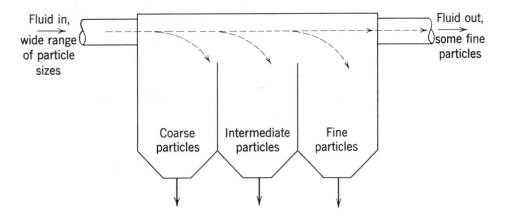

Figure 22.4. Gravity-settling tank.

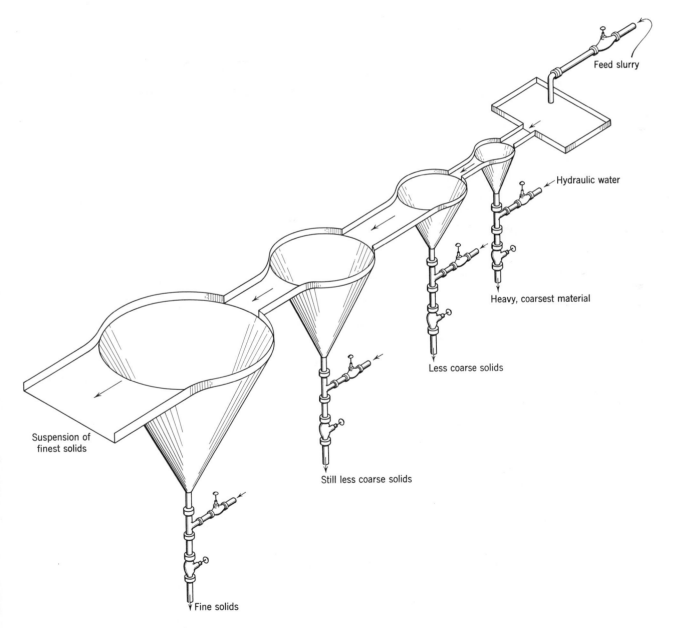

Figure 22.5. Schematic representation of a Spitzkasten.

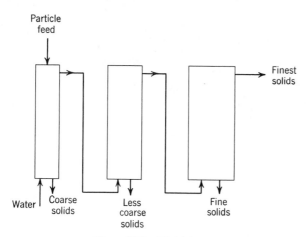

Figure 22.6. Elutriators.

the top of the second vessel, where another separation occurs since the velocity is lower than that in the first vessel. Each succeeding vessel will separate a different range of particles depending upon settling velocities of the particles and upon the fluid velocity in the vessel. In each vessel the velocity of the upflowing fluid is independently controlled to give the desired size range of product from that vessel.

An *elutriator* (Figure 22.6) is a vertical tube through which a fluid passes upwardly at a specific velocity while a solid mixture, whose separation is desired, is fed into the top of the column. The large particles, which settle at a velocity higher than that of the rising fluid, are collected at the bottom of the column, and the smaller particles are carried out of the top of the column with the fluid. Several columns of different diameters in series may be used to bring about a further separation.

A *double-cone classifier* (Figure 22.7) is a conical vessel inside of which is a second cone. The inner cone is slightly larger in angle and is movable in a vertical direction so that a variable annular area is available for flow.

The feed material flows downward through the inner cone and out at a baffle at the bottom of the inner cone. Rising upward through the annular space, fluid is fed at a controlled velocity into the unit in the vicinity of the exit of the inner cone. The solids from the inner cone and fluid are mixed and then flow through the annulus whose cross-sectional area varies. Classification action occurs in this annular space with the small particles leaving with the liquid and the larger particles settling to the bottom for removal.

Another piece of equipment for solid separation whose mechanism is based upon settling velocity is the *rake classifier* (Figure 22.8). The rake classifier is a tank with an inclined bottom in which are provided movable rakes. Feed is introduced near the middle of the tank. The lower end of the tank has a weir overflow over which

the fines that have not settled leave in dilute suspension. The heavy material sinks to the bottom where the rakes scrape it upwards towards the top of the tank. The stroking action of the rakes is such that, when the stroke is completed, the rakes are lifted and returned to the starting position. The slurry is thus kept in continuous agitation. The time of the raking stoke is adjusted so the heavy particles have time to settle while the fines remain near the surface of the slurry in the rake compartment. Thus, the heavy material is moved upwards along the floor of the tank and removed at the top of the apparatus as a dense slurry.

For classifying fine material, a *bowl classifier* (Figure 22.9) is used. It consists of a shallow cylinder with a gentle conical bottom. The feed slurry is fed at the center near the surface, and the liquid flows outward in a radial direction. The solids settle out according to their particle size with the fines overflowing at the bowl wall. The heavy sludge is scraped towards the center for discharge.

Centrifugal Classification. The most widely used of the centrifugal-type separation equipment is the cyclone separator for separating dust or mist from gases (Figure 22.10).

The feed enters the cyclone tangentially near the top and is given a spinning motion as it enters the chamber proper. The tangential velocity of the particles tends to carry them toward the periphery of the chamber. The spiral motion of the fluid results in some inward radial acceleration of the particle, and simultaneously gravitational force imparts downward acceleration. The result is a downward and spiraling path of increasing radius until the particle reaches the boundary. Thereafter the particles continue a spiraling path down the wall; and the gas, freed of solids, moves upward in the central core. At high tangential velocities, the outward force on the particle is many times the force of gravity, hence cyclones accomplish more rapid and more effective separation than gravitational-settling chambers for

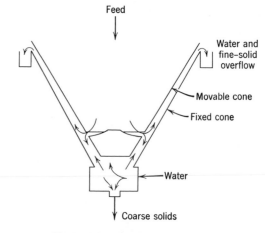

Figure 22.7. Double-cone classifier.

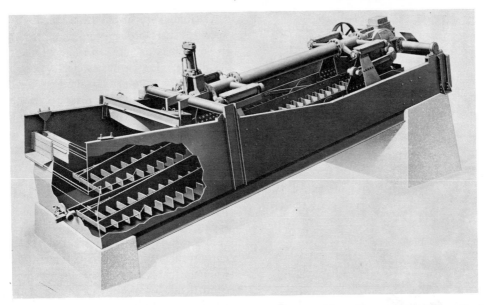

Figure 22.8. Heavy-duty rake classifier. Cutaway shows rake assemblies. (Courtesy Dorr-Oliver, Inc.)

Figure 22.9. 8 ft by 12 ft diameter bowl classifier. (Courtesy Dorr-Oliver, Inc.)

particles of sizes down to a few microns. For very small sizes, the energy represented by the tangential velocity is insufficient to overcome the centripetal force of the rotating fluid, and separation is ineffective.

It is assumed that the particles in a cyclone quickly reach their terminal velocities. Separation of solids from fluids in a cyclone usually involves particle sizes small enough so that Stokes' law may be assumed valid, irrespective of gas velocity. Equations 22.7 and 22.16 give, for a spherical particle obeying Stokes' law,

$$\frac{\pi D_p{}^3}{6} r\omega^2(\rho_s - \rho) = 3\pi D_p v_R \mu \qquad (22.30)$$

where v_R is the radial velocity of the particle. But

$r\omega^2 = v_{\text{tan}}^2/r$ where v_{tan} is the tangential velocity of the particle at radius r. Thus

$$\frac{\pi D_p{}^3}{6} \frac{v_{\text{tan}}^2}{r} (\rho_s - \rho) = 3\pi D_p v_R \mu \qquad (22.30a)$$

Solving Equation 22.30a for v_R gives

$$v_R = \left[\frac{D_p{}^2}{18} \cdot \frac{(\rho_s - \rho)}{\mu}\right] \frac{v_{\text{tan}}^2}{r} \qquad (22.31)$$

but, by Equation 22.15, the bracket term in Equation 22.31 is the gravitational terminal velocity of the particle with the constant, g, omitted. Therefore,

$$v_R = \frac{v_t}{g} \frac{v_{\text{tan}}^2}{r} \qquad (22.32)$$

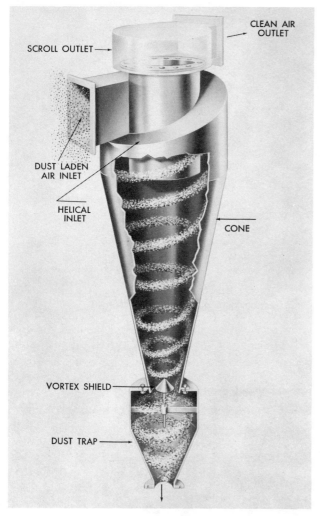

Figure 22.10. Cyclone Separator. (Courtesy The Ducon Co.)

Cyclones are usually designed to operate with a specified pressure drop. The pressure drop through cyclones may be between 1 and 20 times the inlet velocity head (37). Lapple (24) suggests an inlet gas velocity of 50 ft/sec for ordinary installations.

Through the use of plastic cyclones, the effects of size, shape, and air velocity have been studied by several investigators (14, 41). Significant cyclone dimensions are height, diameter, exit-duct diameter, and the ratio of the diameter of the cylinder to that of the exit duct.

Wet Scrubbers. For gases containing very fine particles, liquid scrubbing is sometimes used for effective separation. The dirty gas passes upward through water sprays which tend to wash out the dirt particles and entrain them for removal at the bottom of the scrubber. Figure 22.12 shows a cyclone scrubber in which both centrifugal force and scrubbing action work toward elimination of dirt.

Many variations of this type of scrubbing equipment are available.

Centrifugation. Centrifuges of the solid-bowl type are settling devices that utilize a centrifugal field rather than a gravity field to cause separation of the components of liquid-solid or liquid-liquid systems. Gas-solid systems are separated by the same effects in the cyclone separator discussed earlier. The centrifugal field causes particles of the heavier phase to "fall" through the lighter phase away from the center of rotation. This action is exactly that which occurs under the influence of the gravity field in classification.

Centrifuges. There are three main types of centrifuges distinguishable by the centrifugal force developed, the range of throughputs normally obtained, and the solids concentration that can be handled (15). The first of these is the *tubular-bowl centrifuge*. This centrifuge rotates at high speeds developing centrifugal forces of the order of 13,000 times the force of gravity, but it is built for low capacities in the range between 50 and 500 gal/hr. Since it has no automatic solids removal system, it can handle only small concentrations of solids. The second is the *disk-bowl centrifuge*, which is larger than the tubular-bowl centrifuge and rotates at slower speed, developing a centrifugal force up to 7000 times gravity. This centrifuge may be designed to handle as much as 5000 gal/hr of a feed containing moderate quantities of solids which are discharged continuously in a concentrated stream. Both of these types of centrifuge are primarily designed to separate liquid-liquid systems. However, the disk-bowl centrifuge can be adapted to separate liquid-liquid-solid systems, or liquid-solid systems where the major product is a clarified liquid. The *solid-bowl centrifugal*, which forms the third class, is primarily a solid-liquid separator and operates like a thickener (to be described

Thus the higher the terminal velocity, the greater the radial velocity, and the easier it should be to separate the particle. The evaluation of the radial velocity is complex since it is a function of terminal velocity, tangential velocity, and position from the center of the cyclone. For a given-sized particle, the radial velocity is a minimum near the center of the cyclone and increases towards the wall.

Theoretically, the size of the smallest particle that is retained by the cyclone can be calculated, but experience indicates that agglomeration of particles and entrainment tend to introduce deviations from the calculations. If a cyclone is designed to separate a definite particle size, a cyclone efficiency may be defined as the mass fraction of the particles of that size that is separated in the cyclone. Cyclone efficiency increases as the size of the particles increases. If a cut diameter is defined as that diameter for which half the particles by weight will be separated, the efficiency of a typical cyclone separator is shown in Figure 22.11.

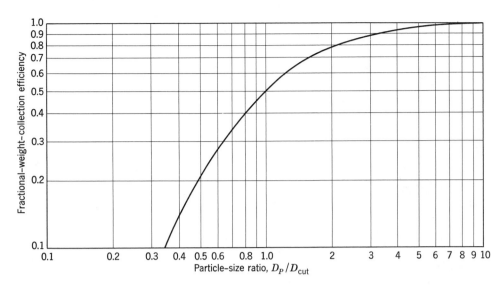

Figure 22.11. Cyclone efficiency.

later). These centrifugals are built to handle solids at rates up to 50 tons/hr.

The *tubular-bowl centrifuge* is shown in Figure 22.13. It consists of a tubular bowl that rotates within a housing. The bowl stands vertically, hung from a thin, flexible, solid shaft supported by a thrust bearing and is belt driven from an electric motor or direct driven with an electric motor or from an air or steam turbine. At the bottom, the bowl is loosely guided by a spring-supported bushing. The feed is delivered to a nozzle at the bottom of the bowl, and jets into the bowl where it is quickly accelerated to bowl speed by the action of light metal vanes loosely fitted into the bowl. The heavy phase collects along the walls of the bowl, the light phase forming a concentric layer on the inside of the heavy phase. Droplets of light liquid in the heavy phase move toward the center of the bowl, and droplets of heavy liquid move toward the wall of the bowl. The 2- to 5-ft length of the bowl affords sufficient residence time for the droplets to reach their proper phase layer. The layers are maintained in the bowl and their separate discharge is controlled by ring dams at the top end of the bowl. There is no arrangement for removing solids, and, if they are present, they usually build up on the walls of the bowl until the unit is stopped and cleaned out.

A cutaway view of a *disk-bowl centrifuge* is shown in Figure 22.14. This type of centrifuge was invented by DeLaval in 1878 and has been widely used since then for such diverse applications as cream separation, catalyst separation, the dehydrating of marine lubricating oils, and the refining of fish oils. In the unit shown, the feed is poured into the open top where it passes through a screen before entering the centrifuge bowl. Feed flows into the centrifuge bowl from the top, downward around the drive spindle, and up through holes in the disks. From these holes, it flows into the spaces between the disks which break the feed into layers that permit the separation of the two liquid phases across a short path and over a large area. The heavy phase flows down the underside of the disks toward the outer reservoir in the bowl, while the light phase flows along the top of the disks toward the center of the bowl. If heavy solids are suspended in the feed, they will flow with the heavy liquid phase and collect at the outer edge of the bowl where twelve nozzles discharge them along with some of the heavy liquid phase. The balance of the heavy phase, like the light phase, discharges over a dam at the top of the centrifuge bowl.

A cross-sectional view of a *solid-bowl centrifugal* is shown in Figure 22.15, and the action of this centrifugal is diagrammed in Figure 22.16. The major parts of this machine are a truncated cone-shaped bowl and an internal screw conveyor for solids that closely fits the cone of the bowl. These parts rotate together, but the screw conveyor rotates at a rate 1 or 2 rpm below the rate of rotation of the bowl. In operation, the feed is admitted through the central screw and enters the bowl about halfway along the side of the cone. The centrifugal action forces both liquid and solid phases to the walls of the cone and down the cone to the large end. The solids, being denser, concentrate against the cone walls and along the bottom of the pool of liquid held in the bowl by the position of the filtrate discharge parts. The solids conveyor, however, has a net rotation toward the small end of the bowl and scrapes the solids from the walls of the cone toward the small end. As the solids move in this direction, they are given a fresh-water wash with wash water entering in the same manner as the feed entered. They are ultimately discharged at the small end of the conical bowl.

These centrifugals are made with maximum bowl

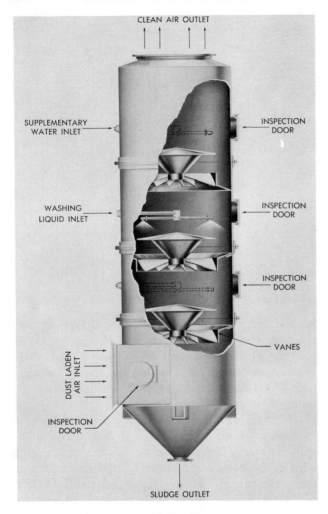

Figure 22.12. Centrifugal washer. (Courtesy The Ducon Co.)

diameters ranging between 4 and 54 in. The 54-in. machine handles up to 50 tons/hr of solids, though this rate must be reduced if the particles are particularly fine or if the liquid phase is viscous. These machines may also be operated as classifiers, in which case the feed rate and bowl speed are adjusted so that satisfactorily small particles will not settle out and leave with the filtrate. This type of operation might occur with the centrifugal accepting the product from a wet-grinding step and feeding the large-particle solids back to the grinder while the filtrate with its fine particles passes on for further processing. These units develop a centrifugal force up to 600 times that of gravity, so that sharp separations can be made in the 1-micron particle-size range.

Similar centrifugals are made with a perforated wall on the cone-shaped bowl. These centrifugals act exactly like filters (to be described later) with the filtrate draining through the cake and bowl wall into a surrounding collector. As with other centrifugal filters, they operate best with coarse-granular or coarse-crystalline, free-flowing solids.

Centrifuge Theory and Calculations—Rate of Separation. The basic force balance around a particle falling in a centrifugal force field was given earlier as

$$\frac{dv}{d\theta} = r\omega^2\left(\frac{\rho_s - \rho}{\rho_s}\right) - \frac{C_D \rho v^2 S}{2m} \qquad (22.7)$$

In the separation of phases by settling either in a gravitational or centrifugal field, the perfection of separation is limited by the rate of fall of the smallest particles present. In most cases, these particles fall at rates low enough so that laminar flow exists and $C_D = 24/N_{Re}$, as shown by Figure 13.6. Making this substitution in Equation 22.7 and considering spherical particles so that

$$\frac{m}{S} = \frac{\rho_s V}{S} = \frac{\frac{\pi}{6} D_p^3 \rho_s}{\frac{\pi}{4} D_p^2} = \frac{4\rho_s D_p}{6}$$

Then Equation 22.7 becomes

$$\frac{dv}{d\theta} = r\omega^2\left(\frac{\rho_s - \rho}{\rho_s}\right) - \frac{18\mu v}{\rho_s D_p^2} \qquad (22.33)$$

As particles move radially in a centrifugal field the field strength changes with their position. Consequently,

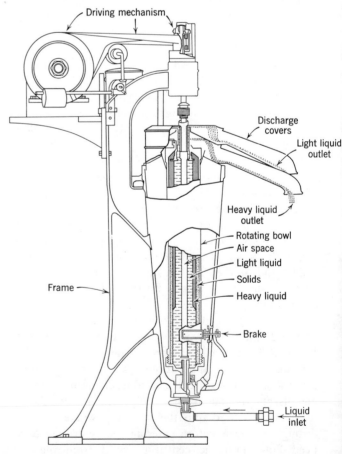

Figure 22.13. Cutaway view of tubular-bowl centrifuge.

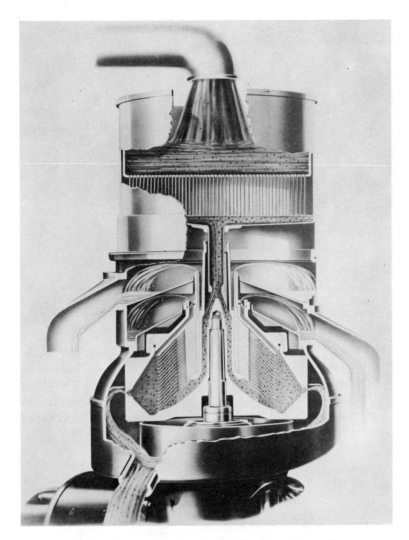

Figure 22.14. Cutaway drawing of a disk-bowl centrifuge with nozzles for continuous solids discharge. (Courtesy DeLaval Separator Company.)

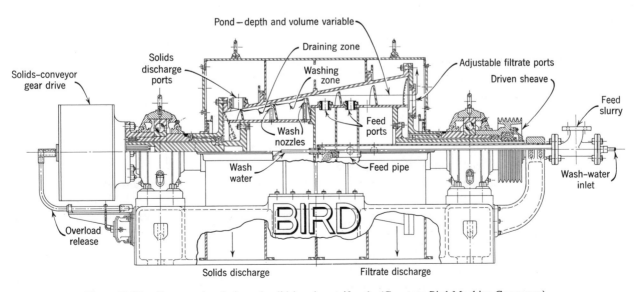

Figure 22.15. Cross-sectional view of solid-bowl centrifugal. (Courtesy Bird Machine Company.)

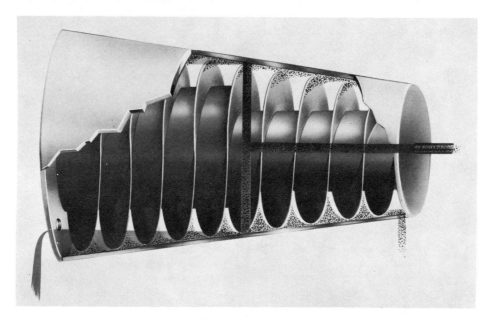

Figure 22.16. Internal separation action of solid-bowl centrifugal. (Courtesy The Sharples Corp.)

the terminal velocity of the particles is a function of radial position. In these developments a particle at any position is considered to move at the unique terminal velocity characteristic of its position. Thus, for any *position*, $dv/d\theta = 0$; but, for any *instant* in the movement of a single particle, dv/dr is positive. Then considering a particular position, $dv/d\theta = 0$, and Equation 22.33 becomes

$$v_R = \frac{r\omega^2(\rho_s - \rho)D_p^{\,2}}{18\mu} \qquad (22.34)$$

where v_R = the terminal falling velocity of spherical particles of diameter D_p at radius r in a centrifugal field rotating at rate ω

The radial distance traveled by the particle may be obtained by multiplying Equation 22.34 by the differential time ($d\theta$).

$$v_R \, d\theta = dr = \frac{r\omega^2(\rho_s - \rho)D_p^{\,2}}{18\mu} \, d\theta \qquad (22.35)$$

which, upon integration, gives

$$\ln\frac{r_2}{r_1} = \frac{\omega^2(\rho_s - \rho)D_p^{\,2}}{18\mu}\,\theta = \frac{\omega^2(\rho_s - \rho)D_p^{\,2}}{18\mu}\cdot\frac{V}{Q} \quad (22.36)$$

where V = volume of material held in the centrifuge
Q = volumetric feed rate to the centrifuge
V/Q = residence time of a particle in the centrifuge

The diameter in Equation 22.36 is that of a particle that falls from r_1 to r_2 during the residence time available in the centrifuge. The significance of this equation is perhaps most readily seen in relation to the parallel equation which holds when the liquid layer within the

centrifuge is very narrow compared to the radius. In this case the centrifugal field is considered constant, and Equation 22.35 may be written directly in terms of the residence time (V/Q).

$$v_R\theta = x = \frac{r\omega^2(\rho_s - \rho)D_p^{\,2}}{18\mu}\cdot\frac{V}{Q} \qquad (22.35a)$$

where x = radial distance traveled by a particle of diameter D_p in the residence time available

If x is taken as half the thickness of the liquid layer $[(r_2 - r_1)/2]$, half the particles of some diameter D_p' will settle to the wall, whereas half of them will still be in suspension when the fluid leaves the centrifuge. D_p', is the "critical diameter." Particles of diameter larger than D_p' will predominantly be settled from the fluid phase, and particles of diameter smaller than D_p' will predominantly remain in solution. Solving Equation 22.35a for D_p', and substituting $(r_2 - r_1)/2$ for x, the critical particle diameter is

$$D_p' = \sqrt{\frac{9\mu Q}{(\rho_s - \rho)\omega^2 V}\cdot\frac{r_2 - r_1}{r}} \qquad (22.37)$$

where $r_2 - r_1$ = the thickness of the liquid layer
D_p' = critical particle diameter

Comparison of this equation with Equation 22.36 shows that for the case where the liquid-layer thickness is great enough so that the variation of centrifugal field with radius must be considered, the effective value of $(r_2 - r_1)/r$ is

$$\left(\frac{r_2 - r_1}{r}\right)_{\text{eff}} = 2\ln\frac{r_2}{r_1} \qquad (22.38)$$

where $\left(\dfrac{r_2 - r_1}{r}\right)_{\text{eff}}$ = the effective value of $(r_2 - r_1)/r$ to be used in Equation 22.37 when $r_2 - r_1$, the thickness of the liquid layer in the centrifuge, is not negligible compared to either r_1 or r_2

A very useful characteristic of a centrifuge can be derived by manipulating Equations 22.37 and 22.15. Equation 22.37 is solved for Q and the gravitational constant inserted.

$$Q = \frac{(\rho_s - \rho)g D_p'^2}{9\mu} \cdot \frac{V\omega^2 r}{g(r_2 - r_1)} = 2v_t \cdot \Sigma \quad (22.39)$$

in which

$$v_t = \frac{(\rho_s - \rho)g D_p'^2}{18\mu} \quad (22.15)$$

where v_t is the terminal settling velocity of a particle in a gravitational field and

$$\Sigma = \frac{V\omega^2 r}{g(r_2 - r_1)} \quad (22.40)$$

where Σ is a characteristic of the centrifuge itself and not of the system being separated. The Σ factor can then be used as a means of comparing centrifuges (1). It is the cross-sectional area of a settler that will remove particles down to the same diameter as those separated in the centrifuge when its volumetric feed rate equals that of the centrifuge. If two centrifuges are to perform the same function,

$$\frac{Q_1}{\Sigma_1} = \frac{Q_2}{\Sigma_2} \quad (22.41)$$

The quantity Σ can be determined for commercial centrifuges, although in some cases the determination requires approximation methods. For the tubular-bowl centrifuge, applying Equation 22.38 gives

$$\Sigma = \frac{\pi\omega^2 l (r_2^2 - r_1^2)}{g \, \ln \dfrac{r_2^2}{r_1^2}} \quad (22.42)$$

where l = the bowl length

For the disk-bowl centrifuge Ambler (1) gives

$$\Sigma = \frac{2n\pi(r_2^3 - r_1^3)\omega^2}{3g \tan \Omega} \quad (22.43)$$

where n = number of spaces between disks in the stack
r_2, r_1 = outer and inner radii of the disk stack
Ω = the conical half angle

Table 22.1 gives values obtained for several types of centrifuges based both on calculations from geometry and on laboratory and plant data. The table shows both the comparative performance of the centrifuges and the effectiveness of the machines in comparison to that which is calculated on the basis of their geometry. Note how much more effective the disk centrifuge is than is any one of the other centrifuges listed. This is the result of the high residence time and short disk-to-disk separating distances built into this machine.

In the development of Equations 22.33 to 22.43 given above, the diameter (D_p) was defined as a particle diameter, with the inference being that the particle would be solid. If a liquid-liquid separation is to be made, the mechanism is no different from that of solid-liquid separations. Droplets of liquid rather than particles of solid migrate, and they migrate across one liquid

Table 22.1. COMPARATIVE PERFORMANCE OF CENTRIFUGES (1)

	Σ Values, sq ft		
	Calculated from Geometry	From Experimental Data Clarifying Ideal Systems	Extrapolation on Commercial Systems (Supercentrifuge Tests)
Laboratory supercentrifuge (tubular bowl $1\frac{3}{4}$ in. I.D. × $7\frac{1}{4}$ in. long)			
(a) 10,000 rpm	582	582	582*
(a) 16,000 rpm	1,485	1,485	
(a) 23,000 rpm	3,070	3,070	1,290
(a) 50,000 rpm	14,520	14,520	not used
No. 16 supercentrifuge (tubular bowl $4\frac{1}{8}$ in. I.D. × 29 in. long)			
(a) 15,000 rpm	27,150	27,150	27,150
No. 2 disk centrifuge, $1\frac{7}{8}$ in. r_1 × $5\frac{3}{4}$ in. r_2 on disks			
1. 52 disks, 35° half angle, 6000 rpm	178,800	98,000	89,400 to 178,800
2. 50 disks, 45° half angle	134,000	72,600	67,900 to 134,000
Super-D-Cantor (solid-bowl centrifugal)			
PN-14 (conical bowl), 3250 rpm (D = 14 in.–8 in., L = 23 in.)	4,750	2,950	2,950*
PY-14 (cylindrical bowl), 3250 rpm (D = 14 in., L = 23 in.)	8,940	5,980	5,980*

* For relatively low throughput rates.

phase and coalesce into the other phase rather than migrating through the fluid phase to the wall. The rate of migration is still calculable through use of Equation 22.37 properly modified for the centrifuge being used.

Illustration 22.3. A liquid-detergent solution of 100-centipoise viscosity and 0.8 gm/cu cm density is to be clarified of fine Na_2SO_4 crystals ($\rho_s = 1.46$ gm/cu cm) by centrifugation. Pilot runs in a laboratory supercentrifuge operating at 23,000 rpm indicate that satisfactory clarification is obtained at a throughput of 5 lb/hr of solution. This centrifuge has a bowl $7\frac{3}{4}$ in. long internally with $r_2 = \frac{7}{8}$ in., and $(r_2 - r_1) = 19/32$ in.

(*a*) Determine the critical particle diameter for this separation.

(*b*) If the separation is to be done in the plant using a No. 2 disk centrifuge with 50 disks at 45° half angle, what production rate could be expected?

SOLUTION. (*a*) The critical particle diameter, (D_p') can be determined directly from Equation 22.37. Since the liquid layer is thick, Equation 22.38 is used to get the effective value of $(r_2 - r_1)/r$.

$$D_p' = \sqrt{\frac{9 \times 100 \times 2.42 \times \left(\dfrac{5}{62.4 \times 0.8}\right) \times 1728}{62.4(1.46 - 0.80) \times (2\pi \times 23,000 \times 60)^2 \, 7.75 \times \pi \, (\overline{0.875^2} - \overline{0.281^2})} \times 2 \ln \frac{0.875}{0.281}}$$

$D_p' = 0.0000033$ ft or 1.08 microns

(*b*) Table 22.1 gives $\Sigma = 1290$ for the laboratory supercentrifuge and $\Sigma = 72,600$ for the No. 2 disk centrifuge. Since these centrifuges are to perform identical functions Equation 22.41 gives

$$Q_2 = \frac{Q_1}{\Sigma_1} \cdot \Sigma_2 = \frac{5}{1290} \times 72,600 = 281 \text{ lb/hr}$$

Centrifuge Theory Calculations—Outlet Dam Settings. In liquid-liquid separations, the position of the outlet

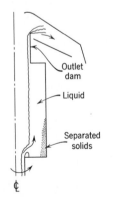

(*a*) tubular-bowl centrifuge arranged for clarifying a liquid of entrained solids.

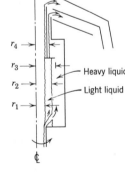

(*b*) tubular-bowl centrifuge arranged for separating two liquid phases.

Figure 22.17. Overflow-dam arrangements in a tubular-bowl centrifuge.

dam becomes more important than it is for solid-liquid separations, for, instead of merely controlling the volumetric hold-up in the centrifuge and the critical particle diameter, the position now also determines whether a separation can be made at all. Figure 22.17 shows the physical situation in a centrifuge arranged for clarifying a liquid phase of entrained solids and that in a centrifuge arranged to separate two liquid phases. In Figure 22.17*b* the distances have the following significances:

$r_1 =$ radius to top of light-liquid layer
$r_2 =$ radius to liquid-liquid interface
$r_3 =$ radius to outside edge of dam
$r_4 =$ radius to surface of heavy-liquid downstream from the dam

The location of the interface is fixed by a balance of forces arising from the hydraulic heads of the two liquid layers. Expressing these forces as pressures gives

$$\int_{r_i}^{r_f} dP = \int_{r_i}^{r_f} \frac{dF}{A} = \int_{r_i}^{r_f} \frac{a \, dm}{g_c A} = \int_{r_i}^{r_f} \frac{\rho(\omega^2 r)(2\pi r l \, dr)}{(2\pi r l) g_c}$$

where the general case is used in which the liquid layer extends from any initial radius (r_i) to any final radius (r_f). Upon simplification

$$P = \int_{r_i}^{r_f} \frac{\rho \omega^2 r \, dr}{g_c} \qquad (22.44)$$

which, upon integration, gives

$$P = \frac{\rho \omega^2}{2g_c} (r_f{}^2 - r_i{}^2) \qquad (22.45)$$

Applying Equation 22.45 to the physical situation of Figure 22.17*b* where the pressure must be the same on either side of the liquid-liquid interface at r_2 results in

$$\frac{\rho_h \omega^2}{2g_c} (r_2{}^2 - r_4{}^2) = \frac{\rho_l \omega^2}{2g_c} (r_2{}^2 - r_1{}^2) \qquad (22.46)$$

or

$$\frac{r_2{}^2 - r_4{}^2}{r_2{}^2 - r_1{}^2} = \frac{\rho_l}{\rho_h} \qquad (22.47)$$

where $\rho_l =$ density of the light phase
  $\rho_h =$ density of the heavy phase

In order for the centrifuge to separate the two liquid phases, the liquid-liquid interface must be located at a radius smaller than r_3 but greater than that of the top of the overflow dam (r_4). Moreover, it is usually found that one of the phases is more difficultly clarified than is the other. To compensate for this, the volume of this phase must be made larger than the volume of the easily clarified phase. This can be done by adjusting the height of the two overflow dams. Note that in the liquid-clarifying centrifuge of Figure 22.17*a* only one dam is

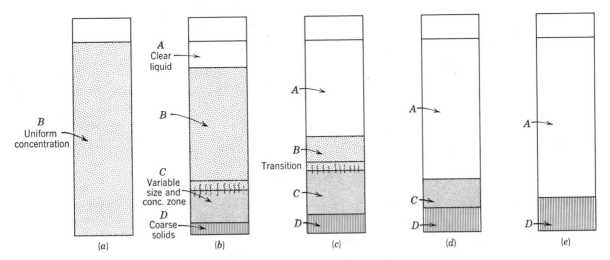

Figure 22.18. Batch sedimentation.

used, and the sole function of this dam is to control the volume of liquid maintained in the centrifuge.

Illustration 22.4. In the primary refining of vegetable oils, the crude oil is partially saponified with caustic and the refined oil separated immediately from the resulting soap stock in a centrifuge. In such a process, the oil has a density of 0.92 gm/cu cm and a viscosity of 20 centipoises, and the soap phase has a density of 0.98 gm/cu cm and a viscosity of 300 centipoises. It is proposed to separate these phases in a tubular-bowl centrifuge with a bowl 30 in. long and 2 in. I.D. rotating at 18,000 rpm. The radius of the dam over which the light phase flows is 0.500 in., whereas that over which the heavy phase flows is 0.510 in.

(a) Determine the location of the liquid-liquid interface within the centrifuge.

(b) If this centrifuge is fed at a rate of 50 gal/hr with feed containing 10 volume percent soap phase, what is the critical droplet diameter of oil held in the soap?

SOLUTION. (a) The interface location can be found directly from Equation 22.47.

$$\frac{\rho_l}{\rho_h} = \frac{r_2^2 - r_4^2}{r_2^2 - r_1^2}; \qquad \frac{0.92}{0.98} = \frac{r_2^2 - \overline{0.510}^2}{r_2^2 - \overline{0.500}^2}$$

$$0.939(r_2^2 - 0.250) = r_2^2 - 0.260$$

$$r_2^2 = 0.426, \ r_2 = 0.654 \text{ in.}$$

(b) The volume of soap phase held in the centrifuge is

$$\frac{30 \times \pi(1^2 - 0.426)}{1728} = 0.0313 \text{ cu ft}$$

This represents a residence time of

$$\frac{0.0313 \times 3600}{50 \times 0.10/7.48} = 169 \text{ sec} = \frac{V}{Q}$$

The critical droplet size, of which 50 per cent will be separated from the soap phase in the 169-sec residence time, can be determined from Equation 22.37.

$$D_p' =$$

$$\sqrt{\frac{9 \times (300 \times 2.42) \times 3600}{(0.98 - 0.92)62.4 \times (2\pi \times 18000 \times 60)^2 \times 169} \, 2 \ln \frac{1.0}{0.654}}$$

$$D_p' = 26.6 \times 10^{-6} \text{ ft or 8.1 microns}$$

Sedimentation. The separation of a dilute slurry by gravity settling into a clear fluid and a slurry of higher solids content is called *sedimentation*. The mechanism of sedimentation may be best described by observation of what occurs during a batch settling test as solids settle from a slurry in a glass cylinder. Figure 22.18a shows a newly prepared slurry of a uniform concentration of solid particles throughout the cylinder. As soon as the process starts, all particles begin to settle and are assumed to approach rapidly the terminal velocities under hindered-settling conditions. Several zones of concentration will be established (Figure 22.18b). Zone D of settled solids will predominantly include the heavier, faster-settling particles. In a poorly defined transition zone above the settled material, there are channels through which fluid must rise. This fluid is forced from zone D as it compresses. Zone C is a region of variable size distribution and nonuniform concentration. Zone B is a uniform-concentration zone, of approximately the same concentration and distribution as initially. At the top of region B is a boundary above which is clear liquid, region A. If the original slurry is closely sized with respect to solids, the line between A and B is sharp.

As sedimentation continues, the heights of each zone vary as indicated in Figure 22.18b, c, d. Note that both A and D grow larger at the expense of B. Eventually, a point is reached where B and C disappear and all the solids appear in D—this is referred to as the *critical settling point* (Figure 22.18e), i.e., the point at which a single distinct interface forms between clear liquid and sediment. The sedimentation process from this point

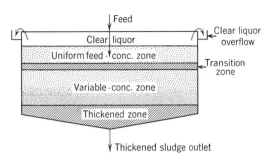

Figure 22.19. Settling zones in continuous thickeners.

on consists of a slow compression of the solids, with liquid being forced upwards through the solids into the clear zone. Settling rates are very slow in this dense slurry. The final phase is an extreme case of hindered settling. Equation 22.22 may be used to estimate settling velocities. It accounts for the effective density and viscosity of the fluid but does not account for agglomeration of particles, so that the calculated settling rate may be in considerable error.

In a batch-sedimentation operation as illustrated, the heights of the various zones vary with time. The same zones will be present in continuously operating equipment. However, once steady state has been reached (where the slurry fed per unit time to the thickener is equal to the rate of sludge and clear liquor removal), the heights of each zone will be constant. The zones are pictured in Figure 22.19 for a *continuous* sedimentation.

Industrial sedimentation operations may be carried out batchwise or continuously in equipment called *thickeners*. The batch thickener operates exactly like the illustration cited above. The equipment is nothing more than a cylindrical tank with openings for a slurry feed and product draw-off. The tank is filled with a dilute slurry, and the slurry is permitted to settle. After a desired period of time, clear liquid is decanted until sludge appears in the draw-off. The sludge is removed from the tank through a bottom opening. Figure 22.20 illustrates a batch thickener.

Continuous thickeners (Figure 22.19 and 22.21) are large-diameter, shallow-depth tanks with slowly revolving rakes for removing the sludge. The slurry is fed at the center of the tank. Around the top edge of the tank is a clear liquid overflow. The rakes serve to scrape the sludge towards the center of the bottom for discharge. The motion of the rake also "stirs" only the sludge layer. This gentle stirring aids in water removal from the sludge.

Continuous-Thickener Calculations The purpose of a continuous thickener is to take a slurry of some initial concentration of solids and through the process of sedimentation produce a slurry of some higher concentration. The calculations necessary for the design of a

continuous thickener are governed by the settling characteristics of the solids in the slurry. The design of a thickener requires a specification of a cross-sectional area and a depth. It is possible through the use of batch-settling information to design a unit to produce a specified product in a continuous manner. The next few paragraphs will indicate the calculation procedures.

In the experimental determination of behavior of slurries for the purpose of determining necessary size of thickeners, experimental measurements introduce a complication in translating batch data to continuous operations. The most satisfactory method of laboratory measurement is the observation of the settling rate of the uniformly mixed slurry in a graduated glass cylinder. Such a settling rate may be observed for dispersions of various amounts of the solid in the liquid. It is found that the settling rate decreases with increasing concentration of solid in the slurry. However, the decrease is less rapid than the increase in solid content at some concentrations. This is important in the translation of batch-settling tests to the design of a continuous thickener for concentrating a particular slurry. In a continuous thickener, conditions at a point are steady with time, whereas each batch-settling test represents the conditions at some particular point in a thickener only at a particular time during the test.

A necessary condition for functioning of a continuous thickener is that the rate at which solids settle through every zone must be at least fast enough to accommodate the solid being delivered to that level. In the upper part of the thickener, the slurry is quite dilute, and settling is quite rapid. In the bottom, the density and solids concentration are extremely high; and, even though the settling velocity of each particle is small, the capacity rate for total solids per unit area is higher than at some intermediate point in the thickener. The velocity of particles has been decreased by the increasing solids concentration. The velocity decreases more rapidly

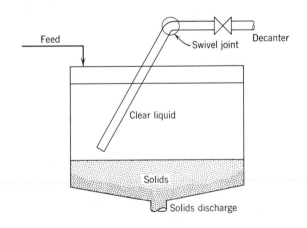

Figure 22.20. Batch thickener.

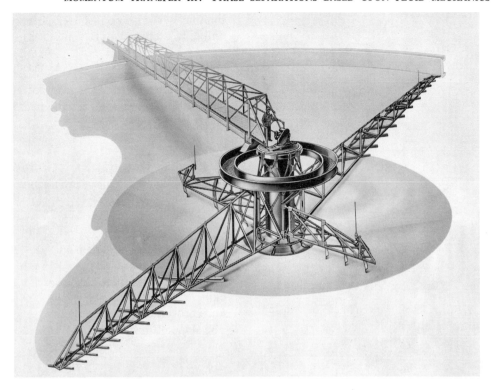

Figure 22.21. Continuous thickener. (Courtesy Dorr-Oliver, Inc.)

than the solids concentration increases in the intermediate levels. Stated another way, the mass velocity decreases, then increases as the solids settle. The result is that there is an intermediate level in the thickener through which the rate of passage of mass of solids is at its lowest value. If the load on the thickener exceeds the capacity at this point, solids will accumulate at this point, and the limiting zone will travel upward through the thickener.

Analysis of thickener performance cannot be based on a model of solid particles settling through a water layer in which downward liquid velocity is uniformly just sufficient to provide the water content of the thickened slurry. A more realistic simplified model recognizes that as a particle settles it is accompanied by an amount of liquid constituting its boundary layer. As the particle reaches the hindered settling zone, and later the compaction zone, the liquid constituting the boundary layers is reduced in quantity by interference and crowding of the boundary layers of all the particles. As liquid is progressively rejected from the boundary layers it must flow upward, resulting in a decrease in the net settling rate of the particles. This upflowing stream can be visualized as zero at the bottom of the thickener, and increasing to a maximum at the level where boundary layers first begin to interfere with each other.

In the design of a thickener for a specified quantity of slurry, the minimum cross-sectional area of the thickener which will allow passage of the solids is found at the limiting intermediate concentration. It is imperative that the cross-sectional area at the limiting level be large enough for solids to pass through at a rate equal to or exceeding the feed rate. If the area is not large enough, the material balance at this level is satisfied only by the accumulation of solids, with resultant travel of the limit zone upward in the thickener.

The determination of minimum area of a thickener requires in succession: (1) the identification of the concentration at which the mass rate of settling of solids is a minimum; (2) the determination of the settling velocity at this concentration; and (3) the evaluation of the amount of water rejected from the slurry between this concentration and that of the product, hence the amount that is flowing upward at the limiting level. The first step is accomplished by mathematical manipulations of settling rates and proper interpretation of the slopes of a batch-settling curve. The second is accomplished by reinterpretation of the same data into a settling rate-versus-concentration curve. The third factor is based on successive evaluations of the upward velocity of fluid at various concentrations

Each of the steps is developed in detail, leading to a value of the minimum thickener area. When the minimum cross section is known, the remaining design variable is the time of retention in the compression zone. Kynch (22) showed that the rate of upward propagation of such a limiting zone is constant if throughput exceeds its capacity and that the rate is a function of the solids

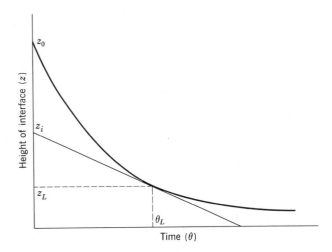

Figure 22.22. Batch-settling results.

concentration. Writing a material balance for constant area and concentration in this zone gives

$$\bar{v}_L = c\frac{dv_L}{dc} - v_L \qquad (22.51)$$

where $\bar{v}_L$ = upward velocity of the capacity-limiting layer

v_L = settling velocity of the particles in this layer

c = solids concentration, mass of solids per unit volume of slurry

However, since it was assumed that velocity was a function only of concentration, that is, $v_L = f(c)$ it follows that

$$\frac{dv_L}{dc} = f'(c)$$

and Equation 22.51 becomes

$$\bar{v}_L = cf'(c) - f(c) \qquad (22.52)$$

Since c is constant for this layer, $f'(c)$ and $f(c)$ are also constant, and $\bar{v}_L$ is, therefore, also constant. The constancy of $\bar{v}_L$ in the rate-limiting zone may be used to determine the concentration of solids at the upper boundary of the layer from a *single batch-settling test*. Let c_0 and z_0 represent the initial concentration and height of the suspended solids in a batch-settling test. The total weight of the solids in the slurry is c_0z_0S, where S is the cross-sectional area of the cylinder in which the test is being performed. If a limiting layer exists, it must first form at the bottom and move upward to the clear-liquid interface. If the concentration of the limiting layer is c_L and the time for it to reach the interface is θ_L, the quantity of solids passing through this layer is $c_LS\theta_L(v_L + \bar{v}_L)$. This quantity must equal the total solids present, since the layer having this

limiting concentration started forming at the bottom and moved upward to the interface. Therefore,

$$c_LS\theta_L(v_L + \bar{v}_L) = c_0z_0S \qquad (22.53)$$

If z_L is the height of the interface at θ_L with $\bar{v}_L$ being constant in accord with Equation 22.52, then

$$\bar{v}_L = \frac{z_L}{\theta_L} \qquad (22.54)$$

Substituting the value of $\bar{v}_L$ from Equation 22.54 into Equation 22.53 and simplifying gives

$$c_L = \frac{c_0z_0}{z_L + v_L\theta_L} \qquad (22.55)$$

The laboratory-test data may be treated by plotting the height of interface as a function of time, as in Figure 22.22. From this plot, the value of v_L is the slope of the curve at $\theta = \theta_L$ as shown by Equation 22.56. The tangent with the curve at θ_L intersects the ordinate at z_i. The slope of this line is

$$\frac{z_i - z_L}{\theta_L} = v_L \qquad (22.56)$$

or

$$z_i = z_L + \theta_Lv_L \qquad (22.57)$$

Combining Equations 22.57 and 22.55 yields

$$c_Lz_i = c_0z_0 \qquad (22.58)$$

It follows, therefore, that z_i is the height which the slurry would occupy if all the solids present were at concentration c_L. In terms of the model postulated above, c_L is the minimum concentration at which boundary layers interfere.

The settling velocity as a function of concentration may also be determined from a single settling test. The procedure is as follows: For several arbitrary θ's,

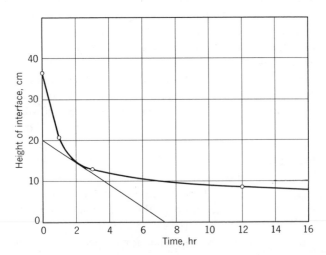

Figure 22.23. Height of interface as a function of time, for Illustration 22.5.

the slope of the tangent and its intercept at $\theta = 0$ is determined from a plot of z as a function of θ. The value of the intercept is used in Equation 22.58 to obtain the corresponding concentration. From this, v_L as a function of c is obtained.

Illustration 22.5. A single batch-settling test was made on a limestone slurry. The interface between clear liquid and suspended solids was observed as a function of time, and the results are tabulated below. The test was made using 236 gm of limestone per liter of slurry. Prepare a curve showing the relationship between settling rate and solids concentration.

Test Data

Time, hr	Height of Interface, cm
0	36.0
0.25	32.4
0.50	28.6
1.00	21.0
1.75	14.7
3.0	12.3
4.75	11.55
12.0	9.8
20.0	8.8

SOLUTION. Using the test data, the height of the interface (z) is plotted as a function of time (θ) (Figure 22.23). From

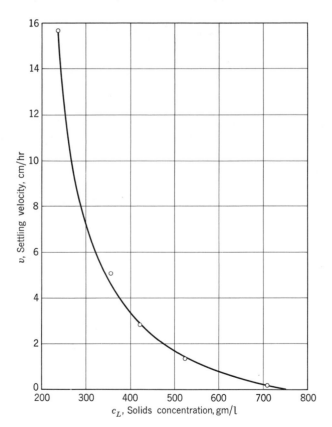

Figure 22.24. Settling-rate–concentration relationship, for Illustration 22.5.

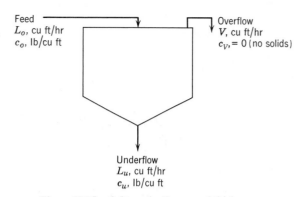

Feed
L_o, cu ft/hr
c_o, lb/cu ft

Overflow
V, cu ft/hr
c_V, = 0 (no solids)

Underflow
L_u, cu ft/hr
c_u, lb/cu ft

Figure 22.25. Schematic diagram of thickener.

the solid concentration of the initial slurry

$$c_0 z_0 = 236 \times 36 = 8500 \text{ gm cm/l}$$

From Equation 22.58,

$$c = \frac{8500}{z_i} \text{ gm/l}$$

The tangent to the curve at $\theta = 2$ hr, is found to have an intercept of $z_i = 20$ cm. The settling velocity at that time is the slope of the curve, $dz/d\theta = v = 2.78$ cm/hr, and $c = 425$ gm/l. Other points are obtained in the same way, tabulated in Table 22.2, and plotted in Figure 22.24.

Table 22.2. SOLUTION TO ILLUSTRATION 22.5

θ, hr	z_i, cm	v, cm/hr	c, gm/l
0.5	36	15.65	236
1.0	36	15.65	236
1.5	23.8	5.00	358
2.0	20	2.78	425
3.0	16.2	1.27	525
4.0	14.2	0.646	600
8.0	11.9	0.158	714

Thickener Area. The required thickener area is fixed at the concentration layer requiring the maximum area to pass a unit quantity of solids. For the thickener shown in Figure 22.25, the material balance for solids is

$$L_0 c_0 = L_u c_u \tag{22.59}$$

$$L_u = \frac{L_0 c_0}{c_u} \tag{22.59a}$$

in which the symbols are identified in Figure 22.25.

An over-all liquid balance gives

$$L_0(\rho_0 - c_0) = V\rho_w + L_u(\rho_u - c_u) \tag{22.60}$$

Eliminating L_u in Equation 22.60 by substitution of Equation 22.59a results in

$$L_0(\rho_0 - c_0) = V\rho_w + L_0 \frac{c_0}{c_u}(\rho_u - c_u) \tag{22.61}$$

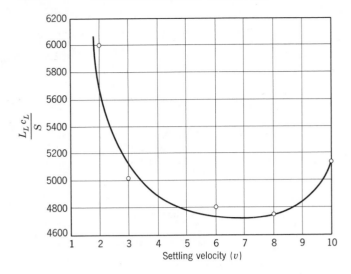

Figure 22.26. Determination of $\left(\dfrac{L_L c_L}{S}\right)_{min}$, Illustration 22.6.

Upon rearrangement Equation 22.61 becomes

$$V = L_0 c_0 \left[\frac{\rho_0}{c_0} - \frac{\rho_u}{c_u}\right] \frac{1}{\rho_w} \qquad (22.62)$$

Dividing both sides of Equation 22.62 by the thickener cross-sectional area (S) and using ρ_{av} for slurries gives

$$\frac{V}{S} = \frac{L_0 c_0}{S} \left[\frac{1}{c_0} - \frac{1}{c_u}\right] \frac{\rho_{av}}{\rho_w} \qquad (22.63)$$

The term V/S is the upward linear velocity of the clarified liquid. Previously, it was mentioned that in order to keep solids from overflowing, the upward velocity of the liquor must be equal to or less than the settling velocity of the solids; therefore, V/S may be replaced by v. Equation 22.63 may be written (40) in terms of the capacity-limiting layer, even though it has not yet been established that a particular value of c_L and corresponding downflow (L_L) represent the true limit; therefore,

$$L_0 c_0 = L_L c_L \qquad (22.64)$$

and

$$\frac{L_L c_L}{S} = \frac{v}{\left[\dfrac{1}{c_L} - \dfrac{1}{c_u}\right] \dfrac{\rho_{av}}{\rho_w}} \qquad (22.65)$$

Equation 22.65 is used with a relation $v = f(c)$ such as that of Illustration 22.5. For various corresponding values of v and c Equation 22.65 is used to calculate $L_L c_L/S$. The lowest value of $L_L c_L/S$ determines the minimum thickener area required.

Illustration 22.6. A limestone–water slurry equivalent to that of Illustration 22.5 is fed to a thickener at a rate of 50 tons of dry solids/hr to produce a thickened sludge of 550 gm limestone per liter. For an initial slurry concentration of 236 gm limestone per liter of slurry, specify the thickener area required.

SOLUTION. Since the slurry conditions entering the thickener are equal to that of the previous illustration, the results of that batch-settling test may be used. Figure 22.24 gives the relationship between v and c_L for this slurry. Using Equation 22.65 and Figure 22.24, Table 22.3 is prepared.

Table 22.3. DATA FOR SOLUTION TO ILLUSTRATION 22.6

v, cm/hr	c_L, gm/l	$\dfrac{1}{c_L}$	$\dfrac{1}{c_L} - \dfrac{1}{c_u}$	$\dfrac{L_L c_L}{S}$
10	265	0.00377	0.00195	5140
8	285	0.00351	0.00169	4740
6	325	0.00307	0.00125	4800
3	415	0.00241	0.00059	5090
2	465	0.00215	0.00033	6060
1	550	0.00182	0	∞

To determine the minimum value of $L_L c_L/S$, the data of Table 22.3 are plotted in Figure 22.26. This plot yields a minimum value of

$$\left(\frac{L_L c_L}{S}\right)_{min} = 4730 \ \frac{\text{cm/hr}}{\text{l/gm}}$$

corresponding to $v = 6.9$ cm/hr, and, from Figure 22.24, $c_L = 310$ gm/l. Since no solids leave in the overflow, a solids material balance (Equation 22.64) gives

$$L_L c_L = 50 \text{ tons/hr} = 100{,}000 \text{ lb solids/hr}$$

Now, $4730 \ \dfrac{\text{cm/hr}}{\text{l/gm}} = 9.68 \ \dfrac{\text{ft/hr}}{\text{cu ft/lb}}$ and

$$S = \frac{100{,}000}{9.68} = 10{,}320 \text{ sq ft}$$

Thickener Depth. Comings (11), in a series of tests on a continuous thickener, determined the effect of underflow rate on thickener capacity. Essentially clear overflow was obtained from all runs. The depth of the thickening zone increased as the underflow rate was decreased. He concluded that the depth of the thickening zone for incompressible slurries is less important than the retention time of the particles within

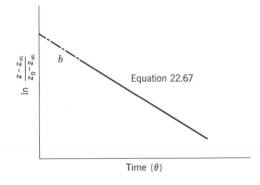

Figure 22.27. Compression-zone variation with time.

the thickener. That is, the area of a thickener may be dependably predicted from the settling limit described above, and the volume of the compression zone determined to give necessary retention time. The retention time may be determined from laboratory data in a batch test.

The shape of the compression curve (Figure 22.22) suggests that the rate of settling as a function of time may be given by (34)

$$\frac{dz}{d\theta} = k(z - z_\infty) \qquad (22.66)$$

where z = the height of the compression zone at time θ
  z_∞ = the height of the compression zone at infinite time
  k = a constant for a particular system

Laboratory data from batch tests give the height-time relationships. Integration of Equation 22.66 gives

$$\ln \frac{z - z_\infty}{z_c - z_\infty} = -k\theta \qquad (22.67)$$

where z_c is the height of the compression zone at the critical concentration at which time θ is taken as zero. The critical concentration is that concentration at which the slurry in the thickening zone begins to compress. The graphical form of Equation 22.67 is indicated in Figure 22.27.

The plotting of Equation 22.67 has to be done by trial and error since z_∞ is unknown. It is done by guessing a value of z_∞ and, with this assumed value of z_∞, plotting Equation 22.67. If the resulting line is not straight, another value of z_∞ is assumed and tried until a linear result is obtained. The slope of the final line equals k.

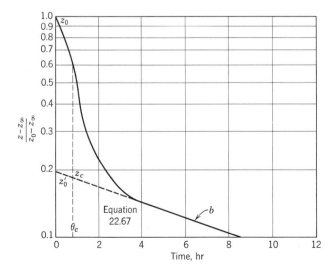

Figure 22.28. Settling characteristics of a slurry.

The critical concentration is not clearly defined. Figure 22.28 shows the complete settling-characteristic curve; settling initially starts at time zero where particles are in a free-falling period for some time. Eventually, a time is reached when settling becomes hindered, and the rate of settling decreases. Somewhere at about this time, the critical concentration is reached. Since those solids reaching the bottom first will perhaps be in compression while those higher are still in free settling, it is clear that a batch test cannot give a true time in the compression zone.

Roberts (34) suggests a way for estimating the critical time. The compression curve (b) in Figure 22.28 is extended to zero time in accord with Equation 22.67 as shown in Figure 22.28, and the critical time is obtained by arithmetically averaging z_0 and z_0'. The critical time is the time at which all the solids go into compression. Actually, part of the solids have entered compression earlier, and part do not go into compression until later. Thus, the retention time of the solids in the compression zone is the difference between the time necessary to reach a desired underflow concentration and the critical time.

If the compression zone is considered as an entity, it may be thought to be moving in the thickener at an average velocity of $V/S\theta_r$, where V is the volume of the compression zone, θ_r is the retention time, and S is the cross-sectional area. This velocity is composed of two components: one, an average settling velocity, and the other, the velocity at which the concentrated sludge is removed from the thickener.

The required volume for the compression zone equals the sum of the volume occupied by the solids and the volume occupied by the associated liquids.

$$V = \frac{L_0 c_0}{\rho_s}(\theta - \theta_c) + \frac{L_0 c_0}{\rho} \int_{\theta_c}^{\theta} \frac{W_l}{W_s} d\theta \qquad (22.68)$$

where V = compression zone volume, cu ft
  $L_0 c_0$ = mass of solids fed per unit time to thickener, lb/hr
  W_l = mass of liquid in compression zone, lb
  W_s = mass of solids in compression zone, lb
  $(\theta - \theta_c)$ = compression zone retention time, hr
  ρ_s = density of solid phase, lb/cu ft
  ρ = density of liquid phase, lb/cu ft.

Equation 22.68 is in keeping with the conclusion of Comings that the retention time within the compression zone is more important than the depth of the zone. The term W_l/W_s is a liquid-concentration term. It can be evaluated as a function of time from such information and methods as that of Illustration 22.5, and the integral term of Equation 22.68 evaluated graphically.

After the volume of the compression zone has been calculated, its depth may be evaluated by dividing the volume by the thickener area calculated by Equation

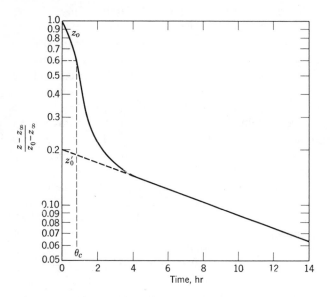

Figure 22.29. Settling characteristics for Illustration 22.7.

22.65. The total depth of the thickener may be estimated (4) by adding to the depth of the compression zone the following depths:

Bottom pitch	1 to 2 ft
Storage capacity	1 to 2 ft
Submergence of feed	1 to 3 ft

Illustration 22.7. Estimate the depth of the thickener required to perform the operation of Illustration 22.6. The batch-settling test indicated a value of $z_\infty = 7.7$ cm. The specific gravity of the limestone is 2.09.

SOLUTION. From the information of Illustration 22.5 and Table 22.2, Table 22.4 may be prepared.

Table 22.4. DATA FOR SOLUTION TO ILLUSTRATION 22.7

Time, hr	$(z - z_\infty)$	$\dfrac{z - z_\infty}{z_0 - z_\infty}$	W_s/W_l, gm solid/l. water	W_l/W_s, l/gm solid
0	28.3	1.0	236	0.00444
0.25	24.7	0.871	236	0.00444
0.50	20.9	0.739	236	0.00444
1.00	13.3	0.470	236	0.00444
1.75	7.0	0.247	392	0.00255
3.0	4.6	0.162	525	0.00191
4.75	3.85	0.136	644	0.00155
12.0	2.1	0.0743	850	0.00118
20.0	1.1	0.0389		
∞	0	0.0		

The critical time is evaluated from Figure 22.29 by determining the time at a value of $(z_0 + z_0')/2$. This time is found to be 0.80 hr. The final concentration is specified by Illustration 22.6 as 550 gm/l. From Table 22.2, a time of about 3.4 hr is required to produce a concentration of solids equal to 550 gm/l. Thus the retention time in the compression zone is (3.4 − 0.8) or 2.6 hr.

Equation 22.68 is used to evaluate the compression zone volume.

$$V = \frac{L_0 c_0}{\rho_s}(\theta - \theta_c) + \frac{L_0 c_0}{\rho}\int_{\theta_c}^{\theta}\frac{W_l}{W_s}\,d\theta$$

Using the data of Table 22.4, the integral of Equation 22.68 is graphically evaluated and found to be 6.89 hr.

$$L_0 c_0 = 100,000 \text{ lb/hr (from Illustration 22.6)}$$
$$\rho_s = 2.09 \times 62.3 = 130 \text{ lb/cu ft}$$
$$(\theta - \theta_c) = 2.6 \text{ hr}$$

$$\text{Therefore, } V = \frac{100,000}{130} \times 2.6 + \frac{100,000}{62.3} \times 6.89$$

$$V = 2000 + 11,000 = 13,000 \text{ cu ft}$$

The thickener area is 10,320 sq ft from Illustration 22.6. Depth of thickener:

Compression zone	13,000/10,320 ≅ 1.3 ft
Bottom pitch	2.0 ft
Storage capacity	2.0 ft
Submergence of feed	2.0 ft
Total depth	= 7.3 ft

APPLICATIONS OF THE MECHANICS OF FLOW OF FLUIDS THROUGH PARTICULATE SOLIDS

In many industrial operations, a fluid phase flows through a particulate-solid phase. Examples include filtration, heat transfer in regenerators and pebble heaters, mass transfer in packed columns, chemical reactions using solid catalysts, adsorption, and flow of oil through the reservoir toward an oil well. In many cases, the solid phase is stationary, as it is in a packed distillation column; in some cases, the bed moves countercurrent to the gas stream, as it does in a pebble heater or in some catalytic reactors. In some cases, the fluid velocity is great enough so that the momentum transferred from the fluid to the solid particles balances the opposing gravitational force on the particles and the bed expands into a fluidlike phase; and, in still other applications, the fluid phase carries the solid phase with it, as it does in pneumatic conveying.

The rate of momentum transfer from the fluid to the solid particles and therefore the pressure drop for flow through the bed are related to the physical mechanisms by which flow occurs. In a packed bed, the flow path is made up of many parallel and interconnecting channels. The channels are not of fixed diameter but widen and narrow repeatedly, and even twist and turn in varying directions as the particles obstruct the passageway. The channels do not even have the same average cross section or total length. In flowing through these passages, the fluid phase is repeatedly accelerated and decelerated and experiences repeated kinetic-energy losses. In addition, the rough surfaces of the particles produce the usual form-drag and skin-friction losses.

The flow through large open channels will be at higher velocity than the flow through parallel narrow constricted channels, because the pressure drop per foot of bed length must be constant regardless of the channel under consideration. For this reason, the transition from laminar to turbulent flow will occur at a much lower bulk flow rate in open passages than it will in restricted channels. Moreover, at the convergence of two channels, eddy currents and turbulence will be promoted because of the inequality of velocity in the two channels The flow behavior in expanded, or fluidized, beds will be very similar to that for packed beds, except that the flow passages will be more open and almost continuously interconnected. In cases where the bed is falling down through the rising fluid or where the fluid is conveying the solid phase, the mechanism is more like that of the hindered settling of particles through the fluid which has been treated above.

It is apparent from the physical mechanism discussed above that the momentum transfer from fluid to particles arises from a form-drag and a kinetic-energy loss, that is,

$$(\tau g_c)_{\text{total}} = (\tau g_c)_{\substack{\text{form drag} \\ \text{and skin friction}}} + (\tau g_c)_{\text{kinetic energy}} \quad (22.69)$$

At low rates of flow through very small passages, the kinetic-energy losses are small compared to the form-drag losses, but for high rates of flow through large passages or fluidized beds, the kinetic-energy losses may completely overshadow the form-drag losses. However, the transition will not be sharp, because the wide variety of parallel passages permits a wide variety of flow conditions.

Flow Through Packed Beds. Fluids are forced to flow through stationary beds of particulate or porous solids in a wide range of practical situations including moisture assimilation by soils, adsorption, ion exchange, and many of the examples noted in the introductory section. In these applications, the form-drag losses can be related to the flow conditions through modification of the relations previously given for friction losses in ducts. For this case,

$$f = \frac{2(-\Delta P)_f g_c D}{\bar{v}^2 \rho L} = \phi(N_{\text{Re}}) = \phi\left(\frac{D\bar{v}\rho}{\mu}\right) \quad (22.70)$$

where $-\Delta P_f$ = pressure drop due to friction and

$$f = \frac{8(\tau_y g_c)_1}{\bar{v}^2 \rho} \quad (13.14)$$

have been previously presented. For noncircular ducts, an equivalent diameter must be used.

$$D_{\text{eq}} = \frac{4S}{b} \quad (13.29)$$

which defines the equivalent diameter of a duct as four times the cross-sectional area divided by the wetted perimeter. In terms of a packed bed, Equation 13.29 can be modified to

$$\frac{4S}{b} \times \frac{L}{L} = \frac{\text{total volume of voids}}{\text{total surface area of particles}} = \frac{4\epsilon N V_p}{(1 - \epsilon) \cdot N A_p} \quad (13.29a)$$

where ϵ = porosity, fraction of total volume that is void

N = number of particles

NV_p = total volume of solid particles

NA_p = total surface area of solid particles

Inserting Equation 13.29a into Equation 22.70,

$$f = \frac{8(-\Delta P)_f g_c \epsilon V_p}{\bar{v}^2 \rho L (1 - \epsilon) A_p} = \phi\left(\frac{4\epsilon V_p \bar{v} \rho}{(1 - \epsilon) A_p \mu}\right) \quad (22.71)$$

It is usually convenient to express this equation in terms of the superficial velocity, that which would occur with the actual mass-flow rate flowing through the container holding the bed if this container were empty. By definition, then, $v_s = \epsilon \bar{v}$, where v_s is the superficial velocity. Also it is common practice to replace the V_p and A_p terms with a "particle diameter." The particle diameter is defined in terms of the diameter of an equivalent sphere.

For a sphere,

$$\frac{A_p}{V_p} = \frac{\pi(D_{\text{sp}})^2}{\pi(D_{\text{sp}})^3/6}$$

from which

$$D_{\text{sp}} = \frac{6}{A_p/V_p} \quad (22.72)$$

where D_{sp} = diameter of a spherical particle

For an irregular particle of known A_p to V_p ratio, there is only one size of sphere having this same ratio. The diameter of this sphere is taken as characteristic of the particle.

$$D_p = \frac{6}{\dfrac{A_p}{V_p}} = D_{\text{sp}} \quad (22.73)$$

Writing Equation 22.71 in terms of v_s and D_p gives

$$f = \frac{8(-\Delta P)_f g_c \epsilon^3 D_p}{6 v_s^2 \rho L (1 - \epsilon)} = \phi\left(\frac{4 D_p v_s \rho}{6(1 - \epsilon)\mu}\right) \quad (22.74)$$

In the region of low flow rate and small particle size or, in other words, at low N_{Re} (laminar flow), Equation 22.74 can be used to express the entire pressure drop, because the kinetic-energy losses are small. Under these conditions, the drag coefficient and hence the friction factor are inversely proportional to the Reynolds number.

$$f = \frac{k_1}{N_{\text{Re}}} \quad (13.16)$$

where k_1 is a proportionality constant.

Combining this equation with Equation 22.74 yields

$$\frac{\epsilon^3(-\Delta P)_f g_c D_p}{(1-\epsilon)v_s^2 \rho L} = k_2 \frac{(1-\epsilon)\mu}{D_p v_s \rho} \qquad (22.75)$$

which rearranges into

$$\frac{(-\Delta P)_f g_c}{L} = k_2 \frac{(1-\epsilon)^2 \mu v_s}{\epsilon^3 D_p^2} \qquad (22.76)$$

where k_1 and k_2 are constants. Equation 22.76 is known as the Carman-Kozeny equation and has been successfully used to calculate pressure drop for laminar flow through packed beds. It was originally derived by Kozeny (21) who used the simplified model of a number of parallel capillary tubes of equal length and diameter to describe the packed bed. If this model were exact, k_1 would be 64 as given by Equation 13.16. Carman (9) applied this equation to experimental results on flow through packed beds and found that $k_2 = 180$.

If the N_{Re} is high, kinetic-energy losses become significant. These losses may be found by modifying the kinetic-energy term in the general relation, Equation 20.16a.

$$\frac{-\Delta P_k}{\rho} = \frac{\bar{v}^2}{2g_c}$$

But, if the energy loss is to occur repeatedly in a unit channel length,

$$\frac{(-\Delta P)_k g_c}{L} = \frac{n}{2}\rho\bar{v}^2 \qquad (22.77)$$

where $n =$ the number of repetitive kinetic-energy losses in a unit length

$-\Delta P_k =$ pressure drop due to kinetic-energy losses

In the channels under consideration, the expansions in channel width probably occur at distances roughly equivalent to the channel diameter. This follows because the particulate bed has been replaced by a model consisting of many parallel, circular ducts. The diameter of these ducts will be proportional to the particle diameter, and the fact of one expansion occurring for each particle is thus approximated. Using n proportional to $1/D_c$ gives

$$\frac{(-\Delta P)_k g_c}{L} = \frac{k_3 \rho \bar{v}^2}{D_c} \qquad (22.78)$$

where k_3 is a constant and D_c is the channel diameter. As with the pressure drop due to form drag, it is convenient to convert this equation to an expression in terms of D_p and v_s. As before,

$$\bar{v} = \frac{v_s}{\epsilon}$$

$$D_p = \frac{6V_p}{A_p} \qquad (22.73)$$

and

$$\frac{A_p}{V_p} = \frac{N_c L \pi D_c}{\dfrac{L\pi D^2}{4}(1-\epsilon)} \qquad (22.79)$$

where $N_c =$ number of channels in the area of the bed
$D =$ bed diameter

In Equation 22.79 the bed has again been likened to a large number of capillary tubes of fixed length and diameter. The particle surface area (A_p) has been taken as equal to the wall surface of these capillaries. Also,

$$\frac{N_c \pi D_c^2}{4} = \frac{\pi D^2}{4}\cdot\epsilon \qquad (22.80)$$

so that

$$D^2 = \frac{N_c D_c^2}{\epsilon} \qquad (22.80a)$$

Combining Equations 22.79 and 22.80a gives

$$\frac{A_p}{V_p} = \frac{N_c D_c}{\dfrac{N_c D_c^2}{4\epsilon}\cdot(1-\epsilon)} = \frac{4\epsilon}{D_c(1-\epsilon)} = \frac{6}{D_p} \qquad (22.81)$$

Equation 22.81 and the superficial velocity, (v_s) may be inserted into Equation 22.78.

$$\frac{(-\Delta P)_k g_c}{L} = k_3 \frac{\dfrac{\rho v_s^2}{\epsilon^2}}{\dfrac{4\epsilon D_p}{6(1-\epsilon)}} = k_4 \frac{\rho v_s^2}{D_p}\cdot\frac{(1-\epsilon)}{\epsilon^3} \qquad (22.82)$$

This relation was first derived by Burke and Plummer (8) to express the pressure drop resulting from turbulent flow through packed beds. Equation 22.76 for the pressure drop caused by form drag and Equation 22.82 for the pressure drop caused by kinetic-energy losses may now be added to obtain the total pressure drop resulting from flow through the bed (13, 26).

$$\frac{(-\Delta P)g_c}{L} = k_2 \frac{(1-\epsilon)^2}{\epsilon^3}\frac{\mu v_s}{D_p^2} + k_4 \frac{(1-\epsilon)}{\epsilon^3}\frac{\rho v_s^2}{D_p} \qquad (22.83)$$

This equation has been developed under the assumption that the velocity is constant throughout the length of the bed. With gas flow at high pressure drop, this would not be the case, and a differential form of Equation 22.83 would have to be written and integrated for the full bed depth. If such an integration is carried out by assuming isothermal expansion of an ideal gas, Equation (22.83) becomes

$$\frac{(-\Delta P)g_c}{L} = k_2 \frac{(1-\epsilon)^2}{\epsilon^3}\frac{\mu v_{sm}}{D_p^2} + k_4 \frac{(1-\epsilon)}{\epsilon^3}\frac{G v_{sm}}{D_p} \qquad (22.84)$$

where $v_{sm} =$ superficial velocity at the average of inlet and outlet pressure

$G =$ mass rate of flow (ρv_s) based on total bed cross section

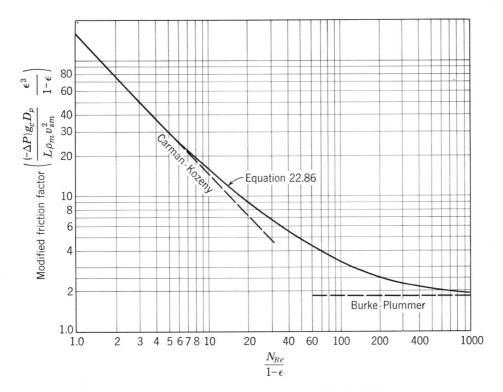

Figure 22.30. Pressure drop for flow through packed beds.

Equation 22.84 can be rearranged into an equation relating a modified friction factor and a N_{Re} as follows (12):

$$\frac{(-\Delta P)g_c}{L}\frac{D_p{}^2}{\mu v_{sm}}\frac{\epsilon^3}{(1-\epsilon)^2} = k_2 + k_4\frac{N_{Re}}{(1-\epsilon)} \quad (22.85)$$

Here, the term on the left side of the equation is an averaged friction factor modified by the inclusion of the porosity terms. Using this relation, the constants k_2 and k_4 can be readily determined from experimental data by plotting

$$\frac{(-\Delta P)}{L}g_c\frac{D_p{}^2}{\mu v_{sm}}\frac{\epsilon^3}{(1-\epsilon)^2}$$

as a function of $N_{Re}/(1-\epsilon)$. Since Equation 22.85 is linear in these groups of variables, straight lines result where the slope is k_4 and the intercept at $N_{Re}/(1-\epsilon) = 0$ is k_2. A large amount of experimental data on flow through beds of granular solids shows that $k_2 = 150$ and $k_4 = 1.75$ (12). The modified friction-factor term of Equation 22.85 is the ratio of the total pressure drop to the pressure drop due to form drag and skin friction. More usually, the friction factor is given as the ratio of the total pressure drop to the kinetic-energy losses. Rearranging Equation 22.85 in this way by dividing by $(D_p v_{sm}\rho_m/\mu)/(1-\epsilon)$ and inserting the values of k_2 and k_4 found experimentally gives

$$\frac{(-\Delta P)g_c}{L}\frac{D_p}{\rho v_{sm}{}^2}\frac{\epsilon^3}{(1-\epsilon)} = 150\frac{(1-\epsilon)}{N_{Re}} + 1.75 \quad (22.86)$$

Both Equation 22.85 and Equation 22.86 are dimensionless, and hence any set of consistent units may be used throughout any complete term of these equations. Typical units might be as follows:

$-\Delta P = $ pressure drop through the packed bed, lb$_f$/sq ft

$L = $ bed length, ft

$D_p = $ particle diameter, ft

$\rho = $ fluid density, lb/cu ft

$v_{sm} = $ superficial velocity at a density averaged between inlet and outlet conditions, ft/sec

$g_c = $ conversion factor, (lb/lb$_f$) (ft/sec^2)

$\epsilon = $ bed porosity, dimensionless

$N_{Re} = $ average Reynolds number based on superficial velocity ($D_p v_{sm}\rho/\mu$), dimensionless

For liquids, v_{sm} may be taken at any point in the bed, but gas-density variations require the use of the average velocity as specified above.

In Figure 22.30 is shown the mean curve obtained when experimental data from several sources (12, 8, 30, 32) are plotted as

$$[(-\Delta P)g_c D_p/L\rho v_{sm}{}^2][\epsilon^3(1-\epsilon)]$$

as a function of $[N_{Re}/(1-\epsilon)]$. The solid line is a plot of Equation 22.86, and the dotted lines marked Carman-Kozeny and Burke-Plummer are plots of Equations 22.76 and 22.82 respectively. The data points themselves have been removed for clarity. They scatter smoothly around the line plotting Equation 22.86 showing no systematic deviation, and a maximum deviation of about 20 per cent if a few erratic points are eliminated. This is excellent agreement in view of the wide range of variables covered by Figure 22.30. Note the similarity in shape between the curve shown here and the drag-coefficient–Reynolds-number plot, Figure 13.5. The gradual transition between laminar and turbulent regions, which covers a range of $N_{Re}/(1 - \epsilon)$ between 5 and 2000, results from the widely differing flow passages through the bed, with the resulting wide variations in actual velocity.

Other relations for the pressure drop through packed beds have been found by empirically correlating experimental data. In this way, Leva and Grummer (26) obtained

$$\frac{(-\Delta P)g_c}{L} = \frac{0.0243G^{1.9}\mu^{0.1}\lambda^{1.1}}{D_p^{1.1}\rho}\frac{(1 - \epsilon)}{\epsilon^3} \quad (22.87)$$

to which they applied different constants for particles of different surface roughness. Here λ is a shape factor obtained by the relation

$$\lambda = 0.205 \frac{A_p}{V_p^{2/3}} \quad (22.88)$$

Leva's experimental data were all in the range of relatively high N_{Re} and relatively large particles. Data on both liquid and gaseous fluids were included in the correlation so that G can be either G_V or G_L.

Brownell and Katz (7) correlated results obtained mainly for very fine particles by using the standard f versus N_{Re} plot but multiplying f and N_{Rc} by factors dependent upon the particle sphericity and the bed porosity.

Illustration 22.8. A bed of $\frac{1}{4}$-in. cubes is to be used as packing for a regenerative heater. The cubes are poured into the cylindrical shell of the regenerator to a depth of 10 ft. If air flows through this bed, entering at 80°F and 100 psia, leaving at 400°F, and flowing at a mass rate of 1000 lb/hr sq ft of free cross section, determine the pressure drop across the bed.

SOLUTION. This problem can be solved by applying Equations 22.85 and 22.86 or Figure 22.30. In any case, the particle diameter (D_p), the bed porosity (ϵ), and the mean superficial velocity must first be determined.

For the $\frac{1}{4}$-in. cubes,

$$D_p = \frac{6V_p}{A_p} = \frac{6 \times \overline{0.25^3}}{6 \times \overline{0.25^2}} = 0.25 \text{ in.}$$

The porosity can be determined using Figure B-12, Appendix B, which gives porosity as a function of sphericity (ψ). By Equation B-26,

$$\psi = \frac{\pi\left(\dfrac{6V_p}{\pi}\right)^{2/3}}{A_p} = \frac{\pi\left(\dfrac{6}{\pi}\right)^{2/3}(\overline{0.25^3})^{2/3}}{6 \times \overline{0.25^2}} = \left(\frac{\pi^{3/2} \cdot 6}{6^{3/2} \cdot \pi}\right)^{2/3} = 0.806$$

From Figure B-12, this gives

$$\epsilon = 0.44$$

Fundamentally, v_{sm} must depend upon the pressure drop, since the gas specific volume is pressure dependent. Here, however, as in many pressure-drop computations, the pressure drop will probably be small relative to the total pressure, so that the effect of this $-\Delta P$ on gas density will be small. With this simplification, v_{s1} and v_{s2} can be fixed directly. Otherwise, a trial-and-error solution would be necessary to find v_2. On this basis,

$$\rho_1 = \frac{PM}{RT_1} = \frac{100 \times 29}{10.73 \times 540} = 0.500 \text{ lb/cu ft}$$

$$\rho_2 = \frac{PM}{RT_2} = \frac{100 \times 29}{10.73 \times 860} = 0.314 \text{ lb/cu ft}$$

and

$$v_{sm} = \frac{\dfrac{1000}{0.500} + \dfrac{1000}{0.314}}{2} = 2595 \text{ ft/hr}$$

Using these values, the group $N_{Re}/(1 - \epsilon)$ can be calculated, and the modified friction factor can be read from Figure 22.30.

$$\frac{N_{Re}}{1 - \epsilon} = \frac{\dfrac{D_pG}{\mu_m}}{1 - \epsilon} = \frac{0.25 \times 1000}{12(1 - 0.44) \times (0.023 \times 2.42)} = 669$$

$$\frac{(-\Delta P)g_c D_p}{L\rho_m v_{sm}^2}\frac{\epsilon^3}{(1 - \epsilon)} = 2.00 \text{ (from Figure 22.30)}$$

This equation may be solved for $(-\Delta P)$. Here, $\rho_m v_{sm}$ are replaced with G.

$$-\Delta P = \frac{2.00 \times 10 \times 1000 \times 2595 \times 0.56}{(32.2 \times \overline{3600^2}) \times \dfrac{0.25}{12} \times \overline{0.44^3}} = 40.4 \text{ lb}_f/\text{sq ft}$$

$$-\Delta P = 0.28 \text{ psi}$$

The assumption of a relatively small $-\Delta P$ has then been justified for this case. Larger relative pressure drops can be expected if the particle diameter is small.

Bed Fluidization. The packed bed expands when the pressure drop due to the upward flow of fluid through a granular unrestricted bed equals the weight of the packing. As the bed expands, it retains its top horizontal surface with the fluid passing through the bed much as it did when the bed was stationary. Now, however, the porosity is much greater, and the individual particles move under the influence of the passing fluid. The bed has many of the appearances of a boiling liquid and is referred to as being "fluidized." Writing the force balance on a section of bed of length L when the pressure drop equals the gravitational force gives

$$\frac{(-\Delta P)g_c}{L} = (1 - \epsilon)(\rho_s - \rho) \cdot g \quad (22.89)$$

where ρ_s = solid-particle density

The extent of bed expansion may be obtained by eliminating $(-\Delta P)g_c/L$ between Equations 22.84 and 22.89.

$$150\frac{(1-\epsilon)}{\epsilon^3}\frac{\mu v_{sm}}{D_p{}^2} + 1.75\frac{\rho_m v_{sm}{}^2}{D_p} = (\rho_s - \rho)g \quad (22.90)$$

Since pressure affects the density and hence the velocity of a gas in the fluid phase, this equation should be solved rigorously by successively inserting v_s varying between inlet and outlet. The bed porosity would show equivalent gradation. However, this effect is usually small, and the use of mean values of ρ and v_s is usually satisfactory. In most industrial applications involving expanded or fluidized beds, the particle diameter is small, and v_{sm} is also small. In these cases, the second term of Equation 22.90 is negligible compared to the first, so that

$$\frac{150\mu v_{sm}}{(\rho_s - \rho)g D_p{}^2} = \frac{\epsilon^3}{1-\epsilon} \quad (22.91)$$

For a given bed, Equation 22.91 may be written for both the unexpanded and the expanded state at the pressure drop required for bed expansion. From the difference in these two states,

$$\frac{150\mu}{g(\rho_s - \rho)D_p{}^2}(v_{sm} - v_{sm}{}^0) = \frac{\epsilon^3}{1-\epsilon} - \frac{\epsilon^{0^3}}{1-\epsilon^0} \quad (22.92)$$

where the superscipt 0 refers to the condition where the fluid velocity is just insufficient to expand the bed. Equation 22.92 relates the bed porosity and therefore the extent of bed expansion to the mean superficial velocity of fluid passing through the bed. It applies only where the pressure drop is constant and balances the force of gravity acting on the solid particles and where D_p is small.

The pressure-drop behavior of a packed bed as the velocity of flow up through it increases is illustrated in Figure 22.31. Between points A and B, the bed is stable, and the pressure drop and Reynolds number are related by Equation 22.86. At point B, the pressure drop essentially balances the bed solids weight. Between points B and C the bed is unstable, and particles adjust their position to present as little resistance to flow as possible. At point C, the loosest possible arrangement is obtained in which the particles are in contact. Beyond this point, the particles begin to move freely but collide frequently so that the motion is similar to that of particles in hindered settling. Point C is referred to as the "point of fluidization." By the time point D is reached, the particles are all in motion, and, beyond this point, increases in N_{Re} result in very small increases in $-\Delta P$ as the bed continues to expand and the particles move in more rapid and more independent motion. Ultimately, the particles will stream with the fluid, and the bed will cease to exist. This occurs at point E.

Two main types of fluidization have been noted experimentally (44). In cases where the fluid and solid densities are not too different, where the particles are small, and therefore where the velocity of flow is low, the bed fluidizes evenly with each particle moving individually through a relatively uniform mean free path. The solid phase has many of the characteristics of a gas. This is called *particulate fluidization*. Where the fluid and solid densities are greatly different or the particles are large, the velocity of flow must be relatively high. In this case, fluidization is uneven, and the fluid passes through the bed mainly in large bubbles. These bubbles burst at the surface spraying solid particles above the bed. Here, the bed has many of the characteristics of a liquid with the fluid phase acting as a gas bubbling through it. This is called *aggregative fluidization*. It appears that particulate fluidization occurs when the Froude number $(v^2/D_p g)$ at the point of fluidization is less than one, whereas aggregative fluidization occurs when the Froude number at the point of fluidization is greater than one. As noted in Chapter 20, the Froude number is the ratio of kinetic to gravitational energy. It is an important criterion whenever the motion of a free surface of fluid is considered. For example, on the surface of lakes and ponds, small waves move at such a rate that the Froude number is unity. In rivers and drainage ditches, flow rates giving $N_{Fr} > 1.0$ exhibit rapid flow characteristics, whereas at $N_{Fr} < 1.0$ tranquil flow characteristics prevail. In rapid flow the flow rate is greater than the rate of travel of a small disturbance, so that disturbances at the downstream end of a passage do not propagate back up the flow path. In tranquil flow a small disturbance moves more rapidly than the bulk flow and hence can travel back against the flow path.

The use of the N_{Fr} as the criterion of aggregative or particulate fluidization is not completely verified experimentally, since the experimental cases of particulate fluidization occurred with a liquid as the fluid phase, and for aggregative fluidization the fluid phase was a gas (44).

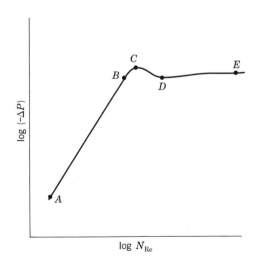

Figure 22.31. Fluidization of a bed of particulate solids (44).

Figure 22.32. A 23,000 bbl/day fluid catalytic cracking unit. The large vessel just to the right of the elevator shaft is the regenerator. The reactor is the next vessel to the right of the regenerator. This is an Esso Model IV unit built at Port Jerome, France. Units of this type first went on stream at the end of 1952 and have been built in sizes from 5000 to 55,000 bbl/day. (Courtesy Esso Research and Engineering Company.)

Industrially, it has been found advantageous to carry out many solid-catalyzed reactions in fluidized beds. The circulation of the bed and uniform agitation within it prevent the occurrence of hot spots and dead regions. It also makes possible the continuous circulation of the catalyst between the reaction vessel and a regeneration vessel. The major disadvantage of this system is that the catalyst is eroded and broken by its constant motion, so that there is continual attrition of the particles requiring continual makeup of fresh catalyst.

The largest-scale industrial application of fluidization is in the fluid catalytic cracking of heavy crude-oil fractions to give gasoline components. This process was developed during the early part of World War II when there was a critical need for high gasoline yields from crude oil and has been very widely applied since then. Figure 22.32 shows a typical fluid-cracking unit, and the catalyst and fluid flow through this unit is diagrammed in Figure 22.33.

Regenerated catalyst at about 1050°F is blown by the vaporized feed into the reactor vessel. The hot catalyst preheats the feed and is conveyed at a rate of about 7 lb of catalyst per pound of feed as required by the system energy balance. Cracking takes place in the reactor at about 900°F with coke depositing on the surface of the activated natural clay or synthetic silica-alumina catalyst. Spent catalyst collects in a standpipe and flows through a valve into an air stream which conveys it back into the regenerator. In the regenerator, the coke is burned off the catalyst, whereupon the regenerated catalyst collects in a second standpipe for return to the reactor. A 25,000 bbl/day unit such as the one shown in Figure 22.32 contains about 500 tons of catalyst of 20 to 400-micron particle diameter circulating at a rate of 20 tons/min and transferring 100×10^6 Btu/hr of heat from the regenerator to the reactor. Only about 1 per cent of the circulating catalyst is entrained by the fluids, and it is separated in cyclones before the fluids leave the contacting chambers (19).

Although the fluid-cracking process is the largest-scale application of fluid-bed reacting, fluidized reactors have been developed for many processes including ore roasting, cement manufacture, extraction of oil from shale and bituminous sand, production of phthalic anhydride by oxidation of naphthalene, and the oxidation of ethylene to form ethylene oxide (36).

Illustration 22.9. A catalyst having spherical particles of $D_p = 50$ microns and $\rho_s = 1.65$ gm/cu cm is to be used to contact a hydrocarbon vapor in a fluidized reactor at 900°F, 1 atm pressure. At operating conditions, the fluid viscosity is 0.02 centipoise and its density is 0.21 lb/cu ft. Determine the superficial gas velocity necessary to fluidize the bed, the velocity at which the bed would begin to flow with the gas, and the extent of bed expansion when the gas velocity is the average of the velocities previously determined. Does aggregative or particulate fluidization occur?

SOLUTION. The mean superficial velocity at the start of fluidization can be obtained directly from Equation 22.91 once the porosity is determined. From Figure B-12 a porosity of 0.42 is shown for loose packing of spheres. Then

$$(v_{sm}^0) = \frac{\epsilon^3}{1 - \epsilon} \frac{(\rho_s - \rho)g D_p^{\,2}}{150\mu}$$

$$= \frac{\overline{0.42^3}}{0.58} \times \frac{(1.65 \times 62.4 - 0.21)32.2 \left(\dfrac{0.050}{25.4 \times 12}\right)^2}{150 \times (0.02 \times 0.000672)}$$

$$= \frac{0.128 \times 102.8 \times 32.2 \times (0.0269 \times 10^{-6})}{0.002014}$$

$$= 0.0056 \text{ ft/sec}$$

is the velocity at which fluidization will begin.

The bed will disintegrate and stream with the flowing gas when the gas velocity equals the velocity of free fall of the particles. For these small particles, the flow is laminar, and the settling velocity can be determined directly from Equation 22.15.

$$v_t = \frac{(\rho_s - \rho)g D_p^{\,2}}{18\mu} = \frac{102.8 \times 32.2 \times (0.0269 \times 10^{-6})}{18 \times (0.02 \times 0.000672)}$$

$$= 0.37 \text{ ft/sec}$$

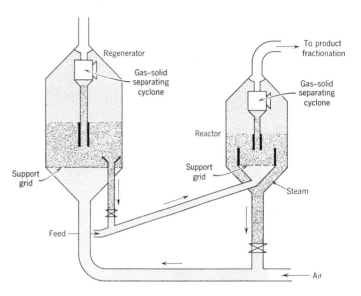

Figure 22.33. Schematic diagram of fluid catalytic-cracking process (26).

The extent of bed expansion at any intermediate flow rate can be determined from Equation 22.92 by solving for ϵ. This was not possible when finding the maximum value of v_{sm} because Equation 22.92 becomes indeterminate when $\epsilon \to 1$. Solving Equation 22.92,

$$\frac{\epsilon^3}{1 - \epsilon} = \frac{150\mu}{g(\rho_s - \rho)D_p^{\,2}} (v_{sm} - v_{sm}^0) + \frac{\epsilon^{0^3}}{1 - \epsilon^0}$$

$$= \frac{0.002014}{32.2 \times 102.8 \times 0.0269 \times 10^{-6}}$$

$$\times \left(\frac{0.37 + 0.0056}{2} - 0.0056\right) + \frac{\overline{0.42^3}}{0.58}$$

$$= 23.5(0.187 + 0.0027) + 0.128 = 4.18 + 0.128$$

$$= 4.308$$

$\epsilon = 0.855$ at $v_{sm} = 0.190$ ft/sec, the average of the previously determined velocities.

The type of fluidization that will probably occur is indicated by the size of the N_{Fr} at the start of fluidization.

$$N_{Fr} = \frac{(v_{sm}^0)^2}{D_p g} = \frac{0.\overline{0056}^2}{0.161 \times 10^{-3} \times 32.2} = \frac{0.314 \times 10^{-4}}{5.15 \times 10^{-3}}$$

$$= 0.0061$$

Thus, particulate fluidization would be expected.

Equations 22.91 and 22.92 were obtained by applying to the Carman-Kozeny equation the restriction of constant pressure drop equivalent to the bed weight (Equation 22.89) and allowing a resulting variation in porosity. Fluidized beds are normally operated with a wide range of particle sizes at flow rates high enough to fluidize all the bed particles. At these flow rates, a large proportion of the fines would be expected to be entrained in the gas phase. For example, cracking catalyst normally ranges between 20 and 500 microns in diameter (28) and is fluidized at superficial velocities of 1 to 2 ft/sec. As seen in Illustration 22.9, these velocities are great enough to entrain particles somewhat greater than 50 microns. However, in operation, relatively little entrainment of solids occurs. This apparent violation of Stokes law probably results from the flocculation of the small particles into agglomerates perhaps held together by static electricity. Such agglomerates have been observed and seem to be particularly large and stable in particles of about 10 microns and smaller diameter. Similar effects were noted by Morse (31) in examining the available experimental data on pressure drop and porosity of fluidized beds. Below N_{Re} of about 10, he found that the modified friction-factor term of Equation 22.86 was lower for fluidized beds than for fixed beds, but, above $N_{Re} = 10$, the modified friction factor for fluidized beds was greater than that for fixed beds. In the low N_{Re} range, this deviation is explained on the basis of flocculation, since the experimental data in this range was obtained by Leva and coworkers (27) with small particles. The data of $N_{Re} > 10$ was obtained by Wilhelm and Kwauk (44) with larger particles. The deviation here was explained as the result of the kinetic energy imparted to the solid particles by the fluid stream. The comparison between fixed- and fluidized-bed pressure-drop data given by Morse is shown in Figure 22.34 (31).

Fluid-Solid Conveying. When the fluid-phase flow rate exceeds the free-settling velocity of the particles, the fluidized bed loses its identity because the solid particles are conveyed in the fluid stream. This method of conveying is frequently used throughout industry; for instance, to unload grain ships, to convey the product from spray dryers, to fill and empty cement silos, etc. It has the advantage of cleanliness, low loss, and the ability to move large quantities of solids rapidly. On the other hand, significant breakage of solid particles may occur, and pipe erosion may be excessive.

In attempting to predict the performance of fluid-solid conveyors, Gasterstadt (16) and Vogt and White (42) used a dimensional analysis to combine the variables that apply. They chose to use the relation

$$\beta = \phi\left[\frac{Dv\rho}{\mu}, Y', \frac{D_p}{D}, \frac{\rho_s}{\rho}, \psi, \frac{\epsilon_p}{D_p}, \frac{\frac{1}{3}(\rho_s - \rho)\rho g D_p{}^3}{\mu^2}\right]$$

(22.93)

where β = frictional pressure drop for conveying solids divided by the frictional pressure drop for fluid flow at the same velocity through the same duct

Y' = mass ratio of solid to fluid phase

ϵ_p/D_p = particle roughness

ψ = sphericity

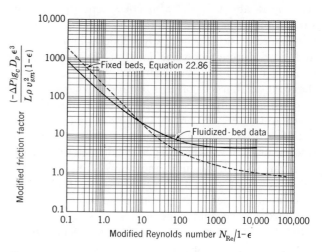

Figure 22.34. Comparison between fixed- and fluidized-bed momentum transfer (31).

The available data gave no information on the effect of ψ or ϵ_p/D_p on β, and the remaining dimensionless groups were related through the equation

$$\beta - 1 = B\left(\frac{D}{D_p}\right)^2\left(\frac{\rho Y' \mu}{\rho_s D v \rho}\right)^k$$

(22.94)

where B and k are functions of the group

$$\frac{\frac{1}{3}(\rho_s - \rho)\rho g D_p{}^3}{\mu^2}$$

and are defined by Figures 22.35 and 22.36.

Note that the resulting relations differ for vertical and horizontal flow. This is because the weight of the solids adds a potential-energy term to those losses which account for the total pressure drop in horizontal flow. In fact, the potential-energy term is so large that, in many cases of vertical fluid-solid flow, the pressure drop can be calculated as accurately by merely accounting for the potential-energy change, as it can be through use of Equation 22.94.

The correlation of Equation 22.94 is an empirical one based on data for gas-solid systems. Its extrapolation to liquid-solid situations should be done with full realization of the resulting uncertainty. Later work (2) has shown that liquid-solid flow can be correlated by a relation of the form of Equation 22.94, though different constants were obtained. Over the very narrow region of available data, k was found to be 1.0 and the constant B to vary from 0.48 to 1.40 as the group

$$\sqrt{\frac{\frac{1}{3}(\rho_s - \rho)\rho g D_p{}^3}{\mu^2}}$$

varied from 0.60 to 1.0. Other work on prediction of pressure drop for liquid-solid flow has used a wide variety of correlating methods, including the determination of the viscosity of the total suspension.

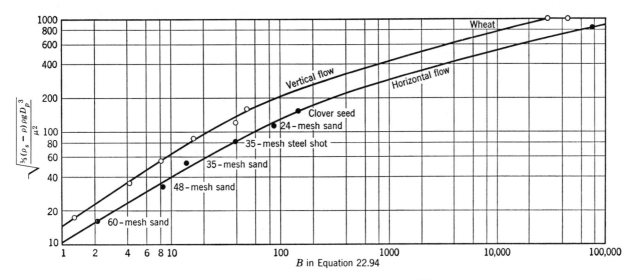

Figure 22.35. Relation between B in Equation 22.94 and the group $\sqrt{\dfrac{\frac{1}{3}(\rho_s - \rho)g\rho D_p^3}{\mu^2}}$ (42). (Courtesy Am. Chem. Soc.)

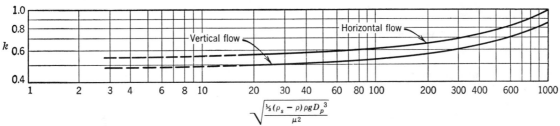

Figure 22.36. Relation between k in Equation 22.94 and the group $\sqrt{\dfrac{\frac{1}{3}(\rho_s - \rho)\rho g D_p^3}{\mu^2}}$ (42). (Courtesy Am. Chem. Soc.)

Illustration 22.10. Wheat is to be pneumatically conveyed from a cargo ship into a dock-side storage bin through a 6-in. tube 400 ft long which may be assumed to be horizontal. If a flow rate of 10 tons/hr of wheat is desired and the weight ratio of wheat to air is 10, what are the capacity and head requirements of the blower?

SOLUTION. Wheat has a density of 1.28 gm/cu cm and a mean particle diameter of 0.158 in. (42). The air flow will be taken to be at 80°F and to be at atmospheric pressure at the pickup end of the conveyor, decreasing in pressure as it moves toward the blower. Gas and solids will be separated in a cyclone just ahead of the blower.

The solution of this problem basically requires solution of the general energy balance in a form like Equation 20.26. The equation must be written and solved for the pickup nozzle where the pressure loss is caused mainly by inertia effects and again for the conveyor tube where frictional effects cause the major pressure loss. Writing Equation 20.26 with the potential energy and work terms eliminated for this case

$$\int \frac{v\,dv}{\alpha g_c} + \int V\,dP + \frac{f\bar{v}^2\,dL}{2g_c D} = 0 \qquad (a)$$

In applying Equation a to the conveyor tube the friction loss term will first be evaluated for air flow alone. The value obtained will be multiplied by the ratio β to get the frictional

losses when solids as well as air flow through the duct. Completing the evaluation of pressure loss is done as shown in Chapter 20 for compressible fluids.

Equation a still must be integrated to eliminate the gas velocity because v varies with pressure along the tube length. Therefore the pressure at the point just inside the pickup nozzle will be calculated and then used as one limit in the integration of Equation a and its application along the tube. In analyzing the pickup nozzle, it will be assumed that the nozzle is frictionless and that the specific volume of gas and solid phases are constant. Then from Equation 20.17 written across the nozzle

$$(w_s + w_g)\frac{\bar{v}_2^2}{2\alpha g_c} + \int w_s V_s\,dP + \int w_g V_g\,dP = 0 \qquad (b)$$

where the subscripts s and g refer to solid and gas phases, respectively.

Taking $\alpha = 1$,

$$v_2 = \frac{412\,(\text{SCFM}) \times \dfrac{540}{492}}{60 \times 0.785 \times 0.25} = 38.4 \text{ ft/sec.};$$

$$V_s = \frac{10}{1.28 \times 62.4} = 0.130 \text{ cu ft/lb air};$$

$$V_g = \frac{379}{29} \times \frac{540}{520} = 13.58 \text{ cu ft/lb air}$$

and inserting these values in Equation b gives

$$\frac{11.0 \times \overline{38.4}^2}{64.4} = -(0.130 + 13.58)\,\Delta P$$

$$\Delta P_{\text{nozzle}} = -\frac{\overline{38.4}^2 \times 11}{64.4 \times 13.71} = -18.4 \text{ lb}_f/\text{sq ft}$$

This value gives the velocity just inside the conveyor as

$$38.4 \times \frac{14.7 \times 144}{14.7 \times 144 - 18.4} = 38.4 \times \frac{2115}{2097.3} = 38.7 \text{ ft/sec}$$

Now the friction loss term of Equation a will be evaluated by assuming the pressure loss is small enough to allow use of a constant v

$$\left(\frac{\Delta P}{\rho}\right)_{f,\text{air}} = \frac{fv^2 L}{2g_c D} \qquad (c)$$

The friction factor f is a function of N_{Re}.

Evaluating N_{Re}

$$N_{\text{Re}} = \frac{0.5 \times 2000 \Big/ \left(\dfrac{\pi}{4} \times \overline{0.5}^2\right)}{0.021 \times 2.42} = 100,000$$

Using the drawn-tubing roughness and the friction factor relations given in Appendix C, $f = 0.018$. Then,

$$\left(\frac{-\Delta P}{\rho}\right)_{f,\text{air}} = \frac{0.018(\overline{38.7}^2)400}{64.4 \times 0.50} = 339. \text{ ft-lb}_f/\text{lb}$$

Having determined the frictional pressure drop for gas flow alone, that for gas-solid flow can be determined using Equation 22.94 and Figures 22.35 and 22.36.

$$\sqrt{\frac{\frac{1}{3}(\rho_s - \rho)\rho g D_p{}^3}{\mu^2}}$$

$$= \sqrt{\frac{\frac{1}{3}(1.28 \times 62.4 - 0.08) \times 0.0795 \times 32.2 \times \left(\dfrac{0.158}{12}\right)^3}{(0.021 \times 0.000672)^2}}$$

$$= 882$$

From Figure 22.35, $B = 90,000$, and, from Figure 22.36, $k = 0.95$. Then

$$\beta - 1 = 90,000\left(\frac{0.5 \times 12}{0.158}\right)^2$$

$$\times \left(\frac{0.0795 \times 10 \times 0.021 \times 0.000672}{79.9 \times 0.5 \times \dfrac{10,200}{3600}}\right)^{0.95}$$

$$= 90,000 \times \overline{38}^2 \times (9.9 \times 10^{-8})^{0.95}$$

$$\beta - 1 = 45.5$$

$$\beta = 46.5$$

Assuming the average pressure is 10 psia:

$$\Delta P_{f,\text{air}+\text{solids}} = 46.5 \times 339 \times \frac{29}{359} \times \frac{492}{540} \times \frac{10}{14.7}$$

$$= 790 \text{ lb}_f/\text{sq ft}$$

Since there are no significant kinetic energy changes in the conveyor tube the frictional pressure loss will be the total loss through the duct. Then

$$\Delta P_{\text{total}} = \Delta P_{\text{nozzle}} + \Delta P_{\text{duct}} = -18.4 - 790$$

$$= -808.4 \text{ lb}_f/\text{sq ft} = 5.6 \text{ psi}$$

This pressure drop should be refined in view of the assumption of constant velocity through the duct and the crude estimate of mean duct pressure used. These refinements can be made using the methods of Chapter 20. It should be noted that the β-value obtained is so large as to seriously limit the accuracy of the result.

Relation between Regions of Fluid-Solid Flow. As evident from the treatments above, flow through packed beds, sedimentation of solids, fluidization, and fluid-solid conveying are all operations depending upon the mechanics of a fluid-solid system. In fact, the flow of droplets and particles through a spray dryer and the behavior of the two liquid phases in a spray-extraction column depend upon the same mechanics. All these operations can be considered in a unified way by focusing attention on the relative, or slip, velocity of particle and fluid (23, 29).

Thus, by definition

$$v_{sl} = v_p - v_{fs}$$

where $\quad v_{sl} =$ relative, or slip velocity in the direction of particle movement

$\qquad v_p =$ velocity of particle

$\qquad v_{fs} =$ fluid free-stream velocity

A single particle is affected by the fluid flowing around it which is displaced from in front of the particle, flows beside it, and passes to the rear. The friction and kinetic-energy changes that result determine the velocity of the solid particle relative to the fluid. If a significant portion of the total stream consists of solid particles, the area for fluid passage around the particles is more sharply restricted than it is with a single particle in a free stream. Moreover, there is a significant interaction between particles. By considering only the flow path around the particles and the effect it has on the separation eddys behind the particles, Equation 22.95 can be derived for spherical particles of fixed diameter spaced in a cubic arrangement (29).

$$\frac{v_{sl,\epsilon}}{v_{sl,\epsilon=1}} = 1 - 1.209(1 - \epsilon)^{2/3} \qquad (22.95)$$

where $\quad v_{sl,\epsilon} =$ slip velocity for particles in a fluid-solid stream of porosity ϵ and infinite extent

$\qquad v_{sl,\epsilon=1} =$ slip velocity for a single particle in a fluid stream

Equation 22.95 is based on the minimum free cross section at the stream section containing the particle

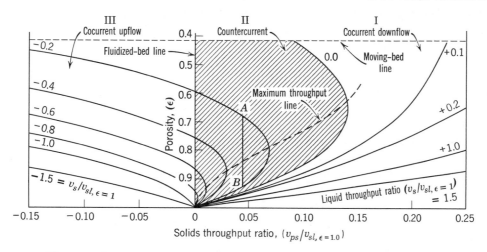

Figure 22.37. Theoretical throughput ratio (29). (Courtesy Am. Inst. Chem. Engrs.)

diameters and hence gives a result that is too low. A similar equation can be derived based on the over-all average free cross section of the stream (29) which gives a result that is too high.

From the definition of slip velocity for a system of identical particles,

$$v_{p,\epsilon} = v_{sl,\epsilon} + v_{fs} \qquad (22.96)$$

where $v_{p,\epsilon}$ = particle velocity in a fluid-solid system of porosity ϵ and infinite extent

$v_{sl,\epsilon}$ = slip velocity of particles in a fluid-solid system of porosity ϵ and infinite extent

v_{fs} = free-stream, or average liquid, velocity

and where the direction of particle movement is taken as positive. Equation 22.96 can be written in terms of the superficial fluid velocity

$$v_{p,\epsilon} = v_{sl,\epsilon} + \frac{v_s}{\epsilon} \qquad (22.97)$$

and then divided by the single-particle slip velocity

$$\frac{v_{p,\epsilon}}{v_{sl,\epsilon=1}} = \frac{v_{sl,\epsilon}}{v_{sl,\epsilon=1}} + \frac{1}{\epsilon}\frac{v_s}{v_{sl,\epsilon=1}} \qquad (22.98)$$

For continuous flow, Equation 22.98 can be written in terms of superficial liquid and solid flow velocities by eliminating $v_{p,\epsilon}$. The *superficial* solid flow rate (v_{ps}) in cubic feet per square foot of cross section per hour is

$$v_{ps} = v_{p,\epsilon}(1 - \epsilon) \qquad (22.99)$$

Placing this equation in Equation 22.98 gives

$$\frac{v_{ps}}{v_{sl,\epsilon=1}} = (1 - \epsilon)\left[\frac{v_{sl,\epsilon}}{v_{sl,\epsilon=1}} + \frac{1}{\epsilon}\frac{v_s}{v_{sl,\epsilon=1}}\right] \qquad (22.100)$$

The term $v_{sl,\epsilon}/v_{sl,\epsilon=1}$ is a function of porosity, as shown by Equation 22.95. Although Equation 22.95 gives too low a result for the slip-velocity ratio, inserting it into

Equation 22.100 will give a result which is qualitatively correct.

$$\frac{v_{ps}}{v_{sl,\epsilon=1}} = (1 - \epsilon)\left[1 - 1.209(1 - \epsilon)^{2/3} + \frac{1}{\epsilon}\frac{v_s}{v_{sl,\epsilon=1}}\right]$$

$$(22.101)$$

Figure 22.37 is a plot of Equation 22.101 (29). There are three major regions in the diagram. At positive values of fluid-throughput ratio ($v_s/v_{sl,\epsilon=1.0}$), cocurrent flow exists. Since this region is also in the region of positive solid-throughput ratio ($v_{ps}/v_{sl,\epsilon=1.0}$), both phases must flow downward. In region II, which lies in the area of negative liquid-throughput ratio but positive solids-throughput ratio, countercurrent flow obtains. This region is the one within which settlers and countercurrent spray dryers and spray extractors operate. In region III, for which the solids-throughput ratio is negative, cocurrent upward flow exists. This is the fluid-solid conveying region. For a fluidized bed, the solid throughput must be zero, but the bed porosity can vary between about 0.42 for a loosely packed bed of spheres and 1.0. At porosities between about 0.42 and about 0.37, the bed is packed and stable. Thus, the packed-bed region is a somewhat elongated point at $\epsilon = 0.4$, $v_{ps}/v_{sl,\epsilon=1} = 0$. The dotted line along the top of the diagram represents moving-bed operation. The maximum-throughput line connects the maximum-solids-throughput-ratio points on the lines of constant-liquid-throughput ratio. This line divides the countercurrent-flow region into two sections. Below the line, increasing the liquid-through-put ratio at constant solids throughput increases the porosity. Above this line, a similar change decreases the porosity. Thus, within this region at any fixed liquid and solid throughput, two values of porosity can be obtained such as points A and B. If a long solids holding time is desired, operation at point A would be preferred over that at point B. In a liquid-extraction column, point A would correspond to operation with the heavy phase continuous, and point *B* would correspond to operation with the light phase continuous. In the operation of a gas-solids device, point *B* would correspond to a falling-particle operation such as in countercurrent spray drying, and operation at point *A* would correspond to a fluidized bed moving downward against an upward-flowing gas stream as is found in some catalytic reactors. Since the maximum-throughput line represents both the maximum solids throughput at fixed liquid throughput and the maximum liquid throughput at fixed-solids rate, it is also the flooding curve.

Figure 22.37 is a highly idealized presentation. The equation on which it is based applies to infinite systems of equal-sized spheres and uses a slip-velocity relation based on minimum free cross section. Therefore, it should be used only for a qualitative understanding of the interrelationship among these operations rather than for quantitative results.

Filtration. Filtration is one of the most common applications of the flow of fluids through packed beds. As carried out industrially, it is exactly analogous to the filtrations carried out in the chemical laboratory using a filter paper in a funnel. The object is still the separation of a solid from the fluid in which it is carried. In every case, the separation is accomplished by forcing the fluid through a porous membrane. The solid particles are trapped within the pores of the membrane and build up as a layer on the surface of this membrane. The fluid, which may be either gas or liquid, passes through the bed of solids and through the retaining membrane.

Industrial filtration differs from laboratory filtration only in the bulk of material handled and in the necessity that it be handled at low cost. Thus, to attain a reasonable throughput with a moderate-sized filter, the pressure drop for flow may be increased, or the resistance to flow may be decreased. Most industrial equipment decreases the flow resistance by making the filtering area as large as possible without increasing the over-all size of the filter apparatus. The choice of filter equipment depends largely on economics, but the economic advantages will vary depending on:

a. Fluid viscosity, density, and chemical reactivity.
b. Solid particle size, size distribution, shape, flocculation tendencies, and deformability.
c. Feed slurry concentration.
d. Amount of material to be handled.
e. Absolute and relative values of liquid and solid products.
f. Completeness of separation required.
g. Relative costs of labor, capital, and power.

Sand Filters. The simplest of industrial filters is the *sand filter*, consisting of layers of rock, gravel, and sand supported by a grating, as shown in Figure 22.38. The feed is pumped onto the top of the sand layer and trickles through the bed by gravity. Sand filters are used only when large flows of very dilute slurry are to be treated, when neither liquid nor solid product has high unit value, and when the solid product is not to be recovered. After a period of operation, the bed is cleaned by backflushing with wash water. Typical applications are found in water- and sewage-treatment plants.

Sand filters may be built from concrete with open tops as is that of Figure 22.38 or may be enclosed to permit pressure operation. Backwashing rates are usually made high enough to fluidize the uniformly sized sand that makes up the actual filter medium.

Flow rate through a sand filter may be calculated using Equation 22.86 for the condition immediately after backwashing when the bed is clean. As solids build up between the sand particles, the porosity decreases and the flow rate drops.

Illustration 22.11. An open sand filter uses a 3-ft-deep bed of −20 + 28 mesh sand as primary filter bed. The sand particles used have an estimated sphericity of 0.9. If the slurry being filtered is essentially water and stands 2 ft deep over the top of the sand, determine the maximum flow rate through the bed which occurs immediately after backwashing.

SOLUTION. The average particle size, as determined from the screen openings, is 0.0280 in. Since the particle area and volume are not known, the particle diameter (D_p) cannot be precisely determined. One method would be to take D_p as equal to 0.0280. An alternate approach is to assume the particles have a volume equal to that of a sphere of $D = 0.0280$ in. Then

$$\text{Surface of sphere} = \frac{\pi \times \overline{0.028}^2}{144} \text{ sq ft}$$

$$A_p = \frac{A_{\text{sphere}}}{\psi} = \frac{\pi \times \overline{0.028}^2}{0.9 \times 144} \text{ and } V_p = \frac{\pi \times \overline{0.028}^3}{6 \times 1728} \text{ cu ft}$$

This gives $D_p = \dfrac{\dfrac{\pi \times \overline{0.028}^3}{6 \times 1728}}{\dfrac{\pi \times \overline{0.028}^2}{0.9 \times 144}} = 0.0021$ ft

Applying Equation 22.86 directly requires a trial-and-error solution for v_s. As a first approximation, Equation 22.76, the Carman-Kozeny equation, will be applied and the answer then adjusted as necessary using Equation 22.86.

$$\frac{(-\Delta P)_f}{L} \cdot g_c = 180 \frac{(1 - \epsilon)^2}{\epsilon^3} \frac{\mu v_s}{D_p{}^2}$$

From Appendix B, Figure B-10, the porosity is estimated at 0.40. Then,

$$\frac{5 \times 62.4}{3} \times 32.2 = 180 \frac{\overline{0.6}^2}{\overline{0.4}^3} \times \frac{(1 \times 0.000672) \times v_s}{\overline{0.0021}^2}$$

$$v_s = 0.0218 \text{ ft/sec}$$

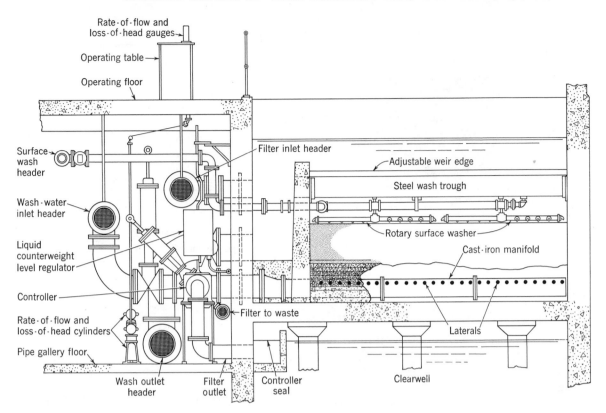

Figure 22.38. Sectional diagram of gravity sand filter showing feed and wash piping systems, graded sand and gravel bed, and clearwell. (Courtesy The Permutit Co.)

Note that in this solution $\rho \, \Delta z \, g/g_c$ has been substituted for ΔP. The Carman-Kozeny equation is correct only at low N_{Re}. Checking the N_{Re},

$$N_{\mathrm{Re}} = \frac{0.0021 \times 0.0218 \times 62.4}{1 \times 0.000672} = 4.25$$

From Figure 22.30, it is apparent that at this N_{Re} the Carman-Kozeny equation is in error by 36.5/35.0 in terms of $1/v_{sm}^2$. Thus, the corrected v_{sm} is

$$v_{sm} = 0.0218 \sqrt{\frac{35}{36.5}} = 0.0214 \text{ ft/sec}$$

The volumetric flow is $0.0214 \times 60 \times 7.48 = 9.6$ gal/sq ft min. This applies immediately after backwashing. The average rate for the entire filter cycle may be less than half this figure.

For filtering a gas-solid material, the *bag*, or *hat*, *filter* is often used. This filter consists of large felt or canvas bags stretched across the openings in a framework built across the gas-flow passageway. Several hundred of these bags may be placed in parallel in this way. The gas passing through the bags deposits the entrained solids on the inside of the bags. Periodically the bags are cleaned by shaking the rack to which they are fastened. The household vacuum cleaner operates on this same principle.

Plate-and-Frame Filter Press. The *filter press* has long been the most common filtering device throughout the chemical industry. Although it is now being replaced in large installations by continuous filter devices it has the advantages of low first cost, very low maintenance, and extreme flexibility. On the other hand, the need for periodic manual disassembly represents a large labor requirement that is often excessive.

The filter press is designed to accomplish a variety of functions, the sequence of which is controlled manually. During filtration the press (a) permits the delivery of feed slurry to the filter surfaces through its own duct, (b) permits the forcing of feed slurry against the filter surfaces, (c) permits filtrate which has passed through the filter surfaces to exit through its own duct, while it (d) retains the solids that were originally in the slurry. During the wash sequence the press (a) permits delivery of wash water to the filtered solids through its own duct, (b) permits the forcing of wash water through the solids retained in the filter, and (c) permits wash water and impurities to leave through a separate duct. Filter design can include four separate ducts as indicated above or can allow for only two ducts where the contamination of the liquid products is not important. After the wash sequence the press is disassembled, and the solids may be collected manually or merely removed and discarded.

The most common filter-press design consists of alternate plates and frames hung on a rack and forced

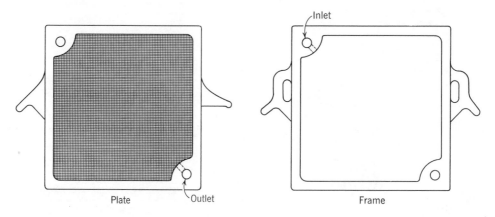

Figure 22.39. Plate-and-frame pair of simple corner-hole nonwashing design with closed discharge and waffle-grid surface. (Courtesy T. Shriver and Company.)

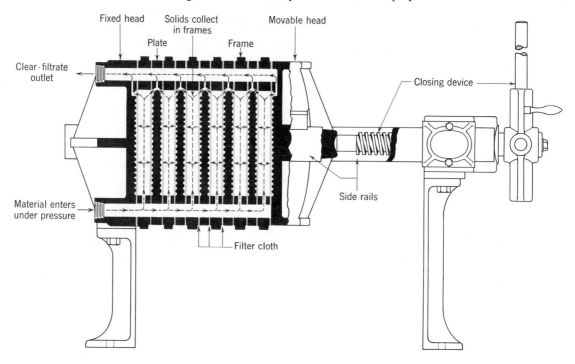

Figure 22.40. Schematic diagram of filter press in operation. (Courtesy T. Shriver and Company.)

tightly together with a screw- or hydraulic-closing mechanism. Figure 22.39 shows a plate-and-frame pair; Figure 22.40 is a diagram of a filter press in operation, and Figure 22.41 is a photograph of a typical filter press. To set up this press, the plates and frames are hung alternately on the side rails of the press using the side lugs on the plates and frames. The filter medium is then hung over the plates extending over both faces of the plate. The filter medium may be canvas or synthetic cloth, filter paper, or woven wire. Holes are cut in the cloth to match the channel holes in the plates and frames. If cloth is used, it may be necessary to preshrink the medium so that the holes will continue to match. When the filter cloths are aligned with the plates and frames, the press is closed with a hand screw

or in large sizes by hydraulic- or electric-closing devices. When the press is closed the filter medium acts as a gasket, sealing the plates and frames and forming a continuous flow channel from the holes in the plates and frames as shown in Figure 22.40. Feed slurry is then pumped to the press under pressure and flows in the press of Figures 22.39 and 22.40 into the bottom-corner duct. This duct has outlets into each of the frames, so the slurry fills the frames in parallel. The solvent, or filtrate, then flows through the filter media while the solids build up in a layer on the frame side of the media. The filtrate flows between the filter cloth and the face of the plate to an outlet duct. As filtration proceeds, the cakes build up on the filter cloths until the cakes being formed on each face of the frame meet in the

Figure 22.41. A filter press operating in a chemical plant separating neutralizing salts from 1,2,6-hexanetriol before distillation. (T. Shriver and Company, Courtesy Am. Chem. Soc.)

center. When this happens, the flow of filtrate, which has been decreasing continuously as the cakes build up, drops off abruptly to a trickle. Usually filtration is stopped well before this occurs.

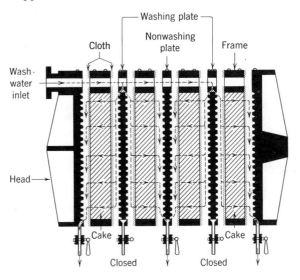

Figure 22.42. Schematic diagram of through-washing in a plate-and-frame filter press with open delivery. Note one-button, two-button, three-button coding on the top edge of the plates and frames. (Courtesy T. Shriver & Co.)

In many cases, it is desirable to wash the filter cake in order to remove the solvent trapped in the cake or dissolve impurities from the cake. In a filter using the plates of Figure 22.39, the washing could be done by feeding wash water into the feed opening, but, if the cake nearly fills the frames, the wash water may be blocked from passage just as is the feed slurry. A better system is afforded by the through-washing plate-and-frame press. In this press, a separate channel is supplied for the wash-water inlet; in closed-delivery presses, a separate exit channel is also supplied. Wash water enters the channel, which has ports opening behind the cloths at every other plate. The wash water then flows through the filter cloths, through the cake built up in an entire frame, through the filter media at the other side of the frames, and out the discharge channel. The flow path is shown in Figure 22.42. In this figure, an open-discharge filter press is pictured. In the closed-discharge press shown in Figure 22.40, the outlet streams would be collected into a common duct like the inlet duct. Note that in this press there are two kinds of plates: those with ducts to admit wash water behind the filter media alternating with those without such ducts. In closed-delivery presses, the alternate plates often have ducts to permit the withdrawal of wash through a

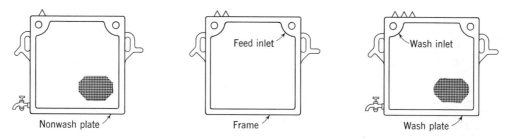

Figure 22.43. Plates and frame for a through-washing open-delivery filter press. (Courtesy D. R. Sperry and Company.)

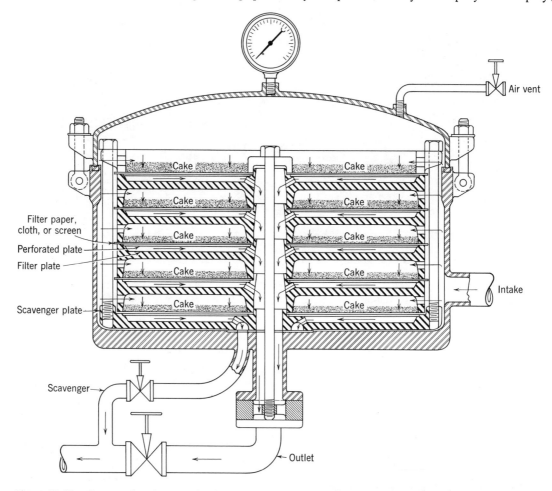

Figure 22.44. Cross-sectional schematic diagram of a horizontal plate filter. Patented scavenger-plate filters last of each batch. The scavenger valve is opened during precoating and closed until the end of the cycle. Then the outlet valve is closed, scavenger valve opened, and remaining liquid is filtered through the scavenger plate by introducing air or gas pressure through the intake. (Courtesy Sparkler Mfg. Company.)

channel separate from the one used to remove filtrate. These various plates and frames are coded with buttons on the top edge. One button signifies a nonwashing plate; two buttons, a frame; and three buttons, a washing plate. Figure 22.43 shows the plates and frame for a through-washing open-delivery filter press.

Filter presses can be made of any construction material desired such as wood, cast iron, rubber, and stainless steel. They can be built for slurry pressures up to 1000 psia. They can handle the filtration of heavy slurries or the "polishing" of a liquid containing only

a faint haze of precipitate. For these reasons, they are used widely throughout the process industries, but, in operations involving a large, continuous throughput, the requirement of manual cleaning and assembling becomes prohibitively expensive.

Other Batch Filters. A large variety of filters is made which, though still batch filters, do not require the complete disassembly for cleaning that is necessary with a plate-and-frame filter press. A few of these are shown in Figures 22.44, 22.45, 22.46 and 22.47. All these filters use varieties of *filter leaves*. The filter leaf is a

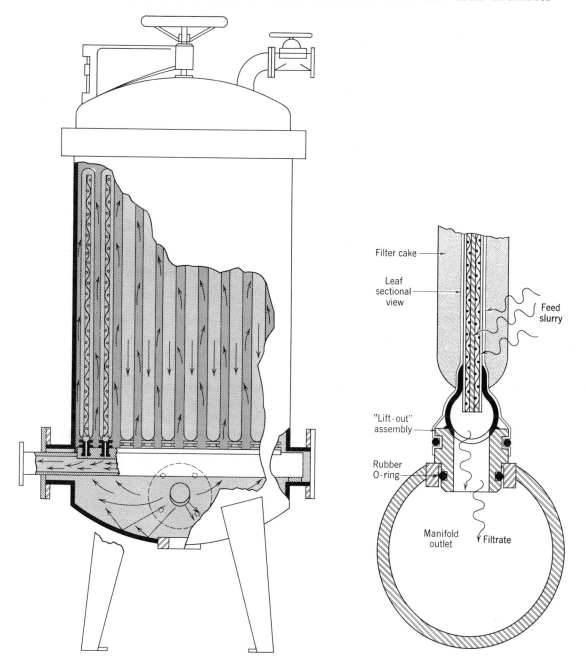

Figure 22.45. Cutaway view of a vertical-leaf filter and sectional diagram showing filter-leaf construction. (Courtesy Industrial Filter & Pump Mfg. Co.)

hollow, internally supported plate as shown in Figure 22.45, which is permanently covered with filter medium. The slurry to be filtered fills the space around the leaf and is forced by pressure on the slurry or vacuum within the leaf to flow through the leaf. Filter cake is built upon the outside of the leaf and filtrate passes from within the leaf to the filtrate-discharge system. When a cake of the desired thickness is built up on the leaves, the filter is opened, and the leaves are either removed for cleaning or are cleaned in place manually or by sluicing away the solids. Of the filters shown the *horizontal-plate filter*, Figure 22.44 is particularly well adapted for the final clarifying of solutions containing minute quantities of solids because of the ease of applying a filter-aid precoat. Filter aids are open-structured incompressible solids that may be deposited on the filter cloths to serve as a high-efficiency filter medium. They are further discussed below. The *horizontal leaf filter* (Figure 22.46) is built in very large sizes and can be opened particularly rapidly for cleaning. The *Sweetland filter*, (Figure 22.47) is made in two half-cylinders. The bottom half opens downward by releasing the

Figure 22.46. A large horizontal-leaf filter open for cleaning. This filter has 800 sq ft of filtering area and a capacity of up to 800 gal/min of water. (Courtesy Niagara Filters Division, American Machine and Metals, Inc.)

quick-opening cam locks to expose vertical disk-shaped leaves which are cleaned in place.

Continuous Filters. Modern, high-capacity processing has made the development of continuous filters mandatory, and several varieties are in common use. In such filters, the slurry is fed continuously, and cake and filtrate are produced continuously.

The *horizontal rotary filter* of Figures 22.48 and 22.49 is particularly well adapted to the filtering of quick-draining crystalline solids. Its horizontal surface prevents the solids from falling off or from being washed off by the wash water, and an unusually heavy layer of solids can be tolerated. This filter consists of a circular horizontal table that rotates around a center axis. The table is made up of a number of hollow pie-piece-shaped segments with perforated or woven metal tops. Each of the sections is covered with a suitable filter medium and is connected to a central valve mechanism that appropriately times the removal of filtrate and wash liquids and the dewatering of the cake during each revolution. Each segment successively receives slurry, is then sprayed with wash liquid in two applications, has its cake dewatered by pulling air through the cake, and has its cake scooped off the surface by the discharge scroll.

The *rotary-disk vacuum filter* is shown in Figure 22.50. This filter gives an especially high filtration rate for a given floor space. The filtering medium is again a wedge-shaped leaf covered with a filter medium. Here, the leaves rotate in a vertical plane around a horizontal axis. The slurry to be filtered fills the filter basin almost

up to the filter axis. As the leaf dips through the slurry, it collects a cake on its surfaces while the filtrate passes through to a central discharge system. The leaf then carries the filter cake through the upper half of its rotation while air pulled through the cake dries it. The cake is scraped off the leaves by doctor knives or is blown off with compressed air fed inside the leaves before they dip again into the slurry. No provision is made for washing the cake. If a filter cloth becomes worn or torn the single segment may be removed and replaced with a new one relatively quickly.

The *rotary-drum vacuum filter* with string discharge is shown schematically in Figure 22.51, and a somewhat different rotary vacuum filter is pictured in Figure 22.52. With either unit, the cycle is very much like that of the horizontal vacuum filter. Filter cake is picked up from a slurry pool by dipping the drum surface and applying vacuum. The cake is then carried around the drum where it is successively washed and dewatered by the continuous application of vacuum to the inside of the drum. The string-discharge system leads the cake away from the drum and over a roll with sharp curvature which causes the cake to drop off. Figure 22.51 shows the agitator in the slurry pool that prevents settling of the slurry, the discharge valve that controls withdrawal of filtrate and wash liquids and the dewatering of the cake, the drain lines that apply vacuum to the drum surface, and the drum surface construction. For coarse particles which settle rapidly and form a porous cake, a feed hopper on top of the drum is more satisfactory than the dipping hopper shown.

The unit shown in Figure 22.52 differs from that in Figure 22.51 mainly in the cake-discharge mechanism. This unit uses a motor-driven doctor knife that moves toward the filter drum at an extremely slow rate. This permits the application of an initial, or precoat, layer

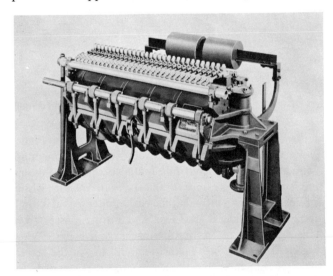

Figure 22.47. A Sweetland pressure filter closed for filtration. (Courtesy Dorr-Oliver, Inc.)

Figure 22.48. Horizontal-rotary vacuum filter showing arrangement of piping for two stages of wash and scroll cake-removal mechanism. One section of filter medium is removed to illustrate deck support and drainage slope. (Courtesy Dorr-Oliver, Inc.)

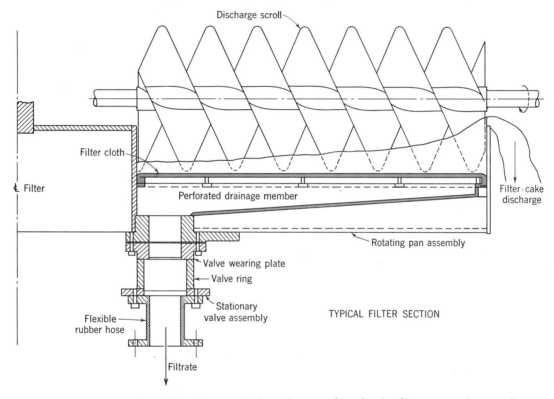

Figure 22.49. Cross-sectional view of a rotary horizontal vacuum filter showing filtrate-removal system, filter cloth, and discharge scroll. (Courtesy Filtration Engineers Division, American Machine and Metals, Inc.)

of filter aid perhaps 1 to 2 in. thick. After this layer is in place the slurry is fed to the dip tank, and filter cake builds up on the precoat layer of filter aid. The doctor knife removes the filter cake and a very thin layer of filter aid. As filtration progresses the filter-aid layer gets progressively thinner until a new layer must be applied. However, the doctor knife moves so slowly that a precoat layer will last as long as a week.

Filter Media and Filter Aids. As mentioned above filter media consisting of cloth, paper, or woven or porous metal may be used. The criteria upon which a filter medium is selected must include ability to remove the solid phase, high liquid throughput for a given pressure drop, mechanical strength, and chemical inertness to the slurry to be filtered and to any wash fluids. Of course, each of these considerations is tempered by the economics involved, so that the filter operator tries to choose a medium that meets the required filtration standards while contributing to the lowest possible over-all filtration cost.

A variety of filter media have been tested for pore-size distribution (18). The tests showed that with a

Figure 22.50. Rotary-disk vacuum filter, 8-ft face by 6 disk, shown from valve and drive end. (Courtesy Dorr-Oliver, Inc.)

woven-cloth medium both interfiber and interyarn pores exist with 30 to 50 per cent of the total pore volume being made up of interfiber pores. The interfiber pores are the spaces between the fibers making up a single thread of the cloth, whereas interyarn pores are the spaces between the threads woven to form the cloth. These interfiber pores were less than 10 microns in radius. The interyarn pores were found to range between 70 and 200 microns in radius. With a felt medium, all the pores are interfiber, but, because of the random fiber orientation, they were found to vary between about 2 and about 180 microns nearly following a logarithmic normal-distribution function with a mean between 10 and 20 microns. The sintered-metal media tested had a much more uniform pore-size distribution with 90 per cent of the pore volume falling between 20 and 50 microns in radius.

In operation, some filter-cake solids usually penetrate the filter medium and fill some of the pores. As filtration continues, these particles are thought to bridge across the pores and cake begins to form on the face of the medium. In normal cases, between 5 and 25 per cent of the pore volume of the filter medium is filled with solids. As a result, the resistance to flow through the medium increases sharply. In some cases, the solids fill the filter medium to such an extent that the filtration rate is seriously reduced.

Filter aids are often used to speed filtration or to make it possible to collect more completely the finest particles held in the slurry. The filter aids are finely divided, hard-structured solids that themselves form an open, noncompressible cake. The most common example is diatomaceous earth, which consists of the skeletons of very small prehistoric marine animals. This material is mined from large surface deposits mainly in California,

Oregon, and Nevada. It is practically pure silica and has a very complicated structure. Applied as a precoat on the filter cloth, the filter aid acts as the primary filter medium and permits complete removal of very fine solid particles from the slurry. Another method of application is to mix the filter aid into the slurry. Here, it distributes throughout the cake, keeping the cake relatively open for flow and continuously supplying a large surface for adhesion of very finely divided solids. This action is particularly valuable when filtering colloidal solids that form a very dense and compressible cake, which is not to be recovered. During 1950, some 180,000 tons of diatomaceous earth were used as filter aid in the United States (20).

Filtration Calculations—General Relations. The flow of filtrate through the filter cake should be describable by any of the general equations for flow through packed beds, such as Equation 22.86. Actually, in almost all practical cases, flow is laminar, and the Carman-Kozeny equation

$$\frac{(-\Delta P)_f g_c}{L} = 180 \frac{(1-\epsilon)^2}{\epsilon^3} \frac{\mu v_s}{D_p^2} \qquad (22.76)$$

applies. This equation relates the pressure drop through the cake to the flow rate, the cake porosity and thickness, and the solid-particle diameter. Some modification of the equation is necessary so that the measureable variables of filtration can be introduced into it.

In its more usual form, it is written in terms of the specific surface area of the particles by incorporating Equation 22.73.

$$D_p = \frac{6}{\dfrac{A_p}{V_p}} = \frac{6}{S_0} \qquad (22.73a)$$

where S_0 = specific surface area of particle, sq ft/cu ft of solid volume

Thus,

$$\frac{(-\Delta P)_f g_c}{L} = \frac{5(1-\epsilon)^2 \mu v_s S_0^2}{\epsilon^3} \qquad (22.102)$$

Solving this equation for the velocity of flow gives

$$v_s = \frac{(-\Delta P)_f g_c \epsilon^3}{5L\mu S_0^2(1-\epsilon)^2} = \frac{1}{A}\left(\frac{dV}{d\theta}\right) \qquad (22.103)$$

where $dV/d\theta$ = the filtration rate, that is, the volume of filtrate passing through the bed per unit time

A = filtration area

In order to integrate Equation 22.103 to obtain a relation usable over the entire process, only two variables may appear in the equation. As written, the quantities V, θ, L, $(-\Delta P)_f$, S_0, and ϵ may all vary. The cake

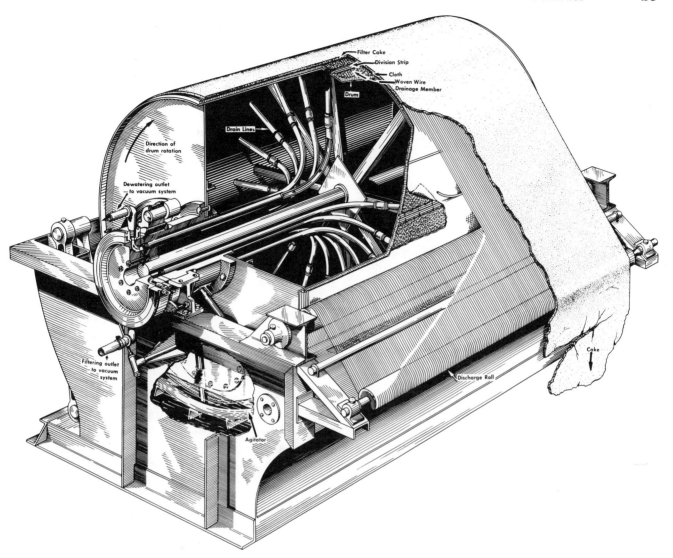

Figure 22.51. Schematic drawing of string-discharge rotary-drum vacuum filter. (Courtesy Filtration Engineers Division, American Machine and Metals, Inc.)

Figure 22.52. Continuous vacuum precoat filter, 5 ft 3 in. diameter × 8 ft face. (Courtesy Dorr-Oliver, Inc.)

thickness (L) may be related to the volume of filtrate by a material balance, since the thickness will be proportional to the volume of feed delivered to the filter.

$$LA(1 - \epsilon)\rho_s = w(V + \epsilon LA) \qquad (22.104)$$

where ρ_s = density of the solids in the cake
w = weight of solids in the feed slurry per volume of liquid in this slurry
V = volume of filtrate which has passed through the filter cake

The final term of Equation 22.104, (ϵLA) represents the volume of filtrate held in the filter cake. It is normally infinitesimal compared to V, the filtrate which has passed through the bed. Assuming this term negligible and combining Equation 22.103 with Equation 22.104 to eliminate L gives

$$\frac{1}{A}\frac{dV}{d\theta} = \frac{(-\Delta P)_f g_c \epsilon^3}{5\frac{wV}{A\rho_s}\mu(1 - \epsilon)S_0^2} = \frac{(-\Delta P)_f g_c}{\frac{\alpha\mu wV}{A}} \quad (22.105)$$

where α is the *specific cake resistance*, defined as

$$\alpha = \frac{5(1 - \epsilon)S_0^2}{\rho_s \epsilon^3} \qquad (22.106)$$

A similar equation in terms of L could also be obtained by eliminating V between Equations 22.103 and 22.104.

Equation 22.105 is the basic filtration equation in terms of the pressure drop across the filter cake alone. The collection of all terms involving the filter-cake properties into the specific cake resistance does not infer that the resistance (α) will be constant for a given feed slurry, regardless of filtering pressure drop or of filter type or size. The specific cake resistance may not even remain constant throughout a given filter operation at constant $(-\Delta P)_f$ because of variations in ϵ and S_0. The void fraction (ϵ) usually varies with variation of the compacting stress applied to the bed. This stress will be directly proportional to $(-\Delta P)/L$, and, since L varies throughout the process, ϵ may also vary. Both ϵ and S_0 are sensitive to the degree of flocculation of the precipitate in the feed. The flocculation may vary with the turbulence of flow of slurry fed to the press and hence may be a function of the filter rate. However, in most constant-pressure filtrations, α is constant except in the initial moments of filtration when the flow rate is very high and the form of the filter cake has not been fixed. In fact, for many filter cakes, α is relatively insensitive to changes in $(-\Delta P)_f$. Such cakes are said to be *incompressible*, though probably no cake is completely incompressible.

Filtration Calculations—Inclusion of Filter-Medium Resistance. Equation 22.105 is expressed in the familiar form of a rate proportional to a driving force divided by a resistance. Here, both driving force and resistance apply to the filter cake alone. However, practically any $(-\Delta P)$ measured will at least include the pressure drop across the filter medium and will probably include the pressure drop of various flow channels before and after the actual filtering area. If such an over-all pressure drop is to be used in an equation like Equation 22.105 the resistance term must also include the flow resistances of the additional parts of the apparatus. Since these resistances are arranged in series, Equation 22.105 becomes

$$\frac{dV}{A\,d\theta} = \frac{(-\Delta P_t)g_c}{\mu\left(\dfrac{\alpha wV}{A} + R_M\right)} \qquad (22.107)$$

where R_M has the units (ft)$^{-1}$ and represents the resistance of filter medium and piping to filtrate flow. For convenience in analyzing filtration-performance data, the resistance of filter cloth and flow channels is usually expressed in terms of an equivalent volume of filtrate

$$\frac{dV}{A\,d\theta} = \frac{(-\Delta P_t)g_c}{\dfrac{\mu\alpha w}{A}(V + V_e)} \qquad (22.108)$$

Here, V_e is the volume of filtrate necessary to build up a fictitious filter cake, the resistance of which is equal to the resistance of the filter medium and the piping between the pressure taps used to measure $(-\Delta P_t)$. The filter-medium resistance of significance here is the resistance of the medium with the pores partially blanked with filter cake and with the initial layer of filter cake, on which the bulk of the filter cake will be built, in place. It is considerably greater than the resistance to water flow of the clean filter cloths.

Filtration Calculations—Integration of Filtration Equations. The integration of Equation 22.108 can now be performed if $(-\Delta P_t)$ and α can be related to the variables of V and θ. As mentioned above, there are many slurries for which α is substantially constant throughout the filter cycle. In the discussion that immediately follows, only those slurries that form incompressible cakes will be considered. Thus, α = constant.

The variation in $(-\Delta P_t)$ through the filter cycle depends upon the operating procedure and the type of filter used. In sand filters, $(-\Delta P_t)$ varies with the liquid head over the filter bed. In rotary vacuum filters, it will be constant at somewhat less than 1 atm. In filter presses and pressure leaf filters, it can be varied at the will of the operator within the equipment limits. The operator may control the pressure at a fixed value throughout the entire run. This may be the easiest method of operation if the slurry is fed from a pressurized tank or by hydrostatic pressure from a storage tank. The operation may be run at a constant feed rate as would be done by delivering the feed through a positive-displacement pump and continuing the run until the filter feed pressure reached a limit. Often, the operation starts at a constant throughput and continues in this way until the pressure reaches a predetermined level, after which the pressure is held constant. This has the advantage of forming a rather loosely knit initial cake and forcing a minimum amount of solids into the pores of the filter medium. It is also a convenient method of operation, for initially the filter will accept the entire throughput of the feed pump at relatively low feed pressure. As the cake builds up, the feed pressure also increases, but the rate of delivery to the filter falls off only slightly even with a centrifugal feed pump. When an economically optimum or maximum safe pressure is reached, the run is continued controlling the feed pressure at this value.

Filtration Calculations for Incompressible Cakes. Performing the integration of Equation 22.108 for incompressible cakes (α = constant) and for operation *at constant* $(-\Delta P_t)$ gives

$$\int_0^V (V + V_e)\, dV = \int_0^\theta \frac{g_c A^2(-\Delta P_t)}{\mu \alpha w}\, d\theta \qquad (22.108a)$$

$$\frac{V^2}{2} + VV_e = \frac{g_c A^2(-\Delta P_t)}{\mu \alpha w}\, \theta \qquad (22.109)$$

or

$$\theta = \frac{\mu \alpha w}{g_c A^2(-\Delta P_t)} \left(\frac{V^2}{2} + V_e V \right) \qquad (22.109a)$$

from which the time necessary to pass any given volume of filtrate can be calculated.

The solution of Equation 22.109a requires evaluation of the two constants α and V_e. The specific cake resistance (α) may perhaps be evaluated from the properties of the filter cake if ϵ and S_0 are known for the particular filtration condition. However, the volume of filtrate equivalent to the filter-medium and piping-flow resistances (V_e) must be determined from pilot-filtration data. For this reason, it is usual practice to evaluate both α and V_e from a pilot-filtration run using the actual slurry to be filtered under conditions as close to those to be employed in the plant as possible. To permit evaluation of these constants from experimental data, Equation 22.108 is inverted to give

$$\frac{d\theta}{dV} = \frac{\mu \alpha w}{g_c A^2(-\Delta P_t)} (V + V_e) \qquad (22.110)$$

This equation is a straight-line relation between $d\theta/dV$ and V if $(-\Delta P_t)$ is held constant. Thus, from actual constant-pressure filtration data, $d\theta/dV$ may be plotted

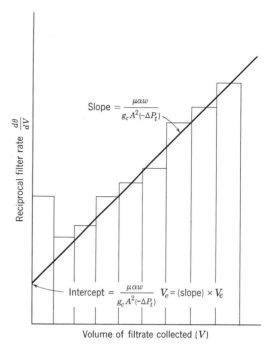

Figure 22.53. Typical filtration-rate data for constant-pressure condition.

as the ordinate as a function of V as the abscissa to give a straight line. Such a plot is shown in Figure 22.53 where measurements have been taken over finite time differences rather than differentials. The bar-graph technique is used to indicate that the value $(\Delta V/\Delta \theta)_i$ represents the average rate during the interval between V and $V + \Delta V$. Therefore in plotting the correlating line a balance should be struck between areas of the bars above the line and areas not in the bars but below the line. The slope of this line is then

$$\alpha \mu w / g_c A^2 (-\Delta P_t)$$

from which α may be determined, and the intercept is

$$[\mu \alpha w / g_c A^2(-\Delta P_t)] V_e.$$

The intercept divided by the slope is V_e. The data plotted in Figure 22.53 are typical in that the initial measuring periods give highly irregular data. This irregularity occurs as the cake begins to form on the filter medium and may be caused partly by higher flow rates giving more tendency toward turbulence and partly by the instability of the filter cake.

If the test run is made under conditions of *constant rate* rather than constant pressure, Equation 22.108 can be rearranged to

$$-\Delta P_t = \frac{\mu \alpha w}{g_c A^2} \left(\frac{dV}{d\theta} \right)(V + V_e) \qquad (22.108b)$$

Equation 22.108b gives a straight line for constant $dV/d\theta$, that is, for constant filtration rate, when $(-\Delta P_t)$ is plotted against V. Here the slope is

$$(\mu\alpha w/g_cA^2)/(dV/d\theta),$$

and the intercept is

$$[(\mu\alpha w/g_cA^2)/(dV/d\theta)]V_e.$$

Again, the unknown constants can be determined from the measured slope and intercept at zero filtrate volume. Once α and V_e are determined, they can be used directly to predict the path of a constant-rate filtration process, but only if the cake is incompressible.

Equations equivalent to Equations 22.108 through 22.110 can be developed in terms of cake thickness (L) rather than filtrate volume. The resulting equations are convenient when dealing with leaf or rotary vacuum filters where it is desirable to set the filtration cycle by the thickness of filter cake built up on the surfaces.

Illustration 22.12. Ruth and Kempe (35) report the results of laboratory filtration tests on a precipitate of $CaCO_3$ suspended in water. A specially designed plate-and-frame press with a single frame was used. The frame had a filtering area of 0.283 sq ft and a thickness of 1.18 in. All tests were conducted at 66°F and with a slurry containing 0.0723 weight fraction $CaCO_3$. The density of the dried cake was 100 lb/cu ft. Test results for one run are given below:

$P = 40$ psi = constant

Volume of Filtrate, l.	Time, sec
0.2	1.8
0.4	4.2
0.6	7.5
0.8	11.2
1.0	15.4
1.2	20.5
1.4	26.7
1.6	33.4
1.8	41.0
2.0	48.8
2.2	57.7
2.4	67.2
2.6	77.3
2.8	88.7

Determine the filtrate volume equivalent in resistance to the filter medium and piping (V_e), the specific cake resistance (α), the cake porosity (ϵ), and the cake specific surface (S_0).

SOLUTION. The filter data given are differenced as shown tabularly below, so that the relation

$$\frac{\Delta\theta}{\Delta V} = \frac{\mu\alpha w}{g_cA^2(-\Delta P_t)}(V + V_e) \qquad (22.110a)$$

may be graphically solved for α and V_e by plotting $\Delta\theta/\Delta V$ as a function of V.

V, l.	θ_{sec},	$\Delta\theta$	$\dfrac{\Delta\theta}{\Delta V}$
0	0		
		1.8	9.0
0.2	1.8		
		2.4	12.0
0.4	4.2		
		3.3	16.5
0.6	7.5		
		3.7	18.5
0.8	11.2		
		4.2	21.0
1.0	15.4		
		5.1	25.5
1.2	20.5		
		6.2	31.0
1.4	26.7		
		6.7	33.5
1.6	33.4		
		7.6	38.0
1.8	41.0		
		7.8	39.0
2.0	48.8		
		8.9	44.5
2.2	57.7		
		9.5	47.5
2.4	67.2		
		10.1	50.5
2.6	77.3		
		11.4	57.0
2.8	88.7		

The plotted data are shown in Figure 22.54. The slope of the line drawn through the differenced data is 18.05 sec/l.², and the intercept is 5.9 sec/l. From these values,

$$V_e = 5.9/18.05 = 0.327 \text{ l.}$$

$$\frac{\alpha\mu w}{g_cA^2(-\Delta P)} = \frac{18.05}{(0.0353)^2} = 14,500 \text{ sec/(cu ft)}^2$$

Solving for $\alpha\mu w$ and for α yields

$$\alpha\mu w = 14,500 \times 32.2 \times (40 \times 144) \times (0.283 \times 2)^2$$

$$= 8.60 \times 10^8 \text{ lb/sec cu ft}$$

$$w = \frac{0.0723}{\dfrac{0.9277}{62.4}} = 4.86 \text{ lb } CaCO_3/\text{cu ft of } H_2O$$

$$\alpha = \frac{8.60 \times 10^8}{1.1 \times 0.000672 \times 4.86} = 2.41 \times 10^{11} \text{ ft/lb}$$

The cake porosity can be directly calculated from the measurement of dry-cake density.

Density of solid $CaCO_3 = 2.93 \times 62.4 = 183$ lb/cu ft $= \rho_s$

$$\epsilon = 1 - \frac{100}{183} = 1 - 0.547 = 0.453$$

The specific cake surface (S_0) can now be determined from the values of α, ϵ, and ρ_s calculated above.

$$\alpha = \frac{5(1-\epsilon)S_0^2}{\rho_s\epsilon^3}$$

$$S_0^2 = \frac{2.41 \times 10^{11} \times 183 \times \overline{0.453}^2}{5 \times 0.547} = 3.35 \times 10^{12}$$

$$S_0 = 1.83 \times 10^6 \text{ sq ft/cu ft of solids}$$

Illustration 22.13. A 30 by 30 in. plate-and-frame filter press with twenty frames 2.50 in. thick is to be used to filter the $CaCO_3$ slurry which was used in the test of Illustration

22.12. The effective filtering area per frame is 9.4 sq ft, and V_e may be assumed to be the same as that found in the test run. If filtration is carried out at constant pressure with $(-\Delta P) = 40$ psi, determine the volume of slurry that will be handled until the frames are full, and the time required for this filtration.

SOLUTION. The required slurry volume can be directly calculated from the volume of frames and the slurry concentration.

$$\text{Volume of frames} = \frac{9.40}{2} \times \frac{2.50}{12} \times 20 = 19.60 \text{ cu ft}$$

$$\text{Solids in cake} = 19.60 \times 100 = 1960 \text{ lb}$$

$$\text{Weight of slurry fed} = \frac{1960}{0.0723} = 27,100 \text{ lb}$$

To calculate the slurry density,

$$\frac{1}{\rho} = \frac{1}{62.4} \times 0.9277 + \frac{1}{183} \times 0.0723 = 0.01485 + 0.00040$$

$$\rho = \frac{1}{0.01525} = 65.5 \text{ lb/cu ft}$$

$$\text{Volume of slurry} = \frac{27,100}{65.5} \times 7.48 = 3100 \text{ gal.}$$

The required filtration time can be found by solving Equation 22.109a, since the constants α and V_e have already been determined.

$$\theta = \frac{\mu \alpha w}{g_c A^2 (-\Delta P_t)} \left(\frac{V^2}{2} + V_e V \right) \qquad (22.109a)$$

In this equation, V is the volume of filtrate, and it can be determined from the material balance based on the cake, Equation 22.104

$$LA(1 - \epsilon)\rho_s = w(V + \epsilon LA) \qquad (22.104)$$

$$19.60(1 - 0.453)183 = 4.86(V + 19.60 \times 0.453)$$

$$V = 395.0 \text{ cu ft}$$

Inserting this value into Equation 22.109a gives

$$\theta = \frac{8.60 \times 10^8}{32.2 \times \overline{188}^2 \times 40 \times 144}$$

$$\times \left(\frac{\overline{395}^2}{2} + 0.327 \times 0.0353 \times 395 \right)$$

$$\theta = 0.131 \times (78,000 + 4.5) = 10,200 \text{ sec} = 2.84 \text{ hr}$$

Filtration Calculations for Compressible Cakes. As mentioned before, most chemical precipitates form compressible filter cakes, in which higher compressive forces deform the solid particles, break up flocculent aggregates, and force the particles closer together.

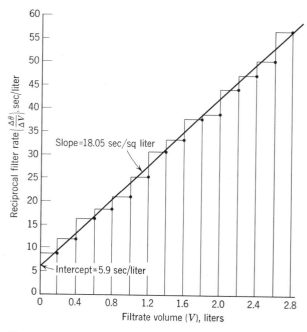

Figure 22.54. Graphical determination of filtration constants for Illustration 22.12.

Empirically, it has been found that at moderate pressures the relation

$$\alpha = \alpha_0 + b(-\Delta P_c)^s \qquad (22.111)$$

where $\alpha_0 =$ the specific cake resistance at zero compressive pressure, a constant

$s =$ the cake compressibility factor, a constant over moderate pressure ranges

$b =$ a constant

$-\Delta P_c =$ pressure drop across the cake

holds for most of these precipitates. The values of α_0, b, and s can be determined from a series of constant-pressure filtration tests.

A simpler equation has been even more generally used:

$$\alpha = \alpha_0'(-\Delta P_c)^{s'} \qquad (22.111a)$$

where α_0' and s' are empirical constants with significance equivalent to α_0 and s of Equation 22.111. This equation is obviously wrong at low values of $-\Delta P_c$, for at $-\Delta P_c = 0$ it predicts a specific cake resistance of zero. It is convenient, however, for the constants are readily determined, and it can be used over restricted pressure ranges.

A more fundamental approach to compressible filter cakes has been made by Grace (17). He notes that the compressive force on a particle in a filter cake depends upon its position in the cake, varying from a maximum at the cloth to a minimum at the cake surface. This gradation even arises for the simple case of pressure filtration on a flat vertical surface, since the hydraulic pressure applied to the surface of the cake equals the sum of the mechanical compressive pressure and the hydraulic pressure at any point within the cake. In this case, the compressive pressure (P_s) varies from zero at the filter-cake–slurry interface to $P_1 - P_2$ at the filter medium, where P_1 is the pressure on the cake surface and P_2 is the pressure at the face of the filter medium. This constancy of hydraulic plus compressive pressures can be shown by making a simple force balance at a point within the filter cake. For more complicated situations, a similar relation exists between bed depth and compressive force. If the relations among ϵ, S_0, and P can be determined, Equation 22.108 can be integrated to give the filtration rate as a function of filtrate volume. This integration requires the writing of Equation 22.108 as a differential function of P.

$$\frac{dV}{A\,d\theta} = \frac{-g_c A \rho_s}{5w\mu V} \int_0^{P_1 - P_2} \frac{\epsilon^3}{(1-\epsilon)S_0{}^2}\,dP \quad (22.112)$$

Alternatively, Equation 22.112 can be simplified by inserting the specific cake resistance, (α_p) applicable at a point in the cake.

$$\frac{dV}{A\,d\theta} = \frac{-g_c A}{wV\mu} \int_0^{P_1 - P_2} \frac{dP}{\alpha_p} \quad (22.113)$$

which is again restricted to pressure filtration on a flat vertical surface. P_1 and P_2 here refer to the pressures at the filter-cake–slurry interface and at the filter-medium surface respectively.

The relations between P and α_p, ϵ, and S_0 can be experimentally determined by "compression-permeability" experiments. The slurry to be tested is enclosed in a cylinder with a porous bottom, and a filter cake is built up on this porous surface by letting the filtrate drain through it. A piston is then put into the cylinder and slowly forced down onto the filter cake. The piston is then loaded with an increasing series of weights. At each piston loading, the porosity is determined by noting the piston position. Filtrate is fed to the filter cake, and the value of α determined by solving Equation 22.113 assuming α to be constant at any given loading. Then, S_0 can be calculated from the values of α and ϵ already determined. Figure 22.55 shows typical compression-permeability data of Grace (17) giving ϵ, S_0, and α_p as functions of ΔP. Grace has shown that the results of such experiments can be used under industrial filtration conditions.

Illustration 22.14. Determine the average specific cake resistance (α) that applies for the filtration of Type B ZnS slurry at a $(-\Delta P_c)$ of 70 psia. Compression permeability data for this slurry are given in Figure 22.55.

SOLUTION. The specific cake resistance at a fixed compressive pressure (α_p) is shown as a function of compressive pressure on line 5 of Figure 22.55a. The data are replotted in arithmetic coordinates on Figure 22.56 with $1/\alpha_p$ as a function of P:

P, psia	α_p, ft/lb	$\dfrac{1}{\alpha_p} \times 10^{-12}$
1	0.85×10^{12}	1.19
10	6.0×10^{12}	0.17
20	10.5×10^{12}	0.095
30	16×10^{12}	0.063
50	29×10^{12}	0.034
70	45×10^{12}	0.022

The specific cake resistance applicable to the entire filter cake can be obtained from Equation 22.113. From this equation

$$\frac{1}{A}\frac{dV}{d\theta} = \frac{-g_c A}{wV\mu} \int_0^{P_1 - P_2} \frac{dP}{\alpha_p} = \frac{-g_c A\,(\Delta P)}{wV\mu}\frac{1}{\alpha}$$

or

$$\alpha = \frac{\Delta P}{\displaystyle\int_0^{P_1 - P_2} \frac{dP}{\alpha_p}}$$

The indicated solution is carried out from Figure 22.56 by graphical integration. In this integration the numerical result is greatly influenced by the value of α_p at $P = 0$. Therefore the data have been replotted with α_p as a function of P. This curve can be more readily extrapolated and gives $\alpha_p = 0.25 \times 10^{12}$ at $P = 0$. Then carrying out the graphical integration

$$\int_0^{P_1 - P_2} \frac{dP}{\alpha_p} = 15.51 \times 10^{-12}$$

and

$$\alpha = \frac{70}{15.51 \times 10^{-12}} = 4.51 \times 10^{12} \text{ ft/lb}$$

Filtration Calculations—Washing and Dewatering. After filtering is completed, the washing and dewatering operations are done with the filter cake in place. Since the cake thickness is unchanged, the wash rate usually varies directly with the pressure drop. If the wash water follows the same path as the slurry and is fed at the same pressure, the wash rate will equal the final filtration rate, as given by Equation 22.108. But in the through-washing filter press, any given volume of wash water passes through only half as many parallel filter-cloth and filter-cake surfaces and twice as thick a layer of filter cake as does the final filtrate. This flow path may be seen by examining Figure 22.42. Therefore, the wash rate in this unit is one-quarter the final filter rate if the filtrate is very nearly water and the pressure drop for washing is the same as the final filtering pressure drop.

In the dewatering part of a filtration cycle, air is drawn through the filter cake, pulling the filtrate or wash water remaining in the pores of the cake out ahead of it and

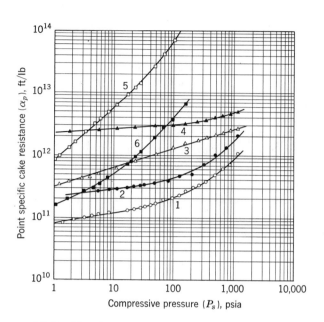

(a) Specific cake resistance.

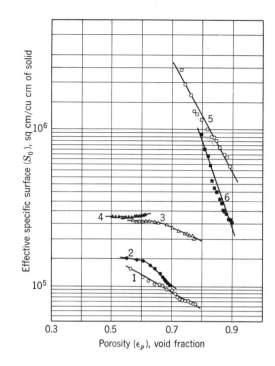

(b) Porosity as a function of specific surface.

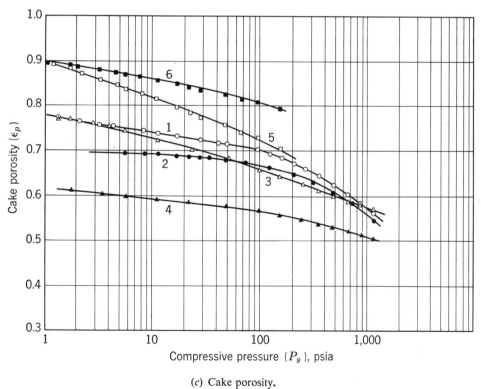

(c) Cake porosity.

Figure 22.55. Compression permeability data (46). (Courtesy Am. Inst. Chem. Engrs.)

1—Superlite $CaCO_3$ (flocculated), pH = 9.8
2—Superlite $CaCO_3$, pH = 10.3
3—R-110 grade TiO_2 (flocculated), pH = 7.8
4—R-110 grade TiO_2, pH = 3.5
5—ZnS, Type B, pH = 9.1
6—ZnS, Type A, pH = 9.1

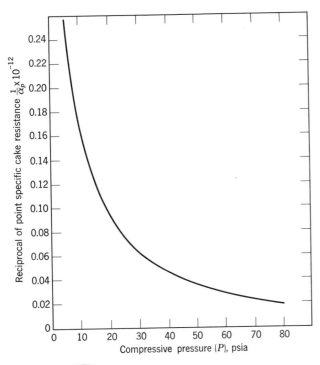

Figure 22.56. Compression-permeability data for ZnS Type B suspension presented as $1/\alpha_p$ as a function of P.

partially through-drying the cake. In part of this operation, two-phase flow occurs. Methods have been developed for calculating the $(-\Delta P_t)$ versus flow rate relation in this period (3, 5, 6, 7). They are particularly valuable in vacuum-filter design because this period of operation places the greatest load on the vacuum system. Such calculations are beyond the scope of this book.

Centrifugal Filtration. A filter operation can be carried out using a centrifugal force rather than the pressure force used in the equipment described above. Filters using centrifugal force are usually used in the filtration of coarse granular or crystalline solids and are available for batch or continuous operation.

Batch centrifugal filters most commonly consist of a basket with perforated sides rotated around a vertical axis, as shown in Figure 22.57. An electric motor, located either above or below the basket, turns it at rates usually below 4000 rpm. Basket diameters may be as great as 48 in. The slurry is fed into the center of the rotating basket and is forced against the basket sides by centrifugal force. There, the liquid passes through the filter medium which is placed around the inside surface of the basket sides and is caught in a shielding vessel, called a curb, within which the basket rotates. The solid phase builds up a filter cake against the filter medium. When this cake is thick enough to retard the filtration to an uneconomical rate or to endanger the balance of the centrifuge, the machine is slowed, and the cake is scraped into a bottom discharge

or is scooped out of the centrifuge. In an underdriven centrifuge, the entire cake and filter medium may be removed and a clean filter cloth inserted.

In the *automatically discharging batch centrifugal filter*, unloading occurs automatically while the centrifuge is rotating, but the filtration cycle is still a batch one. This machine is shown schematically in Figure 22.58. The constant-rotation speed of this machine permits lower power requirements for a given amount of filtrate collected as well as lower labor requirements than would be obtained in a batch centrifugal.

The slurry is fed to the unit through duct A as the drum rotates, until the desired cake is built up in the bowl. Slurry feed then is stopped, and wash water is fed onto the cake through tube G. After washing, the cake is spun dry. When this final spin is completed, the peeler knife (H) is moved up into the cake by an

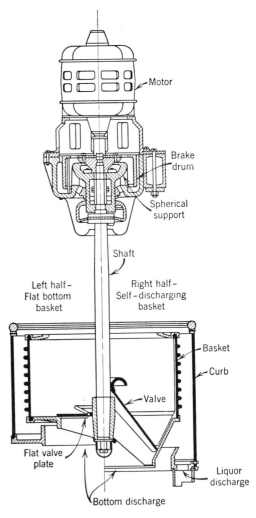

Figure 22.57. A composite cross-sectional view of a suspended-basket centrifuge showing a flat-bottomed and a cone-bottomed discharge on the left and right sides respectively. [From *Chem. & Met. Eng.*, **50**, No. 7, 119 (July 1943) by permission.]

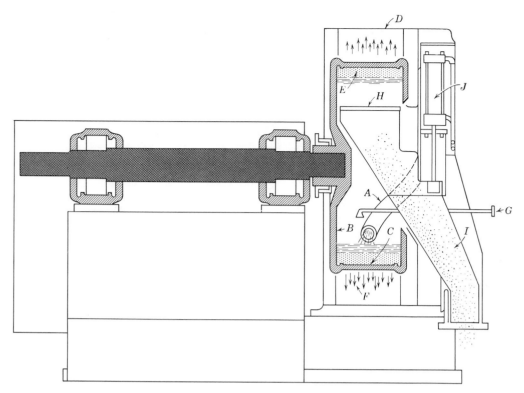

Figure 22.58. Schematic diagram of an automatically discharging batch centrifugal filter. (Courtesy Baker-Perkins, Inc.)

A—feed duct *B*—centrifuge bowl *C*—filter medium *D*—liquor housing *E*—filter cake *F*—filtrate
G—wash pipe *H*—peeler knife *I*—discharge chute *J*—hydraulic cylinder.

hydraulic mechanism peeling it off into the chute (*I*). If desired, the filtrate and wash liquid can be collected separately rather than combined in the housing (*D*).

A *continuous centrifugal filter* is shown in Figure 22.59. In this filter, solids handling is held to a minimum, which permits filtration of fragile solids with the least possible breakage. As with other centrifugal filters, this unit is best suited for handling coarse-granular or coarse-crystalline solids in nonviscous liquids. Units are available with capacities up to 25 tons/hr of solids when handling this type of material.

In operation, the feed slurry is fed through tube *O* to the feed funnel (*F*) of Figure 22.59. The funnel is attached to the pusher (*D*) and rotates at the same speed as the centrifuge drum. In the funnel, the feed is accelerated to drum speed and fed into the back end of the filter drum where the filtrate is forced through the filter screen (*C*), and the solids form a cake on the screen. Intermittently, the cake is pushed toward the discharge end of the centrifuge by the pusher, which then retreats again leaving an open region for the build-up of new cake. In this way, the cake moves across the face of the filter screen until it is pushed off the end into the solids-collector housing (*G*). As it moves, the cake passes through a wash region where wash liquid passes through it into the filtrate housing (*A*). The filtrate and wash

liquid are kept separate by partitions (*H*) in the wet collector.

Other continuous centrifugal filters are built such as the *solid-bowl centrifugal* pictured in Figure 22.15. They, however, would have the bowl wall consisting of a screen through which the filtrate and wash water would drain. Frequently, they have cylindrical bowls rather than the cone-shaped bowl illustrated.

Centrifugal-Filtration Calculations. The mechanism of centrifugal filtration is identical to that of pressure filtration, and the same equations must apply to both cases if the terms within the equation are properly evaluated. Now, however, the driving force is the centrifugal force acting on the fluid, rather than fluid pressure itself. Simple substitution in the pressure-filtration equation (Equation 22.108) of this centrifugal force seems to be indicated. In engineering units,

$$F_c g_c = ma = mr\omega^2 \qquad (22.114)$$

where a = acceleration
  r = radius, ft
  ω = rate of revolution, radians/sec = $2\pi N \times 60$
  N = rate of revolution, rpm
  F_c = centrifugal force, lb_f
  m = mass, lb
  g_c = conversion factor, 32.17 ft/sq sec, $\dfrac{lb}{lb_f}$

Thus the term $\rho r \omega^2/g_c$ is the equivalent of $-\Delta P/L$ when a centrifugal field rather than a pressure field is present.

Careful analysis of centrifugal filtration shows other differences from pressure filtration (17). Primarily, the centrifugal-force field varies with the depth in the filter cake, increasing as r increases. The centrifugal force not only creates an hydraulic pressure at the cake surface but also acts on the filtrate as it flows through the cake and on the solids in the cake, supplementing the hydraulic-pressure head. Finally, in the typical basket filter, the cake forms on the inside vertical walls of the filter. The filter area decreases as the cake thickens, and the kinetic energy of the flowing filtrate also changes. These differences are illustrated in Figure 22.60 which shows the physical situation and the nomenclature to be used in describing it.

A differential pressure balance about a very thin layer of filter cakes gives

$$-dP = -dP_g - dP_k - dP_f \qquad (22.115)$$

where $-dP$ = total effective pressure drop
  $-dP_g$ = hydraulic-pressure gradient developed by liquid in the centrifugal field flowing through the cake
  $-dP_k$ = hydraulic-pressure gradient as a result of kinetic-energy changes in passing through the fluid
  $-dP_f$ = hydraulic-pressure gradient as a result of frictional drag

For this differential cake, the area through which flow passes is

$$A = 2\pi r h \qquad (22.116)$$

A solids-material balance for this differential cake thickness results in

$$w\, dV = \rho_s(1 - \epsilon_p)2\pi h r\, dr \qquad (22.117)$$

where w = the weight of solids per unit volume of original slurry
  dV = the increase in filtrate volume in a differential time span, $d\theta$

Comparing Equation 22.117 with Equation 22.104, which was a similar material balance for the filtration through a flat cake, it is seen that the filtrate volume held in a differential thickness of the filter cake has been neglected here. Finally, a filtrate balance gives

$$\frac{dV}{d\theta} = 2\pi h r v \qquad (22.118)$$

or, upon differentiation,

$$dv = -\left(\frac{dV}{d\theta}\right)\frac{dr}{2\pi h r^2} \qquad (22.119)$$

where v = the linear radial velocity of filtrate flowing through the differential cake thickness, dr

The various terms of Equation 22.115 can now be expanded in terms of the particular nomenclature and

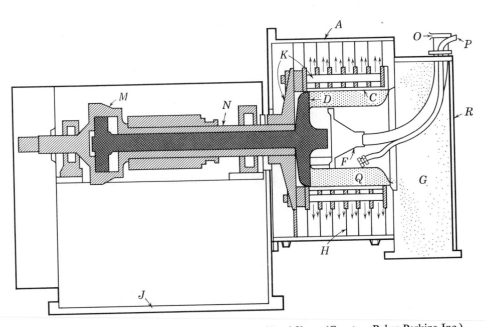

Figure 22.59. Schematic diagram of continuous centrifugal filter. (Courtesy Baker-Perkins, Inc.)
A—filtrate housing *C*—filter screen *D*—cake pusher *F*—feed funnel *G*—cake-discharge chute *H*—wet housing separators *J*—base *K*—basket *M*—hydraulic servo motor *N*—piston rod *O*—slurry feed pipe *P*—wash feed pipe *Q*—filter cake *R*—access door.

physical arrangement of centrifugal filtration as diagrammed in Figure 22.60. The individual terms will then be recombined to give a general equation for centrifugal filtration. The integration of this equation will, of course, depend upon the properties of the particular material and the geometry of the particular equipment involved. The integration will be performed for the simple case of centrifugation in a cylindrical basket of a slurry that forms a noncompressible cake.

The pressure gradient resulting from kinetic-energy variations must be

$$\frac{dP_k}{\rho} = \frac{v\, dv}{g_c}$$

as shown in Chapter 20. Combining this with Equations 22.118 and 22.119 gives

$$dP_k = \frac{-\rho}{g_c}\left(\frac{dV}{d\theta}\right)^2 \frac{dr}{(2\pi h)^2 r^3} \qquad (22.120)$$

where ρ = the density of the filtrate.

The pressure gradient resulting from hydraulic head would be

$$\frac{dP_g}{\rho} = \frac{g}{g_c}\, dz$$

if the head occurs in a gravitational field as shown by Equation 20.5. In the centrifugal field existing here, g must be replaced by $r\omega^2$, and dz must be replaced by dr. As a result,

$$dP_g = \frac{\rho\omega^2 r\, dr}{g_c} \qquad (22.121)$$

for the centrifugal filtration.

Finally, the pressure drop resulting from form drag and skin friction can be found by appropriate alteration of Equation 22.105. For pressure filtration,

$$-dP_f = \frac{\alpha\mu w}{g_c A^2}\left(\frac{dV}{d\theta}\right) dV \qquad (22.122)$$

Substituting Equations 22.116 and 22.117 into Equation 22.122 gives

$$-dP_f = \frac{\rho_s(1-\epsilon_p)\cdot\alpha\mu}{2\pi h g_c}\left(\frac{dV}{d\theta}\right)\frac{dr}{r} \qquad (22.123)$$

Replacing Equations 22.120, 22.121, and 22.123 into Equation 22.115 and integrating results in

$$-\int_{P_2}^{P_3} dP = \frac{-\rho\omega^2}{g_c}\int_{r_2}^{r_3} r\, dr + \frac{\rho}{(2\pi h)^2 g_c}\left(\frac{dV}{d\theta}\right)^2\int_{r_2}^{r_3}\frac{dr}{r^3}$$

$$+ \frac{\rho_s\mu}{2\pi h g_c}\left(\frac{dV}{d\theta}\right)\int_{r_2}^{r_3}\alpha_p(1-\epsilon_p)\frac{dr}{r} \qquad (22.124)$$

where P_2, P_3 = pressures at the filter-cake surface and at the filter-cloth surface respectively

r_2, r_3 = radii to the filter-cake surface and to the filter-cloth surface respectively

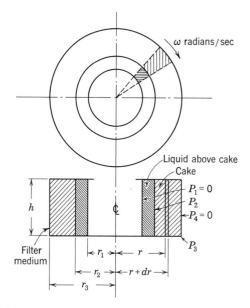

Figure 22.60. Nomenclature and physical arrangement, centrifugal filtration.

ρ, ρ_s = density of filtrate and of solids respectively

ϵ_p = porosity at any point in the bed

α_p = specific cake resistance at any point in the bed

h = height of cylindrical surface on which the cake is being built

$(dV/d\theta)$ = filtration rate

ω = rate of rotation, radians/sec

This equation is the basic rate equation for centrifugal filtration, but, as with Equation 22.105, it only gives the pressure drop across the bed itself. The pressure at the filter-cloth surface (P_3) can be related to the normally measured downstream pressure (P_4) in terms of a filter-medium resistance as was done for pressure filtration.

$$P_3 - P_4 = \frac{\left(\dfrac{dV}{d\theta}\right)\mu R_M{}'}{2\pi r_3 h g_c} \qquad (22.125)$$

where P_4 = pressure at low pressure side of filter cloth

$R_M{}'$ = resistance of the filter medium, analogous to the medium resistance given in Equation 22.107

The pressure at the cake surface (P_2) is

$$P_2 - P_1 = \frac{\rho\omega^2(r_2{}^2 - r_1{}^2)}{2g_c} \qquad (22.126)$$

where r_1 = radius to surface of liquid over the filter cake

Generally Equation 22.124 can be integrated only

if the relations among α_p, ϵ_p, and r can be determined. For a few simplified cases, the integration can be done directly (17). The most useful of the integrations is for a cake that can be considered incompressible. Then, putting Equations 22.125 and 22.126 in Equation 22.124 and integrating with $P_4 = P_1 = 0$ gives

$$\left(\frac{dV}{d\theta}\right)\frac{\mu R_M{}'}{2\pi r_3 h g_c} - \frac{\rho\omega^2(r_2{}^2 - r_1{}^2)}{2g_c} = \frac{\rho\omega^2(r_3{}^2 - r_2{}^2)}{2g_c}$$

$$- \frac{\rho\left(\dfrac{dV}{d\theta}\right)^2}{2g_c(2\pi h)^2}\left(\frac{1}{r_2{}^2} - \frac{1}{r_3{}^2}\right) - \frac{\left(\dfrac{dV}{d\theta}\right)\mu\alpha\rho_s(1-\epsilon)}{2\pi h g_c}\ln\frac{r_3}{r_2}$$

$$(22.127)$$

which rearranges to

$$\frac{dV}{d\theta} = \frac{\pi h\rho\omega^2(r_3{}^2 - r_1{}^2) - \dfrac{\rho(dV/d\theta)^2}{4\pi h}\left(\dfrac{1}{r_3{}^2} - \dfrac{1}{r_1{}^2}\right)}{\mu\left[\alpha\rho_s(1-\epsilon)\ln\dfrac{r_3}{r_2} + \dfrac{R_M{}'}{r_3}\right]}$$

$$(22.128)$$

The second term in the numerator is that coming from the changes in fluid kinetic energy. It is almost always very small. Removing it and writing $\omega = \pi N/30$ gives

$$\frac{dV}{d\theta} = \frac{\pi^3 N^2 \rho h(r_3{}^2 - r_1{}^2)}{(30)^2\mu\left[\alpha\rho_s(1-\epsilon)\ln\dfrac{r_3}{r_2} + \dfrac{R_M{}'}{r_3}\right]} \qquad (22.129)$$

Equation 22.129 can be compared with the standard-pressure filtration equation, Equation 22.107, by considering the total hydraulic driving force to replace $-\Delta P_t$ in the pressure-filtration equation. By integration of Equation 22.121 for the filter cake and the liquid above it,

$$\Delta P_g = \frac{\rho\omega^2}{g_c}\frac{r_3{}^2 - r_1{}^2}{2} \qquad (22.130)$$

Also, by a mass balance on the solids in the cake,

$$\rho_s(1-\epsilon) = \frac{wV}{\pi h(r_3{}^2 - r_2{}^2)} \qquad (22.131)$$

Combining Equations 22.130 and 22.131 with Equation 22.129 gives

$$\frac{dV}{d\theta} = \frac{\Delta P_g g_v}{\mu\left(\dfrac{\alpha wV}{\left(\dfrac{2\pi h(r_3 + r_2)}{2}\right)\left(\dfrac{2\pi h(r_3 - r_2)}{\ln(r_3/r_2)}\right)} + \dfrac{R_M{}'}{2\pi h r_3}\right)}$$

$$(22.132)$$

in which the area terms may be designated

$$\frac{dV}{d\theta} = \frac{\Delta P_g g_c}{\mu\left(\dfrac{\alpha wV}{A_m A_{lm}} + \dfrac{R_M{}'}{A_c}\right)} \qquad (22.133)$$

Thus, for centrifugal filtration, the A^2 term in Equation 22.107 becomes $A_m \cdot A_{lm}$, as defined by Equation 22.132, but A_c (the area of the filter cloth) is correct as a replacement for A when dealing with the resistance of the filter medium.

Illustration 22.15. The $CaCO_3$ slurry described in Illustrations 22.12 and 22.13 is to be filtered in a basket centrifugal filter of 24-in. inside basket diameter and 10-in. basket height rotating at 1200 rpm. Assuming that the cake is incompressible and that the filter medium resistance is negligible, what filtration rate can be expected when the cake is 1 in. thick and the liquid surface corresponds to the filter-cake surface?

SOLUTION. If the cake is incompressible, the values of α and ϵ calculated in Illustration 22.12 can be used here. The rate $(dV/d\theta)$ can then be calculated directly from either Equation 22.128, 22.129, or 22.132. Equation 22.129 will be used since it is more immediately applicable. Thus,

$$\frac{dV}{d\theta} = \frac{\pi^3 \times (1200)^2 \times 62.4 \times \dfrac{10}{12}\left[1^2 - \left(\dfrac{11}{12}\right)^2\right]}{900 \times 1.0 \times 0.000672\left[2.41 \times 10^{11}\right.}$$

$$\left.\times 183(1 - 0.453)\ln\dfrac{12}{11}\right]$$

$$\frac{dV}{d\theta} = 0.000300 \text{ cu ft/sec or } 0.135 \text{ gal/min}$$

Cyclic Operation of Batch Filters. The operation of a batch filter necessarily consists of the steps of filtering, washing and dewatering, and cleaning and reassembly. The free volume within the filter available for cake storage poses the ultimate limit on filtering time during a single cycle. A more practical restriction is often found in the desire of the filter operator to get the maximum amount of material filtered in an operating day. This will occur when the curve of filtrate volume per day as a function of filtering time per cycle reaches a maximum. In other words

$$\frac{d(N_c V)}{d\theta_f} = 0 \qquad (22.134)$$

where N_c = the number of complete operating cycles per day

V = volume of filtrate per cycle

θ_f = time of filtering in each cycle, hr

The number of cycles per day is

$$N_c = \frac{24}{\theta_c} = \frac{24}{\theta_f + \theta_w + \theta_d} \qquad (22.135)$$

where θ_c = time for a complete cycle, hr

θ_w = time for washing and dewatering, hr

θ_d = time for disassembly, cleaning and reassembly, hr

The relations developed earlier allow θ_f to be expressed as a function of V whether pressure or centrifugal filtration is considered, and with compressible or incompressible cakes. In many cases the wash time (θ_w) can also be related to the filtrate volume or filtering time. This, however, is at the discretion of the operator. The disassembly, cleaning, and reassembly time (θ_d) depends on the type of filter, the properties of the material being filtered, and the manpower available. It will usually be a constant, independent of V, for any given installation.

Thus Equation 22.135 can be reduced to a function relating N_c to either θ_f or V. Formal differentiation will then give either $d(NV)/d\theta_f$ or $d(NV)/dV$ which can be set equal to zero to get the desired optimum.

This solution is an example of the formal use of an economic balance to control plant operations. The economic balance can seldom be applied with such mathematical precision, but it must always form the yardstick by which process alternates are chosen and operations are guided.

PARTICLE SEPARATIONS NOT INVOLVING FLUID MECHANICS

Screening is an operation for separating solid materials on the basis of size alone. No fluid mechanics is involved in this process. A mixture of solid particles of various sizes is fed onto a surface provided with suitable openings. Some material passes through the openings, and some remains behind; the resulting portions are now more uniform in size than the original.

The screening equipment may be in the form of stationary or moving bars, perforated steel plate, or woven-wire cloth. The screens may be inclined at an angle so the solids can flow downward by gravity flow, or the screens may be set in some type of vibratory motion.

The effectiveness and analysis of the screening operation depends upon a knowledge of particulate solids, as discussed in Appendix B.

Electrical Precipitation. Gases containing very small particles can be cleaned electrically by passing the gas between two high-voltage electrodes which ionize the gas. The gas ions attach themselves to the solid particles and build up a charge on the particle. The charged particle migrates according to an EMF gradient and is attracted to collecting electrodes—the negatively charged particles to the positive electrode and the positively charged particles to the negative electrode.

There are two general classes of electrical precipitators:

1. Single stage in which ionization and collection are combined.

2. Two stage in which ionization is followed by collection, with each operation occurring in a different part of the apparatus.

Most industrial gases are readily ionized. Electrical precipitators are operated at the highest practical voltage, governed by arcing, since this voltage produces the most rapid migration of particles. Voltages up to 100,000 are used. Collection efficiencies are very good. With low gas velocities, nearly 100 per cent efficiency, on a mass basis, can be obtained, but the economic limit is about 99 per cent.

Magnetic Separation. Materials that have different magnetic attractability may be separated by passing them through a magnetic field. Most often magnetic separation techniques are used to remove iron, steel, or magnetic iron oxide from materials low in magnetic attractability. However the methods have been refined so that they can be used to separate materials that are only slightly reactive to a magnetic field. One common device is simply an electromagnet suspended over a conveyor belt. This might be used to remove tramp iron from the feed to a crusher. Periodically the magnet is unloaded into a bin. Another device employs a magnetized pulley at the end of a conveyor belt. As the material is conveyed over this pulley, magnetically inert material drops off the belt in a normal manner. Magnetic material is held on the belt, however, and finally drops off the belt as it leaves the field of the pulley. Other magnetic separators can be used to remove magnetic material from a slurry.

REFERENCES

1. Ambler, C. M., *Chem. Eng. Progr.*, **48**, 150 (1952).
2. Bhattacharya, A., and A. N. Roy, *Ind. Eng. Chem.*, **47**, 268 (1955).
3. Brown, G. G., and Associates, *Unit Operations*, John Wiley and Sons, New York, 1950, Chapter 18.
4. Brown, G. G., and Associates, *Unit Operations*, John Wiley and Sons, New York, 1950, p. 119.
5. Brownell, L. E., and H. E. Crosier, *Chem. Eng.*, **56**, No. 10, 124 (1949).
6. Brownell, L. E., and G. B. Gudz, *Chem. Eng.*, **56**, No. 9, 112 (1949).
7. Brownell, L. E., and D. L. Katz, *Chem. Eng. Progr.*, **43**, 537, 703 (1947).
8. Burke, S. P., and W. B. Plummer, *Ind. Eng. Chem.*, **20**, 1196 (1928).
9. Carman, P. C., *J. Soc. Chem. Ind.*, London, **57**, 225T (1938).
10. Coe, H. S., and G. H. Clevenger, *Trans. Am. Inst. Mining Met. Engrs.*, **55**, 356 (1916).
11. Comings, E. W., *Ind. Eng. Chem.*, **32**, 663 (1940).
12. Ergun, S., *Chem. Eng. Progr.*, **48**, 89 (1952).
13. Ergun, S., and A. A. Orning, *Ind. Eng. Chem.*, **41**, 1179 (1949).
14. First, M. W., ASME paper 49–A–127 (1949).
15. Flood, J. E., *Chem. Eng.*, No. 6, 217 (1955).
16. Gasterstadt, H., *VDI Forschungsarbeiten*, No. 265 (1924).
17. Grace, H. P., *Chem. Eng. Progr.*, **49**, 303, 367, 427 (1953).

18. Grace, H. P., *A.I.Ch.E. Journal*, **2**, 307 (1956).

19. Gunness, R. C., *Chem. Eng. Progr.*, **49**, 113 (1953).

20. Hull, W. Q., H. Keel, J. Kenney, Jr., and B. W. Gamson, *Ind. Eng. Chem.*, **45**, 256 (1953).

21. Kozeny, J., *Sitzber. Akad. Wiss. Wien, Math-naturw. Kl. Abt. IIa*, **136**, 271–306 (1927).

22. Kynch, G. J., *Trans. Faraday Soc.*, **48**, 166 (1952).

23. Lapidus, L., and J. C. Elgin, *A.I.Ch.E. Journal*, **3**, 63 (1957).

24. Lapple, C. E., in Perry, *Chemical Engineers Handbook*, 3rd ed., McGraw-Hill Book Co., New York, p. 1027.

25. Lapple, C. E., and C. B. Shepherd, *Ind. Eng. Chem.*, **32**, 605 (1940).

26. Leva, M., and M. Grummer, *Chem. Eng. Progr.*, **43**, 549, 633, 713 (1947).

27. Leva, M., M. Grummer, M. Weintraub, and M. Pollchick, *Chem. Eng. Progr.*, **44**, 511, 619 (1948).

28. Matheson, G. L., W. A. Herbst, and P. H. Holt, *Ind. Eng. Chem.*, **41**, 1099 (1949).

29. Mertes, T. S., and H. B. Rhodes, *Chem. Eng. Progr.*, **51**, 429, 517 (1955).

30. Morcom, A. R., *Trans. Inst. Chem. Engrs. (London)*, **24**, 30 (1946).

31. Morse, R. D., *Ind. Eng. Chem.*, **41**, 1117 (1949).

32. Oman, A. O., and K. M. Watson, *National Petrol. News*, **36**, R 795 (1944).

33. Perry, J. H., *Chemical Engineers Handbook*, 3rd ed., McGraw-Hill Book Co., New York, 1950, p. 1026.

34. Roberts, E. J., *Trans. Am. Inst. Mining Met. Engrs.*, **184**, 61 (1949).

35. Ruth, B. F., and L. L. Kempe, *Trans. Am. Inst. Chem. Engrs.*, **33**, 34–83 (1937).

36. Sittig, M., *Petrol. Refiner*, **31**, No. 9, 91 (1952).

37. Shepherd, C. B., and C. E. Lapple, *Ind. Eng. Chem.*, **31**, 972 (1939).

38. Steinour, H. H., *Ind. Eng. Chem.*, **36**, 618, 840, 901 (1944).

39. Stokes, G. G., *Math. and Phys. Papers* (1901); *Trans. Cambridge Phil. Soc.* **9**, Part II, 51 (1851).

40. Talmadge, W. L., and E. B. Fitch, *Ind. Eng. Chem.*, **47**, 38 (1955).

41. Ter Linden, A. J., *Proc. Inst. Mech. Engrs.*, London, **160**, 233 (1949).

42. Vogt, E. G., and R. R. White, *Ind. Eng. Chem.*, **40**, 1731 (1948).

43. Waddel, H., *J. Franklin Inst.*, **217**, 459 (1934).

44. Wilhelm, R. H., and M. Kwauk, *Chem. Eng. Progr.*, **44**, 201 (1948).

PROBLEMS

22.1. Prepare a plot showing the terminal-velocity–particle-diameter relationship for a spherical particle having a specific gravity of 2.0 settling in water at 65°F. Use several diameters between 1 micron and 1000 microns.

22.2. Repeat Problem 22.1 for the particle settling in air at 70°F.

22.3. Calculate the terminal radial velocity of a soluble coffee particle 60 microns in diameter in air at 500°F entering a cyclone 18 in. in diameter. The tangential velocity of the coffee particle is 200 cm/sec, and the specific gravity of the solid is 1.05.

22.4. Square mica plates, $\frac{1}{32}$ in. thick and 0.01 sq in. in area are falling randomly through oil with density of 55 lb/cu ft and with viscosity of 15 centipoises. The specific gravity of the mica is 3.0. What will be the settling velocity? Suppose the area of the mica plate is 1.0 sq in., what would be its settling velocity?

22.5. Glass beads 50 microns in diameter settle in water at 65°F. Specific gravity of glass is 2.6.

(a) If unhindered settling occurs, what is the maximum velocity attained?

(b) If settling occurs such that the mass ratio of water to glass is 2, what is the maximum velocity?

22.6. A laboratory viscosimeter consists of a steel ball and uniform-diameter glass cylinder. The cylinder is filled with the test fluid, and the time for the ball to fall a known distance is recorded. The ball is 0.25 in. in diameter, and the index marks are 8 in. apart. The viscosity of corn syrup having a density of 1.3 gm/cu cm is desired. The measured time interval is 7.32 sec. What is the viscosity of the syrup? Specific gravity of the steel ball is 7.9.

22.7. A mixture of galena and silica is to be elutriated by a water stream flowing at a velocity of 0.02 ft/sec and a temperature of 65°F. The solid feed analyzes 30 weight percent galena, and a screen analysis is indicated in the following table.

Particle dia., microns	20	30	40	50	60	70	80	90	100
Wt. percentage of undersize	33	53	67	77	83	88	91	93	94.5

If the size distribution given applies for both components in the feed, what fraction of the galena fed is in the overhead and bottom products, and what is the weight fraction (dry basis) of galena in these products?

Galena specific gravity = 7.5
Silica specific gravity = 2.65

22.8. Solve Problem 22.7 if carbon tetrachloride is used as the elutriating agent.

22.9. A gravity settling tank is to be used to clean waste water from an oil refinery. The waste stream contains 1 per cent oil by volume (specific gravity of the oil is 0.87) as small drops ranging in size between 10 and 500 microns. The tank is rectangular and measures 10 ft wide by 6 ft deep. Provision is made at the discharge end for the clean water to be continuously removed from the bottom of the tank. Periodic skimming of the liquid surface at the discharge end removes the accumulated oil. If 100,000 gal/min of waste water is to be processed, how long must the settling tank be?

22.10. A mixture of coal and sand in particle sizes smaller than 20 mesh is to be completely separated by screening and then elutriating each of the cuts from the screening operation with water as the elutriating fluid. Recommend a screen size such that the oversize cut can be completely separated into coal and sand fractions by water elutriation. What water velocity will be required? The specific gravity for sand and coal is 2.65 and 1.35 respectively.

22.11. Quartz and pyrites (FeS_2) are separated by continuous hydraulic classifications. The feed to the classifier ranges in size between 10 microns and 300 microns. Three fractions are obtained: a pure-quartz product, a pure-pyrites product, and a mixture of quartz and pyrites. The specific gravity of quartz is 2.65, and that of pyrites is 5.1. What is the size range of the two materials in the mixed fraction for each of the following cases:

(a) The bottoms product to contain the maximum amount of pure pyrites.

(b) The overhead product to contain the maximum amount of quartz.

22.12. The mixed feed of Problem 22.11 is to be separated into two pure fractions of pyrites and quartz by a hindered-settling process. What is the minimum density of the heavy medium that will give this separation?

22.13. A tubular-bowl centrifuge is to be used to separate water from a fish oil. This centrifuge has a bowl 4 in. in diameter by 30 in. long and rotates at 15,000 rpm. The fish oil has a density of 0.94 gm/cu cm and a viscosity of 50 centipoises at 25°C. The radii of the inner and outer overflow dams are 1.246 in. and 1.250 in. respectively. Determine the critical diameter of droplets of oil suspended in water and of droplets of water suspended in oil if the feed rate is 300 gal/hr of a suspension containing 20 weight percent fish oil.

22.14. Determine the Σ value for the centrifuge of Problem 22.13 for oil removal from the water. Note that, with a liquid-liquid separation, Σ is no longer independent of the system used.

22.15. A cylindrical-bowl internal-screw centrifugal is to be used to separate $MgSO_4 \cdot 6H_2O$ crystals from the mother liquor which comes as product from a vacuum crystallizer. The centrifugal has a bowl 14 in. in diameter and 23 in. long and carries liquid to a depth of 3 in. It rotates at 3000 rpm. If no crystals smaller than 5 microns in diameter exist in the slurry, what feed rate to the centrifugal will result in complete removal of the solids? Assume that the internal screw does not lift any of the solid particles back into the liquid or disturb their fall through the liquid. The solution density is 1.21 gm/cu cm, its viscosity is 1.5 centipoises. The crystals have a density of 1.66 gm/cu cm.

22.16. A conical solid-bowl centrifugal is to be used to dewater the liquid product from a classifier in closed circuit with a fine grinder in the preparation of cement rock before feeding the cement kiln. The centrifugal is to remove particles of a size greater than 10 microns in diameter. It has a 40-in. maximum diameter, a 60-in. length, and a 10° angle on the cone wall. Maximum liquid depth is 8 in. Rotation is possible up to a maximum of 1200 rpm. Determine the revolutions per minute to be used if the feed rate to the centrifugal is 300 gal/min.

22.17. On the basis of the ability to separate minute quantities of cottonseed oil at 80°F from a water phase down to droplet diameters of 2 microns, what would be the permissible throughput rate when using a No. 2 disk centrifuge with specifications equal to those of the second unit of this type listed in Table 22.1. Could this centrifuge be operated to remove both oil and water droplets greater than 2 microns in diameter from the other phase when feeding a 50 volume percent oil and water mix?

22.18. Determine outlet-dam heights and throughput that would permit the tubular-bowl centrifuge of Problem 22.13 to separate droplets down to a critical diameter of 1 micron from both oil and water phases.

22.19. A 3 per cent by weight calcium carbonate slurry was subjected to a batch-sedimentation test. The density of the solids in the slurry was 2.63 gm/l., and that of the liquid was 1.0 gm/l. The test results are indicated below.

Time, min	0	20	40	60	80	100	120	140
Height of liquid-solid interface, cm	176	100	74	57	42	34	26	22

Determine the area and depth of a continuous thickener that is to handle 100 tons of solids per day. Initial solids concentration in the feed is 3 per cent by weight and the final concentration is to be 40 per cent solids.

22.20. Waste water from a drinking plant is to be clarified by continuous sedimentation. Feed to the thickener is one million gal. per day containing 1.20 per cent by weight solids. The underflow from the unit analyzes 8 per cent solids. Specify the depth and diameter of the thickener.

A single batch-settling test on the feed material gave the following information:

Specific gravity of solids = 2.00
Specific gravity of solution = 1.00
Concentration of solids in test = 0.12 per cent.

Time, min	0	5	10	20	40	60	180	240	∞
Height of liquid-solid interface, cm	31	21	10	3.2	2.2	2.1	2.0	1.96	1.94

22.21. 40 tons/hr of a metallurgical pulp are to be thickened from 186 gm/l. to 1200 gm/l. in a Dorr Thickener. Batch-settling data

on the feed material gave the following information:

Time, hr	0	0.10	0.25	0.50	0.75	1.0	2.00	4.0
Pulp height, ft	3.0	2.0	1.4	0.9	0.68	0.48	0.25	0.1

Specific gravity of solids = 4.44
Concentration = 186 gm/l.

What depth and diameter thickener is required to accomplish the specified assignment?

22.22. An adsorber bed of "molecular sieves" consists of randomly packed extruded cylinders $\frac{1}{16}$ in. in diameter by $\frac{3}{16}$ in. long held on a sintered metal plate. Oxygen at $-200°F$, 100 psia is to be passed through the bed at a superficial velocity of 1 ft/sec in order to remove impurities such as light hydrocarbons and inert gases. What pressure drop can be expected through a bed 10 ft long? The viscosity of gaseous oxygen under these conditions is 0.0125 centipoise.

22.23. The adsorbent bed of Problem 22.22 is supported by a sintered metal plate $\frac{1}{4}$ in. thick having passages averaging 50 microns in diameter. Estimate the pressure drop through this plate if its porosity is 0.30.

22.24. A sand filter consists of uniform, spherical, particles of $-20 + 28$ mesh size. After backflushing with water, the sand bed has settled to a stable depth of 6 ft and is flooded with water which stands to a depth of 3 ft above the top of the sand. If the drain valves are opened, how long will it take the bed to drain until the water level is even with the top of the bed?

22.25. Boiler feed water is deionized by pumping it downward through a bed of ion-exchange resin. The resin has spherical particles sized to $-8 + 10$ mesh. The deionizer is 4 ft in diameter by 6 ft long, and an operating flow rate of 20 gal/min/cu ft of bed is used. What pressure drop exists through the bed?

22.26. The ion-exchange resin of Problem 22.25 has a particle density of 80 lb/cu ft. It is regenerated by backflushing with 10 per cent brine solution ($\rho = 1.07$ gm/cu cm) at a rate such that the bed is expanded 100 per cent. What rate of brine flow is required?

22.27. In the fluidized roasting of zinc sulfide, flotation concentrate of the sulfide in particle sizes averaging 50 microns is roasted at 1600°F and about 4 psig. Published information [Andersen, T. T., and R. Balduc, *Chem. Eng. Progr.* **49**, 527 (1953)] gives a 1-ft/sec velocity through the bed and a fluidized bed density of 90 lb/cu ft. Using a solid density of 4.00 gm/cu cm for the concentrate and assuming the particles are spherical, locate this data point on the plot showing pressure drop through fixed beds (Figure 22.30).

22.28. Estimate the gas flow necessary to begin fluidization in the roaster of Problem 22.27 and the gas flow necessary to cause conveying of the bed.

22.29. In the manufacture of synthetic detergents, the spray-dried bead is pneumatically conveyed from the bottom of the spray drier to cyclone separators at the top of the manufacturing building. In a plant producing 3000 lb/hr of product the conveyor is a 10-in. I.D. tube running vertically for 60 ft and horizontally for 30 ft. The detergent has a mean particle size of 600 microns and a packaged bulk density of 20 lb/cu ft. The blower pulling air through this conveyor has a rating of 500 std. cu ft/min. What is the pressure at the blower inlet if atmospheric pressure exists at the bottom of the spray drier? Ignore pressure drop through the cyclone and assume the conveyor operates at a constant 80°F.

22.30. Superphosphate fertilizer is to be conveyed to an open storage bin by a pneumatic conveying system. The production rate is 10 tons/hr of product with an average particle size of 200 microns and a particle density of 150 lb/cu ft. The conveyor is to discharge at a point 70 ft above the pick-up location and 500 ft away from it. Find tube diameter, blower capacity, and tube pressure drop such that the product will be satisfactorily conveyed when the blower is located at the feed end of the system.

22.31. Coal is pumped in water suspension from Cadiz, Ohio, to Cleveland, a distance of 108 miles, in a 10-in. I.D. welded pipeline. A 45 weight percent slurry of the $\frac{1}{4}$-in. coal particles is pumped at $3\frac{1}{2}$ mph through the pipeline by high-pressure, positive-displacement pumps located at three pumping stations along the pipeline. Estimate the pump pressure developed and the horse-power required by each pumping station.

22.32. Coffee beans are moved from the wharf to roasters by pneumatic conveying through an 8-in. duct. The duct must convey the beans horizontally 300 ft and 60 ft vertically at a rate of 2000 lb/hr. Specify a blower (capacity and discharge pressure) that can be used if it is located at the feed end of this system. State any assumptions needed.

22.33. The following data were collected in the laboratory on the filtration of a $CaCO_3$ sludge through a washing plate-and-frame filter press. After assembling the press, the data were taken by noting the time required for the collection of succeeding 5-lb batches of filtrate:

Filtrate Collected, lb	Elapsed-Time Interval, min
0	0
5	0.70
10	0.52
15	0.65
20	0.75
25	1.00
30	1.08

Temperature 20°C
Area of filter surface 1.87 sq ft
Feed concentration, 5 per cent $CaCO_3$ in water
Pressure at filter inlet, 6 psig

Determine α and V_e for the filtration of this slurry in this filter press.

22.34. A rotary-disk filter is to be used to collect the solids obtained as flotation-cell product in the feed-preparation system of a cement plant. The filter rotates at $\frac{1}{3}$ rpm with half its surface submerged and an internal pressure of 5 psia. Laboratory experiments run on the same slurry at a constant-pressure differential of 10 psi using a plate-and-frame press with 10 sq ft effective surface gave the following data:

Time, min	Filtrate Volume, cu ft
10	90
20	155
30	210
40	255

What area is required for the disk filter if 95 tons/day of solids are to be filtered from a slurry containing 10 weight percent solids?

22.35. A wet process cement plant grinds their cement rock to -200 mesh, the final grinding being in a water carrier. The resulting slurry is thickened and then filtered with the filter cake being sent to the kilns. The thickened slurry is 30 weight percent solids, with the solids being of 40-micron average diameter. It is planned to filter this slurry through rotary disk vacuum filters. These filters have disks 4 ft in diameter which are submerged in the slurry up to their hubs and which rotate at 1 rpm. The pressure inside the disks is 3 psia, and it is estimated that there will be a 1-psia pressure drop through the filter cloth on the disk face. Laboratory tests with this slurry show that the density of the dried cake is 115 lb/cu ft at any filtration pressure below 40 psig. The solid density of the cement rock is 180 lb/cu ft.

If the filter plant must deliver 50 tons/day of dry solids as filter cake, how many filter disks are required?

22.36. It is desired to increase the capacity of a rotary vacuum drum filter. Give, as accurately as possible, the per cent capacity (weight of product per unit time) increase that results from:

(a) Doubling the rate of rotation.
(b) Doubling the submergence, where submergence is the distance from the slurry pond surface to the bottom of the drum.
(c) Doubling the concentration of solids in the feed.
(d) Doubling the pressure drop.

State qualitatively what effect each of these changes might have on the product quality and on the cost of filtration.

22.37. An inorganic dye is to be filtered in a plate-and-frame filter press having 100 sq ft effective area. The 3 per cent solution will be pumped directly to the filter by a centrifugal pump with the following head-capacity characteristics:

Pressure Developed, psi	Flow Rate, gal/min
1	49
15	46
20	43
25	38
30	27
35	0

Laboratory constant-pressure filtration tests with the same slurry at $(-\Delta P) = 20$ psi give straight lines when $A^2(\Delta\theta/\Delta V)$ is plotted against V. The slopes of the lines are 0.05 min ft^4/sq gal, whereas the intercepts at $V = 0$ are 0.1 min ft^4/gal where the filter medium and piping are as similar to the plant units as can be obtained. Tests at varying $(-\Delta P)$ values indicate that the filter cake is incompressible. If the plant runs are to be stopped when the pump pressure reaches 30 psig, how long will the batch filtrations take?

22.38. Show, for a nonwashing plate-and-frame filter press operating at constant feed pressure with negligible V_e, that the optimum cycle occurs when the time for filtering equals the time lost in opening, dumping, cleaning, and reassembling the press.

22.39. A leaf filter with 20 sq ft of filtering area gave the following data during constant-pressure filtration at $(-\Delta P) = 50$ psi:

Time, min	Filtrate Volume, cu ft
15	100
30	163
45	213
60	258
90	335

The total time required for draining, dumping, and reassembling the filter is 15 min each cycle. The cake is to be washed with a volume of water equal to the filtrate volume.

(a) What volume of filtrate would be collected every 24 hr with the optimum operating cycle using constant-pressure filtration with $(-\Delta P) = 50$ psi?

(b) If the slurry pump can deliver slurry at pressures up to 100 psig and the filter cake is incompressible, what filtration rate is attainable with the above slurry and press in a 60-min constant-rate filtration?

22.40. A 5 per cent slurry of ZnS (Type B), as described in Figure 22.55, in water at 66°F is to be filtered through a plate-and-frame filter having frames 2 in. thick and 3 ft square. A $(-\Delta P)$ of

40 psi is to be used. Under these conditions, it may be assumed that the pressure at the cake surface is 40 psig and that at the filter-cloth surface it is 1 psig. Determine the time required to fill the frames.

22.41. A through-washing plate-and-frame filter press is to be used to filter a slurry containing 2 weight percent of precipitated calcium carbonate, the particles of which average 10 microns in diameter. Laboratory tests show that the cake particles have a sphericity of about 0.8 and that 100 cu cm of cake retain 40 cu cm of filtrate. In operation, the volume of wash water to be used each cycle will be 5 per cent of the volume of filtrate collected. 30 min are required to dump and reassemble the press each cycle.

Assuming that the cake is incompressible and that the filtrate has the same properties as the pure water used in washing and neglecting the resistance of the filter cloth and delivery lines, compute:

(a) The filtering and washing time per cycle to obtain maximum filtrate per day if both are carried out at a constant pressure of 30 psig.

(b) The proper frame thickness under these conditions.

22.42. A leaf-type filter is used for producing a cake that is 50 per cent solids by volume from a 1.0 weight percent slurry of calcium carbonate. The average particle diameter is 5 microns, and the cake is incompressible and dense packed. The filter operates at $(-\Delta P) = 50$ psi, building up a cake 1 in. thick. After filtering, the cake is washed with a volume of water equal to 10 per cent of the volume of filtrate produced. The dumping time for this filter is 25 min. What area is required for the press to produce 50,000 lb of wet (50 per cent solids by volume) cake per 24-hr day? The density of $CaCO_3$ is 2.26 gm/cu cm.

22.43. It is planned to filter flocculated TiO_2 in a plate-and-frame filter press using a constant feed pressure of 200 psia. The press has 20 plates 24 in. by 24 in. with a 3.5 sq ft filtering area per plate face. Compression-permeability measurements on the 5 weight percent slurry gives results as shown in Figure 22.55. How much filtrate will be collected in 1 hr of continuous filtration of such a slurry?

22.44. Using the compression-permeability data for flocculated superlite $CaCO_3$ given in Figure 22.55 determine the compressibility for this material at pressures below 200 psia.

22.45. The $CaCO_3$ slurry described in Problem 22.33 is to be filtered in a batch centrifugal filter 30 in. in diameter by 24 in. deep which rotates at 2400 rpm.

(a) Over what period should the filter be operated in order to build up a 3-in. layer of solids in the filter basket.

(b) If the filter cake is not to be washed, and 15 min are required to stop the centrifuge, empty it, replace a clean filter cloth, and restart it, what is the optimum time of filtration and the cake thickness at the end of this period?

22.46. The $CaCO_3$ slurry described in Problem 22.33 is to be filtered in a continuous centrifugal filter of the intermittent pusher-discharge type. The centrifugal is 24-in. in diameter and rotates at 3000 rpm. The pusher makes 10 strokes/min, each stroke $1\frac{1}{2}$ in. long. Estimate the rate of dry-solids production from this centrifugal filter.

PART III. NOTATION AND NOMENCLATURE

a a constant

a interfacial area per cubic foot of contactor volume, ft^{-1}

a_E external acceleration, ft/sec^2

a_h interfacial area for heat transfer per cubic foot of contactor volume, ft^{-1}

a_m interfacial area for mass transfer per cubic foot of contactor volume, ft^{-1}

A' area projected on the plane normal to a radiant beam, sq ft

A area for filtration, sq ft

A area across which transfer occurs, sq ft

A_B bubble interfacial area

A_c area of filter cloth, sq ft

A_p surface area of a particle, sq ft

b a constant

b wetted perimeter of a duct, ft

B constant in Wien's displacement law, micron °R

B a constant

BPR   boiling point rise, °F

c velocity of the electromagnetic beam, ft/sec

c^* equilibrium equivalent concentration, lb moles/cu ft

c_a heat capacity of component a, Btu/lb mole °F or Btu/lb °F

c_b heat capacity of component b, Btu/lb mole °F or Btu/lb °F

c_h humid heat, Btu/lb mole °F

c_i concentration at the interface, lb moles/cu ft

c_L specific heat of liquid, Btu/lb mole °F

c_P heat capacity, Btu/lb °F

c_s concentration at surface, lb moles/cu ft

C mass of crystals in product magma, lb/hr

C_D drag coefficient

C_o orifice coefficient, dimensionless

C_v venturi coefficient, dimensionless

d the differential operator

D rate of flow of drips, or condensate, lb/hr

D diameter, ft

D' agitator diameter, ft

D_c diameter of a hypothetical channel, ft

D_{eq} equivalent diameter, 4 times the cross-sectional area divided by the wetted perimeter, ft

D_p particle diameter, ft

D_p' critical particle diameter, ft

D_{pm} maximum particle diameter, ft

D_s a characteristic dimension of seed crystal to a crystallizer, ft

D_{sp} diameter of a spherical particle, ft

D_{vs} Sauter-mean particle diameter, ft

$\mathscr{D}$ diffusivity, sq ft/hr

e rate at which thermal energy is absorbed and emitted by a real solid, Btu/hr sq ft

e base of natural logarithm

e_b rate at which thermal energy is absorbed and emitted by a black body, Btu/hr sq ft

e_p particle roughness, ft

E internal energy, Btu/lb

$\mathscr{E}_N$ eddy diffusivity of mass, sq ft/hr

E_r internal energy due to radiation, Btu

E_v Murphree stage efficiency, dimensionless

$E_v°$ point efficiency of a contacting stage, dimensionless

f friction factor $= 8(\tau g_c)_1/v_{fs}^2\rho$, dimensionless

F force, lb_f

F_B buoyant force, lb_f

F_D drag force, lb_f

F_E external force, lb_f

ΣF summation of all frictional (form drag, skin friction, etc.) energy losses, ft lb_f/lb

$F_{1\to2}$ geometric or view factor, the fraction of the total energy emitted from body 1 that strikes body 2, where body 1 is black, dimensionless

$\bar{F}_{1\to2}$ geometric or view factor where the primary surfaces (source and sink) are connected by refractory walls, dimensionless

$\mathscr{F}_{1\to2}$ geometric or view factor where the primary source is an imperfect radiator, dimensionless

g acceleration in the gravity field, ft/sec^2

g_c dimensional constant, 32.174 lb ft/lb_f sec^2

G_F feed rate of solids to a dryer (dry basis), lb (dry solids)/hr sq ft

G_L mass rate of flow of phase L, lb/hr sq ft

G_V mass rate of flow of phase V, lb/hr sq ft

h enthalpy of liquid stream, Btu/lb

h height of a centrifugal filter basket, ft

h_1, h_2 heat-transfer coefficient at surface 1 and 2 respectively, Btu/hr sq ft °F

h_c convective-heat-transfer coefficient, Btu/hr sq ft °F

h_f heat-transfer coefficient for a fin, Btu/hr sq ft °F

h_i heat-transfer coefficient inside tube, Btu/hr sq ft °F

h_o heat-transfer coefficient outside tube, Btu/hr sq ft °F

h_{rb} coefficient for heat transfer by radiation from a black body, Btu/hr sq ft °F

h_V vapor-phase heat-transfer coefficient, Btu/hr sq ft °F

$\bar{h}$ average heat-transfer coefficient Btu/hr sq ft °F

h_t tube-side heat-transfer coefficient Btu/hr sq ft °F

H enthalpy, Btu/lb mole

H enthalpy of gaseous stream, Btu/lb

H_a, H_b enthalpy of components a and b respectively, Btu/lb mole

H^* enthalpy of equilibrium equivalent, Btu/lb mole

H_{dry} enthalpy of dry gas, Btu/lb mole

H_G height of a gas-phase mass-transfer unit, ft

H_L height of a liquid-phase mass-transfer unit, ft

H_L, H_V enthalpy of liquid and vapor phases respectively, Btu/lb mole

H_{OG} height of an over-all gas-phase mass-transfer unit, ft

H_{OL} height of an over-all liquid-phase mass-transfer unit, ft

i a proportionality constant, Btu/hr sq ft

j_q Colburn heat-transfer factor (Equation 13.104)

j_N Colburn mass-transfer factor (Equation 13.100)

k a constant

k thermal conductivity, Btu/hr sq ft (°F/ft)

k_1, k_2, k_3, k_4 constants

k_c, k_c' liquid-phase mass-transfer coefficient based on concentration driving force, lb moles/hr sq ft (lb mole/cu ft)

k_G gas-phase mass-transfer coefficient based on partial-pressure driving force, lb moles/hr sq ft atm

k_L liquid-phase mass-transfer coefficient based on mass-concentration driving force, lb moles/hr sq ft (lb/cu ft)

k_X liquid-phase mass-transfer coefficient based on driving force expressed as mole ratio, lb moles/hr sq ft (mole ratio)

k_x, k_x' liquid-phase mass-transfer coefficient based on driving force expressed as mole fraction, lb moles/hr sq ft (mole fraction)

k_Y gas-phase mass-transfer coefficient based on driving force expressed as a mole ratio, lb moles/hr sq ft (mole ratio)

k_y, k_y' gas-phase mass-transfer coefficient based on driving force expressed as a mole fraction, lb moles/hr sq ft (mole fraction)

K_c, K_c' over-all mass-transfer coefficient based on concentration driving force, lb moles/hr sq ft (lb moles/cu ft)

K_L over-all mass-transfer coefficient based on mass-concentration driving force, lb moles/hr sq ft (lb/cu ft)

K_x, K_x' over-all mass-transfer coefficient based on driving force expressed as liquid-phase mass fraction, lb moles/hr sq ft (mole fraction)

K_y, K_y' over-all mass-transfer coefficient based on driving force expressed as gas-phase mass fraction, lb moles/hr sq ft (mass fraction)

l dryer length, ft, length of centrifuge bowl, ft

l length of radiant beam travel through a gas, ft

L cake thickness, ft

L mass rate of flow of liquid stream, lb/hr or lb moles/hr

L volumetric rate of flow of liquid stream, cu ft/hr

L_{av} average liquid-flow rate, lb moles/hr

L_1, L_2 liquid-flow rate at bottom and top of column respectively, lb moles/hr

m slope of the equilibrium curve, atm/(lb mole/cu ft)

m mass of flowing system, lb

m mass of a photon, lb

m mass of solute deposited per unit time lb/hr

$\bar{M}$ mean molecular weight, lb/lb mole

M_a, M_b molecular weights of condensable and noncondensable components respectively, lb/lb mole

M_a, M_h molecular weight of anhydrous salt and hydrated crystal respectively, lb/lb mole

n frequency of radiant energy, sec^{-1}

n number of spaces between disks in a disk centrifuge, dimensionless

N revolutions per second, sec^{-1}

N number of photons in the system, dimensionless

N number of particles, dimensionless

N number of passes in a heat exchanger, dimensionless

N number of mass-transfer units

N rate of mass transfer, lb moles/hr

N_{Fr} Froude number (v^2/Dg), dimensionless

N_G number of gas-phase mass-transfer units, dimensionless

N_{Gr} Grashof number ($L^3\rho^2 g\beta \Delta T/\mu^2$), dimensionless

N_L number of liquid-phase mass-transfer units, dimensionless

N_{Nu} Nusselt number (hD/k), dimensionless

N_{OG} number of over-all gas-phase mass-transfer units, dimensionless

N_{OL} number of over-all liquid-phase mass-transfer units, dimensionless

N_{Po} Power number ($\mathbf{P}g_c/N^3 D^5 \rho$), dimensionless

N_{Pr} Prandtl number ($c_P \mu/k$), dimensionless

N_{Re} Reynolds number ($Dv\rho/\mu$), dimensionless

N_{Sc} Schmidt number ($\mu/\rho\mathscr{D}$), dimensionless

p partial pressure, atm

p fin perimeter, ft

p^* equilibrium equivalent partial pressure, atm

p_a, p_b partial pressure of components a and b respectively, atm

p_i partial pressure at an interface, atm

$\mathbf{P}$ power, ft lb$_f$/sec

P total pressure, atm

P_a, P_b vapor pressure of components a and b respectively, atm

P° total vapor pressure of pure solvent of T, atm

P_r the pressure effect of radiation, momentum change/(unit time) (unit surface area) atm

$-\Delta P_c$ pressure drop across filter cake, lb/sq ft

$-\Delta P_f$ pressure drop resulting from frictional effects, lb/sq ft

dP_g hydraulic pressure gradient, lb/sq ft

dP_k pressure gradient resulting from kinetic-energy changes, lb/sq ft

q total rate of thermal energy transfer, Btu/hr

q_c, q_k, q_r rates of thermal-energy transfer by the

mechanisms of convection, conduction, and radiation respectively, Btu/hr

q_{rb} rate of radiant emission from a black body, Btu/hr

Q total amount of heat transferred, Btu

Q volumetric feed rate to a centrifuge, cu ft/hr

r radius, ft

r distance from emitter to receiver of radiation, ft

r_1, r_2, r_3, r_4 radii to various points in the centrifuge bowl, ft

R resistance to transfer of heat, hr sq ft °F/Btu

R_c drying rate during the constant-rate period, lb/hr sq ft atm

R_M resistance of the filter medium to flow of filtrate, ft^{-1}

R_M' resistance of the filter medium to flow of the filtrate, centrifugal filtration, ft^{-1}

$R_i, R_p, R_o, R_c,$ $R_w, R_d, R_L, R_T,$ resistance to heat transfer of inside fluid, pipe wall, outside fluid, condensate, tube wall, tube fouling, liquid, and total path respectively, hr sq ft °F/Btu

s filter-cake-compressibility factor, dimensionless

s dryer slope, dimensionless

S cross section available for flow, sq ft

S cross-sectional area of fin, sq ft

S, S_r entropy, and entropy due to radiant energy alone, Btu/°F

T temperature, °F or °R

$T_B, T_B°$ boiling point of solution and of pure solvent at system pressure, °F

T_d dew-point temperature, °F

T_i temperature at the interface, °F

T_L, T_V temperature of liquid and gas phases respectively, °F

T_0 temperature at the base where $H = 0$, °F

T_{1s}, T_{2s} temperature of steam condensing in the steam chests of effects 1 and 2 respectively, °F

T_s surface temperature, °F

T_{sa} adiabatic-saturation temperature, °F

T_w wet-bulb temperature, °F

ΔT a temperature difference $(T_2 - T_1)$, °F

$\Delta T'$ a new assumption of a temperature difference, °F

u velocity component in the x-direction, ft/sec

U over-all heat-transfer coefficient, Btu/hr sq ft °F

U_i, U_o over-all heat-transfer coefficient based on inside and outside areas respectively, Btu/hr sq ft °F

U_0 over-all heat-transfer coefficient at time zero, Btu/hr sq ft °F

U_r internal energy due to radiation per unit volume of system, Btu/cu ft

v velocity in the y-direction, ft/sec

v velocity, ft/sec

v_{fs} free-stream velocity, ft/sec

v_H velocity of hindered settling, ft/sec

v_p particle velocity, ft/sec

v_{ps} superficial particle velocity, ft/sec

$v_{p, \epsilon}$ particle velocity at porosity ϵ, ft/sec

$v_{p, \epsilon=1}$ particle velocity at porosity $\epsilon = 1$, ft/sec

v_R radial velocity, ft/sec

v_s superficial velocity based on empty tower cross section, ft/sec

v_{sl} slip velocity, ft/sec

$v_{sl, \epsilon}$ slip velocity at porosity ϵ, ft/sec

$v_{sl, \epsilon=1}$ slip velocity at porosity $\epsilon = 1$, ft/sec

v_{sm} mean superficial velocity, ft/sec

$v_{sm}°$ mean superficial velocity at the start of fluidization, ft/sec

v_t terminal velocity, ft/sec

v_{tan} tangential velocity, ft/sec

$\bar{v}_L$ velocity of rise of limiting layer in sedimentation, ft/sec

V mass rate of vapor stream, lb moles/hr or lb/hr

V volume, cu ft

V volume of filtrate collected, cu ft

V volume of material held in the filter, cu ft

V' mass rate of flow of noncondensable component, lb moles/hr

V_a, V_b specific volume of components a and b respectively, cu ft/lb mole

V_{dry} specific volume of dry gas, cu ft/lb mole

V_e hypothetical volume of filtrate equivalent in cake resistance to the resistance of filter cloth and piping, cu ft

V_h humid volume, cu ft/lb mole of dry gas

V_0 steam rate to first effect, lb/hr

w velocity in the z-direction, ft/sec

w mass rate of flow, lb/hr

w weight of solids in the feed slurry per volume of liquid in this slurry, lb/cu ft

w weight of condensate per unit time, lb/hr

W_s weight of solids, lb

x a direction, dimensionless

x distance travelled by a particle in a fixed time span, ft

x mole fraction in liquid phase, dimensionless

x wall thickness, ft

x thickness of laminar film, ft

x' mass fraction in liquid phase, dimensionless

x_a, x_b liquid-phase mole fraction of components a and b respectively

x_i, x^* liquid-phase mole fraction at the interface and at equilibrium equivalent concentration respectively

X a temperature ratio (Equation 15.19)

$\mathscr{X}$ volume fraction of liquid in slurry

X mole ratio in liquid phase, dimensionless

X' mass ratio in liquid phase, dimensionless or lb/lb

$\bar{X}'$ moisture content of a solid, dimensionless

$\bar{X}_E', \bar{X}_C'$ equilibrium and critical moisture contents respectively

$\bar{X}_f', \bar{X}_0'$ final and initial moisture contents respectively

y direction of flow, dimensionless

y mole fraction in vapor phase, dimensionless

y' mass fraction in vapor phase, dimensionless

y_a, y_b vapor-phase mole fraction of components a and b respectively

y_i, y^* vapor-phase mole fraction at the interface and at equilibrium equivalent concentration respectively

y_s mole fraction at saturation

Y mole ratio in vapor phase, dimensionless

Y geometric factor for multipass exchangers, dimensionless

Y' mass ratio in vapor phase, dimensionless

Y' mass ratio of solid to fluid phase

z direction of flow

z height, as of a column, or a packed section

Z a temperature ratio (Equation 15.18)

Greek Letters

α absorptivity, proportion of total radiant energy striking a body that is absorbed, dimensionless

α kinetic-energy correction factor, dimensionless

α specific cake resistance, ft/lb

α_p specific cake resistance at a point in the cake, ft/lb

α_0 specific cake resistance at zero compressive pressure, ft/lb

β an angle

β ratio of pressure drop for solid-fluid phase flow to that for fluid-phase flow alone

δ the differential operator when applied to a path function

Δ a finite difference, final minus initial value

ϵ emissivity (e/e_b), dimensionless

ϵ porosity, fraction of total volume not occupied by solids, dimensionless

η fin efficiency (Equation 15.32), dimensionless

θ time, hr

θ_c time of constant rate period, hr

θ_c time of total filtration cycle, hr

θ_d time for disassembly, cleaning, and reassembly of filter, hr

θ_f time of filtering, hr

θ_w time for washing and dewatering a filter cake, hr

λ latent heat of vaporization, Btu/lb mole

λ latent heat of vaporization, Btu/lb

λ wavelength, ft

λ_0 latent heat at T_0, Btu/lb mole

λ fin-efficiency factor, $\sqrt{hp/Sk}$

λ a shape factor defined by Equation 22.88

μ viscosity, lb/ft hr

π pi, 3.141592

ρ density, lb/cu ft

ρ reflectivity, fraction of total incident radiant energy that is reflected from a surface, dimensionless

ρ_s density of solid, lb/cu ft

ρ_B bulk density, lb/cu ft

σ Stefan-Boltzmann constant, 1.73×10^{-9} Btu/hr sq ft °R^4

Σ a summation

Σ a function dependent upon centrifuge geometry, sq ft

τ stress, lb$_f$/sq ft

τ transmissivity, the fraction of incident radiant energy that is transmitted through a body, dimensionless

ϕ an unspecified function

ϕ, ϕ' shape factors in crystal-growth relations

ψ sphericity, area of sphere of same volume as the particle divided by particle area, dimensionless

ω rate of rotation, radians/sec

Ω conical half angle

Subscripts

$0, 1, 2, 3 \ldots n$ stage or effect number, refers to streams leaving stage noted

$1, 2$ positions along flow or transfer path

$a, b, c, \ldots$ component

L, V, F liquid, vapor, and feed streams

s saturation, steam, or solid

0 base condition

f friction

k kinetic energy

appendix A

Dimensions and Units,
Dimensional Analysis, and Model Theory

DIMENSIONS AND UNITS

In this book, the terms *dimension* and *unit* have different meanings. A *dimension* is a description of a particular kind of extent, and its *unit* is a commonly used measure of the same extent. Table A-1 is a list of dimensions and corresponding typical engineering units.

Table A-1.

Dimension	Symbol	Typical Engineering Units
Length[1]	$\mathbf{L}_j$	feet, ft
Time	θ	second, sec
		hour, hr
Mass [2]	$\mathbf{M}$	pound, lb
Force[1]	$\mathbf{F}_j$	pound force, lb$_f$
Thermal energy	$\mathbf{H}$	Btu
Temperature	$\mathbf{T}$	°F, °C
		°R, °K

[1] The subscript j indicates that the quantity cannot be stated completely unless a direction component is included.

[2] In this book, the symbol "lb" without subscript will refer to pounds *mass*.

The six dimensions described in Table A-1 taken alone or in combination are sufficient to describe any engineering phenomenon or condition. In certain instances, relationships that exist among certain of the six dimensions may be used to eliminate dimensions from consideration, but caution must be used in so doing.

Of the six dimensions listed, length ($\mathbf{L}_j$) and force ($\mathbf{F}_j$) are vectors and the direction of application must be stated. As a consequence, the position of a point must be described in terms of three separate dimensions, $\mathbf{L}_x$, $\mathbf{L}_y$, and $\mathbf{L}_z$, when a three-dimensional coordinate system is used, as shown in Figure A-1,

$$\mathbf{L}_j = \sqrt{\mathbf{L}_x{}^2 + \mathbf{L}_y{}^2 + \mathbf{L}_z{}^2} \qquad \text{(A-1)}$$

Length is often considered to be a scalar quantity rather than a vector. Length may be considered a vector quantity when it is defined as the distance *from* a reference point *to* the point of interest. Therefore, as shown in Figure A-1, the length of the line *from* the intersection point of the coordinate axes ($x = 0$, $y = 0$, $z = 0$) *to* the point A is $\mathbf{L}_j$, a vector. Similarly, the length $\mathbf{L}_x$ is measured *from* $x = 0$ *to* $x = \mathbf{L}$ and is

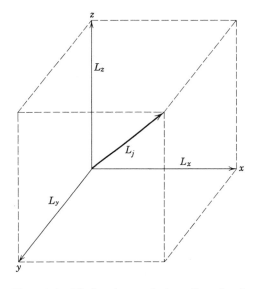

Figure A-1. The length vector in three-dimensional coordinates.

515

therefore a vector in the $+x$-direction. On the other hand when length is considered a scalar quantity, all directions are ignored. The length of the line *between* ($x = 0$, $y = 0$, $z = 0$) and A may be specified simply as $\mathbf{L}$, with no direction indicated. Although the scalar definition of length is satisfactory in many applications, it is often more useful to consider the direction in which the length is measured, i.e., to consider length as a vector. The force at a point must be described by three separate force dimensions in the three-dimensional coordinate system; therefore

$$\mathbf{F}_j = \sqrt{\mathbf{F}_x^2 + \mathbf{F}_y^2 + \mathbf{F}_z^2} \qquad (A\text{-}2)$$

Equations A-1 and A-2 apply in three-dimensional rectilinear coordinates and will be recognized as the addition of vectors as described in elementary mechanics. Similar equations may be written in other coordinate systems.*

Of the six dimensions described in Table A-1, temperature ($\mathbf{T}$) is unique because an absolute-zero value of temperature has been described by thermodynamics. In certain instances of engineering practice, a *difference* in temperature (extent of potential) is important; therefore, the units are not referred to absolute zero, as °C or °F. On the other hand, in calculations with the gas law, the absolute temperature must be used.

Relationships between Force and Mass. The mass of a system is a measure of the quantity of matter present in the system. It takes into account all atomic and subatomic particles of which the system is composed. Mass may be completely characterized by the dimension $\mathbf{M}$, since it is not a vector quantity. The mass of a system is a constant regardless of its location in the universe. Practically, the mass of the system is determined by measurement of its weight, which is the force exerted on the system in the earth's gravitational field. The measurement of mass utilizes Newton's law which states that force is proportional to the product of mass and acceleration, or

$$F \sim ma \qquad (A\text{-}3)$$

Before the units of force and mass are discussed, the relationship (A-3) should be written as an equation rather than as a proportion. An equation relates two combinations of variables, and each combination has the *same dimensions* as well as the same numerical value. Therefore, Equation A-3 must include a proportionality constant.

$$Fg_c = ma$$

where $F =$ the force applied
$m =$ the mass acted upon
$a =$ the acceleration of the mass resulting from the force
$g_c =$ the proportionality constant

A force applied in a direction will have associated with it an acceleration in the same direction. Therefore, the above expression is incomplete in that the direction of the force and acceleration is omitted. Properly, the equation should be

$$F_j g_c = ma_j \qquad (A\text{-}4)$$

where j represents the direction of the applied force and the resulting acceleration. Then, F_j represents the vector sum $\sqrt{F_x^2 + F_y^2 + F_z^2}$, and a_j represents the vector sum $\sqrt{a_x^2 + a_y^2 + a_z^2}$, where F_j and a_j are measured in three-dimensional rectilinear coordinates. In terms of primary dimensions, a_j has the dimensions $\mathbf{L}_j/\boldsymbol{\theta}^2$, and the typical units ft/sec². The dimensions on both sides of Equation A-4 must be the same. Therefore, if the primary dimensions are substituted for F_j, m, and a_j, the dimensions of the constant g_c can be determined.

$$\text{Dimensions of } g_c = \frac{\mathbf{M}\mathbf{L}_j}{\boldsymbol{\theta}^2 \mathbf{F}_j}$$

Several systems of units are in use; they are shown in Table A-2. In American chemical engineering practice, it is customary to measure forces in pounds force. For example, pressures are usually expressed as pounds force per square inch, lb_f/sq in. On the other hand, quantities

---

* In this book, cylindrical coordinates are written as $\mathbf{L}_y$, the axial direction, and $\mathbf{L}_r$, the radial direction. The general designation $\mathbf{L}_r$ represents a point in the x-z plane in rectilinear coordinates. For example, in the figure below, which is a portion of the x-z plane, with origin designated at 0 at the center of a cylindrical

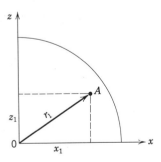

shape, A is any point with rectilinear coordinates z_1 and x_1 at a radial distance r_1 from the origin; then, by the Pythagoras theorem,

$$r_1 = \sqrt{x_1^2 + z_1^2}$$

From this equation, it is evident that r_1 has dimensions $\mathbf{L}_r = \sqrt{\mathbf{L}_x^2 + \mathbf{L}_y^2}$. In this book, systems involving cylindrical geometry have identical properties and conditions at any point on the periphery of a circle of any radius r about the origin. In this case, there is no basis for logical designation of the axes z and x; therefore, unless otherwise stated, the dimension $\mathbf{L}_r$ can have only the general dimensions $\sqrt{\mathbf{L}_x^2 + \mathbf{L}_z^2}$.

of matter in process are measured in pounds of mass. This practice is consistent with the English engineering system of units (see Table A-2). Since one pound force represents the force exerted on one pound mass in the standard gravitational field,* g_c may be defined using Equation A-4.

$$g_c = \frac{ma_j}{F_j} = \frac{(1 \text{ lb mass})(32.174 \text{ ft/sec}^2)}{(1 \text{ lb force})}$$

$$= 32.174 \frac{\text{lb ft}}{\text{sec}^2 \text{ lb}_f} \cong 32.2 \frac{\text{lb ft}}{\text{sec}^2 \text{ lb}_f}$$

Thus, one pound force may be defined as the force necessary to impart an acceleration of 32.2 ft/sec² to

Table A-2.

System of Units	Force	Mass	Acceleration	g_c
English engineering	pound force	pound	ft/sec²	32.2 $\frac{\text{lb ft}}{\text{sec}^2 \text{ lb}_f}$
English absolute	poundal	pound	ft/sec²	1 $\frac{\text{lb ft}}{\text{sec}^2 \text{ poundal}}$
English gravitational	pound force	slug	ft/sec²	1 $\frac{\text{slug-ft}}{\text{sec}^2 \text{ lb}_f}$
Metric centimeter-gram-second (cgs)	dyne	gram	cm/sec²	1 $\frac{\text{gm cm}}{\text{sec}^2 \text{ dyne}}$
Special metric*	gram force	gram	cm/sec²	981 $\frac{\text{gm cm}}{\text{sec}^2 \text{ gm}_f}$

* Occasionally in chemical engineering literature, the gram force is encountered.

one pound of mass. The constant g_c is a force-mass conversion factor.

The values of g_c given in Table A-2 are universal constants. Although they are defined by the standard gravitational field and are sometimes numerically equal to but dimensionally different from the gravitational acceleration, they can be used for all forces regardless of origin.

Equation A-4 is often written without the dimensional conversion factor (g_c).

$$F_j = ma_j \qquad (A-5)$$

This equation assumes that g_c is equal to unity, which is correct only in certain sets of units where the unit of force or of mass has been specially defined. For example, in the English Gravitational System (Table A-2), a special unit, the slug, is defined for mass. The *slug* is defined as the mass which will be accelerated 1 ft/sec² by 1 lb_f. Application of Equation A-4 gives

$$(1 \text{ lb}_f)\left(32.174 \frac{(\text{lb ft})}{(\text{sec}^2 \text{ lb}_f)}\right) = (1 \text{ slug})(1 \text{ ft/sec}^2)$$

or

$$1 \text{ slug} = 32.174 \text{ lb}$$

* As specified by the International Committee on Weights and Measures, the standard acceleration of gravity is 32.174 ft/sec².

Use of Equation A-5 gives

$$1 \text{ lb}_f = 1 \text{ slug} (1 \text{ ft/sec}^2)$$

For dimensional consistency, it is necessary to insert $g_c = 1 \text{ slug-ft/sec}^2 \text{ lb}_f$. The *dyne* and the *poundal* are special units of force defined in a similar manner.

Equation A-4 may be used to eliminate either mass or force from an equation and consequently to reduce by one the number of dimensions required to describe any system in which mass and force are both present. Whether force or mass is chosen as the primary dimension is a matter of convenience. The chemical engineer is concerned with chemical reactions, with masses of reactants and products to be processed and sold. It would seem then that the dimension mass would be of primary consideration to the chemical engineer. This restriction applies in this book, which uses the English engineering system. When a force is considered, the force must be expressed in terms including mass using Equation A-4.

On the other hand, civil and mechanical engineers are concerned with forces and reactions. The mass of the structure is of no consequence per se; only the force that is exerted by that mass is of consequence. Therefore, to the civil engineer, force is considered the primary dimension, and mass is defined in terms of an equivalent force.

In certain areas, civil, mechanical, and chemical engineers have a common interest; and therefore, when reading the literature of other engineering disciplines, the differences in point of view must be recognized.

Energy Relationships. Mechanical and thermal energy may be encountered in engineering calculations. Since energy can presumably be converted from one form to another, an initial assumption might be that one energy dimension (with an appropriate set of units) might suffice for all kinds of energy. However, a single unit is not possible. Thermal energy, usually expressed as an enthalpy content, is a scalar quantity. The enthalpy content is the result of random motion of all molecules in a body, and therefore no specific direction can be assigned. On the other hand, a change in mechanical energy, which is expressed as a product of force and distance, must have associated with it the direction of the force-distance product, with consequent dimensions $F_j L_j$ or, since the direction of the force and distance are the same, $(FL)_j$. Kinetic energy (one half the product of mass and square of velocity) has dimensions ML_j^2/θ^2 with velocity appearing as a vector term. Thus, the mechanical forms of energy require a directional designation. The second law of thermodynamics admits the conversion of a fraction of the thermal energy of a system to mechanical energy within certain exact limits for a given set of conditions. It should be intuitively evident that the conversion of the random motions of thermal energy to the directed motions and forces of mechanical energy cannot be unrestricted in a practical sense.

If a recognized fraction of the thermal energy of a system can be converted into mechanical energy, an appropriate equation would be

$$H = \frac{1}{J_j}(F_j L_j) \qquad (A\text{-}6)$$

where H = thermal energy which is converted to mechanical energy (**H**)
   $(F_j L_j)$ = mechanical energy (**FL**)$_j$
   J_j = proportionality constant with dimensions, $\mathbf{F_j L_j / H}$

The constant J_j is the mechanical-energy equivalent of heat and is equal to 778 ft lb$_f$/Btu.   The direction of the force and distance must be included in a rigorous definition.

A similar equation must exist for conversion of thermal energy to kinetic energy.

$$H = \frac{1}{J_j g_c}(\tfrac{1}{2}m v_j^2) \qquad (A\text{-}7)$$

where H = the thermal energy which is converted to kinetic energy (**H**)
   g_c = force-mass conversion constant ($\mathbf{ML_j/F\theta^2}$)
   v_j = velocity, with direction j specified ($\mathbf{L_j/\theta}$)
   m = mass (**M**)

Equations A-6 and A-7 may be used to express that part of the thermal energy that is converted to mechanical or kinetic energy in appropriate systems of units.

Thermal energy can also be related to temperature, through the constant pressure equation

$$H = m c_p T \qquad (A\text{-}8)$$

where m = mass of the system, with dimension **M**
   c_p = thermal energy capacity of the material making up the system, with dimensions **H/MT**
   T = temperature, with dimension **T**

with the use of Equation A-8, the dimension of either thermal energy or temperature can be eliminated from equations describing the system.

Notes on Dimensions and Units.   Chemical engineering practice has infrequently recognized the vectorial nature of length and force.   As a consequence, the *units* used to describe a system differ from the true dimensions of a system.   As an example, consider the viscosity, defined in Chapter 9, as

$$\tau_y g_c = -\mu \left(\frac{dv}{dx}\right)$$

where τ_y = stress on a plane of fluid, in the y-direction, per unit area in a y-z plane

v = velocity in the y-direction
x = distance in the x-direction
g_c = force-mass conversion constant with dimensions and units consistent with τ_y.

The dimensions of each of the above terms are

$$\text{Dimensions of } \tau_y = \frac{\mathbf{F}_y}{\mathbf{L}_y \mathbf{L}_z}$$

$$\text{Dimensions of } v = \frac{\mathbf{L}_y}{\boldsymbol{\theta}}$$

$$\text{Dimensions of } x = \mathbf{L}_x$$

$$\text{Dimensions of } g_c = \frac{\mathbf{ML}_y}{\boldsymbol{\theta}^2 \mathbf{F}_y}$$

The dimensions of each term may be substituted for the variables.   Dimensions of μ, the absolute viscosity, are obtained by substituting the known dimensions into the viscosity equation.

$$\frac{\mathbf{F}_y}{\mathbf{L}_y \mathbf{L}_z} \cdot \frac{\mathbf{ML}_y}{\boldsymbol{\theta}^2 \mathbf{F}_y} = \left(\frac{\mathbf{L}_y}{\boldsymbol{\theta}}\right)\left(\frac{1}{\mathbf{L}_x}\right)(\text{dimensions of } \mu)$$

By solving the above for μ, the dimensions of μ may be determined

$$\text{Dimensions of } \mu = \frac{\mathbf{ML}_x}{\mathbf{L}_y \mathbf{L}_z \boldsymbol{\theta}}$$

Note that $\mathbf{L}_x$, $\mathbf{L}_y$, and $\mathbf{L}_z$ are different dimensions.   Past chemical-engineering practice has tended to ignore the different direction components of lengths, and therefore $\mathbf{L}_x$, $\mathbf{L}_y$, and $\mathbf{L}_z$ have been canceled indiscriminately.   If this procedure is followed,

$$\mu = \frac{\mathbf{M}}{\mathbf{L\theta}}$$

In chemical engineering data sources, the units of absolute viscosity are generally stated as lb/ft sec, which indicates no recognition of the vectorial nature of the dimension of length.   Fortunately, this practice has not led to serious difficulty because, in the simple systems, the direction of forces and lengths are well known.   The nomenclature in this book is written with the vectorial *dimensions* and with the common nonvectorial *units*.

Because a number of systems of units are used in industrial practice, it is desirable to list the conversion factors which are commonly used.   These are given in Table A-3.

DIMENSIONAL ANALYSIS*

Chemical engineering practice sometimes depends upon empirical correlations that are equations made up of dimensionless groups of variables raised to various

---

* In this book empirical equations are usually developed by a mechanism-ratio analysis as described and demonstrated in Chapter 13.   An alternate procedure is presented here.

Table A-3. UNITS AND CONVERSION FACTORS

Mass, Length, (**M, L**)

1 cubic foot	= 28.316 liters
	= 28,316 cubic centimeters
	= 7.481 gallons (U.S.)
1 foot	= 30.480 centimeters
1 gallon	= 3.7853 liters
	= 231 cubic inches
1 inch	= 2.540 centimeters
1 mile (U.S.)	= 1.60935 kilometers
1 pound (avoirdupois)	= 453.5924 grams
	= 7000 grains
1 slug	= 32.174 pounds
1 square foot	= 929.034 square centimeters
1 ton (short)	= 2000 pounds
1 ton (long)	= 2240 pounds
1 ton (metric)	= 1000 kilograms
	= 2204.62 pounds

Temperature (**T**)

1 Centigrade degree = 1.8 Fahrenheit degree
Temperature, Kelvin = $T\,°C + 273.18$
Temperature, Rankine = $T\,°F + 459.7$
Temperature, Fahrenheit = $9/5\ T\,°C + 32$
Temperature, Centigrade = $5/9\ (T\,°F - 32)$

Force (**F**)

1 dyne = 1 gm cm/sec^2
1 gram force = 981 dynes
1 pound force = 32.174 poundals

Pressure (**F/L^2**)

1 atm = 760 mm Hg at 0°C (density 13.5951 gm/cu cm)
= 29.921 inches Hg at 32°F
= 14.696 pounds force/square inch
= 33.899 feet of water at 39.1°F
= 1.01325×10^6 dynes/square centimeter

Density (**M/L^3**)

1 gram/cubic centimeter = 62.43 pounds/cubic foot

Energy (**H** or **FL**)

1 British thermal unit = 251.98 gram-calories
= 777.97 foot-pounds force
= 10.409 liter-atmospheres
= 1054.8 joules
= 0.555 pound-centigrade unit (pcu)
= 0.2930 watt-hour

Diffusivity (**L^2/θ**)

1 sq cm/sec = 3.87 sq ft/hr

Viscosity (**M/Lθ**)

1 centipoise = 0.01 poise
= 0.01 dyne sec/sq cm
= 0.01 gm/cm sec
= 0.000672 lb/ft sec
= 2.42 lb/ft hr

Thermal Conductivity [**H/θL^2 (T/L)**]

1 Btu/hr sq ft (°F/ft) = 0.00413 cal/sec sq cm (°C/cm)

Heat Capacity (**H/MT**)

1 Btu/lb °F = 1 cal/gm °C = 1 pcu/lb °C

Gas Constant

1.987 Btu/lb mole °R = 1.987 cal/gm mole °K
= 1.987 pcu/lb mole °K
= 82.057 atm cu cm/gm mole °K
= 0.7302 atm cu ft/lb mole °R
= 10.73 (lb$_f$/sq in.) (cu ft)/lb mole °R
= 1545 (lb$_f$/sq ft) (cu ft)/lb mole °R
= 1545 ft-lb$_f$/lb mole °R

---

powers. Means of derivation, application, and the significance of these dimensionless groups will be examined here. Note that the vectorial nature of length is ignored.

Derivation: The Rayleigh Method. Any equation, to be complete, must be made up of variables and constants of proportionality in the form of two equal terms (the "left" and the "right" sides of the equation) *that have the same dimensions.* For example, earlier it was shown that

$$Fg_c = ma \quad (A-4)$$

is a dimensionally consistent equation. If the dimensions are substituted for the variables, Equation A-4 *as written* will have the dimensions

$$(F)\left(\frac{ML}{\theta^2 F}\right) = \frac{ML}{\theta^2} \quad (A-9)$$

or with cancellation $ML/\theta^2 = ML/\theta^2$

If Equation A-4 is rearranged to

$$F = ma/g_c \quad (A-4a)$$

and the dimensions substituted, then, after cancellation within terms,

$$\mathbf{F = F}$$

Thus, in both cases, the conditions for a dimensionally consistent equation are satisfied.

This principle can be useful in handling complex physical situations involving a large number of dimensional variables. If all the proper dimensional variables can be assumed, an *equation form* can be deduced based entirely upon the condition that an equation must be dimensionally consistent. For example, consider the complex phenomenon of fluid flow in tubes. Assume that no prior knowledge of fluid flow is available but that a certain set of variables are believed to be significant.

Using this assumption, the following functional relationship may be written:

$$(-\Delta P)g_c^* = \Phi(D, \bar{v}, \rho, \mu, L) \qquad \text{(A-10)}$$

where $(-\Delta P)$ = pressure loss due to friction
g_c = force-mass conversion constant
D = geometric factor, tube diameter
$\bar{v}$ = mean velocity
ρ = fluid density
μ = absolute viscosity
L = geometric factor, tube length

Equation A-10 is a statement of the assumed significant variables. Almost any mathematical function may be expressed as a power series with a sufficient number terms. Therefore, *assuming* that Equation A-10 may be written as a power series gives

$$(-\Delta P)g_c = C_1(D^a \bar{v}^b \rho^c \mu^d L^e)$$
$$+ C_2(D^{a'} \bar{v}^{b'} \rho^{c'} \mu^{d'} L^{e'}) + \dots \quad \text{(A-11)}$$

where a, b, c, d, and e are constant exponents and C_1 and C_2 are constants. The constants in Equation A-11 are dimensionless by this definition; therefore, to be dimensionally consistent, each term in the series must have the same dimensions as the term on the left side of the equation. Since the dimensions of each term in the series are identical, only the dimensions of the first term of Equation A-11 need be considered for dimensional homogeneity

$$(-\Delta P)g_c = C_1(D^a \bar{v}^b \rho^c \mu^d L^e) \qquad \text{(A-11a)}$$

Note that distances measured in different directions (D and L) are both included in the analysis. The only force $(-\Delta P)$ is undirectional. In this simple system, some partial recognition of the vectorial nature of distance is evident because two length terms are included for lengths measured in different directions. The directional nature of g_c is consistent with the directional nature of $-\Delta P$.

For each variable in Equation A-11a, substitution of the appropriate dimensions gives

$$\frac{F}{L^2}\frac{ML}{\theta^2 F} = \frac{M}{L\theta^2} = C_1(L)^a \left(\frac{L}{\theta}\right)^b \left(\frac{M}{L^3}\right)^c \left(\frac{M}{L\theta}\right)^d (L)^e \text{ (A-11b)}$$

This is an "equation" made up of three dimensions. To be dimensionally consistent, the sum of the exponents on each dimension must be the same on both sides of the equation. The three dimensions can be separated into equations for **M**, **L**, and **θ**, and the exponents on each side of Equation A-11b can be equated.

---

* Combining g_c with $(-\Delta P)$ eliminates the need for force as a dimension in the analysis. Only mass, length, and time remain because viscosity is expressed in mass units.

Sum of the exponents for **M**: $\quad 1 = c + d$
Sum of the exponents for **L**: $\quad -1 = a + b - 3c$
$\qquad\qquad\qquad\qquad\qquad\qquad - d + e$
Sum of the exponents for **θ**: $\quad -2 = -b - d$

Since there are three equations in five unknowns, they can be solved for three of the unknowns in terms of the other two. Solving for a, b, and c in terms of d and e gives

$$\begin{aligned} c &= 1 - d \\ b &= 2 - d \\ a &= -d - e \end{aligned} \qquad \text{(A-11c)}$$

Equation A-11a may be written substituting for a, b, and c the corresponding values in terms of d and e so determined.

$$(-\Delta P)g_c = C_1(D^{-d-e})(v^{2-d})(\rho^{1-d})(\mu^d)(L^e) \quad \text{(A-12)}$$

The terms may be collected in groups bearing the exponents d, e, and unity.

$$(-\Delta P)g_c = C_1(\bar{v}^2\rho)^1 \left(\frac{\mu}{D\bar{v}\rho}\right)^d \left(\frac{L}{D}\right)^e$$

or:

$$\frac{(-\Delta P)g_c}{\bar{v}^2\rho} = C_1 \left(\frac{D\bar{v}\rho}{\mu}\right)^{-d} \left(\frac{L}{D}\right)^e \qquad \text{(A-12a)}$$

Experimental evidence indicates that, for fully developed turbulence, $e = 1$; therefore Equation A-12 becomes

$$\frac{(-\Delta P)g_c D}{\bar{v}^2 L\rho} = C_1 \left(\frac{D\bar{v}\rho}{\mu}\right)^{-d} \qquad \text{(A-13)}$$

Equation A-13 is the familiar friction-factor-Reynolds-number equation.

Equation A-12a is in a form made up of several groups of variables. Each group of variables is in itself dimensionless. The constant C_1 was defined as dimensionless; consequently Equation A-12a is dimensionally consistent. The equation is a restatement of Equation A-11a with the necessary condition of dimensional consistency applied. The condition of dimensional consistency automatically brings into being a relationship among the exponents a, b, c, d, and e (Equation A-11b) which must be satisfied in *any* dimensionally consistent equation. Since Equation A-11a is one term of a series expression, the entire series may be written in the same form as Equation A-12a

$$\frac{-\Delta P g_c}{\bar{v}^2\rho} = C_1 \left(\frac{D\bar{v}\rho}{\mu}\right)^{-d} \left(\frac{L}{D}\right)^e + C_2 \left(\frac{Dv\rho}{\mu}\right)^{-d'} \left(\frac{L}{D}\right)^{e'} + \dots$$
$$\text{(A-14)}$$

This polynomial series is made up of the sum of dimensionless terms and is therefore also dimensionally consistent. During the derivation, no conditions other than Equation A-11b were applied to the exponents. Therefore, after dimensional analysis, the remaining exponents may be constant or variable, real or imaginary, positive or negative and may be functions of any of the dimensionless groups in the equation.

Although Equation A-11 has been modified by this dimensional analysis, the resulting Equation A-14 sheds no further light upon the mechanism of the phenomenon. The constants of the equation may be evaluated *only from experimental data*. The advantage of dimensional analysis lies in the ease of correlation of experimental data. In Equation A-11a, the constant and five exponents need evaluation. These data may be obtained experimentally only by holding four variables constant, varying the fifth, and observing the sixth. Contrast Equation A-12a in which two exponents and a constant must be evaluated, so that one *group* is held constant, one *group* varied, and the third observed. The actual number of experiments required to describe a phenomenon is reduced many fold. This advantage is possible only if the variables that actually describe the system are the same as the variables that are assumed in the analysis.

Dimensional analysis does not indicate extraneous, omitted, or redundant variables. All the foregoing information is available *only from experimental data*. Dimensional analysis may offer some hints as to error in reasoning. For example, consider the assumption

$$(-\Delta P)g_c = (D)^a(\rho)^b(L)^c$$

This equation will not respond to dimensional analysis because time (θ) as a dimension appears only in the left-hand term. Every basic dimension must appear at least twice before the analysis can be written. Since the analysis cannot be written, a dimensionally consistent equation cannot exist, and therefore a term is omitted, or a term is extraneous. This information is useful but, in the more complicated analyses, the basic dimensions almost invariably appear frequently enough so that the analysis of erroneous assumptions can be completed only with experimental data. When the exponents and constants are evaluated from experimental data, exponents on the groups containing extraneous or redundant variables go to zero, or equal exponents appear on several groups. When equal exponents appear, the groups may be combined, and the redundant or extraneous variable will cancel from the equation. If a significant variable has been omitted, the exponents and constants cannot be evaluated from experimental data.

A dimensional analysis has certain inherent properties. For a given set of assumptions, the number of dimensionless groups resulting from analysis is usually equal to the number of variables minus the number of dimensions. For the example given earlier, six variables were assumed, and three dimensions were required for the analysis; therefore three dimensionless groups should remain after

analysis. At no time in dimensional analysis may universal constants such as g_c or J be thought of as variables. In general, g_c, J, or the product Jg_c should be combined with the assumed variables to minimize the number of dimensions required to completely define the system. In the example above, g_c was combined with force terms to eliminate force as a dimension necessary to define the system.

If a dimensional analysis results in three or more groups, several solutions are possible. The solutions are not fundamentally different but have a different form. In the analysis presented above, several solutions are possible depending upon the exponents chosen as independent variables in the solution of the simultaneous set of exponent equations. In the example cited above, the independent variables were d and e, the exponents on μ and L, respectively. Note that in Equation A-12a, μ and L each appear in *single* separate dimensionless groups. If the exponents d and a, the exponents on μ and D, respectively were chosen as independent variables that is, b, c, and e were expressed in terms of d and a, the dimensional analysis would be

$$\frac{(-\Delta P)g_c}{\bar{v}^2\rho} = C_1\left(\frac{L\bar{v}\rho}{\mu}\right)^{-d}\left(\frac{L}{D}\right)^a \qquad \text{(A-15)}$$

In Equation A-15, μ and D each appear in single dimensionless groups. If b and e, the exponents on $\bar{v}$ and L respectively, were chosen as independent variables, the dimensional analysis would yield

$$\frac{(-\Delta P)\rho g_c D^2}{\mu^2} = C_1\left(\frac{D\bar{v}\rho}{\mu}\right)^b\left(\frac{L}{D}\right)^e \qquad \text{(A-16)}$$

in which case $\bar{v}$ and L each appear in single separate dimensionless groups. Although these equations look different, they are identical. This can be shown by multiplying Equation A-16 by $(D\bar{v}\rho/\mu)^{-2}$.

$$\left(\frac{(-\Delta P)\rho g_c D^2}{\mu^2}\right)\left(\frac{D\bar{v}\rho}{\mu}\right)^{-2} = C_1\left(\frac{D\bar{v}\rho}{\mu}\right)^{b-2}\left(\frac{L}{D}\right)^e$$

After cancellation of like terms in the left side,

$$\frac{(-\Delta P)g_c}{\bar{v}^2\rho} = C_1\left(\frac{D\bar{v}\rho}{\mu}\right)^{b-2}\left(\frac{L}{D}\right)^e \qquad \text{(A-16a)}$$

From Equation A-11c, note that $-d = b - 2$, so that Equation A-16a is identical to Equation A-12a. Therefore, for every combination of independent variables, a different but equivalent equation form will result after dimensional analysis. This property is useful for the experimentalist who chooses to isolate one or several variables.

A dimensional analysis may also be written in which L_x, L_y, and L_z and F_x, F_y, and F_z are used as separate dimensions, recognizing the vectorial nature of force and distance.

A dimensionless group is unique because any consistent set of units used in the variables within the group will result in the same numerical value for the group. This is demonstrated in Illustration A.1.

Illustration A.1. Calculate $D\bar{v}\rho/\mu$ for water flowing through a 2-in. pipe at 7 ft/sec, using consistent units of the metric system and of the English Engineering System.

The data are μ is 1 centipoise and density is 1 gm/cu cm.

SOLUTION. The centipoise as such is a special measure of viscosity. It cannot be used in calculations of dimensionless groups because its dimensions are not stated. By definition, 1 centipoise = 0.01 dyne sec/sq cm. Multiplication by $g_c = 1$ gm cm/sec² dyne (Table A-2) gives 1 centipoise = 0.01 gm/cm sec. Conversion to the English system gives 0.000672 lb/ft sec. Specific gravity is a ratio of density of a substance to the density of a reference substance, and, as such, it is a dimensionless ratio. The ratio cannot be used in situations involving statements of dimension. Since water is the reference substance, the *density* of water = 1 gm/cu cm. The latter term is stated in dimensions and can be used in dimensionless group calculations. A table is set up of each variable expressed in metric and English Engineering units:

Variable	Metric	English
D	5.08 cm	2/12 ft
$\bar{v}$	214 cm/sec	7 ft/sec
ρ	1 gm/cu cm	62.4 lb/cu ft
μ	0.01 gm/cm sec	0.000672 lb/ft sec

In metric units,

$$\frac{D\bar{v}\rho}{\mu} = \frac{(5.08 \text{ cm})\left(\frac{214 \text{ cm}}{\text{sec}}\right)\left(\frac{1 \text{ gm}}{\text{cu cm}}\right)}{\left(\frac{0.01 \text{ gm}}{\text{cm sec}}\right)}$$

$$= 108,500$$

In English units,

$$\frac{D\bar{v}\rho}{\mu} = \frac{(2/12 \text{ ft})\left(\frac{7 \text{ ft}}{\text{sec}}\right)\left(\frac{62.4 \text{ lb}}{\text{cu ft}}\right)}{\left(\frac{0.000672 \text{ lb}}{\text{ft sec}}\right)}$$

$$= 108,500$$

Note that in the metric system, the units used consistently are grams, centimeters, and seconds, and, in the English engineering system, the units that are used consistently are pounds, feet, and seconds.

MODEL THEORY

In many cases, the behavior of an operating unit or device can be predicted by test procedures using a conveniently sized scale model. The use of less expensive models and less time-consuming tests with small test equipment has increased immeasurably the store of available performance characteristics of equipment. The interpretation of test data from scale-model tests and application to full-sized equipment depends upon model theory, the fundamentals of which are set down here. The scaling-up or scaling-down procedure for existing equipment is also based upon model theory.

Model theory is dependent upon several criteria of similarity. *Geometric similarity* is established in elementary geometry courses as existing between two geometrical figures if all counterpart length dimensions bear a constant ratio. *Kinematic similarity* exists in a geometrically similar system of models of different size if all velocities at counterpart positions bear a constant ratio. Geometric similarity must exist, otherwise counterpart positions would not exist. *Dynamic similarity* exists in a geometrically similar system of two models if all forces at counterpart positions bear a constant ratio.

Consider two lengths of smooth tubing, one of which is 1 in. I.D. and 12 in. long, and the other 1 ft I.D. and 12 ft long. The two sections are geometrically similar since any two counterpart linear dimensions are in the ratio 1/12. Next consider the velocity distributions in the two lengths of pipe. If the point velocities at the midpoint and the velocities at all other geometrically counterpart points between the center and the wall bear a constant ratio, the model and prototype are kinematically similar. If well-developed turbulence exists throughout the entire length of both models, the point velocity is independent of axial position. If the ratio of kinetic energy per unit volume (inertial forces) to fluid stress (viscous forces) is constant and no other forces (for example, surface tension or gravitational forces) exist to significant degree, the two models are dynamically similar. Note that the ratio of forces is a statement of the Reynolds number as described in Chapter 13. Therefore, any two tubes, regardless of size, that operate at the same Reynolds number are dynamically similar. This does not mean that D, $\bar{v}$, ρ, and μ are related in any way in the two pipes but that the product $(D\bar{v}\rho/\mu)$ is the same in both pipes. For geometrically similar systems, any combinations of D, $\bar{v}$, ρ, and μ that result in the same Reynolds number describe dynamically similar systems. It is the latitude in variables in the preceding statement that illustrates the usefulness of model theory. If any other forces exist, new ratios exist. The Froude number ($\bar{v}^2/Dg$), where g is gravitational acceleration, is the ratio of inertial forces to gravitational forces; the Weber number $(D^2\bar{v}\rho/\gamma g_c)$, where γ is surface tension, is the ratio of inertial forces to surface forces; and the Euler number $((-\Delta P)g_c/\bar{v}^2\rho)$, is a ratio of pressure forces to inertial forces.

Model theory may be stased as follows: If two models are geometrically, kinematically, and dynamically similar, all velocities and forces are in a constant relationship at counterpart positions.

Illustration A.2. A proposal exists to construct an airplane. The "drag," or surface stress, due to air resistance must be evaluated. If the proposed airplane is to cruise at 75 mph and has a wingspan of 25 ft, describe the conditions for a model test using a 3-ft-wingspan scale model in (*a*) a wind tunnel at 70°F and 1 atm and (*b*) a towing-tank test, in which the model is towed under water, at 70°F.

SOLUTION. Since a scale model is to be used, all linear dimensions are in a 3:25 ratio, and the system is geometrically similar. In order to establish some criterion for comparison, a dimensional analysis can be written that includes all forces and velocities that exist or that defines kinematic and dynamic conditions.

$$\tau_y g_c = \phi(L, \bar{v}, \rho, \mu)$$

With the exponential-series assumptions, a solution of the dimensional analysis is

$$\frac{\tau_y g_c}{\bar{v}^2\rho} = \phi\left(\frac{L\bar{v}\rho}{\mu}\right)^n + \phi'\left(\frac{L\bar{v}\rho}{\mu}\right)^{n'} + \ldots$$

where $\tau_y g_c$ = drag
$\quad L$ = wingspan
$\quad \rho$ = fluid density
$\quad \mu$ = viscosity
$\quad \bar{v}$ = velocity
$\quad \tau_y g_c/\bar{v}^2\rho$ = drag coefficient
$\quad L\bar{v}\rho/\mu$ = a form of Reynolds number based on wingspan
$\quad \phi$ and n = unknown functions

For purposes of test, if dynamic similarity exists, the two models operate at the same Reynolds number. If the Reynolds number is constant, the drag coefficient is constant regardless of the nature of ϕ, n, and n'. The solution lies in establishing the conditions for equal Reynolds number for both models, or

$$\left(\frac{L\bar{v}\rho}{\mu}\right)_1 = \left(\frac{L\bar{v}\rho}{\mu}\right)_2 \qquad (A-17)$$

Subscript 1 = small-scale model
Subscript 2 = standard-sized unit

The conditions, except velocity, are the same in atmospheric air and in the wind tunnel; therefore, $(\rho/\mu)_1 = (\rho/\mu)_2$. Substituting into Equation A-17 gives

$$3 \times \bar{v}_1 \times \left(\frac{\rho}{\mu}\right)_1 = 25 \times 75 \times \left(\frac{\rho}{\mu}\right)_2$$

$$v_1 = 625 \text{ mph}$$

This velocity is very close to the sonic barrier and consequently would be of dubious value for experiment. If the towing tank were used, $(\rho/\mu)_1 = 62.4/(1 \times 6.72 \times 10^{-4}) = 9.3 \times 10^4$; $(\rho/\mu)_2 = 0.075/(0.0182 \times 6.72 \times 10^{-4}) = 6.13 \times 10^3$. These values can be substituted into Equation A-17.

$$v_1 = \frac{(25)(75)(6.13 \times 10^3)}{(3)(9.3 \times 10^4)} = 41 \text{ mph}$$

This velocity is quite reasonable for towing tanks, and therefore would be the recommended procedure.

After the scale model is tested, during which time the stress is determined, the stress for the full-scale model could be estimated. At (N_{Re}) = constant,

$$\left(\frac{\tau_y g_c}{\bar{v}^2\rho}\right)_1 = \left(\frac{\tau_y g_c}{\bar{v}^2\rho}\right)_2$$

$$(\tau_y g_c)_2 = \frac{(\tau_y g_c)_1(\bar{v}^2\rho)_2}{(\bar{v}^2\rho)_1}$$

$$= (\tau_y g_c)_1 \frac{(75)^2 \, 0.075}{(41)^2 \, 62.4}$$

$$(\tau_y g_c)_2 = 4.02 \times 10^{-3} (\tau_y g_c)_1$$

Or, in words, the stress on the full-scale plane would be 4.02×10^{-3} times the stress on the small-scale model. Since the stress on the small-scale model is so much greater than the stress on the full-scale plane, the experimental procedure can be carried out without specialized instruments.

PROBLEMS

A-1. Consider a cube in rectangular coordinates. (*a*) Write an expression for the *dimensions* of the area of each of the six faces of the cube, considering length a vector.
(*b*) Write an expression for the dimensions of the volume of the cube.

A-2. What force must be exerted to accelerate a 1-lb mass at 41 ft/sec²? Express the answer in pounds force, in poundals, and in dynes.

A-3. What force must be applied to a body with a mass of 1.23 slugs to give it an acceleration of 11 ft/sec²? Express the answer in pounds force, in poundals, and in dynes.

A-4. It is believed that the heat-transfer coefficient (*h*) in a circular pipe depends on the following physical variables of the system.
$\quad k$ = thermal conductivity
$\quad c_P$ = heat capacity
$\quad \mu$ = viscosity
$\quad \rho$ = density
$\quad \bar{v}$ = mean velocity of fluid
$\quad D$ = diameter of pipe
Derive an equation of dimensionless group by dimensional analysis which relates these variables.

A-5. An attempt to correlate data using the equation developed in Problem A-4 shows different exponents on one dimensionless group for heating and for cooling. Further consideration indicates that the viscosity at the wall temperature (μ_1), may be influential. Using this variable and those listed in Problem A-4, derive an expression of dimensionless groups to be used in correlation.

A-6. In natural convection from horizontal tubes, the heat-transfer coefficient (h) depends upon the pipe diameter and the fluid thermal conductivity, viscosity, heat capacity, density, and the coefficient of thermal expansion (β, T^{-1}). In addition, the acceleration of gravity [g, (L/θ^2)] and the temperature difference between the pipe and fluid (ΔT) are influential. Using dimensional analysis, derive an equation of dimensionless groups relating these variables.

A-7. The mass-transfer coefficient (k_c) in a circular pipe is believed to be some function of the following variables.

$\mathscr{D}$ = mass diffusivity
μ = viscosity of fluid
ρ = density of fluid
$\bar{v}$ = mean velocity of fluid
D = diameter of pipe

Derive an equation of dimensionless groups to be used in correlating mass transfer coefficients.

A-8. A spherical particle is falling through a fluid. (*a*) List the physical variables that are likely to influence the terminal velocity (v_t) of the particle. (*b*) Develop an equation of dimensionless groups which relates the terminal velocity to these variables.

A-9. By dimensional analysis, develop an expression relating the pressure drop for a fluid flowing through a packed bed to other pertinent variables. The bed consists of particles of uniform size and shape.

A-10. A wind tunnel is to be used to test a scale model of a submarine. The model is one-tenth of full size. What air velocity would be equivalent to a velocity of 25 mph for the submarine in water? Air is at 75°F and 1 atm, and water is at 50°F.

A·11. What would be a satisfactory diameter and velocity for an aluminum sphere in air at 70°F which is to be dynamically similar to a lead sphere of 1-in. diameter moving at a velocity of 1 ft/min in mercury at 100°F?

appendix B

Description of Particulate Solids

The problem of characterizing a particulate solid and of predicting its characteristics is one that pervades all engineering. Civil engineers deal in concrete aggregates, soils, river silting. Mining and metallurgical engineers face this problem whenever an ore is handled. Chemical engineers meet particulate solids in carrying out many unit operations, as for example crushing, drying, filtering, crystallization, solid-fluid reacting, dust collecting, as a part of any process that produces a solid product, and as catalysts in many industrially important chemical reactions. In this book, such operations are treated, but the significance of terms describing the solid phase has not been dealt with. For example, when considering the flow of fluid relative to a solid particle, the solid has been treated as a group of identical particles of fixed "diameter" and face area. Also, when considering spray-drying operations, the statistical average of particle diameters and measures of particle-size distribution are used. In this appendix, a particulate-solid phase will be described, the methods of measuring it will be treated, and the use of mathematical techniques to describe the measurements will be discussed.

In chemical engineering, solid particles ranging in size between quarried rock and smoke are of interest. The size range covered is from about 10^5 to 1 microns (1 micron = 1/1000 mm) (3). In this range, the particles are larger than most individual molecules and larger than particles in colloids; hence, they are unaffected by Brownian-motion forces. On the other hand, they are small enough so that they are usually found in large numbers. Mathematically, these facts lead naturally to the use of summation arithmetic and statistical expressions in describing properties of particulate solids.

The properties of such solids fall into two classes: those belonging to the individual particle and those belonging to the solids-voids phase of particles. In the first class are the size and shape of the particle, its volume, surface area, and mass. Properties of the bulk solid material which are retained by the small particles must also be considered in this class. Examples are thermal conductivity, solid density, specific heat, hardness, and hygroscopic tendency. In the second class are the void fraction of the mass, the effective density of the aggregate mass of solids and voids, surface area per cubic foot, and many secondary characteristics such as effective thermal conductivity, permeability (which in this case is a measure of the pressure drop due to fluid flow through the mass), and angle of repose (the steepness of the sides of a poured pyramid). Obviously, the properties of the solids-voids phase must depend upon the properties of the particles, but the phase properties must now be expressed so that the effect of all the particles present is considered—no problem at all when the particles are all identical, but a very difficult problem when the particles exist in a range of sizes, a variety of shapes, and even sometimes a difference in composition.

METHODS OF PARTICLE-SIZE MEASUREMENT

The methods of expressing particle size depend upon the measuring device used. The commonest of them is the standard sieve. Here, the solid phase is placed on top of a series of screens. Each screen has smaller openings than the one above, usually in $2^{1/n}$ series. As the sieves are shaken, the particles fall through them until a screen is reached in which the openings are too small for the particle to pass. The size of particles found on any screen is expressed as an appropriate mean length between the openings in the screen above and that on which the particle rests.

A second method of size measurement is that of

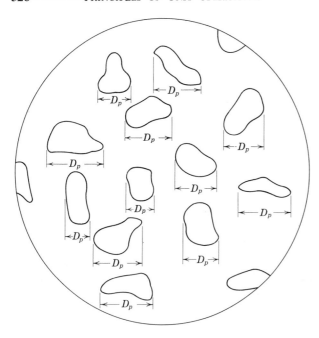

Figure B-1. Particle-size measurement with a micrometer.

counting the particles and actually measuring them. If the particles are small, a sample of the material is put under a microscope, and each particle within the field of vision is measured by an optical micrometer. With irregularly shaped particles, a large variety of dimensions might be used. For instance, the longest or shortest dimension or an average of two dimensions might be used. However, it is normal practice to choose a direction of measurement and take the longest distance across the particle in this direction. This method is illustrated in Figure B-1. The optical micrometer usually consists of a cross hair moved across the microscope field by a vernier dial. The readings taken are then dial readings. They may be converted to inches or microns by noting the difference in dial reading while traversing a known distance. The measured distances are then treated as particle diameters in determining the average particle diameter. The accuracy of the mean diameter so obtained depends on the use of a statistically large number of particles. For such a sample, the random orientation of particles will eliminate any effect of direction. This method is, of course, extremely laborious, but the answers obtained are not dependent upon the perfection of a screen, the uncertainty caused by use of screen sizes arranged in finite differences or on theories of particle motion in a fluid. Moreover, this method can be used where others cannot. For example, liquid droplets can be sized with this technique by spraying them into a catching liquid in which they are insoluble or by photographing them and then measuring the photographic image. This method recognizes visually any agglomeration of particles which may go undetected by other methods.

A third method of size measurement is that of separating the sample according to particle settling velocities. The theory behind these methods is presented in Chapter 22. The result is that the settling velocity is a function of both the particle and fluid densities and the particle projected area (see Equation 22.10). Hence, for a uniform-density material, separation is by projected area. In terms of the micrometer measurements, this area would be some combination of maximum and minimum measurements. For convenience, the dimension is given as the diameter of a sphere that would fall with a velocity equal to the observed falling velocity of the particle. If the particle is markedly nonspherical this dimension may not be representative of the particle.

Particle-Size Measurement by Screen Analysis. Although the three methods of particle-size determination do not give identical answers, especially in the small-size range, there seems to be no clear-cut choice between methods. Therefore, most particle-size determinations are made by screen analysis when the particles are within the size range that can be measured by screens. Standard screens of either the Tyler series or U.S. sieve series are used with almost equal frequency throughout U. S. industry. These screen series use identical 200-mesh screens but differ very slightly in other sizes. The characteristics of the Tyler series are given in Appendix C-8, in which the length of one side of the square opening is given as the hole size. In each of the screen series, the wire size and mesh, or number of openings per linear inch, are adjusted so that the ratio of hole size of any two consecutive screens in the series is constant giving a

Figure B-2. Fisher-Wheeler sieve shaker with nested screens in place. (Courtesy Fisher Scientific Co.)

geometric ratio of hole sizes. As shown in Appendix C-8 for the Tyler series, enough screens are available in each of the series so that the ratio can be $\sqrt[4]{2}$. Usually, it is unnecessary to determine the particle-size distribution using such narrow intervals, and every other screen in the $\sqrt[4]{2}$ series is used to give a new series with openings progressing by the $\sqrt{2}$. In some cases, every fourth screen in the $\sqrt[4]{2}$ series may be used to give a progression with a factor of 2 between screens. The 200-mesh screen is usually the finest one used, although screens are available down to 400-mesh. The finer screens are expensive and extremely delicate. The clogging of the screen openings with sample particles (screen blinding) is a much more serious problem in these sizes, and the screen wires are easily displaced to give uneven large and small openings.

In making a screen analysis, screens are nested one above another, arranged so that each screen has openings larger than the one below. Under the bottom screen is put a solid pan. The sample is then put onto the top screen, a lid put on, and the stack of screens clamped into a shaker. The shaking action influences the efficiency of the screen set, so that a reproducible shaking motion is desirable. The Fisher-Wheeler sieve shaker, which is shown in Figure B-2, gives the screens a horizontal rotary motion while tapping them on the bottom. This motion has been found to give higher efficiency than most other shaker types, where

$$\text{Efficiency} = \frac{\text{percentage material actually passing}}{\text{percentage material capable of passing}}$$

Shaking is continued for a fixed time, usually 10 to 20 minutes; the sieves are removed, and the material held on each of the sieves is collected and weighed. The

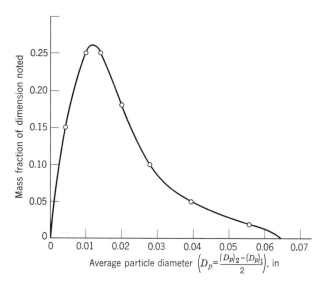

Figure B-4. Fractional-distribution plot for screen analysis of Table B-1.

efficiency of sieves is found to depend upon the material and its size. Blinding of screens increases, and efficiency decreases, as the sample size increases, and particle size decreases. Blinding is also more severe for light, sticky solids or for those with a large concentration of particles almost identical to the openings size of one screen. Some materials can only be analyzed satisfactorily by suspending them in a nonsolvent liquid and washing them through the screens in a shaker. With this method, analyses of very high efficiency can be made.

The results of such a screen analysis are initially presented tabularly. A typical screen analysis is shown in Table B-1.

Table B-1. RESULTS OF TYPICAL SCREEN ANALYSIS

Size Range (Tyler Mesh)	Average Particle Diameter $(\bar{D}_p)$, in.	Mass Fraction Retained, weight percent
−10 + 14	0.0555	2
−14 + 20	0.0394	5
−20 + 28	0.0280	10
−28 + 35	0.0198	18
−35 + 48	0.0140	25
−48 + 65	0.0099	25
−65	0.0041	15

The designation −10 + 14 means particles smaller than 10 mesh but greater than 14 mesh. Alternate methods of designation would be 10/14 or "through 10 mesh, on 14 mesh." The results could be presented as a *histogram*, such as that of Figure B-3. Here, the vertical axis would be the mass fraction of the total sample

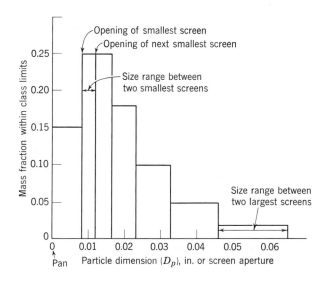

Figure B-3. Histogram presentation of screen analysis of Table B-1.

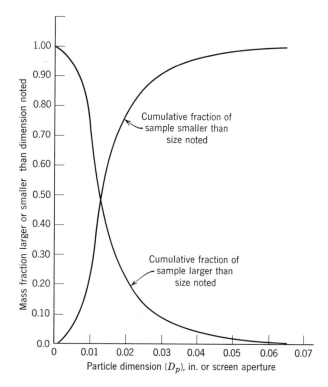

Figure B-5. Cumulative-distribution plot for screen analysis of Table B-1.

between two screens. The results could also be presented as fractional or cumulative distribution curves. These curves are usually made by assuming the material between two screens to have a particle diameter that is the arithmetic average of the two screen openings. Such plots are shown in Figures B-4 and B-5. In many cases, it is more convenient to plot the distribution curves on semilogarithmic coordinates so that the appropriate mass fraction is plotted as a function of the logarithm of the particle size. The fractional-distribution plot of Figure B-4 is replotted on semilogarithmic coordinates in Figure B-6. It can be seen that the crowding of most of the data into a small section of the diagram and the skewing of the distribution curve toward the small-particle-diameter end of the diagram that occurred in Figure B-4 is avoided by the semilogarithmic plot.

Screen analyses cannot be made to determine particle-size distribution below 0.0015 in. and are unsatisfactory below 0.003 in. For this reason, calculational methods have been developed to estimate the fine-particle-size distribution. Empirically, it has been found that for the fine tailings of product fresh from a size-reduction step, the fractional distribution as determined by a screen analysis is usually an exponential function of the particle diameter (4). Thus,

$$-\frac{dx}{d\bar{D}_p} = a(\bar{D}_p)^k \tag{B-1}$$

where a and k = constants
x = mass fraction of sample of average diameter ($\bar{D}_p$) as determined by screen analysis

Integrating gives

$$x = \frac{-a}{k+1} \bar{D}_p^{k+1} + b \tag{B-2}$$

so that a logarithmic fractional-distribution plot should show a straight-line relationship in the fine-particle region. Equation B-2, however, is an empirical result only and is restricted to crushed products. Moreover, this equation applies to the results of a screen analysis. Therefore, it is important that any fractional-distribution values obtained by Equation B-2 be read using the same interval of successive screen openings as was used in the screen analysis from which the constants a and b were obtained. Thus, if a $\sqrt{2}$ series of screens was used, the logarithmic-fractional-distribution-plot extrapolation must be read at intervals of $\bar{D}_p$ each $\sqrt{2}$ smaller than the preceding one.

Illustration B-1. A sample of crushed ilmenite as removed from a ball mill has the following screen analysis:

Screen Size (Tyler Mesh)	Weight Fraction	$\bar{D}_p$, in.
−10 + 14	0.075	0.0555
−14 + 20	0.136	0.0394
−20 + 28	0.158	0.0280
−28 + 35	0.154	0.0198
−35 + 48	0.133	0.0140
−48 + 65	0.106	0.0099
−65 + 100	0.082	0.0070
−100 + 150	0.056	0.0050
−150 + 200	0.043	0.0035
−200	0.057	

Complete the particle-size-distribution information by fixing the size distribution of the −200-mesh fraction.

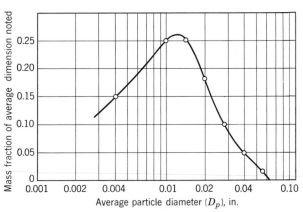

Figure B-6. Fractional-distribution plot for screen analysis of Table B-1 on semilogarithmic coordinates.

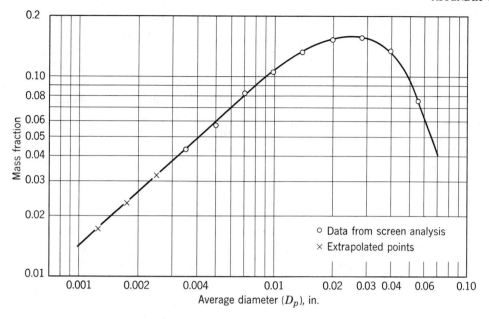

Figure B-7. Logarithmic plot of screen analysis of ilmenite from ball mill, Illustration B-1.

SOLUTION. This analysis is plotted on logarithmic coordinates in Figure B-7. The data points are shown as circles. The data points for the four smallest fractions obtained fall on a straight line which is extrapolated into the fine-size region. According to Equation B-2, this line must also give the fractional analysis of the fines at intervals of $\sqrt{2}$ in diameter extrapolated into the fine-particle-size range. The extrapolation is continued until the sum of the weight fraction so obtained equals the weight fraction of fines, which in this case is that portion through the 200 mesh screen, or 0.057. The results are

$\bar{D}_p$, in.	Weight Fraction		Adjusted Weight Fraction
0.0025	0.032	→	0.032
0.00175	0.023	→	0.023
0.00124	0.017	→	0.002
	0.072		0.057

The excess weight fraction obtained through extrapolation must be removed and has here been arbitrarily subtracted from the smallest size fraction.

In making screen analyses, care is taken that the screens are in good repair and that the sample is shaken long enough so that the fines have ample opportunity to pass through the screen. Even so, some oversize material passes through the screen, and some fine material stays on top of the screen. This is not usually a serious problem, and the screen effectiveness, defined as the fractional recovery of desired material times the fractional removal of undesired material for either the oversize or undersized fraction, is nearly unity. The primary standard here would be a new and perfect testing screen.

In industrial screening, the effectiveness is not so high, and quantitative measurements of effectiveness are more useful. From the definition,

$$\text{Effectiveness} = \text{recovery} \times \text{rejection}$$

$$\text{Recovery} = \frac{Px_P}{Fx_F} \qquad \text{(B-3)}$$

where P, F = mass of product and feed respectively, where either the oversize or undersize cut may be considered as product

x_P, x_F = mass fraction of desired-size-range material in feed and product, respectively

and Rejection = 1 − recovery of undesired material

$$= 1 - \frac{P(1 - x_P)}{F(1 - x_F)}$$

Then,

$$\text{Effectiveness} = \frac{Px_P}{Fx_F}\left[1 - \frac{P(1 - x_P)}{F(1 - x_F)}\right] \qquad \text{(B-4)}$$

This definition requires both a screen analysis of feed and product, and a weight measurement of feed and product for a determination of effectiveness. The weight measurement can be eliminated by combining a material balance with this equation. Thus,

$$Fx_F = Px_P + Rx_R$$

where R = reject mass

x_R = mass fraction of desired product in reject

$$Fx_F = Px_P + (F - P)x_R$$
$$F(x_F - x_R) = P(x_P - x_R)$$
$$\frac{F}{P} = \frac{(x_P - x_R)}{(x_F - x_R)}$$

and

$$\text{Effectiveness} = \frac{x_P(x_F - x_R)}{x_F(x_P - x_R)} \left(1 - \frac{(x_F - x_R)(1 - x_P)}{(x_P - x_R)(1 - x_F)}\right)$$

(B-5)

which can be solved on the basis of screen analysis data only.

The effectiveness of any screening operation will decrease as the screen gets blinded or as it wears to give uneven hole sizes. The effectiveness will also be reduced if the screening is done too rapidly, if the sample size is too large, or if the solids become moist so that they tend to agglomerate. In these cases some appropriate particles will have no opportunity to pass through a hole. Conversely, too long a shaking interval may cause incorrect results because of particle attrition.

Sampling. Proper sampling techniques are necessary in the study of any material, especially if the material is a particulate-solid phase. Errors arising because of poor sampling can be greater by many fold than those that occur because of inefficient screening, poor calibration of a microscope, or insufficient care in timing the fall of a particle in a liquid phase. Moreover, sampling errors are extremely common and are important whether particle size or shape, phase density or porosity, or any other property is under study. The problem is aggravated in the case of particulate solids because of the tendency of these materials to classify themselves. In a bin or pile of solid particles, the fine ones and the more dense ones tend to migrate toward the bottom of the bin. If the heavy particles are different in composition from the light ones, the analysis of the material in the bin would vary with location of the sample. In any case, the particle size, the bulk density, and the phase porosity would vary. If the sample were a spray-dried product, the moisture content and the caking tendency would vary through the bin for similar reasons. The problem of self-classification constantly plagues producers of packaged mixed solids, such as dry food products.

Unlike a liquid phase, a solid phase cannot always be remixed by agitation because of the delicacy of the solid particles. It can, of course, be tumbled in a dry blender until it again becomes uniform. This method is expensive in terms of handling costs and produces a pile, or bin, of solids which again must be sampled. The likelihood of a poor sample is reduced but not entirely eliminated.

The most generally accepted method of solid sampling is by splitting, which can be done manually or mechanically. Manually, the entire batch to be sampled is piled in a uniform pyramid. The pile is then quartered by drawing a cross through the peak of the pile and separating the four segments. The process is repeated with any one of the four segments by starting again with a uniform pyramid. Successive steps finally result in a sample of the desired size. Mechanically, the sampling is done with the help of a splitter. This device is made of many parallel, equally spaced, thin metal walls. The walls form the sides of chutes that alternately discharge to the right and left sides of the splitter. A material poured into the top of a splitter will form two separate and probably identical piles. One of these piles is then discarded, and the other retained. Successive splitting will result in a uniform small test sample. If desired, splitters can be arranged in series so that a solid phase fed into the top splitter will result in a representative sample and the rejected bulk of the product issuing separately from the bottom of the splitter train.

ANALYTICAL REPRESENTATION OF PARTICLE-SIZE DISTRIBUTION

In efforts to relate particle size of a solid phase to the characteristics of the equipment that generates it and the material from which it is made, the graphical representations mentioned above are clumsy at best. For instance, in spray drying, the droplet-size distribution in terms of a distribution curve would be difficult to relate to the fluid properties and nozzle construction. Therefore, it would be desirable to find as few parameters as possible that will express the particle-size distribution so that the parameters may be related mathematically to the rest of the system variables.

Mean-Diameter Expressions. The simplest method of expressing particle size is merely to use some mean diameter as typical of all the particles in a sample. Even here, difficulties arise because the engineer must decide in what characteristic he wishes the diameter to be typical. Thus, if the important characteristic is particle mass, he must choose a mean diameter such that a particle of mean diameter will have a mean mass. The particle will not, except by chance, also have a mean surface area, a mean projected area, or a mean linear size. In mass-transfer operations, as for instance in packed distillation columns, the important characteristic is the surface area per unit volume. A similar case is that of catalyst effectiveness. On the other hand, in studying the distribution of mass throughout the spray from a nozzle, the particle volume is of prime interest.

As stated in Chapter 10, the true mean value of any property can be expressed by the equation

$$\bar{y}(x_2 - x_1) = \int_{x_1}^{x_2} y \, dx \tag{10.3}$$

or

$$\bar{y} = \frac{\int_{x_1}^{x_2} y \, dx}{x_2 - x_1} \tag{10.3a}$$

where y = a property the mean value of which is sought

$\bar{y}$ = the appropriate mean value over the interval $x_2 - x_1$

x = a second property upon which y is *solely* dependent

This general definition will be used in the determination of mean diameters. It must, of course, be specialized for the case at hand.

The analytical determination of mean diameters must be based on an analytical expression for the particle-size distribution. In functional notation, the equation

$$\frac{dm}{dD_p} = f_m(D_p) \qquad (B\text{-}6)$$

where dm/dD_p = the differential change in sample mass for a differential change in particle dimension

applies to the cumulative distribution curve based on mass-fraction analysis. If this curve were based on a particle-size count, the function would be

$$\frac{dn}{dD_p} = f(D_p) \qquad (B\text{-}7)$$

where dn/dD_p = the number of specimens of dimension lying between the limits D_p and $(D_p + dD_p)$

If these functions properly express the particle-size distribution of a representative sample,

$$\int_0^\infty \frac{dm}{dD_p}\, dD_p = \Sigma m = \int_0^\infty f_m(D_p)\, dD_p \qquad (B\text{-}8)$$

and

$$\int_0^\infty \frac{dn}{dD_p}\, dD_p = \Sigma n = \sigma = \int_0^\infty f(D_p)\, dD_p \qquad (B\text{-}9)$$

where Σm and σ are the total sample mass and the total number of individual particles in the sample respectively.

Equations B-6, B-7, B-8, and B-9 are closely related to the distribution plots presented earlier. The relation is most easily seen by reference to the histogram, Figure B-3. The ordinate on this plot may be written as

$$\Delta m = \frac{\Delta m}{\Delta D_p}\,\Delta D_p$$

If the size range between screens is continuously reduced until it becomes differential, this expression becomes

$$dm = \frac{dm}{dD_p}\, dD_p$$

Such a plot would consist of an infinite number of blocks, each differentially wide and high. It would not retain the shape of the histogram or approach the shape of the fractional distribution plot, Figure B-4. On the other hand, if the ordinate of Figure B-3 were mass fraction divided by screen opening size range, $\Delta m/\Delta D_p$, reducing the ΔD_p to differential size would not reduce the value of $\Delta m/\Delta D_p$. Ultimately the ordinate would become dm/dD_p. This is the function $f_m(D_p)$ of Equation B-6. Similar reasoning based on a particle count analysis leads to $f(D_p)$ as in Equation B-7.

Equations B-8 and B-9 are statements of the fact that the area under such an altered differential fractional distribution plot must equal the total sample mass or total number of particles counted, respectively. If the group limits are different by finite amounts, these equations become difference equations as will be discussed below. In all screen analyses, these group limits are screen openings and are spaced at finite intervals.

Using the analytical distribution function based on number of particles, as given by Equations B-7 and B-9, the summation of all the particle diameters must be

$$\Sigma D_p = \sigma \bar{D}_p = \int_0^\infty D_p f(D_p)\, dD_p \qquad (B\text{-}10)$$

The summation of all the particle surface areas must be

$$\Sigma A_p = \sigma \bar{A}_p = \int_0^\infty \psi_A D_p{}^2 f(D_p)\, dD_p \qquad (B\text{-}11)$$

and the summation of all the particle volumes must be

$$\Sigma V_p = \sigma \bar{V}_p = \int_0^\infty \psi_V D_p{}^3 f(D_p)\, dD_p \qquad (B\text{-}12)$$

In these relations, $\bar{D}_p$ = the mean particle dimension

$\bar{A}_p$ = the mean particle area

$\bar{V}_p$ = the mean particle volume

ψ_A = a shape factor for area

ψ_V = a shape factor for volume

Equations B-10, B-11, and B-12 are identical to Equation 10.3 in form but have been written for the specific mean values of interest here. Equations B-8 and B-9 give the range of variable over which the integration is to be carried out. For example σ of Equation B-9 is equivalent to $(x_2 - x_1)$ as used in Equation 10.3a.

In order to integrate Equations B-11 and B-12, it may be necessary to assume that all the particles have the same shape. Then, ψ_A and ψ_V can be taken as constants. For example, if the particles are spheres, the particle volume $V_p = (\pi/6)D_p{}^3$ so that $\psi_V = \pi/6$, whereas again for spherical particles the particle area $A_p = \pi D_p{}^2$ so that $\psi_A = \pi$. In this example, D_p is the diameter of the spherical particles. In cases where the particle shape varies with particle size, it is sometimes possible to relate shape and size so that $\psi = f_\psi(D_p)$ can be inserted into Equations B-11 and B-12. If a mean particle diameter based, for instance, on surface area is desired,

$$\Sigma A_p = \sigma \bar{A}_p = \sigma \cdot \bar{\psi}_A \bar{D}_{pA}{}^2 = \int_0^\infty \psi_A D_p{}^2 f(D_p)\, dD_p \qquad (B\text{-}13)$$

where $\bar{D}_{pA}$ = the dimension of a particle having the mean particle area

$\bar{\psi}_A$ = a mean-area shape factor applicable over the entire size range of the sample

If the shape factor is a constant, Equation B-13 can be simplified to give

$$\bar{D}_{pA} = \sqrt{\frac{\int_0^\infty D_p^2 f(D_p)\, dD_p}{\sigma}} \qquad \text{(B-15)}$$

This relation and others giving mean particle dimensions based on other desired particle characteristics are given in Table B-2 and Equations B-14, B-15, B-16, and B-17, which are written as part of the table. Several other such mean dimensions can be developed and are sometimes found to be convenient (7).

Of the mean diameter equations shown in Table B-2, the length mean given by Equation B-14 gives a dimension that is the average for the entire sample. It is simply the summation of all the particle dimensions measured divided by the total number of measurements. Similarly, Equation B-16 for the volume mean dimension gives the dimension of that particle having a volume that is the average of the volumes of all the particles in the sample. It could be obtained by measuring the total solid volume of the sample, perhaps by a liquid displacement method, and dividing this volume by the total number of particles in the sample. As indicated in the third column of Table B-2, the volume mean would be the most-significant average particle dimension to use when investigating the proportion of the total mass of a spray that exists in various size ranges.

Perhaps the mean dimension of most wide application in chemical engineering is the Sauter mean given by Equation B-17. This dimension is that of the particle which has the average volume per unit surface among those particles sampled. It could be obtained by directly measuring the volume and area of the entire sample and taking the ratio of them. This procedure would, however, be difficult, and the Sauter mean is usually obtained by particle-size measurement. This mean has application whenever the surface area per unit volume of a solid phase is important. Examples could be found in adsorption work where the surface area available in a bed of fixed volume is important, in studies on the effectiveness of a solid catalyst, in determining the rate of solution of crystals in a solvent, and in calculations on packed-column distillation where the effective interface between gas and liquid phases depends on the packing surface per unit of column volume. In this book, the Sauter mean has been used in characterizing droplet-size distribution from spray nozzles. Other mean values could be used for this application, but the Sauter mean is the significant one if the power required for atomization is of interest since this power is theoretically used to generate new surface as well as to accelerate the fluid. It is also the significant mean if the drying time is of interest, for the drying time depends upon the surface area per unit volume of material as well as on other factors such as driving forces for transfer, liquid- and gas-phase resistance to transfer, and time of exposure.

Usually, the sample distribution is given in discrete segments rather than in differential terms. Then, summation arithmetic is preferable to integral calculus as a summation form. The distribution function becomes

$$\frac{\Delta n}{\Delta D_p} = \phi(D_{pi}) \qquad \text{(B-7a)}$$

where $\Delta n / \Delta D_p$ is the number of particles in the dimension class bounded by D_p and $D_p + \Delta D_p$. The functional notation $\phi(D_{pi})$ indicates that the distribution is taken over discrete intervals rather than over differential intervals as was done in obtaining $f(D_p)$. In the limit as ΔD_p approaches zero, $\Delta n / \Delta D_p = dn/dD_p$ and $f(D_p) = \phi(D_{pi})$. The total number of particles in the sample is then

$$\Sigma n = \sigma = \sum_{i=1}^k \phi(D_{pi})\, \Delta D_{pi} \qquad \text{(B-9a)}$$

and the total of all particle volumes will be

$$\Sigma V_p = \sigma \bar{V}_p' = \sum_{i=1}^k \psi_V D_{pi}^3 \phi(D_{pi})\, \Delta D_{pi} \qquad \text{(B-12a)}$$

which gives the dimension of a particle having a mean volume as

$$\bar{D}_{pV} = \sqrt[3]{\frac{\sum_{i=1}^k D_{pi}^3 \phi(D_{pi})\, \Delta D_{pi}}{\sigma'}} \qquad \text{(B-16a)}$$

This mean-dimension equation is based on a count of the individual particles distributed within finite segments. Others giving mean dimensions with different characteristics are also listed in Table B-2 as Equations B-14a, B-15a, and B-17a.

However, screen analyses are given in terms of mass distribution rather than numerical distribution. To alter these equations to fit this case, the relation between Equation B-6 and Equation B-7 must be considered. From Equations B-6 and B-7,

$$dm = f_m(D_p)\, dD_p = dn \cdot m_p = f(D_p)\, dD_p \rho \psi_V D_p^3 \qquad \text{(B-18)}$$

where m_p = mass of an individual particle = $\rho V_p = \rho \psi_V D_p^3$ so that

$$f(D_p) = \frac{f_m(D_p)}{\rho \psi_V D_p^3} \qquad \text{(B-19)}$$

or, if finite class sizes are used,

$$\phi(D_{pi}) = \frac{\phi_m(D_{pi})}{\rho \psi_V D_{pi}^3} \qquad \text{(B-19a)}$$

Table B-2. MEAN DIAMETERS BASED ON PARTICLE-COUNT ANALYSIS

Type	Equations Based on Continuous Variation	Equations Based on Discrete Distribution	Possible Uses
Length mean	$$\bar{D}_p = \frac{\int_0^\infty D_p f(D_p)\,dD_p}{\sigma} \quad \text{(B-14)}$$	$$\bar{D}_p = \frac{\sum_{i=1}^{k} D_{pi}\phi(D_{pi})\,\Delta D_{pi}}{\sigma} \quad \text{(B-14}a\text{)}$$	Comparisons of droplet evaporation
Surface mean	$$\bar{D}_{pA} = \sqrt{\frac{\int_0^\infty D_p^2 f(D_p)\,dD_p}{\sigma}} \quad \text{(B-15)}$$	$$\bar{D}_{pA} = \sqrt{\frac{\sum_{i=1}^{k} D_{pi}^2\phi(D_{pi})\,\Delta D_{pi}}{\sigma}} \quad \text{(B-15}a\text{)}$$	Adsorption, crushing, light diffusion
Volume mean	$$\bar{D}_{pV} = \sqrt[3]{\frac{\int_0^\infty D_p^3 f(D_p)\,dD_p}{\sigma}} \quad \text{(B-16)}$$	$$\bar{D}_{pV} = \sqrt[3]{\frac{\sum_{i=1}^{k} D_{pi}^3\phi(D_{pi})\,\Delta D_{pi}}{\sigma}} \quad \text{(B-16}a\text{)}$$	Distribution of mass in a spray
Volume surface mean or Sauter mean	$$\bar{D}_{pVA} = \frac{\int_0^\infty D_p^3 f(D_p)\,dD_p}{\int_0^\infty D_p^2 f(D_p)\,dD_p} \quad \text{(B-17)}$$	$$\bar{D}_{pVA} = \frac{\sum_{i=1}^{k} D_{pi}^3\phi(D_{pi})\,\Delta D_{pi}}{\sum_{i=1}^{k} D_{pi}^2\phi(D_{pi})\,\Delta D_{pi}} \quad \text{(B-17}a\text{)}$$	Efficiency studies, mass transfer, catalytic reactions

Note: All equations presented here are based on distribution function by number of individuals.

Applying this relation to Equation B-16a gives

$$\bar{D}_{pV} = \sqrt[3]{\frac{\sum_{i=1}^{k} \dfrac{D_{pi}^3\phi_m(D_{pi})\,\Delta D_{pi}}{\rho\psi_V D_{pi}^3}}{\sum_{i=1}^{k} \dfrac{\phi_m(D_{pi})\,\Delta D_{pi}}{\rho\psi_V D_{pi}^3}}}$$

If particle density and shape may be assumed to be independent of particle size,

$$\bar{D}_{pV} = \sqrt[3]{\frac{\sum_{i=1}^{k} \phi_m(D_{pi})\,\Delta D_{pi}}{\sum_{i=1}^{k} \dfrac{\phi_m(D_{pi})\,\Delta D_{pi}}{D_{pi}^3}}} \quad \text{(B-20}a\text{)}$$

Equation B-20a is exactly analogous to Equations B-16a of Table B-2. Here, the volume mean dimension is given on the basis of a typical mass-fraction screen analysis. Data on the mass fraction of the total sample lying between the size limits of D_p and $D_p + \Delta D_p$ can be used directly in Equation B.20a to obtain the volume mean diameter. This is perhaps more clearly shown if Equation B-20a is expanded.

$$D_{pV} = \sqrt[3]{\frac{x_{-14+20} + x_{-20+28} + x_{-28+25} + \cdots}{\dfrac{x_{-14+20}}{(\bar{D}_{p-14+20})^3} + \dfrac{x_{-20+28}}{(\bar{D}_{p-20+28})^3} + \dfrac{x_{-28+35}}{(\bar{D}_{p-28+35})^3} + \cdots}}$$

$$\text{(B-20}a\text{)}$$

In this expansion, the terms are given in relation to a screen analysis using 14-, 20-, 28-, 35-, etc., mesh screens. The term x_{-14+20} is the mass fraction of sample resting on the 20-mesh screen, and $\bar{D}_{p-14+20}$ is the average diameter for the $-14+20$-mesh cut.

Similar equations can be readily developed for length, surface, and volume-surface mean dimensions that are analogous to Equations B-14a, B-15a, and B-17a.

Illustration B.2. Determine the surface mean diameter of the crushed-ilmenite sample analyzed in Table B-1.

SOLUTION. From Equation B-15a as altered by the insertion of Equation B-19a,

$$\bar{D}_{pA} = \sqrt{\frac{\sum_{i=1}^{k} \dfrac{\phi_m(D_{pi})\,\Delta D_{pi}}{D_{pi}}}{\sum_{i=1}^{k} \dfrac{\phi_m(D_{pi})\,\Delta D_{pi}}{D_{pi}^3}}} \quad \text{(B-21}a\text{)}$$

Substituting the screen analysis information of Table B-1 gives

$$D_{pA} = \left[\frac{0.02/0.0555 + 0.05/0.0394 + 0.10/0.0280 + 0.18/0.0198 + \cdots}{0.02/\overline{0.0555}^3 + 0.05/\overline{0.0394}^3 + 0.10/\overline{0.0280}^3 + 0.18/\overline{0.0198}^3 + \cdots}\right]^{1/2}$$

$$D_{pA} = \left[\frac{0.360 + 1.27 + 3.57 + 9.09 + 17.85 + 25.25 + 36.58}{(0.012 + 0.082 + 0.455 + 2.315 + 9.10 + 25.78 + 217.5)10^4}\right]^{1/2}$$

$$D_{pA} = 0.00615$$

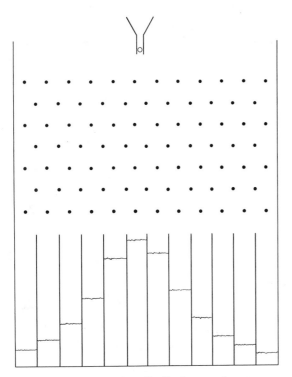

Figure B-8. Random-distribution experiment.

Comparing this result with the screen analysis shows the very heavy contribution of the fine material to the total surface area.

Analytical Distribution Expressions. The mean dimensions discussed above do not adequately describe the entire particulate phase. Identical mean diameters could be obtained for a bin of marbles $\frac{1}{2}$ in. in diameter and for a bin of random-sized marbles or even of random-sized crushed limestone. Even if the shapes are held constant, as in the case of the marbles, a measure of the distribution of particle sizes within the bin is necessary. Basically, such a representation exists in a screen analysis or particle-size count, but, as mentioned before, these forms of result are not readily incorporated into correlations involving particle size.

The expression of $f(D_p)$ in analytical terms has been attempted in various ways, though empirical equations have generally been more successful than the functions based on simplified concepts of the factors contributing to the size distribution (7). Nonetheless, one of the simplified expressions deserves mention because it is so generally applied to large samples. The normal-distribution function expresses the frequency of occurrence of a value as a function of its deviation from a most common value when the deviations result from purely random effects (1).

The most common demonstration of the sort of factors that result in a normal-, or Gaussian-distribution function is shown in Figure B-8. This device consists of a slanted board such as on a pin-ball machine in which pins are placed in a regular pattern above a series of evenly spaced slots. Marbles are dropped one-by-one onto the top row of pins at the center. The marbles bounce from pin to pin in a purely random way and enter the bottom slots. If a large number of marbles are dropped the slots will fill in the pattern shown, which can be related by the normal-probability function. Note that the effects producing this function are entirely random. Note also that adding or subtracting rows of pins should not change the general shape of the distribution pattern formed by the marbles in the slots, though it might decrease or increase the height of the central peak in relation to the heights of the marbles in the edge slots. In other words, it might increase or decrease the dispersion of the marbles.

The normal-probability function is usually expressed as

$$\frac{dn}{dD_p} = f(D_p) = \frac{1}{S_n\sqrt{2\pi}}\, e^{-(D_p - \bar{D}_p)^2/2S_n^2} \quad \text{(B-22)}$$

where the distribution of particle diameters is of concern, and where S_n = standard deviation, a measure of the dispersion of the values

$\qquad\qquad S_n^2$ = variance

Equation B-22 is the equation of a perfectly symmetrical bell-shaped curve. The larger the value of S_n the flatter is the bell, but the basic symmetry is unchanged.

The standard deviation is defined as the square root of the mean-squared deviation from the mean. It is sometimes called the root-mean-square deviation and is expressed mathematically as

$$S_n = \sqrt{\frac{\Sigma(D_p - \bar{D}_p)^2}{n}} \quad \text{(B-23)}$$

where $\quad \bar{D}_p$ = the mean dimension

$\qquad D_p$ = the dimension of one particle

$\qquad n$ = the total number of particles for which $(D_p - \bar{D}_p)$ is computed

In terms of the area under the fractional-distribution plot, which is the fraction of the total particles found between the chosen limits, S_n is the difference between $\bar{D}_p$ and the value of D_{ps} for which

$$\int_0^{D_{ps}} f(D_p)\, dD_p = 0.8413 \int_0^{\infty} f(D_p)\, dD_p \quad \text{(B-24)}$$

The qualitative understanding of S_n is aided by realizing that about two-thirds of the total number of measurements lie within one standard deviation of the mean value; about 95 per cent of the observations lie less than two standard deviations away from the mean; and about 99 per cent of the observations lie within three standard deviations of the mean.

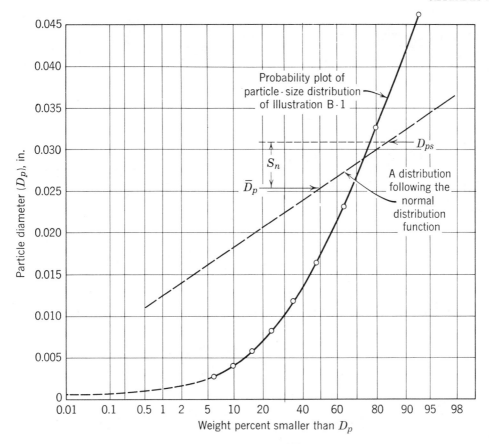

Figure B-9. Particle-size distribution on probability coordinates.

If the distribution of particles can be fit by Equation B-22, the function $f(D_p)$ is completely defined by the two parameters $\bar{D}_p$ and S_n. Even though this distribution is generally unsatisfactory for chemical-engineering application, particle-size-distribution relations are often given in these terms or are plotted on probability coordinates to show by the curvature of the resulting line how the distribution differs from normal.

Figure B-9 shows the distribution of particle size of the sample analyzed in Illustrations B-1 and B-2 plotted on normal-probability coordinates. The data on particle sizes below 200 mesh are not plotted. The curvature obtained results from higher concentrations of large particles than would be predicted by the normal-probability function. Also plotted on Figure B-9 is a distribution fitting the normal-distribution function. Values of $\bar{D}_p$, D_{ps}, and S_n are shown for this line.

SHAPE FACTORS

The shape of particles of solids may be as important as the particle-size distribution. In the development above, it was assumed that the shape did not vary with particle size, but no further concern was given to the shape. The definition of the shape factor will depend on whether particle surface, volume, or a linear particle dimension is important to the application being considered, as was indicated in Equations B-11 and B-12. A normal way of expressing the shape factor is to make it the ratio of the particle property to the property of a sphere having a diameter equal to the measured particle dimension. Thus a volume-based shape factor becomes: particle volume/volume of a sphere of same diameter or

$$\psi_V = \frac{\psi_V D_p{}^3}{\frac{\pi}{6} D_p{}^3} = \frac{\psi_V}{\frac{\pi}{6}} \tag{B-25}$$

The most commonly used of the shape factors is the sphericity (ψ), a surface-volume shape factor. ψ is defined as the ratio of the surface area of a sphere of volume equal to that of the particle, to the surface area of the particle. Thus,

$$\psi_V{}' = \frac{A_0}{A_p} = \frac{\pi D_0{}^2}{A_p} = \frac{\pi \left(\frac{6V_p}{\pi}\right)^{2/3}}{A_p} \tag{B-26}$$

where A_0, A_p = surface area of the equivalent sphere and of the particle respectively
D_0 = diameter of the equivalent sphere
V_p = particle volume

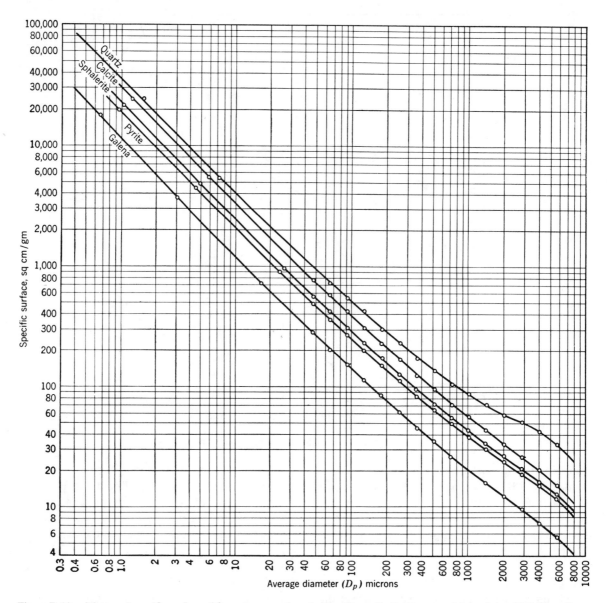

Figure B-10. Measured specific surface of five common minerals (5). (From G. G. Brown and Associates, *Unit Operations*, John Wiley and Sons, New York, 1950, by permission.)

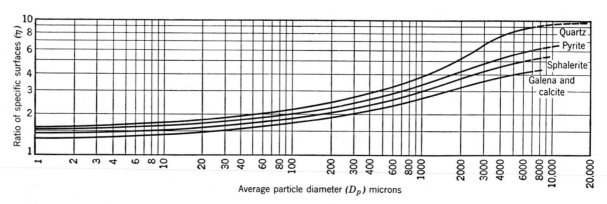

Figure B-11. Ratio of specific surface, η, as a function of average particle diameter for the minerals of Figure B-10 (5). (From G. G. Brown and Associates, *Unit Operations*, John Wiley and Sons, New York, 1950, by permission.)

In terms of the shape factors previously given,

$$\psi = \frac{\pi \left(\dfrac{6}{\pi}\right)^{2/3} \psi_V^{2/3} D_p^{\ 2}}{\psi_A D_p^{\ 2}} = \frac{\pi \left(\dfrac{6}{\pi}\right)^{2/3} \psi_V^{2/3}}{\psi_A} \qquad \text{(B-27)}$$

Still another shape factor that may be used is the *ratio* of specific surfaces (η). This ratio is that of the surface of the particle to the surface of a sphere of the same "diameter."

$$\eta = \frac{\text{Specific surface (sq cm/gm)}}{\dfrac{6}{\rho \bar{D}_p}} \qquad \text{(B-28)}$$

The "diameter" is usually taken as the mean screen opening. The specific surface is defined as the surface area per unit mass of material. Figure B-10 shows the specific surface of several common minerals as functions of particle size. The data were obtained by experimental determination of rates of solution (5). The same data are shown as a ratio of specific surfaces (η) in Figure B-11. The major advantage of representing particle shape as a ratio of specific surfaces is that materials within rather broad class ranges, as for example the crushed ores of Figure B-11, have values of η that fall within a relatively narrow range. Thus, the specific surface of a material for which there are no data may be roughly estimated from the ratio of specific surfaces of a similar material. For a solid phase consisting of a range of particle sizes, the value of η must be obtained from a summation process. The ratio of specific surfaces is basically a function of particle size, so that the total surface can only be obtained by summing the surfaces for each small size using the correct η for each fraction.

$$\text{Total surface} = \frac{6\eta_1 m_1}{\rho (\bar{D}_p)_1} + \frac{6\eta_2 m_2}{\rho (\bar{D}_p)_2} + \cdots$$
$$+ \frac{6\eta_i m_i}{\rho (\bar{D}_p)_i} = \frac{6}{\rho} \sum_{i=1}^{k} \frac{\eta_i m_i}{(\bar{D}_p)_i} \qquad \text{(B-29)}$$

The average specific surface will be the total surface divided by the total mass.

$$\text{Average specific surface} = \frac{\dfrac{6}{\rho} \displaystyle\sum_{i=1}^{k} \dfrac{\eta_i m_i}{(\bar{D}_p)_i}}{\Sigma m_i} = \frac{6}{\rho} \sum_{i=1}^{k} \frac{\eta_i x_i}{(\bar{D}_p)_i}$$
$$\text{(B-30)}$$

BED POROSITY

The particle size and shape, which have been discussed above, are basically properties of single particles, though this is not true of particle-size distribution. Usually, the particulate-solid phase must be dealt with rather than the particles that exist in it. This phase may exist as a stationary bed, a fluidized bed, or a fog. In any case, one of its most important characteristics is

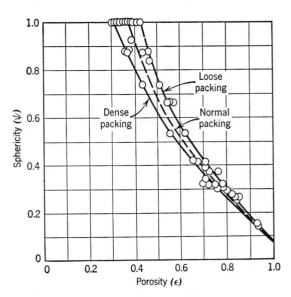

Figure B-12. Sphericity as a function of porosity for random-packed beds of uniformly sized particles (2). (From G. G. Brown and Associates, *Unit Operations*, John Wiley and Sons, New York, 1950, by permission.)

its fraction void volume, or porosity. The porosity influences the pressure drop for flow through the phase, the electrical resistivity of the phase, its effective thermal conductivity, the reactive surface area, and in fact any property of the bulk phase.

The importance of porosity has long been recognized, and a great deal of experimental work has been done to try to relate porosity to the properties of the individual particles. These works have shown that porosity of a static bed depends upon the particle size and size distribution, the particle shape and surface roughness, the method of packing, and the size of container relative to the particle diameter. Of the variables, the method of packing is the only one that can be eliminated from consideration. Vessels are usually packed by dumping the particles into the empty vessel, though it may be done by dumping them into the vessel filled with water and then emptying the water. Though the water-fill method initially gives a more porous packing, vibration of the vessel and the effect of liquid or gas flow through it ultimately compacts the bed. As a result, filling method has little effect on ultimate bed porosity, except for stacked shapes which form a rigid bed (8).

Particle shape is a much more important variable in porosity determination than is surface roughness, though both of the variables act in the same way. The lower the particle sphericity the more open is the bed. Particles settle across each other and pack with pointed ends against each other, preventing a close packing. The dependence of porosity upon sphericity for beds of uniform-sized granular particles is shown in Figure B-12. In this figure, a range of values is given depending

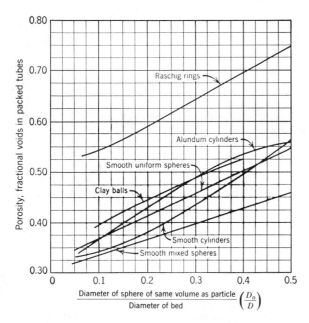

Figure B-13. Porosity as a function of the ratio of particle diameter to bed diameter (6). (Courtesy Am. Inst. Chem. Engrs.)

upon packing method. In practical design, the dense packing must be assumed for pressure-drop prediction and for most other applications, though the assumption of loose packing would be safe when considering the effective thermal conductivity.

The presence of fine and coarse particles results in a bed of lower porosity than would be obtained with uniform particles, for the fine particles slide down between the larger ones filling the interstices. This has been repeatedly demonstrated, but none of the data has been collected in such a way that it can be correlated with the quantitative measurements of particle-size dispersion mentioned above.

The particle size and vessel size are interrelated in their influence upon porosity. The presence of the container wall interrupts the pattern of particle-to-particle contacts and hence makes for a larger fraction voids at the wall. This has prompted some engineers to consider porous beds as consisting of an annular ring just inside the wall having a high porosity, and the central core of the bed with a lower porosity. In such calculations as that of pressure drop for flow through the bed, the flow is apportioned between those two regions in such a way that the pressure drop through the two sections is balanced. Another approach, which is simpler but which corresponds less closely to the physical picture, is to consider porosity as a function of the diameter ratio, D_p/D. The result of this type of analysis is shown in Figure B-13 (6). Note that the curves for all the shapes tested tend to come together at low D_p/D ratio except for the Raschig rings. Note also that the particulate

beds tested in obtaining Figure B-13 were made up of particles very large in relation to the total bed diameter. In common practice, D_p/D is so close to zero that the wall effect is negligible in its influence on total porosity. Still, the large porosity at the bed wall is important even with very small D_p/D values. For example, in commercial packed distillation columns, the liquid collects in the region of the wall, leaving the bulk of the packing relatively low in reflux. To prevent this, elaborate systems for redistributing the liquid every few feet down the column must be used.

In fluidized beds, fogs, and suspensions, porosity depends upon the motion of the liquid phase as well as the properties of the solid phase. For this reason it is discussed in the chapters dealing with fluid–solid operations.

REFERENCES

1. Bennett, C. A., and N. L. Franklin, *Statistical Analysis in Chemistry and the Chemical Industry*, John Wiley and Sons, New York, 1954.
2. Brown, G. G., and Associates, *Unit Operations*, John Wiley and Sons, New York, 1950.
3. Dalla Valle, J. M., *Micromeritics*, 2nd ed., Pitman Publishing Corp., New York, 1948.
4. Gaudin, A. M., *Principles of Mineral Dressing*, McGraw-Hill Book Co., New York, 1939.
5. Gross, J., "Crushing and Grinding," *U.S. Bur. Mines Bull.* 402 (1938).
6. Leva, M., *Chem. Eng. Progr.*, **43**, 549 (1947).
7. Marshall, W. R., Jr., "Atomization and Spray Drying," *Chem. Eng. Progr. Monograph Ser.*, **50** No. 2.
8. Martin, J. J., Thesis, *Carnegie Institute of Technology* (1942).

PROBLEMS

B-1. Figures 18.43 and 18.44 are photomicrographs of spray dried products. Each has been enlarged 200 times. Using one of these figures:

(*a*) Measure the size of the particles shown, if possible using a vernier caliper, and present this information in the form of a histogram.

(*b*) Present this information as a cumulative-distribution plot.

(*c*) Determine the length-mean dimension of the particles.

(*d*) Determine the Sauter-mean dimension of the particles.

B-2. A spray-dried detergent has the following screen analysis:

Tyler Screen Mesh	Weight Fraction
−10 + 14	0.02
−14 + 20	0.17
−20 + 28	0.34
−28 + 35	0.22
−35 + 48	0.14
−48 + 65	0.06
−65 + 100	0.03
−100	0.02

(a) Present this information as a histogram, a fractional-distribution plot, and as a cumulative-distribution plot.

(b) Determine the surface-mean dimension and the volume-mean dimension of this material.

B-3. Crushed galena from a ball mill has the following screen analysis:

Tyler Screen Mesh	Weight Fraction
−28 + 35	0.150
−35 + 48	0.200
−48 + 65	0.171
−65 + 100	0.134
−100 + 150	0.104
−150 + 200	0.080
−200	0.161

(a) Present this information in the form of a histogram, as a plot of cumulative weight fraction greater than given size versus size, as a logarithmic fractional-distribution plot, and as a plot on arithmetic probability coordinates.

(b) Complete the screen analysis by determining the size distribution in the −200-mesh cut.

(c) Determine the average particle diameter based on:
1. Length.
2. Mass.
3. Surface per unit volume.

(d) Determine the total surface area of a 100-gm sample of the crushed ore.

(e) Determine the sphericity of one of the cuts given above.

(f) If the −65 + 100 mesh cut were packed in a beaker what would be the net weight in a 300 cu cm filled beaker?

B-4. What sort of mean particle diameter would you consider most significant for describing the solid phase in the following applications:

(a) An explosive powder in which the rate of explosion is the phenomenon to be studied.

(b) Pebbles in a pebble heater. In this device, hot pebbles travel slowly countercurrent to a rising gas stream thus heating the gas. The heat-transfer coefficient per unit heater volume is to be studied.

(c) Sands in an oil reservoir. Here, the characteristic of interest will be the rate of flow of oil toward the well tip at fixed pressure drop.

B-5. Determine the sphericity of:

(a) A spherical bead $\frac{1}{4}$ in. in diameter with a $\frac{1}{16}$ in. hole drilled through it.

(b) A Lessing ring which is 1 in. O.D. by $\frac{3}{4}$ in. I.D. by 1 in. high with a web as thick as the ring itself which crosses the ring diameter and makes a quarter turn from one end of the ring to the other.

B-6. If college-entrance test scores have a mean of 500 and a standard deviation of 100, what fraction of the students taking any one of the tests have a score above 660?

B-7. Show that the ratio of specific surfaces (η) and the sphericity (ψ) may be related by the equation

$$\psi = \frac{\bar{D}_p}{D_0} \times \frac{1}{\eta}$$

Equipment Design Data

Appendix C-1. RELATIVE ROUGHNESS AS A FUNCTION OF DIAMETER FOR PIPE OF VARIOUS MATERIALS. [Moody, L. F., *Trans. ASME*, **66,** 671–84 (1944).]

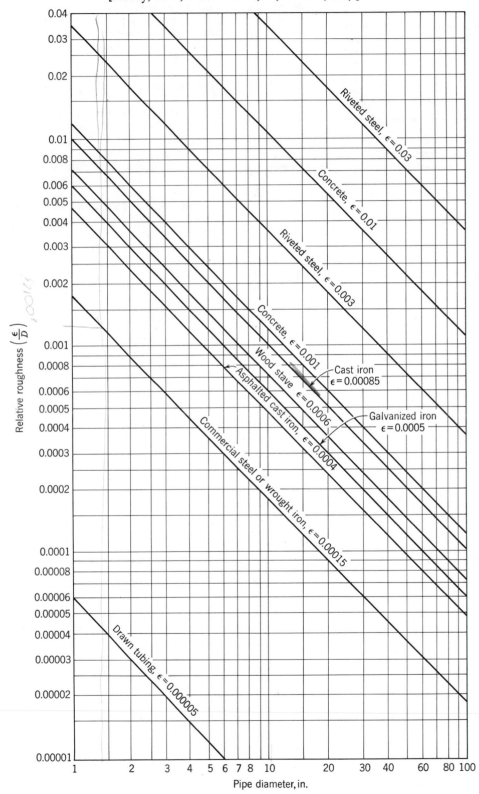

Appendix C-2a. REPRESENTATIVE EQUIVALENT LENGTH IN PIPE DIAMETERS (L/D) OF VARIOUS VALVES AND FITTINGS (CRAN

Description	Equivalent Le in Pipe Diame (L/D)
Globe valves	
Conventional	
With no obstruction in flat, bevel, or plug type seat—Fully open	340
With wing or pin guided disk—Fully open	450
Y-pattern	
(No obstruction in flat, bevel, or plug type seat)	
With stem 60 degrees from run of pipe line—Fully open	175
With stem 45 degrees from run of pipe line—Fully open	145
Angle valves	
Conventional	
With no obstruction in flat, bevel, or plug type seat—Fully open	145
With wing or pin guided disk—Fully open	200
Gate valves	
Conventional wedge disk, double disk, or plug disk	
Fully open	13
Three-quarters open	35
One-half open	160
One-quarter open	900
Pulp stock	
Fully open	17
Three-quarters open	50
One-half open	260
One-quarter open	1200
Conduit pipe line—Fully open	3*
Check valves	
Conventional swing—0.5†—Fully open	135
Clearway swing—0.5†—Fully open	50
Globe lift or stop—2.0†—Fully open	Same as globe
Angle lift or stop—2.0†—Fully open	Same as angle
In-line ball—2.5 vertical and 0.25 horizontal†—Fully open	150
Foot valves with strainer	
With poppet lift-type disk—0.3†—Fully open	420
With leather-hinged disk—0.4†—Fully open	75
Butterfly valves (6-inch and larger)—Fully open	20
Cocks	
Straight-through	
Rectangular plug port area equal to 100% of pipe area—Fully open	18
Three-way	
Rectangular plug port area equal to 80% of pipe area (fully open)	
Flow straight through	44
Flow through branch	140

* Exact equivalent length is equal to the length between flange faces or welding ends.
† Minimum calculated pressure drop (psi) across valve to provide sufficient flow to lift disk fully.

Appendix C-2a.—*Continued.*

Description	Equivalent Length in Pipe Diameters (L/D)
Fittings	
90-degree standard elbow	30
45-degree standard elbow	16
90-degree long radius elbow	20
90-degree street elbow	50
45-degree street elbow	26
Square corner elbow	57
Standard tee	
With flow through run	20
With flow through branch	60
Close pattern return bend	50

Appendix C-2b. RESISTANCE DUE TO SUDDEN ENLARGEMENTS AND CONTRACTIONS. (CRANE CO.)

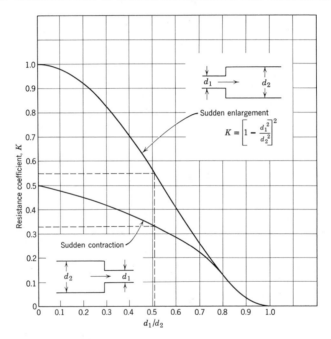

$$K = \left[1 - \frac{d_1^2}{d_2^2}\right]^2$$

Appendix C-2c. RESISTANCE DUE TO PIPE ENTRANCE AND EXIT. (CRANE CO.)

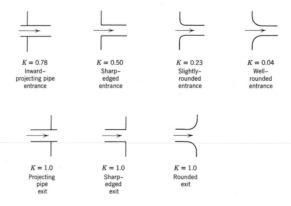

$K = 0.78$
Inward–projecting pipe entrance

$K = 0.50$
Sharp–edged entrance

$K = 0.23$
Slightly–rounded entrance

$K = 0.04$
Well–rounded entrance

$K = 1.0$
Projecting pipe exit

$K = 1.0$
Sharp–edged exit

$K = 1.0$
Rounded exit

Appendix C-2d. EQUIVALENT LENGTHS L AND L/D AND RESISTANCE COEFFICIENT K. (CRANE CO.)

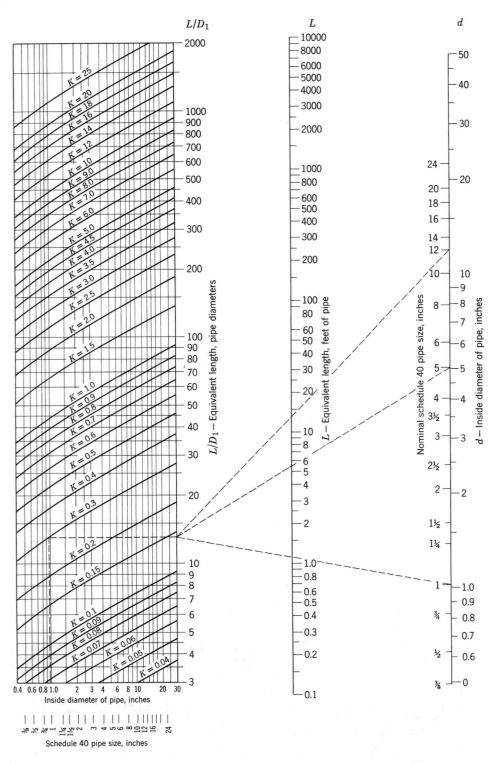

Problem: Find the equivalent length in pipe diameters and feet of Schedule 40 pipe, and the resistance factor K for 1-, 5-, and 12-inch fully opened gate valves.

Solution				
Valve Size	1″	5″	12″	*Refer to*
Equivalent length, pipe diameters	13	13	13	App. C-2a
Equivalent length, feet of Sch. 40 pipe	1.1	5.5	13	Dotted lines
Resist. factor K, based on Sch. 40 pipe	0.30	0.20	0.17	on chart

Appendix C-3. FRICTION FACTOR AS A FUNCTION OF REYNOLDS NUMBER WITH RELATIVE ROUGHNESS AS A PARAMETER. [Moody, L. F., *Trans. ASME*, **66**, 671–84 (1944).]

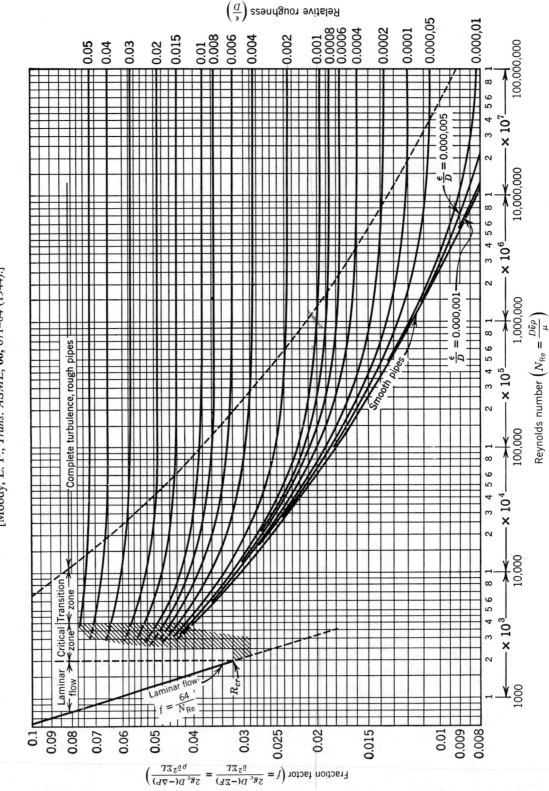

Appendix C-4. FRICTION FACTOR AS A FUNCTION OF KÁRMÁN NUMBER.
(From G. G. Brown and Associates, *Unit Operations*, John Wiley and Sons, New York, 1950.)

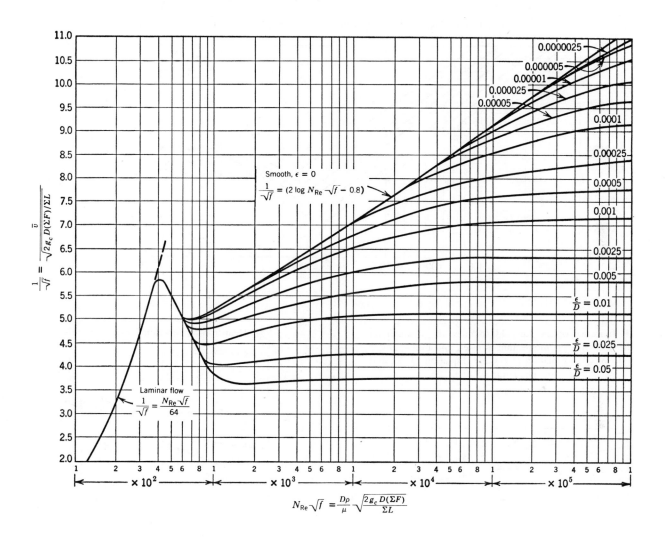

Appendix C-5. CORRELATIONS FOR HEAT AND MASS TRANSFER TO SMOOTH TUBES
(From G. G. Brown and Associates, *Unit Operations*, John Wiley and Sons, 1950.)

(*a*) Influence of Reynolds number (low range) on the heat-transfer function, j_q.

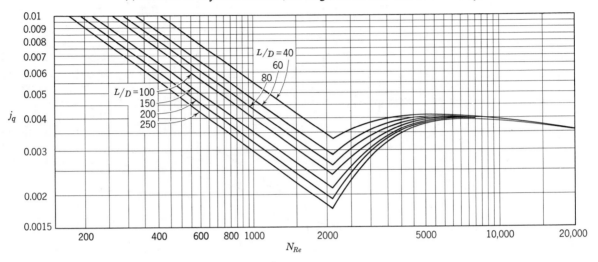

(*b*) Influence of the Reynolds number (high range) on the heat-transfer, momentum-transfer, and mass-transfer functions; j_q (Equation 13.97), $F/8$ (Equation 13.98), and j_N' (Equation 13.101).

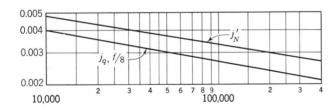

(*c*) Heat-transfer function for liquid metals.

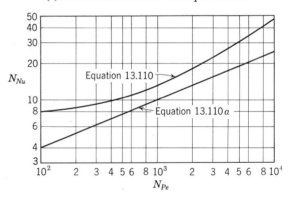

Appendix C-6. DIMENSIONS OF STANDARD STEEL PIPE (ASA STANDARDS B36.10–1939)

Nominal Pipe Size, in.	Outside Diameter, in.	Schedule No.	Wall Thickness, in.	Inside Diameter, in.	Cross-Sectional area of Metal, sq in.	Inside Sectional Area, sq ft	Circumference, ft, or Surface, sq ft/ft of length	
							Outside	Inside
$\frac{1}{8}$	0.405	40	0.068	0.269	0.072	0.00040	0.106	0.0705
		80	0.095	0.215	0.093	0.00025	0.106	0.0563
$\frac{1}{4}$	0.540	40	0.088	0.364	0.125	0.00072	0.141	0.0954
		80	0.119	0.302	0.157	0.00050	0.141	0.0792
$\frac{3}{8}$	0.675	40	0.091	0.493	0.167	0.00133	0.177	0.1293
		80	0.126	0.423	0.217	0.00098	0.177	0.1110
$\frac{1}{2}$	0.840	40	0.109	0.622	0.250	0.00211	0.220	0.1630
		80	0.147	0.546	0.320	0.00163	0.220	0.1430
$\frac{3}{4}$	1.050	40	0.113	0.824	0.333	0.00371	0.275	0.2158
		80	0.154	0.742	0.433	0.00300	0.275	0.1942
1	1.315	40	0.133	1.049	0.494	0.00600	0.344	0.2745
		80	0.179	0.957	0.639	0.00499	0.344	0.2505
$1\frac{1}{4}$	1.660	40	0.140	1.380	0.669	0.01040	0.435	0.362
		80	0.191	1.278	0.881	0.00891	0.435	0.335
$1\frac{1}{2}$	1.900	40	0.145	1.610	0.799	0.01414	0.498	0.422
		80	0.200	1.500	1.068	0.01225	0.498	0.393
2	2.375	40	0.154	2.067	1.075	0.02330	0.622	0.542
		80	0.218	1.939	1.477	0.02050	0.622	0.508
$2\frac{1}{2}$	2.875	40	0.203	2.469	1.704	0.03322	0.753	0.647
		80	0.276	2.323	2.254	0.02942	0.753	0.609
3	3.500	40	0.216	3.068	2.228	0.05130	0.917	0.804
		80	0.300	2.900	3.016	0.04587	0.917	0.760
$3\frac{1}{2}$	4.000	40	0.226	3.548	2.680	0.06870	1.047	0.930
		80	0.318	3.364	3.678	0.06170	1.047	0.882
4	4.500	40	0.237	4.026	3.173	0.08840	1.178	1.055
		80	0.337	3.826	4.407	0.07986	1.178	1.002
5	5.563	40	0.258	5.047	4.304	0.1390	1.456	1.322
		80	0.375	4.813	6.112	0.1263	1.456	1.263
6	6.625	40	0.280	6.065	5.584	0.2006	1.734	1.590
		80	0.432	5.761	8.405	0.1810	1.734	1.510
8	8.625	40	0.322	7.981	8.396	0.3474	2.258	2.090
		80	0.500	7.625	12.76	0.3171	2.258	2.000
10	10.75	40	0.365	10.020	11.90	0.5475	2.814	2.620
		80	0.593	9.564	18.92	0.4989	2.814	2.503
12	12.75	40	0.406	11.938	15.77	0.7773	3.338	3.13
		80	0.687	11.376	26.03	0.7058	3.338	2.98

Appendix C-7. DIMENSIONS OF HEAT-EXCHANGER TUBES*

Outside Diameter, in.	Wall Thickness		Inside Diameter, in.	Inside Cross-Sectional Area, sq ft	Circumference, ft, or Surface, sq ft/ft of length	
	BWG	in.			Outside	Inside
$\frac{1}{4}$	14	0.083	0.084	0.000039	0.0654	0.0219
	16	0.065	0.120	0.000079	0.0654	0.0314
	18	0.049	0.152	0.000126	0.0654	0.0397
$\frac{1}{2}$	14	0.083	0.334	0.000608	0.1309	0.0874
	16	0.065	0.370	0.000747	0.1309	0.0969
	18	0.049	0.402	0.000882	0.1309	0.1052
$\frac{3}{4}$	14	0.083	0.584	0.00186	0.1963	0.1529
	16	0.065	0.620	0.00210	0.1963	0.1623
	18	0.049	0.652	0.00232	0.1963	0.1707
1	14	0.083	0.834	0.00379	0.2618	0.2183
	16	0.065	0.870	0.00413	0.2618	0.2277
	18	0.049	0.902	0.00444	0.2618	0.2361
$1\frac{1}{4}$	14	0.083	1.084	0.00641	0.3271	0.2839
	16	0.065	1.120	0.00684	0.3271	0.2932
	18	0.049	1.152	0.00724	0.3271	0.3015
$1\frac{1}{2}$	14	0.083	1.334	0.00971	0.3925	0.3492
	16	0.065	1.370	0.0102	0.3925	0.3587
	18	0.049	1.402	0.0107	0.3925	0.3670
2	14	0.083	1.834	0.0183	0.5233	0.4801
	16	0.065	1.870	0.0191	0.5233	0.4896
	18	0.049	1.902	0.0197	0.5233	0.4979

* A more complete listing may be found in the *Chemical Engineers' Handbook*.

Appendix C-8.   TYLER STANDARD SCREEN SIZES

Interval = $\sqrt[4]{2}$

Standard Interval = $\sqrt{2}$, Aperture, in.	Aperture, in.	Aperture, mm	Mesh Number	Wire Diameter, in.
1.050	1.050	26.67	...	0.148
	0.883	22.43	...	0.135
0.742	0.742	18.85	...	0.135
	0.624	15.85	...	0.120
0.525	0.525	13.33	...	0.105
	0.441	11.20		0.105
0.371	0.371	9.423	...	0.092
	0.312	7.925	$2\frac{1}{2}$	0.088
0.263	0.263	6.680	3	0.070
	0.221	5.613	$3\frac{1}{2}$	0.065
0.185	0.185	4.699	4	0.065
	0.156	3.962	5	0.044
0.131	0.131	3.327	6	0.036
	0.110	2.794	7	0.0326
0.093	0.093	2.362	8	0.032
	0.078	1.981	9	0.033
0.065	0.065	1.651	10	0.035
	0.055	1.397	12	0.028
0.046	0.046	1.168	14	0.025
	0.0390	0.991	16	0.0235
0.0328	0.0328	0.833	20	0.0172
	0.0276	0.701	24	0.0141
0.0232	0.0232	0.589	28	0.0125
	0.0195	0.495	32	0.0118
0.0164	0.0164	0.417	35	0.0122
	0.0138	0.351	42	0.0100
0.0116	0.0116	0.295	48	0.0092
	0.0097	0.248	60	0.0070
0.0082	0.0082	0.208	65	0.0072
	0.0069	0.175	80	0.0056
0.0058	0.0058	0.147	100	0.0042
	0.0049	0.124	115	0.0038
0.0041	0.0041	0.104	150	0.0026
	0.0035	0.088	170	0.0024
0.0029	0.0029	0.074	200	0.0021
	0.0024	0.061	230	0.0016
0.0021	0.0021	0.053	270	0.0016
	0.0017	0.043	325	0.0014
0.0015	0.0015	0.038	400	0.0010

appendix D

Physical Data

The data given here are of use in solving the problems included in this book. More complete tables of physical data may be found in such sources as the *Chemical Engineers' Handbook* (1) and the *International Critical Tables* (2).

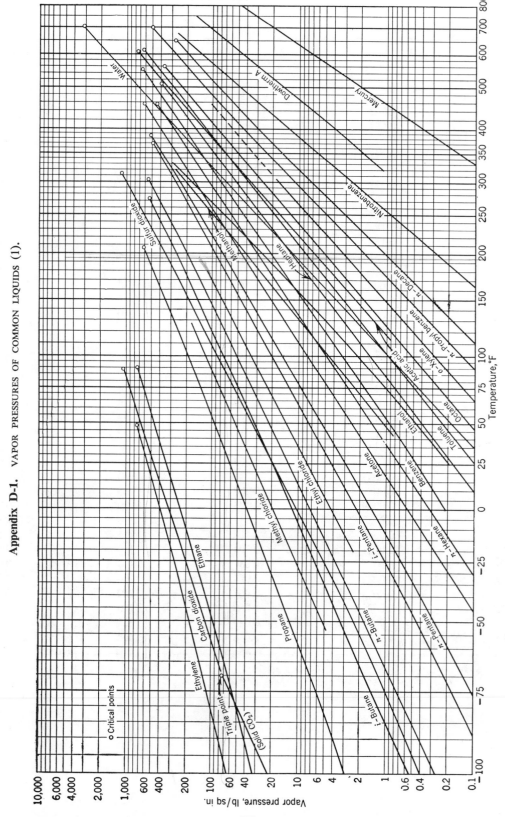

Appendix D-1. VAPOR PRESSURES OF COMMON LIQUIDS (1).

Appendix D-2. VAPOR–LIQUID EQUILIBRIUM CONSTANTS FOR HYDROCARBONS (1).

(*a*) At 44.7 lb/sq in.

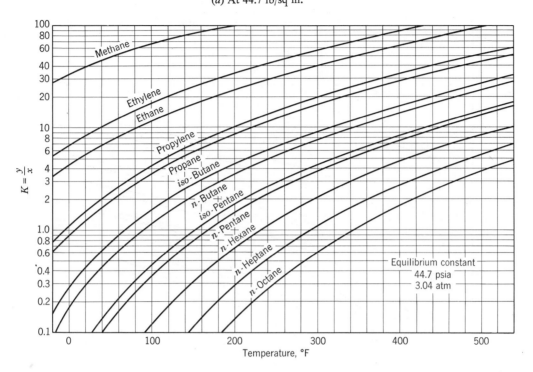

(*b*) At 215 lb/sq in.

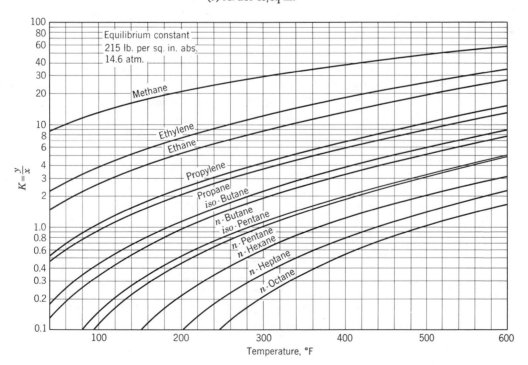

Appendix D-2.—*Continued.*

(*c*) At 465 lb/sq in.

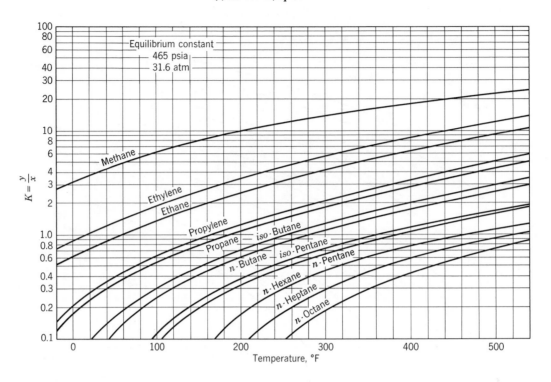

Appendix D-3. HENRY'S LAW CONSTANTS FOR VARIOUS GASES IN WATER (2)

$$p_a = H_a x_a$$

where p_a = partial pressure of the solute a in the gas phase, atm
x_a = mole fraction of solute a in the liquid phase, mole fraction
H_a = Henry's law constant, atm/mole fraction

$$H_a \times 10^{-4}, \text{ atm/mole fraction}$$

T, °C	Air	CO_2	CO	C_2H_6	H_2	H_2S	CH_4	NO	N_2	O_2
0	4.32	0.0728	3.52	1.26	5.79	0.0268	2.24	1.69	5.29	2.55
10	5.49	0.104	4.42	1.89	6.36	0.0367	2.97	2.18	6.68	3.27
20	6.64	0.142	5.36	2.63	6.83	0.0483	3.76	2.64	8.04	4.01
30	7.71	0.186	6.20	3.42	7.29	0.0609	4.49	3.10	9.24	4.75
40	8.70	0.233	6.96	4.23	7.51	0.0745	5.20	3.52	10.4	5.35
50	9.46	0.283	7.61	5.00	7.65	0.0884	5.77	3.90	11.3	5.88
60	10.1	0.341	8.21	5.65	7.65	0.103	6.26	4.18	12.0	6.29
70	10.5		8.45	6.23	7.61	0.119	6.66	4.38	12.5	6.63
80	10.7		8.45	6.61	7.55	0.135	6.82	4.48	12.6	6.87
90	10.8		8.46	6.87	7.51	0.144	6.92	4.52	12.6	6.99
100	10.7		8.46	6.92	7.45	0.148	7.01	4.54	12.6	7.01

Appendix D-4. SOLUBILITY DATA FOR GASES THAT DO NOT FOLLOW HENRY'S LAW IN WATER

[Sherwood, T. K., *Ind. Eng. Chem.*, **17**, 745 (1925).]

(a) Ammonia
Partial pressure NH$_3$, mm Hg

Mass NH$_3$ per 100 Masses H$_2$O	0°C	10°C	20°C	30°C	40°C	50°C	60°C
100	947						
90	785						
80	636	987					
70	500	780					
60	380	600	945				
50	275	439	686				
40	190	301	470	719			
30	119	190	298	454	692		
25	89.5	144	227	352	534	825	
20	64	103.5	166	260	395	596	834
15	42.7	70.1	114	179	273	405	583
10	25.1	41.8	69.6	110	167	247	361
7.5	17.7	29.9	50.0	79.7	120	179	261
5	11.2	19.1	31.7	51.0	76.5	115	165
4		16.1	24.9	40.1	60.8	91.1	129.2
3		11.3	18.2	29.6	45.0	67.1	94.3
2			12.0	19.3	30.0	44.5	61.0
1					15.4	22.2	30.2

(b) Sulfur Dioxide
Partial pressure of SO$_2$, mm Hg

Mass SO$_2$ per 100 Masses H$_2$O	0°C	7°C	10°C	15°C	20°C	30°C	40°C	50°C
20	646	657						
15	474	637	726					
10	308	417	474	567	698			
7.5	228	307	349	419	517	688		
5.0	148	198	226	270	336	452	665	
2.5	69	92	105	127	161	216	322	458
1.5	38	51	59	71	92	125	186	266
1.0	23.3	31	37	44	59	79	121	172
0.7	15.2	20.6	23.6	28.0	39.0	52	87	116
0.5	9.9	13.5	15.6	19.3	26.0	36	57	82
0.3	5.1	6.9	7.9	10.0	14.1	19.7		
0.1	1.2	1.5	1.75	2.2	3.2	4.7	7.5	12.0
0.05	0.6	0.7	0.75	0.8	1.2	1.7	2.8	4.7
0.02	0.25	0.3	0.3	0.3	0.5	0.6	0.8	1.3

Appendix D-5. ELASTIC-SPHERE EQUIVALENT DIAMETER

[E. H. Kennard, *Kinetic Theory of Gases*, McGraw-Hill Book Co., New York, 1938, p.149.]

Gas	σ, 10^{-8} cm	Gas	σ, 10^{-8} cm
H$_2$	2.74	C$_2$H$_6$	5.30
He	2.18	O$_2$	3.61
CH$_4$	4.14	HCl	4.46
NH$_3$	4.43	A	3.64
H$_2$O	4.60	CO$_2$	4.59
Ne	2.59	Kr	4.16
N$_2$	3.75	Xe	4.85
C$_2$H$_4$	4.95		

Appendix D-6. CONSTANTS FOR THE LENNARD-JONES (6-12) POTENTIAL*

(Hirschfelder, Curtiss, and Bird, *Molecular Theory of Gases and Liquids*, John Wiley and Sons, New York, 1954).

D-6a Force Constants Evaluated from Viscosity Data

Gas	ϵ/k, °K	σ, 10^{-8} cm	Gas	ϵ/k, °K	σ, 10^{-8} cm
Air	97	3.617	$CHCl_3$	327	5.430
Ar	124	3.418	CO	110	3.590
Br_2	520	4.268	CO_2	190	3.996
CCl_4	327	5.881	CS_2	488	4.438
CH_4	137	3.882	D_2	39.3	2.948
C_2H_2	185	4.221	F_2	112	3.653
C_2H_4	205	4.232	H_2	38	2.915
C_2H_6	230	4.418	HCl	360	3.305
C_3H_8	254	5.061	HI	324	4.123
$n\text{-}C_4H_{10}$	410	4.997	He	10.22	2.576
$i\text{-}C_4H_{10}$	313	5.341	Hg	851	2.898
$n\text{-}C_5H_{12}$	345	5.769	I_2	550	4.982
$n\text{-}C_6H_{14}$	413	5.909	Kr	190	3.61
$n\text{-}C_8H_{18}$	320	7.451	N_2	79.8	3.749
$n\text{-}C_9H_{20}$	240	8.448	NO	91.0	3.599
Cyclohexane	324	6.093	N_2O	237	3.816
C_6H_6	440	5.270	Ne	27.5	2.858
CH_3OH	507	3.585	O_2	88.0	3.541
C_2H_5OH	391	4.455	SO_2	252	4.290
CH_3Cl	855	3.375	$SnCl_4$	1550	4.540
CH_2Cl_2	406	4.759			

* Estimation of the Lennard-Jones constants for gases not given in this table may be made from the following relationships (2):

$$\epsilon/k = 0.77 T_c$$

and $\quad \sigma = 0.841(V_c)^{1/3} = 2.44\left(\dfrac{T_c}{P_c}\right)^{1/3}$

or $\quad \epsilon/k = 1.15 T_b$

and $\quad \sigma = 1.17(V_b)^{1/3}$

where T_c = critical temperature, °K
V_c = molar volume at critical temperature, cu cm/gm mole
P_c = critical pressure, atm
T_b = normal boiling point, °K
V_b = molar volume at the boiling point (Appendix D-7)

D-6b. COLLISION INTEGRALS, Ω_1, Ω_2.

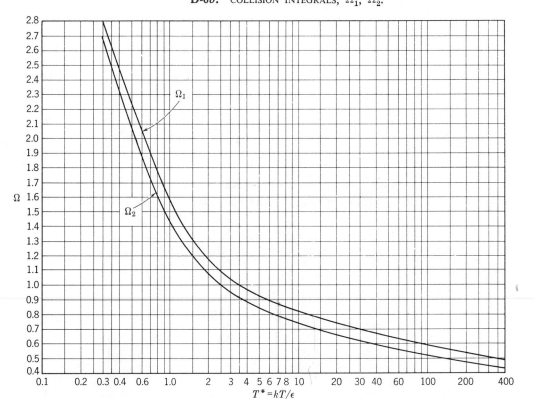

Appendix D-7. ATOMIC VOLUMES FOR USE IN CALCULATING THE MOLAR VOLUME AT THE NORMAL BOILING POINT

(LeBas, *The Molecular Volumes of Liquid Chemical Compounds*, Longmans, London, 1915.)

Element	Atomic Volume, cu cm/gm atom
Air	29.9
Antimony	34.2
Arsenic	30.5
Bismuth	48.0
Bromine	27.0
Carbon	14.8
Chlorine, terminal as R—Cl	21.6
medial as R—CHCl—R	24.6
Chromium	27.4
Fluorine	8.7
Germanium	34.5
Hydrogen, in compounds	3.7
as hydrogen molecule	7.15
Iodine	37.0
Lead	46.5–50.1
Mercury	19.0
Nitrogen	15.6
in primary amines	10.5
in secondary amines	12.0
Oxygen, doubly bound as —C=O	7.4
Coupled to two other elements:	
In aldehydes and ketones	7.4
In methyl ethers	9.9
In methyl esters	9.1
In higher ethers and esters	11.0
In acids	12.0
In union with S, P, N	8.3
Phosphorus	27.0
Silicon	32.0
Sulfur	25.6
Tin	42.3
Titanium	35.7
Vanadium	32.0
Water	18.8
Zinc	20.4
For 3-membered ring, as in ethylene oxide	−6
For 4-membered ring, as in cyclobutane	−8.5
For 5-membered ring, as in furan	−11.5
For six-membered ring, as in benzene, cyclohexane	−15
For naphthalene ring	−30
For anthracene ring	−47.5

Appendix D-8. CORRELATION FOR LIQUID MASS DIFFUSIVITIES
[Wilke, C. R., *Chem. Eng. Progr.* **45,** 218 (1949).]

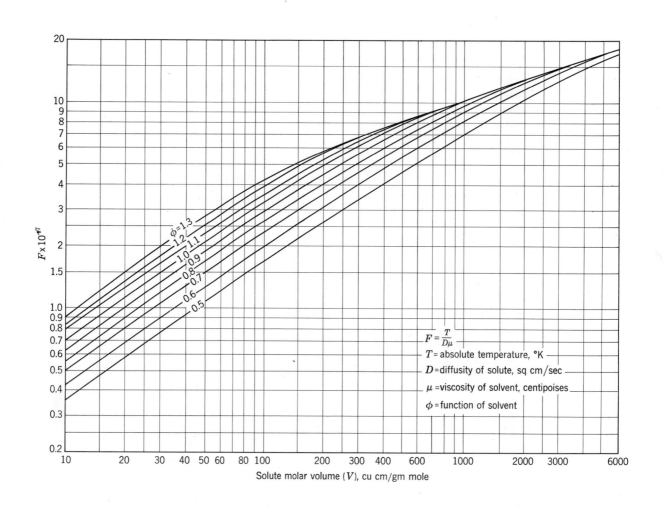

Appendix D-9. VISCOSITIES OF GASES AND LIQUIDS AS A FUNCTION OF TEMPERATURE AT 1 ATM (3)

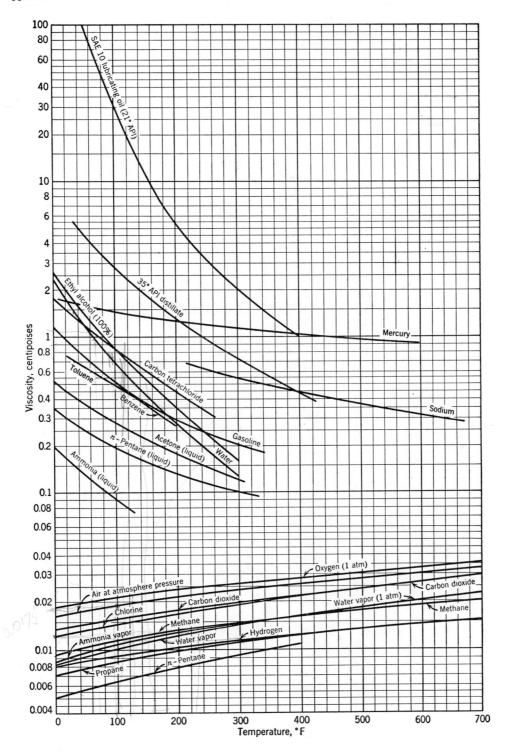

Appendix D-10. THERMAL CONDUCTIVITY OF VARIOUS MATERIALS (1)

(a) Gases and Vapors
$k = $ Btu/hr sq ft (°F/ft)

Substance	T, °F	k	Substance	T, °F	k
Acetone	32	0.0057	Ethylene	−96	0.0064
	115	0.0074		32	0.0101
	212	0.0099		122	0.0131
	363	0.0147		212	0.0161
Acetylene	−103	0.0068	Heptane (*n*-)	212	0.0103
	32	0.0108		392	0.0112
	122	0.0140	Hexane (*n*-)	32	0.0072
	212	0.0172		68	0.0080
Air	−148	0.0095	Hydrogen	−148	0.065
	32	0.0140		−58	0.083
	212	0.0183		32	0.100
	392	0.0226		122	0.115
	572	0.0265		212	0.129
Ammonia	−76	0.0095		572	0.178
	32	0.0128	Methane	−148	0.0100
	122	0.0157		−58	0.0145
	212	0.0185		32	0.0175
Benzene	32	0.0052		122	0.0215
	115	0.0073	Methyl alcohol	32	0.0083
	212	0.0103		212	0.0128
	363	0.0152	Methyl chloride	32	0.0053
	413	0.0176		115	0.0072
Butane (*n*-)	32	0.0078		212	0.0094
	212	0.0135		363	0.0130
Carbon dioxide	−58	0.0068		413	0.0148
	32	0.0085	Nitrogen	−148	0.0095
	212	0.0133		32	0.0140
	392	0.0181		122	0.0160
	572	0.0228		212	0.0180
Carbon tetrachloride	115	0.0041	Oxygen	−148	0.0095
	212	0.0052		−58	0.0119
	363	0.0065		32	0.0142
Chlorine	32	0.0043		122	0.0164
Dichlorodifluoromethane	32	0.0048		212	0.0185
	122	0.0064	Pentane (*n*-)	32	0.0074
	212	0.0080		68	0.0083
	302	0.0097	Propane	32	0.0087
Ethane	−94	0.0066		212	0.0151
	−29	0.0086	Sulfur dioxide	32	0.0050
	32	0.0106		212	0.0069
	212	0.0175	Water vapor, zero pressure*	32	0.0132
Ethyl alcohol	68	0.0089		200	0.0159
	212	0.0124		400	0.0199
Ethyl ether	32	0.0077		600	0.0256
	115	0.0099		800	0.0306
	212	0.0131		1000	0.0495
	363	0.0189			
	413	0.0209			

* For saturated vapor:

Lb/sq in. abs	250	500	1000	1500	2000
T, °F	401	467	545	596	636
k	0.0248	0.0299	0.0395	0.0486	0.0578

Appendix D-10.—*Continued.*

(b) Liquids

$k = $ Btu/hr sq ft (°F/ft)

Liquid	T, °F	k	Liquid	T, °F	k
Acetic acid, 100%	68	0.099	Kerosene	68	0.086
50%	68	0.20		167	0.081
Acetone	86	0.102	Mercury	82	4.83
	167	0.095	Methyl alcohol, 100%	68	0.124
Ammonia	5–86	0.29		122	0.114
Benzene	86	0.092	Methyl chloride	5	0.111
	140	0.087		86	0.089
Carbon tetrachloride	32	0.107	Octane (n-)	86	0.083
	154	0.094		140	0.081
Dichlorodifluoromethane	20	0.057	Pentane (n-)	86	0.078
	60	0.053		167	0.074
	100	0.048	SAE 10 oil (21° API) (3)	0	0.074
	140	0.043		100	0.071
	180	0.038		200	0.068
Ethyl alcohol, 100%	68	0.105		300	0.066
	122	0.087	Sodium	212	49
Ethyl ether	86	0.080		410	46
	167	0.078	Sulfuric acid, 90%	86	0.21
Ethylene glycol	32	0.153	Sulfur dioxide	5	0.128
Glycerol, 100%	68	0.164		86	0.111
	212	0.164	Toluene	86	0.086
Heptane (n-)	86	0.081	Water	32	0.343
	140	0.079		100	0.363
Hexane (n-)	86	0.080		200	0.393
	140	0.078		300	0.395
				420	0.376
				620	0.275

(c) Solids

$k = $ Btu/hr sq ft (°F/ft)

Metals	Density, lb/cu ft	T, °C	k	Metals	Density, lb/cu ft	T, °C	k
Aluminum	165	0	117	Nickel	537	0	36
		100	119			100	34
		200	124			200	33
		300	133			300	32
		400	144	Steel, mild	489	100	26
		500	155			200	26
Copper	556	0	224			300	25
		100	218			400	23
		200	215			500	22
		300	212	Tin	459	0	36
		400	210			100	34
		500	207			200	33
Iron, cast	450	54	27.6	Zinc	440	0	65
		102	26.8			100	64
Lead	710	0	20			200	62
		100	19			300	59
		200	18			400	54
		300	18				

Appendix D-10.—*Continued.*

(c) *continued*

Miscellaneous

$k = $ Btu/hr sq ft (°F/ft)

Miscellaneous	Apparent Density (ρ), lb/cu ft at room temperature	T, °C	k	Miscellaneous	Apparent Density (ρ), lb/cu ft at room temperature	T, °C	k
Asbestos	29.3	−200	0.043	Calcium carbonate, natural	162	30	1.3
	29.3	0	0.090	Carbon stock	94	−184	0.55
	36	0	0.087			0	3.6
	36	100	0.111	Cardboard, corrugated	...	...	0.037
	36	200	0.120	Celluloid	87.3	30	0.12
	36	400	0.129	Coke, petroleum	...	100	3.4
	43.5	−200	0.090			500	2.9
	43.5	0	0.135	Concrete (cinder)	...	...	0.20
Asphalt	132	20	0.43	(stone)	...	...	0.54
Bricks:				(1:4 dry)	...	...	0.44
Alumina (92–99% Al_2O_3 by wt.) fused	...	427	1.8	Cork board	10	30	0.025
Alumina (64–65% Al_2O_3 by wt.)	...	1315	2.7	Cork (regranulated)	8.1	30	0.026
	115	800	0.62	(ground)	9.4	30	0.025
	115	1100	0.63	Diatomaceous earth powder,			
Building brick work	...	20	0.4	coarse	20.0	38	0.036
Carbon	96.7	...	3.0		20.0	871	0.082
Chrome brick (32% Cr_2O_3 by wt.)	200	200	0.67	fine	17.2	204	0.040
	200	650	0.85		17.2	871	0.074
	200	1315	1.0	molded pipe covering	26.0	204	0.051
Diatomaceous earth,					26.0	871	0.088
natural, across strata	27.7	204	0.051	4 vol. calcined earth and			
	27.7	871	0.077	1 vol. cement, poured			
Diatomaceous earth, natural,				and fired	61.8	204	0.16
parallel to strata	27.7	204	0.081		61.8	871	0.23
	27.7	871	0.106	Glass	...	...	0.2–0.73
Diatomaceous earth,				Borosilicate type	139	30–75	0.63
molded and fired	38	204	0.14	Window glass	...	...	0.3–0.61
	38	871	0.18	Soda glass	...	...	0.3–0.44
Diatomaceous earth and				Granite	...	...	1.0–2.3
clay, molded and fired	42.3	204	0.14	Graphite, longitudinal	...	20	95
	42.3	871	0.19	powdered, through 100			
Diatomaceous earth, high				mesh	30	40	0.104
burn, large pores	37	200	0.13	Gypsum (molded and dry)	78	20	0.25
	37	1000	0.34	Ice	57.5	0	1.3
Fire clay (Missouri)	...	200	0.58	Kapok	0.88	20	0.020
		600	0.85	Leather, sole	62.4	...	0.092
		1000	0.95	Limestone (15.3 vol. % H_2O)	103	24	0.54
		1400	1.02	Magnesia (powdered)	49.7	47	0.35
Kaolin insulating brick	27	500	0.15	Magnesia (light carbonate)	13	21	0.034
	27	1150	0.26	Magnesium oxide (compressed)	49.9	20	0.32
Kaolin insulating firebrick	19	200	0.050	Marble	...	...	1.2–1.7
	19	760	0.113	Mineral wool	9.4	30	0.0225
Magnesite (86.8% MgO,					19.7	30	0.024
6.3% Fe_2O_3, 3% CaO,				Paper	...	...	0.075
2.6% SiO_2 by wt.)	158	204	2.2	Porcelain	...	200	0.88
	158	650	1.6	Portland cement, see concrete	...	90	0.17
	158	1200	1.1		...	21–66	0.14

Appendix D-10.—*Continued.*

(c) (*continued*)

Miscellaneous

$k = $ Btu/hr sq ft (°F/ft)

Miscellaneous	Apparent Density (ρ), lb/cu ft at room temperature	T, °C	k	Miscellaneous	Apparent Density (ρ), lb/cu ft at room temperature	T, °C	k
Rubber (hard)	...	0	0.087	Wood (across grain):			
(soft)	...	21	0.075–0.092	Balsa	7–8	30	0.025–0.03
Sand (dry)	94.6	20	0.19	Oak	51.5	15	0.12
Sandstone	140	40	1.06	Maple	44.7	50	0.11
Sawdust	12	21	0.03	Pine, white	34.0	15	0.087
Slag, blast furnace	...	24–127	0.064	Teak	40.0	15	0.10
Slag wool	12	30	0.022	White fir	28.1	60	0.062
Slate	...	94	0.86	Wood (parallel to grain):			
Snow	34.7	0	0.27	Pine	34.4	21	0.20
Wall board, insulating type	14.8	21	0.028	Wool, animal	6.9	30	0.021
Wall board, stiff paste board	43	30	0.04				

Appendix D-11. MASS DIFFUSIVITIES (1)

(a) Gases at 25°C, 1 atm in air

Substance	$\mathscr{D}$, sq cm/sec	$(\mu/\rho\mathscr{D})$*
Ammonia	0.229	0.67
Carbon dioxide	0.164	0.94
Hydrogen	0.410	0.22
Oxygen	0.206	0.75
Water	0.256	0.60
Carbon disulfide	0.107	1.45
Ethyl ether	0.093	1.66
Methanol	0.159	0.97
Ethyl alcohol	0.119	1.30
Propyl alcohol	0.100	1.55
Butyl alcohol	0.090	1.72
Amyl alcohol	0.070	2.21
Hexyl alcohol	0.059	2.60
Formic acid	0.159	0.97
Acetic acid	0.133	1.16
Propionic acid	0.099	1.56
i-Butyric acid	0.081	1.91
Valeric acid	0.067	2.31
i-Caproic acid	0.060	2.58
Diethyl amine	0.105	1.47
Butyl amine	0.101	1.53
Aniline	0.072	2.14
Chlorobenzene	0.073	2.12
Chlorotoluene	0.065	2.38
Propyl bromide	0.105	1.47
Propyl iodide	0.096	1.61
Benzene	0.088	1.76
Toluene	0.084	1.84
Xylene	0.071	2.18
Ethylbenzene	0.077	2.01
Propylbenzene	0.059	2.62
Diphenyl	0.068	2.28
n-Octane	0.060	2.58
Mesitylene	0.067	2.31

* The group $(\mu/\rho\mathscr{D})$ in the above table is evaluated for mixtures composed largely of air.

(b) Liquids at 20°C, dilute solutions

Solute	Solvent	$\mathscr{D} \times 10^5$ (sq cm/sec) $\times 10^5$	$\left(\dfrac{\mu}{\rho\mathscr{D}}\right)$*
O_2	Water	1.80	558
CO_2	Water	1.77	670
N_2O	Water	1.51	665
NH_3	Water	1.76	570
Cl_2	Water	1.22	824
Br_2	Water	1.2	840
H_2	Water	5.13	196
N_2	Water	1.64	613
HCl	Water	2.64	381
H_2S	Water	1.41	712
H_2SO_4	Water	1.73	580
HNO_3	Water	2.6	390
Acetylene	Water	1.56	645
Acetic acid	Water	0.88	1140
Methanol	Water	1.28	785
Ethanol	Water	1.00	1005
Propanol	Water	0.87	1150
Butanol	Water	0.77	1310
Allyl alcohol	Water	0.93	1080
Phenol	Water	0.84	1200
Glycerol	Water	0.72	1400
Pyrogallol	Water	0.70	1440
Hydroquinone	Water	0.77	1300
Urea	Water	1.06	946
Resorcinol	Water	0.80	1260
Urethane	Water	0.92	1090
Lactose	Water	0.43	2340
Maltose	Water	0.43	2340
Glucose	Water	0.60	. . .
Mannitol	Water	0.58	1730
Raffinose	Water	0.37	2720
Sucrose	Water	0.45	2230
Sodium chloride	Water	1.35	745
Sodium hydroxide	Water	1.51	665
CO_2	Ethanol	3.4	445
Phenol	Ethanol	0.8	1900
Chloroform	Ethanol	1.23	1230
Phenol	Benzene	1.54	479
Chloroform	Benzene	2.11	350
Acetic acid	Benzene	1.92	384
Ethylene dichloride	Benzene	2.45	301

* Based on $\mu/\rho = 0.01005$ sq cm/sec for water, 0.00737 for benzene, and 0.01511 for ethanol, all at 20°C. Applies only for dilute solutions.

Appendix D-12. DENSITIES OF VARIOUS MATERIALS (1)

(*a*) Gases at 1 atm and 0°C

Gas	Formula	Molecular Weight	Density, lb/cu ft
Acetylene	C_2H_2	26.02	0.0732
Air	–	–	0.0808
Ammonia	NH_3	17.03	0.0482
Argon	Ar	39.91	0.1114
Butane	C_4H_{10}	58.08	0.1623
Carbon dioxide	CO_2	44.00	0.1235
Chlorine	Cl_2	70.91	0.2011
Ethane	C_2H_6	30.05	0.0848
Ethylene	C_2H_4	28.03	0.0783
Helium	He	4.00	0.0111
Hydrogen	H_2	2.016	0.0056
Methane	CH_4	16.03	0.0448
Methyl chloride	CH_3Cl	50.48	0.1440
Nitrogen	N_2	28.022	0.0782
Oxygen	O_2	32.00	0.0892
Sulfur dioxide	SO_2	64.06	0.1828

(*b*) Liquids at 20°C

	Density, lb/cu ft
Acetic acid	65.4 (−79°C)
Acetone	49.4
Ammonia	51.0
Benzene	54.8
Carbon tetrachloride	99.5
Ethyl ether	44.2
Ethyl alcohol	49.2
Methyl alcohol	49.4
Mercury	849
Pentane (*n*)	39.2
SAE 10 oil (21° API)	57.8
Sodium	58.0 (100°C)
	54.7 (600°F)
Toluene	54.0
Water	62.3

(*c*) Solids*

Metals	Density, lb/cu ft
Aluminum	165
Copper	556
Iron, cast	450
Lead	710
Nickel	537
Steel, mild	489
Tin	459
Zinc	440

* For other solids see Appendix D-10*c*.

Appendix D-13. HEAT CAPACITIES (3)

(*a*) Gases and Liquids at 1 atm

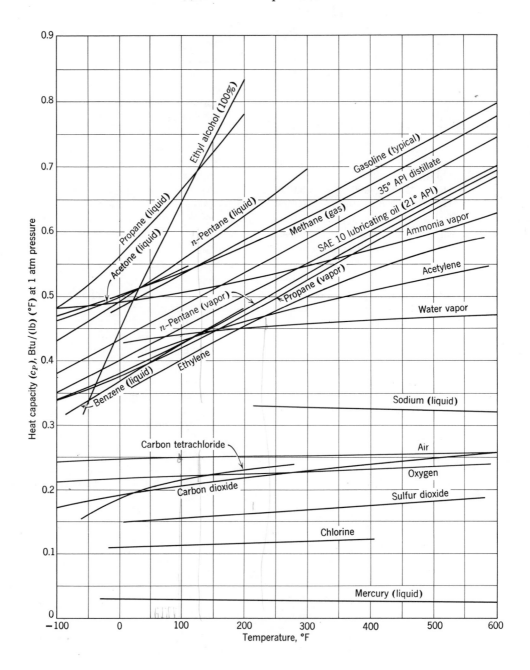

Appendix D-13.—*Continued.*

(*b*) Ratio of Heat Capacities at 1 atm (C_p/C_v)

Compound	Formula	Temperature, °C	Ratio of Specific Heats $\gamma = C_p/C_v$	Compound	Formula	Temperature, °C	Ratio of Specific Heats $\gamma = C_p/C_v$
Acetylene	C_2H_2	15	1.26	Ethylene	C_2H_4	100	1.18
		−71	1.31			15	1.255
Air	...	925	1.36			−91	1.35
		17	1.403	Helium	He	−180	1.660
		−78	1.408	Hexane (*n-*)	C_6H_{14}	80	1.08
		−118	1.415	Hydrogen	H_2	15	1.410
Ammonia	NH_3	15	1.310			−76	1.453
Argon	Ar	15	1.668			−181	1.597
		−180	1.76	Methane	CH_4	600	1.113
		0–100	1.67			300	1.16
Benzene	C_6H_6	90	1.10			15	1.31
						−80	1.34
Carbon dioxide	CO_2	15	1.304			−115	1.41
		−75	1.37	Methyl alcohol	CH_4O	77	1.203
		−180	1.41	Nitrogen	N_2	15	1.404
Chlorine	Cl_2	15	1.355			−181	1.47
Dichlorodifluormethane	CCl_2F_2	25	1.139	Oxygen	O_2	15	1.401
Ethane	C_2H_6	100	1.19			−76	1.415
		15	1.22			−181	1.45
		−82	1.28	Pentane (*n-*)	C_5H_{12}	86	1.086
Ethyl alcohol	C_2H_6O	90	1.13				
Ethyl ether	$C_4H_{10}O$	35	1.08	Sulfur dioxide	SO_2	15	1.29
		80	1.086				

(*c*) Solids

Metals	Heat Capacity, Btu/lb °F
Aluminum	0.374 (0°C), 0.405 (100°C)
Copper	0.164 (0°C), 0.169 (100°C)
Iron, cast	0.214 (20–100°C)
Lead	0.0535 (0°C), 0.0575 (100°C)
Nickel	0.0186 (0°C), 0.206 (100°C)
Steel	0.12
Tin	0.096 (0°C), 0.104 (100°C)
Zinc	0.164 (0°C), 0.172 (100°C)

Miscellaneous	Heat Capacity, Btu/lb °F
Alumina	0.2 (100°C), 0.274 (1500°C)
Asbestos	0.25
Brickwork	About 0.2
Carbon (mean values)	0.168 (26–76°C)
	0.314 (40–892°C)
	0.387 (56–1450°C)
Cellulose	0.32
Cement, Portland	
Clinker	0.186
Charcoal (wood)	0.242
Chrome brick	0.17
Clay	0.224
Coal	0.26 to 0.37

Miscellaneous	Heat Capacity Btu/lb °F
Coke (mean values)	0.265 (21–400°C)
	0.359 (21–800°C)
	0.403 (21–1300°C)
Concrete	0.156 (70–312°F), 0.219 (72–1472°F)
Fireclay brick	0.198 (100°C), 0.298 (1500°C)
Fluorspar	0.21 (30°C)
Glass (crown)	0.16 to 0.20
(flint)	0.117
(pyrex)	0.20
(silicate)	0.188 to 0.204 (0–100°C)
	0.24 to 0.26 (0–700°C)
wool	0.157
Graphite	0.165 (26–76°C), 0.390 (56–1450°C)
Gypsum	0.259 (16–46°C)
Limestone	0.217
Magnesia	0.234 (100°C), 0.188 (1500°C)
Magnesite brick	0.222 (100°C), 0.195 (1500°C)
Marble	0.21 (18°C)
Quartz	0.17 (0°C), 0.28 (350°C)
Sand	0.191
Stone	About 0.2
Wood (oak)	0.570
Most woods vary between	0.45 and 0.65

Appendix D-14. HUMIDITY CHART FOR THE AIR–WATER SYSTEM.

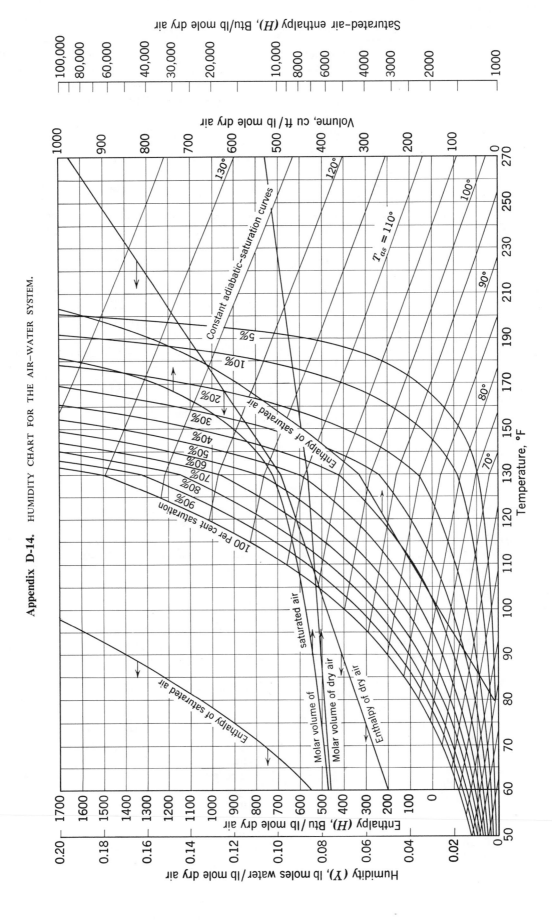

Appendix D-15. TOTAL EMISSIVITIES OF VARIOUS SOLID MATERIALS

[McAdams, W. H., *Heat Transmission*, 3rd ed., McGraw-Hill Book Co., New York, 1954.]

Surface	Temperature, °F	Emissivity ϵ	Surface	Temperature, °F	Emissivity ϵ
I METALS			**II BUILDING MATERIALS**		
Aluminum			Asbestos board	74	0.96
Highly polished plate	440–1070	0.039–0.057	Brick		
Oxidized at 1110°F	390–1110	0.11–0.19	Red, rough	70	0.93
			Building	1832	0.45
Brass			Fireclay	1832	0.75
Highly polished (73–27)	476–674	0.028–0.031			
Polished	100–600	0.10	Lampblack, other blacks	122–1832	0.96
Dull plate	120–660	0.22	Concrete tiles	1832	0.63
			Enamel, white, fused on iron	66	0.90
Copper			Glass, Pyrex, lead, and soda	500–1000	~0.95–0.85
Polished	242	0.023	Oak, planed	70	0.90
Plate heated to 1110°F	390–1110	0.57			
			Paints		
Iron			Black or white lacquer	100–200	0.80–0.95
Electrolytic, highly polished	350–440	0.052–0.064	Flat black lacquer	100–200	0.96–0.98
Cast iron, newly turned	72	0.44	Oil paints, 16 different, all colors	212	0.92–0.96
Smooth sheet iron	1650–1900	0.55–0.60	Aluminum, varying age and Al content	212	0.27–0.67
Steel, oxidized at 1100°F	390–1110	0.79			
Steel plate, rough	100–700	0.94–0.97			
Molten, mild steel	2910–3270	0.28	Plaster, rough lime	50–190	0.91
Stainless, Type 304 after heating	420–914	0.44–0.36	Rubber, soft, gray, rough (reclaimed)	76	0.86
Mercury	32–212	0.09–0.12	Water	32–212	0.95–0.963
Silver, polished	100–700	0.022–0.031	Snow		~1.0
Tungsten, filament	6000	0.39	Ice, solid		0.63
Zinc, galvanized sheet iron, gray	75	0.28			

REFERENCES

1. *Chemical Engineers' Handbook*, J. H. Perry, ed., 3rd ed., McGraw-Hill Book Co., New York, 1950.
2. *International Critical Tables*, McGraw-Hill Book Co., New York, 1929.
3. Brown, G. G., and Associates, *Unit Operations*, John Wiley and Sons, New York, 1950.

Appendix D-16. EQUILIBRIUM DATA

Propane-Oleic Acid-Cottonseed Oil System
(98.5°C, 625 lb/in²abs)

[Hixson, A. W. and J. B. Bockelmann, *Trans. Am. Inst. Chem. Eng.* **38,** 891 (1942)]

Equilibrium Tie Line Data, Weight Percent

Upper Phase			Lower Phase		
Oleic Acid	Cottonseed Oil	Propane	Oleic Acid	Cottonseed Oil	Propane
0	2.3	97.7	0	63.5	36.5
0.25	2.2	97.6	0.25	63.0	36.8
0.5	2.1	97.4	1.0	62.1	36.9
0.75	2.0	97.3	2.5	60.2	37.3
1.0	1.9	97.1	5.0	57.1	37.9
1.5	1.7	96.8	10.0	50.9	39.1
2.5	1.4	96.1	15.0	44.9	40.1
5.0	1.0	94.0	20.0	38.8	41.2
6.0	0.9	93.1	25.0	32.7	42.3
7.2	0.8	92.0	30.0	26.0	44.0
6.5	0.5	93.0	35.0	18.3	46.7
6.0	0.4	93.6	40.0	7.0	53.0
5.5	0.2	94.3	42.5	0.9	56.6

Styrene-Ethyl Benzene-Diethylene Glycol System

[Boobar, M. G. et. al., *Ind. Eng. Chem.* **43,** 2922 (1951)]

Equilibrium Compositions Weight Percent, at 25°C

Ethylbenzene Layer		Diethylene Glycol Layer	
Styrene	Diethylene Glycol	Styrene	Diethylene Glycol
8.63	0.81	1.64	88.51
18.67	0.93	3.49	87.20
28.51	1.00	5.48	85.80
37.98	1.09	7.45	84.48
45.84	1.20	9.49	83.20
57.09	1.40	12.54	81.40
76.60	1.80	18.62	77.65

Isopropyl Ether-Acetic Acid-Water System

Othmer, D. F., R. E. White and E. Trueger, *Ind. Eng. Chem.* **33,** 1240 (1941)

Solubility Data, Weight Percent, at 20°C

Ipe	50.3	31.1	16.7	13.3	41.85	3.5	4.6	5.12	60.2	48.4	11.1
Acid	36.8	45.1	48.4	48.1	47.3	37.6	40.0	41.5	30.7	35.0	48.0
Water	13.0	23.8	34.9	38.6	10.85	58.8	55.4	53.4	9.1	11.6	40.9

Distribution Data, Weight Percent Acetic Acid in:

Ipe Layer	6.1	9.4	16.75	30.2	39.0
Water Layer	16.1	21.9	33.5	46.3	42.1

Index